Signals and Communication Technology

More information about this series at http://www.springer.com/series/4748

Gianfranco Cariolaro

Quantum Communications

 Springer

Gianfranco Cariolaro
Department of Information Engineering
University of Padova
Padova
Italy

ISSN 1860-4862   ISSN 1860-4870   (electronic)
Signals and Communication Technology
ISBN 978-3-319-34375-4   ISBN 978-3-319-15600-2   (eBook)
DOI 10.1007/978-3-319-15600-2

Springer Cham Heidelberg New York Dordrecht London

Springer International Publishing AG Switzerland is part of Springer Science+Business Media
(www.springer.com)

To David, Shu-Ning, Gabriele, and Elena

Foreword

The birth of the original idea of Quantum Communications might be dated back to the same age when Claude E. Shannon formulated the mathematical theory of communications in 1948. In 1950, Dennis Gabor wrote a seminal paper on how to revise Information Theory by considering Quantum Physics, introducing the term "quantum noise". Actually, as the carrier frequency goes up to few tens of a terahertz, quantum noise rapidly becomes more dominant than thermal noise. In 1960, Theodore H. Maiman succeeded in producing the first beam of laser light, whose frequency was at a few hundred terahertz. For a long time, it was a crucial trigger for full-scale studies on Quantum Communications. It was not, however, a straightforward task at all for researchers to establish the unification of the para-doxical aspects of Quantum Mechanics with the landmarks of Communications Theory. It was only recently that the core of Quantum Communications, that is, the theory of capacity for a lossy quantum-limited optical channel, was established. Until now, many new ideas and schemes have been added to the original standard scheme of Quantum Communications, represented by quantum key distribution, quantum teleportation, and so on. Realizing a new paradigm of Quantum Communications is now an endeavor in science and technology, because it requires a grand sum of not only the latest Quantum Communications technologies but also the basics of Information Theory and Signal Detection and Processing technologies. Therefore, it is not easy for students and researchers to learn all the necessary knowledge, to acquire techniques to design and implement the system, and to operate it in practice. These tasks usually take a long time through a variety of courses, and by reading many papers and several books.

This book is meant to achieve this very purpose. The author, Professor Gianfranco Cariolaro, has been working for a long time in the fields of Communications and Image Processing technologies, Deep Space Communications, and Quantum Communications. From this book, readers can track a history of Quantum Communications and learn its core concepts and very practical techniques. For this decade, commercial applications of quantum key distribution have been taking place, and in 2013, lunar laser communication was successfully demonstrated by the National Aeronautics and Space Administration, where a novel photon counting method was employed. This means that an era of Quantum Communications in practice is around the corner. I am very excited to have this book at such a time. Through this book, readers will also be able to see a future Communications Technology on the shoulder of a long history of Quantum Communications.

Tokyo, Japan, September 2014

Masahide Sasaki
Director of Quantum ICT Laboratory
National Institute of Information
and Communications Technology

Preface

Quantum Mechanics represents one of the most successful theories in the history of science. Born more than a hundred years ago, for several decades Quantum Mechanics was confined to a revolutionary interpretation of Physics and related fields, like Astronomy. Only in the last decades, after the discovery of laser with the possibility of producing coherent light, did Quantum Mechanics receive a strong interest in the area of information, with very innovative and promising applications (Quantum Computer, Quantum Cryptography, and Quantum Communications).

In particular, the original ideas of Quantum Communications were developed by Helstrom [10] and by scientists from MIT [11, 13] proving the superiority of quantum systems with respect to classic optical systems. However, the research in this specific field did not obtain the same spectacular expansion as the other fields of quantum information. In our personal opinion, the reason is twofold. One is the difficulty in the implementation of quantum receivers, which involves sophisticated optical operations. The other reason, perhaps the most relevant, was due to the advent of optical fibers, whose tremendous capacity annihilated the effort on the improvement of performances of the other transmission systems. This may explain the concentration of interest in the other fields of quantum information. Nevertheless, Quantum Communications deserve a more adequate attention for us to be prepared for the future developments, being confident that a strong progress in quantum optics will be surely achieved.

There is another motivation for considering Quantum Communications, especially for educational purposes in Information Engineering. In fact, continuing with our personal viewpoint, Quantum Mechanics is a discipline that cannot be ignored in the future curriculum of information engineers (electronics, computer science, telecommunications, and automatic control). On the other hand, Quantum Mechanics is a difficult discipline for its mathematical and also philosophical impact, and cannot be introduced at the level of Physics and Mathematical Physics because the study burden in information engineering is already quite heavy. However, we realized (with some surprise) that the notions of Quantum Mechanics needed for Quantum Communications may be easily tackled by information engineering students. In fact, the notions needed at this level (vector spaces and probability theory) are already known to these students and require only an ad hoc recall. Following these ideas, six years ago the author introduced a course on Quantum Communications in the last year of the Telecommunications degree (master level) at the Faculty of Engineering of the University of Padova, and, as confirmed by students and colleagues, the conclusion was that the teaching experiment has proved very successful.

At the same time, experience shows that the majority of students, who join quantum optics and quantum information community after taking courses in quantum mechanics with concentration on elementary particles and high-energy physics, have very little feeling for the real notion of information transfer and manipulation as it is known in practical telecommunications. The comprehensive consideration of Quantum Communication concepts presented in this book serves to establish this missing conceptual link between the formal Quantum Mechanics theory formulated originally for particles and the quantum optical information manipulation utilizing quantum mechanics along with optics and telecommunications tools.

It is difficult to predict in what direction quantum information will evolve or when the quantum computer will arrive, but it will surely have a strong impact in the future. Students and researchers that will have learned Quantum Communications, having acquired the methodology and language, will be open to any other application in the field of Quantum Information.

Organization of the Book

The book is organized into three parts and 13 chapters.

Chapter 1 (Introduction) essentially describes the evolution of Quantum Mechanics in the previous century, with special emphasis on the last part of the evolution in the area of Quantum Information, with its promising and exciting applications.

Part I: Fundamentals

Chapter 2 collects the mathematical background needed in the formulation and development of Quantum Mechanics: mainly notions of linear vector spaces and Hilbert spaces, with special emphasis on the eigendecomposition of linear operators.

Chapter 3 introduces the fundamentals of Quantum Mechanics, in four postulates. Postulate 1 is concerned with the environment of Quantum Mechanics: a Hilbert space. Postulate 2 formulates the evolution of a quantum system, according to Schrödinger's and Heisenberg's visions. Postulate 3 is concerned with the quantum measurements, which prescribes the possibility of extracting information from a quantum system. Finally, Postulate 4 deals with the combination of two or more interacting quantum systems. A particular emphasis is given to Postulate 3, because it manages the information in a quantum system and will be the basis of Quantum Communications and Quantum Information consideration.

Part II: Quantum Communications Systems

Chapter 4 deals with the general foundations of telecommunications systems and the difference between Classical and Quantum Communications systems. In the second part of the chapter the foundations of optical classical communications, which is the necessary prologue to optical quantum communications, are developed.

Chapter 5 develops the concept of optimal quantum decision, which establishes the best criterion to perform the measurements of Postulate 3 in a quantum system to extract information. Here a nontrivial effort is made to express the results within the language of telecommunications, where the quantum decision is applied to the receiver.

Chapter 6 develops suboptimization in quantum decision. Since optimization is very difficult, and exact solutions are only known in few cases, suboptimization techniques are considered, the most important of which is called square-root measurements (SRM).

Chapter 7 deals with the general formulation of quantum communication systems, where the transmitter (Alice) prepares and launches the information in a quantum channel and the receiver (Bob) extracts the information by applying the quantum decision rules. Although, in principle, the transmission of analog information would be possible, according to the lines of present-day technology, only digital information (data) is considered. In any case, we will refer to optical communications, in which the information is conveyed through a coherent radiation produced by a laser. The quantum formulation of coherent radiation is expressed according to the universal and celebrated Glauber's theory.

In the second part of the chapter, these basic ideas are applied to most popular quantum communication systems, each one characterized by a specific modulation format (OOK, PPM, PSK, and QAM). The performance of each specific system is compared to that of the corresponding classical optical system, where the decision is based on a simple photon counting. The comparisons will clearly state the superiority of the quantum systems.

Chapter 8 reconsiders the analysis of Chap. 7 with the introduction of thermal noise, in order to get a more realistic evaluation of the performance. Technically speaking, the analysis in the absence of thermal noise is carried out using the description of the system status made in terms of pure states, whereas the presence of thermal noise requires a description in terms of density operators. Consequently, the analysis becomes much more complicated (but challenging).

Chapter 9 deals with the implementation of coherent quantum communication systems. The few implementations available in the literature and the difficulties encountered in the realization are described. Also, some original ideas for an improved implementation of quantum communication systems are described.

Part III: Quantum Information

Chapter 10 begins by dealing with Quantum Information, which exhibits two forms, discrete and continuous. Discrete quantum information is based on discrete variables, the best known example of which is the quantum bit or qubit. Continuous quantum information is based on continuous variables, the best known example of which is provided by the quantized harmonic oscillator. An important remark is that most of the operations in quantum information processing can be carried out both with discrete and continuous variables (this last possibility is a quite recent discovery).

Chapter 11, Quantum Mechanics fundamentals of Chap. 3 are confined to the basic notions (relatively few) necessary to the development of Quantum Communications systems in Part II. In this chapter, for a full development of Quantum Information, the above fundamentals are extended to continuous quantum variables, to include Gaussian states and Gaussian transformations.

Chapter 12 deals with Information Theory, starting from Classical Shannon's Information Theory and then extending the concepts to Quantum Information Theory. The latter is a relatively new discipline, which is based on quantum mechanical principles and in particular on its intriguing resources, such as entanglement.

Chapter 13 deals with the applications of Quantum Information, as quantum random number generation, quantum key distribution, and teleportation. These applications are developed with both discrete and continuous variables.

Suggested Paths

For the choice of the path one should bear in mind that the book is a combination of Quantum Mechanics and Telecommunications, and perhaps students and researchers in the area of Information Engineering have no preliminary knowledge of Quantum Mechanics, whereas students and researchers in the area of Physics may have no preliminary knowledge of Telecommunications (for which we recommend reading Chap. 4 on Telecommunications fundamentals).

As said above, the mathematics needed for the comprehension of the book is confined to Linear Vector Spaces, as developed in Chap. 2. Hilbert spaces are introduced for completeness, but they are not really used. The other mathematical requirement is Probability Theory (probability fundamentals and random variables, sometimes extended to random processes). These preliminaries must be known at a good, but not too sophisticated level.

The book could be used by both graduate students (meaning people who have no knowledge of Quantum Mechanics) and researchers (meaning people who have a good knowledge of Quantum Mechanics, but not of classical Telecommunications) following two different paths.

In the Introduction we will indicate in detail two different paths for "students" and for "researchers".

Manuscript Preparation

To prepare the manuscript we used LATEX, supplemented with a personal library of macros. The illustrations too are composed with LATEX, sometimes with the help of `Mathematica`©.

Padova, October 2014 Gianfranco Cariolaro

Acknowledgments

As an expert in traditional Telecommunications, in the twilight of my life, I decided to tread on the unknown territory of Quantum Information. This required the help of many people, without whom this book could never have been written. First of all, I would like to mention Tommaso Occhipinti and Federica Fongher, students, with whom I outlined the main ideas of the project.

I owe special thanks to Gianfranco Pierobon, who made fundamental contributions to many subjects in the book. In addition to sharing the same name, Gianfranco and I shared a similar enthusiastic wonder toward Quantum Mechanics, as we were both newcomers to the discipline. I still remember the trepidation and scepticism with which we submitted our first paper to an international journal (on the performance of Quantum Communications systems based on square root measurements). But it turned out to be a success. Incidentally, I would like to mention that the topic and the methodology of that paper were inspired by the work of Professor Masahide Sasaki and his collaborators. Gianfranco's help was so valuable that I repeatedly offered him to co-author the book, but he always refused, and, knowing his stubbornness, I had to give up. However, I hope he will accept next time, with the next book.

Roberto Corvaja was very helpful by re-reading the manuscript, over and over again, and integrating it with essential numeric computations. Nicola Laurenti provided invaluable assistance in the development of Part III of the book, concerning the applications of Quantum Information, in particular, by helping me to find my way in the jungle of this rapidly growing subject, as well as by proposing alternative arguments. Tomaso Erseghe was kind enough to learn Quantum Mechanics for the sole purpose of helping me, and he did so with great competence.

Several other dedicated readers offered me numerous, detailed, and insightful suggestions: Antonio Assalini, Luigi Bellato, Cesare Barbieri, Gianpaolo Naletto, Ezio Obetti, Stefano Olivares, Silvano Pupolin, Edi Ruffa, Lorenzo Sartoratti, Giovanna Sturaro, Francesco Ticozzi, Paolo Villoresi, and the young students: Nicola Dalla Pozza, Alberto Dall'Arche, Davide Marangon, and Giuseppe Vallone.

I am particularly indebted to Nino Trainito, perhaps the only one to actually read the whole manuscript!, who made several comments and considerably improved the language.

I would like to mention that I was forced to interrupt my work for several months due to an unfortunate, tragic event that affected my family. On that occasion, three people in particular were crucial in helping me to cope with the situation and to return to normal life, namely Consul Vincenzo De Luca, Prof. Renato Scienza, and my friend Stefano Gastaldello. Other friends were very close to me in this difficult period, namely, Cesare Barbieri, Peter Kraniauskas, Umberto Mengali, Marina Munari, Silvano Pupolin, Romano Valussi, and Guido Vannucchi. I take the liberty to mention all this, an unusual subject for an acknowledgments section, because without the support of all these friends the book would have never been finished.

To all these people I owe a great debt of gratitude and offer heartfelt thanks.

Padova, October 2014   Gianfranco Cariolaro

Contents

Chapter 1
Introduction

1.1 A Brief History of Quantum Mechanics

A Few Milestones in Quantum Mechanics

1900: Black body radiation law (Max Planck)

1905: Postulation of photons to explain photoelectric effect (Albert Einstein)

1909: Interference experiments (Geoffrey Ingram Taylor)

1913: Quantization of angular momentum of hydrogen (Niels Bohr)

1923: Compton effect (Arthur Holly Compton)

1924: Wave–particle duality extended to incorporate matter (Louis de Broglie)

1925: Matrices as basis for Quantum Mechanics (Werner Heisenberg)

1926: Probabilistic interpretation of the wavefunction (Max Born)

1926: Gilbert Lewis coined the word photon

1926: Wave equation to explain the hydrogen atom (Erwin Schrödinger)

1927: Uncertainty principle (Werner Heisenberg)

1927: Copenhagen interpretation (Niels Bohr)

1928: First solution of Quantum Mechanics explaining spin (Paul Dirac)

1930: Principles of Quantum Mechanics (Paul Dirac)

1930: Interference, how quantized light interacts with atoms (Enrico Fermi)

1932: Mathematical foundations of Quantum Mechanics (John von Neumann)

1935: EPR paradox (Einstein, Podolsky, and Rosen)

1950s: Theory of photon statistic and counting (Hanbury Brown, and Twiss)

1960s: Quantum theory of coherence (Glauber, Wolf, Sudarshan, and others)

1970: (early 1970s) Tunable lasers

© Springer International Publishing Switzerland 2015

G. Cariolaro, *Quantum Communications*, Signals and Communication Technology,

DOI 10.1007/978-3-319-15600-2_1

1.1.1 The Dawn

In the last decade of the nineteenth century Newton's mechanics, Maxwell's electromagnetic theory, and Boltzmann's statistical mechanics seemed capable of exhaustively explaining any relevant physical phenomenon. However, some phenomena, initially deemed as marginal, did not completely fit in the structure of these *classic* disciplines. It all began with the discoveries of a Physics student called **Max Planck** (1858–1947).[1] Planck's research was triggered by the study of the emission and absorption of light by physical bodies. At that time, the founding theory of radiation emission by a black body was based on classical electromagnetism. Applying this theory, the phenomenon was well explained for relatively low frequencies of the emitted radiation (visible or near infrared and downwards); however, for high frequencies (ultraviolet and upwards) classical theory would predict an infinite increase in the energy of the emitted radiation, which, as matter of fact, does not happen in reality. To overcome such a problem, Planck formulated the hypothesis that the radiating energy could only exist in the form of discrete quantities, or "packets", which he called **quanta**. To set the framework of Planck's problem, we must recall the previous research of the physicist **J.W. Strutt Lord Rayleigh** (1842–1919), who studied the radiation of the black body from a classical point of view, modeling it as a collection of electromagnetic oscillators, and considering the presence of the radiation at frequency ν as the consequence of the excitation of the oscillator at such frequency. With some contribution by **Sir James Hopwood Jeans** (1877–1946), he arrived at the formulation of the **Rayleigh-Jeans Law**, given by the expression

$$E(\nu) = \frac{8\pi kT \nu^4}{c^4} = \frac{8\pi kT}{\lambda^4}, \tag{1.1}$$

which gives the value $E(\nu)$ of energy density per frequency unit emitted by a black body at frequency ν. In (1.1) $k = 1.38 \; 10^{-23} \mathrm{JK}^{-1}$ is Boltzmann's constant, T is the absolute temperature of the black body, c is the speed of light, and $\lambda = c/\nu$ is the wavelength. This law shows that the energy density irradiated by a black body increases linearly with temperature and with the fourth power of the frequency of the emitted radiation. Experimental measurements demonstrate that this law is perfectly adequate at low frequencies: in fact, it is well known that, with increasing temperature, the irradiated energy increases proportionally, at least up to the infrared. However, measurements carried out at higher frequencies, for example in the ultraviolet range, clearly show that the emitted energy values diverge considerably from those foreseen by the theory. In addition, from a careful analysis of Eq. (1.1), one can see that the expected result in this spectral interval has no physical meaning. In fact, this equation states that, with increasing frequency, energy density increases indefinitely. As a consequence, the equation asserts that the high-frequency oscillators (very low wavelength, corresponding to the ultraviolet radiation, to the X-rays, and to

---

[1] On December 14, 1900, Planck publishes his first paper on Quantum Theory in *Verh. Deut. Phys. Ges. 2,237–45.*

the γ-rays) should be excited even at room temperature. Such absurd result, which posits the emission of a large amount of energy in the high-frequency region of the electromagnetic spectrum, went under the name of *ultraviolet catastrophe*.

The solution of the problem was in fact due to Max Planck, who tackled it in mathematical terms. Instead of integrating the energies of the "elementary oscillators" (that is, in practice, of the electrons "oscillating" around the nucleus) considering them as continuous quantities, he performed a summation of the energies, hypothesizing that they could assume only discrete values, proportional to the characteristic oscillation frequency ν of the electrons, by an appropriate constant h

$$E = h\nu. \tag{1.2}$$

The relation discovered by Planck for the energy density per frequency unit of the black body turns out to be (*Planck's relation*)

$$E(\nu) = \frac{8\pi}{c^3} \frac{h\nu^3}{e^{h\nu/kT} - 1}$$

and it appears to be in perfect agreement with the experimental distribution for each temperature, assuming $h = 6.63 \ 10^{-34}$ Js; h is known as *Planck's constant*.

Planck's theoretical discovery on quanta became accepted by the classical physicists only when **Albert Einstein** (1879–1955)[2] succeeded in explaining the photoelectric effect, speculating that light radiation was constituted by energy packets, subsequently called "photons". Einstein showed that, thanks to quanta, other physical phenomena could be explained, in addition to the black body emission proposed by Planck, and at that point the discrete nature of electromagnetic radiation became a fundamental and generally accepted concept.

Another problem that could not be explained by classical mechanics was the regularity of the emission spectrum of an atom, that is, the fact that it always appeared as formed by the same characteristic frequencies, independently of its origin and of possible excitation processes it had undergone, a fact that could not be convincingly explained by the model proposed by **Ernest Rutherford** (1871–1937) in 1911. The first one to address the problem in mathematical terms was **Niels Bohr** (1885–1962) in 1913. Bohr hypothesized that the lines of an atomic spectrum were originated by the transition of an electron between two discrete states of an atom. This theory correctly interpreted, for the first time, the emission and absorption properties of an atom of hydrogen.

The next step in the development of Quantum Mechanics was due to **Louis-Victor Pierre Raymond de Broglie**[3] (1892–1987), who extended to the particles with mass the *wave–particle duality* that had been evidenced for electromagnetic

---

[2] In 1905 he published on the *Annalen der Physik* three articles, the first on light quanta, the second on Brownian motion, which would definitely confirm the atomicity of matter, the third on the foundations of restricted relativity.

[3] After publishing a few papers, he developed in full form this original idea in his Ph.D. thesis (1924): *Recherches sur la théorie des quanta*.

radiations. Louis de Broglie surmised that not only would light, generally modeled as a wave, sometimes behave as a particle, but also electrons, usually modeled as particles, could at times behave as waves. De Broglie suggested that the key for the description of electrons in terms of wave–particle could be given by the relation

$$\lambda = \frac{h}{mv} \tag{1.3}$$

where λ is the wavelength of the wave associated to the electron, and m e v are, respectively, the mass and the velocity of the electron itself. For example, a wave is associated to an electron moving along a closed orbit around the atomic nucleus. In this particular case, the wave is stationary and its wavelength is linked to mass and velocity by relation (1.3).

We can say that de Broglie's contribution marks the end of the pioneering phase of Quantum Mechanics, whose various phenomena were examined and explained individually, without attempting to formulate a general theory.

1.1.2 The Maturity of Quantum Mechanics

Quantum Mechanics reached maturity in the 1920s and in the 1930s, moving from Quantum Theory to Quantum Mechanics, thanks to the work of Schrödinger, Heisenberg, Dirac, Pauli, and others.

Shortly after de Broglie's conjecture, almost simultaneously, Quantum Mechanics was presented by **Erwin Schrödinger** (1887–1961) and **Werner Heisenberg** (1902–1976).[4] Among the greatest physicists of the century, Schrödinger, stated the fundamental equation of Undulatory Mechanics, known nowadays as Schrödinger's equation

$$H\psi = E\,\psi, \tag{1.4}$$

where ψ is an eigenfunction describing the state of the system, H is an operator, called *Hamiltonian*, and E is the eigenvalue accounting for the system's energy.[5] This equation, stated for non relativistic energies, is the basis for the description of the various phenomena of molecular, atomic, and quantum nuclear physics.

Heisenberg, instead, introduced into Physics the uncertainty of physical entities. His *Uncertainty Principle*, in fact, asserts that it is impossible to know, simultaneously and exactly, couples of physical entities, like position and velocity of a particle. In essence, the more precisely we know the position of a particle, the less information we have on momentum, and vice versa, according to:

---

[4] In 1927, he published on *Zeitschrift fur Physik* his famous paper on the uncertainty principle, entitled: *Über den anschaulichen Inhalt der quanten theoretischen Kinematik und Mechanik.*

[5] Equation (1.4) is Schrödinger's time-independent equation, where ψ is an eigenfunction. Schrödinger's equation can also include the time to take into account system evolution.

$$\Delta x \, \Delta p \geq \frac{h}{4\pi}. \tag{1.5}$$

This principle is of general validity, but it is particularly appreciable at the atomic or subatomic scale.

The statistical laws related to the concept of probability became a reality: uncertainty is a fundamental fact, and the relations connected to the principle evidence an insuperable limit to our knowledge of nature.

> The more precisely the position is determined, the less precisely the momentum is known in this instant, and vice versa. (Heisenberg, *Uncertainty Paper*, 1927)

To conclude this historical note, we find it appropriate to mention the fundamental contribution, albeit indirect, given by the mathematician **David Hilbert** (1862–1943), since the modern version of Quantum Mechanics requires a Hilbert space as mathematical context.

1.2 Revolutionary Concepts of Quantum Mechanics

In describing reality, Quantum Mechanics presents a few concepts that appear revolutionary with respect to Classical Physics, and even seem in contrast with common sense. These concepts will be briefly summarized below.

1.2.1 Randomness

The fundamental difference between *Classical Mechanics* and Quantum Mechanics lies in the fact that, while Classical Mechanics is a deterministic theory, Quantum Mechanics envisages and formalizes indeterminate aspects of reality.

In the mathematical models of Classical Mechanics, once the initial state of a system is known, and so are the forces acting on it, the system's evolution is perfectly predictable and *deterministically measurable*. Resort to probabilistic models is then justified exclusively by the need to account for lack of information on entities characterizing the system.

In Quantum Mechanics, instead, randomness is an intrinsic element of the theory. In fact, it states that the measurements performed on a system, starting from exactly the same initial conditions, may produce different results. This is not due to measurement imprecision, but rather to the fact that the result of any measurement is intrinsically random and must be dealt with the Theory of Probability.

Randomness in Quantum Mechanics is expressed by the fact that the measure of an entity is described by a complex function (the *wave function*), whose squared modulus gives the probability density of the result (intended as a random variable).

1.2.2 Indeterminacy

Another peculiar aspect of Quantum Mechanics is that in any experiment the measurement procedure interferes with the system, altering it. In Classical Physics there is no such problem, because measurement errors can be acknowledged and estimated, but the measurement itself, if accurately performed, does not modify the system. In Quantum Mechanics this is not possible any more, because, as established by the above-mentioned Heisenberg's principle, the accuracy in the knowledge of one quantity (e.g., the position of a particle) inhibits an equal accuracy in the knowledge of another quantity (e.g., velocity). This should be interpreted not only in the sense that two quantities cannot be measured simultaneously with an arbitrary degree of accuracy. As we shall see, they are conceptually undetermined with an uncertainty whose lower bound is given by Heisenberg's inequality.

1.2.3 Complementarity

The above example of position and momentum is a typical case of *conjugate* or *complementary* entities. This corresponds to a distinctive feature of Quantum Mechanics, whose fundamental example is the case of the wave function $\psi(x)$ of position and the wave function of momentum $\tilde{\psi}(p)$: there exists no wave function $\psi(x, p)$ providing a joint statistical description of both entities. The same applies to other couples of complementary variables.

1.2.4 Quantization

Differently from Classical Mechanics, in Quantum Mechanics, the states of a quantum system can only correspond to discrete energy levels. In other words, the granular nature of matter can be extended to energy.

This fact is in good agreement with the requirements of telecommunications, where digital information is represented by quantities that can assume a finite number of values.

1.2.5 Linearity and Superposition

Paradoxically, the states of a quantum system, although characterized by discrete energy levels, have a continuous nature, in the sense that wave functions are continuous functions. In addition, if $\psi(x)$ and $\phi(x)$ are two possible wave functions, also their linear combination $a\,\psi(x) + b\,\phi(x)$, with a and b complex numbers, is still a wave function.

Linearity is then another feature of Quantum Mechanics. The algebraic structure in which its models are represented is constituted by Hilbert spaces, that are linear spaces, and Schrödinger's equation, which governs the evolution of the state, is a linear differential equation.

Linearity and superposition, very simple mathematical concepts, are actually the basis of *Quantum Information and Computation* and have practical consequences of great importance.

1.2.6 Entanglement

The *entanglement* is a phenomenon of Quantum Mechanics in blatant contradiction with physical intuition, as Classical Physics would suggest, to the point that its meaning itself is still open to discussion.

Two particles emitted from the same source, when in the entanglement condition, show strictly correlated characteristics that are preserved even when they move away from each other. And when the state of one of them is measured, the state of the other changes immediately with a "spooky action at a distance," in total contrast with common sense.

1.3 Quantum Information

The natural field of application of Quantum Mechanics is within Physics. However, in the last 20 years (starting from the 1980s) it has exceptionally expanded into the area of Information science and technologies. The main ideas come from the *Postulates* of Quantum Mechanics, which in the last 100 years have never been disproved, and, after a substantial reformulation, envisage extremely innovative applications, like the *quantum computer*, *quantum coding*, *quantum cryptography*, and *quantum communications*. Many of these innovations, consequences of the Postulates, have already had experimental verification and are the subject of a frenzied research activity.

It is worthwhile to introduce these innovations by adding some more historical notes.

1.3.1 The Discovery of Laser and the Theory of Quantum TLC

In the 1960s, after the discovery of laser, **Ronny J. Glauber** of Harvard University formulated the quantum theory of optical coherence [1, 2]. The possibility of producing *coherent* light led Helstrom [3], and other scientists from the Massachusetts

Institute of Technology (MIT), to formulate the Theory of Quantum Telecommunications, that is, a theory where the information is related to quantum states and the analysis and design is based on the rules of Quantum Mechanics. This theory, which we will develop in Chaps. 7 and 8, aimed to realize optical transmissions in free air, as optical fibers were not yet available at that time; unfortunately it did not generate appreciable applications, because the technology was not mature enough, and mostly because the appearance of optical fibers, with their enormous throughput, obscured the interest toward quantum telecommunications. Nevertheless, these pioneering investigations may be considered the beginning of Quantum Information.

Recently, the QTLCs (Quantum Telecommunications) have been vigorously revived at the **Jet Propulsion Laboratory** (JPL) of NASA, where the Deep Space Network is in operation, and, in fact, it is in the area of deep space transmissions that Quantum Communications are expected to play a crucial role. We are dealing, for the time being, with niche applications, but it should be remembered that other fields, like the application of error-correction codes, started precisely at JPL, and they led to fully fledged applications many years later.

1.3.2 *Quantum Information Based on Discrete Quantum Variables. The Qubit*

To understand the motivations that, in the early 1980s, led to studying information in the context of Quantum Mechanics, we can start from *Moore's Law* of electronic circuit technology. As we know, this law, stated by Gordon Moore in 1965, asserts that the complexity of electronic circuits (chips), at equal size, doubles approximately every18 months, and this prediction has been substantially confirmed in the last 50 years. However, it assumes an indefinite reduction in the size of components, down to the limit of atomic dimensions, where quantum effects become predominant. At this point, a natural development is to try to reformulate Information Theory in the framework of Quantum Mechanics. Following this line of thought, **Benioff**, **Manin**, and **Feynman** postulated the idea of a Quantum Computer, for the simulation of Quantum Systems. Differently from the classical computer, which, as is well known, is a power-consuming device, the quantum computer, in theory, does not require power consumption (this theoretical possibility had already been demonstrated by **Charles Bennett** within IBM). Subsequently, in 1985 **David Deutsch** proved that a Quantum Computer can naturally operate in parallel mode (quantum parallelism), in the sense that it makes it possible to evaluate any function $f(x)$, for every value of x, in a single step. With this parallelism, the theoretical superiority of the quantum computer with respect to the conventional one was demonstrated.

Still around those times, **Charles Bennett** and **Gilles Brassard** explored the possibility of secure information transmission based on the laws of Quantum Mechanics. The principle is related to *quantum measurements* (Postulate 3 of Quantum Mechanics) according to which, if the information is intercepted, the receiver is automatically and securely alerted. This marks the birth of *Quantum Cryptography*. On the other hand, in 1991 **Arthur Eckert** proposes another form of secure transmission based on entanglement, a phenomenon predicted by Postulate 4 of Quantum Mechanics.

In any case, Quantum Cryptography, as a *quantum key distribution*, is one of the most concrete applications of Quantum Mechanics in the information area, in that it already shows significant implementations.

The phenomenon of entanglement, typical of quantum mechanics, and totally unforeseen by the classical theory, gave origin to another research thread: *Superdense Coding*, according to which, by sending a single bit of quantum information (qubit), two bits of classical information can be transmitted. This originated a very promising new field, *Quantum Coding*, steadily growing, as witnessed by the numerous papers appearing on the IEEE Trans. on Information Theory. It should be noticed that, in this context, Shannon's Information Theory is being reviewed, giving way to *Quantum Information Theory*. Superdense Coding was invented by Bennett and Wiesner [4] and experimentally implemented by Mattle et al. [5].

Bennett et al. [6] found another use of entanglement, *quantum teleportation*, in which separate experiments sharing two halves of entangled systems can make use of entanglement to transfer a quantum state from one to another using only classical communications. Teleportation was later experimentally realized by Boschi et al. [7] using optical techniques and by Bouwmeester et al. [8] using photon polarization.

Going back to Quantum Computing, we must mention the milestone achieved by **Peter Shor** of AT&T in 1994, who demonstrated that a Quantum Computer can decompose an integer number into prime factors with polynomial complexity, whereas it is conjectured that the classic computer requires exponential complexity. It is an alarming discovery, because the majority of current cryptographic security systems are based on the (exponential) difficulty of prime factor decomposition. On the other hand, this confirms the importance of investing in ideas and resources on Quantum Cryptography.

The above history (1990–2010) on quantum computers, quantum cryptography, and quantum teleportation refers to the manipulation of individual quanta of information, known as quantum bits or *qubits*; in other words, based on *discrete* quantum variables.

1.3.3 Quantum Information Based on Continuous Quantum Variables

Very recently it was realized that the use of *continuous* quantum variables, instead of qubits, represents a powerful alternative to quantum information processing [9]. In this context, on the theoretical side, simple analytical tools are available (Gaussian states, Gaussian operators, and Gaussian measurements) and, on the practical side, the corresponding laboratory implementation is readily available. Hence, the continuous state approach opens the way to a variety of tasks and applications, in competition with the discrete state approach. Furthermore, these new possibilities provide a new challenge to the implementation of Quantum Communications systems.

In conclusion, Quantum Information comes in two forms, discrete and continuous. From a historic viewpoint, the continuous form was developed in pioneering works for Quantum Communications systems (1970) and the discrete form in the last two decades, but now continuous and discrete forms are in competition.

1.4 Content of the Book

This book is a collection of ideas for an "educational experiment" on the teaching of Quantum Information and particularly Quantum Telecommunications to students of the Departments of Engineering and Physics, hence with the twofold objective of opening a cross-disciplinary field of study and possibly providing a common background for scientific collaboration.

Part I: Fundamentals

Chapter 2: Hilbert Spaces

This chapter contains the mathematical foundations required to understand Quantum Mechanics, which develops over Hilbert spaces on complex numbers. Many notions (vector spaces, and inner product vector spaces) are already known to students, others, like the spectral decomposition of a Hermitian operator, are less known and represent a fundamental subject in Quantum Measurements.

In any case, the collection provides a run-through and a symbolism acquisition, useful to come to grips with the subsequent subjects.

Chapter 3: Elements of Quantum Mechanics

The formulation of these elements is presented following in sequence the four *Postulates* of Quantum Mechanics. The development is partly parallel to Nielsen and Chuang's book [10]. However, herein to the four postulates are given different emphasis; in particular, Postulate 3 on Quantum Measurements is developed in great detail, as it represents the most interesting part with respect to Quantum Communications.

Part II: Quantum Communications

Chapter 4: Introduction to Quantum Communications

The general foundations of telecommunications systems are introduced and the difference between Classical and Quantum Communications systems is explained.

In the second part of the chapter we introduce the foundations of *optical classical communications*, which is the necessary prologue to *optical quantum communications* developed in the subsequent chapters. The mathematical framework is given by Poisson processes, and more specifically by doubly stochastic Poisson processes.

Chapter 5: Quantum Decision Theory: Analysis and Optimization

Only data transmission is considered, starting from the description and analysis of a general scheme, shown in Fig. 1.1. For a general K-ary system, the rules are given to calculate the transition probabilities and the error probabilities, obviously in terms of quantum parameters. Then we develop, in a fully general way, the best choice of quantum measurements that minimize the error probability (optimization).

Two important topics are also introduced: the geometrically uniform symmetry (GUS) of a constellation of states and the compression of quantum states.

Chapter 6: Quantum Decision Theory: Suboptimization

Optimization in quantum decision is very difficult, and exact solutions are only known in few cases. To overcome such a difficulty *suboptimization* is considered. In quantum communications the most important suboptimal decision is called square-root measurement (SRM), because its solution is based on the square root of an operator. Particularly attractive is the SRM combined with the GUS of quantum states.

Chapter 7: Quantum Communications Systems

In this chapter, the general Quantum Decision Theory is applied to systems in which digital information is carried by the monochromatic radiation produced by a laser (*coherent states*). As a preliminary, *classic* optical telecommunications systems are outlined in order to provide the background and the inspiration for the transition from classic to *quantum* optical telecommunication systems. The quantum version differs mainly at the receiver, where the analysis and the design are carried out using the Postulates of Quantum Mechanics.

Then the theory is explicitly applied to the more popular systems, like OOK (on off keying), PSK (phase shift keying), PPM (pulse position modulation), and QAM (quadrature amplitude modulation). Anyhow, we shall eventually demonstrate the *net gain in terms of performance that can be obtained by the quantum versions compared to the classic schemes*.

Chapter 8: Quantum Communications Systems in the Presence of Thermal Noise

In the analysis of Chap. 7, the background noise (or thermal noise) is neglected, and the uncertainty of the result (the message) is only due to the randomness arising in

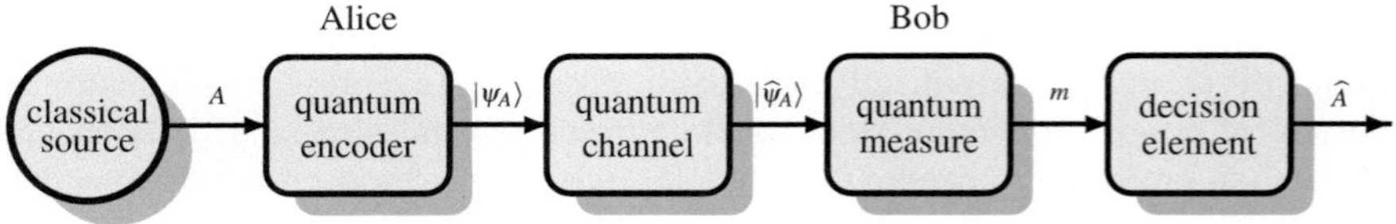

Fig. 1.1 Quantum Communications system for digital transmission. A symbol to be transmitted, $|\psi_A\rangle$ quantum state prepared by Alice, $|\widehat{\psi}_A\rangle$ received quantum state, m outcome of the quantum measurement, and $\widehat{A}$ decided symbol

quantum measurements. In this chapter, the analysis of the main quantum transmission systems, developed in Chap. 7, takes into account background noise, which is always present in real-world systems.

Chapter 9: Implementation of Quantum Communications Systems

While Quantum Communications theory is reaching a steady state, the implementation of the corresponding systems is still at an early stage. The chapter describes the implementations realized so far around the world, a few promising ideas, and some open problems.

Part III: Quantum Information

Chapter 10: Introduction to Quantum Information

Quantum Information exhibits two forms, discrete and continuous. Discrete quantum information is based on *discrete variables*, the best known example of which is the quantum bit or, briefly, *qubit*. Continuous quantum information is based on *continuous variables*, the best known example of which is provided by the quantized harmonic oscillator, which represents the fundamental tool in quantum optics and is the basis for the introduction of coherent states and more generally of Gaussian states.

An important remark is that most of the operations in quantum information processing can be carried out both with discrete and continuous variables (this last possibility is a quite recent discovery).

Chapter 11: Fundamentals of Quantum Continuous Variables

In Quantum Mechanics formulation of Chaps. 2 and 3 we have considered some fundamentals, as bases, eigendecompositions, measurements, and operators, in the *discrete* case. Specifically, we assumed the bases consisting of finite or enumerable sets of vectors, the operator eigendecompositions having a finite or enumerable spectrum, and quantum measurements having a finite set (alphabet) of possible outcomes. This formulation was sufficient because in the subsequent chapters we limit ourselves to the development of *digital* Quantum Communications.

In this chapter, for a full development of Quantum Information, we extend the above fundamentals to the continuous case, where the sets become a continuum. A particular relevance is given to Gaussian states and Gaussian transformations.

Chapter 12: Quantum Information Theory

Information Theory was born in the field of Telecommunication in 1948 with the revolutionary ideas developed by Shannon [11]. Its purpose is mainly: (1) to define *information* mathematically and quantitatively, (2) to represent information in an efficient way (data compression) for storage and transmission, and (3) to ensure information protection (encoding) in the presence of noise and other impairments. Recently, with the interest in quantum information processing, Information Theory was extended to Quantum Mechanics. Of course, Quantum Information Theory, is based on quantum mechanical principles and in particular on its intriguing phenomena, like entanglement.

The chapter provides an overview of Quantum Information Theory starting from Classical Information Theory, which represents a necessary preliminary. Thus, each of the three items listed above are developed in the framework of Quantum Mechanics, starting from the classical case.

Chapter 13: Applications of Quantum Information

The list of topics that will be developed in this chapter is:

- quantum random number generation,
- quantum key distribution,
- teleportation,

considered with both discrete and continuous variables.

1.5 Suggested Paths

As mentioned in the Preface, the book is meant to address readers from Physics and Telecommunications, both graduate students and researchers, providing that they are familiar with Linear Vector Spaces and Probability Theory. In order to account for the different backgrounds and academic levels, two different paths through the book are suggested, as illustrated in Fig. 1.2, with the indication of the difficulties[6] probably encountered in each chapter.

Graduate students should begin by checking their mathematical background while studying carefully Chap. 2, and solving some specific exercises to get familiarity and confidence with the topic. In the study of Part III, they can skip, at least at the first reading, the description of quantum systems in terms of *density operators*. In fact, the formulation in terms of *pure states* is adequate to tackle the essence of Quantum Communications and the comparison with classic optical systems. Therefore Chap. 8 can be completely omitted (the content of this chapter may be regarded as a very advanced topic). Once completed the comprehension of Part II, students will have reached a reasonable and adequate mastering level on the subject. But they might as well consider moving on to the more advanced topics of Part III, if they have enough time and spirit of inquiry.

Researchers could avoid the study of Part I (or they could quickly browse through it to acquire the symbolism and references for the next chapter). They will have to study simultaneously the developments based on pure-state and density-operator representations. In particular, Chap. 8, which is very advanced, may offer them stimulating hints for original research. Eventually, they will complete their path with the last three chapters.

---

[6] Of course, the difficulty scale strongly depends on the preparation and on the personality of the reader.

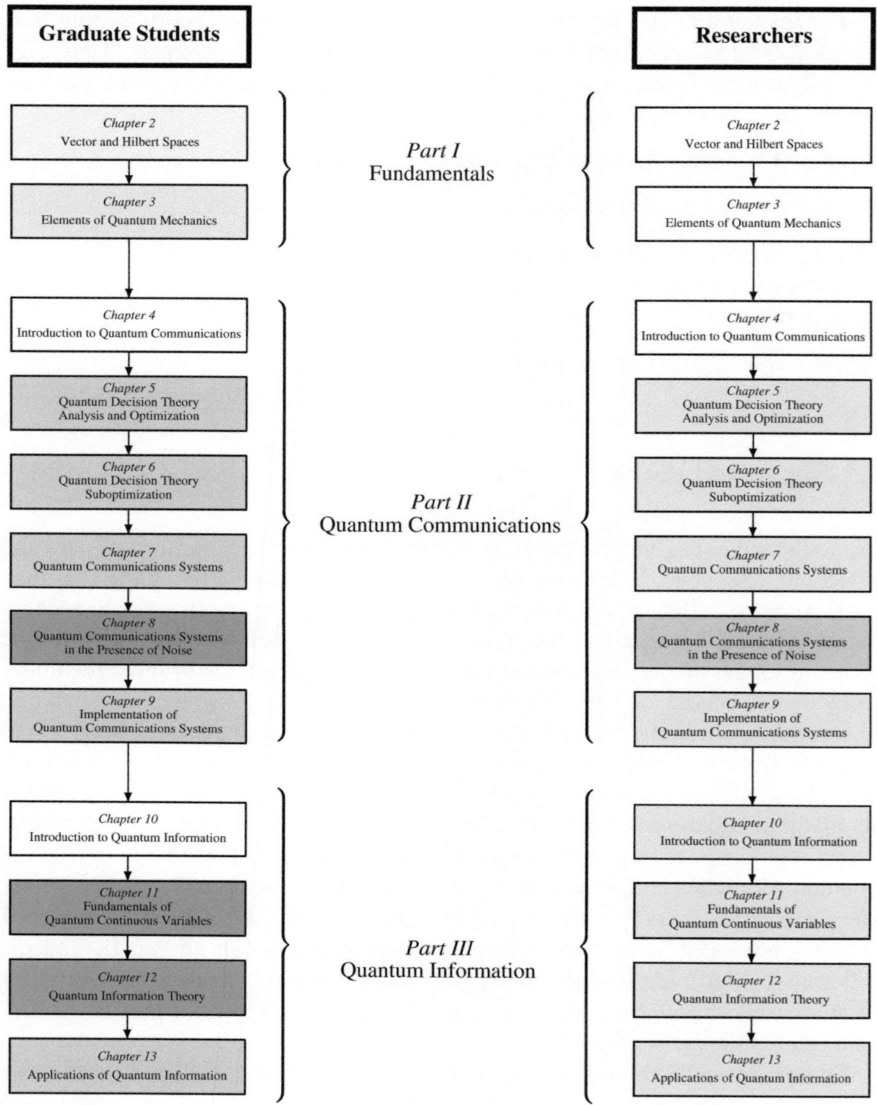

Fig. 1.2   The two suggested paths with the difficulties indicated by a gray level in the blocks

1.6 Conventions on Notation

Sections of advanced topics that can be omitted at the first reading are marked by
$\Downarrow$. Problems are marked by asterisks indicating difficulty (* = easy, ** = medium,
*** = difficult). Sections and problems marked with the symbol ∇ require notions
that are developed further on.

Throughout the book, notations are explicitly specified at the first use and are frequently recalled. Matrices and operators are denoted by uppercase letters, e.g., A, Ψ. For quantum states, Dirac's notation *bra* and *ket* is used, as $\langle x|$ and $|x\rangle$.

List of Symbols

$:=$	Equal by definition
$\otimes$	Tensor product
$\oplus$	Direct sum
$\mathrm{Tr}[\cdot]$	Trace
$\mathrm{Tr}_A[\cdot]$	Trace over the subsystem A
$\mathrm{E}[\cdot]$	Expectation (of a random variable)
$\mathrm{P}[\cdot]$	Probability (of an event)
$q_i := \mathrm{P}[A = i]$	Source probabilities
$p_c(j|i)$ or $p(j|i)$	Transition probabilities
$\mathcal{V}$	Vector space
$\mathcal{H}$	Hilbert space
$\mathbb{Z}$	Set of integer numbers
$\mathbb{R}$	Set of real numbers
$\mathbb{C}$	Set of complex numbers
$\mathcal{A}$	Alphabet (source)
$\mathcal{M}$	Alphabet of a quantum measurement
$|x\rangle$	Ket
$\langle x|$	Bra
$\langle x|y\rangle$	Inner product of vectors $|x\rangle$ and $|y\rangle$
$|x\rangle\langle y|$	Outer product of vectors $|x\rangle$ and $|y\rangle$
$|x\rangle \perp |y\rangle$	$|x\rangle$ and $|y\rangle$ are orthogonal ($\langle x|y\rangle = 0$)
$\|x\|$	Norm of vector $|x\rangle$
$[x_{ij}]$	Matrix with entries x_{ij}
$[A, B], \{A, B\}$	Commutator and anticommutator of operators A and B
$I_{\mathcal{H}}$	Identity operator of $\mathcal{H}$
I_n	Identity matrix of size n
$|z|$	Absolute value of complex number z
$|\mathcal{A}|$	Dimension of set $\mathcal{A}$
z^*	Conjugate of complex number z
A^*	Adjoint of operator A or conjugate transpose of matrix A
A^{T}	Transpose of matrix A
a, a^*	Annihilator and creation operators
q, p	Quadrature operators
δ_{ij}	Kronecker's symbol
$\delta(x)$	Dirac delta function
h	Planck's constant

$\hbar := h/(2\pi)$ Reduced Planck's constant
k Boltzmann's constant
$W_N := \mathrm{e}^{\mathrm{i}\,2\pi/N}$ Nth radix of unity
$W_{[N]}$ DFT matrix of order N

- for the list of symbols on Continuous Variables, see the beginning of Chap. 11
- for the list of symbols on Information Theory, see the beginning of Chap. 12

List of Acronyms

A/D	Analog-to-digital
D/A	Digital-to-analog
CFT	Complex Fourier transform
CSP	Convex semidefinite programming
DFT	Discrete Fourier transform
EID	Eigendecomposition
EPR	Einstein-Podolsky-Rosen
FT	Fourier transform
GUS	Geometrically uniform symmetry
IID	Independent Identically Distributed
LMI	Linear matrix inequality
OOK	On–off keying
POVM	Positive Operator-Valued Measurements
PSD	Positive semidefinite
PSK	Phase shift keying
BPSK	Binary PSK
PPM	Pulse position modulation
QAM	Quadrature amplitude modulation
QKD	Quantum Key Distribution
SNR	Signal to noise ratio
SRM	Square root measurement
SVD	Singular-value decomposition
TLC	Telecommunications

References

1. K.E. Cahill, R.J. Glauber, Ordered expansions in Boson amplitude operators. Phys. Rev. **177**, 1857–1881 (1969)
2. R.J. Glauber, The quantum theory of optical coherence. Phys. Rev. **130**, 2529–2539 (1963)
3. C.W. Helstrom, J.W.S. Liu, J.P. Gordon, Quantum-mechanical communication theory. Proc. IEEE **58**(10), 1578–1598 (1970)
4. C.H. Bennett, F. Bessette, G. Brassard, L. Salvail, J. Smolin, Experimental quantum cryptography. J. Cryptol. **5**(1), 3–28 (1992)
5. K. Mattle, H. Weinfurter, P.G. Kwiat, A. Zeilinger, Dense coding in experimental quantum communication. Phys. Rev. Lett. **76**, 4656–4659 (1996)

6. C.H. Bennett, G. Brassard, C. Crépeau, R. Jozsa, A. Peres, W.K. Wootters, Teleporting an unknown quantum state via dual classical and Einstein-Podolsky-Rosen channels. Phys. Rev. Lett. **70**, 1895–1899 (1993)
7. D. Boschi, S. Branca, F. De Martini, L. Hardy, S. Popescu, Experimental realization of teleporting an unknown pure quantum state via dual classical and Einstein-Podolsky-Rosen channels. Phys. Rev. Lett. **80**, 1121–1125 (1998)
8. D. Bouwmeester, J.W. Pan, K. Mattle, M. Eibl, H. Weinfurter, A. Zeilinger, Experimental quantum teleportation. Nature **390**, 575–579 (1997)
9. C. Weedbrook, S. Pirandola, R. Garcí a Patrón, N.J. Cerf, T.C. Ralph, J.H. Shapiro, S. Lloyd, Gaussian quantum information. Rev. Mod. Phys. **84**, 621–669 (2012)
10. M.A. Nielsen, I.L. Chuang, *Quantum Computation and Quantum Information* (Cambridge University Press, Cambridge, 2000)
11. C.E. Shannon, A mathematical theory of communication. Bell Syst. Tech. J. **27**(3), 379–423 (1948)

Part I
Fundamentals

Chapter 2
Vector and Hilbert Spaces

2.1 Introduction

The purpose of this chapter is to introduce Hilbert spaces, and more precisely the *Hilbert spaces* on the field of complex numbers, which represent the abstract environment in which Quantum Mechanics is developed.

To arrive at Hilbert spaces, we proceed gradually, beginning with spaces mathematically less structured, to move toward more and more structured ones, considering, in order of complexity:

(1) *linear* or *vector* spaces, in which the points of the space are called *vectors*, and the operations are the *sum* between two vectors and the *multiplication by a scalar*;
(2) *normed vector spaces*, in which the concept of *norm* of a vector x is introduced, indicated by $||x||$, from which one can obtain the distance between two vectors x and y as $d(x, y) = ||x - y||$;
(3) *vector spaces with inner product*, in which the concept of *inner product* between two vectors x, y is introduced, and indicated in the form (x, y), from which the norm can be obtained as $||x|| = (x, x)^{1/2}$, and then also the distance $d(x, y)$;
(4) *Hilbert spaces*, which are vector spaces with inner product, with the additional property of *completeness*.

We will start from vector spaces, then we will move on directly to vector spaces with inner product and, eventually, to Hilbert spaces. For vectors, we will initially adopt the standard notation $(x, y,$ etc.), and subsequently we will switch to Dirac's notation, which has the form $|x\rangle$, $|y\rangle$, etc., universally used in Quantum Mechanics.

© Springer International Publishing Switzerland 2015

G. Cariolaro, *Quantum Communications*, Signals and Communication Technology,

DOI 10.1007/978-3-319-15600-2_2

2.2 Vector Spaces

2.2.1 Definition of Vector Space

A vector space on a field $\mathbb{F}$ is essentially an Abelian group, and therefore a set provided with the addition operation $+$, but completed with the operation of multiplication by a scalar belonging to $\mathbb{F}$.

Here we give the definition of vector space in the field of complex numbers $\mathbb{C}$, as it is of interest to Quantum Mechanics.

Definition 2.1 A *vector space* in the field of complex numbers $\mathbb{C}$ is a nonempty set $\mathcal{V}$, whose elements are called *vectors*, for which two operations are defined. The first operation, *addition*, is indicated by $+$ and assigns to each pair $(x, y) \in \mathcal{V} \times \mathcal{V}$ a vector $x + y \in \mathcal{V}$. The second operation, called multiplication by a scalar or simply *scalar multiplication*, assigns to each pair $(a, x) \in \mathbb{C} \times \mathcal{V}$ a vector $ax \in \mathcal{V}$. These operations must satisfy the following properties, for $x, y, z \in \mathcal{V}$ and $a, b \in \mathbb{C}$:

(1) $x + (y + z) = (x + y) + z$ (associative property),
(2) $x + y = y + x$ (commutative property),
(3) $\mathcal{V}$ contains an identity element 0 with the property $0 + x = x$, $\forall x \in \mathcal{V}$,
(4) $\mathcal{V}$ contains the opposite (or inverse) vector $-x$ such that $-x + x = 0$, $\forall x \in \mathcal{V}$,
(5) $a(x + y) = ax + ay$,
(6) $(a + b)x = ax + bx$. $\square$

Notice that the first four properties assure that $\mathcal{V}$ is an *Abelian group* or commutative group, and, globally, the properties make sure that every linear combination

$$a_1 x_1 + a_2 x_2 + \cdots + a_n x_n \qquad a_i \in \mathbb{C}, \ x_i \in \mathcal{V}$$

is also a vector of $\mathcal{V}$.

2.2.2 Examples of Vector Spaces

A first example of a vector space on $\mathbb{C}$ is given by $\mathbb{C}^n$, that is, by the set of the n-tuples of complex numbers,

$$x = (x_1, x_2, \ldots, x_n) \quad \text{with} \quad x_i \in \mathbb{C}$$

where scalar multiplication and addition must be intended in the usual sense, that is,

$$ax = (ax_1, ax_2, \ldots, ax_n), \qquad \forall a \in \mathbb{C}$$

$$x + y = (x_1 + y_1, x_2 + y_2, \ldots, x_n + y_n).$$

A second example is given by the sequence of complex numbers

$$x = (x_1, x_2, \ldots, x_i, \ldots) \quad \text{with} \quad x_i \in \mathbb{C}.$$

In the first example, the vector space is *finite dimensional*, in the second, it is *infinite dimensional* (further on, the concept of dimension of a vector space will be formalized in general).

A third example of vector space is given by the class of continuous-time or discrete-time, and also multidimensional, signals (complex functions). We will return to this example with more details in the following section.

2.2.3 Definitions on Vector Spaces and Properties

We will now introduce the main definitions and establish a few properties of vector spaces, following Roman's textbook [1].

Vector Subspaces

A nonempty subset S of a vector space $\mathcal{V}$, itself a vector space provided with the same two operations on $\mathcal{V}$, is called a *subspace* of $\mathcal{V}$. Therefore, by definition, S is *closed* with respect to the linear combinations of vectors of S.

Notice that $\{0\}$, where 0 is the identity element of $\mathcal{V}$, is a subspace of $\mathcal{V}$.

Generator Sets and Linear Independence

Let S_0 be a nonempty subset of $\mathcal{V}$, not necessarily a subspace; then the set of all the linear combinations of vectors of S_0 *generates a subspace* S of $\mathcal{V}$, indicated in the form

$$S = \text{span} \, (S_0) = \{a_1 x_1 + a_2 x_2 + \cdots + a_n x_n \mid a_i \in \mathbb{C}, \; x_i \in S_0\}. \tag{2.1}$$

In particular, the generator set S_0 can consist of a single point of $\mathcal{V}$. For example, in $\mathbb{C}^2$, the set $S_0 = \{(1, 2)\}$ consisting of the vector $(1, 2)$, generates $S = \text{span} \, (S_0) = \{a(1, 2) \mid a \in \mathbb{C}\} = \{(a, 2a) \mid a \in \mathbb{C}\}$, which represents a straight line passing through the origin (Fig. 2.1); it can be verified that S is a subspace of $\mathbb{C}^2$. The set $S_0 = \{(1, 2), (3, 0)\}$ generates the entire $\mathbb{C}^2$, that is,[1]

$$\text{span} \, ((1, 2), (3, 0)) = \mathbb{C}^2.$$

The concept of linear independence of a vector space is the usual one. A set $S_0 = \{x_1, x_2, \ldots, x_n\}$ of vectors of $\mathcal{V}$ is *linearly independent*, if the equality

---

[1] If S_0 is constituted by some points, for example $S_0 = \{x_1, x_2, x_3\}$, the notation $\text{span}(S_0) = \text{span}(\{x_1, x_2, x_3\})$ is simplified to $\text{span}(x_1, x_2, x_3)$.

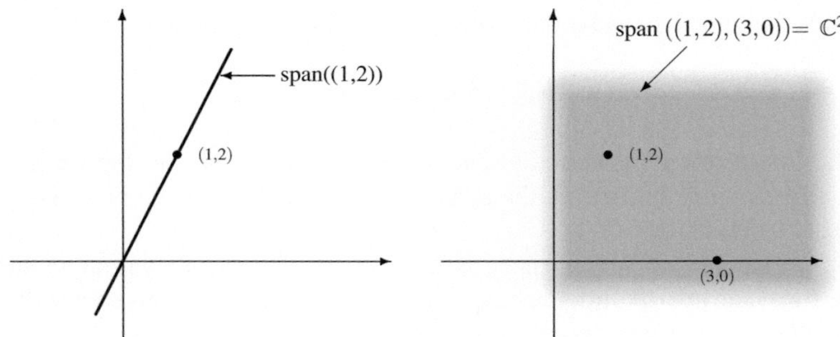

Fig. 2.1 The set $\{(1, 2)\}$ of $\mathbb{C}^2$ generates a straight line trough the origin, while the set $\{(3, 0), (1, 2)\}$ generates $\mathbb{C}^2$ (for graphical reason the representation is limited to $\mathbb{R}^2$)

$$a_1 x_1 + a_2 x_2 + \cdots + a_n x_n = 0 \tag{2.2}$$

implies

$$a_1 = 0, \; a_2 = 0, \; \ldots, \; a_n = 0.$$

Otherwise, the set is *linearly dependent*. For example, in $\mathbb{C}^2$ the set $\{(1, 2), (0, 3)\}$ is constituted by two linear independent vectors, whereas the set $\{(1, 2), (2, 4)\}$ is linearly dependent because

$$a_1(1, 2) + a_2(2, 4) = (0, 0) \quad \text{for} \quad a_1 = 2 \; \text{e} \; a_2 = -1.$$

2.2.4 Bases and Dimensions of a Vector Space

A subset $\mathcal{B}$ of a vector space $\mathcal{V}$ constituted by linearly independent vectors is *a basis of* $\mathcal{V}$ if $\mathcal{B}$ generates $\mathcal{V}$, that is, if two conditions are met:

(1) $\mathcal{B} \subset \mathcal{V}$ is formed by linearly independent vectors,
(2) span $(\mathcal{B}) = \mathcal{V}$.

It can be proved that [1, Chap.1]:

(a) Every vector space $\mathcal{V}$, except the degenerate space $\{0\}$, admits a basis $\mathcal{B}$.
(b) If $b_1, b_2, \ldots, b_n$ are vectors of a basis $\mathcal{B}$ of $\mathcal{V}$, the linear combination

$$a_1 b_1 + a_2 b_2 + \cdots + a_n b_n = x \tag{2.3}$$

is unique, i.e., the coefficients $a_1, a_2, \ldots, a_n$, are uniquely identified by x.
(c) All the bases of a vector space have the same *cardinality*. Therefore, if $\mathcal{B}_1$ and $\mathcal{B}_2$ are two bases of $\mathcal{V}$, it follows that $|\mathcal{B}_1| = |\mathcal{B}_2|$.

The property (c) is used to define the *dimension* of a vector space $\mathcal{V}$, letting

$$\dim \mathcal{V} := |\mathcal{B}|. \tag{2.4}$$

Then the dimension of a vector space is given by the common cardinality of its bases. In particular, if $\mathcal{B}$ is finite, the vector space $\mathcal{V}$ is of *finite dimension*; otherwise $\mathcal{V}$ is of *infinite dimension.*

In $\mathbb{C}^n$ the *standard basis* is given by the n vectors

$$(1, 0, \ldots, 0), \quad (0, 1, \ldots, 0), \quad \ldots, \quad (0, 0, \ldots, 1). \tag{2.5}$$

Therefore, $\dim \mathbb{C}^n = n$. We must observe that in $\mathbb{C}^n$ there are infinitely many other bases, all of cardinality n.

In the vector space consisting of the sequences $(x_1, x_2, \ldots)$ of complex numbers, the standard basis is given by the vectors

$$(1, 0, 0, \ldots), \quad (0, 1, 0, \ldots), \quad (0, 0, 1, \ldots), \quad \ldots \tag{2.6}$$

which are infinite. Therefore this space is of infinite dimension.

2.3 Inner-Product Vector Spaces

2.3.1 Definition of Inner Product

In a vector space $\mathcal{V}$ on complex numbers, the *inner product*, here indicated by the symbol $\langle \cdot, \, \cdot \rangle$, is a function

$$\langle \cdot, \, \cdot \rangle \; : \; \mathcal{V} \times \mathcal{V} \; \rightarrow \; \mathbb{C}$$

with the following properties, for $x, y, z \in \mathcal{V}$ and $a, b \in \mathbb{C}$:

(1) it is a positive definite function, that is,

$$\langle x, x \rangle \geq 0 \quad \text{and} \quad \langle x, x \rangle = 0 \quad \text{if and only if} \quad x = 0;$$

(2) it enjoys the Hermitian symmetry

$$\langle x, y \rangle = \langle y, x \rangle^*;$$

(3) it is linear with respect to the first argument

$$\langle ax + by, z \rangle = a\langle x, z \rangle + b\langle y, z \rangle. \qquad \qquad \square$$

From properties (2) and (3) it follows that with respect to the second argument the so-called *conjugate linearity* holds, namely

$$\langle z, ax + by \rangle = a^* \langle z, x \rangle + b^* \langle z, y \rangle.$$

We observe that within the same vector space $\mathcal{V}$ it is possible to introduce different inner products, and the choice must be made according to the application of interest.

2.3.2 Examples

In $\mathbb{C}^n$, the standard form of inner product of two vectors $x = (x_1, x_2, \ldots, x_n)$ and $y = (y_1, y_2, \ldots, y_n)$ is defined as follows:

$$\langle x, y \rangle = x_1 y_1^* + \cdots + x_n y_n^* = \sum_{i=1}^{n} x_i y_i^* \tag{2.7a}$$

and it can be easily seen that such expression satisfies the properties (1), (2), and (3). Interpreting the vectors $x \in \mathbb{C}^n$ as column vectors ($n \times 1$ matrices), and indicating with y^* the conjugate transpose of y ($1 \times n$ matrix), that is,

$$x = \begin{bmatrix} x_1 \\ \vdots \\ x_n \end{bmatrix}, \qquad y^* = \begin{bmatrix} y_1^*, & \ldots, & y_n^* \end{bmatrix} \tag{2.7b}$$

and applying the usual matrix product, we obtain

$$\langle x, y \rangle = y^* x = x_1 y_1^* + \cdots + x_n y_n^* \tag{2.7c}$$

a very handy expression for algebraic manipulations.

The most classic example of infinite-dimensional inner-product vector space, introduced by Hilbert himself, is the space ℓ_2 of the square-summable complex sequences $x = (x_1, x_2, \ldots)$, that is, with

$$\sum_{i=1}^{\infty} |x_i|^2 < \infty \tag{2.8}$$

where the standard inner product is defined by

$$\langle x, y \rangle = \sum_{i=1}^{\infty} x_i y_i^* = \lim_{n \to \infty} \sum_{i=1}^{n} x_i y_i^*.$$

The existence of this limit is ensured by Schwartz's inequality (see (2.12)), where (2.8) is used.

Another example of inner-product vector space is given by the continuous functions over an interval $[a, \ b]$, where the standard inner product is defined by

$$\langle x, \ y \rangle = \int_a^b x(t)\, y^*(t)\, \mathrm{d}t.$$

2.3.3 Examples from Signal Theory

These examples are proposed because they will allow us to illustrate some concepts on vector spaces, in view of the reader's familiarity with the subject.

We have seen that the class of signals $s(t)$, $t \in I$, defined on a domain I, form a vector space. If we limit ourselves to the signals $\mathcal{L}_2(I)$, for which it holds that [2]

$$\int_I \mathrm{d}t \, |s(t)|^2 < \infty, \tag{2.9}$$

we can obtain a space with inner product defined by

$$\langle x, \ y \rangle = \int_I \mathrm{d}t \, x(t)\, y^*(t) \tag{2.10}$$

which verifies conditions (1), (2), and (3).

A first concept that can be exemplified through signals is that of a *subspace*. In the space $\mathcal{L}_2(I)$, let us consider the subspace $\mathcal{E}(I)$ formed by the *even* signals. Is $\mathcal{E}(I)$ a subspace? The answer is yes, because every linear combination of even signals is an even signal: therefore $\mathcal{E}(I)$ is a subspace of $\mathcal{L}_2(I)$. The same conclusion applies to the class $\mathcal{O}(I)$ of odd signals. These two subspaces are illustrated in Fig. 2.2.

2.3.4 Norm and Distance. Convergence

From the inner product it is possible to define the norm $||x||$ of a vector $x \in \mathcal{V}$ through the relation

$$||x|| = \sqrt{\langle x, \ x \rangle}. \tag{2.11}$$

Intuitively, the norm may be thought of as representing the *length* of the vector. A vector with unit norm $||x|| = 1$, is called *unit vector* (we anticipate that in Quantum Mechanics only unit vectors are used). In terms of inner product and norm, we can

---

[2] To proceed in unified form, valid for all the classes $\mathcal{L}_2(I)$, we use the Haar integral (see [2]).

Fig. 2.2 Examples of *vector subspaces* of the signal class $\mathcal{L}_2(I)$

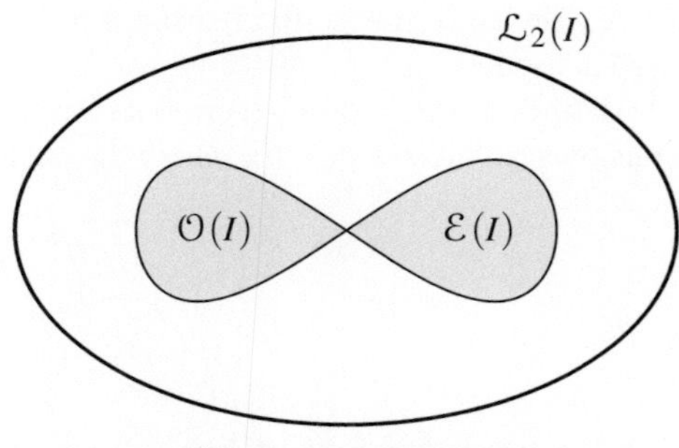

$\mathcal{E}(I)$: class of even signals

$\mathcal{O}(I)$: class of odd signals

write the important *Schwartz's inequality*

$$|\langle x, \, y \rangle| \leq ||x|| \, ||y|| \tag{2.12}$$

where the equal sign holds if and only if y is proportional to x, that is, $y = kx$ for an appropriate $k \in \mathbb{C}$.

From (2.11) it follows that an inner-product vector space is also a normed space, with the norm introduced by the inner product.

In an inner-product vector space we can also introduce the distance $d(x, y)$ between two points $x, y \in \mathcal{V}$, through the relation

$$d(x, y) = ||x - y|| \tag{2.13}$$

and we can verify that this parameter has the properties required by distance in metric spaces, in particular the triangular inequality holds

$$d(x, y) \leq d(x, z) + d(y, z). \tag{2.14}$$

So an inner-product vector space is also a metric space.

Finally, the inner product allows us to introduce the concept of *convergence*. A sequence $\{x_n\}$ of vectors of $\mathcal{V}$ *converges* to the vector x if

$$\lim_{n \to \infty} d(x_n, x) = \lim_{n \to \infty} ||x_n - x|| = 0. \tag{2.15}$$

Now, suppose that a sequence $\{x_n\}$ has the property (*Cauchy's sequence* or *fundamental sequence*)

$$d(x_m, x_n) \to 0 \quad \text{for} \quad m, n \to \infty.$$

In general, for such a sequence, the limit (2.15) is not guaranteed to exist, and, if it exists, it is not guaranteed that the limit x is a vector of $\mathcal{V}$. So, an inner-product vector space in which all the Cauchy sequences converge to a vector of $\mathcal{V}$ is said to be *complete*. At this point we have all we need to define a Hilbert space.

2.4 Definition of Hilbert Space

Definition 2.2 A Hilbert space is a *complete* inner-product vector space.

It must be observed that a *finite* dimensional vector space is always complete, as it is closed with respect to all its sequences, and therefore it is always a Hilbert space. Instead, if the space is *infinite* dimensional, the completeness is not ensured, and therefore it must be added as a hypothesis, in order for the inner-product vector space to become a Hilbert space.

At this point, we want to reassure the reader: the theory of optical quantum communications will be developed at a level that will not fully require the concept of a Hilbert space, but the concept of inner-product vector space will suffice. Nonetheless, the introduction of the Hilbert space is still done here for consistency with the Quantum Mechanics literature.

From now on, we will assume to operate on a Hilbert space, but, for what we just said, we can refer to an inner-product vector space.

2.4.1 Orthogonality, Bases, and Coordinate Systems

In a Hilbert space, the basic concepts, introduced for vector spaces, can be expressed by using orthogonality.

Let $\mathcal{H}$ be a Hilbert space. Then two vectors $x, y \in \mathcal{H}$ are *orthogonal* if

$$\langle x, \ y \rangle = 0. \tag{2.16}$$

Extending what was seen in Sect. 2.2, we have that a Hilbert space admits orthogonal bases, where each basis

$$\mathcal{B} = \{b_i, i \in I\} \tag{2.17}$$

is formed by pairwise orthogonal vectors, that is,

$$\langle b_i, b_j \rangle = 0 \qquad i, j \in I, \ i \neq j$$

and furthermore, $\mathcal{B}$ generates $\mathcal{H}$

$$\mathrm{span}(\mathcal{B}) = \mathcal{H}.$$

The set I in (2.17) is finite, $I = \{1, 2, \ldots, n\}$, or countably infinite, $I = \{1, 2, \ldots\}$, and may even be a continuum (but not considered in this book until Chap. 11).

Remembering that a vector b is a *unit vector* if $||b||^2 = \langle b, b \rangle = 1$, a basis becomes *orthonormal*, if it is formed by unit vectors. The *orthonormality condition* of a basis can be written in the compact form

$$\langle b_i, b_j \rangle = \delta_{ij}, \tag{2.18}$$

where δ_{ij} is Kronecker's symbol, defined as $\delta_{ij} = 1$ for $i = j$ and $\delta_{ij} = 0$ for $i \neq j$. In general, a Hilbert space admits infinite orthonormal bases, all, obviously, with the same cardinality.

For a fixed orthonormal basis $\mathcal{B} = \{b_i, i \in I\}$, every vector x of $\mathcal{H}$ can be uniquely written as a linear combination of the vectors of the basis

$$x = \sum_{i \in I} a_i \, b_i \tag{2.19}$$

where the coefficients are given by the inner products

$$a_i = \langle x, b_i \rangle. \tag{2.20}$$

In fact, we obtain

$$\langle x, b_j \rangle = \left\langle \sum_i a_i \, b_i, b_j \right\rangle = \sum_i a_i \, \langle b_i, b_j \rangle = a_j$$

where in the last equality we used orthonormality condition (2.18).

The expansion (2.19) is called *Fourier expansion* of the vector x and the coefficients a_i the *Fourier coefficients* of x, obtained with the basis $\mathcal{B}$.

Through Fourier expansion, every orthonormal basis $\mathcal{B} = \{b_i, i \in I\}$ defines a *coordinate system* in the Hilbert space. In fact, according to (2.19) and (2.20), a vector x uniquely identifies its Fourier coefficients $\{a_i, i \in I\}$, which are the *coordinates* of x obtained with the basis $\mathcal{B}$. Of course, if the basis is changed, the coordinate system changes too, and so do the coordinates $\{a_i, i \in I\}$. Sometimes, to remark the dependence on $\mathcal{B}$, we write $(a_i)_{\mathcal{B}}$.

For a Hilbert space $\mathcal{H}$ with finite dimension n, a basis and the corresponding coordinate system establish a one-to-one correspondence between $\mathcal{H}$ and $\mathbb{C}^n$: the vectors x of $\mathcal{H}$ become the vectors of $\mathbb{C}^n$ composed by the Fourier coefficients of a_i, that is,

$$x \in \mathcal{H} \quad \xrightarrow{\text{coordinates}} \quad x_\mathcal{B} = \begin{bmatrix} a_1 \\ a_2 \\ \vdots \\ a_n \end{bmatrix} \in \mathbb{C}^n. \tag{2.21}$$

Example 2.1 (Periodic discrete signals) Consider the vector space $\mathcal{L}_2 = \mathcal{L}_2(\mathbb{Z}(T)/\mathbb{Z}(NT))$ constituted by periodic discrete signals (with spacing T and period NT); $\mathbb{Z}(T) := \{nT \,|\, n \in \mathbb{Z}\}$ is the set of multiples of T. A basis for this space is formed by the signals

$$b_i = b_i(t) = \frac{1}{T}\delta_{\mathbb{Z}(T)/\mathbb{Z}(NT)}(t - iT), \qquad i = 0,\, 1,\, \ldots,\, N - 1,$$

where $\delta_{\mathbb{Z}(T)/\mathbb{Z}(NT)}$ is the periodic discrete impulse [2]

$$\delta_{\mathbb{Z}(T)/\mathbb{Z}(NT)}(t) = \begin{cases} 1/T & t \in \mathbb{Z}(NT) \\ 0 & t \notin \mathbb{Z}(NT) \end{cases} \qquad t \in \mathbb{Z}(T).$$

This basis is orthonormal because

$$\langle b_i, b_j \rangle = \int_{\mathbb{Z}(T)/\mathbb{Z}(NT)} \mathrm{d}t \; b_i(t)\, b_j^*(t) = \delta_{ij}.$$

A first conclusion is that this vector space has finite dimension N.

For a generic signal $x = x(t)$, coefficients (2.20) provide

$$a_i = \langle x, b_i \rangle = \int_{\mathbb{Z}(T)/\mathbb{Z}(T_p)} \mathrm{d}t \; x(t)\, b_i^*(t) = \frac{1}{T}x(iT),$$

and therefore the signal coordinates are given by a vector collecting the values in one period, divided by T.

2.4.2 Dirac's Notation

In Quantum Mechanics, where systems are defined on a Hilbert space, vectors are indicated with a special notation, introduced by Dirac [3]. This notation, although apparently obscure, is actually very useful, and will be adopted from now on.

A vector x of a Hilbert space $\mathcal{H}$ is interpreted as a *column vector*, of possibly infinite dimension, and is indicated by the symbol

$$|x\rangle \tag{2.22a}$$

which is called *ket*. Its transpose conjugate $|x\rangle^*$ should be interpreted as a *row vector*, and is indicated by the symbol

$$\langle x| = |x\rangle^* \tag{2.22b}$$

which is called *bra*.[3] As a consequence, the inner product of two vectors $|x\rangle$ and $|y\rangle$ is indicated in the form

$$\langle x|y\rangle. \tag{2.22c}$$

We now exemplify this notation for the Hilbert space $\mathbb{C}^n$, comparing it to the standard notation

$$x = \begin{bmatrix} x_1 \\ \vdots \\ x_n \end{bmatrix} \quad \text{becomes} \quad |x\rangle = \begin{bmatrix} x_1 \\ \vdots \\ x_n \end{bmatrix} \tag{2.23}$$

$$x^* = \begin{bmatrix} x_1^*, & \ldots, & x_n^* \end{bmatrix} \quad \text{becomes} \quad \langle x| = |x\rangle^* = \begin{bmatrix} x_1^*, & \ldots, & x_n^* \end{bmatrix}$$

$$\langle x, y \rangle = y^*x \quad \text{becomes} \quad \langle y|x\rangle = x_1 \, y_1^* + \cdots + x_n \, y_n^*.$$

Again, to become familiar with Dirac's notation, we also rewrite some relations, previously formulated with the conventional notation. A linear combination of vectors is written in the form

$$|x\rangle = a_1|x_1\rangle + a_2|x_2\rangle + \cdots + a_n|x_n\rangle.$$

The norm of a vector is written as $||x|| = \sqrt{\langle x|x\rangle}$. The orthogonality condition between two vectors $|x\rangle$ and $|y\rangle$ is now written as

$$\langle x|y\rangle = 0,$$

and the orthonormality of a basis $\mathcal{B} = \{|b_i\rangle, i \in I\}$ is written in the form

$$\langle b_i|b_j\rangle = \delta_{ij}.$$

The Fourier expansion with a finite-dimensional orthonormal basis $\mathcal{B} = \{|b_i\rangle | i = 1, \ldots, n\}$ becomes

$$|x\rangle = a_1|b_1\rangle + \cdots + a_n|b_n\rangle \tag{2.24}$$

where

$$a_i = \langle b_i|x\rangle, \tag{2.24a}$$

and can also be written in the form

$$|x\rangle = (\langle b_1|x\rangle) \, |b_1\rangle + \cdots + (\langle b_n|x\rangle) \, b_n). \tag{2.25}$$

---

[3] These names are obtained by splitting up the word "bracket"; in the specific case, the brackets are $\langle \, \rangle$.

Schwartz's inequality (2.12) becomes

$$|\langle x|y\rangle|^2 \le \langle x|x\rangle\langle y|y\rangle \quad \text{or} \quad \langle x|y\rangle\langle y|x\rangle \le \langle x|x\rangle\langle y|y\rangle. \tag{2.26}$$

Problem 2.1 $\star$ A basis in $\mathcal{H} = \mathbb{C}^2$ is usually denoted by $\{|0\rangle, |1\rangle\}$. Write the standard basis and a nonorthogonal basis.

Problem 2.2 $\star\star$ An important basis in $\mathcal{H} = \mathbb{C}^n$ is given by the columns of the Discrete Fourier Transform (DFT) matrix of order n, given by

$$|w_i\rangle = \frac{1}{\sqrt{n}}\left[1, W_n^{-i}, W_n^{-2i}, \ldots, W_n^{-i(n-1)}\right]^{\mathrm{T}}, \qquad i = 0, 1, \ldots, n-1 \tag{E1}$$

where $W_n := \exp(\mathrm{i}2\pi/n)$ is the nth root of 1. Prove that this basis is orthonormal.

Problem 2.3 $\star$ Find the Fourier coefficients of ket

$$|x\rangle = \begin{bmatrix} 1 \\ \mathrm{i} \\ 2 \end{bmatrix} \in \mathbb{C}^3$$

with respect to the orthonormal basis (E1).

Problem 2.4 $\star$ Write the Fourier expansion (2.24) and (2.25) with a general orthonormal basis $\mathcal{B} = \{|b_i\rangle | i \in I\}$.

2.5 Linear Operators

2.5.1 Definition

An *operator* A from the Hilbert space $\mathcal{H}$ to the same space $\mathcal{H}$ is defined as a function

$$A : \mathcal{H} \to \mathcal{H}. \tag{2.27}$$

If $|x\rangle \in \mathcal{H}$, the operator A returns the vector

$$|y\rangle = A|x\rangle \quad \text{with} \quad |y\rangle \in \mathcal{H}. \tag{2.28}$$

To represent graphically the operator A, we can introduce a block (Fig. 2.3) containing the symbol of the operator, and in (2.28) $|x\rangle$ is interpreted as *input* and $|y\rangle$ as *output*.

The operator $A : \mathcal{H} \to \mathcal{H}$ is *linear* if the superposition principle holds, that is, if

$$A(a_1|x_1\rangle + a_2|x_2\rangle) = a_1 A|x_1\rangle + a_2 A|x_2\rangle$$

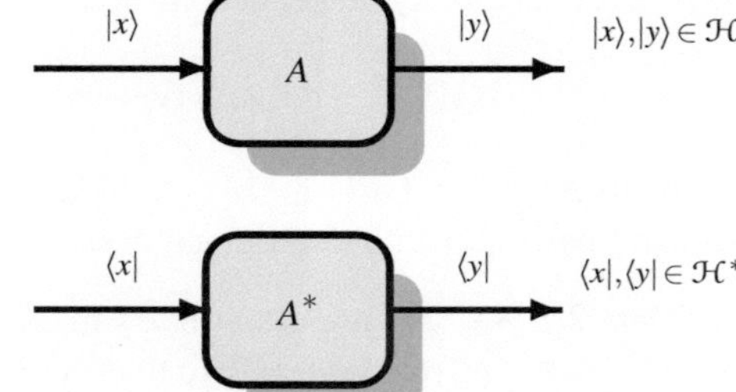

Fig. 2.3 Graphical representation of a linear operator

Fig. 2.4 The linear operator for "bras"; A^* is the *adjoint* of A

for every $|x_1\rangle$, $|x_2\rangle \in \mathcal{H}$ and $a_1, a_2 \in \mathbb{C}$.

A trivial linear operator is the *identity operator* $I_{\mathcal{H}}$ on $\mathcal{H}$ defined by the relation $I_{\mathcal{H}}\,|x\rangle \equiv |x\rangle$, for any vector $|x\rangle \in \mathcal{H}$. Another trivial linear operator is the *zero operator*, $0_{\mathcal{H}}$, which maps any vector onto the zero vector, $0_{\mathcal{H}}|x\rangle \equiv 0$.

In the interpretation of Fig. 2.3 the operator A acts on the kets (column vectors) of $\mathcal{H}$: assuming as input the ket $|x\rangle$, the operator outputs the ket $|y\rangle = A|x\rangle$. It is possible to associate to $\mathcal{H}$ a Hilbert space $\mathcal{H}^*$ (*dual* space) creating a correspondence between each ket $|x\rangle \in \mathcal{H}$ and its bra $\langle x|$ in $\mathcal{H}^*$. In this way, to each linear operator A of $\mathcal{H}$ a corresponding A^* of $\mathcal{H}^*$ can be associated, and the relation (2.28) becomes (Fig. 2.4)

$$\langle y| = \langle x|A^*.$$

The operator A^* is called the *adjoint*[4] of A. In particular, if $A = [a_{ij}]_{i,j=1,\ldots,n}$ is a square matrix, it results that $A^* = [a_{ji}^*]_{i,j=1,\ldots,n}$ is the *conjugate transpose*.

2.5.2 Composition of Operators and Commutability

The composition (*product*)[5] AB of two linear operators A and B is defined as the linear operator that, applied to a generic ket $|x\rangle$, gives the same result as would be obtained from the successive application of B followed by A, that is,

$$\{AB\}|x\rangle = A\{B|x\rangle\}. \tag{2.29}$$

In the graphical representation the product must be seen as a cascade of blocks (Fig. 2.5).

In general, like for matrices, the commutative property $AB = BA$ does not hold. Instead, to account for noncommutativity, the **commutator** between two operators

---

[4] This is not the ordinary definition of adjoint operator, but it is an equivalent definition, deriving from the relation $|x\rangle^* = \langle x|$ (see Sect. 2.8).

[5] We take for granted the definition of *sum $A + B$* of two operators, and of *multiplication of an operator by a scalar kA*.

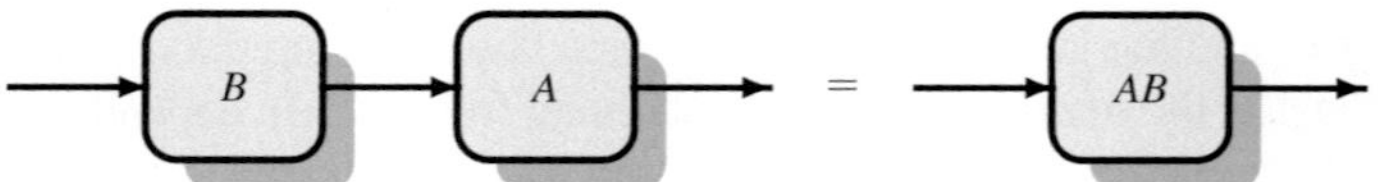

Fig. 2.5 Cascade connection of two operators

A and B is introduced, defined by

$$[A, \ B] := AB - BA. \tag{2.30}$$

In particular, if two operators commute, that is, if $AB = BA$, the commutator results in $[A, \ B] = 0$. Also an **anticommutator** is defined as

$$\{A, \ B\} := AB + BA. \tag{2.31}$$

Clearly, it is possible to express the product between two operators A and B in terms of the commutator and anticommutator

$$AB = \frac{1}{2}[A, \ B] + \frac{1}{2}\{A, \ B\}. \tag{2.31a}$$

In Quantum Mechanics, the commutator and the anticommutator are extensively used, e.g., to establish Heisenberg's uncertainty principle (see Sect. 3.9). Since most operator pairs do not commute, specific *commutation relations* are introduced through the commutator (see Chap. 11).

2.5.3 *Matrix Representation of an Operator*

As we have seen, a linear operator has properties very similar to those of a square matrix and, more precisely, to the ones that are obtained with a linear transformation of the kind $y = Ax$, where x and y are column vectors, and A is a square matrix, and it can be stated that linear operators are a generalization of square matrices. Also, it is possible to associate to each linear operator A a square matrix $A_{\mathcal{B}}$ of appropriate dimensions, $n \times n$, if the Hilbert space has dimension n, or of infinite dimensions if $\mathcal{H}$ has infinite dimension.

To associate a matrix to an operator A, we must fix an orthonormal basis $\mathcal{B} = \{|b_i\rangle, \ i \in I\}$ of $\mathcal{H}$. The relation

$$a_{ij} = \langle b_i|A|b_j\rangle, \qquad |b_i\rangle, |b_j\rangle \in \mathcal{B} \tag{2.32}$$

allows us to define the elements a_{ij} of a complex matrix $A_{\mathcal{B}} = [a_{ij}]$. In (2.32), the expression $\langle b_i|A|b_j\rangle$ must be intended as $\langle b_i|\{A|b_j\rangle\}$, that is, as the inner product of the bra $\langle b_i|$ and the ket $|A|b_j\rangle$ that is obtained by applying the operator A to the ket $|b_j\rangle$. Clearly, the matrix $A_{\mathcal{B}} = [a_{ij}]$, obtained from (2.32), depends on the basis chosen, and sometimes the elements of the matrix are indicated in the form $a_{ij\,\mathcal{B}}$ to stress such dependence on $\mathcal{B}$. Because all the bases of $\mathcal{H}$ have the same cardinality, all the matrices that can be associated to an operator have the same dimension.

From the matrix representation $A_{\mathcal{B}} = [a_{ij\,\mathcal{B}}]$ we can obtain the operator A using the *outer product* $|b_i\rangle\langle b_j|$, which will be introduced later on. The relation is

$$A = \sum_i \sum_j a_{ij\,\mathcal{B}} |b_i\rangle\langle b_j| \tag{2.33}$$

and will be proved in Sect. 2.7.

The matrix representation of an operator turns out to be useful as long as it allows us to interpret relations between operators as relations between matrices, with which we are usually more familiar. It is interesting to remark that an appropriate choice of a basis for $\mathcal{H}$ can lead to an "equivalent" matrix representation, simpler with respect to a generic choice of the basis. For example, we will see that a Hermitian operator admits a diagonal matrix representation with respect to a basis given by the eigenvectors of the operator itself.

In practice, as previously mentioned, in the calculations we will always refer to the Hilbert space $\mathcal{H} = \mathbb{C}^n$, where the operators can be interpreted as $n \times n$ square matrices with complex elements (the dimension of $\mathcal{H}$ could be infinite), keeping in mind anyhow that matrix representations with different bases correspond to the usual basis changes in normed vector spaces.

2.5.4 Trace of an Operator

An important parameter of an operator A is its *trace*, given by the sum of the diagonal elements of its matrix representation, namely

$$\boxed{\mathrm{Tr}[A] = \sum_i \langle b_i|A|b_i\rangle.} \tag{2.34}$$

The operation $\mathrm{Tr}[\cdot]$ appears in the formulation of the third postulate of Quantum Mechanics, and it is widely used in quantum decision.

The trace of an operator has the following properties, which will be often used in the following:

(1) The trace of A is independent of the basis with respect to which it is calculated, and therefore it is a characteristic parameter of the operator.
(2) The trace has the *cyclic* property

$$Tr[AB] = Tr[BA] \tag{2.35}$$

which holds even if the operators A and B are not commutable; such property holds also for rectangular matrices, providing that the products AB and BA make sense.

(3) The trace is *linear*, that is,

$$Tr[aA + bB] = a\,Tr[A] + b\,Tr[B], \qquad a, b \in \mathbb{C}. \tag{2.36}$$

For completeness we recall the important identity

$$\boxed{\langle u|A|u\rangle = Tr[A|u\rangle\langle u|]} \tag{2.37}$$

where $|u\rangle$ is an arbitrary vector, and $|u\rangle\langle u|$ is the operator given by the outer product, which will be introduced later.

2.5.5 *Image and Rank of an Operator*

The *image* of an operator A of $\mathcal{H}$ is the set

$$im(A) := A\mathcal{H} = \{A|x\rangle \mid |x\rangle \in \mathcal{H}\}. \tag{2.38}$$

It can be easily proved that $im(A)$ is *a subspace of* $\mathcal{H}$ (see Problem 2.6).

The dimension of this subspace defines the *rank* of the operator

$$rank(A) = \dim im(A) = |A\,\mathcal{H}|. \tag{2.39}$$

This definition can be seen as the extension to operators of the concept of rank of a matrix. As it appears from (2.38), to indicate the image of an operator of $\mathcal{H}$ we use the compact symbol $A\mathcal{H}$.

Problem 2.5 ★ Prove that the image of an operator on $\mathcal{H}$ is a subspace of $\mathcal{H}$.

Problem 2.6 ★ Define the 2D operator that inverts the entries of a ket and write its matrix representation with respect to the standard basis.

Problem 2.7 ★★ Find the matrix representation of the operator of the previous problem with respect to the DFT basis.

Fig. 2.6 Interpretation of eigenvalue and eigenvector of a linear operator A

2.6 Eigenvalues and Eigenvectors

An *eigenvalue* λ of a given operator A is a complex number such that a vector $|x_0\rangle \in \mathcal{H}$ exists, different from zero, satisfying the following equation:

$$A|x_0\rangle = \lambda|x_0\rangle \qquad |x_0\rangle \neq 0 \ . \tag{2.40}$$

The vector $|x_0\rangle$ is called *eigenvector corresponding to the eigenvalue* λ.[6] The interpretation of relation (2.40) is illustrated in Fig. 2.6.

The set of all the eigenvalues is called *spectrum of the operator* and it will be indicated by the symbol $\sigma(A)$.

From the definition it results that the eigenvector $|x_0\rangle$ associated to a given eigenvalue is not unique, and in fact from (2.40) it results that also $2|x_0\rangle$, or $i|x_0\rangle$ with i the imaginary unit, are eigenvectors of λ. The set of all the eigenvectors associated to the same eigenvalue

$$\mathcal{E}_\lambda = \{|x_0\rangle \ | \ A|x_0\rangle = \lambda|x_0\rangle\} \tag{2.41}$$

is always a subspace,[7] which is called *eigenspace* associated to the eigenvalue λ.

2.6.1 Computing the Eigenvalues

In the space $\mathcal{H} = \mathbb{C}^n$, where the operator A can be interpreted as an $n \times n$ matrix, the eigenvalue computation becomes the procedure usually followed with complex square matrices, consisting in the evaluation of the solutions to the *characteristic equation*

$$c(\lambda) = \det[A - \lambda \, I_{\mathcal{H}}] = 0$$

where $c(\lambda)$ is a polynomial. Then, for the fundamental theorem of Algebra, the number $r \leq n$ of *distinct* solutions is found: $\lambda_1, \lambda_2, \ldots, \lambda_r$, forming the spectrum of A

$$\sigma(A) = \{\lambda_1, \lambda_2, \ldots, \lambda_r\}.$$

---

[6] The eigenvector corresponding to the eigenvalue λ is often indicated by the symbol $|\lambda\rangle$.

[7] As we assume $|x_0\rangle \neq 0$, to complete $\mathcal{E}_\lambda$ as a subspace, the vector 0 of $\mathcal{H}$ must be added.

The solutions allow us to write $c(\lambda)$ in the form

$$c(\lambda) = a_0(\lambda - \lambda_1)^{p_1}(\lambda - \lambda_2)^{p_2} \ldots (\lambda - \lambda_r)^{p_r}, \qquad a_0 \neq 0$$

where $p_i \geq 1$, $p_1 + p_2 + \cdots + p_r = n$, and p_i is called the *multiplicity* of the eigenvalue λ_i. Then we can state that the characteristic equation has always n solutions, counting the multiplicities.

As it is fundamental to distinguish whether we refer to distinct or to multiple solutions, we will use different notations in the two cases

$$\begin{aligned} &\lambda_1, \lambda_2, \ldots, \lambda_r && \text{for distinct eigenvalues} \\ &\tilde{\lambda}_1, \tilde{\lambda}_2, \ldots, \tilde{\lambda}_n && \text{for eigenvalues with repetitions} \end{aligned} \qquad (2.42)$$

Example 2.2 The 4×4 complex matrix

$$A = \frac{1}{4}\begin{bmatrix} 7 & -1+2\,\mathrm{i} & -1 & -1-2\,\mathrm{i} \\ -1-2\,\mathrm{i} & 7 & -1+2\,\mathrm{i} & -1 \\ -1 & -1-2\,\mathrm{i} & 7 & -1+2\,\mathrm{i} \\ -1+2\,\mathrm{i} & -1 & -1-2\,\mathrm{i} & 7 \end{bmatrix} \qquad (2.43)$$

has the characteristic polynomial

$$c(\lambda) = 6 - 17\lambda + 17\lambda^2 - 7\lambda^3 + \lambda^4$$

which has solutions $\lambda_1 = 1$ with multiplicity 2, and $\lambda_2 = 2$ and $\lambda_3 = 3$ with multiplicity 1. Therefore, the distinct eigenvalues are $\lambda_1 = 1$, $\lambda_2 = 2$ and $\lambda_3 = 3$, whereas the eigenvalues with repetition are

$$\tilde{\lambda}_1 = 1, \qquad \tilde{\lambda}_2 = 1, \qquad \tilde{\lambda}_3 = 2, \qquad \tilde{\lambda}_4 = 3.$$

The corresponding eigenvectors are, for example,

$$|\tilde{\lambda}_1\rangle = \begin{bmatrix} 1+\mathrm{i} \\ \mathrm{i} \\ 0 \\ 1 \end{bmatrix} \quad |\tilde{\lambda}_2\rangle = \begin{bmatrix} -\mathrm{i} \\ 1-\mathrm{i} \\ 1 \\ 0 \end{bmatrix} \quad |\tilde{\lambda}_3\rangle = \begin{bmatrix} -1 \\ 1 \\ -1 \\ 1 \end{bmatrix} \quad |\tilde{\lambda}_4\rangle = \begin{bmatrix} -\mathrm{i} \\ -1 \\ \mathrm{i} \\ 1 \end{bmatrix}. \qquad (2.44)$$

As we will see with the spectral decomposition theorem, it is possible to associate different eigenvectors to coincident, or even orthogonal, eigenvalues. In (2.44) $|\tilde{\lambda}_1\rangle$ and $|\tilde{\lambda}_2\rangle$ are orthogonal, namely, $\langle \tilde{\lambda}_1 | \tilde{\lambda}_2 \rangle = 0$.

What was stated above for $\mathbb{C}^n$ can apply to any finite dimensional space n, using matrix representation. For an infinite dimensional space the spectrum can have infinite cardinality, but not necessarily. In any case it seems that no general procedures exist to compute the eigenvalues for the operators in an infinite dimensional space.

Trace of an operator from the eigenvalues   It can be proved that the sum of the eigenvalues with coincidences gives the trace of the operator

$$\sum_{i=1}^{n} \tilde{\lambda}_i = \sum_{i=1}^{r} p_i\, \lambda_i = \mathrm{Tr}[A]. \tag{2.45}$$

It is also worthwhile to observe that the product of the eigenvalues $\tilde{\lambda}_i$ gives the determinant of the operator

$$\tilde{\lambda}_1 \tilde{\lambda}_2 \dots \tilde{\lambda}_n = \lambda_1^{p_1} \lambda_2^{p_2} \dots \lambda_r^{p_r} = \det(A) \tag{2.46}$$

and that the rank of A is given by the sum of the multiplicities of the $\tilde{\lambda}_i$ different from zero.

2.7 Outer Product. Elementary Operators

The *outer product* of two vectors $|x\rangle$ and $|y\rangle$ in Dirac's notation is indicated in the form

$$|x\rangle\langle y|,$$

which may appear similar to the inner product notation $\langle x|y\rangle$, but with factors inverted. This is not the case: while $\langle x|y\rangle$ is a complex number, $|x\rangle\langle y|$ is an operator. This can be quickly seen if $|x\rangle$ is interpreted as a column vector and $\langle y|$ as a row vector, referring for simplicity to the space $\mathbb{C}^n$, where

$$|x\rangle = \begin{bmatrix} x_1 \\ \vdots \\ x_n \end{bmatrix}, \qquad \langle y| = \begin{bmatrix} y_1^*, \dots, y_n^* \end{bmatrix}.$$

Then, using the matrix product, we have

$$|x\rangle\langle y| = \begin{bmatrix} x_1 \\ \vdots \\ x_n \end{bmatrix} \begin{bmatrix} y_1^*, \dots, y_n^* \end{bmatrix} = \begin{bmatrix} x_1 y_1^* & \cdots & x_1 y_n^* \\ \vdots & & \vdots \\ x_n y_1^* & \cdots & x_n y_n^* \end{bmatrix}$$

that is, $|x\rangle\langle y|$ is an $n \times n$ square matrix.

The outer product[8] makes it possible to formulate an important class of linear operators, called *elementary operators* (or rank 1 operators) in the following way

---

[8] Above, the outer product was defined in the Hilbert space $\mathbb{C}^n$. For the definition in a generic Hilbert space one can use the subsequent (2.48), which defines $C = |c_1\rangle\langle c_2|$ as a linear operator.

Fig. 2.7 Interpretation of the
linear operator $C = |c_1\rangle\langle c_2|$

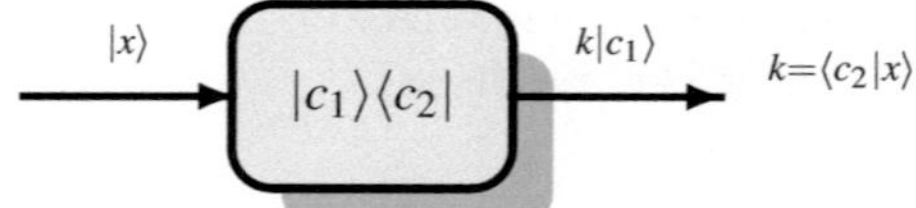

$$C = |c_1\rangle\langle c_2| \qquad (2.47)$$

where $|c_1\rangle$ and $|c_2\rangle$ are two arbitrary vectors of the Hilbert space $\mathcal{H}$. To understand
its meaning, let us apply to $C = |c_1\rangle\langle c_2|$ an arbitrary ket $|x\rangle \in \mathcal{H}$ (Fig. 2.7), which
results in

$$C|x\rangle = (|c_1\rangle\langle c_2|)|x\rangle = (\langle c_2|x\rangle)|c_1\rangle, \qquad \forall\, |x\rangle \in \mathcal{H}, \qquad (2.48)$$

namely, a vector proportional to $|c_1\rangle$, with a proportionality constant given by the
complex number $k = \langle c_2|x\rangle$.

From this interpretation it is evident that the image of the elementary operator C
is a straight line through the origin identified by the vector $|c_1\rangle$

$$\mathrm{im}(|c_1\rangle\langle c_2|) = \{h\, |c_1\rangle\,|\, h \in \mathbb{C}\}$$

and obviously the elementary operator has unit rank.

Within the class of the elementary operators, a fundamental role is played, especially in Quantum Mechanics, by the operators obtained from the outer product of a
ket $|b\rangle$ and the corresponding bra $\langle b|$, namely

$$B = |b\rangle\langle b|. \qquad (2.49)$$

For these elementary operators, following the interpretation of Fig. 2.7, we realize
that B transforms an arbitrary ket $|x\rangle$ into a ket proportional to $|b\rangle$. As we will see,
if $|b\rangle$ is unitary, then $|b\rangle\langle b|$ turns out to be a *projector*.

2.7.1 Properties of an Orthonormal Basis

The elementary operators allow us to reinterpret in a very meaningful way the prop-
erties of an orthonormal basis in a Hilbert space $\mathcal{H}$. If $\mathcal{B} = \{|b_i\rangle, i \in I\}$ is an
orthonormal basis on $\mathcal{H}$, then $\mathcal{B}$ identifies $k = |I|$ elementary operators $|b_i\rangle\langle b_i|$, and
their sum gives the identity

$$\boxed{\sum_{i \in I} |b_i\rangle\langle b_i| = I_{\mathcal{H}} \qquad \text{for every orthonormal}\ \ \mathcal{B} = \{|b_i\rangle, i \in I\}.} \qquad (2.50)$$

In fact, if $|x\rangle$ is any vector of $\mathcal{H}$, its Fourier expansion (see (2.24)), using the basis $\mathcal{B}$ (see (2.24)), results in

$$\sum_i |b_i\rangle\langle b_i|x\rangle = \sum_i |b_i\rangle \sum_j a_i \langle b_i|b_j\rangle$$
$$= \sum_i |b_i\rangle a_i = |x\rangle$$

and, recalling that $\langle b_i|b_j\rangle = \delta_{ij}$, we have

$$\sum_i |b_i\rangle\langle b_i|x\rangle = \sum_i a_i|b_i\rangle = |x\rangle.$$

In other words, if we apply to the sum of the elementary operators $|b_i\rangle\langle b_i|$ the ket $|x\rangle$, we obtain again the ket $|x\rangle$ and therefore such sum gives the identity. The property (2.50), illustrated in Fig. 2.8, can be expressed by stating that the elementary operators $|b_i\rangle\langle b_i|$ obtained from an orthonormal basis $\mathcal{B} = \{|b_i\rangle, i \in I\}$ *give a resolution of the identity $I_{\mathcal{H}}$ on $\mathcal{H}$.*

The properties of an orthonormal basis $\mathcal{B} = \{|b_i\rangle, i \in I\}$ on the Hilbert space $\mathcal{H}$ can be so summarized:

(1) $\mathcal{B}$ is composed of *linearly independent* and orthonormal vectors

$$\langle b_i|b_j\rangle = \delta_{ij};$$

(2) the cardinality of $\mathcal{B}$ is, by definition, equal to the dimension of $\mathcal{H}$

$$|\mathcal{B}| = \dim \mathcal{H};$$

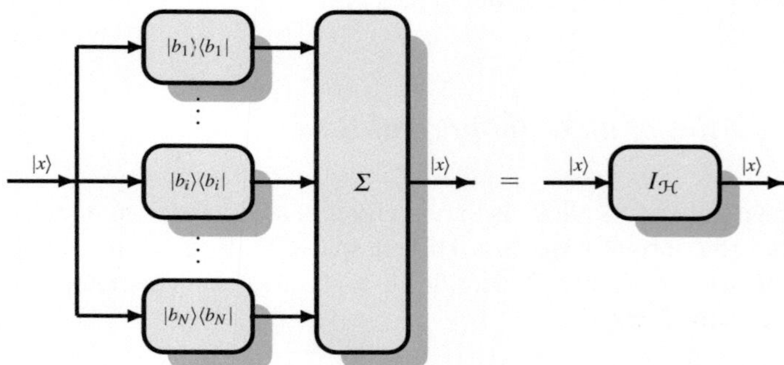

Fig. 2.8 The elementary operators $|b_i\rangle\langle b_i|$ obtained from an orthonormal basis $\mathcal{B} = \{|b_1\rangle, \ldots, |b_N\rangle\}$ provide a *resolution of the identity $I_{\mathcal{H}}$*

(3) the basis $\mathcal{B}$, through its elementary operators $|b_i\rangle\langle b_i|$, gives a *resolution of the identity* on $\mathcal{H}$, as stated by (2.50);

(4) $\mathcal{B}$ makes it possible to develop every vector $|x\rangle$ of $\mathcal{H}$ in the form (Fourier expansion)

$$|x\rangle = \sum_i a_i |b_i\rangle \quad \text{with} \quad a_i = \langle b_i|x\rangle. \tag{2.51}$$

Continuous bases Above we have implicitly assumed that the basis consists of an enumerable set of kets $\mathcal{B}$. In Quantum Mechanics also *continuous bases*, which consist of a continuum of eigenkets, are considered. This will be seen in the final chapters in the context of Quantum Information (see in particular Sect. 11.2).

2.7.2 Useful Identities Through Elementary Operators

Previously, we anticipated two identities requiring the notion of elementary operator. A first identity, related to the trace, is given by (2.37), namely

$$\langle u|A|u\rangle = \mathrm{Tr}[A|u\rangle\langle u|]$$

where A is an arbitrary operator, and $|u\rangle$ is a vector, also arbitrary. To prove this relation, let us consider an orthonormal basis $\mathcal{B} = \{|b_i\rangle, i \in I\}$ and let us apply the definition of a trace (2.34) to the operator $A|u\rangle\langle u|$. We obtain

$$\begin{aligned}
\mathrm{Tr}[A|u\rangle\langle u|] &= \sum_i \langle b_i|A|u\rangle\langle u|b_i\rangle \\
&= \sum_i \langle u|b_i\rangle\langle b_i|A|u\rangle = \langle u| \sum_i |b_i\rangle\langle b_i|A|u\rangle \\
&= \langle u|I_{\mathcal{H}}A|u\rangle = \langle u|A|u\rangle
\end{aligned}$$

where we took into account the fact that $\sum_i |b_i\rangle\langle b_i|$ coincides with the identity operator $I_{\mathcal{H}}$ on $\mathcal{H}$ (see (2.50)).

A second identity is (2.33)

$$A = \sum_i \sum_j a_{ij} |b_i\rangle\langle b_j|$$

which makes it possible to reconstruct an operator A from its matrix representation $A_{\mathcal{B}} = [a_{ij}]$ obtained with the basis $\mathcal{B}$. To prove this relation, let us write A in the form $I_{\mathcal{H}}AI_{\mathcal{H}}$, and then let us express the identity $I_{\mathcal{H}}$ in the form (2.50). We obtain

$$A = I_{\mathcal{H}} A I_{\mathcal{H}} = \sum_i |b_i\rangle\langle b_i| A \sum_j |b_j\rangle\langle b_j|$$

$$= \sum_i \sum_j |b_i\rangle\langle b_i|A|b_j\rangle\langle b_j|$$

where (see (2.32)) $\langle b_i|A|b_j\rangle = a_{ij}$.

2.8 Hermitian and Unitary Operators

Basically, in Quantum Mechanics only *unitary* and *Hermitian* operators are used. Preliminary to the introduction of these two classes of operators is the concept of an *adjoint* operator.

The definition of adjoint is given in a very abstract form (see below). If we want to follow a more intuitive way, we can refer to the matrices associated to operators, recalling that if $A = [a_{ij}]$ is a complex square matrix, then:

- A^* indicates the conjugate transpose matrix, that is, the matrix with elements a_{ji}^*,
- A is a *Hermitian* matrix, if $A^* = A$,
- A is a *normal* matrix, if $AA^* = A^*A$,
- A is a *unitary* matrix, if $AA^* = I$, where I is the identity matrix.

Note that the class of normal matrices includes as a special cases both Hermitian and unitary matrices (Fig. 2.9). It is also worthwhile to recall the conjugate transpose rule for the product of two square matrices

$$(AB)^* = B^*A^*. \tag{2.52}$$

2.8.1 The Adjoint Operator

The adjoint operator A^* was introduced in Sect. 2.5.1 as the operator for the bras $\langle x|$, $\langle y|$, whereas A is the operator for the kets $|x\rangle$, $|y\rangle$ (see Figs. 2.3 and 2.4). But the standard definition of adjoint is the following.

Fig. 2.9 The class of *normal matrices* includes *Hermitian matrices* and unitary matrices

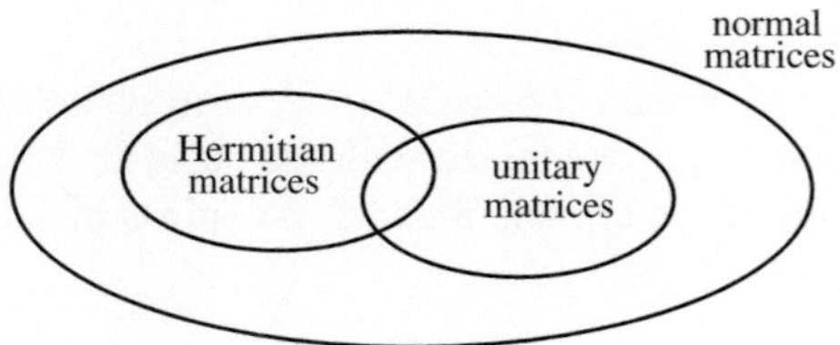

Given an operator $A : \mathcal{H} \rightarrow \mathcal{H}$, the *adjoint (or Hermitian adjoint)* operator A^* is defined through the inner product from the relation[9]

$$(A|x\rangle, |y\rangle) = \left(|x\rangle, A^*|y\rangle\right), \qquad |x\rangle, |y\rangle \in \mathcal{H}. \tag{2.53}$$

It can be proved that the operator A^* verifying such relation exists and is unique and, also, if $A_\mathcal{B} = [a_{ij}]$ is the representative matrix of A, the corresponding matrix of A^* is the *conjugate transpose* of $A_\mathcal{B}$, namely the matrix $A^*_\mathcal{B} = [a^*_{ji}]$.

In addition, between two operators A and B and their adjoints the following relations hold:

$$\begin{aligned}
(A^*)^* &= A \\
(A + B)^* &= A^* + B^* \\
(AB)^* &= B^* A^* \\
(aA)^* &= a^* A^* \qquad a \in \mathbb{C}
\end{aligned} \tag{2.54}$$

that is, exactly the same relations that are obtained interpreting A and B as complex matrices.

2.8.2 Hermitian Operators

An operator $A : \mathcal{H} \rightarrow \mathcal{H}$ is called *Hermitian* (or self-adjoint) if it coincides with its adjoint, that is, if
$$A^* = A.$$

As a consequence, every representative matrix of A is a Hermitian matrix.

A fundamental property is that the spectrum of a Hermitian operator is composed of *real eigenvalues*. To verify this property, we start out by observing that for each vector $|x\rangle$ it results

$$\langle x|A|x\rangle \in \mathbb{R}, \qquad \forall |x\rangle \in \mathcal{H}, \qquad A \text{ Hermitian.} \tag{2.55}$$

In fact, the conjugate of such product gives $(\langle x|A|x\rangle)^* = \langle x|A^*|x\rangle = \langle x|A|x\rangle$. Now, if λ is an eigenvalue of A and $|x_0\rangle$ the corresponding eigenvector, it results that $\langle x_0|A|x_0\rangle = \langle x_0|\lambda x_0\rangle = \lambda\langle x_0|x_0\rangle$. Then $\lambda = \langle x_0|A|x_0\rangle / \langle x_0|x_0\rangle$ is real, being a ratio between real quantities.

Another important property of Hermitian operators is that the *eigenvectors corresponding to distinct eigenvalues are always orthogonal*. In fact, from $A|x_1\rangle = \lambda_1|x_1\rangle$ and $A|x_2\rangle = \lambda_2|x_2\rangle$, remembering that λ_1 and λ_2 are real, it follows that $\lambda_2\langle x_1|x_2\rangle = \langle x_1|A|x_2\rangle = \langle \lambda_1 x_1|x_2\rangle = \lambda_1\langle x_1|x_2\rangle$. Therefore, if $\lambda_1 \neq \lambda_2$, then nec-

---

[9] In most textbooks the adjoint operator is indicated by the symbol $A^\dagger$ and sometimes by A^+.

essarily $\langle x_1 | x_2 \rangle = 0$. This property can be expressed in terms of *eigenspaces* (see (2.41)) in the form: $\mathcal{E}_\lambda = \{ |x_0\rangle \mid A|x_0\rangle = \lambda |x_0\rangle \}$, $\lambda \in \sigma(A)$, that is, the eigenspaces of a Hermitian operator are orthogonal and this is indicated as follows:

$$\mathcal{E}_\lambda \perp \mathcal{E}_\mu, \qquad \lambda \neq \mu.$$

Example 2.3 The matrix 4×4 defined by (2.43) is Hermitian in $\mathbb{C}^4$. As $\sigma(A) = \{1, 2, 3\}$, we have three eigenspaces $\mathcal{E}_1$, $\mathcal{E}_2$, $\mathcal{E}_3$. For the eigenvalues indicated in (2.44) we have

$$|\tilde{\lambda}_1\rangle, |\tilde{\lambda}_2\rangle \in \mathcal{E}_1, \qquad |\tilde{\lambda}_3\rangle \in \mathcal{E}_2, \qquad |\tilde{\lambda}_4\rangle \in \mathcal{E}_3.$$

It can be verified that orthogonality holds between eigenvectors belonging to different eigenspaces. For example,

$$\langle \lambda_1 | \lambda_4 \rangle = [1 - i, -i, 0, 1] \begin{bmatrix} -i \\ -1 \\ i \\ 1 \end{bmatrix} = (1 - i)(-i) + (-i)(-1) + 1 = 0.$$

2.8.3 Unitary Operators

An operator $U : \mathcal{H} \to \mathcal{H}$ is called unitary if

$$U U^* = I_{\mathcal{H}} \tag{2.56}$$

where $I_{\mathcal{H}}$ is the identity operator. Both unitary and Hermitian operators fall into the more general class of *normal operators*, which are operators defined by the property $AA^* = A^*A$.

From definition (2.56) it follows immediately that U is *invertible*, that is, there exists an operator U^{-1} such that $UU^{-1} = I_{\mathcal{H}}$, given by

$$U^{-1} = U^*. \tag{2.57}$$

Moreover, it can be proved that the spectrum of U is always composed of eigenvalues λ_i *with unit modulus*.

We observe that, if $\mathcal{B} = \{ |b_i\rangle, i \in I \}$ is an orthonormal basis of $\mathcal{H}$, all the other bases can be obtained through unitary operators, according to $\{ U |b_i\rangle, i \in I \}$.

An important property is that the unitary operators preserve the inner product. In fact, if we apply the same unitary operator U to the vectors $|x\rangle$ and $|y\rangle$, so that $|u\rangle = U|x\rangle$ and $|v\rangle = U|y\rangle$, from $\langle u| = \langle x|U^*$, we obtain

$$\langle u | v \rangle = \langle x | U^* U | y \rangle = \langle x | y \rangle.$$

Example 2.4 A remarkable example of unitary operator in $\mathcal{H} = \mathcal{L}_2(I)$ is the operator/matrix F which gives the discrete Fourier transform (DFT)

$$F = \left(1/\sqrt{N}\right)\ [W_N^{-(r-s)}]_{r,s=0,1,\dots,N-1} \tag{2.58}$$

where $W_N = \mathrm{e}^{\mathrm{i}2\pi/N}$. Then F is the DFT matrix. The inverse matrix is

$$F^{-1} = F^* = \left(1/\sqrt{N}\right)\ [W_N^{r-s}]_{r,s=0,1,\dots,N-1}. \tag{2.59}$$

The columns of F, like for any unitary matrix, form an orthonormal basis of $\mathcal{H} = \mathbb{C}^N$.

Problem 2.8 $\star$ Classify the so-called *Pauli matrices*

$$\sigma_0 = I = \begin{bmatrix} 1 & 0 \\ 0 & 1 \end{bmatrix}, \quad \sigma_x = \begin{bmatrix} 0 & 1 \\ 1 & 0 \end{bmatrix}, \quad \sigma_y = \begin{bmatrix} 0 & -\mathrm{i} \\ \mathrm{i} & 0 \end{bmatrix}, \quad \sigma_z = \begin{bmatrix} 1 & 0 \\ 0 & -1 \end{bmatrix} \tag{E2}$$

which have an important role in quantum computation.

2.9 Projectors

Orthogonal projectors (briefly, projectors) are Hermitian operators of absolute importance for Quantum Mechanics, since quantum measurements are formulated with such operators.

2.9.1 Definition and Basic Properties

A projector $P : \mathcal{H} \to \mathcal{H}$ is an *idempotent* Hermitian operator, that is, with the properties

$$P^* = P, \qquad P^2 = P \tag{2.60}$$

and therefore $P^n = P$ for every $n \geq 1$.

Let $\mathcal{P}$ be the image of the projector P

$$\mathcal{P} = \mathrm{im}(P) = P\,\mathcal{H} = \{P|x\rangle \mid |x\rangle \in \mathcal{H}\}, \tag{2.61}$$

then, if $|s\rangle$ is a vector of $\mathcal{P}$, we get

$$P|s\rangle = |s\rangle, \qquad |s\rangle \in \mathcal{P}. \tag{2.62}$$

In fact, as a consequence of idempotency, if $|s\rangle = P|x\rangle$ we obtain $P|s\rangle = P(P|x\rangle) = P|x\rangle = |s\rangle$. Property (2.62) states that the subspace $\mathcal{P}$ is *invariant* with respect to the operator P.

Expression (2.62) establishes that each $|s\rangle \in \mathcal{P}$ is an eigenvector of P with eigenvalue $\lambda = 1$; the spectrum of P can contain also the eigenvalue $\lambda = 0$

$$\sigma(P) \subset \{0, 1\}. \tag{2.63}$$

In fact, the relation $P|x\rangle = \lambda|x\rangle$, multiplied by P gives

$$P^2|x\rangle = \lambda P|x\rangle = \lambda^2|x\rangle \quad \rightarrow \quad P|x\rangle = \lambda^2|x\rangle = \lambda|x\rangle.$$

Therefore, every eigenvalue satisfies the condition $\lambda^2 = \lambda$, which leads to (2.63).

Finally, (2.63) allows us to state that projectors are *nonnegative* or *positive semidefinite* operators (see Sect. 2.12.1). This property is briefly written as $P \geq 0$.

2.9.2 Why Orthogonal Projectors?

To understand this concept we must introduce the *complementary* projector

$$P_c = I - P \tag{2.64}$$

where $I = I_{\mathcal{H}}$ is the identity on $\mathcal{H}$. P_c is in fact a projector because it is Hermitian, and also $P_c^2 = I^2 + P^2 - IP - PI = I - P = P_c$. Now, in addition to the subspace $\mathcal{P} = P\mathcal{H}$, let us consider the *complementary* subspace $\mathcal{P}_c = P_c\mathcal{H}$. It can be verified that (see Problem 2.9):

(1) all the vectors of $\mathcal{P}_c$ are orthogonal to the vectors of $\mathcal{P}$, that is,

$$\langle s^\perp|s\rangle = 0 \quad |s\rangle \in \mathcal{P}, \ |s^\perp\rangle \in \mathcal{P}_c \tag{2.65}$$

and then we write $\mathcal{P}_c = \mathcal{P}^\perp$.

(2) the following relations hold:

$$P|s^\perp\rangle = 0, \ |s^\perp\rangle \in \mathcal{P}_c \qquad P_c|s\rangle = 0, \ |s\rangle \in \mathcal{P}. \tag{2.66}$$

(3) the decomposition of an arbitrary vector $|x\rangle$ of $\mathcal{H}$

$$|x\rangle = |s\rangle + |s^\perp\rangle \tag{2.67}$$

is uniquely determined by

$$|s\rangle = P|x\rangle, \qquad |s^\perp\rangle = P_c|x\rangle. \tag{2.67a}$$

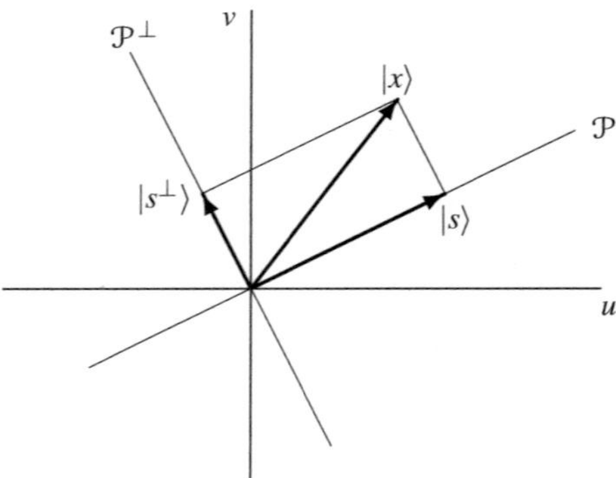

Fig. 2.10 Projection of the vector $|x\rangle$ *on* $\mathcal{P}$ along $\mathcal{P}^{\perp}$

Property (3) establishes that *the space $\mathcal{H}$ is given by the direct sum of the subspaces $\mathcal{P}$ and $\mathcal{P}^{\perp}$*, and we write $\mathcal{H} = \mathcal{P} \oplus \mathcal{P}^{\perp}$. According to properties (1) and (2), the projector P "projects the space $\mathcal{H}$ on $\mathcal{P}$ along $\mathcal{P}^{\perp}$".

Example 2.5 Consider the space $\mathcal{H} = \mathbb{R}^2$, which is "slightly narrow" for a Hilbert space, but sufficient to graphically illustrate the above properties. In $\mathbb{R}^2$, let us introduce a system of Cartesian axes u, v (Fig. 2.10), and let us indicate by $|x\rangle = \begin{bmatrix} u \\ v \end{bmatrix}$ the generic point of $\mathbb{R}^2$.

Let P_h be the real matrix

$$P_h = \frac{1}{1 + h^2} \begin{bmatrix} 1 & h \\ h & h^2 \end{bmatrix} \tag{2.68}$$

where h is a real parameter. It can be verified that $P_h^2 = P_h$, therefore P_h is a projector. The space generated by P_h is

$$\mathcal{P} = \left\{ P_h \begin{bmatrix} u \\ v \end{bmatrix} \Big| (u, v) \in \mathbb{R}^2 \right\} = \left\{ \begin{bmatrix} u \\ hu \end{bmatrix} \Big| u \in \mathbb{R} \right\}$$

This is a straight line passing through the origin, whose slope is determined by h. We can see that the complementary projector

$$P_h^{(c)} = I - P_h$$

has the same structure as (2.68) with the substitution $h \to -1/h$, and therefore $\mathcal{P}^{\perp}$ is given by the line through the origin orthogonal to the one above. The conclusion is that the projector P_h projects the space $\mathbb{R}^2$ onto the line $\mathcal{P}$ along the line $\mathcal{P}^{\perp}$.

It remains to verify that the geometric orthogonality here implicitly invoked, coincides with the orthogonality defined by the inner product. If we denote by $|s\rangle$ the generic point of $\mathcal{P}$, and by $|s^{\perp}\rangle$ the generic point of $\mathcal{P}^{\perp}$, we get

$$|s\rangle = \begin{bmatrix} u \\ hu \end{bmatrix} \qquad |s^\perp\rangle = \begin{bmatrix} u_1 \\ (-1/h)u_1 \end{bmatrix}$$

for given u and u_1. Then

$$\langle s^\perp | s \rangle = \begin{bmatrix} u_1, & (-1/h)u_1 \end{bmatrix} \begin{bmatrix} u \\ hu \end{bmatrix} = 0.$$

Finally, the decomposition $\mathbb{R}^2 = \mathcal{P} \oplus \mathcal{P}^\perp$ must be interpreted in the following way (see Fig. 2.10): each vector of $\mathbb{R}^2$ can be uniquely decomposed into a vector of $\mathcal{P}$ and a vector of $\mathcal{P}^\perp$.

Example 2.6 We now present an example less related to the geometric interpretation, a necessary effort if we want to comprehend the generality of Hilbert spaces.

Consider the Hilbert space $\mathcal{L}_2 = \mathcal{L}_2(I)$ of the signals defined in I (see Sect. 2.3, Examples from Signal Theory), and let $\mathcal{E}$ be the subspace constituted by the *even signals*, that is, those verifying the condition $s(t) = s(-t)$. Notice that $\mathcal{E}$ is a subspace because every linear combination of even signals gives an even signal. The orthogonality condition between two signals $x(t)$ and $y(t)$ is

$$\int_I \mathrm{d}t \; x(t) \, y^*(t) = 0.$$

We state that the orthogonal complement of $\mathcal{E}$ is given by the class $\mathcal{O}$ of the *odd* signals, those verifying the condition $s(-t) = -s(t)$. In fact, it can be easily verified that if $x(t)$ is even and $y(t)$ is odd, their inner product is null (this for sufficiency; for necessity, the proof is more complex). Then

$$\mathcal{E}^\perp = \mathcal{O}.$$

We now check that a signal $x(t)$ of $\mathcal{L}_2$, which in general is neither even nor odd, can be uniquely decomposed into an even component $x_p(t)$ and an odd component $x_d(t)$. We have in fact (see Unified Theory [2])

$$x(t) = x_p(t) + x_d(t)$$

where

$$x_p(t) = \frac{1}{2}[x(t) + x(-t)], \qquad x_d(t) = \frac{1}{2}[x(t) - x(-t)].$$

Then (Fig. 2.11)

$$\mathcal{L}_2 = \mathcal{E} \oplus \mathcal{O}.$$

Notice that $\mathcal{E} \cap \mathcal{O} = \{0\}$, where 0 denotes the identically null signal.

Fig. 2.11 The space $\mathcal{L}_2$ is obtained as a direct sum of the subspaces $\mathcal{E}$ and $\mathcal{O} = \mathcal{E}^\perp$

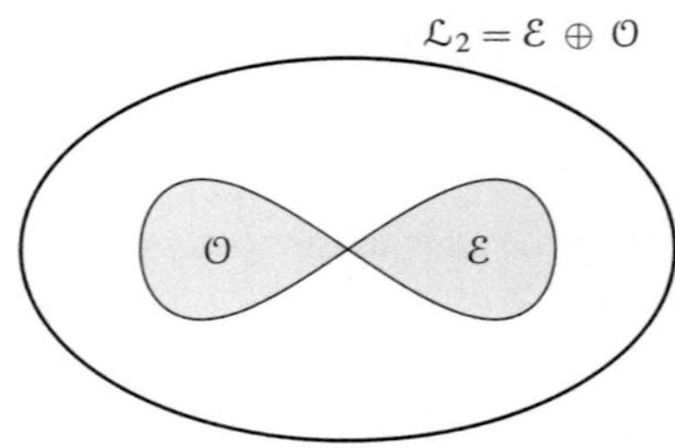

$\mathcal{E}$: class of even signals

$\mathcal{O}$: class of odd signals

2.9.3 General Properties of Projectors

We summarize the general properties of a projector P:

(1) the spectrum is always $\sigma(P) = \{0, 1\}$,
(2) the multiplicity of eigenvalue 1 gives the rank of P and the dimension of the subspace $\mathcal{P} = P\mathcal{H}$,
(3) $P \geq 0$: is a positive semidefinite operator (see Sect. 2.12),
(4) $\text{Tr}[P] = \text{rank}(P)$: the trace of P gives the rank of the projector.

(1) has already been seen. (3) is a consequence of (1) and of Theorem 2.6. (4) follows from (2.45).

2.9.4 Sum of Projectors. System of Projectors

The sum of two projectors P_1 and P_2

$$P = P_1 + P_2$$

is not in general a projector. But if two projectors are *orthogonal* in the sense that

$$P_1 P_2 = 0,$$

the sum results again in a projector, as can be easily verified. An example of a pair of orthogonal projectors has already been seen above: P and the complementary projector P_c verify the orthogonality condition $PP_c = 0$, and their sum is

$$P + P_c = I,$$

where $I = I_{\mathcal{H}}$ is the identity on $\mathcal{H}$, which is itself a projector.

The concept can be extended to the sum of several projectors:

$$P = P_1 + P_2 + \cdots + P_k \tag{2.69}$$

where the addenda are pairwise orthogonal

$$P_i P_j = 0, \qquad i \neq j. \tag{2.69a}$$

To (2.69) one can associate $k + 1$ subspaces

$$\mathcal{P} = P\mathcal{H} \quad \text{and} \quad \mathcal{P}_i = P_i\mathcal{H}, \ i = 1, \ldots, k$$

and, generalizing what was previously seen, we find that every vector $|s\rangle$ of $\mathcal{P}$ can be uniquely decomposed in the form

$$|s\rangle = |s_1\rangle + |s_2\rangle + \cdots + |s_k\rangle \quad \text{with} \quad |s_i\rangle = P_i|s\rangle. \tag{2.70}$$

Hence $\mathcal{P}$ is given by the *direct sum* of the subspaces $\mathcal{P}_i$

$$\mathcal{P} = \mathcal{P}_1 \oplus \mathcal{P}_2 \oplus \cdots \oplus \mathcal{P}_k.$$

If in (2.69) the sum of the projectors yields the identity I

$$P_1 + P_2 + \cdots + P_k = I \tag{2.71}$$

we say that the projectors $\{P_i\}$ *provide a resolution of the identity on* $\mathcal{H}$ and form a *complete* orthogonal class of projectors. In this case the direct sum gives the Hilbert space $\mathcal{H}$

$$\mathcal{P}_1 \oplus \mathcal{P}_2 \oplus \cdots \oplus \mathcal{P}_k = \mathcal{H}, \tag{2.72}$$

as illustrated in Fig. 2.12 for $k = 4$, where each point $|s\rangle$ of $\mathcal{P}$ is uniquely decomposed into the sum of 4 components $|s_i\rangle$ obtained from the relations $|s_i\rangle = P_i|s\rangle$, also shown in the figure.

For later use we find it convenient to introduce the following definition:

Definition 2.3 A set of operators $\{P_i, i \in I\}$ of the Hilbert space $\mathcal{H}$ constitutes a complete system of orthogonal projectors, briefly **projector system**, if they have the properties:

(1) the P_i are projectors (Hermitian and idempotent),
(2) the P_i are pairwise orthogonal ($P_i P_j = 0$) for $i \neq j$,
(3) the P_i form a resolution of the identity on $\mathcal{H}$ ($\sum_i P_i = \mathcal{I}_\mathcal{H}$).

The peculiarities of a projector system have been illustrated in Fig. 2.12.

Rank of projectors A projector has always a reduced rank with respect to the dimension of the space, unless P coincides with the identity I

$$\mathcal{P}_i = P_i\,\mathcal{H}$$

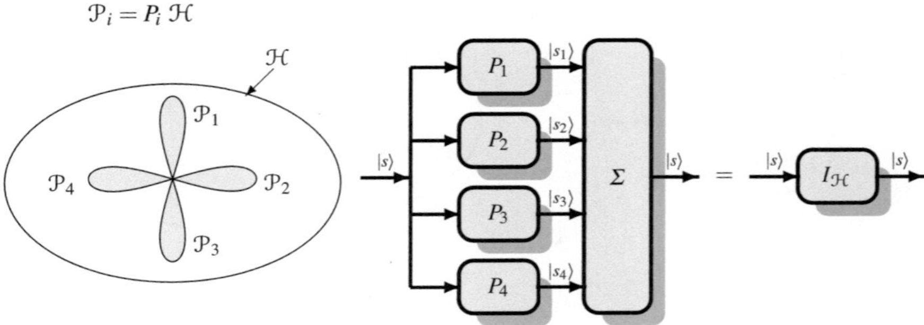

Fig. 2.12 The subspaces $\mathcal{P}_i$ give the space $\mathcal{H}$ as a *direct sum*. The projectors P_i give a *resolution of the identity* $I_{\mathcal{H}}$ and form a **projector system**

$$\mathrm{rank}(P) = \dim\mathcal{P} < \dim\mathcal{H} \qquad (P \neq I).$$

In the decomposition (2.71) it results as

$$\mathrm{rank}(P_1) + \cdots + \mathrm{rank}(P_k) = \mathrm{rank}(I) = \dim\mathcal{H},$$

and therefore in the corresponding direct sum we have that

$$\dim\mathcal{P}_1 + \cdots + \dim\mathcal{P}_k = \dim\mathcal{H}.$$

2.9.5 Elementary Projectors

If $|b\rangle$ is any unitary ket, the elementary operator

$$B = |b\rangle\langle b| \quad \text{with} \quad ||b|| = 1 \tag{2.73}$$

is Hermitian and verifies the condition $B^2 = |b\rangle\langle b|b\rangle\langle b| = B$, therefore it is a *unit rank projector*. Applying to B any vector $|x\rangle$ of $\mathcal{H}$ we obtain

$$B|x\rangle = |b\rangle\langle b|x\rangle = k\,|b\rangle \quad \text{with } k = \langle b|x\rangle.$$

Hence B projects the space $\mathcal{H}$ on the straight line through the origin identified by the vector $|b\rangle$ (Fig. 2.13).

If two kets $|b_1\rangle$ and $|b_2\rangle$ are orthonormal, the corresponding elementary projectors $B_1 = |b_1\rangle\langle b_1|$ and $B_2 = |b_2\rangle\langle b_2|$ verify the orthogonality (for operators)

$$B_1 B_2 = |b_1\rangle\langle b_1|b_2\rangle\langle b_2| = 0$$

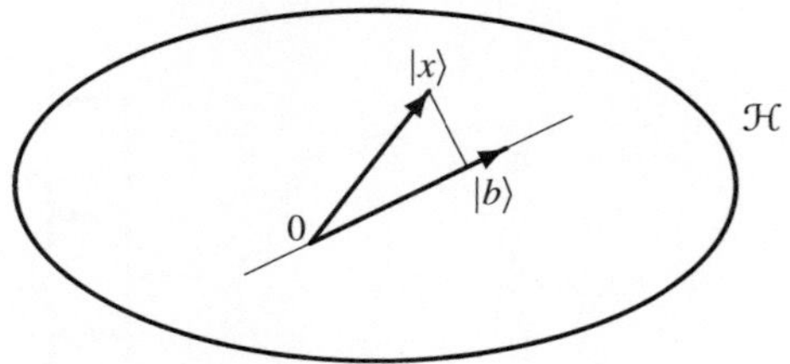

Fig. 2.13 The elementary projector $|b\rangle\langle b|$ *projects* an arbitrary ket $|x\rangle$ of $\mathcal{H}$ on the line of the ket $|b\rangle$

and therefore their sum

$$B_1 + B_2 = |b_1\rangle\langle b_1| + |b_2\rangle\langle b_2|$$

is still a projector (of rank 2).

Proceeding along this way we arrive at the following:

Theorem 2.1 *If* $\mathcal{B} = \{|b_i\rangle, i \in I\}$ *is an orthonormal basis of* $\mathcal{H}$*, the elementary projectors* $B_i = |b_i\rangle\langle b_i|$ *turn out to be orthogonal in pairs, and give the identity resolution*

$$\sum_{i \in I} B_i = \sum_{i \in I} |b_i\rangle\langle b_i| = I_{\mathcal{H}}.$$

In conclusion, through a generic basis of a Hilbert space $\mathcal{H}$ of dimension n it is always possible to "resolve" the space $\mathcal{H}$ through n elementary projectors, which form a projector system.

Problem 2.9 ⋆ Prove properties (2.65), (2.66) and (2.67) for a projector and its complement.

Problem 2.10 ⋆ Prove that projectors are positive semidefinite operators.

2.10 Spectral Decomposition Theorem (EID)

This theorem is perhaps the most important result of Linear Algebra because it sums up several previous results and opens the door to get so many interesting results. It will appear in various forms and will be referred to in different ways, for example, as *diagonalization* of a matrix and also as eigendecomposition or EID.

2.10.1 Statement and First Consequences

Theorem 2.2 *Let A be a Hermitian operator (or unitary) on the Hilbert space* $\mathcal{H}$*, and let* $\{\lambda_i\}, i = 1, 2, \ldots, k$ *be the distinct eigenvalues of A. Then A can be uniquely*

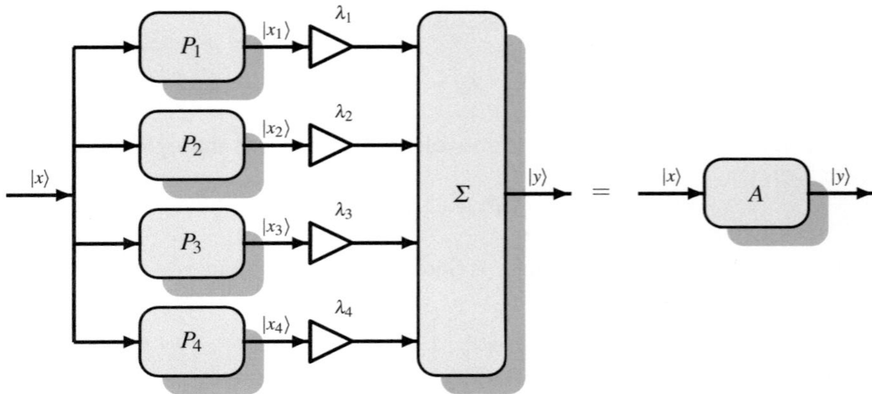

Fig. 2.14 Spectral decomposition of an operator A with four distinct eigenvalues

decomposed in the form

$$A = \sum_{i=1}^{k} \lambda_i P_i \tag{2.74}$$

where the $\{P_i\}$ form a projector system.

The spectral decomposition is illustrated in Fig. 2.14 for $k = 4$.

The theorem holds both for Hermitian operators, in which case the spectrum of the operator is formed by *real* eigenvalues λ_i, and for unitary operators, the eigenvalues λ_i have *unit modulus*.

We observe that, if p_i is the multiplicity of the eigenvalue λ_i, the rank of the projector P_i is just given by p_i

$$\operatorname{rank}(P_i) = p_i.$$

In particular, if the eigenvalues have all unit multiplicity, that is, if A is *nondegenerate*, the projectors P_i have unit rank and assume the form

$$P_i = |\lambda_i\rangle\langle\lambda_i|$$

where $|\lambda_i\rangle$ is the eigenvector corresponding to eigenvalue λ_i. In this case the eigenvectors define an orthonormal basis for $\mathcal{H}$.

Example 2.7 The 4×4 complex matrix considered in Example 2.1

$$A = \frac{1}{4}\begin{bmatrix} 7 & -1+2\,i & -1 & -1-2\,i \\ -1-2\,i & 7 & -1+2\,i & -1 \\ -1 & -1-2\,i & 7 & -1+2\,i \\ -1+2\,i & -1 & -1-2\,i & 7 \end{bmatrix} \tag{2.75}$$

is Hermitian. It has been found that the distinct eigenvalues are

$$\lambda_1 = 1, \qquad \lambda_2 = 2, \qquad \lambda_3 = 3$$

with λ_1 of multiplicity 2. Then it is possible to decompose A through 3 projectors, in the form

$$A = \lambda_1 P_1 + \lambda_2 P_2 + \lambda_3 P_3$$

with P_1 of rank 2 and P_2, P_3 of rank 1. Such projectors result in

$$P_1 = \frac{1}{4}\begin{bmatrix} 2 & 1-i & 0 & 1+i \\ 1+i & 2 & 1-i & 0 \\ 0 & 1+i & 2 & 1-i \\ 1-i & 0 & 1+i & 2 \end{bmatrix} \qquad P_2 = \frac{1}{4}\begin{bmatrix} 1 & -1 & 1 & -1 \\ -1 & 1 & -1 & 1 \\ 1 & -1 & 1 & -1 \\ -1 & 1 & -1 & 1 \end{bmatrix}$$

$$P_3 = \frac{1}{4}\begin{bmatrix} 1 & i & -1 & -i \\ -i & 1 & i & -1 \\ -1 & -i & 1 & i \\ i & -1 & -i & 1 \end{bmatrix}.$$

We leave it to the reader to verify the idempotency, the orthogonality, and the completeness of these projectors. In other words, to prove that the set $\{P_1, P_2, P_3\}$ forms a projector system.

2.10.2 Interpretation

The theorem can be interpreted in two ways:

- as *resolution* of a given Hermitian operator A, which makes it possible to identify a projector system $\{P_i\}$, as well as the corresponding eigenvalues $\{\lambda_i\}$,
- as *synthesis*, in which a projector system $\{P_i\}$ is known, and, for so many fixed distinct real numbers λ_i, a Hermitian operator can be built based on (2.74), having the λ_i as eigenvalues.

It is very interesting to see how the spectral decomposition acts on the input $|x\rangle$ and on the output $|y\rangle$ of the operator, following Fig. 2.14. The parallel of projectors decomposes in a unique way the input vector *into orthogonal components* (see (2.70)).

$$|x\rangle = |x_1\rangle + |x_2\rangle + \cdots + |x_k\rangle \tag{2.76}$$

where $|x_i\rangle = P_i|x\rangle$. In fact, as a consequence of the orthogonality of the projectors, we have

$$\langle x_i|x_j\rangle = \langle x|P_iP_j|x\rangle = 0, \qquad i \neq j$$

while (2.76) is a consequence of completeness. In addition, *each component* $|x_i\rangle$ *is an eigenvector of A with eigenvalue* λ_i. In fact, we have

$$A|x_i\rangle = \sum_j \lambda_j P_j |x_i\rangle = \lambda_i |x_i\rangle$$

where we have taken into account that $P_j|x_i\rangle = P_j P_i |x\rangle = 0$ if $i \neq j$.

Yet from Fig. 2.14, it appears that the decomposition of the input (2.76) is followed by the decomposition of the output in the form

$$|y\rangle = \lambda_1 |x_1\rangle + \lambda_2 |x_2\rangle + \cdots + \lambda_k |x_k\rangle$$

namely, as a sum of eigenvectors multiplied by the corresponding eigenvalues.

Going back to the decomposition of the input, as each component $|x_i\rangle = P_i |x\rangle$ belongs to the eigenspace $\mathcal{E}_{\lambda_i}$, and as $|x\rangle$ is an arbitrary ket of the Hilbert space $\mathcal{H}$, it results that

$$P_i \, \mathcal{H} = \mathcal{E}_{\lambda_i}$$

is the corresponding eigenspace. Furthermore, for completeness we have (see (2.72))

$$\mathcal{E}_{\lambda_1} \oplus \mathcal{E}_{\lambda_2} \oplus \cdots \oplus \mathcal{E}_{\lambda_k} = \mathcal{H}.$$

The reader can realize that the Spectral Decomposition Theorem allows us to refine what we saw in Sect. 2.9.4 on the sum of orthogonal projectors.

2.10.3 Decomposition via Elementary Projectors

In the statement of the theorem the eigenvalues $\{\lambda_i\}$ are assumed distinct, so the spectrum of the operator A is

$$\sigma(A) = \{\lambda_1, \lambda_2, \ldots, \lambda_k\} \quad \text{with} \quad k \leq n$$

where k may be smaller than the space dimension.

As we have seen, if $k = n$ all the eigenvalues have unit multiplicity and the decomposition (2.74) is done with elementary projectors. We can obtain a decomposition with elementary projectors even if $k < n$, that is, not all the eigenvalues have unit multiplicity. In this case we denote with

$$\tilde{\lambda}_1, \tilde{\lambda}_2, \ldots, \tilde{\lambda}_n$$

the eigenvalues with repetition (see (2.42)). Then the spectral decomposition (2.74) takes the form

$$A = \sum_{i=1}^{n} \tilde{\lambda}_i \, |b_i\rangle\langle b_i| \tag{2.77}$$

where now the projectors $B_i = |b_i\rangle\langle b_i|$ are all elementary, and form a complete system of projectors.

To move from (2.74) to (2.77), let us consider an example regarding the space $\mathcal{H} = \mathbb{C}^4$, with A a 4×4 matrix. Suppose now that

$$\sigma(A) = \{\lambda_1, \lambda_2, \lambda_3\} \qquad p_1 = 2, \; p_2 = 1, \; p_3 = 1.$$

Then Theorem 2.2 provides the decomposition

$$A = \lambda_1 P_1 + \lambda_2 P_2 + \lambda_3 P_3$$

with

$$P_1 \quad \text{of rank 2}, \qquad P_2 = |\lambda_2\rangle\langle\lambda_2|, \qquad P_3 = |\lambda_3\rangle\langle\lambda_3|.$$

Consider now the subspace $\mathcal{P}_1 = P_1\mathcal{H}$ having dimension 2, which in turn is a Hilbert space, and therefore with basis composed of two orthonormal vectors, say $|c_1\rangle$ and $|c_2\rangle$. Choosing such a basis, the sum of the corresponding elementary projectors yields (see Theorem 2.1)

$$P_1 = |c_1\rangle\langle c_1| + |c_2\rangle\langle c_2|.$$

In this way we obtain (2.77) with $(\tilde{\lambda}_1, \tilde{\lambda}_2, \tilde{\lambda}_3, \tilde{\lambda}_4) = (\lambda_1, \lambda_1, \lambda_2, \lambda_3)$ and $|b_1\rangle = |c_1\rangle$, $|b_2\rangle = |c_2\rangle$, $|b_3\rangle = |\lambda_2\rangle$, $|b_4\rangle = |\lambda_3\rangle$. Notice that the $|c_i\rangle$ and the $|\lambda_i\rangle$ are independent (orthogonal) because they belong to different eigenspaces.

Example 2.8 In Example 2.1 we have seen that the eigenvalues with repetition of the matrix (2.75) are

$$\tilde{\lambda}_1 = 1, \qquad \tilde{\lambda}_2 = 1, \qquad \tilde{\lambda}_3 = 2, \qquad \tilde{\lambda}_4 = 3$$

and the corresponding eigenvectors are

$$|\tilde{\lambda}_1\rangle = \begin{bmatrix} 1+i \\ i \\ 0 \\ 1 \end{bmatrix} \quad |\tilde{\lambda}_2\rangle = \begin{bmatrix} -i \\ 1-i \\ 1 \\ 0 \end{bmatrix} \quad |\tilde{\lambda}_3\rangle = \begin{bmatrix} -1 \\ 1 \\ -1 \\ 1 \end{bmatrix} \quad |\tilde{\lambda}_4\rangle = \begin{bmatrix} -i \\ -1 \\ i \\ 1 \end{bmatrix}. \tag{2.78}$$

These vectors are not normalized (they all have norm 2). The normalization results in

$$|b_i\rangle = \frac{1}{2}|\tilde{\lambda}_i\rangle \tag{2.79}$$

and makes it possible to build the elementary projectors $Q_i = |b_i\rangle\langle b_i|$. Therefore the spectral decomposition of matrix A via elementary projectors becomes

$$A = \tilde{\lambda}_1 Q_1 + \tilde{\lambda}_2 Q_2 + \tilde{\lambda}_3 Q_3 + \tilde{\lambda}_4 Q_4. \tag{2.80}$$

We encourage the reader to verify that with choice (2.79) the elementary operators Q_i are idempotent and form a system of orthogonal (elementary) projectors.

2.10.4 Synthesis of an Operator from a Basis

The Spectral Decomposition Theorem in the form (2.77), revised with elementary projectors, identifies an orthonormal basis $\{|b_i\rangle\}$.

The inverse procedure is also possible: given an orthonormal basis $\{|b_i\rangle\}$, a Hermitian (or unitary) operator can be built choosing an n-tuple of real eigenvalues $\tilde{\lambda}_i$ (or with unit modulus to have a unitary operator). In this way we obtain the synthesis of a Hermitian operator in the form (2.77).

Notice that with synthesis we can also obtain non-elementary projectors, thus arriving at the general form (2.74) established by the Spectral Decomposition Theorem. To this end, it suffices to choose some $\tilde{\lambda}_i$ equal. For example, if we want a rank 3 projector, we let

$$\tilde{\lambda}_1 = \tilde{\lambda}_2 = \tilde{\lambda}_3 = \lambda_1$$

and then

$$\tilde{\lambda}_1|b_1\rangle\langle b_1| + \tilde{\lambda}_2|b_2\rangle\langle b_2| + \tilde{\lambda}_3|b_3\rangle\langle b_3| = \lambda_1 P_1$$

where $P_1 = |b_1\rangle\langle b_1| + |b_2\rangle\langle b_2| + |b_3\rangle\langle b_3|$ is actually a projector, as can be easily verified.

2.10.5 Operators as Generators of Orthonormal Bases

In Quantum Mechanics, orthonormal bases are usually obtained from the EID of operators, mainly Hermitian operators. Then, considering a Hermitian operator B, the starting point is the *eigenvalue relation*

$$B|b\rangle = b\,|b\rangle \tag{2.81}$$

where $|b\rangle$ denotes an eigenket of B and b the corresponding eigenvalue; B is given, while b and $|b\rangle$ are considered unknowns. The solutions to (2.81) provide the spectrum $\sigma(B)$ of B and also an orthonormal basis

$$\mathcal{B} = \{|b\rangle, b \in \sigma(B)\}$$

where $|b\rangle$ are supposed to be normalized, that is, $\langle b|b\rangle = 1$. Note the economic notation (due to Dirac [3]), where a single letter (b or B) is used to denote the operator, the eigenkets, and the eigenvalues.

The bases obtained from operators are used in several ways, in particular to represent kets and bras through the Fourier expansion and operators through the matrix representation. A systematic application of these concepts will be seen in Chap. 11 in the context of Quantum Information (see in particular Sect. 11.2).

2.11 The Eigendecomposition (EID) as Diagonalization

In the previous forms of spectral decomposition (EID) particular emphasis was given to projectors because those are the operators that are used in quantum measurements. Other forms are possible, or better, other interpretations of the EID, evidencing other aspects.

Relation (2.77) can be written in the form

$$\boxed{A = U\,\tilde{\Lambda}\,U^*} \tag{2.82}$$

where U is the $n \times n$ matrix having as columns the vectors $|b_i\rangle$, and $\tilde{\Lambda}$ is the diagonal matrix with diagonal elements $\tilde{\lambda}_i$, namely

$$U = [|b_1\rangle, |b_2\rangle, \ldots, |b_n\rangle], \qquad \tilde{\Lambda} = \mathrm{diag}\,[\tilde{\lambda}_1, \tilde{\lambda}_2, \ldots, \tilde{\lambda}_n]\,, \tag{2.82a}$$

and U^* is the conjugate transpose of U having as rows the bras $\langle b_i|$. As the kets $|b_i\rangle$ are orthonormal, the product UU^* gives identity

$$UU^* = I_{\mathcal{H}}. \tag{2.82b}$$

Thus U is a unitary matrix.

The decomposition (2.82) is a consequence of the Spectral Decomposition Theorem and is interpreted as *diagonalization* of the Hermitian matrix A. The result also holds for unitary matrices and more generally for normal matrices. Furthermore, from diagonalization one can obtain the spectral decomposition, therefore the two decompositions are equivalent.

Example 2.9 The diagonalization of the Hermitian matrix A, defined by (2.75), is obtained with

$$U = \frac{1}{2}\begin{bmatrix} 1+i & -i & -1 & -i \\ i & 1-i & 1 & -1 \\ 0 & 1 & -1 & i \\ 1 & 0 & 1 & 1 \end{bmatrix}, \qquad \tilde{\Lambda} = \begin{bmatrix} 1 & 0 & 0 & 0 \\ 0 & 1 & 0 & 0 \\ 0 & 0 & 2 & 0 \\ 0 & 0 & 0 & 3 \end{bmatrix} \tag{2.83}$$

2.11.1 *On the Nonuniqueness of a Diagonalization*

Is the diagonalization $A = U \Lambda U^*$ unique? A first remark is that the eigenvalues and also their multiplicity are unique. Next, consider two diagonalizations of the same matrix

$$A = U \Lambda U^*, \qquad A = U_1 \Lambda U_1^*. \tag{2.84}$$

By a left multiplication by U^* and a right multiplication by U, we find

$$U^* U \Lambda U^* U = \Lambda = U^* U_1 \Lambda U_1^* U.$$

Hence, a sufficient condition for the equivalence of the two diagonalizations is

$$U^* U_1 = I_{\mathcal{H}} \tag{2.85}$$

which reads: if the unitary matrices U and U_1 verify the condition (2.85), they both diagonalize the same matrix A.

But, we can also permute the order of the eigenvalues in the diagonal matrix Λ, combined with the same permutation of the eigenvectors in the unitary matrix, to get a new diagonalization. These, however, are only formal observations. The true answer to the question is given by [4]:

Theorem 2.3 *A matrix is uniquely diagonalizable, up to a permutation, if and only if its eigenvalues are all distinct.*

2.11.2 *Reduced Form of the EID*

So far, in the EID we have not considered the rank of matrix A. We observe that the rank of a linear operator is given by the number of nonzero eigenvalues (with multiplicity) $\tilde{\lambda}_i$. Then, if the $n \times n$ matrix A has the eigenvalue 0 with multiplicity p_0, the rank results in $r = n - p_0$. Sorting the eigenvalues with the null ones at the end, (2.77) and (2.82) become

$$\boxed{A = \sum_{i=1}^{r} \tilde{\lambda}_i \, |b_i\rangle \langle b_i| = U_r \, \tilde{\Lambda}_r \, U_r^*} \tag{2.86}$$

where

$$U_r = [|b_1\rangle, |b_2\rangle, \ldots, |b_r\rangle], \qquad \tilde{\Lambda}_r = \text{diag} \, [\tilde{\lambda}_1, \tilde{\lambda}_2, \ldots, \tilde{\lambda}_r]. \tag{2.86a}$$

Therefore, U_r has dimensions $n \times r$ and collects as columns only the eigenvectors corresponding to nonzero eigenvalues, and the $r \times r$ diagonal matrix $\tilde{\Lambda}_r$ collects such eigenvalues. In the form (2.82) the eigenvectors with null eigenvalues are not relevant (because they are multiplied by 0), whereas in (2.86) all the eigenvectors are relevant. These two forms will be often used in quantum decision and, to distinguish them, the first will be called *full* form and the second *reduced* form.

2.11.3 Simultaneous Diagonalization and Commutativity

The diagonalization of a Hermitian operator given by (2.82), that is,

$$A = U \tilde{\Lambda} U^* \tag{2.87}$$

is done *with respect to the orthonormal basis* constituted by the columns of the unitary matrix U. The possibility that another operator B be diagonalizable with respect to the same basis, namely

$$B = U \tilde{\Lambda}_1 U^* \tag{2.88}$$

is bound to the commutativity of the two operators. In fact:

Theorem 2.4 *Two Hermitian operators A and B are commutative, $BA = AB$, if and only if they are simultaneously diagonalizable, that is, if and only if they have a common basis made by eigenvectors.*

The sufficiency of the theorem is immediately verified. In fact, if (2.87) and (2.88) hold simultaneously, we have

$$BA = U \tilde{\Lambda}_1 U^* U \tilde{\Lambda} U^* = U \tilde{\Lambda}_1 \tilde{\Lambda} U^*$$

where the diagonal matrices are always commutable, $\tilde{\Lambda}_1 \tilde{\Lambda} = \tilde{\Lambda} \tilde{\Lambda}_1$, thus $BA = AB$. Less immediate is the proof of necessity (see [5, p. 229]).

Commutativity and non-commutativity of Hermitian operators play a role in Heisenberg's Uncertainty Principle (see Sect. 3.9).

2.12 Functional Calculus

One of the most interesting applications of the Spectral Theorem is Functional Calculus, which allows for the introduction of arbitrary functions of an operator A, such as

$$A^m, \quad e^A, \quad \cos A, \quad \sqrt{A}.$$

We begin by observing that from the idempotency and the orthogonality, in decomposition (2.74) we obtain

$$A^m = \sum_k \lambda_k^m P_k.$$

Thus a polynomial $p(\lambda)$, $\lambda \in \mathbb{C}$ over complex numbers is extended to the operators in the form

$$p(A) = \sum_k p(\lambda_k) P_k.$$

This idea can be extended to an arbitrary complex function $f : \mathbb{C} \to \mathbb{C}$ through

$$f(A) = \sum_k f(\lambda_k) P_k. \tag{2.89}$$

An alternative form of (2.89), based on the compact form (2.82), is given by

$$\boxed{f(A) = U f(\tilde{\Lambda}) U^*} \tag{2.89a}$$

where

$$f(\tilde{\Lambda}) = \operatorname{diag} [f(\tilde{\lambda}_1), \ldots, f(\tilde{\lambda}_n)]. \tag{2.89b}$$

The following theorem links the Hermitian operators to the unitary operators [4]

Theorem 2.5 *An operator U is unitary if and only if it can be written in the form*

$$U = e^{iA} \qquad \text{with } A \text{ Hermitian operator,}$$

where, from (2.89),

$$e^{iA} = \sum_k e^{i\lambda_k} P_k. \tag{2.90}$$

For a proof, see [4]. As a check, we observe that, if A is Hermitian, its eigenvalues λ_k are real. Then, according to (2.90), the eigenvalues of U are $e^{i\lambda_k}$, which, as it must be, have unit modulus.

Importance of the exponential of an operator   In Quantum Mechanics a fundamental role is played by the exponential of an operator, in particular in the form e^{A+B}, where A and B do not commute. This topic will be developed in Sect. 11.6.

2.12.1 Positive Semidefinite Operators

We first observe that if A is a Hermitian operator, the quantity $\langle x|A|x\rangle$ is always a real number. Then a Hermitian operator is:

- *nonnegative* or *positive semidefinite* and is written as $A \geq 0$ if

$$\langle x|A|x\rangle \geq 0 \qquad \forall\, |x\rangle \in \mathcal{H} \tag{2.91}$$

- *positive definite* and is written as $A > 0$ if

$$\langle x|A|x\rangle > 0 \qquad \forall\, |x\rangle \neq 0. \tag{2.92}$$

From the Spectral Theorem it can be proved that [1, Theorem 10.23]:

Theorem 2.6 *In a finite dimensional space, a Hermitian operator A is positive semidefinite (positive) if and only if its eigenvalues are nonnegative (positive).*

Remembering that the spectrum of a projector P is $\sigma(P) = \{0, 1\}$, we find, as anticipated in Sect. 2.9.3:

Corollary 2.1 *The projectors are always positive semidefinite operators.*

2.12.2 Square Root of an Operator

The square root of a positive semidefinite Hermitian operator $A \geq 0$ is introduced starting from its spectral resolution

$$A = \lambda_1 P_1 + \cdots + \lambda_k P_k \qquad \lambda_j \geq 0$$

in the following way:

$$\sqrt{A} = \sqrt{\lambda_1}\, P_1 + \cdots + \sqrt{\lambda_k}\, P_k \tag{2.93a}$$

or in equivalent form (see (2.89a))

$$\sqrt{A} = U \sqrt{\tilde{\Lambda}}\, U^* \tag{2.93b}$$

where $\sqrt{\tilde{\Lambda}} = \mathrm{diag}\,[\sqrt{\tilde{\lambda}_1}, \ldots, \sqrt{\tilde{\lambda}_n}]$. The definition of $\sqrt{A}$ is unique and it can be soon verified that from (2.93) it follows that

$$\left(\sqrt{A}\right)^2 = A.$$

The square root of a Hermitian operator will find interesting applications in Quantum Communications starting from Chap. 6.

2.12.3 Polar Decomposition

These decompositions regard arbitrary operators and therefore not necessarily Hermitian or unitary [4]

Theorem 2.7 (Polar decomposition) *Let A be an arbitrary operator. Then there always exists a unitary operator U and two positive definite Hermitian operators J and K such that*

$$A = UJ = KU$$

where J and K are unique with

$$J = \sqrt{A^*A} \quad and \quad K = \sqrt{AA^*}. \tag{2.94}$$

The theorem can be considered as an extension to square matrices of the polar decomposition of complex numbers: $z = |z|\,\exp(\mathrm{i}\arg z)$.

2.12.4 Singular Value Decomposition

So far we have considered operators of the Hilbert space, which in particular become complex square matrices. The singular value decomposition (SVD) considers more generally rectangular matrices.

Theorem 2.8 *Let A be an $m \times n$ complex matrix. Then the singular value decomposition of A results in*

$$A = UDV^*, \tag{2.95}$$

where

- *U is an $m \times m$ unitary matrix,*
- *V is an $n \times n$ unitary matrix,*
- *D is an $m \times n$ diagonal matrix with real nonnegative values on the diagonal.*

The positive values d_i of the diagonal matrix D are called the *singular values* of A. It can be proved that the SVD of a matrix A is strictly connected to the EIDs of the Hermitian matrices AA^* and A^*A (see [4] and Chap. 5).

If the matrix has rank r, the positive values d_i are r and a more explicit form can be given for the decomposition

$$A = U_r D_r V_r^* = \sum_{i=1}^{r} d_i \, |u_i\rangle\langle v_i| \tag{2.96}$$

where

- $U_r = [|u_1\rangle \cdots |u_r\rangle]$ is an $m \times r$ matrix,
- $V_r = [|v_1\rangle \cdots |v_r\rangle]$ is an $n \times r$ matrix,
- D is an $r \times r$ diagonal matrix collecting on the diagonal the singular values d_i.

The form (2.96) will be called the *reduced* form of the SVD. Both forms play a fundamental role in the theory of quantum decision.

Example 2.10 Consider the 4×2 matrix

$$A = \frac{1}{12\sqrt{2}} \begin{bmatrix} 5 & 1 \\ 3 - 2\,\mathrm{i} & 3 + 2\,\mathrm{i} \\ 1 & 5 \\ 3 + 2\,\mathrm{i} & 3 - 2\,\mathrm{i} \end{bmatrix}.$$

The SVD UDV^* of A results in

$$U = \begin{bmatrix} \frac{1}{2} & \frac{1}{2} & \frac{1}{2} & \frac{1}{2} \\ \frac{1}{2} & \frac{-\mathrm{i}}{2} & -\frac{1}{2} & \frac{\mathrm{i}}{2} \\ \frac{1}{2} & -\frac{1}{2} & \frac{1}{2} & -\frac{1}{2} \\ \frac{1}{2} & \frac{\mathrm{i}}{2} & -\frac{1}{2} & \frac{-\mathrm{i}}{2} \end{bmatrix}, \quad V = \begin{bmatrix} \frac{1}{\sqrt{2}} & \frac{1}{\sqrt{2}} \\ \frac{1}{\sqrt{2}} & -\frac{1}{\sqrt{2}} \end{bmatrix}, \quad D = \begin{bmatrix} \frac{1}{2} & 0 \\ 0 & \frac{1}{3} \\ 0 & 0 \\ 0 & 0 \end{bmatrix}.$$

Therefore the singular values are $d_1 = 1/2$ and $d_2 = 1/3$. The reduced form $U_r\, D_r\, V_r^*$ becomes

$$U_r = \begin{bmatrix} \frac{1}{2} & \frac{1}{2} \\ \frac{1}{2} & \frac{-\mathrm{i}}{2} \\ \frac{1}{2} & -\frac{1}{2} \\ \frac{1}{2} & \frac{\mathrm{i}}{2} \end{bmatrix}, \quad V_r = V = \begin{bmatrix} \frac{1}{\sqrt{2}} & \frac{1}{\sqrt{2}} \\ \frac{1}{\sqrt{2}} & -\frac{1}{\sqrt{2}} \end{bmatrix}, \quad D_r = \begin{bmatrix} \frac{1}{4} & 0 \\ 0 & \frac{1}{9} \end{bmatrix}.$$

We leave it to the reader to verify that, carrying out the products in these decompositions, one obtains the original matrix A.

2.12.5 Cholesky's Decomposition

Another interesting decomposition for Hermitian matrices is given by Cholesky's decomposition [4].

Theorem 2.9 (Cholesky's decomposition) *Let A be an $n \times n$ positive semidefinite Hermitian matrix, then there exists an upper triangular matrix U, of dimensions $n \times n$, with nonnegative elements on the main diagonal, such that*

$$A = U^*U.$$

If A is positive semidefinite then the matrix U is unique, and the elements of its main diagonal are all positive.

For a given matrix A it can turn out to be useful as an alternative to the EID for the factor decomposition of the density operators (see Chap. 5).

Example 2.11 Consider again the Hermitian matrix of Example 2.1

$$A = \frac{1}{4}\begin{bmatrix} 7 & -1+2i & -1 & -1-2i \\ -1-2i & 7 & -1+2i & -1 \\ -1 & -1-2i & 7 & -1+2i \\ -1+2i & -1 & -1-2i & 7 \end{bmatrix}.$$

Cholesky's decomposition, obtained with `Mathematica`, is specified by the triangular factor

$$U = \begin{bmatrix} \frac{\sqrt{7}}{2} & -\frac{\frac{1}{2}-i}{\sqrt{7}} & -\frac{1}{2\sqrt{7}} & -\frac{\frac{1}{2}+i}{\sqrt{7}} \\ 0 & \sqrt{\frac{11}{7}} & -\frac{2-3i}{\sqrt{77}} & -\frac{1+i}{\sqrt{77}} \\ 0 & 0 & \sqrt{\frac{17}{11}} & -\frac{3-4i}{\sqrt{187}} \\ 0 & 0 & 0 & 2\sqrt{\frac{6}{17}} \end{bmatrix} = \begin{bmatrix} 1.32 & -0.19+i0.38 & -0.19 & -0.19-i0.38 \\ 0 & 1.25 & -0.23+i0.34 & -0.11-i0.11 \\ 0 & 0 & 1.24 & -0.22+i0.29 \\ 0 & 0 & 0 & 1.19 \end{bmatrix}.$$

The decomposition is unique and has positive elements on the main diagonal, as predicted by Theorem 2.9.

Problem 2.11 $\star\star\star$ Let A be an arbitrary operator of the Hilbert space $\mathcal{H}$. Show that the operator AA^* is always positive semidefinite.

Hint: use diagonalization of A.

2.13 Tensor Product

The *tensor product* makes it possible to combine two or more vector spaces to obtain a larger vector space. In Quantum Mechanics it is used in Postulate 4 to combine quantum systems.

Before giving the definition, we introduce the symbolism that will be used. If $\mathcal{H}_1$ and $\mathcal{H}_2$ are two Hilbert spaces (on complex numbers), their tensor product is indicated in the form

$$\mathcal{H} = \mathcal{H}_1 \otimes \mathcal{H}_2$$

and, as we will see, the new Hilbert space $\mathcal{H}$ has dimension

$$\dim(\mathcal{H}_1 \otimes \mathcal{H}_2) = \dim(\mathcal{H}_1)\dim(\mathcal{H}_2).$$

If $|x\rangle \in \mathcal{H}_1$ and $|y\rangle \in \mathcal{H}_2$ the kets and the bras of $\mathcal{H}$ are indicated, respectively, in the form

$$|x\rangle \otimes |y\rangle \qquad \langle x| \otimes \langle y| , \tag{2.97}$$

which is sometimes simplified as

$$|x\rangle|y\rangle \qquad \langle x|\langle y|. \tag{2.97a}$$

If A is an operator of $\mathcal{H}_1$ and B is an operator of $\mathcal{H}_2$, the operator of $\mathcal{H}$ is indicated in the form

$$A \otimes B.$$

We now list the abstract definitions of the vectors and of the operators that are obtained through the tensor product. However, as these definitions are very abstract, or better, scarcely operational, they can be skipped and the reader may move to the next section, where the tensor product is developed for matrices and is more easily understood.

2.13.1 Abstract Definition ⇓

We want to combine two Hilbert spaces $\mathcal{H}_1$ and $\mathcal{H}_2$ using the tensor product $\otimes$, and let us denote by $|x\rangle$, $|x_1\rangle$, $|x_2\rangle$ arbitrary kets of $\mathcal{H}_1$, and by $|y\rangle$, $|y_1\rangle$, $|y_2\rangle$ arbitrary kets of $\mathcal{H}_2$. Then, by definition, the tensor product of kets must have the following properties:

(1) homogeneity

$$a(|x\rangle \otimes |y\rangle) = (a|x\rangle) \otimes |y\rangle = |x\rangle \otimes a(|y\rangle) \tag{2.98a}$$

(2) linearity with respect to the first factor

$$(|x_1\rangle + |x_2\rangle) \otimes |y\rangle = |x_1\rangle \otimes |y\rangle + |x_2\rangle \otimes |y\rangle \tag{2.98b}$$

(3) linearity with respect to the second factor

$$|x\rangle \otimes (|y_1\rangle + |y_2\rangle) = |x\rangle \otimes |y_1\rangle + |x\rangle \otimes |y_2\rangle. \tag{2.98c}$$

Once defined the tensor products between the kets, imposing the above conditions, the tensor product between bras can be obtained as follows:

$$\langle x| \otimes \langle y| = (|x\rangle)^* \otimes (|y\rangle)^*.$$

Then we can move on to define the tensor product of two operators in this way:

$$(A \otimes B)(|x\rangle \otimes |y\rangle) = (A|x\rangle) \otimes (B|y\rangle) \tag{2.99}$$

where on the right-hand side we find the tensor product of two kets, which has already been defined.

Finally, the inner product on $\mathcal{H} = \mathcal{H}_1 \otimes \mathcal{H}_2$ is defined by the inner products on $\mathcal{H}_1$ and $\mathcal{H}_2$, through the relation

$$((\langle x_1| \otimes \langle y_1|)(|x_2\rangle \otimes |y_2\rangle) = \langle x_1|x_2\rangle \, \langle y_1|y_2\rangle. \tag{2.100}$$

2.13.2 Kronecker's Product of Vector and Matrices

Consider two row vectors written in standard notation

$$x = [x_0, \ldots, x_{m-1}], \qquad y = [y_0, \ldots, y_{n-1}]$$

and suppose we want to form a "product" containing all possible products between the elements of two vectors

$$x_i \, y_j, \qquad i = 0, 1, \ldots, m - 1, \; j = 0, 1, \ldots, n - 1$$

The natural procedure would be to build the $m \times n$ matrix

$$[x_i \, y_j]$$

but with the Kronecker product we want to build *a vector* containing all the mn products. The problem is that of passing from a bidimensional configuration (2D), like the matrix $[x_i \, y_j]$, to a 1D configuration, as a vector is. But, while in 1D there is a natural order of the indexes, namely 0, 1, 2, ..., in 2D such order does not exist for the indexes (i, j). We must then introduce a conventional ordering to establish, for example, whether $(1, 2)$ comes before or after $(2, 1)$. A solution to the problem is given by the *lexicographical order*[10] obtained as follows: in the pair of indexes (i, j) we fix the first index starting from $i = 0$ and let run the second index j along

---

[10] This name comes from the order given to words in the dictionary: a word of k letters, $a = (a_1, \ldots, a_k)$ appears in the dictionary before the word $b = (b_1, \ldots, b_k)$, symbolized $a < b$, *if and only if the first a_i which is different from b_i comes before b_i in the alphabet*. In our context the

its range, obtaining $(0, 0)$, $(0, 1), \ldots, (1, n - 1)$, we then move to the value $i = 1$, until $i = m - 1$. In this way we associate the pair of integer indexes to a single index given by

$$h = j + (i - 1)\,n, \qquad j = 0, 1, \ldots, n - 1 \quad i = 0, 1, \ldots, m - 1$$

which gives the required 1D ordering. The resulting vector can be written in the compact form:

$$x \otimes y = [x_1 y, x_2 y, \ldots, x_m y] \tag{2.101}$$

where the form $x_i y$ indicates the n-tuple $(x_i y_1, \ldots, x_i y_n)$. Relation (2.101) defines Kronecker's product of two vectors x and y.

Next we consider two matrices

$$A = [a_{ir}], \qquad i = 1, \ldots, m, \ r = 1, \ldots, p$$
$$B = [b_{is}], \qquad j = 1, \ldots, n, \ s = 1, \ldots, q$$

where the dimensions are respectively $m \times p$ and $n \times q$. To collect all the products of the entries of the two matrices we would have to form a 4D matrix

$$[a_{ir}\, b_{js}]$$

but, if we want a standard 2D matrix, we apply the lexicographical order to the pairs of indexes (i, j) and (r, s), given by the integers

$$h = j + i(p - 1), \qquad k = s + r(q - 1)$$

where h goes from 1 to mp and k from 1 to nq. In this way we build an $A \otimes B$ matrix of dimension $mn \times pq$.

The compact form for $A \otimes B$ is

$$A \otimes B = \begin{bmatrix} a_{11}B & \ldots & a_{1p}B \\ \vdots & \ddots & \vdots \\ a_{m1}B & \ldots & a_{mp}B \end{bmatrix} \tag{2.102}$$

where on the left-hand side we notice the "blocks" $a_{ij}B$, which are $n \times q$ matrices. Once we expand these blocks, we can see that on the left-hand side the resulting matrix has dimensions $mn \times pq$.

(Footnote 10 continued)
alphabet is given by the set of integers. Then we find, e.g., that $(1, 3) < (2, 1)$, $(0, 3, 2) < (1, 0, 1)$ and $(1, 1, 3) < (1, 2, 0)$.

It can be verified that (2.102) falls into the abstract definition of tensor product, based on the previous "abstract" conditions (1), (2), and (3).

Relation (2.102) extends to matrices in the compact form (2.101) seen for vectors and represents the definition of the *Kronecker product* for matrices. It includes the case of vectors, provided that vectors are regarded as matrices.

Example 2.12 If

$$
|a\rangle = \begin{bmatrix} a_1 \\ a_2 \end{bmatrix}, \qquad |b\rangle = \begin{bmatrix} b_1 \\ b_2 \\ b_3 \end{bmatrix}
$$

the tensor product gives

$$
|a\rangle \otimes |b\rangle = \begin{bmatrix} a_1 b \\ a_2 b \end{bmatrix} = \begin{bmatrix} a_1 b_1 \\ a_1 b_2 \\ a_1 b_3 \\ a_2 b_1 \\ a_2 b_2 \\ a_2 b_3 \end{bmatrix},
$$

which is a 6×1 vector. In particular,

$$
|a\rangle = \begin{bmatrix} 1+i \\ 2+i \end{bmatrix}, \qquad |b\rangle = \begin{bmatrix} 1+i \\ 2+2i \\ 3+2i \end{bmatrix} \qquad \rightarrow \qquad |a\rangle \otimes |b\rangle = \begin{bmatrix} 2i \\ 4i \\ 5+i \\ 1+3i \\ 2+6i \\ 4+7i \end{bmatrix}.
$$

If

$$
A = \begin{bmatrix} a_{11} & a_{12} \\ a_{21} & a_{22} \end{bmatrix}, \qquad B = \begin{bmatrix} b_{11} & b_{12} \\ b_{21} & b_{22} \\ b_{31} & b_{32} \end{bmatrix}
$$

are two matrices of dimensions, respectively, 2×2 and 3×2, the tensor product yields

$$
A \otimes B = \begin{bmatrix} a_{11} B & a_{12} B \\ a_{21} B & a_{22} B \end{bmatrix}
$$

which is a 6×4 matrix. In particular, if

$$
A = \begin{bmatrix} i & 3 \\ 2+i & 1 \end{bmatrix}, \quad B = \begin{bmatrix} 2i & 1 \\ 2 & i \\ i & 3 \end{bmatrix} \; \rightarrow \; A \otimes B = \begin{bmatrix} -2 & i & 6i & 3 \\ 2i & -1 & 6 & 3i \\ -1 & 3i & 3i & 9 \\ -2+4i & 2+i & 2i & 1 \\ 4+2i & -1+2i & 2 & i \\ -1+2i & 6+3i & i & 3 \end{bmatrix}.
$$

For Kronecker's product (2.102) the following rules can be established. The transpose and the conjugate transpose simply result in

$$
(A \otimes B)^{\mathrm{T}} = A^{\mathrm{T}} \otimes B^{\mathrm{T}}, \qquad (A \otimes B)^* = A^* \otimes B^* \tag{2.103}
$$

and also the important rule holds (valid if the dimensions are compatible with ordinary products)

$$
\boxed{(A \otimes B)(C \otimes D) = (AC) \otimes (BD)} \tag{2.104}
$$

which contains both the Kronecker product and the ordinary matrix product and will be called **mixed-product law**. In addition, for two *invertible square* matrices we have

$$
(A \otimes B)^{-1} = A^{-1} \otimes B^{-1}. \tag{2.105}
$$

2.13.3 *Properties of the Tensor Product*

Kronecker's product of matrices allows us now to interpret and verify the definition and the properties of the tensor product on Hilbert spaces. This is done directly when $\mathcal{H}_1 = \mathbb{C}^m$ an $\mathcal{H}_2 = \mathbb{C}^n$ and it turns out that $\mathcal{H}_1 \otimes \mathcal{H}_2 = \mathbb{C}^{mn}$, but using the matrix representation it can be done for arbitrary Hilbert spaces of finite dimension (and with some effort even of infinite dimension).

Then, if $\mathcal{H}_1$ and $\mathcal{H}_2$ have dimensions, respectively, m and n, and if $|x\rangle \in \mathcal{H}_1$, $|y\rangle \in \mathcal{H}_2$, it results that:

- $|x\rangle \otimes |y\rangle$ is a ket of dimension mn (column vector)
- $\langle x| \otimes \langle y|$ is a bra of dimension mn (row vector).

For example, given the two kets $|x\rangle \in \mathcal{H}_1 = \mathbb{C}^2$ and $|y\rangle \in \mathcal{H}_2 = \mathbb{C}^3$

$$
|x\rangle = \begin{bmatrix} x_1 \\ x_2 \end{bmatrix}, \qquad |y\rangle = \begin{bmatrix} y_1 \\ y_2 \\ y_3 \end{bmatrix}
$$

the tensor product gives

$$|x\rangle \otimes |y\rangle = \begin{bmatrix} x_1 y_1 \\ x_1 y_2 \\ x_1 y_3 \\ x_2 y_1 \\ x_2 y_2 \\ x_2 y_3 \end{bmatrix}, \qquad \langle x| \otimes \langle y| = \left[x_1^* y_1^*, x_1^* y_2^*, x_1^* y_3^*, x_2^* y_1^*, x_2^* y_2^*, x_2^* y_3^* \right].$$

If A is an operator of $\mathcal{H}_1$ and B is an operator of $\mathcal{H}_2$, then

- $A \otimes B$ is an operator to which an $mn \times mn$ square matrix must be associated.

The following general properties can also be established:

(1) If $\{|b_i\rangle, i \in I\}$ is a basis for $\mathcal{H}_1$ and $\{|c_j\rangle, j \in J\}$ is a basis for $\mathcal{H}_2$, then

$$\{|b_i\rangle \otimes |c_j\rangle, i \in I, j \in J\} \tag{2.106}$$

is a basis for $\mathcal{H}_1 \otimes \mathcal{H}_2$.
(2) $\dim(\mathcal{H}_1 \otimes \dim \mathcal{H}_2) = \dim \mathcal{H}_1 \ \dim \mathcal{H}_2$.
(3) If $\{\lambda_i, i \in I\}$ is the spectrum of A and $\{\mu_j, j \in J\}$ is the spectrum of B, the spectrum of $A \otimes B$ results in

$$\sigma(A \otimes B) = \{\lambda_i \mu_j, \ i \in I, j \in J\}. \tag{2.107}$$

Analogously, the eigenvalues of $A \otimes B$ are given by $\{|\lambda_i\rangle \otimes |\mu_i\rangle\}$.
(4) If A and B are (unitary) Hermitian operators, also $A \otimes B$ is a (unitary) Hermitian operator.
(5) If A and B are positive definite Hermitian operators, also $A \otimes B$ is a positive definite Hermitian operator.
(6) For the trace, the simple rule holds that

$$\mathrm{Tr}[A \otimes B] = \mathrm{Tr}[A]\,\mathrm{Tr}[B]. \tag{2.108}$$

Final Comment on Tensor Product

As mentioned, the tensor product appears in Postulate 4 of Quantum Mechanics. The properties of this product, albeit with a rather heavy symbolism, seem natural enough at first glance. However, just these apparently "intuitive" properties lead to paradoxical consequences, which are at the foundations of very interesting applications, as we will see at the end of the following chapter.

Problem 2.12 ⋆ Prove that if A and B are Hermitian operators, also $A \otimes B$ is a Hermitian operator.

Problem 2.13 ⋆⋆ Establish the compatibility conditions for the dimensions of the matrices in the mixed-product law (2.104).

Problem 2.14 ⋆⋆ Prove property (2.107) of the Kronecker product and, more specifically, prove that, if λ is an eigenvalue of A with eigenvector $|\lambda\rangle$ and μ is an eigenvalue of B with eigenvector $|\mu\rangle$, then $\lambda\mu$ is an eigenvalue of $A \otimes B$ with eigenvector $|\lambda\rangle \otimes |\mu\rangle$.

Problem 2.15 ⋆⋆⋆ The mixed-product law can be extended in several ways. In particular,

$$(A_1 \otimes A_2)(B_1 \otimes B_2)(C_1 \otimes C_2) = (A_1 B_1 C_1) \otimes (A_2 B_2 C_2). \tag{E5}$$

Prove this relation using (2.104).

Problem 2.16 ⋆⋆ Prove that, if the matrices A_1 and A_2 have, respectively, the diagonalizations (see (2.87))

$$A_1 = U_1 \Lambda_1 U_1^*, \qquad A_2 = U_2 \Lambda_2 U_2^*$$

then

$$A_1 \otimes A_2 = (U_1 \otimes U_2)(\Lambda_1 \otimes \Lambda_2)(U_1^* \otimes U_2^*) \tag{E6}$$

is a diagonalization of $A_1 \otimes A_2$.

2.14 Other Fundamentals Developed Throughout the Book

This chapter developed the essential fundamentals necessary for the comprehension of the elements of Quantum Mechanics that will be used in the next chapter, which in turn are indeed required in the study of Quantum Communications systems developed in Part II.

On the other hand, the mathematics encountered in the field of Quantum Mechanics is very extensive and a further development of fundamentals is out of the scope of this book. Considering our philosophy of introducing the needed preliminaries in a gradual form, a few fundamentals, which will be needed in Part III on Quantum Information, will be introduced just before describing the applications. We mention in particular the EID with a continuous spectrum and the partial trace.

References

1. S. Roman, *Advanced Linear Algebra* (Springer, New York, 1995)
2. G. Cariolaro, *Unified Signal Theory* (Springer, London, 2011)
3. P.A.M. Dirac, *The Principles of Quantum Mechanics* (Oxford University Press, Oxford, 1958)
4. R.A. Horn, C.R. Johnson, *Matrix Analysis* (Cambridge University Press, Cambridge, 1998)
5. M.A. Nielsen, I.L. Chuang, *Quantum Computation and Quantum Information* (Cambridge University Press, Cambridge, 2000)

Chapter 3
Elements of Quantum Mechanics

3.1 Introduction

Quantum Mechanics will be formulated assuming four *postulates*:

- **Postulate 1**: gives the universal model of any physical system: a Hilbert space on the field of complex numbers.
- **Postulate 2**: models the temporal evolution of a *closed* physical system that is not influenced by other physical systems.
- **Postulate 3**: regards the *information* that can be extracted (through a *quantum measurement*) from a quantum system at a given time instant.
- **Postulate 4**: formalizes the interaction among physical systems through a combination of multiple Hilbert spaces into a single Hilbert space.

In this formulation we will partly follow Nielsen and Chuang's textbook [1], which appears as one of the most complete and up to date. The variations with respect to such textbook, which is mainly concerned with *Quantum Computing* and *Quantum Information*, come mostly from the fact that our main objective is *Quantum Communications*. Therefore, some aspects of Quantum Mechanics will not be further expanded, such as the consequences of Postulate 2 on the evolution of a quantum system, while other points, in particular *quantum measurements* (Postulate 3) will be exhaustively investigated.

Clearly, the postulates of Quantum Mechanics are completely abstract, and give no indications on how to associate to a given physical system a corresponding Hilbert space. However, this is a common aspect of all the models of reality, in which the match between the mathematical model and the physical reality to be described must be done in a "reasonable" way, often with the help of Classical Mechanics. The success of the choice will depend on the consequences and the results that will be obtained. This does not rule out the existence of well-established choices in particular domains, such as the model of an atomic or subatomic particle, or the model of the electromagnetic radiation produced by a laser. An example of real-world model, again based on few axioms, and certainly familiar to the reader, is

© Springer International Publishing Switzerland 2015

G. Cariolaro, *Quantum Communications*, Signals and Communication Technology,
DOI 10.1007/978-3-319-15600-2_3

given by Probability Theory which forms the framework for many disciplines, with undoubtedly useful consequences and results.

The fundamentals of Quantum Mechanics formulated in this chapter through four postulates are adequate for the development of Quantum Communications in Part II (which will be confined to the digital transmission of information). But a further in-depth analysis will be necessary in the final chapters (Part III) for the development of Quantum Information.

3.2 The Environment of Quantum Mechanics

The first postulate of Quantum Mechanics defines the environment in which any physical system must be described (Fig. 3.1).

Postulate 1 To each closed (or isolated) physical system, a Hilbert space $\mathcal{H}$ of appropriate dimension on the field $\mathbb{C}$ of complex numbers, called *state space*, must be associated. At each time instant of its evolution, the system is completely specified by a *state* $|\psi\rangle$, given by a *unit* vector of $\mathcal{H}$. □

Clearly, this postulate, like the other postulates of Quantum Mechanics, is completely abstract, and does not specify the nature of the state $|\psi\rangle$, except for the mathematical detail that $|\psi\rangle$ must be unitary, that is, it must verify the normalization condition

$$\langle\psi|\psi\rangle = 1. \tag{3.1}$$

An important consequence on the possible states of a quantum system, deriving from the *linearity* of Hilbert spaces, is **state superposition**: if $|\psi_1\rangle, |\psi_2\rangle, \ldots, |\psi_n\rangle$ are states of $\mathcal{H}$, also their linear combination

$$|\psi\rangle = a_1|\psi_1\rangle + a_2|\psi_2\rangle + \cdots + a_n|\psi_n\rangle, \qquad a_i \in \mathbb{C} \tag{3.2}$$

Fig. 3.1 Representation of a quantum system in the state $|\psi\rangle$

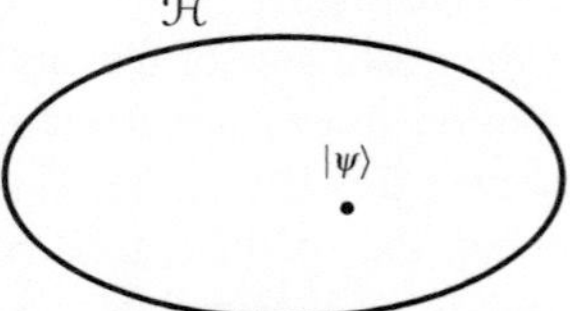

$\mathcal{H}$: Hilbert space on complex numbers

$|\psi\rangle$: state of the system

is a state of $\mathcal{H}$. The complex coefficients a_i must verify the normalization condition (3.1), given by

$$\sum_i \sum_j a_i^* a_j \langle \psi_i | \psi_j \rangle = 1 . \tag{3.2a}$$

The state superposition has several consequences, sometimes surprising, and represents a remarkable difference of Quantum Mechanics with respect to Classic Mechanics.

On the phase of a quantum state. As an additional mathematical detail, it must be pointed out that a quantum state is defined *modulus a phase rotation*, and precisely all the vectors obtained by multiplying a ket $|\psi\rangle$ by an arbitrary *phasor* $e^{i\varphi}$ are representing the same state.

3.2.1 An Elementary Example of a Quantum System: The Qubit

The most elementary example of a quantum system is the *qubit*, which must be seen as a bidimensional Hilbert space, substantially $\mathcal{H} = \mathbb{C}^2$. In such space, a basis is provided by two orthonormal vectors, indicated in the form $|0\rangle$ and $|1\rangle$. A generic state of a qubit system can be expressed in the form

$$|\psi\rangle = a\,|0\rangle + b\,|1\rangle \tag{3.3}$$

where a and b are complex numbers. As the normalization condition must hold, a and b are not arbitrary, but must verify the condition

$$|a|^2 + |b|^2 = 1.$$

For instance

$$|\psi_1\rangle = \frac{1}{\sqrt{2}}\,|0\rangle + \frac{1}{\sqrt{2}}\,|1\rangle, \qquad |\psi_2\rangle = \frac{i}{\sqrt{3}}\,|0\rangle + \frac{i+1}{\sqrt{3}}\,|1\rangle$$

are possible states of this bidimensional system. In particular, $|\psi_2\rangle$ is the superposition of the state $|0\rangle$, with amplitude $i/\sqrt{3}$, and of the state $|1\rangle$, with amplitude $(i+1)/\sqrt{3}$.

The qubit, intended as a quantum system, can be compared to the *bit*, intended as a two-state binary system (and not as the unit of measure of information). To the state $|0\rangle$ $(|1\rangle)$ of the qubit it corresponds the state 0 (1) of the bit; but the qubit can be in every state given by the superposition of $|0\rangle$ and $|1\rangle$ according to (3.3) and therefore has an infinity of possible states, whereas the bit presents only two states: state 0 and state 1, with no intermediate states. Other more remarkable differences between the qubit and the bit will be seen with the other postulates. Considering the importance, we must examine the qubit more thoroughly. This will be done in Sect. 3.12 of this chapter.

3.2.2 A More Elaborate Example: The Laser Radiation

The quantum system describing a monochromatic coherent radiation emitted by a laser at a certain optical frequency v, and observed within a certain time interval, is formulated by an infinite dimensional Hilbert space $\mathcal{H}$. In this system the basis is given by a set $\{|n\rangle,\ n = 0, 1, \ldots\}$ of orthonormal states in which the parameter n is a natural number representing the number of photons contained in the state $|n\rangle$; for this reason, the states $|n\rangle$ are called *number states* (or *Fock states*). Once fixed this basis, the monochromatic radiation of a laser is represented by a *coherent state* $|\alpha\rangle$, which is given by the linear combination of number states, according to the expression

$$|\alpha\rangle = \mathrm{e}^{-\frac{1}{2}|\alpha|^2} \sum_{n=0}^{\infty} \frac{\alpha^n}{\sqrt{n!}} \, |n\rangle \, . \tag{3.4}$$

Here α is a complex parameter characterizing the state $|\alpha\rangle$. Note that in this expression $|\alpha\rangle$ is a quantum state, a point of the infinite dimensional Hilbert space $\mathcal{H}$ with basis the number states $|n\rangle$, while α is a complex number. In symbols $|\alpha\rangle \in \mathcal{H}$ and $\alpha \in \mathbb{C}$. Also, $|n\rangle \in \mathcal{H}$, while n is a non negative integer.

As we will see, using the postulate on quantum measurements, it can be proved that α has the meaning given by

$$|\alpha|^2 = \text{average number of photons in state } |\alpha\rangle. \tag{3.4a}$$

By fixing α, e.g., $\alpha = 0.6 - \mathrm{i}2.4$, one gets from (3.4) the coherent state $|0.6 - \mathrm{i}2.4\rangle$, having $|\alpha|^2 = 6.12$ as average number of photons. In particular, for $\alpha = 0$ we obtain the state $|0\rangle$, containing no photons (ground state). According to (3.4), this state coincides with the state $|0\rangle$ of the number states $|n\rangle$, that is,

$$|\alpha\rangle_{\alpha=0} = |n\rangle_{n=0} \, . \tag{3.4b}$$

We can guess from the structure of (3.4) that the random variables bound to the coherent states are related to the Poisson regime.

Coherent states will play a fundamental role in optical communications formulated according to Quantum Mechanics [2, 3]. They will be further investigated in Chap. 7 and subsequent, and mainly in Chap. 11 in the context of Gaussian states.

Example 3.1 Let us prove that the state given by (3.4) verifies the normalization condition $\langle \alpha | \alpha \rangle = 1$, as required by Postulate 1. The bra $\langle \alpha |$ corresponding to the ket (3.4) results in

$$\langle \alpha | = \mathrm{e}^{-\frac{1}{2}|\alpha|^2} \sum_{m=0}^{\infty} \frac{(\alpha^*)^m}{\sqrt{m!}} \langle m |$$

and therefore, remembering that the number states $|n\rangle$ are orthonormal, that is, $\langle m | n \rangle = \delta_{mn}$, we obtain

$$\begin{aligned}
\langle \alpha | \alpha \rangle &= \mathrm{e}^{-|\alpha|^2} \sum_{m=0}^{\infty} \sum_{n=0}^{\infty} \frac{(\alpha^*)^m \alpha^n}{\sqrt{m!n!}} \langle m | n \rangle \\
&= \mathrm{e}^{-|\alpha|^2} \sum_{m=0}^{\infty} \frac{|\alpha|^{2m}}{m!} \\
&= \mathrm{e}^{-|\alpha|^2} \mathrm{e}^{|\alpha|^2} = 1 \ .
\end{aligned}$$

Notice that the normalization applies also to the ground state $|0\rangle$, which then has unit amplitude (and should not be confused with the null element of the Hilbert space).

3.3 On the Statistical Description of a Closed Quantum System

The first postulate of Quantum Mechanics establishes that a closed (or isolated) quantum system, formalized through a Hilbert space $\mathcal{H}$, at each time instant of its evolution, is *completely described by a unit vector of* $\mathcal{H}$, called *state*. We must insist[1] on the fact that at each preset instant t, the *system is at one precise point of* $\mathcal{H}$ expressed by a ket.

3.3.1 Pure States and Mixed States

We first remark that, once we have decided to represent a physical system by a Hilbert space of a given dimension, any point of $\mathcal{H}$ is an *admissible state*. For instance, in a qubit system, where $\mathcal{H} = \mathbb{C}^2$, every pair of complex numbers, constrained by normalization, is an admissible state.

---

[1] This insistence is justified by the fact that in the literature one finds expressions like "the system is in a mixture of states (mixed states)," which suggests the idea of a simultaneous presence of many states, whereas the first postulate of Quantum Mechanics establishes that an isolated system is in just one well-defined state.

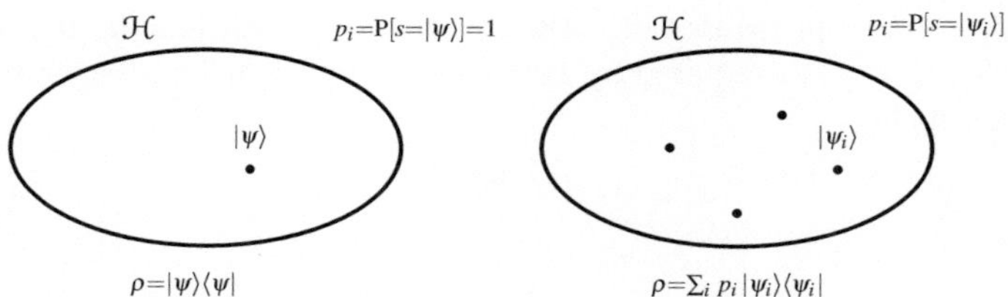

Fig. 3.2 Quantum system $\mathcal{H}$ in condition of certain state (*pure state*) and of random state (*mixed states*)

The following classification (pure and mixed states) is concerned with the knowledge one has about the state of the system at a given time instant. The degree to which the state of a quantum system *is known* depends **on the point of view of the observer**, who may either fully know the system's state, or only know its statistical description. In other words, two cases must be considered:

(1) the observer knows the state of the system, say $s = |\psi\rangle \in \mathcal{H}$, with certainty,
(2) the observer knows that the state of the system belongs to a subset of $\mathcal{H}$, say $\mathcal{S} = \{|\psi_1\rangle, |\psi_2\rangle, \ldots\}$, but knows the specific state only probabilistically, through the probabilities

$$p_i := \mathrm{P}[s = |\psi_i\rangle]\,, \tag{3.5}$$

obviously with $p_i \geq 0$ and $\sum_i p_i = 1$.

The two situations are schematically illustrated in Fig. 3.2.

If we want to describe these two points of view in a single probabilistic model, in case (1) the state $\{s = |\psi\rangle\}$ represents the certain event, i.e. with $\mathrm{P}[s = |\psi\rangle] = 1$, while in case (2) the state s must be expressed as a random entity, that is, as a random variable, or, more specifically, as a **random state**. Then, according to Probability Theory, the random state is completely described by the **ensemble**

$$\mathcal{E} = (\mathcal{S}, p), \qquad \text{with} \quad \mathcal{S} = \{|\psi_1\rangle, |\psi_2\rangle, \ldots\} \quad \text{and} \quad p_i = \mathrm{P}[s = |\psi_i\rangle]\,. \tag{3.6}$$

An alternative way of describing the random state s, commonly adopted in Quantum Mechanics, is given by the **density operator**, defined as follows

$$\boxed{\rho = \sum_i p_i\, |\psi_i\rangle\langle\psi_i|} \tag{3.7}$$

and is therefore given by a linear combination of the *elementary operators* $|\psi_i\rangle\langle\psi_i|$, weighed by the respective probabilities.

The density operator also allows us to describe a system in a pure state $|\psi\rangle$ through the degenerate form

$$\rho = |\psi\rangle\langle\psi| \tag{3.8}$$

where $P[s = |\psi\rangle] = 1$.

As will be seen in the coming paragraphs, from the density operator ρ we can obtain all the statistical descriptions relevant to Quantum Mechanics, because ρ encompasses in its structure both the states and their probabilities.

Pure states and mixture of states. In Quantum Mechanics, to indicate that the state of the quantum system is known, or certain, it is said to be in a *pure state*. On the other hand, if the state is probabilistically known, and, as we have seen, it is a random variable, the system is said to be in a *mixture of states* (mixed states), characterized by the density operator ρ [4, 5]. Nielsen and Chuang [1] to indicate a mixture of states use the terminology *ensemble* $\{p_i, |\psi_i\rangle\}$, which appears more appropriate. The terms pure state and mixed states are so consolidated in the literature that they will be used also in this book. Note that we often encounter expression of the form: "the quantum system is in state ρ" to indicate the presence of a mixed state characterized by the density operator ρ (not excluding that ρ be a pure state with the degenerate form (3.8)).

3.3.2 Properties of the Density Operator

As seen above, the density operator derives from two entities: the probabilities $\{p_i\}$ and the states $\{|\psi_i\rangle\}$. In general, no restriction is imposed on the states, apart from the fact that they must be unitary, $\langle\psi_i|\psi_i\rangle = 1$. Thus, in general, the states of the ensemble $\{p_i, |\psi_i\rangle\}$ are not orthogonal.

It can be verified that the density operator has the following properties:

(1) $\rho = \rho^*$, it is a Hermitian operator,
(2) $\rho \geq 0$, it is a positive semidefinite operator,
(3) $\mathrm{Tr}(\rho) = 1$, it has unitary trace,
(4) $\mathrm{Tr}(\rho^2) \leq 1$ and in particular $\mathrm{Tr}(\rho^2) = 1$ if and only if the system is in a pure state.

To prove property (2), consider an arbitrary state $|\varphi\rangle \in \mathcal{H}$. We obtain

$$\langle\varphi|\rho|\varphi\rangle = \sum_i p_i \langle\varphi|\psi_i\rangle\langle\psi_i|\varphi\rangle$$

$$= \sum_i p_i |\langle\varphi|\psi_i\rangle|^2 \geq 0$$

and then, by definition (see (2.91)), $\rho \geq 0$. For property (3) we use the linearity of the trace

$$\mathrm{Tr}[\rho] = \sum_i p_i \mathrm{Tr}[|\psi_i\rangle\langle\psi_i|] = \sum_i p_i = 1$$

where $\mathrm{Tr}[|\psi_i\rangle\langle\psi_i|] = 1$ because $|\psi_i\rangle\langle\psi_i|$ is an elementary projector (see Sect. 2.9). Property (4), which will be proved in Problem 3.2 and also verified on some examples, makes it possible to check whether the system is in a certain state (for the observer) or in a state only statistically known.

Example 3.2 Consider a qubit in a generic pure state $|\psi\rangle = a|0\rangle + b|1\rangle$ with $|a|^2 + |b|^2 = 1$, where $\{|0\rangle, |1\rangle\}$ is the basis. The density operator results in

$$\rho = |\psi\rangle\langle\psi| = (a|0\rangle + b|1\rangle)(a^*\langle 0| + b^*\langle 1|)$$
$$= |a|^2|0\rangle\langle 0| + |b|^2|1\rangle\langle 1| + ab^*|0\rangle\langle 1| + ba^*|1\rangle\langle 0| .$$

In terms of matrices, with

$$|0\rangle = \begin{bmatrix} 1 \\ 0 \end{bmatrix}, \qquad |1\rangle = \begin{bmatrix} 0 \\ 1 \end{bmatrix}$$

it results

$$|0\rangle\langle 0| = \begin{bmatrix} 1 & 0 \\ 0 & 0 \end{bmatrix}, \qquad |1\rangle\langle 1| = \begin{bmatrix} 0 & 0 \\ 0 & 1 \end{bmatrix}, \qquad |0\rangle\langle 1| = \begin{bmatrix} 0 & 1 \\ 0 & 0 \end{bmatrix}, \qquad |1\rangle\langle 0| = \begin{bmatrix} 0 & 0 \\ 1 & 0 \end{bmatrix}$$

and therefore

$$\rho = \begin{bmatrix} |a|^2 & ab^* \\ a^*b & |b|^2 \end{bmatrix} = \begin{bmatrix} a \\ b \end{bmatrix} [a^* \ b^*] .$$

We observe that

$$\mathrm{Tr}[\rho] = |a|^2 + |b|^2 = 1 ,$$

while we leave it to the reader to verify that $\mathrm{Tr}[\rho^2] = 1$.

Suppose now that the qubit is in a mixed state, namely in the state $|0\rangle$ with probability $\frac{1}{3}$ and in the state $|1\rangle$ with probability $\frac{2}{3}$. Then, the qubit is described by the density operator

$$\rho = \frac{1}{3}|0\rangle\langle 0| + \frac{2}{3}|1\rangle\langle 1|$$

which in term of matrices yields

$$\rho = \begin{bmatrix} \frac{1}{3} & 0 \\ 0 & \frac{2}{3} \end{bmatrix} \quad \rightarrow \quad \rho^2 = \begin{bmatrix} \frac{1}{9} & 0 \\ 0 & \frac{4}{9} \end{bmatrix} .$$

Now $\mathrm{Tr}[\rho] = 1$ while $\mathrm{Tr}[\rho^2] = \frac{5}{9} < 1$.

Example 3.3 (*Thermal noise*) In a cavity at thermal equilibrium at a certain absolute temperature T_0 a chaotic radiation takes place, known as *thermal noise* or *background noise*. Through the statistical theory of thermodynamics we can formulate such noise in the framework of the same Hilbert space seen for coherent states, where the orthonormal basis is formed by the number states $|n\rangle$. It can be shown [5] that the corresponding density operator results in

$$\rho = (1 - \varepsilon) \sum_{n=0}^{\infty} \varepsilon^n |n\rangle\langle n|, \qquad \varepsilon := e^{-h\nu/kT_0} \tag{3.9}$$

where ν is the frequency of the radiation mode and T_0 the absolute temperature (h and k are Planck's and Boltzmann's constants respectively). It can be proved (see Problem 3.3) that such operator verifies the condition

$$\mathrm{Tr}[\rho^2] = \frac{1 - \varepsilon}{1 + \varepsilon} < 1,$$

which establishes that the cavity is not in a pure state, but in a mixture of states (as expected, because the observer doesn't know the states of the system).

3.3.3 Nonunicity of the Density Operator Decomposition

Given a density operator ρ, with properties (1) to (4), the decomposition (3.7) is not unique. This means that two or more ensembles $\mathcal{E} = (\mathcal{S}, p)$ can give the same ρ. For instance, in the qubit system of Example 3.2 we have seen that the ensemble

$$|\psi_1\rangle = |0\rangle, \ p_1 = \frac{1}{3}, \qquad |\psi_2\rangle = |1\rangle, \ p_2 = \frac{2}{3} \tag{3.10}$$

gives the density operator $\rho = \frac{1}{3}|0\rangle\langle 0| + \frac{2}{3}|1\rangle\langle 1|$. But the same density operator is obtained with the ensemble (see Problem 3.4)

$$|\psi_1\rangle = \sqrt{\frac{1}{3}}|0\rangle + \sqrt{\frac{2}{3}}|1\rangle, \ p_1 = \frac{1}{2}, \qquad |\psi_2\rangle = \sqrt{\frac{1}{3}}|0\rangle - \sqrt{\frac{2}{3}}|1\rangle, \ p_2 = \frac{1}{2} \tag{3.11}$$

and also with the ensemble of four states

$$|\psi_1\rangle = -\frac{1}{\sqrt{3}}|0\rangle - \frac{2i}{\sqrt{6}}|1\rangle, \ p_1 = \frac{1}{4}, \qquad |\psi_2\rangle = +\frac{1}{\sqrt{3}}|0\rangle - \frac{2}{\sqrt{6}}|1\rangle, \ p_2 = \frac{1}{4}$$

$$|\psi_3\rangle = -\frac{1}{\sqrt{3}}|0\rangle + \frac{2i}{\sqrt{6}}|1\rangle, \ p_3 = \frac{1}{4}, \qquad |\psi_4\rangle = +\frac{1}{\sqrt{3}}|0\rangle + \frac{2}{\sqrt{6}}|1\rangle, \ p_4 = \frac{1}{4}. \tag{3.12}$$

In general, to a density operator one can associate infinitely many ensembles.

The topic of the multiplicity of the ensembles of a density operator will be systematically developed in Sect. 3.11 at the end of this chapter.

3.3.4 Role of Density Operators

The density operators play a fundamental role in Quantum Mechanics, to the point that its very postulates can be formulated exclusively in terms of density operators [1, Chap. 2].

In Quantum Communications, developed in Part II, at the transmission side, the modulator (or encoder) chooses a pure state among an alphabet of possible states; the density operator is then of the type $|\psi\rangle\langle\psi|$. In reception, one can still refer to a pure state, if the thermal noise in neglected, but if we consider the thermal noise, the state is not pure anymore, and we must proceed with density operators.

In Quantum Information, developed in Part III, a more sophisticated use is made of a density operator, through the representation in the phase space by the *characteristic function* and by the *Wigner function*. These functions will allow for the introduction of the important class of *Gaussian states*.

Problem 3.1 ⋆   Prove that the density operator ρ of a quantum system in a pure state is *idempotent*.

Problem 3.2 ⋆ ⋆ ∇ Prove that, if and only if $\mathrm{Tr}[\rho^2] = 1$, the density operator ρ represents a pure state.
Hint: see Proposition 3.5.

Problem 3.3 ⋆⋆   Prove relation (3.9), which states that a cavity at thermal equilibrium is in a mixed state.

Problem 3.4 ⋆⋆   Verify that the ensembles (3.10)–(3.12) give the same density operator.

3.4 Dynamical Evolution of a Quantum System

Above we considered the state of a quantum system at a fixed time and now we consider its temporal evolution. The assumption is that the system is *closed* as in Postulate 1, that is, "left to itself", and in particular such that it is not affected by measurement instruments. To describe this temporal evolution, two equivalent visions are available: one is *Schrödinger's picture*, the other is *Heisenberg's picture*. Both are described by a Hamiltonian operator H, a Hermitian operator corresponding to the total energy of the physical system.[2] The Hamiltonian generates the unitary

---

[2] The Hamiltonian is often called *observable*, for a reason we shall see in the context of quantum measurements (see Sect. 3.6).

evolution U. In the Schrödinger's picture the quantum state evolves according to U and the observable is fixed (constant), while in the Heisenberg's picture the state is fixed and the observable evolves in time.

3.4.1 Postulate 2 of Quantum Mechanics

Let us indicate by $|\psi(t)\rangle \in \mathcal{H}$ the state of the system at time t, which in this context is called *wave function*.

Postulate 2 The evolution of a closed quantum system is described by a **Hamiltonian operator** $H(t)$ through Schrödinger's equation

$$i\hbar \frac{\partial}{\partial t}|\psi(t)\rangle = H(t)\,|\psi(t)\rangle \tag{3.13}$$

where $|\psi(t)\rangle$ is the wave function and $\hbar = h/(2\pi)$ is the reduced Planck's constant.

As an alternative: The evolution of a closed quantum system is described by a **unitary operator** U. If $|\psi(t_0)\rangle$ is the state of the system at time t_0, the state of the system at time t becomes

$$|\psi(t)\rangle = U(t_0, t)\,|\psi(t_0)\rangle \qquad t > t_0 \tag{3.14}$$

where $U = U(t_0, t)$ depends only on t_0 and t. $\qquad\qquad\square$

According to this postulate, the dynamics of the quantum system is described by the *Hamiltonian operator* $H(t)$, or by the *temporal evolution operator* $U = U(t_0, t)$, in the sense that once we know $H(t)$, we can solve the differential equation (3.13), and then compute the temporal evolution (3.14).

It is important to remark that the postulate must be completed by some **noncommutativity conditions** (see (3.21)). As a matter of fact, it will be the consequent noncommutative algebra that will make the difference between Quantum Mechanics and Classical Mechanics. A second remark is that the restriction of the temporal evolution of a quantum system to unitary operators has the important consequence that not all temporal evolutions are possible. An example of an *impossible* action is the copy, or clonation, of information (**No-cloning Theorem**). This will be seen at the end of the chapter because it requires Postulate 4.

The equivalence of the two formulations of Postulate 2 is based on the following statement.

Proposition 3.1 *The Hermitian property of the Hamiltonian, $H^*(t) = H(t)$, implies the unitary property of the temporal evolution operator, $U(t_0, t)\,U^*(t_0, t) = I_{\mathcal{H}}$.*

We limit the proof to the case in which the Hamiltonian is independent of time t. In the case in which the Hamiltonian contains explicitly the time t, the conclusion is the same, but the proof is cumbersome and can be found in [6].

If the Hamiltonian H is not time-dependent, that is, $H(t) = H$, the solution to Schrödinger's equation is immediate and given by

$$|\psi(t)\rangle = \exp\left[-\frac{i}{\hbar}H(t-t_0)\right]|\psi(t_0)\rangle, \qquad t > t_0 \tag{3.15}$$

thus the temporal evolution results in

$$U(t-t_0) = \exp\left[-\frac{i}{\hbar}H(t-t_0)\right]. \tag{3.16}$$

This solution is the same as the one we would obtain by considering H as a scalar, but the exponential of an operator must be interpreted according to Functional Analysis (see Sect. 2.12.3). Now we can easily check that relation (3.16) actually defines a unitary operator, $U U^* = I_{\mathcal{H}}$, as soon as we impose that H be Hermitian. This conclusion is in agreement with Theorem 2.5, which claims that an operator U is unitary if and only if it can be expressed in the form $U = \exp(i H)$, with H Hermitian.

From the explicit solution (3.16) some important properties can be verified, valid for any other solution of Schrödinger's equation. In particular (see Problem 3.5), the norm of the wave function $|\psi(t)\rangle$ at time t remains of unit length, as it must be from Postulate 1. Moreover, the inner product of two wave functions $|\psi_1(t)\rangle$ and $|\psi_2(t)\rangle$ doesn't change during the evolution; in particular, if two kets are orthogonal at time t_0, they stay orthogonal at each $t > t_0$. In other words, in the geometry of quantum evolution, lengths and angles between states are preserved.

Remark In Postulate 2 the wave function is expressed as a pure state $|\psi(t)\rangle$, but it is easy to express the evolution in terms of a density operator. Starting from definition (3.7), which can now be rewritten showing the temporal dependency

$$\rho(t) = \sum_i p_i |\psi_i(t)\rangle\langle\psi_i(t)|$$

and keeping in mind that for each state we have $|\psi_i(t)\rangle = U(t_0, t)|\psi_i(t_0)\rangle$, we obtain the relation

$$\rho(t) = U^*(t_0, t)\rho(t_0)U(t_0, t), \qquad t > t_0 \tag{3.17}$$

which expresses the density operator at time t as a function of its initial value $\rho(t_0)$.

3.4.2 An Explicit Form of Hamiltonian. Reference System

The Hamiltonian operator H is associated to the energy of the system. For many systems H can be obtained from the classic expression of energy in terms of coordinates and moments, substituting these with the corresponding quantum operators.

A common *reference system* is given by a particle of mass m constrained to move in one direction in a potential $V(q)$, where q is the coordinate of the particle, which may be any real number. We know from Classical Mechanics that the total energy of a system is given by the sum of kinetic and potential energies

$$E = \frac{1}{2} m v^2 + V(q) = \frac{1}{2} \frac{p^2}{m} + V(q) \tag{3.18}$$

where $p = m v$ is the momentum and $V(q)$ is a real function of the coordinate q. In particular, in the *harmonic oscillator* the potential has the form $V(q) = \frac{1}{2} m \omega^2 q^2$, where ω is the pulsation of the oscillator. In any case, the energy is expressed in terms of two continuous *dynamical variables*: the momentum p and the coordinate (position) q.

To treat this system according to the rules of Quantum Mechanics, **we have to replace the dynamical variables** p **and** q **with Hermitian operators** (observables).[3] Then the Hamiltonian reads as

$$H = \frac{1}{2} \frac{p^2}{m} + V(q) \quad \text{(reference system)} . \tag{3.19}$$

It is remarkable that also the Hamiltonian H is a Hermitian operator, as can be easily verified in the general case (3.19), and in particular in the case of the harmonic oscillator, where

$$H = \frac{(p^2 + \omega^2 q^2)}{2m} \quad \text{(harmonic oscillator)} . \tag{3.20}$$

As noted before, the specification of the operators describing the physical system, as p, q, and H, is not sufficient in Quantum Mechanics, and should be completed by the indication of the algebra the operators must obey. In the specific case the algebra is non-commutative with the following *commutation relation*

$$[q, p] = i \hbar I_{\mathcal{H}} \tag{3.21}$$

where $[\cdot, \cdot]$ denotes the *commutator* (see Sect. 2.5), which in this case reads $[q, p] = q p - p q$. This condition will lead to the quantization of the energy and represents a remarkable difference with respect to Classical Mechanics, where p and q commute. The commutation relation provides a link (Correspondence Principle) between Quantum Mechanics and Classical Mechanics: if the Planck constant $h \to 0$, the operators p and q commute and a quantum solution should coincide with a classical solution.

For an application of these concepts see the harmonic oscillator developed in Sect. 11.3.

---

[3] We use the same notation, p, q for the dynamic variables and operators. In the literature the operators are usually marked with a hat, that is, $\hat{p}, \hat{q}$.

3.4.3 The Schrödinger and the Heisenberg Pictures

In the Schrödinger picture the Hermitian operators[4] as p_S, q_S, and H_S are **time-independent** (fixed or stationary). Also, the bases are assumed as time-independent and act as coordinate systems in Classical Mechanics. On the other hand, the state describing the dynamical behavior of the system is time-dependent: if $|\psi(t_0)\rangle$ is the initial state, the state at time t is given by $|\psi_S(t)\rangle = U(t, t_0)|\psi(t_0)\rangle$, as postulated by (3.14). In the Heisenberg picture the Hermitian operators $p_H(t)$, $q_H(t)$, and $H_H(t)$, and also the bases, are moving, while the state is fixed at the initial value $|\psi(t_0)\rangle$. To summarize

	Schrödinger picture	Heisenberg picture
Operators	Fixed	Moving
Bases	Fixed	Moving
State	Moving	Fixed

The two modes (pictures) of formulating Quantum Mechanics are physically equivalent and this should not appear to be strange. For analogy, consider a moving object in Classical Mechanics: observing from a fixed coordinate system, one sees the position of the object moving in time with a given time-dependent law, say $q(t)$, with initial value $q(t_0)$. On the other hand, observing from a coordinate system anchored to the object, one sees a fixed position $q(t_0)$.

To establish the equivalence we have to link the state and the operators of the two pictures. The states are related by (3.14), rewritten in the form

$$|\psi_S(t)\rangle = U(t, t_0)\,|\psi_H(t_0)\rangle \tag{3.22}$$

where $\psi_H(t_0)\rangle$ is fixed and $|\psi_S(t)\rangle$ is time-dependent. If A_S is a (fixed) Hermitian operator in the Schrödinger picture, in the Heisenberg picture it must be defined as

$$A_H(t) = U^*(t, t_0)\, A_S\, U(t, t_0) \tag{3.23}$$

which provides the time-dependence of the operator.

But, for the equivalence, the commutation relations should be the same. In fact, consider a general commutation relation, which in the Schrödinger picture has the form

$$[A_S, B_S] = \mathrm{i}\, C_S \tag{3.24a}$$

where A_S, B_S, and C_S are observables. Then, using (3.23) for the three observables one gets (see Problem 3.6)

---

[4] The subscripts S and H refer to Schrödinger's and Heisenberg's picture, respectively.

$$[A_H(t), B_H(t)] = \mathrm{i}\, C_H(t) \tag{3.24b}$$

which states that the commutation relation is the same in the two pictures. This invariance holds in particular for the commutation relation of position and momentum operators, given by (3.21), that is,

$$[q_S, p_S] = [q_H(t), p_H(t)] = \mathrm{i}\, I_{\mathcal{H}}$$

where $q_H(t) = U^*(t, t_0)\, q_S\, U(t, t_0)$ and $p_H(t) = U^*(t, t_0)\, p_S\, U(t, t_0)$.

Problem 3.5 ⋆ Prove that if the temporal evolution operator $U = U(t, t_0)$ is unitary, as assumed in Postulate 2, then the norm of the wave function $|\psi(t)\rangle$ at time t remains of unit length, as it must be from Postulate 1. Moreover, prove that the inner product of two wave functions $|\psi_1(t)\rangle$ and $|\psi_2(t)\rangle$ doesn't change during the evolution.

Problem 3.6 ⋆ Suppose that A_S, B_S, and C_S are three observables in the Schrödinger picture that verify the commutation condition

$$[A_S, B_S] = \mathrm{i}\, C_S \,.$$

Prove that in the Heisenberg picture the commutation condition becomes

$$[A_H(t), B_H(t)] = \mathrm{i}\, C_H(t) \,.$$

3.5 Quantum Measurements

The third postulate of Quantum Mechanics regards *quantum measurements*, that is, the methods for extracting information from a quantum system $\mathcal{H}$, described in general by a density operator ρ, and in particular by a pure state $|\psi\rangle$. In Quantum Communications, *quantum measurements* are performed to obtain the information required for decision, and therefore play a fundamental role.

It must be observed that the outcome of a quantum measurement is intrinsically a random, or unpredictable, quantity, in the sense that if we prepare a set of identical quantum systems, for which the same measurement technique is used, the results are in general different. This is a fundamental difference from Classical Mechanics (see Sect. 1.2).

The model of a quantum measurement is formulated on the basis of appropriate Hermitian operators. The standard formulation is based on projectors, and was introduced by von Neumann (this is why they are referred to as *projective* or *von Neumann measurements*), but other equivalent formulations are found in the literature, and also various generalizations. Here the presentation of quantum measurements will follow the approach recently proposed by Eldar and Forney [7]. From a speculative viewpoint, the subject of quantum measurements is highly debated.

Given their importance for Quantum Communications, the quantum measurements will be discussed and amply exemplified following a constructive order, i.e., we will start from a restricted form of Postulate 3, introducing increasing degrees of generalization. Alternatively, one could proceed the other way around, starting from the most general case, and gradually developing the special cases, but we believe that the order we chose will be more effective at first reading (see Sect. 3.8 for an overview).

Finally, we observe that the objective of quantum measurements typically regards the identification of the state of a physical system, but also, more in general, other attributes of the system.

3.5.1 The Third Postulate of Quantum Mechanics

Let $\{\Pi_i, i \in \mathcal{M}\}$ be a **projector system** in the Hilbert space $\mathcal{H}$, as defined in Sect. 2.9, Definition 2.3, i.e., with the properties

$$\Pi_i^* = \Pi_i, \qquad \Pi_i \geq 0 \tag{3.25a}$$

$$\Pi_i^2 = \Pi_i, \qquad \Pi_i \Pi_j = 0_{\mathcal{H}} \quad i \neq j \tag{3.25b}$$

$$\sum_{i \in \mathcal{M}} \Pi_i = I_{\mathcal{H}} \tag{3.25c}$$

where $0_{\mathcal{H}}$ is the null operator, $I_{\mathcal{H}}$ is the identity operator on $\mathcal{H}$, and $\mathcal{M}$ is a finite or enumerable alphabet. In particular, the last property establishes that the projectors give a *resolution of the identity* on $\mathcal{H}$.

Postulate 3 A measurement on a quantum system, in the framework of a Hilbert space $\mathcal{H}$ (state space), is obtained through a projector system $\{\Pi_i, i \in \mathcal{M}\}$. The alphabet $\mathcal{M}$ provides the possible outcomes of the measurements. If immediately before the measurement the system is in the state $s = |\psi\rangle$, the probability that the outcome is $m = i \in \mathcal{M}$ is given by

$$\boxed{p_m(i|\psi) := \mathrm{P}[m = i | s = |\psi\rangle] = \langle \psi | \Pi_i | \psi \rangle, \qquad i \in \mathcal{M}.} \tag{3.26}$$

If the outcome is $m = i$, after the measurement the system collapses to the state

$$|\psi_{\mathrm{post}}^{(i)}\rangle = \frac{\Pi_i |\psi\rangle}{\sqrt{\langle \psi | \Pi_i | \psi \rangle}} = \frac{\Pi_i |\psi\rangle}{\sqrt{p_m(i|\psi)}} . \qquad \square \tag{3.27}$$

Postulate 3 is illustrated in Fig. 3.3. We note that in (3.27) the denominator is due to the normalization, which imposes that $\langle \psi_{\mathrm{post}}^{(i)} | \psi_{\mathrm{post}}^{(i)} \rangle = 1$. Since the alphabet $\mathcal{M}$ is assumed enumerable, the outcome of the measurement m must be modeled

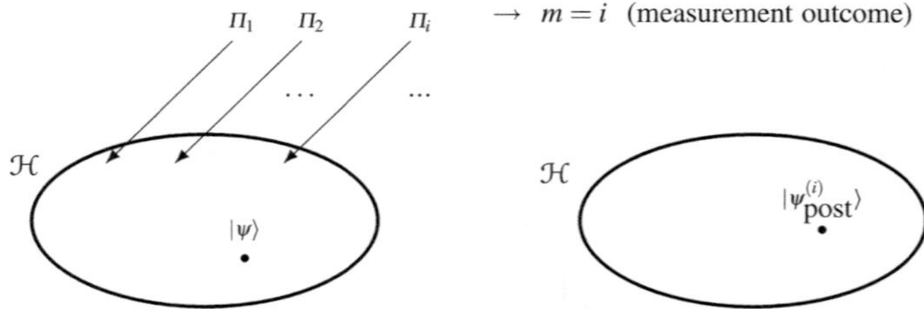

Fig. 3.3 Illustration of the third postulate of Quantum Mechanics. In a quantum system, which is set in the state $|\psi\rangle$, a projector system $\{\Pi_i\}$ is applied and the outcome m of the measurement is a random variable, described by the probability distribution $p_m(i|\psi) := P[m = i|s = |\psi\rangle] = \langle\psi|\Pi_i|\psi\rangle$. After the measurement the system moves to the state $|\psi^{(i)}_{\text{post}}\rangle = \Pi_i|\psi\rangle/\sqrt{p_m(i|\psi)}$

as a *discrete random variable* with alphabet $\mathcal{M}$ and (conditioned) probability distribution[5] $p_m(i|\psi)$ given by (3.26), where the condition is that the system is in the state $s = |\psi\rangle$.

One can easily see that $p_m(i|\psi)$ verifies the standard conditions of a probability distribution ($p_m \geq 0$ and normalization). The condition $p_m \geq 0$ is ensured by the property of the projectors to be positive semidefinite, that is, $\langle\psi|\Pi_i|\psi\rangle \geq 0, \forall\, |\psi\rangle \in \mathcal{H}$ (see Sect. 2.12.1). The normalization condition for the distribution $p_m(i \mid \psi)$ is ensured by the completeness condition (3.25b). In fact

$$\sum_{i\in\mathcal{M}} p_m(i|\psi) = \langle\psi| \sum_{i\in\mathcal{M}} \Pi_i|\psi\rangle = \langle\psi|I_{\mathcal{H}}|\psi\rangle$$
$$= \langle\psi|\psi\rangle = 1,$$

where we used the fact that a state is always a unitary vector. Note that in the detailed symbolism $p_m(i \mid \psi)$ the "condition" is explicitly indicated to stress that the outcome of the measurement depends on the state of the quantum system ($|\psi\rangle$ is shortened to ψ).

It remains to investigate the meaning of the postulate and its connection with the physical reality. The very abstract formulation of Quantum Mechanics leads to a very difficult interpretation. The dynamics described by Postulate 3 is limited to "immediately before" and "right after the measurement", and this leads us to think that the measurement is *instantaneous* and that the temporal evolution of the measurement is not described. The expression "the measurement ... is obtained by means of a projector system" stresses the instantaneity and induces us to interpret

---

[5] In Probability Theory a function as $p_m(i|\psi)$, which describes a discrete random variable, is called "mass probability distribution", while the term "probability distribution" refers to the integral of a probability density. In Quantum Mechanics the term "mass" is usually omitted and in this book we will follow this convention.

the measurement through a system having at the input all the projectors Π_i and at the output the measurement outcome m. But how is this achieved?

For the moment the author has found a partial answer given by the receiver of Barnett and Riis [8], where a binary projector system is involved.

3.5.2 Measurements with Elementary Projectors

A very important case of projective measurements is obtained with *elementary* projectors (see Sect. 2.9.5). Let $\mathcal{H}$ be an M-dimensional Hilbert space and let

$$\mathcal{A} = \{|a_1\rangle, |a_2\rangle, \ldots, |a_M\rangle\}$$

be an orthonormal basis. Then, through the outer products

$$\Pi_i = |a_i\rangle\langle a_i|, \qquad i = 1, \ldots, M \tag{3.28}$$

a projector system is obtained (see Theorem 2.1 of Sect. 2.9).

The *elementary* projectors (3.28), which are all rank-one, when applied to a generic state $|\psi\rangle$, define the vectors

$$|b_i\rangle = \Pi_i|\psi\rangle = |a_i\rangle\langle a_i|\psi\rangle = k_i|a_i\rangle$$

where $k_i = \langle a_i|\psi\rangle$. Hence, $\Pi_i = |a_i\rangle\langle a_i|$ projects the state $|\psi\rangle$ onto the one-dimensional subspace of $\mathcal{H}$ generated by the ket $|a_i\rangle$. The vectors $|a_i\rangle$ forming the projectors Π_i are called *measurement vectors* [7].[6]

We now apply Postulate 3 using elementary projectors (3.28). From (3.26) it results that the probability that the measurement yield the outcome $m = i$, when the system sits in the state $|\psi\rangle$, is simply

$$\boxed{P[m = i|\psi] = |\langle\psi|a_i\rangle|^2} \tag{3.29}$$

and therefore it is given by the squared modulus between the state $|\psi\rangle$ and the measurement vector $|a_i\rangle$. Immediately after the measurement the state of the system becomes

$$|\psi_{\text{post}}^{(i)}\rangle = \frac{\Pi_i|\psi\rangle}{|\langle\psi|a_i\rangle|} = \frac{|a_i\rangle\langle a_i|\psi\rangle}{|\langle a_i|\psi\rangle|} = |a_i\rangle\frac{\langle a_i|\psi\rangle}{|\langle a_i|\psi\rangle|} \tag{3.30}$$

---

[6] The term "measurement vector" is not consolidated. More often the term "rank-one measurement operator" is used to indicate a projector with the form (3.28).

where the last fraction is a complex number with a unitary modulus (phasor) and therefore it can be neglected. Hence, if the measurement outcome is $m = i$, the quantum system collapses to the state specified by the ket $|a_i\rangle$

$$\boxed{|\psi_{\text{post}}^{(i)}\rangle = |a_i\rangle .} \tag{3.31}$$

3.5.3 Postulate 3 in Terms of Density Operators

Postulate 3 tacitly supposed that the state $|\psi\rangle$ in which the system is "prepared" before the measurement were known (pure state). The density operator corresponding to the pure state $|\psi\rangle$ is given by $\rho = |\psi\rangle\langle\psi|$ and then the probability (3.26) can be rewritten in the form

$$P[m = i|\psi] = \text{Tr}[\rho\, \Pi_i], \qquad i \in \mathcal{M} \tag{3.32}$$

where $\text{Tr}[.]$ denotes the trace. To get this result it is sufficient to apply to (3.26) the important identity on the trace (2.34), which in this case becomes

$$\langle\psi|\Pi_i|\psi\rangle = \text{Tr}[|\psi\rangle\langle\psi|\Pi_i].$$

More generally, if the system state is statistically known through the density operator

$$\rho = \sum_j p_j\, |\psi_j\rangle\langle\psi_j| \tag{3.33}$$

where p_j is the probability that the system be in the state $|\psi_j\rangle$, we can still apply Postulate 3, provided that we specify in which of the possible states the quantum system is before the measurement. In other words, it is required that, in the underlying probability space, the *event* measurement be conditioned by the given state $s = |\psi_j\rangle$ and the outcome of the measurement be formulated as a *random variable conditioned by the given state*. Then one can apply (3.26) in the form

$$P[m = i|s = |\psi_j\rangle] = \langle\psi_j|\Pi_i|\psi_j\rangle = \text{Tr}[|\psi_j\rangle\langle\psi_j|\Pi_i] \tag{3.34}$$

which gives the probability that the measurement outcome be $m = i$, under the condition that the system state before the measurement is $s = |\psi_j\rangle$. To get the unconditioned probability one must compute the average with respect to the probabilities of the condition, that is,

$$P[m = i] = \sum_j P[s = \psi_j]P[m = i|s = \psi_j] .$$

Hence

$$P[m = i] = \sum_j p_j \text{Tr}[|\psi_j\rangle\langle\psi_j|\Pi_i] = \text{Tr}[\sum_j p_j|\psi_j\rangle\langle\psi_j|\Pi_i]$$

$$= \text{Tr}[\rho \Pi_i] \tag{3.35}$$

where the linearity of the trace has been used.

The result is expressed in the form:

Proposition 3.2 *In a quantum system where the states are described by the density operator ρ, the probability that the outcome of the measurement, obtained with the projector system $\{\Pi_i, i \in \mathcal{M}\}$, be $m = i$ is given by*

$$\boxed{P[m = i|\rho] = Tr[\rho \Pi_i], \quad i \in \mathcal{M}.} \tag{3.36}$$

If the outcome is $m = i$, the ensemble of states after the measurement is described by the density operator

$$\rho_{post}^{(i)} = \frac{\Pi_i \, \rho \, \Pi_i}{Tr[\rho \Pi_i]} = \frac{\Pi_i \, \rho \, \Pi_i}{P[m = i|\rho]} \,. \tag{3.37}$$

In the symbolism in (3.36) we write explicitly the "condition" ρ to stress the fact that the system states are statistically described by the operator ρ. The proof of (3.36) is given in Appendix section "Probabilities and Random Variables in a Quantum Measurement" and involves the statistical description of the random quantities: (1) the state of the system s before the measurement, (2) the outcome of the measurement m, and (3) the state of the system s_{post} after the measurement.

With elementary projectors, $\Pi_i = |a_i\rangle\langle a_i|$, relation (3.36) becomes

$$P[m = i|\rho] = \text{Tr}[\rho|a_i\rangle\langle a_i|] = \langle a_i|\rho|a_i\rangle \,. \tag{3.38}$$

Proposition 3.2 gives the general result since it is comprehensive of the measurements performed in a quantum system set in a pure state, where $\rho = |\psi\rangle\langle\psi|$, as well as the measurements performed in a quantum system set in a "mixture of states", describes by the density operator ρ given by (3.33). As we shall see in the next chapters, (3.36) will be the fundamental relation in the *decision* of Quantum Communications systems.

Example 3.4 (*Measurement in a qubit system with pure states*) Consider a qubit system, where $\mathcal{H} = \mathbb{C}^2$, and the projector system is given by

$$\Pi_1 = \frac{1}{2}\begin{bmatrix} 1 & i \\ -i & 1 \end{bmatrix}, \qquad \Pi_2 = \frac{1}{2}\begin{bmatrix} 1 & -i \\ i & 1 \end{bmatrix} \,. \tag{3.39}$$

The projective measurement is performed assuming that the system is in the state

$$|\psi\rangle = \frac{1}{\sqrt{5}} \begin{bmatrix} 1 \\ 2i \end{bmatrix} .$$

The alphabet of the measurement outcome m is $\mathcal{M} = \{1, 2\}$ and the probabilities evaluated from (3.26) are given by

$$p_m(1|\psi) = \langle\psi|\Pi_1|\psi\rangle = \frac{1}{10}, \qquad p_m(2|\psi) = \langle\psi|\Pi_2|\psi\rangle = \frac{9}{10} .$$

The post-measurement states, evaluated from (3.27), are respectively

$$|\psi_{\text{post}}^{(1)}\rangle = \frac{1}{\sqrt{2}} \begin{bmatrix} -1 \\ i \end{bmatrix}, \qquad |\psi_{\text{post}}^{(2)}\rangle = \frac{1}{\sqrt{2}} \begin{bmatrix} 1 \\ i \end{bmatrix} .$$

Example 3.5 (Measurement in a qubit system with mixed states) Consider a qubit system, where the projector system $\{\Pi_1, \Pi_2\}$ is given by (3.39). The projective measurement is performed assuming that the system is in a mixed state described by the following ensemble

$$p(|\psi_1\rangle) = \frac{1}{3}, \qquad |\psi_1\rangle = \begin{bmatrix} \frac{1}{\sqrt{5}} \\ \frac{2i}{\sqrt{5}} \end{bmatrix}, \qquad p(|\psi_2\rangle) = \frac{2}{3}, \qquad |\psi_2\rangle = \begin{bmatrix} \frac{1}{\sqrt{2}} \\ \frac{-i}{\sqrt{2}} \end{bmatrix} .$$

The corresponding density operator is

$$\rho = \frac{1}{5} \begin{bmatrix} 2 & -i \\ i & 3 \end{bmatrix} .$$

The alphabet of the measurement outcome m is $\mathcal{M} = \{1, 2\}$ and the probabilities evaluated from (3.35) are given by

$$\mathrm{P}[m = 1|\rho] = \mathrm{Tr}[\rho\Pi_1] = \frac{3}{10}, \qquad \mathrm{P}[m = 2|\rho] = \mathrm{Tr}[\rho\Pi_2] = \frac{7}{10} .$$

The post-measurement density operators are obtained from (3.37) and read

$$\rho_{\text{post}}^{(1)} = \frac{\Pi_1 \rho \Pi_1}{\mathrm{Tr}[\rho\Pi_1]} = \frac{1}{2} \begin{bmatrix} 1 & i \\ -i & 1 \end{bmatrix}, \qquad \rho_{\text{post}}^{(2)} = \frac{\Pi_2 \rho \Pi_2}{\mathrm{Tr}[\rho\Pi_2]} = \frac{1}{2} \begin{bmatrix} 1 & -i \\ i & 1 \end{bmatrix} .$$

Problem 3.7 ⋆ Apply Postulate 3 to a quantum system "prepared" in a pure state $|\psi\rangle$, when the measurement is obtained by a set of orthonormal *measurement vectors* $\{|a_0\rangle, |a_1\rangle, \ldots, |a_{M-1}\rangle\}$. Find the probability distribution of the measure m *when the state of the system is one of the measurement vectors.* Which is the state of the system after the measurement?

3.6 Measurements with Observables

Postulate 3 can also be formulated through a single Hermitian operator A, which, in this context, is called *observable*. An observable combines both the projectors and the alphabet of the measurement, with some advantages concerning the evaluation of the statistical averages on the measurements. This possibility is ensured by the *spectral decomposition theorem*, seen in the previous chapter and now recalled for its importance.

Theorem 3.1 *Let A be an observable and let $\{a_i, i \in \mathcal{M}\}$ be the distinct eigenvalues of A. Then A may be uniquely decomposed in the form*

$$A = \sum_{i \in \mathcal{M}} a_i P_i \tag{3.40}$$

where the $P_i, i \in \mathcal{M}$ form a **projector system**.

In particular, if all the eigenvalues have unitary multiplicity, the decomposition is given by elementary projectors

$$A = \sum_{i \in \mathcal{M}} a_i \, |a_i\rangle\langle a_i| \tag{3.40a}$$

with $|a_i\rangle$ the eigenkets of A.

Then an observable A provides a projector system, which allows us to apply Postulate 3 for a quantum measurement, as depicted in Fig. 3.4. The only modification is the alphabet, which is now given by the spectrum of the operator, that is, $\mathcal{M} = \sigma(A)$. Consequently, in a measurement with an observable **the outcome is always an eigenvalue of the observable**.

To stress the equivalence between a measurement based on projectors and a measurement made with an observable, it is easy to show that also Postulate 3 implies an observable. In fact, given the projector system $\{\Pi_i, i \in \mathcal{M}\}$, we can define the operator

$$A = \sum_{i \in \mathcal{M}} i \, \Pi_i \; . \tag{3.41}$$

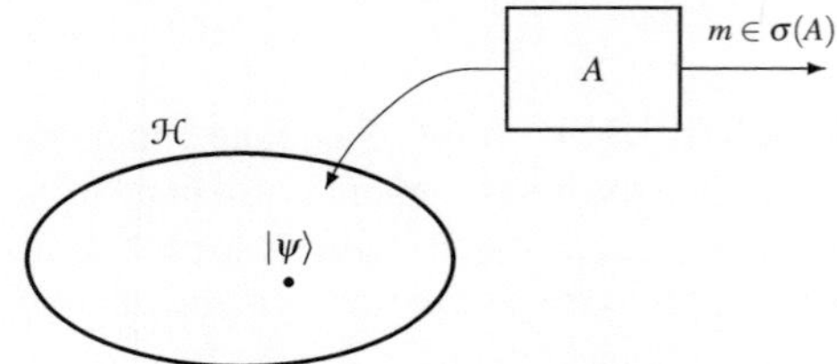

Fig. 3.4 Measurement by means of an observable A in a quantum system $\mathcal{H}$ set in a *pure state*. The outcome of the measurement is always an eigenvalue of A

Next, recalling that the decomposition in Theorem 3.1 is unique, one sees by inspection that A is an observable (Hermitian) with eigenvalues i and projectors Π_i.

3.6.1 Expectation and Moments in a Quantum Measurement

Postulate 3 gives the complete statistical description of the random variable m, the measurement outcome, through the probability distribution (3.26). Here we show that the moments of m, and in particular the mean and the variance, can be obtained from the observable A in a concise form.

We recall from Probability Theory that the *expectation* (or mean or average) of a discrete random variable can be defined from its probability distribution $p_m(i|\psi)$ as

$$E[m|\psi] = \sum_{i \in \mathcal{M}} i \; p_m(i|\psi) \; . \tag{3.42}$$

This definition is extended to the expectation of an arbitrary function $f(m)$ of m as

$$E[f(m)|\psi] = \sum_{i \in \mathcal{M}} f(i) \; p_m(i|\psi) \; . \tag{3.42a}$$

Here the random variable is conditioned by a generic "condition" ψ and (3.42) are called *conditional expectations*.

Coming back to a quantum measurement, we get:

Proposition 3.3 *Let m be the outcome of a quantum measurement obtained with an observable A, when the system is in the state $|\psi\rangle$. Then the expectation of m is given by*

$$\boxed{E[m|\psi] = \langle \psi|A|\psi\rangle \; .} \tag{3.43}$$

In fact, using the relation $p_m(i|\psi) = \langle\psi|\Pi_i|\psi\rangle$ and (3.41) in definition (3.42), one gets

$$E[m|\psi] = \sum_{i \in \mathcal{M}} i \, \langle\psi|\Pi_i|\psi\rangle = \langle\psi| \sum_{i \in \mathcal{M}} i\Pi_i |\psi\rangle = \langle\psi|A|\psi\rangle \; .$$

From A it is also possible to get the quadratic mean and then the variance. Considering property (3.25a), the square of A results in

$$A^2 = \sum_i i^2 \, \Pi_i$$

from which

$$E[m^2|\psi] = \sum_i i^2 p_m(i|\psi) = \langle\psi| \sum_i i^2 \Pi_i |\psi\rangle$$

$$= \langle\psi| A^2 |\psi\rangle \,. \tag{3.44}$$

Analogously, we can calculate the higher order moments $E[m^k|\psi]$ for $k > 2$.

3.6.2 Quantum Expectation with Mixed States

The quantum expectation can be evaluated from the observable A even when the system state is not known, but it is described by a density operator. Then, the distribution of m is given by (3.36) and the quantum expectation results in

$$\boxed{E[m|\rho] = \mathrm{Tr}[\rho A] := \langle A\rangle \,.} \tag{3.45}$$

In fact,

$$E[m|\rho] = \sum_i i\, \mathrm{P}[m = i|\rho] = \sum_i i\, \mathrm{Tr}[\rho \Pi_i]$$

from which one gets (3.45) using the linearity of the trace. Expression (3.45) generalizes (3.43).

3.6.3 Combinations of Projectors

Starting from a system of M projectors $\{P_i, i \in \mathcal{M}\}$, we can build a system with a reduced number of projectors, $N < M$, in the following way. We partition the alphabet $\mathcal{M}$ into N parts

$$\mathcal{I}_1, \mathcal{I}_2, \dots, \mathcal{I}_N \tag{3.46}$$

with

$$\bigcup_{k=1}^{N} \mathcal{I}_k = \mathcal{M} \quad \text{and} \quad \mathcal{I}_h \bigcap \mathcal{I}_k = \emptyset,\, h \neq k$$

and then we define the N Hermitian operators

$$\Pi_k = \sum_{i \in \mathcal{I}_k} P_i, \qquad k = 1, 2, \dots, N. \tag{3.47}$$

We verify immediately from the properties of the projectors $\{\Pi_i\}$:

(1) the idempotency: $\Pi_k^2 = \Pi_k$,
(2) the orthogonality: $\Pi_h \Pi_k = 0_{\mathcal{H}}$,
(3) the identity resolution: $\sum_{k=1}^N \Pi_k = I_{\mathcal{H}}$.

Therefore the operators (3.47) form a new projector system $\{\Pi_k, k \in \mathcal{K}\}$ with $\mathcal{K} = \{1, \ldots, N\}$ and then, according to Postulate 3, they can be used to perform a measurement on $\mathcal{H}$ in which the result can take N possible values.

Once the new projectors Π_k are obtained, a new observable can be formulated representing them, given by

$$A_{\mathcal{J}} = \sum_{k=1}^N i\, \Pi_k$$

which depends on the partition (3.46).

3.6.4 Continuous Observables

Above we assumed that the possible results of the measurements belong to a finite or enumerable set $\mathcal{M}$, called *alphabet*. This is the case of interest for Quantum Communications, but there are important cases in which the possible results belong to a continuum, typically the set of real numbers $\mathbb{R}$. For instance the position operator q and the momentum operator p, introduced in Sect. 3.4, are observables with a continuous spectrum.

The topic of quantum measurements, where the outcome is a continuous random variable, described by a probability density instead of probability distribution, will be seen in Chap. 11 in the context of Quantum Information with continuous variables.

3.6.5 Remarks on Terminology

In Quantum Mechanics the expression "the outcome of the measurement associated to the operator A" is simplified as "the observable A", that is, measurement and operator are identified. Moreover, "mean of the outcome obtained with the observable A" is simplified in "mean of the observable A". In agreement with these simplified expressions, instead of $\mathrm{E}[m|\psi]$ we should write $\mathrm{E}[A]$, so that (3.43) becomes $\mathrm{E}[A] = \langle \psi | A | \psi \rangle$. However, in the author's opinion, these expressions are not convenient and the interpretation of a mean of a measurement must be done according to (3.43), instead of referring to the expectation of A.[7] To avoid confusion it is convenient to use a notation different from $\mathrm{E}[A]$, e.g.,

---

[7] Strictly speaking, in the framework of *Probability Theory* an operator A is a fixed (non random) object and its expectation would coincide with A itself.

$$\langle A \rangle = \langle \psi | A | \psi \rangle, \qquad \langle A^2 \rangle = \langle \psi | A^2 | \psi \rangle . \tag{3.48}$$

With this notation the variance of m results in

$$\sigma_m^2(\psi) = \langle A^2 \rangle - (\langle A \rangle)^2 . \tag{3.49}$$

Having specified the probabilistic meaning of the expectation in the measurement with an observable, in the following, expressions of the form $\langle A \rangle$, $\langle A^2 \rangle$, etc., will be called **quantum expectations**.

Problem 3.8 $\star$ Consider the Hermitian operator

$$H = \frac{1}{2} \begin{bmatrix} 3 & -i \\ i & 3 \end{bmatrix}$$

and use it as an observable for the measurement in a qubit system prepared in the pure state

$$|\psi\rangle = \frac{1}{\sqrt{5}} \begin{bmatrix} 1 \\ 2i \end{bmatrix} .$$

Evaluate the probability of the measurement outcome m and the post-measurement states.

Problem 3.9 $\star\star$ Let A be an observable with spectrum $\sigma(A)$. Show that the *moments* of a measurement m obtained with the observable when the state $|\psi\rangle$ is set to an eigenket $|a\rangle$ of A, are simply given by

$$\mathrm{E}[m^k | a] = a^k, \qquad k = 1, 2, \ldots \tag{E2}$$

where $a \in \sigma(A)$ is the eigenvalue corresponding to the eigenket $|a\rangle$. Explain why.

3.7 Generalized Quantum Measurements (POVM)

The generalized quantum measurements are carried out through a set of Hermitian operators, which are not necessarily projectors, and are called POVM (positive operator-valued measurements).

3.7.1 Definition of POVM

A system of general measurement operators (POVM) $\{Q_i, i \in \mathcal{M}\}$ is defined imposing the following conditions to the operators Q_i:

(1) they are Hermitian operators, $Q_i^* = Q_i$,
(2) they are positive semidefinite: $Q_i \geq 0$,
(3) they resolve the identity: $\sum_i Q_i = I_{\mathcal{H}}$.

Condition (3) is illustrated in Fig. 3.5.

From the above conditions, the POVM systems constitute a broader class than the one of the projector systems, because the POVMs are not required to enjoy idempotency and orthogonality. However, the above properties on POVMs ensure that the probabilities calculated according to Postulate 3

$$P[m = i|\psi] = \langle \psi|Q_i|\psi \rangle \qquad (3.50)$$

respect the conditions relative to a probability distribution, that is,

$$p_m(i|\psi) \geq 0, \qquad \sum_i p_m(i|\psi) = 1.$$

Furthermore, starting from (3.50), the "probability calculus" can be extended to systems placed in a mixture of states described by a given density operator ρ, that is,

$$p_m(i|\rho) = P[m = i|\rho] = \mathrm{Tr}[\rho \, Q_i] \, . \qquad (3.51)$$

We arrive at this result in exactly the same way as with the projective measurements (see (3.32)).

However, with the POVMs, Postulate 3 cannot be fully applied to know the system's state immediately after the measurement. On the other hand, this knowledge is irrelevant in many applications and in particular in quantum communications.

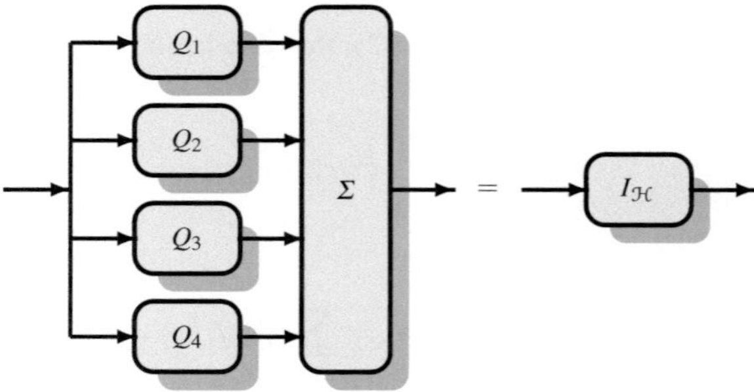

Fig. 3.5 Resolution of the identity of a POVM system with a quaternary alphabet $\mathcal{M}$

According to many authors, for example Helstrom [5], and Eldar and Forney [9], through the **POVM measurements** from a quantum system more useful results can be obtained with respect to von Neumann's **projective measurements**, based on projectors. As will be seen in Chap. 5 in quantum decision theory, to improve generality and simplicity, it is convenient to work out the formulation considering POVMs, but often, due to contextual constraints, we will arrive at the conclusion that such operators turn out to be projectors.

3.7.2 Elementary POVMs

A very interesting class of POVMs has the elementary form

$$Q_i = |\mu_i\rangle\langle\mu_i|$$

where $|\mu_i\rangle$ are vectors of $\mathcal{H}$ not necessarily orthogonal, and not even normalized. The elementary measurement operators are always Hermitian and also verify the condition $Q_i \geq 0$, because $\langle\psi|\mu_i\rangle\langle\mu_i|\psi\rangle = |\langle\psi|\mu_i\rangle|^2 \geq 0$. Instead, the condition that must be imposed is completeness, which assumes the form

$$\sum_{i\in\mathcal{M}} |\mu_i\rangle\langle\mu_i| = I_{\mathcal{H}} \,. \tag{3.52}$$

This condition can be modified when it is specified that the state $|\psi\rangle$ of the quantum system belongs to a subspace $\mathcal{U}$ of the Hilbert space $\mathcal{H}$ and then becomes

$$\sum_{i\in\mathcal{M}} Q_i = \sum_{i\in\mathcal{M}} |\mu_i\rangle\langle\mu_i| = P_{\mathcal{U}} \tag{3.53}$$

where $P_{\mathcal{U}}$ is a projector on $\mathcal{U}$. In fact, this condition guarantees, equally well as (3.52), the normalization condition of the measurement probabilities. We recall in fact from Sect. 2.9 (see (2.62)) that if $|\gamma\rangle$ is a vector of $\mathcal{U}$, the application of $P_{\mathcal{U}}$ does not modify $|\gamma\rangle$, that is,

$$P_{\mathcal{U}}|\gamma\rangle = |\gamma\rangle, \qquad \forall|\gamma\rangle \in \mathcal{U} \,.$$

We then have

$$\sum_i P[m = i|\gamma] = \sum_i \langle\gamma|Q_i|\gamma\rangle = \langle\gamma|P_{\mathcal{U}}|\gamma\rangle = \langle\gamma|\gamma\rangle = 1 \,.$$

3.7.3 The POVMs as Projective Measurements

The POVMs do not really fall into Postulate 3. However, it has been demonstrated (Neumark's theorem) that, given the Hilbert space $\mathcal{H}$ in which the measurement operators satisfying the conditions (1), (2) and (3) seen in Sect. 3.7.1 are applied, an auxiliary quantum system can always be found (*ancilla*) $\mathcal{H}_a$, which, combined with $\mathcal{H}$ (in the sense of Postulate 4, as will be seen), forms an extended quantum system, in which the POVM measurements reappear as standard (von Neumann) projective measurements.

A simple demonstration of Neumark's theorem is found in the paper by Eldar and Forney [7].

3.8 Summary of Quantum Measurements

Quantum measurements, as developed in the previous sections from Postulate 3, can be categorized in various ways.

A first categorization reflects the hypotheses made on the quantum system on which the measurements are carried out, and precisely:

(a) quantum system in a *pure state* $|\psi\rangle$,
(b) quantum system in a *mixture of states* specified by a density operator ρ.

These two hypotheses have been extensively discussed in Sect. 3.3, where it has been seen that (a) is a special case of (b) with density operator $\rho = |\psi\rangle\langle\psi|$.

A second categorization regards the measurement operators, and precisely:

(1) *projective measurements* or *von Neumann measurements*, performed with a *projector system* $\{\Pi_i, i \in \mathcal{M}\}$;
(2) *measurements with an observable*, performed with a single Hermitian operator A;
(3) *generalized measurements*, performed with a POVM system $\{Q_i, i \in \mathcal{M}\}$.

We have already seen the perfect equivalence of (1) and (2), as from the observable A, through the Spectral Decomposition Theorem, a projector system can be obtained, and vice versa, from this an observable can be built. The measurements with POVM actually represent a generalization, because the operators $\{Q_i, i \in \mathcal{M}\}$ are not necessarily orthogonal. On the other hand, it has been mentioned that, introducing an "appropriately expanded" Hilbert space, the POVMs can be viewed as projectors, falling thus back into projective measurements.

Conceptually, we can reverse the line followed so far, in which we started from the projective measurements applied to a quantum system in a pure state. Instead, we can start from the generalized measurements on system in a mixture of states. Then Postulate 3 can be reformulated as follows:

Postulate 3 (**reformulation**). A measurement on a quantum system defined on a Hilbert space $\mathcal{H}$ (state space) is obtained applying a POVM system $\{Q_i, i \in \mathcal{M}\}$, where the alphabet $\mathcal{M}$ provides the possible results of the measurement. If before the measurement the system is in a mixture of states specified by the density operator ρ, the probability that the measurement outcome in $m = i \in \mathcal{M}$ is given by

$$\boxed{\mathrm{P}[m = i|\rho] = \mathrm{Tr}[\rho Q_i], \qquad i \in \mathcal{M}.} \qquad \Box \qquad (3.54)$$

We leave it to the reader to verify that, starting from the postulate reformulated in the more general way, all the other cases considered can be obtained.

Finally, a third classification regards the rank of the measurement operators, and precisely:

 (i) operators with generic rank,
(ii) operators with unit rank (elementary operators).

We recall in particular the elementary operators. In the projective measurements the elementary projectors are obtained from an orthonormal basis , $\mathcal{A} = \{|a_i\rangle, i \in \mathcal{M}\}$, with $\langle a_i|a_j\rangle = \delta_{ij}$, through the outer products $\Pi_i = |a_i\rangle\langle a_i|$. In the measurements with observable A, it must be assumed that A be non-degenerate (with distinct eigenvalues) to make sure that the spectral decomposition be made in terms of projectors with unit rank. In the generalized measurements the elementary POVMs are obtained starting from a set of vectors $\mathcal{B} = \{|b_i\rangle, i \in \mathcal{M}\}$ of the Hilbert space, still through the outer products $Q_i = |b_i\rangle\langle b_i|$, but with vectors $|b_i\rangle$ not necessarily orthogonal.

Terminology Conventions

 In this book we use the expressions **measurement operators** for the generalized measurements (POVM) and **system of measurement operators** to stress the fact that the measurement operators verify the completeness. So "measurement operators" will be equivalent to "POVM". In particular, the measurement operators can be pairwise orthogonal **projectors**, and a complete set of the same is called a **projector system**.

The measurement operators of unit rank, that is, of the kind $Q_i = |b_i\rangle\langle b_i|$, are called **elementary measurement operators** and their factors $|b_i\rangle$ **measurement vectors**.

3.9 Combined Measurements

We now consider a combination of measurements on the same quantum system. A first case is the measurement on the same entity, repeated with the same procedure, that is, with the same projector system. A second case regards two measurements of different entities, performed simultaneously, and will open the way to the well-known Heisenberg's uncertainty principle.

In these considerations we refer to Postulate 3, in the "projective" version of Sect. 3.5, because it requires the knowledge of the system state after the measurement.

3.9.1 Sequential Repetition of the Same Measurement

Consider a quantum system $\mathcal{H}$ in the state $|\psi\rangle$, where a measurement is performed with a projector system $\{\Pi, i \in \mathcal{M}\}$, and the same projector system is applied again for a second measurement (Fig. 3.6).

The first measurement yields $m = i$ with probability $p_i = \langle\psi|\Pi_i|\psi\rangle$ and the system collapses to the state

$$|\psi_{\text{post}}^{(i)}\rangle = \frac{\Pi_i|\psi\rangle}{\sqrt{p_i}} .$$

In the second measurement the system is initially in the state $|\psi_{\text{post}}^{(i)}\rangle$ and from Postulate 3 we obtain the result $m' = j$ with probability

$$p'_j = \text{P}[m' = j|\psi_{\text{post}}^{(i)}] = \langle\psi_{\text{post}}^{(i)}|\Pi_j|\psi_{\text{post}}^{(i)}\rangle = \frac{\langle\psi|\Pi_i\Pi_j\Pi_i|\psi\rangle}{p_i} .$$

Now, if $i = j$, we get $\Pi_i\Pi_j\Pi_i = \Pi_i$ and then

$$p'_i = \text{P}[m' = i|\psi_{\text{post}}^{(i)}] = \frac{\langle\psi|\Pi_i|\psi\rangle}{p_i} = \frac{p_i}{p_i} = 1 .$$

Instead, if $i \neq j$ and $p_j \neq 0$, for the orthogonality of the projectors, we get $\Pi_i\Pi_j\Pi_i = 0$. So the result of the repetition of the measurement (with the same projectors) is $m' = i$ with probability 1, that is, with the repetition of the measurement we get no new information on the system, and the repetition is useless.

It can be verified (see Problem 3.10), applying (3.27), that the state after the second measurement remains the same as it was after the first measurement, namely

$$|\psi_{\text{post,post}}^{(i)}\rangle = |\psi_{\text{post}}^{(i)}\rangle . \tag{3.55}$$

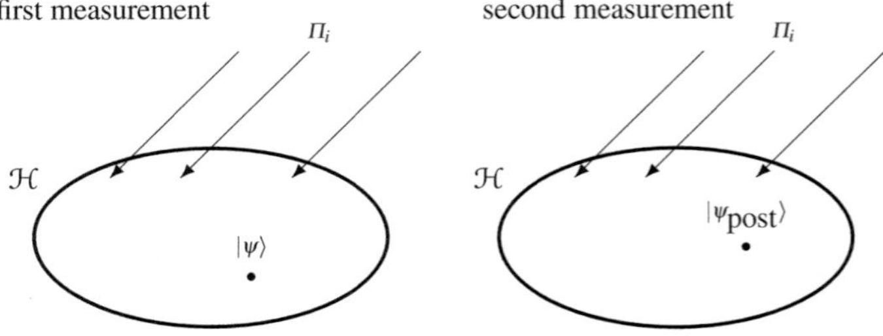

Fig. 3.6 Repetition of a quantum measurement with the same projectors

3.9.2 Exact Simultaneous Measurements

We examine the possibility of performing *simultaneous and exact measurements* on two (or more) entities of a quantum system (Fig. 3.7).

For the first measurement $m \in \mathcal{M}$ we use a projector system $\{\Pi_i^A, i \in \mathcal{M}\}$, and for the second measurement $n \in \mathcal{N}$ another system of projectors $\{\Pi_j^B, j \in \mathcal{N}\}$. The two projector systems can be summarized respectively by the observables

$$A = \sum_{i \in \mathcal{M}} i \, \Pi_i^A \quad \text{and} \quad B = \sum_{j \in \mathcal{N}} j \, \Pi_j^B \,.$$

The problem is the compatibility of the two measurements, because the system, starting from the state $|\psi\rangle$, after two simultaneous measurements, must collapse to a state $|\psi_{\text{post}}\rangle$ compatible with both procedures, that is,

$$|\psi_{\text{post}}\rangle = \frac{\Pi_i^A |\psi\rangle}{\sqrt{p_i^A}} = \frac{\Pi_j^B |\psi\rangle}{\sqrt{p_j^B}} \,.$$

For simplicity, let us assume that the projectors be elementary in both cases

$$\Pi_i^A = |a_i\rangle\langle a_i|, \qquad \Pi_j^B = |b_j\rangle\langle b_j|,$$

then the observables become

$$A = \sum_{i \in \mathcal{M}} i \, |a_i\rangle\langle a_i|, \qquad B = \sum_{j \in \mathcal{N}} j \, |b_j\rangle\langle b_j| \,.$$

Consequently, the compatibility condition becomes (see (3.31))

$$|a_i\rangle = |b_j\rangle \,, \quad \forall i \in \mathcal{M}, \ \forall j \in \mathcal{N} \,.$$

The conclusion is that the two observables must have the same set of eigenvectors. On the other hand, we recall (see Theorem 2.4 of Sect. 2.11)) that two Hermitian

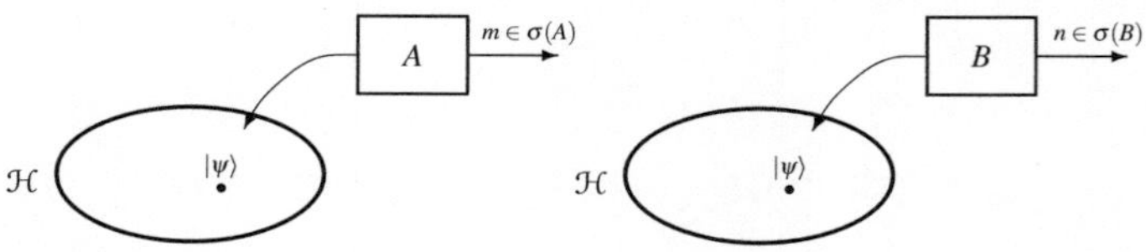

Fig. 3.7 Simultaneous quantum measurements with two different observables, A and B, when the system is in the state $|\psi\rangle$ of the Hilbert space $\mathcal{H}$. The possible outcomes are two random variables m and n

operators A and B admit a common basis if and only if the two operators commute, that is,

$$AB = BA \ .$$

The same condition is required for the projectors, in the sense that A and B commute if and only if the corresponding projectors Π_i^A and Π_j^B commute (see [10], pp. 157 and 170). The conclusion is that two distinct measurements on a quantum system are possible *if and only if the two projector systems, or the corresponding observables, are commutable.*

3.9.3 Heisenberg's Uncertainty Principle

If the two observables A and B do not commute there is an uncertainty between the variances of the outcomes. In order to quantify such uncertainties, we consider the variances of the two outcomes m and n in the two measurements (see (3.48) and (3.49))

$$\sigma_m^2 = \langle \psi | A^2 | \psi \rangle - (\langle \psi | A | \psi \rangle)^2, \qquad \sigma_n^2 = \langle \psi | B^2 | \psi \rangle - (\langle \psi | B | \psi \rangle)^2 \qquad (3.56)$$

Such formulation of the measurement and of the variance in terms of observables gives origin, in an elegant manner, to an important result known as **Heisenberg's Uncertainty Principle**.

Remembering that A and B are two Hermitian operators, we use the identity (2.31a), that is,

$$2AB = [A,\ B] + \{A,\ B\}$$

where $[A,\ B] = AB - BA$ and $\{A,\ B\} = AB + BA$ are respectively the commutator and the anticommutator of A and B. Letting $\langle \psi | AB | \psi \rangle = x + iy$, with x and y real, we have

$$\langle \psi | [A,\ B] | \psi \rangle = 2iy, \qquad \langle \psi | \{A,\ B\} | \psi \rangle = 2x$$

and so

$$4|\langle \psi | AB | \psi \rangle|^2 = |\langle \psi | [A,\ B] | \psi \rangle|^2 + |\langle \psi | \{A,\ B\} | \psi \rangle|^2$$
$$\geq |\langle \psi | [A,\ B] | \psi \rangle|^2 \ . \qquad (3.57)$$

On the other hand, Cauchy-Schwartz 's inequality (see (2.26)) gives

$$|\langle \psi | AB | \psi \rangle|^2 \leq \langle \psi | A^2 | \psi \rangle \langle \psi | B^2 | \psi \rangle \qquad (3.58)$$

which, combined to (3.57), yields

$$|\langle\psi|[A, B]|\psi\rangle|^2 \leq 4\langle\psi|A^2|\psi\rangle\langle\psi|B^2|\psi\rangle. \tag{3.59}$$

In the right-hand side of (3.59) there appear the mean squares $\mathrm{E}[m^2]$ and $\mathrm{E}[n^2]$ of the random variables m ed n. If we want to find an inequality for the variances σ_m^2 and σ_n^2, all it takes is to substitute A with $A - \langle A\rangle$ and B with $B - \langle B\rangle$. We obtain

$$\boxed{\sigma_m\sigma_n \geq \frac{1}{2}\left|\langle\psi|[A, B|\psi\rangle\right|} \tag{3.60}$$

which represents Heisenberg's uncertainty principle.

We illustrate the principle with a "historical" application in which we consider a particle with one degree of freedom and the operators A and B are observables given by the *position* operator q and the *momentum* operator p of the particle. (see the Reference System introduced in Sect. 3.4). In this case, the two operators do not commute and their commutation relation is given by

$$[q, p] = \mathrm{i}\,\hbar\, I_{\mathcal{H}}$$

where $\hbar = h/(2\pi)$ is the reduced Planck constant. Then

$$\langle\psi|[q - \langle q\rangle, p - \langle p\rangle]|\psi\rangle = \langle\psi|[q, p]|\psi\rangle = \langle\psi|\mathrm{i}\,\hbar\, I_{\mathcal{H}}|\psi\rangle = \mathrm{i}\hbar$$

and (3.60) gives

$$\boxed{\sigma_m\sigma_n \geq \frac{1}{2}\hbar\,.} \tag{3.61}$$

The interpretation of (3.61) must be done considering a large number N of identical systems all sitting in the quantum state $|\psi\rangle$. In some of these systems the position m is measured, in others, the momentum n. With increasing N the estimates of the average $\mathrm{E}[m]$ and $\mathrm{E}[n]$, and of the variances σ_m^2 and σ_n^2, settle around values depending on the state $|\psi\rangle$ of the N (identical) systems. The uncertainty principle asserts that in no state $|\psi\rangle$ the product $\sigma_m\sigma_n$ can be less than $\frac{1}{2}\hbar$.

Problem 3.10 ⋆⋆ Prove that the state after the second measurement with the same projector system remains the same as the one in which the system was after the first measurement, as stated by (3.55).

3.10 Composite Quantum Systems

The last postulate of Quantum Mechanics concerns the model of systems consisting of subsystems, such as a pair of particles, or a system combining both the system on which we want to perform the measurement and the measurement apparatus. In any case we must arrive at a (larger) overall isolated and closed system, in which two (or more) subsystems are identifiable. The formulation of the postulate is based on the tensor product seen in Sect. 2.13.

Postulate 4 A system composed by two subsystems $\mathcal{H}_1$ and $\mathcal{H}_2$ must be treated in the framework of a Hilbert space $\mathcal{H}$ given by the *tensor product* of the two component subsystems

$$\mathcal{H} = \mathcal{H}_1 \otimes \mathcal{H}_2$$

and therefore, if $|\psi_1\rangle$ is a state of $\mathcal{H}_1$ and $|\psi_2\rangle$ is a state of $\mathcal{H}_2$, then

$$|\psi\rangle = |\psi_1\rangle \otimes |\psi_2\rangle \tag{3.62}$$

is a state of the composite system. This procedure extends in a straightforward way to the composition of an arbitrary number of subsystems. $\qquad\square$

The postulate is schematically illustrated in Fig. 3.8.

The formulation of a composite system is extremely simple and essential, and this seems to suggest obvious consequences. On the contrary, composite systems conceal surprising and even disconcerting aspects, like the *entanglement*, which lead to revolutionary applications, especially in Quantum Information and in Quantum Computing. Surprising and paradoxical aspects were already evidenced at the very dawn of Quantum Mechanics, like the universally renowned *Einstein, Podolski and Rosen's paradox* [11], but only in the last two decades did research on the subject develop substantially. Here we will point out only a few of these aspects.

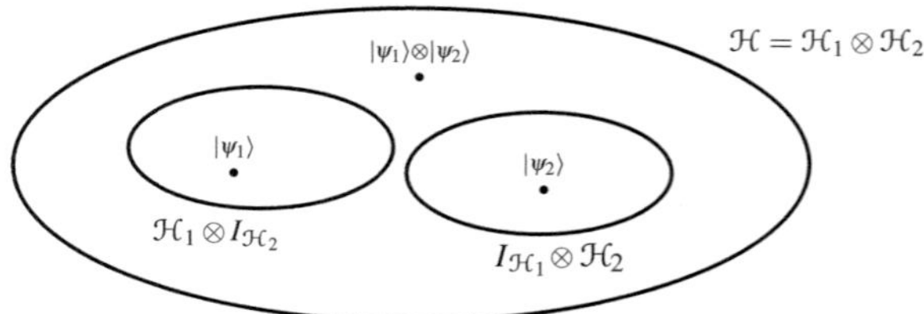

Fig. 3.8 The system composed by two subsystems $\mathcal{H}_1$ and $\mathcal{H}_2$ is given by the *tensor product* $\mathcal{H} = \mathcal{H}_1 \otimes \mathcal{H}_2$. In the composite system $\mathcal{H}_1 \otimes \mathcal{H}_2$ the original subspaces appear as $\mathcal{H}_1 \otimes I_{\mathcal{H}_2}$ and $I_{\mathcal{H}_1} \otimes \mathcal{H}_2$

3.10.1 Marginal Rules of a Composite Space

We now consider the rules that could be labeled as "obvious", because they derive directly from the rules of tensor product and of Kronecker's product seen in Sect. 2.13.

(1) Composition of states. If $|\psi_1\rangle \in \mathcal{H}_1$ and $|\psi_2\rangle \in \mathcal{H}_2$, in the composite space $\mathcal{H} = \mathcal{H}_1 \otimes \mathcal{H}_2$ the state becomes

$$|\psi\rangle = |\psi_1\rangle \otimes |\psi_2\rangle \tag{3.63}$$

as anticipated in (3.62).

(2) Evolution. If U_1 is a unitary operator in $\mathcal{H}_1$, the corresponding unitary operator in $\mathcal{H}$ is $U_1 \otimes I_2$, where I_2 is the identity operator on $\mathcal{H}_2$. The action of $U_1 \otimes I_2$ is the following

$$(U_1 \otimes I_2)(|\psi_1\rangle \otimes |\psi_2\rangle) = (U_1|\psi_1\rangle) \otimes |\psi_2\rangle. \tag{3.64}$$

Introducing the initial time t_0 and the generic time $t > t_0$, as done in Sect. 3.4, from (3.64) we deduce that if the initial state is $|\psi(t_0)\rangle = |\psi_1(t_0)\rangle \otimes |\psi_2(t_0)\rangle$ the final state is

$$|\psi(t)\rangle = |\psi_1(t)\rangle \otimes |\psi_2(t)\rangle$$

where

$$|\psi_1(t)\rangle = U_1|\psi_1(t_0)\rangle, \qquad |\psi_2(t)\rangle = |\psi_2(t_0)\rangle$$

with the obvious interpretation that the operator $U_1 \otimes I_2$ evolves only the states of the subsystem $\mathcal{H}_1$ of $\mathcal{H}_1 \otimes \mathcal{H}_2$. Similar conclusions can be drawn for a unitary operator U_2 of $\mathcal{H}_2$ in the composite system, which becomes $I_1 \otimes U_2$, where I_1 is the identity of $\mathcal{H}_1$.

(3) Measurements. If A_1 and A_2 are two *observables* respectively of $\mathcal{H}_1$ and of $\mathcal{H}_2$, then the corresponding observables in $\mathcal{H}$ are $A_1 \otimes I_2$ and $I_1 \otimes A_2$. Notice that $A_1 \otimes I_2$ and $I_1 \otimes A_2$ *are always commuting* and therefore the measurements on the two subsystems can be performed simultaneously (see Sect. 3.9.2).

As you can see, borrowing the language of Probability Theory, the above rules can be seen as *marginal laws*.

3.10.2 Quantum Measurements in a Composite Hilbert Space

We reconsider in more detail the quantum measurements in a composite Hilbert space, confining the considerations to projective measurements. Then, if $\{\Pi_i, i \in \mathcal{M}\}$ is a

projector system in the Hilbert space $\mathcal{H} = \mathcal{H}_1 \otimes \mathcal{H}_2$, and the quantum system is in the state $|\psi\rangle \in \mathcal{H}$, we can apply integrally Postulate 3. Thus, the probability that the outcome is $m = i \in \mathcal{M}$ is given by

$$p_m(i|\psi) := \mathrm{P}[m = i | s = |\psi\rangle] = \langle\psi|\Pi_i|\psi\rangle, \qquad i \in \mathcal{M}. \tag{3.65}$$

If the system is in a separable state, that is,

$$|\psi\rangle = |\psi_1\rangle \otimes |\psi_2\rangle := |\psi_1\rangle|\psi_2\rangle$$

we can consider a separable projector system

$$\Pi_{i_1 i_2} = \Pi^1_{i_1} \otimes \Pi^2_{i_2}, i \in \mathcal{M}_1, i_2 \in \mathcal{M}_2$$

where $\Pi^1_{i_1}$ and $\Pi^2_{i_2}$ are projectors of $\mathcal{H}_1$ and of $\mathcal{H}_2$, respectively. In this case the outcome of the measurement is a pair of random variables $m = (m_1, m_2)$ and we can apply (3.65) with

$$i = (i_1, i_2), \qquad m = (m_1, m_2), \qquad |\psi\rangle = |\psi_1\rangle \otimes |\psi_2\rangle, \qquad \Pi_i = \Pi^1_{i_1} \otimes \Pi^2_{i_2}$$

to get

$$\begin{aligned}
p_{m_1 m_2}(i_1, i_2|\psi_1\psi_2) &:= \mathrm{P}[m_1 = i_1, m_2 = i_2 | s = |\psi\rangle_1 \otimes |\psi_2\rangle] \\
&= \langle\psi_1| \otimes \langle\psi_2|\Pi^1_{i_1} \otimes \Pi^2_{i_2}||\psi_1\rangle \otimes |\psi_2\rangle. \tag{3.66}
\end{aligned}$$

The separability (and the intuition) suggests that the random variables may be statistically independent. To see this we apply the mixed product law

$$(A \otimes B)(C \otimes D)(E \otimes F) = (A\,C\,E) \otimes (B\,D\,F)$$

in (3.66), to get

$$\langle\psi_1| \otimes \langle\psi_2|\Pi^1_{i_1} \otimes \Pi^2_{i_2}|\psi_1\rangle \otimes |\psi_2\rangle = (\langle\psi_1|\Pi^1_{i_1}|\psi_1\rangle) \otimes (\langle\psi_2|\Pi^2_{i_2}|\psi_2\rangle)$$

where

$$\langle\psi_1|\Pi^1_{i_1}|\psi_1\rangle = \mathrm{P}[m_1 = i_1|\psi_1], \qquad \langle\psi_2|\Pi^2_{i_2}|\psi_2\rangle = \mathrm{P}[m_2 = i_2|\psi_1]$$

which are scalars so that $\otimes$ can be dropped. Hence

$$\mathrm{P}[m_1 = i_1, m_2 = i_2 | s = |\psi\rangle_1 \otimes |\psi_2\rangle] = \mathrm{P}[m_1 = i_1|\psi_1]\mathrm{P}[m_2 = i_2|\psi_1]$$

and the random variables turn out to be independent.

The conclusion is that a quantum measurement with separable projectors when the composite Hilbert space is in a separable state leads to independent random variables. If the state is entangled, in general the random variables turn out to be correlated.

As an example of application we consider the photon counting in a two-mode coherent states $|\alpha_1\rangle \otimes |\alpha_2\rangle$ using the elementary projectors given by the number states of the two modes, that is,

$$\Pi^1_{i_1} = |i_1\rangle_1 \, {}_1\langle i_1|, \qquad \Pi^2_{i_2} = |i_2\rangle_2 \, {}_2\langle i_2| \, .$$

The outcome (m_1, m_2) is a pair of independent Poisson random variables.

Example 3.6 (*Two qubit system*) Consider the composition of two binary systems, $\mathcal{H}_1 = \mathcal{H}_2 = \mathbb{C}^2$, that is, the composition of two qubits. Indicating for clarity the basis of $\mathcal{H}_1$ by $|b_0\rangle$, $|b_1\rangle$ (instead of $|0\rangle$ and $|1\rangle$ as done at the beginning of the chapter), the generic state of $\mathcal{H}_1$ results in

$$|\psi_1\rangle = u_0|b_0\rangle + u_1|b_1\rangle, \qquad |u_0|^2 + |u_1|^2 = 1 \, .$$

Analogously, indicating by $|c_0\rangle$, $|c_1\rangle$ the basis of $\mathcal{H}_2$ we have

$$|\psi_2\rangle = v_0|c_0\rangle + v_1|c_1\rangle, \qquad |v_0|^2 + |v_1|^2 = 1 \, .$$

The composite system $\mathcal{H} = \mathcal{H}_1 \otimes \mathcal{H}_2$ has dimension 4 (it is isomorphic to $\mathbb{C}^4$) and one of its basis is given by (see (2.106))

$$\{|b_0\rangle \otimes |c_0\rangle, \, |b_0\rangle \otimes |c_1\rangle, \, |b_1\rangle \otimes |c_0\rangle, \, |b_1\rangle \otimes |c_1\rangle\} \, . \tag{3.67}$$

The state of $\mathcal{H}$ obtained as tensor product of $|\psi_1\rangle$ and $|\psi_2\rangle$ results in

$$\begin{aligned}
|\psi\rangle = |\psi_1\rangle \otimes |\psi_2\rangle = {}&u_0v_0|b_0\rangle \otimes |c_0\rangle + u_0v_1|b_0\rangle \otimes |c_1\rangle \\
&+ u_1v_0|b_1\rangle \otimes |c_0\rangle + u_1v_1|b_1\rangle \otimes |c_1\rangle \, .
\end{aligned} \tag{3.68}$$

It can be verified that the length of $|\psi\rangle$ is unitary

$$|u_0v_0|^2 + |u_0v_1|^2 + |u_1v_0|^2 + |u_1v_1|^2 = (|u_0|^2 + |u_1|^2)(|v_0|^2 + |v_1|^2) = 1 \, .$$

3.10.3 States Not Covered by the Tensor Product. Entanglement

The most relevant surprises of a composite space come from the fact that not all the states, not all the evolutions, and not all the measurements can be expressed as tensor product of factors belonging to the component subsystems. Let us start out with an example.

Example 3.7 We assert that in a **two qubit system** $\mathcal{H}_1 \otimes \mathcal{H}_2$ not all the states are given by the tensor product developed in (3.68). For example, the state

$$|\psi\rangle = \frac{1}{\sqrt{2}}|b_0\rangle \otimes |c_1\rangle + \frac{1}{\sqrt{2}}|b_1\rangle \otimes |c_0\rangle \tag{3.69}$$

is a state of $\mathcal{H}_1 \otimes \mathcal{H}_2$ because it is a particular linear combination of the vectors of the basis (3.67). To reach this state from (3.68) we should have

$$u_0 v_0 = 0, \qquad u_0 v_1 = \frac{1}{\sqrt{2}}, \qquad u_1 v_0 = \frac{1}{\sqrt{2}}, \qquad u_1 v_1 = 0$$

which has no solutions (the second and the third require u_0, v_1, u_1, v_0 different from zero). So, the state (3.69) cannot be obtained as a product and must be classified as *entangled* state.

Going back to the general case:

Definition 3.1 In a composite Hilbert space $\mathcal{H}_1 \otimes \mathcal{H}_2$ any state that is not given by the tensor product of two states of the component systems, that is,

$$|\psi\rangle \neq |\psi_1\rangle \otimes |\psi_2\rangle \quad \forall |\psi_1\rangle \in \mathcal{H}_1 , \ \forall |\psi_2\rangle \in \mathcal{H}_2 . \tag{3.70}$$

is called *entangled state*. If $|\psi\rangle = |\psi_1\rangle \otimes |\psi_2\rangle$ the state is called *separable*.

For each of these states it is not possible to identify in the two component subsystems two states that generate it as a product. It is, so to speak, a state that achieves its existence only when the two systems are combined.

To see with a systemic eye that in a space $\mathcal{H} = \mathcal{H}_1 \otimes \mathcal{H}_2$ the "majority of the states are entangled", we observe that all the states of $\mathcal{H}_1 \otimes \mathcal{H}_2$ can be generated, like for any other Hilbert space, as a linear combination of a basis. In this case the basis is

$$\{|b_i\rangle \otimes |c_j\rangle, i \in I, j \in J\}$$

and therefore all the states of $\mathcal{H}_1 \otimes \mathcal{H}_2$ are given by

$$|x\rangle = \sum_{i \in I} \sum_{j \in J} t_{ij}|b_i\rangle \otimes |c_j\rangle, \qquad \forall t_{ij} \in \mathbb{C} \tag{3.71}$$

with the normalization constraint $\sum_{ij} |t_{ij}|^2 = 1$.

On the other hand, the states of $\mathcal{H}_1$ and $\mathcal{H}_2$ are given respectively by

$$|x_1\rangle = \sum_i u_i|b_i\rangle, \qquad |x_2\rangle = \sum_j v_j|c_j\rangle$$

with the normalization constraint $\sum_i |u_i|^2 = 1$ and $\sum_j |v_j|^2 = 1$, so the states given by the tensor product result in

$$|x_1\rangle \otimes |x_2\rangle = \sum_i \sum_j u_i v_j |b_i\rangle \otimes |c_j\rangle, \qquad \forall u_i, v_j \in \mathbb{C}. \tag{3.72}$$

These states can be obtained from (3.71) with

$$t_{ij} = u_i v_j, \tag{3.73}$$

but the opposite is not true, in the sense that it is not always possible to express the states of $\mathcal{H}_1 \otimes \mathcal{H}_2$ in the form (3.72) of tensor product of a state of $\mathcal{H}_1$ and of a state of $\mathcal{H}_2$.

This fact is better understood with spaces of finite dimensions: if $\mathcal{H}_1$ and $\mathcal{H}_2$ have, respectively, dimensions $m_1 = 4$ and $m_2 = 3$, it turns out that $\mathcal{H}$ has dimensions $m_1 m_2 = 12$. Then, setting in (3.73) the 12 values of t_{ij} we have a system of 12 equations with 7 unknowns $u_1 u_2 u_3 u_4$, $v_1 v_2 v_3$, which in general has no solutions.

The problem of the separability will be seen systematically in Sect. 10.3 with the Schmidt decomposition.

The consequences of the entanglement are numerous and they would deserve a deeper insight, which will be seen in Part III.

3.10.4 No-Cloning Theorem

We describe another surprising consequence that takes place in the combination of quantum systems, due to the fact that Postulate 2 requires that the evolutions of a system be always governed by a *unitary* operator, for which some evolutions are *impossible*. One of these impossible evolutions is the copy (clonation) of a quantum state.

Consider a quantum system $\mathcal{H}$ in a state $|\psi\rangle$ and suppose we wanted, through an appropriate evolution, to transfer (copy) this state to another system $\mathcal{H}_c$, which at a certain initial time is in whatever state; and it is not restrictive to indicate this state by $|0\rangle$. Then, in the combination of the two systems $\mathcal{H} \otimes \mathcal{H}_c$ we want to move from the initial state $|\psi\rangle \otimes |0\rangle$ to the state $|\psi\rangle \otimes |\psi\rangle$, in which even the system $\mathcal{H}_c$ reaches the state $|\psi\rangle$. For this to happen, there must exist a unitary operator of $\mathcal{H} \otimes \mathcal{H}_c$ such that

$$U(|\psi\rangle \otimes |0\rangle) = |\psi\rangle \otimes |\psi\rangle \tag{3.74a}$$

for every state $|\psi\rangle$. Because $|\psi\rangle$ is arbitrary, the same relation must hold even for a state $|\phi\rangle \neq |\psi\rangle$

$$U(|\phi\rangle \otimes |0\rangle) = |\phi\rangle \otimes |\phi\rangle. \tag{3.74b}$$

Now, if such operator existed, due to the linearity of the tensor product with respect to the argument (see Sect. 2.13), we would have

$$U(\alpha|\psi\rangle + \beta|\varphi\rangle) \otimes |0\rangle = \alpha|\psi\rangle \otimes |\psi\rangle + \beta|\varphi\rangle \otimes |\varphi\rangle$$
$$\neq (\alpha|\psi\rangle + \beta|\varphi\rangle) \otimes (\alpha|\psi\rangle + \beta|\varphi\rangle) \,. \tag{3.75}$$

Therefore no unitary transformation U exists allowing the copy according to (3.74). This result is known as the **no-cloning theorem**: *quantum information, differently from classical, cannot be copied.*

Also, this important topic will be reconsidered in the final chapters.

Problem 3.11 ⋆⋆ Consider the non normalized state

$$|\psi'\rangle = 2\,|0\,0\rangle + i\,|0\,1\rangle + 3\,|0\,1\rangle$$

of a two-qubit system with basis $\mathcal{B} = \{|0\,0\rangle, |0\,1\rangle, |1\,0\rangle, |1\,1\rangle\}$ (here $|0\,0\rangle$ stands for $|0\rangle \otimes |0\rangle$, etc.). Find the normalized form, $\langle\psi|\psi\rangle = 1$, and prove that the two qubits $|\psi\rangle$ are entangled.

3.11 Nonunicity of the Density Operator Decomposition ⇓

In an n-dimensional Hilbert space $\mathcal{H}$ a (discrete) density operator is defined starting from an *ensemble* $\mathcal{E} = (\mathcal{S}, p)$, where $\mathcal{S} = \{|\psi_1\rangle, \ldots, |\psi_k\rangle\}$ is a set of normalized states and p is a probability distribution over $\mathcal{S}$. The resulting density operators is (see (3.7))

$$\rho = \sum_{i=1}^{k} p_i\,|\psi_i\rangle\langle\psi_i| \,. \tag{3.76}$$

In this section we discuss the multiplicity of the ensembles that generate the same density operator.

3.11.1 Matrix Representation of an Ensemble

An ensemble $\mathcal{E} = (\mathcal{S}, p)$ can be represented by an $n \times k$ matrix

$$\widehat{\Psi} = [\sqrt{p_1}\,|\psi_1\rangle, \ldots, \sqrt{p_k}\,|\psi_k\rangle] = [|\widehat{\psi}_1\rangle, \ldots, |\widehat{\psi}_k\rangle] \tag{3.77}$$

where

$$|\widehat{\psi}_i\rangle = \sqrt{p_i}\,|\psi_i\rangle, \qquad i = 1, \ldots, k \,. \tag{3.77a}$$

are called *weighted states*. From the matrix $\widehat{\Psi}$ the density operator (3.76) is obtained in the form

$$\rho_{n \times n} = \widehat{\Psi}_{n \times k} \widehat{\Psi}^*_{k \times n} = [|\widehat{\psi}_1\rangle|, \ldots, |\widehat{\psi}_k\rangle] \begin{bmatrix} \langle\widehat{\psi}_1| \\ \vdots \\ \langle\widehat{\psi}_k| \end{bmatrix} \tag{3.78}$$

where $\widehat{\Psi}^*$ is the conjugate transpose of $\widehat{\Psi}$. Then $\widehat{\Psi}$ is called k-factor or simply a *factor* of the given density operator.

The factor $\widehat{\Psi}$ bears the full information of the ensemble $\mathcal{E}$ as well as of the density operator.

Proposition 3.4 *Factors and ensembles are in one-to-one correspondence, and therefore generate the same density operator.*

In fact, given an ensemble $\mathcal{E} = (\mathcal{S}, p)$, one gets the corresponding factor $\widehat{\Psi}$ from (3.77). Given a factor $\widehat{\Psi}$, one gets the probabilities using (3.77a) as $p_i = \langle\widehat{\psi}_i|\widehat{\psi}_i\rangle$ and the normalized states of $\mathcal{S}$ as $|\psi_i\rangle = (1/\sqrt{p_i})|\widehat{\psi}_i\rangle$.

Example 3.8 Consider the density operator obtained from the normalized states of $\mathcal{H} = \mathbb{C}^4$

$$|\psi_1\rangle = \begin{bmatrix} \frac{1}{2} \\ -\frac{i}{2} \\ -\frac{1}{2} \\ \frac{i}{2} \end{bmatrix}, \qquad |\psi_2\rangle = \begin{bmatrix} \frac{2+\sqrt{2}}{2\sqrt{6}} \\ \frac{-2i+\sqrt{2}}{2\sqrt{6}} \\ \frac{-2+\sqrt{2}}{2\sqrt{6}} \\ \frac{2i+\sqrt{2}}{2\sqrt{6}} \end{bmatrix}, \qquad |\psi_3\rangle = \begin{bmatrix} \frac{-1+\sqrt{2}}{2\sqrt{3}} \\ -\frac{2i+\sqrt{2}}{2\sqrt{6}} \\ -\frac{2+\sqrt{2}}{2\sqrt{6}} \\ -\frac{-2i+\sqrt{2}}{2\sqrt{6}} \end{bmatrix}$$

with probabilities $p_1 = 1/4$, $p_2 = 3/8$ and $p_3 = 3/8$. The expression of ρ is therefore

$$\rho = \frac{1}{4}|\psi_1\rangle\langle\psi_1| + \frac{3}{8}|\psi_2\rangle\langle\psi_2| + \frac{3}{8}|\psi_3\rangle\langle\psi_3| = \begin{bmatrix} \frac{1}{4} & \frac{1}{16}+\frac{3i}{16} & -\frac{1}{8} & \frac{1}{16}-\frac{3i}{16} \\ \frac{1}{16}-\frac{3i}{16} & \frac{1}{4} & \frac{1}{16}+\frac{3i}{16} & -\frac{1}{8} \\ -\frac{1}{8} & \frac{1}{16}-\frac{3i}{16} & \frac{1}{4} & \frac{1}{16}+\frac{3i}{16} \\ \frac{1}{16}+\frac{3i}{16} & -\frac{1}{8} & \frac{1}{16}-\frac{3i}{16} & \frac{1}{4} \end{bmatrix}.$$

A factor is

$$\widehat{\Psi} = \frac{1}{4} \begin{bmatrix} 1 & 1+\frac{1}{\sqrt{2}} & 1-\frac{1}{\sqrt{2}} \\ -i & -i+\frac{1}{\sqrt{2}} & -i-\frac{1}{\sqrt{2}} \\ -1 & -1+\frac{1}{\sqrt{2}} & -1-\frac{1}{\sqrt{2}} \\ i & i+\frac{1}{\sqrt{2}} & i-\frac{1}{\sqrt{2}} \end{bmatrix} \tag{3.79}$$

and, in fact, we can check that $\widehat{\Psi}\,\widehat{\Psi}^* = \rho$. From the columns of $\widehat{\Psi}$ we can obtain both the normalized states and the probabilities. For instance, from the first weighted state one gets

$$
|\widehat{\psi}_1\rangle = \frac{1}{4}\begin{bmatrix} 1 \\ -i \\ -1 \\ i \end{bmatrix} \quad\rightarrow\quad p_1 = \langle\psi_1|\psi_1\rangle = \frac{1}{4}, \quad |\psi_1\rangle = \frac{1}{\sqrt{p_1}}|\widehat{\psi}_1\rangle = \begin{bmatrix} \frac{1}{2} \\ -\frac{i}{2} \\ -\frac{1}{2} \\ \frac{i}{2} \end{bmatrix}.
$$

Note that, while a factor (or an ensemble) uniquely determines a density operator, for a given density operator we may find infinitely many factors, according to the paradigm

$$
\text{ensemble} \quad\Longleftrightarrow\quad \text{factor} \quad\Longrightarrow\quad \text{density operator}.
$$

This problem was considered by Hugston, Josa and Wootters in a letter [12] and here it is reconsidered in a new form, completely based on matrix analysis. For instance, in [12] the compact form (3.78) and the consequent application of the singular value decomposition (SVD) are not considered. Note that factors of a density operator play a fundamental role in quantum detection (see Chap. 5, Sect. 5.7, and Chap. 8) and also in Quantum Information Theory (see Chap. 12).

Before proceeding we refine a few definitions. For a k-factor it is easy to see that the minimum value of k is given by the rank r of ρ, which is also the rank of any factor of ρ, but the value of k may be arbitrarily large. An r-factor, with $r = \text{rank}(\rho)$, will be called a *minimum factor of ρ*. An ensemble $\mathcal{E} = (\mathcal{S}, p)$, where $\mathcal{S} = \{|\psi_1\rangle, \ldots, |\psi_k\rangle\}$ consists of orthonormal states, that is, $\langle\psi_i|\psi_j\rangle = \delta_{ij}$, is called *orthonormal*, and so is for the corresponding factor, where the orthonormality condition is stated by

$$
\widehat{\Psi}^*\,\widehat{\Psi} = \text{diag}\,\{\langle\widehat{\psi}_1|\widehat{\psi}_1\rangle, \ldots, \langle\widehat{\psi}_k|\widehat{\psi}_k\rangle\} = \text{diag}\,\{p_1, \ldots, p_k\}. \tag{3.80}
$$

An orthonormal k-factor $\widehat{\Psi}$ is necessarily minimum. In fact, the k orthonormal columns of $\widehat{\Psi}$ are linearly independent and therefore $\widehat{\Psi}$ has rank k, but $\text{rank}\widehat{\Psi} = \text{rank}\rho = r$. In the previous example, where $k = 3$, the rank of ρ is $r = 2$ and $\widehat{\Psi}^*\widehat{\Psi}$ is not a diagonal matrix, so $\widehat{\Psi}$ is neither minimum nor orthonormal.

3.11.2 Minimum Factor from the EID

Now, we are ready to get the minimum factor from a density operator ρ, using the eigendecomposition (EID).

Proposition 3.5 *Let ρ be a density operator in an n-dimensional Hilbert space $\mathcal{H}$ and let $r = \text{rank}\rho$. The* **reduced EID** *of ρ has the form (see Sect. 2.11)*

$$
\rho = \sum_{i=1}^{r} \sigma_i^2\,|u_i\rangle\langle u_i| = U\,\Sigma^2\,U^* \tag{3.81}
$$

where σ_i^2 are the r positive eigenvalues of ρ, $|u_i\rangle$ are the corresponding orthonormal eigenvectors, $U = [|u_1\rangle, \ldots, |u_r\rangle]$ and $\Sigma^2 = diag\{\sigma_1^2, \ldots, \sigma_r^2\}$. Then,

$$\widehat{\Psi}_0 = U\,\Sigma = [|\widehat{\psi}_{01}\rangle, \ldots, |\widehat{\psi}_{0r}\rangle] \quad with \quad |\widehat{\psi}_{0i}\rangle = \sigma_i\,|u_i\rangle \qquad (3.82)$$

is a minimum orthonormal factor of ρ.

In terms of an ensemble, from the EID we have the orthonormal ensemble $\mathcal{E}_0 = (\mathcal{S}_0, p_0)$, where $\mathcal{S}_0 = \{|u_0\rangle, \ldots, |u_k\rangle\}$ and $p_{0i} = \sigma_i$. $\square$

Proof Clearly $\widehat{\Psi}_0\widehat{\Psi}_0^* = \rho$, that is, $\widehat{\Psi}_0$ is a factor of ρ. Since ρ is Hermitian and positive semidefinite (PSD), its nonzero eigenvalues σ_i^2 are positive. Moreover, $\mathrm{Tr}[\rho] = 1$ and the trace is given by the sum of the eigenvalues. Hence, the σ_i^2 form a probability distribution. Finally note that in (3.81) U collects r orthonormal eigenvectors, so that $\widehat{\Psi}_0$ represents a minimum orthonormal factor of ρ.

Example 3.9 Reconsider the density operator of the previous example, which has rank $r = 2$. The EID is given by $\rho = U\,\Sigma^2\,U^*$ with

$$U = \begin{bmatrix} -\frac{1}{2} & -\frac{1}{2} \\ \frac{i}{2} & -\frac{1}{2} \\ \frac{1}{2} & -\frac{1}{2} \\ -\frac{1}{2} & -\frac{1}{2} \end{bmatrix}, \qquad \Sigma^2 = \begin{bmatrix} \frac{3}{4} & 0 \\ 0 & \frac{1}{4} \end{bmatrix} \quad \rightarrow \quad \Sigma = \begin{bmatrix} \sqrt{\frac{3}{4}} & 0 \\ 0 & \frac{1}{2} \end{bmatrix}.$$

Hence,

$$\widehat{\Psi}_0 = U\,\Sigma = \begin{bmatrix} -\frac{\sqrt{3}}{4} & -\frac{1}{4} \\ \frac{i\sqrt{3}}{4} & -\frac{1}{4} \\ \frac{\sqrt{3}}{4} & -\frac{1}{4} \\ -\frac{\sqrt{3}}{4} & -\frac{1}{4} \end{bmatrix} \qquad (3.83)$$

is a minimum orthonormal factor of ρ. We can check that $U^*\,U = I_2$ and $\widehat{\Psi}_0\widehat{\Psi}_0^* = \rho$.

The factor multiplicity is further investigated in problems. Problem 3.12 establishes how to get a minimum factor from an arbitrary factor, using the singular-value decomposition (SVD). Problem 3.13 establishes how to get the whole class of factors of a given ρ.

Problem 3.12 $\star\star$ Minimum factor from an arbitrary factor. Let $\widehat{\Psi}$ be an arbitrary k-factor of ρ. The **reduced SVD** of the $n \times k$ matrix $\widehat{\Psi}$ has the form (see Sect. 2.12)

$$\widehat{\Psi} = \sum_{i=1}^{r} \sigma_i\,|u_i\rangle\langle v_i| = U\,\Sigma\,V^* \qquad (3.84)$$

where the σ_i are the square roots of the r positive eigenvalues σ_i^2 of $\widehat{\Psi}\,\widehat{\Psi}^* = \rho$, $\Sigma = \mathrm{diag}\,\{\sigma_1, \ldots, \sigma_r\}$, $|u_i\rangle$ and U are the same as in the EID of (3.81), $|v_i\rangle$ are orthonormal vectors of length k, and $V = [|v_1\rangle, \ldots, |v_r\rangle]$. Prove that a minimum orthonormal factor of ρ is given by

$$\widehat{\Psi}_0 = U\,\Sigma\,. \tag{3.85}$$

Problem 3.13 ★★★ Generation of all possible factors of a density operator. Let $\widehat{\Psi}$ be a k-factor of ρ, that is, $\widehat{\Psi}\,\widehat{\Psi}^* = \rho$, and let A be an arbitrary $k \times p$ complex matrix that verifies the condition $A\,A^* = I_k$. Prove that

$$\Phi = \widehat{\Psi}\,A \tag{3.86}$$

is a p-factor of ρ. This relation allows us to generate all the possible factors of a given density operator

Problem 3.14 ★★ Find the reduced SVD of the factor (3.79) and show that it gives the same minimum factor $\widehat{\Psi}_0$ obtained with the EID of ρ.

Problem 3.15 ★★ Consider the minimum factor given by (3.83) and find a 2×3 matrix to generate a 3-factor. Also, apply the 2×8 matrix

$$A = \frac{1}{2\sqrt{2}} \begin{bmatrix} 1 & 1 & 1 & 1 & 1 & 1 & 1 & 1 \\ 1 & e^{-\frac{i\pi}{4}} & -i & e^{-\frac{3i\pi}{4}} & -1 & e^{\frac{3i\pi}{4}} & i & e^{\frac{i\pi}{4}} \end{bmatrix}$$

to generate an 8-factor.

3.12 Revisiting the Qubit and Its Description

The qubit has been introduced in Sect. 3.2 as the simplest quantum system. It was then considered in Example 3.2 in the context of quantum measurements. Due to the relevance of such quantum system for applications, the qubit is now considered in further detail.

We recall the main difference between the bit and the qubit. Regardless of its physical realization, a bit is always understood to be either a 0 or a 1. An analogy to this is a light switch, with the off position representing 0 and the on position representing 1. A qubit has a few similarities to a classical bit, but, overall, it is very different. Like a bit, a qubit can have two possible values, 0 and 1. The difference is that, whereas a bit must be either 0 or 1, a qubit can be 0, 1, or a superposition of both.

3.12.1 Representation of Pure Qubit States on the Bloch sphere

We reconsider the expression (3.3) of a generic qubit state

$$|\psi\rangle = a\,|0\rangle + b\,|1\rangle \tag{3.87}$$

which is a linear combination of the elements of the bases $\{|0\rangle, |1\rangle\}$ with a and b complex numbers. This representation is redundant because a and b are not arbitrary, but are constrained by the normalization condition

$$|a|^2 + |b|^2 = 1. \tag{3.88}$$

We can thus move from a representation with four degrees of freedom (dependence on four real numbers, the real and the imaginary parts of a and b, respectively) to three degrees of freedom. The redundancy can be further reduced to two degrees of freedom. In fact, by letting $a = e^{i\gamma}\cos\frac{1}{2}\theta$ and $b = e^{i(\gamma+\phi)}\sin\frac{1}{2}\theta$, where the normalization is verified, we have

$$|\psi\rangle = e^{i\gamma}\left(\cos\frac{1}{2}\theta\,|0\rangle + e^{i\phi}\sin\frac{1}{2}\theta\,|1\rangle\right).$$

The first exponential represents a global phase, and can thus be ignored because it has no observable effects, as explained in Sect. 3.2. Hence, up to the irrelevant phase, a general normalized qubit vector can be written as

$$|\psi\rangle = \cos\frac{1}{2}\theta\,|0\rangle + e^{i\phi}\sin\frac{1}{2}\theta\,|1\rangle, \tag{3.89}$$

in dependence of the two real parameters θ and ϕ. The possible states for a qubit can then be put in correspondence, and thus visualized, as the points of the surface of a unitary sphere, where a point is determined be the angles θ and ϕ, as shown in Fig. 3.9. This representation is called *Bloch sphere representation*.

In order to develop some familiarity with the Bloch sphere representation, let us provide the representation for some noteworthy states. For example, for $\phi = 0$ and $\theta = 0$ ("North Pole") one obtains the basis ket $|0\rangle$, while for $\phi = 0$ and $\theta = \pi$ ("South Pole") one gets the basis ket $|1\rangle$. The points on the Equator correspond to $\theta = \frac{1}{2}\pi$. It is easily seen that two qubit states $|\psi_1\rangle$ and $|\psi_2\rangle$ are orthogonal if they are antipodal in the Bloch sphere, that is, if $\theta_1 + \theta_2 = \pi$.

The Bloch sphere representation allows us to illustrate geometrically the difference between the classical bit and the qubit. With the classical bit the representation is limited to two points: the North Pole and the South Pole, while the qubit may be located at any point of the surface, which is a two-dimensional manifold.

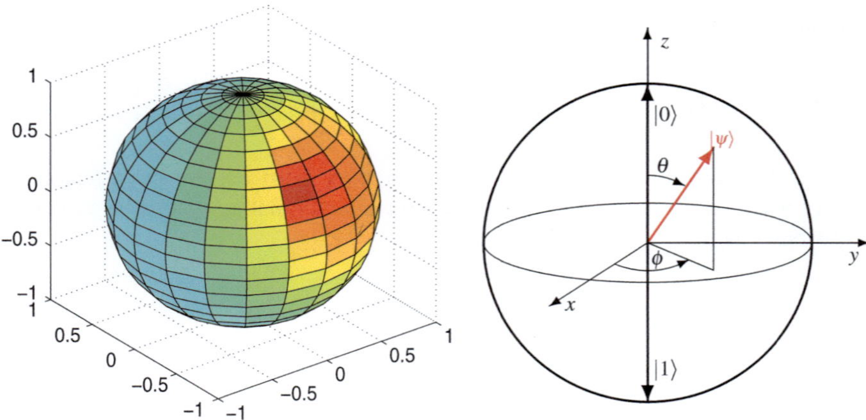

Fig. 3.9 Representation of a qubit through the *Bloch sphere*

3.12.2 Representation of Mixed Qubit States in the Bloch Ball

It is possible to extend the Bloch sphere representation to visualize also mixed states for qubits. These will correspond to points inside the Bloch sphere. In general, a mixed qubit state can be written as a statistical combination of two orthogonal pure qubit states, say

$$\rho = p|\psi_1\rangle\langle\psi_1| + q|\psi_2\rangle\langle\psi_2|, \qquad p, q \geq 0,\ p + q = 1,\ \langle\psi_1|\psi_2\rangle = 0$$

where $|\psi_i\rangle = a_i|0\rangle + b_i|1\rangle$ with $|a_i|^2 + |b_i|^2 = 1$.

To our aim, however, a more efficient representation is obtained by recalling that a density operator acting on $\mathcal{H} = \mathbb{C}^2$ can also be represented as a PSD 2×2 Hermitian matrix having unitary trace. Then, without restrictions, we have that the general form is given by

$$\rho = \frac{1}{2}\begin{bmatrix} 1 + r_z & r_x - ir_y \\ r_x + ir_y & 1 - r_z \end{bmatrix} = \frac{1}{2}I_{\mathcal{H}} + \frac{1}{2}\begin{bmatrix} r_z & r_x - ir_y \\ r_x + ir_y & -r_z \end{bmatrix} \tag{3.90}$$

where r_x, r_y, and r_z are real numbers. The PSD condition of ρ can be translated to the condition $r_x^2 + r_y^2 + r_z^2 \leq 1$. In this way, the 3D vector $\mathbf{r} := (r_x, r_y, r_z)$, with $\|\mathbf{r}\| \leq 1$, allows for the representation of the density operator as a point inside the Bloch sphere.

It can be seen that the state represented by (3.90) is pure if and only if $\|\mathbf{r}\| = 1$ (see Problem 3.16). Note also that the center of the sphere obtained with $\mathbf{r} = \mathbf{0}$ gives the completely chaotic mixed state $\rho = \frac{1}{2}I_{\mathcal{H}}$.

Relation (3.90) can be expressed in an elegant form by using *Pauli's matrices*

$$\sigma_0 = I = \begin{bmatrix} 1 & 0 \\ 0 & 1 \end{bmatrix}, \quad \sigma_x = \begin{bmatrix} 0 & 1 \\ 1 & 0 \end{bmatrix}, \quad \sigma_y = \begin{bmatrix} 0 & -i \\ i & 0 \end{bmatrix}, \quad \sigma_z = \begin{bmatrix} 1 & 0 \\ 0 & -1 \end{bmatrix}. \tag{3.91}$$

These matrices form a basis for the 2×2 complex matrices, and in particular for density operators. In fact, from (3.80) we have directly that any ρ can be expressed in this basis as

$$\rho = \frac{1}{2} \left(\sigma_0 + r_x \sigma_x + r_y \sigma_y + r_z \sigma_z \right) . \tag{3.92}$$

3.12.3 Operations with a Qubit

We recall from Postulate 2 that the evolution of a system in a Hilbert space is provided by a unitary operator U, which acts in the form $|\psi\rangle \rightarrow U |\psi\rangle$ for pure states and $\rho \rightarrow U \rho U^*$ for mixed states. Now all Pauli's matrices are unitary and can be considered as unitary operators for the space $\mathcal{H} = \mathbb{C}^2$, that is, for a qubit system.

Now we suppose that the basis of the qubit is given explicitly by the vector

$$|0\rangle = \begin{bmatrix} 1 \\ 0 \end{bmatrix}, \qquad |1\rangle = \begin{bmatrix} 0 \\ 1 \end{bmatrix}$$

and we consider the evolution when the input is one of the basis kets and the unitary operator is one of the Pauli matrices. Of course, with $U = \sigma_0 = I_2$ with have no evolution. With the other Pauli's matrix we find:

- application of σ_x: **bit-flip**

$$\sigma_x |0\rangle = |1\rangle, \qquad \sigma_x |1\rangle = |0\rangle$$

- application of σ_z: **bit-flip**

$$\sigma_z |0\rangle = |0\rangle, \qquad \sigma_z |1\rangle = -|1\rangle$$

- application of σ_y: **bit-phase-flip**

$$\sigma_y |0\rangle = i|1\rangle, \qquad \sigma_y |1\rangle = -i|0\rangle .$$

Here we have indicated the names given to the transformations in the fields of Quantum Computation and Quantum Information, where they are quite fundamental.

3.12.4 The Qubit in Nature

The qubit represents a wonderful theoretical discovery[8] and nowadays it is the fundamental tool of Quantum Information and Quantum computation (with discrete variables). But the question is: do qubits exist in Nature? The first concrete experiment was conceived by Stern in 1921 and carried out by Gerlach in 1922 (today known as the Stern–Gerlach experiment) by beaming silver atoms with a magnetic field.

At the present time there is an enormous amount of experiments that set a system into a qubit state with simpler methods related to the two different polarizations of photons, to the alignment of a nuclear spin, and the two states of electrons orbiting in a single atom [1].

3.12.5 Quantum Measurements with a Qubit

A quantum measurement allows us to gain information about the state of the qubit. We recall from Sect. 3.5.2 that a quantum measurement performed with a system of elementary projectors $\{\Pi_i = |a_i, i \in \mathcal{M}\}$ in a quantum system prepared in the state $|\psi\rangle$ gives the outcome $m = i$ with probability $\mathrm{P}[m = i|\psi] = |\langle a_i|\psi\rangle|^2$. After the measurement the system moves to the state $|\psi_{\mathrm{post}}\rangle = |a_i\rangle$.

We can apply these statements to the qubit $|\psi\rangle = a|0\rangle + b|1\rangle$, using the projectors $\Pi_0 = |0\rangle\langle 0|$ and $\Pi_1 = |1\rangle\langle 1|$ obtained from the basis $\{|0\rangle, |1\rangle\}$. We find that the result of the measurement will be either $|0\rangle$, with probability $|a|^2$, or $|1\rangle$, with probability $|b|^2$. The measurement alters the state of the qubit: if the outcome is $m = 0$, the qubit collapses to the state $|\psi_{\mathrm{post}}\rangle = |0\rangle$, while if $m = 1$ the qubit collapses to the state $|\psi_{\mathrm{post}}\rangle = |1\rangle$.

This quantum measurement allows us to reflect about the amount of information contained in a qubit. Considering the representation in the Bloch sphere, the cardinality of a qubit is the same as the one of a two-dimensional real space and therefore a qubit could contain an infinite amount of information. But, if we try to extract this information with a quantum measurement, the qubit system collapses into either state $|0\rangle$ or state $|1\rangle$. Hence with a measurement one gets only a single bit of information, as with a classical bit. Then, the question is: why does the qubit, potentially containing an infinite amount of information, does collapse into a pure binary form? According to Nielsen and Chang [1, p.15]: "nobody knows".

But even more interesting and intriguing questions arise with multiple qubits for the possibility of the entanglement.

---

[8] The introduction of the term "qubit" is attributed to Schumacher [13].

3.12.6 Multiple Qubits

With two qubit systems $\mathcal{H}_1$ and $\mathcal{H}_2$ we can construct a composite systems by the tensor product, $\mathcal{H} = \mathcal{H}_1 \otimes \mathcal{H}_2$. If the bases of the component systems are $\{|0_1\rangle, |1_1\rangle\}$ and $\{|0_2\rangle, |1_2\rangle\}$ a basis of $\mathcal{H}$ is given by

$$\{|0_1\rangle \otimes |0_2\rangle, |0_1\rangle \otimes |1_2\rangle, |1_1\rangle \otimes |0_2\rangle, |1_1\rangle \otimes |1_2\rangle\} \qquad (3.93a)$$

which is usually abbreviated in the form

$$\{|00\rangle, |01\rangle, |10\rangle, |11\rangle\} . \qquad (3.93b)$$

Thus, in particular, with the bases, we can form the four two-qubits $|00\rangle$, $|01\rangle$, $|10\rangle$, and $|11\rangle$, exactly as in the classical case of four two-bits 00, 01, 10, and 11.

In general a two-qubit state is given by a linear combination of the basis vectors, namely

$$|\psi\rangle = a_{00} |00\rangle + a_{01} |01\rangle + a_{10} |10\rangle + a_{11} |11\rangle \qquad (3.94)$$

where the normalization condition is $\sum_i \sum_j |a_{ij}|^2 = 1$.

It is interesting to find conditions on the coefficient a_{ij} for the presence of entanglement. This can be done using Schmidt's decomposition developed in Chap. 10, which is essentially a singular value decomposition of the coefficient matrix. The result is (see Problem 3.17):

Proposition 3.6 *The two-qubit state* (3.94) *is* **separable** *(non entangled) if and only if the coefficients verify the condition*

$$a_{01}a_{10} = a_{00}a_{11} . \qquad (3.95)$$

Then, e.g., we find that the *Bell state*

$$|\psi\rangle = \frac{1}{\sqrt{2}}(|00\rangle + |11\rangle) \qquad (3.96)$$

is entangled.

It would be interesting to see the information in a two-qubit system, in particular in the case of entanglement, but this will be seen in the final part of the book.

Problem 3.16 $\star\star$ Prove that the mixed state qubit expressed in the form (3.90) represents a pure state if and only if the vector $\mathbf{r}$ has unit length. Under this condition, from (3.90) find the corresponding pure state.

Problem 3.17 $\star\star\nabla$ Using Schmidt's decomposition given in Chap. 10, prove Proposition 3.6.

Problem 3.18 $\star\star$ Prove Proposition 3.6 using the considerations of Sect. 3.10.2, in particular relations (3.71) to (3.73).

Appendix

Probabilities and Random Variables in a Quantum Measurement

In this appendix we develop in detail the statistical description of the random quantities one finds in a quantum measurement, starting from the case of pure states and then considering the general case of mixed states.

Measurements with Pure States

The random quantity is given by the outcome of the measurement m, which is described by the conditioned probability distribution given by (3.26), that is, $p_m(i|\psi) := P[m = i|s = |\psi\rangle] = \langle\psi|\Pi_i|\psi\rangle$, where the condition $s = |\psi\rangle$ is not random if the state of the system is known before the measurement. Also, the state of the system after the measurement given by

$$s_{\text{post}} = \frac{\Pi_i|\psi\rangle}{\sqrt{\langle\psi|\Pi_i|\psi\rangle}} = \frac{\Pi_i|\psi\rangle}{\sqrt{p_m(i|\psi)}} \, .$$

This state is not random because it is uniquely determined by the original state $s = |\psi\rangle$.

Measurements with Mixed States

The scenario is more complicated because the random quantities become:

(1) the state of the system s before the measurement,
(2) the outcome of the measurement m,
(3) the state of the system s_{post} after the measurement.

The statistical description of the state s in encoded in the density operator

$$\rho = \sum_{|\psi\rangle\in\mathcal{S}} p_s(\psi)|\psi\rangle\langle\psi|$$

which gives the ensemble (see (3.6)) $\mathcal{E} = (\mathcal{S}, p_s)$, where $\mathcal{S} = \{|\psi_1\rangle, |\psi_2\rangle, \ldots\}$ is the alphabet of the states and $p_s(\psi) = P[s = |\psi\rangle]$ is the probability distribution over $\mathcal{S}$. The statistical description of the measurement outcome m is given by the conditional probability distribution

$$p_{m|s}(i|j) := P[m = i|s = |\psi_j] = \langle\psi_j|\Pi_i|\psi_j\rangle = \text{Tr}(|\psi_j\rangle\langle\psi_j|\Pi_i) \qquad (3.97)$$

with absolute probability distribution (see (3.35))

$$p_m(i) := \mathrm{P}[m = i] = \mathrm{Tr}[\Pi_i\,\rho]\,. \tag{3.98}$$

The relations to get the reverse conditional probability are

$$p_{s|m}(j|i) := \mathrm{P}[s = |\psi_j\rangle|m = i] = \mathrm{P}[m = i, s = |\psi_j\rangle]/\mathrm{P}[m = i]$$
$$= \mathrm{P}[m = i|s = |\psi_j\rangle]\mathrm{P}[s = |\psi_j\rangle]/\mathrm{P}[m = i]$$

and, considering (3.97) and (3.98),

$$p_{s|m}(j|i) := \mathrm{P}[s = |\psi_j\rangle|m = i] = \frac{\mathrm{Tr}[\Pi_i|\psi_j]\,p_s(j)]}{\mathrm{Tr}[\Pi_i\,\rho]}\,. \tag{3.99}$$

Finally we consider the state of the system after the measurement, s_{post}. This state can be evaluated from (3.27), for a given original state $s = |\psi_j\rangle$, that is,

$$|\psi_{\mathrm{post},\,j}^{(i)}\rangle = \frac{\Pi_i|\psi_j\rangle}{\sqrt{\langle\psi_j|\Pi_i|\psi_j\rangle}} = \frac{\Pi_i|\psi_j\rangle}{\sqrt{\mathrm{Tr}[|\psi_j\rangle\langle\psi_j|\Pi_i]}}\,. \tag{3.100}$$

Here we consider that $m = i$ is given, then also Π_i is given, and therefore, as $s = |\psi_j\rangle \in \mathcal{S}$, (3.100) generates the alphabet of the post-measurement states

$$\mathcal{S}_{\mathrm{post}}^{(i)} = \left\{ |\psi_{\mathrm{post},\,j}^{(i)}\rangle\,|\psi_j \in \mathcal{S} \right\}\,.$$

The randomness of s_{post} depends only on the randomness of the original state $s = |\psi_j\rangle$ and the related probability distribution reads as

$$p_{s_{\mathrm{post}}}(j|i) = \mathrm{P}[s_{\mathrm{post}} = |\psi_{\mathrm{post},\,j}^{(i)}\rangle \,\big|\, m = i]\,. \tag{3.101}$$

Now we get the density operator $\rho_{\mathrm{post}}^{(i)}$ describing the system after the measurement from the ensemble $\mathcal{E}_{\mathrm{post}}^{(i)} = (\mathcal{S}_{\mathrm{post}}^{(i)}, p_{s_{\mathrm{post}}})$, that is,

$$\rho_{\mathrm{post}}^{(i)} = \sum_j p_{s_{\mathrm{post}}}(j|i)|\psi_{\mathrm{post},\,j}^{(i)}\rangle\langle\psi_{\mathrm{post},\,j}^{(i)}|.$$

For the explicit evaluation we remark that (3.100) establishes a one-to-one correspondence between the random states s and s_{post}, so that the probability distribution (3.100) coincides with the probability distribution (3.99). Hence, using (3.99) and (3.100) we get

$$\rho_{\mathrm{post}}^{(i)} = \sum_j \frac{\mathrm{Tr}[\Pi_i|\psi_j\rangle\langle\psi_j|]\,p_s(j)}{\mathrm{Tr}[\Pi_i\,\rho]}\frac{\Pi_i|\psi_j\rangle\langle\psi_j|\Pi_i}{\mathrm{Tr}[\Pi_i|\psi_j\rangle\langle\psi_j|]} = \sum_j \frac{p_s(j)\Pi_i|\psi_j\rangle\langle\psi_j|\Pi_i}{\mathrm{Tr}[\Pi_i\,\rho]}$$

and (3.37) follows.

References

1. M.A. Nielsen, I.L. Chuang, *Quantum Computation and Quantum Information* (Cambridge University Press, Cambridge, 2000)
2. C.W. Helstrom, *Quantum detection and estimation theory*, Mathematics in Science and Engineering vol. 123, (Academic Press, New York, 1976)
3. R.J. Glauber, The quantum theory of optical coherence. Phys. Rev. **130**, 2529–2539 (1963)
4. P.A.M. Dirac, *The Principles of Quantum Mechanics* (Oxford University Press, Oxford, 1958)
5. C.W. Helstrom, J.W.S. Liu, J.P. Gordon, Quantum-mechanical communication theory. Proc. IEEE **58**(10), 1578–1598 (1970)
6. W.H. Louisell, *Radiation and Noise in Quantum Electronics* (McGraw-Hill, New York, 1964)
7. Y.C. Eldar, G.D. Forney, Optimal tight frames and quantum measurement. IEEE Trans. Inf. Theory **48**(3), 599–610 (2002)
8. S.M. Barnett, E. Riis, Experimental demonstration of polarization discrimination at the Helstrom bound. J. Mod. Opt. **44**(6), 1061–1064 (1997)
9. Y.C. Eldar, G.D. Forney, On quantum detection and the square-root measurement. IEEE Trans. Inf. Theory **47**(3), 858–872 (2001)
10. P.R. Halmos, *Finite-Dimensional Vector Spaces* (Van Nostrand, New York, 1958)
11. A. Einstein, B. Podolsky, N. Rosen, Can quantum-mechanical description of physical reality be considered complete? Phys. Rev. **47**, 777–780 (1935)
12. L.P. Hughston, R. Jozsa, W.K. Wootters, A complete classification of quantum ensembles having a given density matrix. Phys. Lett. A **183**(1), 14–18 (1993)
13. B. Schumacher, Quantum coding. Phys. Rev. A **51**, 2738–2747 (1995)

Part II
Quantum Communications

Chapter 4
Introduction to Part II: Quantum Communications

This second part of the book is concerned with Quantum Communications, and more specifically, with Quantum Telecommunications, as our scenario will be the transmission of information from a sender to a user, located at a certain distance. The starting point will be Classical Telecommunications, whose fundamentals lie upon Classical Physics and this is due to several reasons. First, historical reasons, in that Classical Telecommunication originated a hundred years before Quantum Telecommunications (see below the fundamental dates), but mainly because the two types of Telecommunications often share the same goal and therefore will be crucial to compare their performance.

Organization of this Chapter

In this chapter we outline the foundations of Telecommunications Systems and introduce the two basic approaches, Classical and Quantum Communications, and then formally describe their differences.

In the second part of the chapter we introduce the foundations of *optical classical communications*, which is the necessary prologue to *optical quantum communications* developed in Chaps. 5–9. The mathematical framework for optical classical communications is given by Poisson processes, which allow us to represent adequately the optical power and its processing inside the system, and also the electrical current in photodetection. The theory of Poisson processes will be developed in Sect. 4.5 and then applied to photodetection.

Organization of Part II

We now summarize the content of Part II, which consists of five chapters.

In Chap. 5, *Quantum Detection Theory: Analysis and Optimization*, the fundamentals of Quantum Mechanics of Chap. 3 are applied to develop Quantum Detection, which represents the essential part of quantum receivers. Quantum Detection is based on quantum measurements and its goal is to extract the information from the incoming quantum states. The choice of the projectors or POVMs for the quantum measurements becomes the key strategy to design the best quantum receiver.

© Springer International Publishing Switzerland 2015
G. Cariolaro, *Quantum Communications*, Signals and Communication Technology,
DOI 10.1007/978-3-319-15600-2_4

Chapter 6, *Quantum Detection Theory: Suboptimization*. In general, optimization does not give explicit results and thus suboptimal detections are investigated. The most important technique of suboptimization is based on square root measurements (SRM), which can be applied to every quantum communication system and achieves "pretty good" estimation of the system performance.

Chapter 7, *Quantum Communications Systems*, develops the fundamentals of Quantum Communications Systems, where the information carrier is based on coherent states (representing the laser radiation at an optical frequency). These fundamentals are then applied to examine the most popular quantum communication systems, based on PAM, QAM, PSK, and PPM modulations, and for each system the performance is compared with the ones of the classical counterpart. In this chapter it is assumed that the information carrier is given by pure states. This is useful as a first simplified formulation because it allows for a better understanding of the essential parts of a quantum communications system.

In Chap. 8, *Quantum Communications Systems in the Presence of Thermal Noise*, the assumption of pure states is abandoned and mixed states are assumed at reception. This allows us to obtain a more realistic formulation where thermal noise, always present in a real-world system, is taken into account. Of course, the formulation with mixed states, represented by density operators, becomes more difficult. Also, with this more accurate formulation, the most popular quantum communication systems are examined and compared to the corresponding classical systems.

In Chap. 9, *Implementation of Quantum Communications Systems*, significant realizations of quantum communications systems are developed. The main solutions proposed in the literature are described and also some original ideas are outlined.

A Few Milestones in Telecommunications

1838: First commercial telegraph (Cooke and Wheatstone)

1854: Antonio Meucci invented a telephone-like device

1858: First communication over a transatlantic telegraphic cable

1876: Alexander Graham Bell patented the first practical telephone

1896: Guglielmo Marconi patented the radio

1901: First wireless signal across the Atlantic Ocean

1926: First public demonstration of a television set by John Baird

1927: Opening of the first transatlantic telephone service

1931: First detection of radio astronomy waves coming from the deep space

1940: First TV transmissions over coaxial cable

1948: "A Mathematical Theory of Communications" (Claude Shannon)

1959: The word laser is coined (Gordon Gould)

1962: First commercial telecommunications satellite (Telstar)

1966: First fiber optical communications (Kao and Hockam)

1969: ARPANET sends the first packet: it is the beginning of computer networks

1973: First modern era mobile cellular phone (Motorola)

1974: First release of the TCP/IP protocol (the foundation of Internet)

1975: First commercial fiber-optic network

1979: First mobile cellular phone network (NTT)

1980: Release of the Ethernet protocol

1982: Release of the SMTP protocol for e-mail

1983: 1st January: ARPANET changes from NCP to TCP/IP

1990: 23rd July: First HDTV transmission (Telettra & RAI, Italy)

1990: December: First web site developed by Tim Berners Lee

2003: August: Skype is released.

4.1 A General Scheme of a Telecommunications System

An essential scheme of a communication system, which is valid for both Classical
and Quantum Communications, is depicted in Fig. 4.1. A source of information emits
a *message*, say m, or $m(t)$ to emphasize the time dependence, which must be sent
to a user located at a certain distance. The message may be of several kinds and of
any degree of complexity, as voice, music, still images, images with motion (cinema
and television), written text, numerical data, and so on. The transmitter converts
the message to a state or *signal* $v_m(t)$ representing a physical quantity, e.g., an
electromagnetic microwave or an optical radiation emitted by a laser. Then the signal
has the role of *information carrier*. The signal is sent to the desired distance using
a physical channel, e.g., an optical fiber, the free space, or a satellite link, which
delivers to the receiver a corrupted version $\widehat{v}_m(t)$ of the original signal $v_m(t)$. Finally,
the receiver extracts from the received signal $\widehat{v}_m(t)$ an approximate replica $\widehat{m}$ of the
original message.

In dependence of the context, the term "transmitter" is sometimes replaced by
"encoder" (and also by "modulator") and the "receiver" by "decoder" ("demodulator"). In Quantum Mechanics it is customary to humanize the transmitter/encoder as
Alice and the receiver/decoder as Bob.

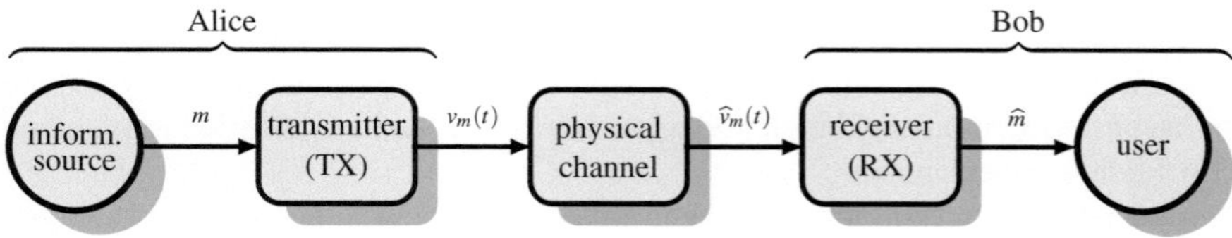

Fig. 4.1 Essential scheme of a telecommunications system

4.1.1 Analog and Digital Messages

A fundamental classification is based on the nature of the message, which may
be continuous, as most of "natural" messages are (voice, images) or discrete, as
data. Correspondingly, the signals have a continuous-time evolution or a discrete
time evolution, respectively. Historically, communications were implemented for
more than a century for continuous messages (apart from the rudimentary telegraph
system), with the technical jargon of *analog communications*, but in the last four
decades all communications are implemented in the form of *digital communications*.
In fact, this is what is commonly referred to as digital revolution.

The digital revolution was started by a wonderful paper by Oliver et al. entitled
"The Philosophy of PCM," published by the *Proceedings of IRE* [1] (now *IEEE*) in
1948 (the same *annus mirabilis* in which Shannon published the other masterpiece,
on Information Theory [2]). In essence, the authors proved that all continuous information can be converted into digital form with an arbitrarily prefixed and then
controlled fidelity. In other words, they claimed that all messages can be converted
to a sequence of 0s and 1s, and, from this sequence the original message can be
recovered with the desired accuracy (see the next section). This possibility had a
tremendous advantage in terms of unification in the system implementation and design. Nowadays this conclusion seems to be trivial, but it was not in 1948. However,
the implementation of this idea came 30 years later because it needed Very Large
Scale Integration of electronic components. Nowadays all information is digitalized
and, ultimately, this motivates our choice of dealing only with digital communications in this book.

Note that the same scheme of Fig. 4.1 can also be adapted to the storage of information. The physical channel becomes the storage medium and the receiver becomes
the player that restores the message from the medium. The main difference is that a
communication system works in "real time," whereas a recorder works in "deferred
time."

4.1.2 Mathematical Representation of Messages and Signals

Messages and signals in a communications system are always modeled as *random*
or *stochastic processes*, and therefore are statistically described according to Probability Theory. The *realizations* of the random process are conveniently interpreted
as possible messages or signals that the communications system can convey and the
source of information may be viewed as an urn from which a realization is extracted
with a random mechanism.

An analog source of information is modeled as continuous-time continuous-
amplitude random process $s(t)$. Figure 4.2 shows a few realizations of such kind
of process.

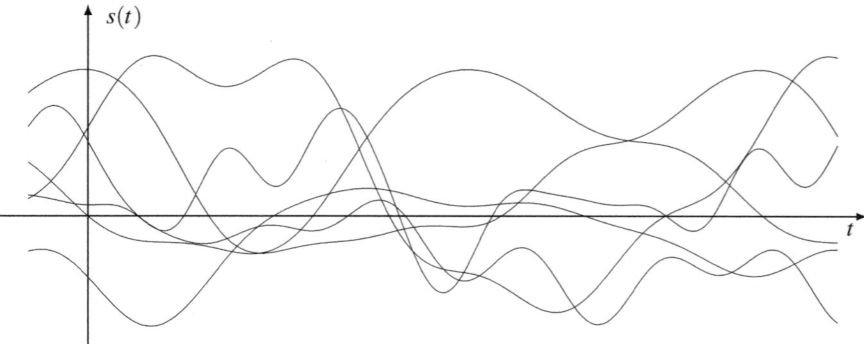

Fig. 4.2 Realizations of a continuous-time continuous-amplitude random process

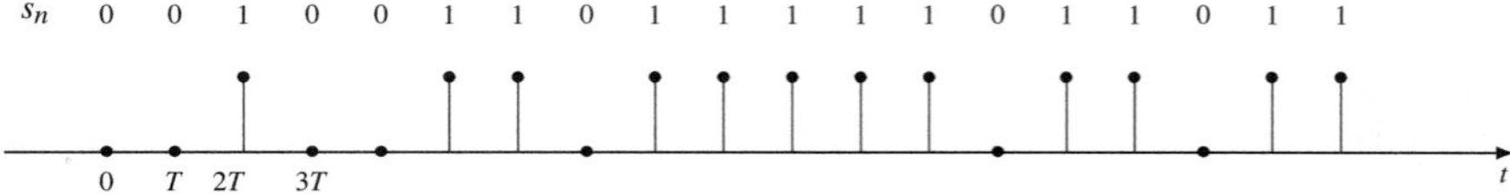

Fig. 4.3 Realization of a binary digital signal $s_n = s(nT)$

A digital source of information is modeled as a discrete-time discrete-amplitude random process. In this case the realizations are written in the form $s(nT)$, where the time t consists of equally T-spaced instants and the amplitude takes its values from a finite-size alphabet $\mathcal{A}$, which often is the binary alphabet $\mathcal{A} = \{0, 1\}$. Figure 4.3 shows a realization of a binary process. The parameters of a digital source are: the symbol period T (in seconds), or equivalently the symbol frequency $f_c = 1/T$ (in symbols per second), and the nominal rate

$$R_0 = f_c \log_2 K \quad \text{bits/symbol} \quad \text{with} \quad K = |\mathcal{A}| = \text{size of } \mathcal{A}. \qquad (4.1)$$

The effective rate of the source R is evaluated through the entropy (see Chap. 12) and is equal to R_0 only when the random process consists of equally likely and statistically independent symbols. In all the other cases $R < R_0$, because the information source is redundant.

4.2 Essential Performances of a Communication System

The analysis of a communication system is usually subdivided into several parts (blocks), where the functionality of each block is specified, and combined in order to evaluate the overall performance. The accuracy in the specification of the blocks depends on the target of the analysis. In general, to understand the principles upon which the system works, a simplified description is the most convenient approach. On the opposite side, if the target is the computer simulation of the system before its realization, a very detailed description is necessary. The same considerations hold for the specification of the global performance that the system should realize, which may either be limited to the essential or very detailed.

In this section we discuss the essential performance.

4.2.1 Analog Systems

For reasons of completeness we begin with an analog system, where the target is the transmission of a class of continuous-time continuous-amplitude signals. If the transmitted signal $s(t)$ is a member of this class, in the ideal case the received signal $\widehat{s}(t)$ should coincide with $s(t)$, but also the form $\widehat{s}(t) = A_0 s(t - t_0)$, where A_0 is an amplification or an attenuation and t_0 is a delay, is accepted as a uncorrupted version of $s(t)$. More generally, an equivalent scheme of an analog system is given by a linear system, specified by the impulse response $h(t)$, followed by an additive noise $n(t)$ (Fig. 4.4). Then, the received signal can be decomposed in the form

$$\widehat{s}(t) = s_u(t) + n(t)$$

where $s_u(t)$ is the output of the linear system (useful signal) and $n(t)$ is the noise. A global parameter to quantify this impairment is usually given by the **signal-to-noise ratio** (SNR)

$$\Lambda = \frac{\mathrm{E}[s_u^2(t)]}{\mathrm{E}[n^2(t)]} \tag{4.2}$$

where $\mathrm{E}[\cdot]$ denotes expectation (or statistical mean).

Another fundamental parameter of an analog system is the **bandwidth** B, given by the maximum frequency of the signals that the system can transmit ensuring a given SNR Λ. For instance, we say that an analog system has a bandwidth of $B = 1$ MHz with an SNR of 60 dB.

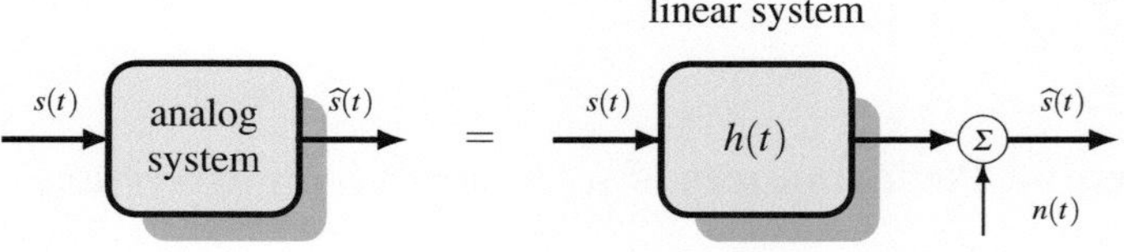

Fig. 4.4   Analog communication system and equivalent scheme with additive noise

4.2.2 Digital Systems

The parameters giving the performance of a digital system are specified in terms of probabilities, and therefore very differently with respect to the analog case. We consider the specification in the simplest case, where: (1) the source emits *stationary* and *statistically independent* symbols A_n, belonging to a finite size alphabet, say $\mathcal{A} = \{0, 1, \ldots, K-1\}$, and (2) the global digital channel is *memoryless* and *permanent*, as the transmission of A_n is not influenced (statistically) by the transmission of the other symbols and is independent of n. The size K of the symbol alphabet $\mathcal{A}$ characterizes the system as a K-ary digital system, so we have a binary system for $K = 2$, a ternary system for $K = 3$, and so on.

The source sequence $\{A_n\}$ is specified by the symbol probabilities

$$p_A(a) := P[A_n = a], \qquad a \in \mathcal{A} \tag{4.3}$$

often called *a priori probabilities*. The function $p_A(a)$ forms a probability distribution, that is, it verifies the conditions $p_A(a) \geq 0$ and $\sum_{a \in \mathcal{A}} p_A(a) = 1$.

Let $\{\widehat{A}_n\}$ be the sequence of symbols at the output of the system. Then the global channel is completely specified by the *transition probabilities*

$$p_c(b|a) := P[\widehat{A}_n = b | A_n = a], \qquad a, b \in \mathcal{A}. \tag{4.4}$$

The probabilities $p_A(a)$ and $p_c(b|a)$ are usually represented by a graph, as shown Fig. 4.5 for a ternary digital system.

Note that the channel should be regarded as a random system, because, given the input, say $A_n = 1$, the output is not uniquely determined and may be $\widehat{A}_n = 1$ with probability $p_c(1|1)$, but also $\widehat{A}_n = 0$ with probability $p_c(0|1)$, etc.

From the a priori probabilities (4.3) and from the transition probabilities (4.4) one can evaluate all the probabilities in the system, such as the joint probabilities given by $P[\widehat{A}_n = b, A_n = a] = p_c(b|a)p_A(a)$. A global parameter, which corresponds to the SNR of the analog case, is given by the **correct decision probability**

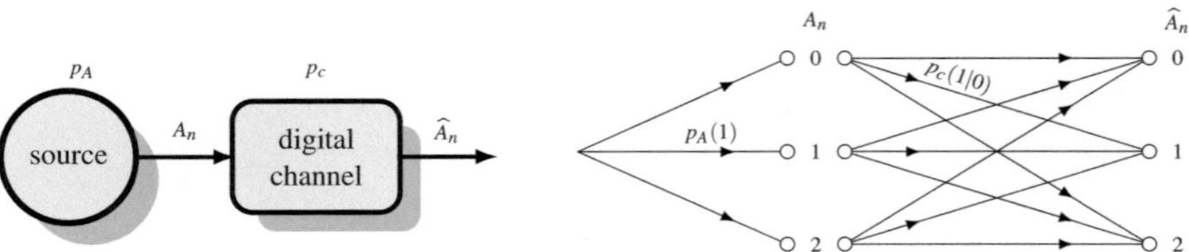

Fig. 4.5 Statistical description of a ternary digital system. The source is specified by the *prior probabilities* $p_A(a)$ and the channel by the *transition probabilities* $p_c(b|a)$

$$P_c := \mathrm{P}\left[\widehat{A}_n = A_n\right] = \sum_{a \in \mathcal{A}} \mathrm{P}[A_n = a]\, \mathrm{P}\left[\widehat{A}_n = a \mid A_n = a\right]$$

$$= \sum_{a \in \mathcal{A}} p_A(a)\, p_c(a|a) \tag{4.5}$$

or by the complementary **error probability**

$$P_e := \mathrm{P}\left[\widehat{A}_n \neq A_n\right] = 1 - P_c. \tag{4.6}$$

The parameter P_c, or P_e, gives the reliability of the transmission. Other important parameters are concerned with the amount of information in the system. The source is characterized by the nominal rate R_0 defined by (4.1). Often, the system performance is specified by the nominal rate and by the error probability, e.g., a digital communication system with a rate $R_0 = 10$ Mbits/s where an error probability $P_e \leq 10^{-10}$ is ensured.

A more sophisticated parameter of a digital communication system is given by the **mutual information** $I(A_n; \widehat{A}_n)$, also called **accessible information** in Quantum Information Theory (see Chap. 12). It gives the average information transmitted by the channel and has the following expression:

$$I(A_n; \widehat{A}_n) = \sum_{a,b} p_A(a)\, p_c(b|a)\, \log_2 \frac{p_c(b|a)}{\sum_{a'} p_A(a') p_c(b|a')}. \tag{4.7}$$

It depends both on the source, through the a priori probabilities, and on the channel, through the transition probabilities. Now, for a given channel, we can vary $I(A; \widehat{A})$ by changing the a priori probabilities $p_A(a)$. The maximum that one obtains gives the **capacity of the channel**:

$$C := \max_{p_A} I(A_n; \widehat{A}_n) \tag{4.8}$$

These information parameters will be discussed in detail in Chap. 12.

4.2.3 Analog Systems Through Digital Systems (Digital Revolution)

For the transmission of an analog message we have two possibilities: using directly an analog system, with the performance established above, or using a digital system with appropriate interfaces, an A/D conversion in transmission and a D/A conversion in reception, as shown in Fig. 4.6, having in mind to obtain the same performance. Considering the importance of this second solution, which was at the core of the

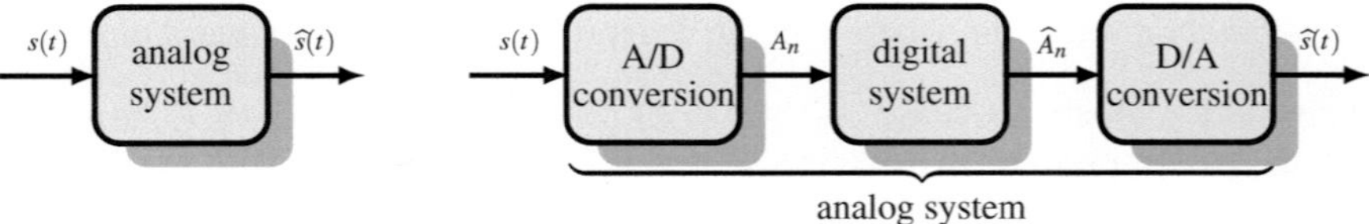

Fig. 4.6 *Direct* analog transmission and analog transmission *through a digital transmission*

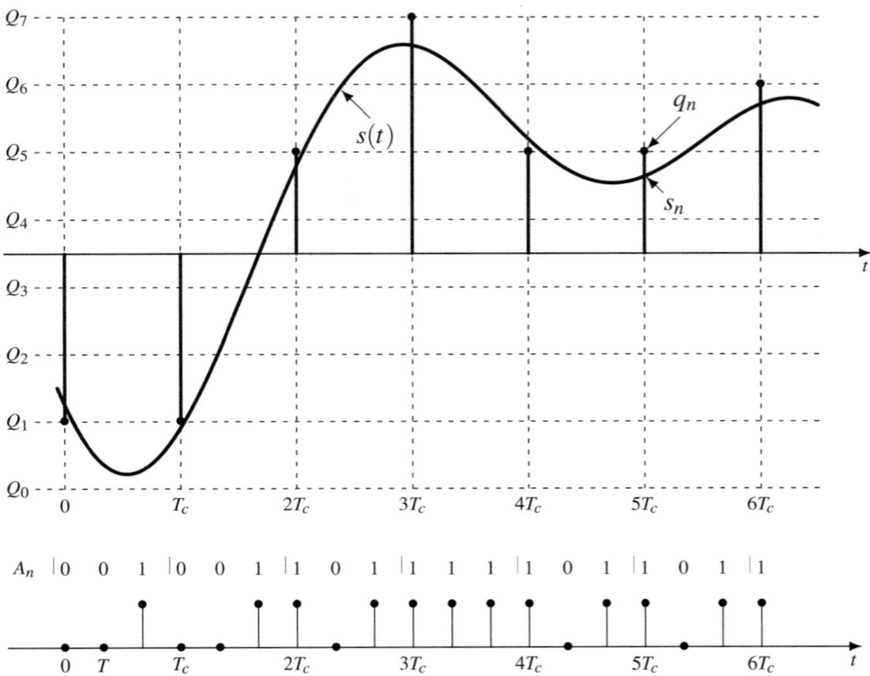

Fig. 4.7 A/D conversion of an analog signal $s(t)$ to a binary sequence $A_n = A(nT)$. The signal $s(t)$ is first sampled to get the discrete-time sequences s_n and quantized into L levels. The quantized sequence q_n is finally converted to binary words of length m. In the illustration $L = 8$ and $m = 3$

digital revolution, we now see it in detail, thus reviewing the "philosophy" of Oliver, Pierce, and Shannon as cited above.

In the A/D conversion we start from a continuous-time continuous-amplitude signal $s(t)$ and we want to represent it through a binary sequence. To this end, three operations are needed (Fig. 4.7).

(1) A *sampling*, to get a discrete-time signal $s_n = s(nT_c)$, where the sampling frequency F_c is chosen according to the Sampling Theorem. Specifically, as $F_c = 2B$, where B is the signal bandwidth.

(2) A *quantization*, which approximates an amplitude $s_n = s(nT_c)$ by an amplitude q_n chosen within a set of $L = 2^m$ levels, $\{Q_0, Q_1, \ldots, Q_{L-1}\}$,

(3) A *binary conversion*, where each Q_i is represented by a binary word of m bits.

The resulting binary sequence $A_n = A(nT)$ has symbol period $T = T_c/m$. Then, after the A/D conversion, we obtain a source of digital information with nominal rate

$$R_0 = f_c = 2m\,B \quad \text{bits.} \tag{4.9}$$

The sampling does not introduce any error and the only error is in quantization. Then the performance of the A/D conversion is established by the *signal-to-quantization error ratio* $\Lambda_q = \mathrm{E}[s_n^2]/\mathrm{E}[e_n^2]$, where $e_n = s_n - q_n$ is the quantization error. A reference (simplified) formula for Λ_q is given by [3]

$$\Lambda_q = L^2 = 2^{2m}. \tag{4.10}$$

For instance, using $m = 10$ bits/sample one gets $\Lambda = 10^6 = 60.2\,\text{dB}$.

In the D/A conversion the quantized levels are recovered from the binary sequence. Some quantized level may be wrong for the error introduced by the digital system, but if the error probability P_e is sufficiently small ($P_e \ll 1/L^2$), the channel errors have a negligible influence and the global performance is determined by the quantization, that is, $\Lambda \simeq \Lambda_q$.

Finally, we note that an analog system must be designed "ad hoc" for the given analog signal, usually with stringent criteria. On the other hand, the digital system can be used for all analog signals (converted to digital form), and this is just the key factor that determined the digital revolution.

4.2.4 Simplified and Detailed Schemes

A simplification usually done in the analysis of digital systems is the reduction of a message to a single symbol, although in reality a sequence of symbols with their time evolution should be considered. The question is: to what extent we lose generality in considering **the transmission of a single symbol**? The theory developed for a single symbol leads to results valid for a sequence $\{A_n\}$ of symbols under the following conditions: (1) the sequence $\{A_n\}$ is *stationary* and with *independent* symbols, (2) the channel is *permanent* and *memoryless*, and (3) absence of interference between symbols (intersymbol interference).

In the following we will often work under this simplification, and the "single symbol" will be denoted by A_0 or simply by A.

Problem 4.1 ⋆ A still image (photo) is quantized in 800×800 pixels with 8 bit/pixel and transmitted by a digital channel with nominal rate $R_0 = 100$ kbits/s. Find (1) the signal-to-quantization error Λ_q, (2) the error probability P_e of the digital channel such that the channel error is negligible, and (3) the time needed to transmit the photo. Note that the global SNR is given by [3]

$$\Lambda = \Lambda_q / (1 + P_e L^2).$$

Problem 4.2 ⋆ A video signal (produced by a TV camera) has bandwidth $B = 5\,\text{MHz}$. Evaluate the A/D conversion parameters that ensure $\Lambda_q = 60\,\text{dB}$, and in particular the nominal rate of the digital channel.

4.3 Classical and Quantum Communications Systems

In this section we introduce other important classifications, which allow us to evidence the distinction between classical and quantum information and between classical and quantum communications systems.

4.3.1 Classical and Quantum Information

The information contained in a message, as a written text, speech, music, images, strings of data, is strongly related to its randomness or uncertainty: if a message is known in advance, it does not bring any information and, on the opposite end, when a message is completely uncertain it brings the maximum of information. As we shall see in Information Theory of Chap. 12, the mathematical tool for evaluating the amount of information of a message is Probability Theory, where a symbol becomes a random variable and a message evolving in time becomes a random process.

In any case, the information is classified as **classical information** or **quantum information**, the distinction being not easy in these preliminary considerations. Broadly speaking, *classical* is the information related to random symbols and *quantum* is the information related to random quantum states (in the sense explained in Sect. 3.3). The distinction may also regard the methodology used in the context: a message of classical information is dealt with using the tools of **Probability Theory**, and a message of quantum information is dealt with using **Probability Theory, but through the laws of Quantum Mechanics**.

4.3.2 Classical and Quantum Communications Systems

The distinction will become clearer in the comparison between Classical and Quantum Communications, as depicted in Fig. 4.8 in the simplest case: the transmission of a message consisting of a single symbol A. Then in both systems the purpose is the transmission of classical information.

Classical communication system

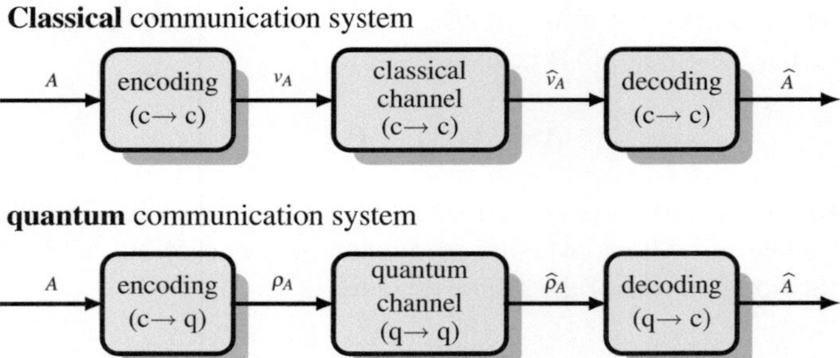

Fig. 4.8 Comparison of *classical and quantum communications systems* for the transmission of a classical symbol A

In the **Classical** Communications system the initial encoding is formulated as a classical–classical $(c \to c)^1$ mapping $A \to v_A(t)$, $A \in \mathcal{A}$, where the symbol A in converted to a physical quantity $v_A(t)$ to be sent to the channel. Note that $v_A(t)$ should belong to a set with $|\mathcal{A}|$ distinct waveforms. The classical channel provides a $c \to c$ mapping $v_A(t) \to \widehat{v}_A(t)$, where $\widehat{v}_A(t)$ is a corrupted version of $v_A(t)$. Finally, the decoding provides again a $c \to c$ mapping $\widehat{v}_A(t) \to \widehat{A}$, where $\widehat{A}$ has the original format, $\widehat{A} \in \mathcal{A}$, but may be different from A for the reasons explained above. In conclusion, in a Classical Communications system all the operations are performed in the classical domain according to $c \to c$ mappings.

In the **Quantum** Communications system the classical information given by the symbol A is transmitted through quantum states. Then the initial encoding becomes a classical–quantum $(c \to q)$ mapping $A \to \rho_A$, where ρ_A is the quantum state to be associated to the symbol A. Alice should be able to prepare $|\mathcal{A}|$ different quantum states. The quantum channel provides a $q \to q$ mapping $\rho_A \to \widehat{\rho}_A$, where $\widehat{\rho}_A$ is a corrupted version of ρ_A. Finally, the decoding provides a $q \to c$ mapping $\widehat{\rho}_A \to \widehat{A}$.

To get a further insight into the differences between the two types of systems, we consider a specific case (one of the most important in applications): a binary data transmission using a laser radiation at optical frequencies using the 2-PSK (phase-shift keying) modulation. We still remain in a situation where the system is simplified with respect to the real-world system (see below), but we arrive at some important general conclusions.

In the **classical** 2-PSK (or BPSK) system the encoder (or better the modulator) provides a laser radiation at an optical frequency v with waveforms in the interval $[0, T)$, $v_0(t) = V_0 \cos(2\pi v t)$ when $A = 0$ and $v_1(t) = V_0 \cos(2\pi v t + \pi)$ when $A = 1$ (Fig. 4.9). The radiation is sent to the optical channel, which may be an optical fiber. At the receiver, the signal, in simplified conditions where the fiber attenuation is compensated by an amplification at the receiver front end, has the form $\widehat{v}_A(t) =$

---

^{1} Here we follow the formalism $c \to c$, $c \to q$, etc., used by Holevo and Giovannetti in a recent paper [4].

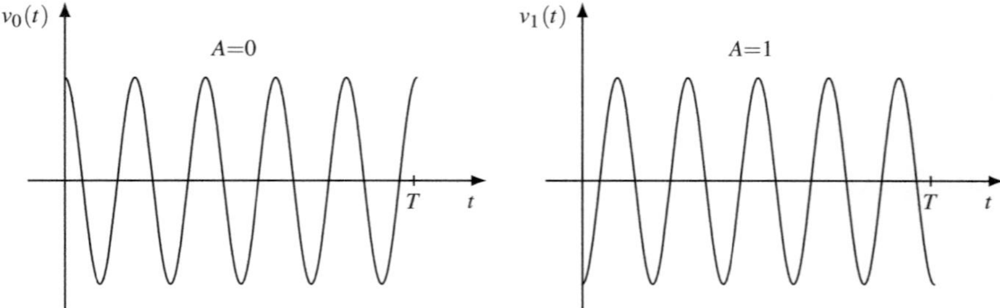

Fig. 4.9   The two modulation waveforms in the 2-PSK system for the transmission of the symbol $A = 0$ and $A = 1$

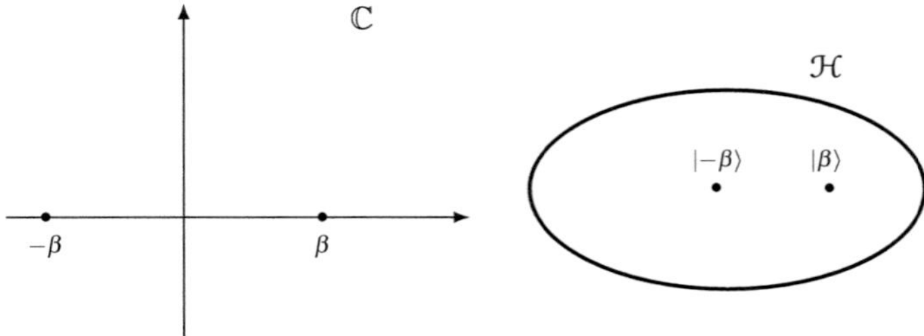

Fig. 4.10   States in 2-PSK modulation. For the transmission of symbol $A = 0$ Alice prepares the system in the coherent state $|\beta\rangle$ and for the transmission of the symbol $A = 1$ the coherent state $|-\beta\rangle$. A coherent state is in general determined by a complex number; in this case $\pm\beta$ are real

$v_A(t) + n(t)$, where $n(t)$ is an additive Gaussian noise, and a classical demodulator extracts from $\widehat{v}_A(t)$ the transmitted symbol with a certain error probability due to the Gaussian noise. As we will see in detail in Chap. 7, the mathematics needed to evaluate the performances is simply given by Probability Theory.

In the **quantum** 2-PSK system the encoder (Alice) prepares two coherent states of the form $|\pm\beta\rangle$, where β is a real amplitude (see Sect. 7.2). She sends to the fiber $|\beta\rangle$ when $A = 0$ and $|-\beta\rangle$ when $A = 1$ (Fig. 4.10). Continuing with simplified conditions, the fiber followed by an amplification may be considered as an ideal quantum channel that produces at the output the state sent by Alice. Finally, the decoder (Bob) performs a quantum measurement with a POVM system to extract the transmitted symbol. In this case the error probability is determined by the law of Quantum Mechanics related to quantum measurements.

We can further simplify the comparison to find what is the essential difference between the two types of systems. At the transmitter and at the channel the two systems (in simplified conditions!) do not exhibit a substantial difference, apart from the mathematical formulation. The core of the difference lies in the receiver, where the optimization to achieve the best performance is carried out with different approaches: according to the Classical Detection Theory (substantially working with Probability

Theory) in the classical case and according to the theory of Quantum Detection in the quantum case. As we shall see in the next chapters, this will really make the difference in favor of the quantum system.

4.4 Scenarios of Classical Optical Communications

In this second part of the chapter we give the fundamentals of *classical* communications working at optical frequencies. Optical frequencies have a wide range, going from ultraviolet to infrared, as shown in Table 4.1. In practice, the choice of a specific frequency ν depends mainly on the propagation and on the availability of the components working at that frequency. So far, the main applications have been implemented in the infrared range in the band $200\,\mathrm{THz} \div 430\,\mathrm{THz}$.

In presenting the scenarios, we consider essentially the budget of the optical power, referring to the **average** optical power $P(t)$, while the **instantaneous** optical power $p(t)$, which exhibits explicitly the presence of photons as energy quanta, will be considered after the theory of Poisson processes of the next section. At a given frequency ν the optical power $P(t)$ consists of energy quanta $h\,\nu$, whose intensity is given by

$$\lambda(t) = \frac{P(t)}{h\,\nu} \tag{4.11}$$

and has the meaning of *average number of photons per second*. For instance, with a power $P(t) = 1$ mW at the optical frequency $\nu = 300$ THz the quantum has energy $h\,\nu = 1.9878\ 10^{-19}$ J and the intensity is $\lambda(t) \simeq 0.5\ 10^{16}$ photons/s.

4.4.1 General Scheme

A general scheme of Classical Optical Communications consists of several parts, as shown in Fig. 4.11, where the *digital* transmission of a sequence of symbols $\{A_n\}$ is

Table 4.1 Where optical frequencies are located

Name	Wavelength λ	Frequency ν	Energy quantum $h\,\nu$
Gamma ray	Less than 0.01 nm	More than 10 EHz	
X-ray	0.01 nm $\div$ 10 nm	30 EHz $\div$ 30 PHz	9.878 fJ $\div$ 9.878 aJ
Ultraviolet	10 nm $\div$ 380 nm	30 PHz $\div$ 790 THz	523.454 zJ $\div$ 19.878 aJ
Visible	380 nm $\div$ 700 nm	790 THz $\div$ 430 THz	19.878 aJ $\div$ 284.918 zJ
Infrared	700 nm $\div$ 1 mm	430 THz $\div$ 300 GHz	284.918 zJ $\div$ 1198.78 zJ
Microwave	1 mm $\div$ 1 m	300 GHz $\div$ 300 MHz	1198.78 zJ $\div$ 1198.78 yJ
Radio	More than 1 mm	Less than 300 MHz	

$$y = 10^{-24} \quad z = 10^{-21} \quad a = 10^{-18} \quad f = 10^{-15} \quad P = 10^{15} \quad E = 10^{18}$$

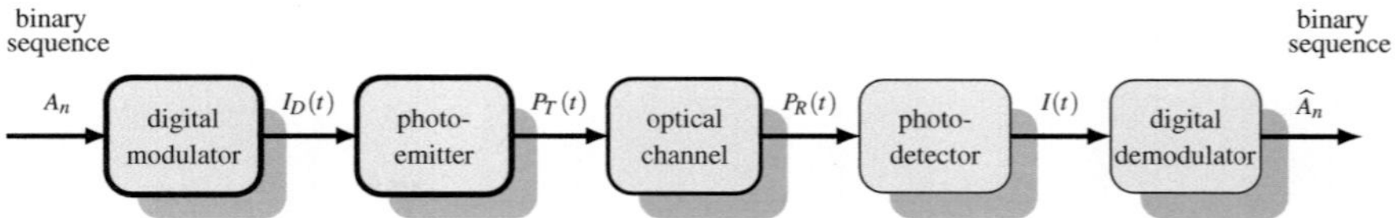

Fig. 4.11 Digital transmission over an optical channel

considered. The first block is a digital modulator, which converts the digital sequence to a continuous time signal $I_D(t)$, an electrical current. This current drives a photon emitter (laser or LED) to produce an optical power $P_T(t)$, which is sent over a distance through an optical channel. At reception, a photodetector (pin diode or APD) converts the received power $P_R(t)$ to an electrical current $I(t)$, which drives a demodulator to get a replica of the original digital sequence.

The central part of the scheme, the optical channel, will be discussed below. It modifies the incoming optical power $P_T(t)$ in the form

$$P_R(t) = A_c\, P_T(t - t_c)$$

where A_c is the attenuation and t_c is the delay produced by the channel.

A fundamental classification of optical communication is into incoherent transmissions and coherent transmissions.

4.4.2 Incoherent Transmissions

In incoherent transmissions the information (data) is conveyed as intensity of the optical radiation, that is, as the average power $P(t)$. The most popular modulation formats are the PAM (pulse amplitude modulation) and the PPM (pulse position modulation), which are illustrated in Fig. 4.12.

In the PAM, the symbols, usually binary 0 and 1, are encoded into the amplitude of the pulses in the form $A_n\, g(t - nT)$; in practice, the full pulse $g(t - nT)$ when $A_n = 1$ and the absence of the pulse when $A_n = 0$. In the PPM format the symbols are encoded into the position of the pulse in the form $g(t - nT - A_n \Delta T)$, where $\Delta T = T/M$ with M the size of the alphabet ($M = 3$ in Fig. 4.12). In any case, the pulses must satisfy the constraint of being nonnegative because they physically represent a power. Both laser and LED can be used in incoherent transmissions. It is worth remarking that incoherent PAM, which is not very efficient but extremely robust, is the most used format worldwide in connections with optical fibers, reaching the capacity rate of ten terabits/s in a single fiber. Only very recently, coherent transmission (more efficient) has been implemented with optical fibers.

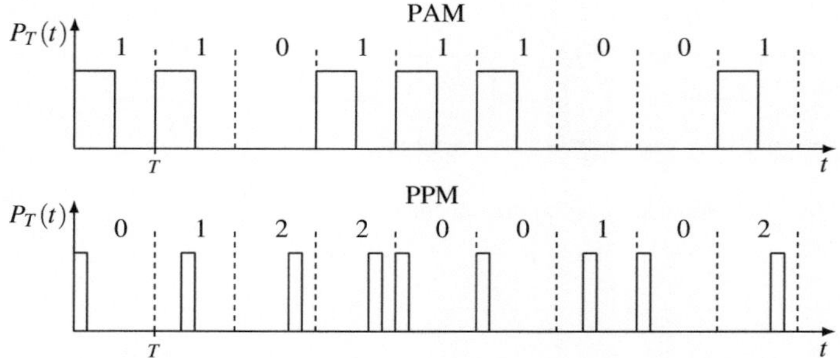

Fig. 4.12 Optical power $P_T(t)$ obtained with a binary PAM modulation, where the symbols are encoded into the pulse amplitudes and with a ternary PPM modulation, where the symbols are encoded into the pulse positions

4.4.3 Coherent Transmissions

Coherent transmissions exploit the fact that the radiation at frequency ν produced by a laser has a sinusoidal waveform

$$v_0(t) = V_0 \cos(2\pi \nu t + \phi_0) \tag{4.12}$$

where the amplitude, when appropriately normalized, gives the optical power as $P = V_0^2$. The waveform (4.12) can be modulated in several forms, exactly as at radio frequencies. The general form of a modulated signal is

$$v(t) = V(t) \cos(2\pi \nu t + \phi(t)) \tag{4.13}$$

where the amplitude $V(t)$ and the phase $\phi(t)$ are modulated according to the information signal. The power becomes time-dependent as $P(t) = V^2(t)$.

Then we have a lot of modulation formats, as OOK (on–off keying), PSK (phase-shift keying), FSK (frequency-shift keying), and QAM (quadrature amplitude modulation). These formats will be examined in the next chapters for a comparison with the corresponding quantum communications.

Coherent transmissions are more efficient than incoherent transmissions, but they require the recovery of the sinusoidal carrier by a local laser at reception.

4.4.4 Guided Optical Channels (Optical Fibers)

A very important optical channel is provided by an optical fiber which "guides" the optical power over a given distance D. A fiber is characterized by an attenuation, which exponentially decreases with D, as

$$A_F = \exp(-\alpha D)$$

where the attenuation coefficient α depends on the type of the fiber, but also on the optical frequency ν. The attenuation is usually specified by the attenuation per unit length $A_{F1} = 10\alpha/\log 10$, expressed in dB/km. This parameter had an important and amazing story since its value passed from tens of dB/km in 1972 to less then 1 dB/km in the 1980s, thus permitting transoceanic connections.

From the attenuation we can calculate the received power as a function of the transmitted power as

$$P_R(t) = A_F \, P_T(t - t_A).$$

Another important optical component that can be introduced in the optical channel is the optical amplifier, which in practice is given by an optical fiber of few meters, appropriately doped. In particular, an optical amplifier is used in transmission as a "booster" to increase the power emitted by the photoemitter. The typical use is to compensate the attenuation of the fiber according to the relation $G_A A_F = 1$, an optical amplifier is characterized by the gain G_A with an input/output relation of the form

$$P_A(t) = G_A \, P(t - t_A) + P_0$$

where P_0 is a spontaneous contribution of the amplifier.

Ordinary fibers and optical amplifiers can be combined in several ways to realize optical channels, with the scenarios illustrated in Fig. 4.13.

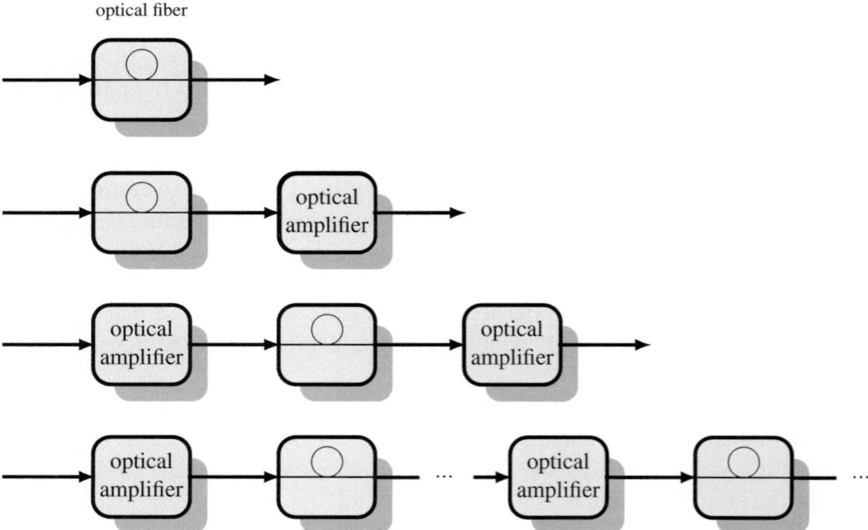

Fig. 4.13 Examples of optical channels obtained with the connection of optical fibers and optical amplifiers

In the evaluation of the performance (error probability), in addition to the power budget, the evaluation of the noise amount is necessary. The noise has two components: the *shot noise*, caused by the granularity of the instantaneous power $p(t)$, and the *thermal noise* generated in the passive components of the receiver circuitry. The shot noise will be considered after the illustration of the theory of Poisson processes. The amount (variance) of the thermal noise can be evaluated with the standard methods of electronics through the noise figure or the noise temperature.

Example 4.1 (*transatlantic connection*) In a transatlantic connection with optical fibers the global distance is $D_g = 5000$ km. The transmitter power, obtained by a laser and a booster amplifier, is $P_T = 100$ mW. The fiber attenuation is $A_{F1} = 1$ dB/km. We want to find the number of links consisting of optical amplifiers and fibers needed with the following constraints: (1) the optical power be greater than $P_{\min} = 1$ nW in all the connection, and (2) the optical amplifiers have a gain G_A of 60 dB.

We first evaluate the distance D_1 obtained with the first link, passing from $P_T = 100$ mW to $P_{\min} = 1$ mW. The attenuation is $A_F = 10^{-9}/10^{-1} = 10^{-8} = 80$ dB. Then, $D_1 = 80$ km. The first amplifier increases the power to $P_1 = G_A P_{\min} = 10^6 \, 10^{-9} = 1$ mW. The second link has an attenuation of $A_2 = P_{\min}/P_1 = 10^{-9}/10^{-3} = 60$ dB and therefore it reaches the distance $D_2 = 60$ km. At this point the connection becomes periodic with a distance $D_n = 60$ km. Then the number of amplifier–fiber links is $5000/60 \simeq 84$.

Example 4.2 (*transatlantic connection: practical implementation*) In transoceanic connections, typical cable lengths are approximately 6000–7000 km for crossing the Atlantic Ocean, and about 9000–11,000 km for crossing the Pacific Ocean. The distance of 6000 km is generally taken as reference for "transoceanic" systems. Modern submarine transmission systems can provide Tb/s capacity per fiber, using wavelength division multiplexing (WDM). One of the most technologically advanced systems installed is the Tata Global Network (TGN) in the Pacific, which provides connectivity between North America and Asia. Each of the submarine cables of the TGN Pacific section has eight pairs of fibers.

Undersea systems use erbium-doped fiber amplifiers (EDFAs) to counteract the attenuation in the optical fiber cable. Thus the undersea amplified line can be seen as a chain of single-mode fibers and optical amplifiers. For a 6000 km system the repeaters (i.e., the amplification stages) are positioned every 50–90 km to balance the cable attenuation. The gain needed to propagate a signal across that system is about (0.2 dB/km) × (6000 km), namely 1200 dB, just to compensate for the loss in the cable.

The noise generated in each EDFA limits the performance of a transmission system and finally determines the spacing between amplifiers. So WDM systems require the use of gain equalization filters in order to expand and/or flatten the intrinsic gain shape of the amplifier chain.

The end-to-end gain change over the optical bandwidth of the transmission line should be limited to about 5–10 dB for suitable performance. If a 10-dB gain error were equally distributed across a 6000 km system (75 amplifiers spaced 80 km apart) the gain equalization would need to be accurate to within 0.1 dB per amplifier stage. This level of gain equalization is routinely achieved for installed systems with optical bandwidths of about 28 nm.

4.4.5 Free-space Optical Channels

An optical power can be transmitted through the free space. We refer to a *directive transmission* (Fig. 4.14), where the propagation is essentially determined by geometrical considerations. The power $P_T(t)$ produced by a laser is first focalized to form a beam and sent to the free space by an antenna (or telescope). Then at a given distance D a new antenna (or telescope) collects and focalizes a fraction $P_R(t)$ of the optical power, which is conveyed to a photodetector. In the simplest case, the received power $P_R(t)$ can be related to the transmitted power as (Friis' formula)

$$P_R(t) = G_T \left(\frac{\lambda}{4\pi D} \right)^2 G_R \, P_T(t - t_0) \tag{4.14}$$

where G_T and G_R are the gains of the transmitting and of the receiving antennas, respectively, and λ is the wavelength of the laser radiation. The expressions of the gains are

$$G_T = (4d_T/\lambda)^2, \qquad G_R = (4\pi/\lambda^2) \, A_R$$

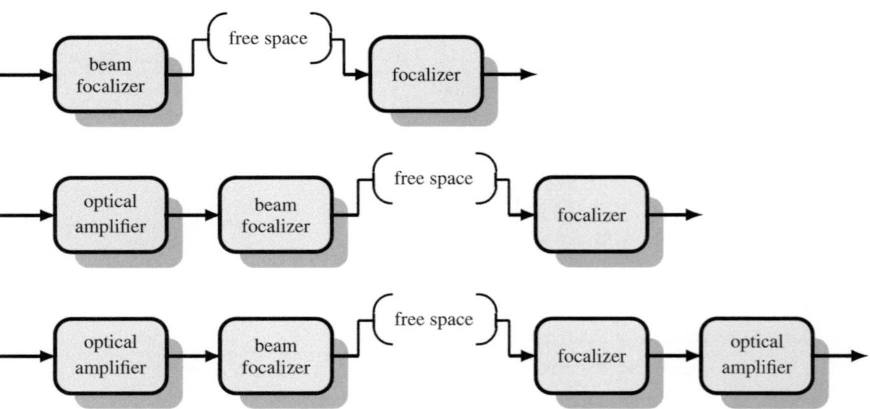

Fig. 4.14 Examples of optical channels in the free space

where d_T is the diameter of the transmitting antenna and A_R is the area of the receiving antenna. Note that (4.14) requires that the aperture of the beam at the distance D, given by $D\,\lambda/d_T$, is greater than the diameter d_R of the receiving antenna.

Example 4.3 We compare the performance at radio frequencies and at optical frequencies, choosing

$$\lambda = 10\,\text{mm} \quad \rightarrow \quad \nu = 30\,\text{GHz} \quad \text{(in radio band, band Ka)}$$
$$\lambda = 1\mu\text{m} \quad \rightarrow \quad \nu = 300\,\text{THz} \quad \text{(optical band, infrared)}$$

and in both cases the transmitting power and the antenna diameter are

$$P_T = 1\,\text{W}, \qquad d_T = 100\,\text{mm}.$$

The comparison is illustrated in Table 4.2. Note in particular the huge gain in the optical case, which enhances the power of 1 W to the apparent power of 160 MW. The comparison considers three distances D: few meters (indoor), few kilometers (outdoor), and the deep space distance of 5 AU. The astronomical unit is the average distance of the Sun from the Earth (1 AU $=$ 150 Gm), 5 AU is the distance of Jupiter from the Sun, the reference distance for missions to outer planets). Note that in a transmission from Jupiter to the Earth the beam diameter is half an astronomical unit (25 thousands times the Earth diameter) at radio frequencies and the power collected by the receiving antenna is extremely small. At optical frequencies the beam diameter is only of 7500 km and the received power is 10^8 greater.

The previous formula establishes the power budget of the system. As regards the noise, in the free-space communications at the receiver front end we find an additional noise, called *sky noise*, which is due to the fact that the receiving antenna

Table 4.2 Performance comparison between radio and optical frequencies in free-space transmission

	Parameter	Radio frequency	Optical frequency
	Transmission gain G_T	31 dB	111.2 dB
	Apparent power $G_T P_T$	1.6 kW	160 MW
Indoor			
$D = 20\,\text{m}, d_R = 10\,\text{mm}$	Beam aperture L_D	2 m	0.2 mm
	Received power P_R	62.4 μW	6.24 kW*
Outdoor			
$D = 2\,\text{km}, d_R = 100\,\text{mm}$	Beam aperture L_D	200 m	20 mm
	Received power P_R	624 nW	62.4 W*
Deep space			
$D = 5\,\text{AU}, d_R = 10\,\text{m}$	Beam aperture L_D	0.5 AU	7.5 Gm
	Received power P_R	$1.77\,10^{-21}$ W	0.77 pW

* Virtual values (condition $d_R < L_D$ not verified)

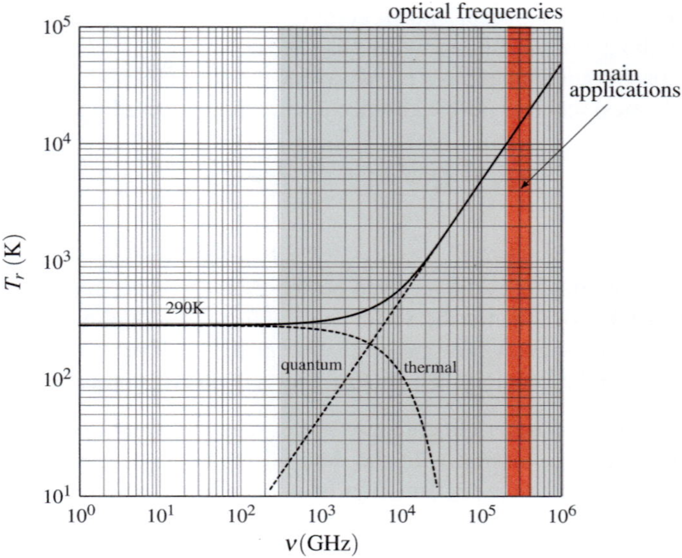

Fig. 4.15 Noise temperature T_r versus the frequency ν due to the combination of thermal noise and quantum noise, with a space temperature $T_a = 290$ K

"sees the sky" and collects extra radiations. This noise is conveniently evaluated by the noise temperature, expressed in absolute degrees (Kelvin) and has two main components. One component is the thermal noise, which is due to the fact that the receiving antenna sees a solid angle of the space at the absolute temperature T_a. The corresponding noise temperature is given by

$$T_{\text{thermal}} = \frac{h\nu}{k\left[\exp(h\nu/kT_a)\right]}$$

where ν is the frequency and k is Boltzmann's constant ($k = 1.3805\ 10^{-23}$ J/K). As shown in Fig. 4.15, $T_{\text{thermal}} \simeq T_a$ for the frequencies up to the terahertz and quickly decreases. The second component is due to the granularity of the radiation and the corresponding noise temperature has the simple expression

$$T_{\text{shot}} = h\nu/k.$$

It is negligible up to the terahertz, but becomes dominant at the optical frequencies.

We realize that sky noise is huge at the optical frequencies: for $\nu \simeq 300$ THz, $T_{\text{sky}} \simeq T_{\text{shot}}$ is in the order of 10,000 K. This leads to think that optical frequencies represent a "forbidden band" for free-space optical transmissions. But, this is not the case because at these frequencies directivity makes it possible to reach a large antenna gain and, on balance, optical communication has a superior budget with respect to radio communications, as we shall see now.

In free-space communications it is easy to find the error probability achieved by an optimum classical receiver (in the absence of encoding), according to the important relation

$$P_e = Q(\sqrt{\Lambda}), \qquad \Lambda = \frac{P_R}{kT_r R} \tag{4.15}$$

where $Q(x) := 1 - (1/\sqrt{2\pi}) \int_{-\infty}^{x} \exp(-y^2/2)\,dy$ is the complementary normalized Gaussian distribution, P_R is the received power, T_r is the noise temperature (in K), R is the rate (in bit/s), and Λ is the SNR. The first trivial but important remark about this formula is that it allows for the evaluation of the achievable rate with a given received power and given noise temperature, for a fixed error probability.

Example 4.4 We reconsider the comparison between radio and optical frequencies for a deep space transmission with the data of Table 4.2. A reasonable value of the error probability for the transmission of images from Jupiter to Earth may be $P_e = 10^{-4}$, which corresponds to a SNR $\Lambda = 16.0$.

With a radio frequency of 30 GHz the received power is $P_R = 1.77\ 10^{-21}$ W; and considering that cryogenic components are used at reception, the noise temperature may be $T_r = 16$ K. Then the available rate is[2]

$$R = \frac{P_R}{kT_r \Lambda} = \frac{1.77\ 10^{-21}}{2.2\ 10^{-22}\ 16} = 0.5\ \text{bit/s}.$$

With an optical frequency of 300 THz the received power is $P_R = 1.77\ 10^{-13}$ W, while the noise temperature becomes $T_r \simeq T_{\text{shot}} = h\nu/k = 1\,600$ K. Then

$$R = \frac{P_R}{kT_r \Lambda} = \frac{1.77\ 10^{-13}}{2.13\ 10^{-20}\ 16} = 508\ \text{kbit/s}.$$

In conclusion, although at the optical frequencies the noise temperature is a thousand times higher than at the radio frequencies, the available rate is a hundred thousands times higher. To emphasize the difference we note that for the transmission of an image of $H = 5$ Mbits, at radio frequencies the time needed is $T = H/R = 10\,\text{Ms} = 116\,\text{days!}$, while at an optical frequency the time is reduced to about 10 s. Incidentally, note that with quantum optical communications the performance is an order of magnitude better than with classical optical communications (see Chap. 7) and the time for an image would be further reduced.

Problem 4.3 ⋆ Write the expressions of a PAM optical power $P(t)$, with a generic fundamental pulse $g(t)$, valid for all $t \in \mathbb{R}$.

---

[2] The rate of NASA Voyager 2 at Jupiter in 1979 mission was of 115.2 kbit/s. This was achieved acting on several factors: increasing the transmitting and the receiving antenna diameters and also increasing the transmitter power (see Problem 4.4).

Problem 4.4 ⋆   The physical parameters of the transmitter (on spacecraft board) and of the receiver (at Earth, Goldstone, California) of NASA Voyager 2 mission at Jupiter (1979) were: radio frequency $\nu = 8.9\,\mathrm{GHz}$, transmitted power $P_T = 24\,\mathrm{W}$, transmitter's antenna diameter $d_T = 3.660\,\mathrm{m}$, receiver's antenna diameter $d_R = 64\,\mathrm{m}$, noise temperature $T_r = 14\,\mathrm{K}$, accepted error probability $P_e = 10^{-3}$. Find the available rate. Repeat the evaluation at the optical frequency $\nu = 300\,\mathrm{THz}$.

4.5 Poisson Processes

In the previous section we have seen the scenarios of optical classical communications and how the *average* optical power $P(t)$ is handled in the different systems. But for performance evaluation we need to consider the *instantaneous* power $p(t)$, where the energy quanta appear explicitly, as well as the instantaneous current composed by elementary charges. In both cases "granularity" causes the shot noise. To formulate "granularity" the fundamental mathematical tool is provided by Poisson processes which are briefly introduced in this section. For a general and exhaustive theory a good reference on Poisson processes is given by Parzen's book [5].

Poisson processes represent a relevant subclass of the family of *point* processes, which describes random time instants at which some events happen. For instance, the time at which a photon crosses a given section of an optical fiber (but also the arrivals of cars at a highway entrance and the arrivals of telephone calls). A realization of point processes is given by a discrete set of real numbers, $\ldots, t_{-1}, t_0, t_1, t_2, \ldots$, whose elements are called *arrival times* or simply *arrivals* (Fig. 4.16). A convenient way to represent a realization of a point process is given by a train of delta functions located at arrival times (Fig. 4.16)

$$x_\delta(t) = \sum_i \delta(t - t_i). \tag{4.16}$$

This signal allows us to count the number of arrivals within a given interval, say $(s, t]$, with $t > s$, as

$$n(s, t] := \int_s^t x_\delta(u)\,\mathrm{d}u. \tag{4.17}$$

This defines the *counting process* associated to the given point process, as shown in Fig. 4.16.

4.5.1 Definition of Poisson Process

A Poisson process is defined as a point process that verifies the following axiomatic properties:

(I) The countings $n(s_i, t_i]$ in disjoint intervals are independent random variables.

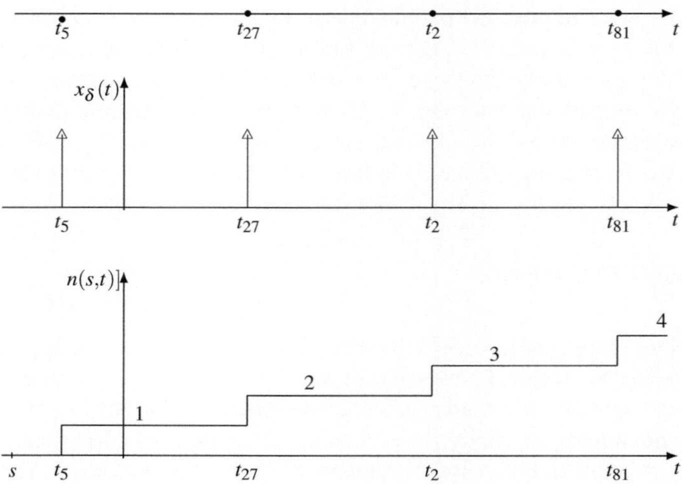

Fig. 4.16 Example of realization of a point process and of the corresponding impulsive process $x_\delta(t)$ and counting process $n(s, t]$

(II) For every $s \in \mathbb{R}$, the limit

$$\lim_{h \to 0^+} \frac{\mathrm{P}[n(s, \, s+h] = 1]}{h} = \lambda(s) \tag{4.18}$$

exists and is finite. The function $\lambda(t)$ is called **intensity** of the Poisson process.

(III) For every $s \in \mathbb{R}$,

$$\lim_{h \to 0^+} \frac{\mathrm{P}[n(s, \, s+h] > 1]}{h} = 0. \tag{4.19}$$

Conditions (II) and (III) are somewhat technical. Broadly speaking, condition (II) states that the probability that a single arrival happen in an infinitesimal interval $(t, t+dt]$ is given by $\lambda(t)\,dt$, and condition (III) states that the probability of simultaneous arrivals is zero.

As we shall see, the statistical description of a Poisson process is completely specified by its intensity $\lambda(t)$. The intensity allows for the classification into *homogeneous* Poisson processes if $\lambda(t)$ is independent of t, and *nonhomogeneous* Poisson processes if $\lambda(t)$ depends on t.

4.5.2 Statistical Description of a Poisson Process

From the above axioms it is possible to evaluate the statistics of the number of arrivals in an arbitrary interval [5].

Theorem 4.1 *In a Poisson process with intensity $\lambda(t)$ the number of arrivals $n(s, t]$ in the interval $(s, t]$ is a Poisson random variable with mean*

$$\Lambda = \mathrm{E}[n(s, t]] = \int_s^t \lambda(a)\,\mathrm{d}a. \tag{4.20}$$

Hence the probability distribution of the number of arrivals is given by

$$p(k; s, t) := \mathrm{P}\left[n(s, t] = k\right] = \mathrm{e}^{-\Lambda}\frac{\Lambda^k}{k!} \tag{4.21}$$

and is completely specified by the mean Λ. It is important to recall that a Poisson random variable has the peculiarity that the variance of the number of arrivals $n = n(s, t]$ coincides with the mean

$$\sigma_n^2 = m_n = \Lambda. \tag{4.22}$$

We also recall that the characteristic function of $n = n(s, t]$ is given by

$$\Psi_n(z) := \mathrm{E}[e^{inz}] = \mathrm{e}^{\Lambda[\exp(iz)-1]}. \tag{4.23}$$

The Poisson distribution is illustrated in Fig. 4.17 for a few values of the mean Λ (it is also illustrated in Fig. 4.26 in the range of quantum limit).

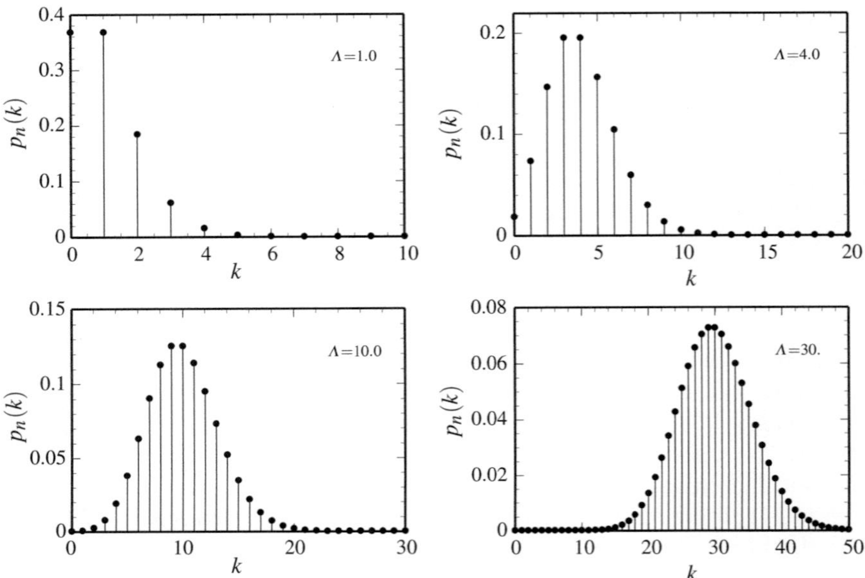

Fig. 4.17 Poisson distribution $p_n(k)$ for some values of the average Λ

From (4.20) one can obtain the meaning of the intensity $\lambda(t)$. Considering that $E[n(0, t]] = \int_0^t \lambda(u)du$, one gets $\lambda(t) = \frac{d}{dt} E[n(0, t]]$. Hence the intensity $\lambda(t)$ is the derivative of the average (expectation) of the counting process and therefore represents the *average number of arrivals per second*.

If the Poisson process is homogeneous with $\lambda(t) = \lambda$, the average is simply given by $\Lambda = (t - s)\lambda$, and consequently the distribution of the number of arrivals $n(s, t]$ in the interval $(s, t]$ does not depend on s, but only on the duration $t - s$. This defines the *stationarity of a homogeneous Poisson process*.

Problem 4.5 ⋆⋆ Consider a Poisson random variable n. Prove that the variance σ_n^2 is equal to the mean $m_n = E[n]$ and that the characteristic function is given by (4.23).

Problem 4.6 ⋆ In the technique of **single photon** the following probabilities are of interest

$$p_0 = P[n = 0], \qquad p_1 = P[n = 1], \qquad p_{>1} = P[n > 1].$$

Assuming that the arrivals are described by a Poisson process, write and plot these probabilities. Moreover, find the average of photon arrivals such that $p_{>1} = 0.1 p_1$

Problem 4.7 ⋆ In the technique of **single photon** the optical power is attenuated to realize the condition of the arrival of a single photon in a given symbol period $(0, T]$. Assuming that the power produced by the laser be $P_0 = 10\,\text{mW}$ at the frequency $\nu = 300\,\text{THz}$ and that the symbol period be $T = 10\,\text{ns}$, find the attenuation A needed to ensure that the condition of a single photon is verified in the 15 % of the symbol periods.

4.6 Filtered Poisson Processes

In several applications a signal is obtained as the superposition of equal pulses (signals of small duration) generated in correspondence with the arrivals of a Poisson process, that is,

$$y(t) = \sum_i h(t - t_i) \tag{4.24}$$

where $h(t)$ is the *fundamental pulse* (Fig. 4.18). The resulting signal is called *filtered Poisson process*. The term "filtered" is justified by the fact that $y(t)$ can be obtained by filtering the impulsive process $x_\delta(t)$ defined by (4.16), with a filter having impulse response $h(t)$. As we shall see, this model is useful in the representation of optical powers and electrical currents in photodetection.

In (4.24) all the pulses have the same form, being a shifted replica of $h(t)$. A generalization of the model is obtained by amplifying each pulse by a random gain g_i in the form

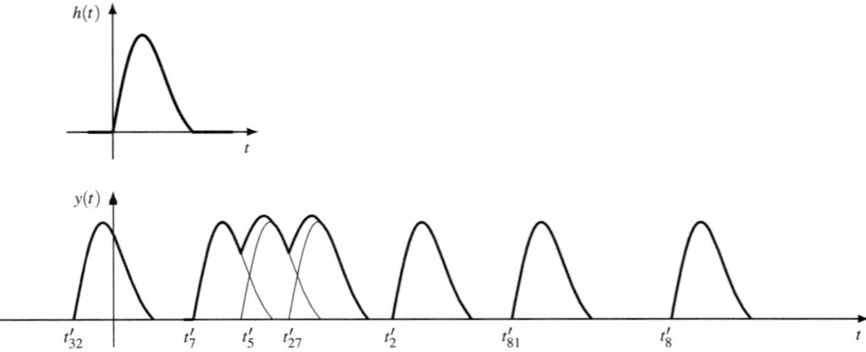

Fig. 4.18 Realization of a filtered Poisson process $y(t)$ with fundamental pulse $h(t)$

$$y(t) = \sum_i g_i\, h(t - t_i).\tag{4.25}$$

This model is called *marked and filtered Poisson process*, where the marks represent the gains. It finds applications to photodetection with the APD and in optical amplification.

Note that in (4.24) the randomness of the filtered process is only due to the presence of the point process $\{t_i\}$, since $h(t)$ is a deterministic (nonrandom) waveform. In (4.25) we have two kind of randomness: the point process $\{t_i\}$ and the sequence of gains $\{g_i\}$.

4.6.1 Campbell's Theorems

The statistical description of filtered Poisson process $y(t)$ is established by a sequel of Campbell's theorems, which are formulated here in order of increasing complexity. The theorems give the complete statistical description of the random process $y(t)$ through the hierarchy of the characteristic functions from which one gets the hierarchy of the probability densities. This in principle, but the actual evaluation is very difficult. Here we limit ourselves to the evaluation of the mean and of the variance or the covariance, which is simpler.

Filtered Poisson Processes with Constant Intensity.

The first theorem is concerned with a filtered (not marked) Poisson process with a constant intensity. In this case the specification of the process is given by the intensity λ and by the pulse shape $h(t)$.

Theorem 4.2 *A filtered Poisson process*

$$y(t) = \sum_i h(t - t_i),\tag{4.26}$$

with a constant intensity λ is stationary. Its mean and covariance are, respectively,

$$m_y = \lambda \int_{-\infty}^{+\infty} h(t)\, \mathrm{d}t, \qquad k_y(\tau) = \lambda \int_{-\infty}^{+\infty} h(t)\, h(t+\tau)\, \mathrm{d}t. \tag{4.27}$$

Marked and Filtered Poisson Processes with Time-Dependent Intensity.

In this case the specification of the process requires also the statistical description of the gains. The assumption is that the gains $\{g_i\}$ are statistically independent and uniformly distributed, and also independent of the arrivals $\{t_i\}$.

Theorem 4.3 *A marked and filtered Poisson process*

$$y(t) = \sum_i g_i\, h(t - t_i), \tag{4.28}$$

with a time-dependent intensity $\lambda(t)$ is not stationary. The mean and the variance are given by

$$\boxed{\begin{aligned} m_y(t) &= G \int_{-\infty}^{+\infty} \lambda(a)\, h(t-a)\, \mathrm{d}a \\ \sigma_y^2(t) &= G_2 \int_{-\infty}^{+\infty} \lambda(a)\, h^2(t-a)\, \mathrm{d}a \end{aligned}} \tag{4.29}$$

where $G = \mathrm{E}[g_i]$ and $G_2 = \mathrm{E}[g_i^2]$ are the moments of the first two orders of the gains g_i.

4.6.2 Doubly Stochastic Poisson Processes

In the above formulation we assumed that the intensity $\lambda(t)$ of the Poisson processes was a given (nonrandom) function of time. This is not the case with Optical Communications, where the information is often encoded into the intensity $\lambda(t)$ of a Poisson process, so that $\lambda(t)$ becomes itself a random process. This leads to the term *doubly stochastic* Poisson processes.

The approach to deal with such processes is to work **under the condition of a given intensity** $\lambda(t)$. Technically, this is achieved with appropriate conditional expectations, which allow us to use all the previous results on ordinary Poisson processes (Poisson distribution of the arrivals and Campbell theorems). For instance, the average number of arrivals in an interval $(s, t]$, given by (4.20), becomes *the mean number of arrivals, conditioned by a given intensity in such interval*, and is given by

$$\mathrm{E}_\lambda[n(s,t)\mid \lambda(a),\ a \in (s,t]] = \int_s^t \lambda(a)\mathrm{d}a. \tag{4.30}$$

Correspondingly, $n(s, t]$ becomes a *conditioned Poisson random variable*, where the condition is "a given intensity." From the conditioned mean (4.30) one can evaluate the unconditioned mean, by taking the expectation with respect to the intensity. Note that in general the unconditioned random variable $n(s, t]$ will be no more a Poisson random variable. A detailed application of these ideas will be described in Sect. 4.7.

4.6.3 *Instantaneous Power as a Filtered Poisson Process*

A monochromatic optical power at the frequency v consists of *energy quanta hv*, where h is Planck's constant ($h = 6.6256 \ 10^{-34}$ J/Hz). The contribution of a quantum that crosses a section at the time t_i can be represented as a power of the form [6]

$$(hv) \ \delta(t - t_i). \tag{4.31}$$

In fact, integrating this impulse one gets just the energy hv; the delta function indicates that the energy is concentrated around the "arrival" time t_i. The instantaneous optical power crossing the prefixed section has the expression (Fig. 4.19)

$$p(t) = \sum_i (hv) \ \delta(t - t_i), \tag{4.32}$$

where $\{t_i\}$ are the arrival times in that section.

The model, universally accepted for the arrival times, is a Poisson point process and therefore is specified by an intensity $\lambda(t)$, which represents the *average number of*

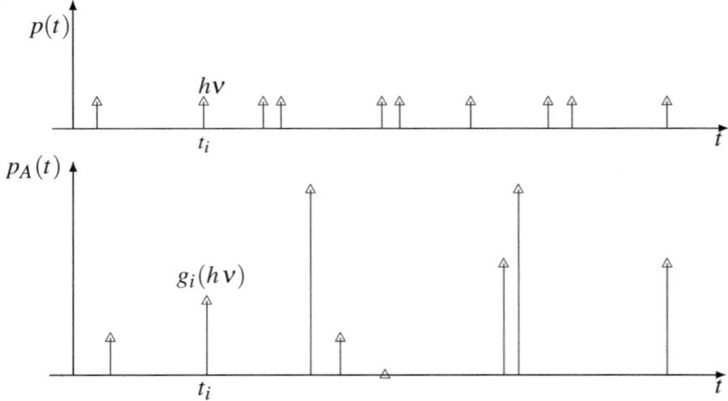

Fig. 4.19 Examples of an instantaneous optical power $p(t)$, where the delta impulses have area $h\,v$, and of an instantaneous *amplified* optical power $p_A(t)$, where the delta impulses have area $g_i(h\,v)$, with g_i random gains

photons per second. According to the theory developed above, Eq. (4.32) represents a *filtered* Poisson process with fundamental pulse $(h\nu)\,\delta(t)$. From Campbell's theorem (Theorem 4.3), the statistical average of the instantaneous power results in

$$P(t) = \mathrm{E}[p(t)] = h\nu\,\lambda(t). \tag{4.33}$$

Note that from the average power $P(t)$, one gets the intensity as $\lambda(t) = P(t)/h\nu$.

Equation (4.32) may represent the optical power produced by a laser and also the power in a generic cross section of the optical medium (fiber or free space) and in particular at the detection. In an *amplified* optical power the quantum $h\nu$ is replaced by a *packet of quanta* and (4.3) becomes

$$g_i\,(h\nu)\,\delta(t - t_i) \tag{4.34}$$

where the gains g_i are positive integer random variables having a geometric distribution [7, 8]. Then an amplified optical power is written in the form (Fig. 4.19)

$$p_A(t) = \sum_i g_i\,(h\nu)\,\delta(t - t_i) \tag{4.35}$$

which represents a *marked and filtered* Poisson process with fundamental pulse $(h\nu)\,\delta(t)$. The statistical average of $p_A(t)$ is given by (from Theorem 4.3)

$$P_A(t) = G_A(h\nu)\,\lambda(t) \tag{4.36}$$

where $G_A = \mathrm{E}[g_i]$ is the average gain of the amplified power (G_A gives also the nominal gain of the amplifier).

4.6.4 Instantaneous Current in a Photodiode

Also the instantaneous current in a photodiode can be represented by a filtered Poisson process in that the current is due to the motion of electrons, or better, electron/hole pairs in a pin photodiode (pin = positive intrinsic negative). In fact, the instantaneous current can be written in the form (Fig. 4.20)

$$i(t) = \sum_k e\,\delta(t - t_k) \tag{4.37}$$

where e is the electron charge ($e = 1.6022\,10^{-19}$ C) and $\{t_k\}$ are the arrival times, which can be modeled as a point Poisson process.

The statistical average of $i(t)$ results in

$$I(t) = \mathrm{E}[i(t)] = e\,\lambda(t) \tag{4.38}$$

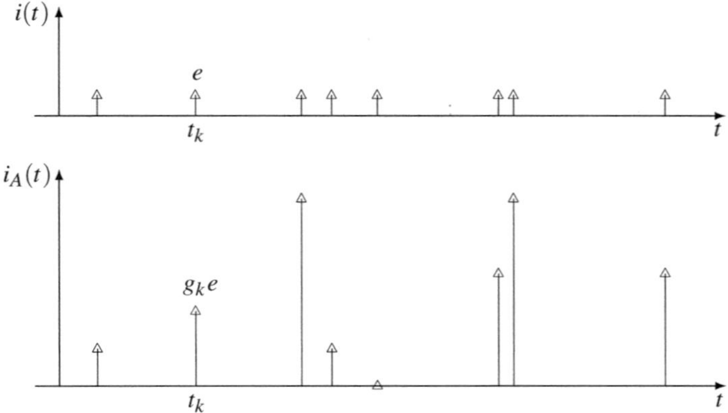

Fig. 4.20 Examples of an instantaneous electrical current $i(t)$ in a pin diode, where the delta impulses have area e, and of an instantaneous *amplified* electrical current $i_A(t)$ in an APD, where the delta impulses have area $g_i\, e$, with g_i random gains

and represents the current usually considered in the photodiode at a macroscopic level. The difference $i_s := i(t) - I(t)$ is the so-called *shot noise*.

An APD with the presence of random gains produces the instantaneous current

$$i_A(t) = \sum_k (g_k\, e)\, \delta(t - t_k) \tag{4.39}$$

where the gains g_k are positive integers with a complicated probability distribution [9]. The corresponding average current results in

$$I_A(t) = G_A\, e\, \lambda(t) \tag{4.40}$$

$G_A = \mathrm{E}[g_k]$ is the average gain (G_A gives also the nominal gain of the APD).

In (4.37) the impulses are ideal and this is an approximation. In reality the fundamental pulse has a shape $i_0(t)$ which depends on the geometry of the photodiode and on the electrical field inside. Then a more accurate model is given by

$$i(t) = \sum_k i_0(t - t_k) \tag{4.41}$$

which is still a filtered Poisson process. Similar considerations hold for the current in APD. Figure 4.21 shows the instantaneous currents in an APD. Note that in any case the area of $i_0(t)$ is given by the elementary charge e.

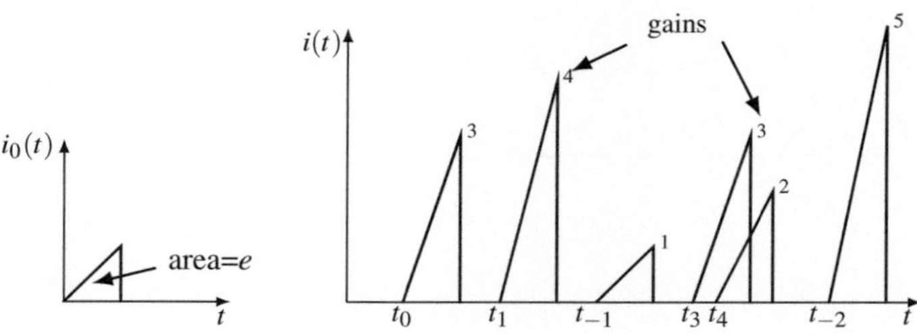

Fig. 4.21 Examples of current pulses in an APD

4.6.5 About the Gaussianity of Filtered Poisson Processes

Campbell's theorems provide the complete statistical description of filtered Poisson processes through the characteristic functions. In particular the first-order characteristic function of the filtered Poisson process (4.37) results in

$$\psi_i(z, t) := \mathrm{E}[\exp(iz\, i(t))]$$

$$= \exp\left\{-\int_{-\infty}^{+\infty} \lambda(a)[1 - z\, i_0(t - a)]\mathrm{d}a\right\} \tag{4.42}$$

from which one gets the probability density $f_i(a, t)$ (by the Fourier transform). This probability density is not Gaussian, but it may be well approximated by the Gaussian density

$$f_i(a, t)_{\mathrm{Gauss}} = \frac{1}{\sqrt{2\pi}\,\sigma_i(t)} \exp\left(-\frac{(a - m_i(t))^2}{2\sigma_i^2(t)}\right) \tag{4.43}$$

where $m_i(t) := \mathrm{E}[i(t)]$ is the mean and $\sigma_i(t)^2$ is the variance, which can be calculated by Theorem 4.3. The goodness of the Gaussian approximation is ensured by the *central limit theorem*, which states that a random variable given by superposition of several independent random variables tends to Gaussianity, and this is the case with the current $i(t)$ given by (4.41). A further trend to Gaussianity is provided when the Poisson current $i(t)$ is added to a thermal noise component $i_\eta(t)$, which is itself Gaussian. This will be seen in detail when dealing with photodetection in the next section.

Problem 4.8 ⋆⋆ Evaluate the mean $m_y(t)$ and the variance $\sigma_y(t)^2$ of a filtered Poisson process $y(t)$, where the intensity is constant $\lambda(t) = \lambda_0$, and the fundamental pulse is rectangular of amplitude h_0 in $(0, T]$.

Problem 4.9 ⋆⋆⋆ Evaluate the mean $m_y(t)$ and the variance $\sigma_y(t)^2$ of a marked and filtered Poisson process $y(t)$, where the intensity is constant $\lambda(t) = \lambda_0$, the fundamental pulse is rectangular of amplitude h_0 in $(0, T]$, and the gains have the geometrical distribution. $p_g(k) = (1 - a)a^k$, $k = 0, 1, 2, \ldots$ with a a positive constant.

Problem 4.10 ⋆⋆ To illustrate the conditioning of a double stochastic Poisson process we consider a (simplified) binary modulation, where the intensity of the optical power in $(0, T]$ has the values $\lambda_0 = 10^9$ photon/s with the symbol $A = 0$ and $\lambda_1 = 4\ 10^9$ photon/s with the symbol $A = 1$. Find the *conditioned* distribution of the number of arrivals

$$p_n(k|A = 0), \qquad p_n(k|A = 1)$$

and the *unconditioned* distribution $p_n(k)$, assuming $\mathrm{P}[A = 1] = 1/4$ and $T = 1$ ns.

4.7 Optical Detection: Semiclassical Model

The *semiclassical* model of optical detection treats the electromagnetic field in classical form, that is, according to Maxwell's undulatory theory, and derives the photoemission current through a probabilistic scheme (theory of doubly stochastic Poisson processes) [10–12]. We use the term "semiclassical" because the final evaluation is done on the count of the photons, and therefore the electromagnetic field is considered quantized into energy quanta.

The model assumes that the surface A of the photodetector is hit by a field of intensity $I(x, t)$ (measured in W/m^2), and emits a current density $J(x, t)$ (measured in A/m^2). We suppose that in every neighborhood of point x and in every neighborhood of time t an electron is emitted with *birth* probability

$$\lambda(x, t)\, dx\, d t.$$

The intensity $\lambda(x, t)$ is linked to the intensity of the electromagnetic field on the surface of the photoemitter by the relation

$$\lambda(x, t) = \frac{\mathcal{R}}{e} I(x, t)$$

where e is the electron's charge and $\mathcal{R} = e/(h\nu)$ is the *responsivity* of the photodetector, which we assume of unitary emittance (one photon per electron). It should be noted that the intensity $\lambda(x, t)$ may be deterministic, but may also be random, depending on the nature of the field. We assume that the electron emission from the surface of the photodetector produces a current density

$$J(x, t) = \sum_n e\, \delta(x - x_n)\delta(t - t_n), \tag{4.44}$$

that is, an impulsive process, with $\{t_n, x_n\}$ a *point process* of time instants and photoemission points of intensity (of the point process) $\lambda(x, t)$.

The photoemission process is then a *doubly stochastic* Poisson process. This means that, with a given field intensity $I(x, t)$, the electron count in the time interval $[0, T)$

$$n = \frac{1}{e} \int_0^T \int_A I(x, t)\, dx\, \mathrm{d}t$$

is a Poisson random variable with random parameter

$$\bar{n} = \int_0^T \int_A \lambda(x, t)\, dx\, \mathrm{d}t = \frac{\mathcal{R}}{e} \int_0^T \int_A I(x, t)\, dx\, \mathrm{d}t$$

which represents the (conditioned) average of the number of arrivals n. The intensity of the field is given by

$$I(x, t) = \frac{1}{Z} |\mathbf{e}(x, t)|^2$$

where $\mathbf{e}(x, t)$ is the *complex envelope* (see below) of the electric field and Z is the wave impedance.

4.7.1 Simplified Model

The previous theory becomes simpler when the electric field is *separable* in the form

$$\mathbf{e}(x, t) = \mathbf{e}_1(x)\, e_2(t) \tag{4.45}$$

which allows us to substitute relation (4.44) with a simple temporal process (as soon as the integration with respect to x is carried out).

In this model we can evaluate the count statistics directly from the optical power, instead of the current. Then, considering a quasimonochromatic radiation around the optical frequency v, and indicating with hv the corresponding energy quantum, the (instantaneous) optical power has the expression

$$p(t) = \sum_k (hv)\, \delta(t - t_k) \tag{4.46}$$

where the arrival instants $\{t_k\}$ must be considered as a *doubly stochastic Poisson process* with random intensity $\lambda(t)$. For this kind of processes we can use integrally the theory of ordinary Poisson processes, providing we operate under the condition of

a given intensity. In particular, the *average power*, conditioned by a given intensity $\lambda(t)$, is obtained through Campbell's theorem according to

$$P(t) = \mathrm{E}[p(t)|\{\lambda(\cdot)\}] = (h\nu)\,\lambda(t), \tag{4.47}$$

where $\mathrm{E}[\cdot|\{\lambda(\cdot)\}]$ denotes the conditional expectation.

The number of arrivals in the reference interval $[0, T)$ is obtained by integrating the instantaneous power

$$n = \frac{1}{h\nu} \int_0^T p(t)\,\mathrm{d}\,t \tag{4.48}$$

and has a random average given by

$$\bar{n}_\lambda = \mathrm{E}[n|\{\lambda(\cdot)\}] = \int_0^T \lambda(t)\,\mathrm{d}\,t = \frac{1}{h\nu} \int_0^T P(t)\,\mathrm{d}\,t. \tag{4.49}$$

Our observations on doubly stochastic Poisson processes imply that the statistics of the number of arrivals in an interval is Poisson, *under the condition of a given intensity*. Therefore,

$$p_n(k|\lambda(\cdot)) = \mathrm{P}[n = k|\bar{n}_\lambda] = \mathrm{e}^{-\bar{n}_\lambda} \frac{\bar{n}_\lambda^k}{k!}, \qquad k = 0, 1, 2, \ldots \quad . \tag{4.50}$$

It remains to relate the average power, and therefore the intensity of the point process, to the *signal* (typically a modulated signal) $v(t)$ present in the optical transmission system in the form of electric field. From the previous general theory, using the hypothesis of separability (4.45) and expressing $v(t)$ through its **complex envelope** $c_v(t)$ in the form

$$v(t) = \Re\, c_v(t)\, \mathrm{e}^{\mathrm{i}\,2\pi\,\nu t},$$

we find that the average power is proportional to $|c_v(t)|^2$ and, by appropriate normalization of the electric field, we can directly write [12]

$$\boxed{P(t) = |c_v(t)|^2.} \tag{4.51}$$

Then, from the statistics of the complex process $c_v(t)$, we can evaluate the statistics of the average power $P(t)$, and therefore of the intensity $\lambda(t)$, that conditions the point process.

What is the complex envelope? In Modulation Theory a real signal $v(t)$, whose frequency content is around a frequency ν (bandpass signal), is efficiently represented by the complex envelope [13], which is obtained as follows. With the ordinary Fourier transform representation a bandpass signal has two modes, $v_+(t)$ with frequency content around the frequency ν, and $v_-(t)$ with frequency content around the specular frequency $-\nu$. Then the complex envelope of $v(t)$ is defined as

$$c_V(t) := 2\,v_+(t)\,\mathrm{e}^{-\mathrm{i}\,2\pi\,\nu\,t}$$

where the exponential shifts the frequency content around the zero frequency. The complex envelope represents the original real signal as

$$v(t) = \Re c_V(t)\,\mathrm{e}^{\mathrm{i}\,2\pi\,\nu\,t}.$$

For instance, the sinusoidal signal $v(t) = V\,\cos(2\pi t + \phi)$ is decomposed through Euler's formula as $v(t) = v_+(t) + v_-(t)$, where $v_\pm(t) = \frac{1}{2}V\,\mathrm{e}^{\pm\mathrm{i}(2\pi\,\nu\,t+\phi)}$. Then, its complex envelope is $c_V(t) = V\,\mathrm{e}^{\mathrm{i}\phi}$.

4.7.2 Complex Envelope and Instantaneous Power Duality

In the analysis of an optical system, at every point of the system, we have the "signal" $v(t)$, conveniently represented by its complex envelope $c_V(t)$. This is the *classical viewpoint*, the same considered at radio frequencies. But at the same point of the optical system we have the instantaneous power $p(t)$, containing the quanta as stated by (4.46), and therefore related to Quantum Mechanics. For this reason the joint formulation of signal/complex envelope and instantaneous power is sometimes called *semiclassical model*, rather than classical model.

Both $c_V(t)$ and $p(t)$ must be considered for an exhaustive analysis of the system. The fundamental relation linking the two quantities is given by (4.51). Note that the average power can be obtained from the complex envelope, but not the converse, because in $P(t)$ the phase information is lost. On the other hand, the complex envelope does not give any direct information about the granular nature of the instantaneous power.

The duality intrinsic in the semiclassical model is summarized in Fig. 4.22.

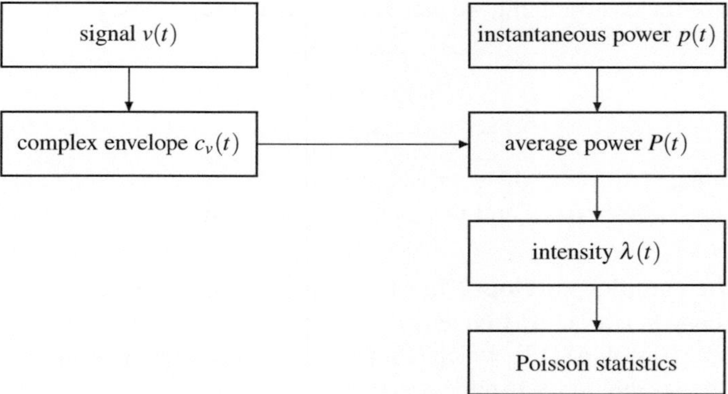

Fig. 4.22 Complex envelope and instantaneous power duality

We develop two examples. The first is the representation of a monochromatic radiation produced by a laser at frequency ν. The signal and complex envelope are, respectively,

$$v_0(t) = V_0 \cos(2\pi \nu t + \phi_0), \qquad c_{v_0} = V_0\, e^{i\phi_0}$$

while the average power and the intensity are, respectively,

$$P_{v_0} = V_0^2, \qquad \lambda_0 = V_0^2/(h\nu).$$

The instantaneous power $p_0(t)$ has the general form (4.46) with the arrivals $\{t_k\}$ having the constant intensity λ_0.

As a second example we consider a modulated signal obtained with the carrier $v_0(t)$. The general form is

$$v(t) = V(t) \cos[2\pi t + \phi(t)], \qquad c_v(t) = V(t)\, e^{i\phi(t)} \tag{4.52}$$

where $V(t) \geq 0$ represents the amplitude modulation and $\phi(t)$ the phase modulation. The average power and the intensity are, respectively,

$$P(t) = V^2(t), \qquad \lambda(t) = V^2(t)/(h\nu). \tag{4.53}$$

The instantaneous power $p_(t)$ has the form (4.46) where now the arrivals $\{t_k\}$ have a time-dependent intensity $\lambda(t)$, which is also random if the modulated signal contains an amplitude modulation. Note that in the passage to the average power, $P(t) = |c_v(t)|^2$, the phase information is lost.

4.7.3 Digital Modulation

Relation (4.52) includes digital modulation but with a specific structure. The target is the transmission of a sequence of complex symbols $\{C_n\}$ through a finite number of waveforms $\{\gamma_0(t), \gamma_1(t), \ldots, \gamma_{K-1}(t)\}$. The expression of the complex envelope is

$$c_v(t) = \sum_{n=-\infty}^{+\infty} \gamma_{C_n}(t - nT), \qquad -\infty < t < +\infty$$

meaning that C_n is transmitted in the symbol period $(nT, nT+T]$ with the waveform $\gamma_{C_n}(t - nT)$. For simplicity we refer to the symbol C_0 and we suppose that its waveform $\gamma_{C_0}(t)$ is confined to the interval $(0, T]$, so that the complex envelope is simply given by

$$c_v(t) = \gamma_{C_0}(t), \qquad 0 < t \leq T. \tag{4.54}$$

In practice (4.54) represents the signal in the absence of thermal noise, but if we want to take into account an additive thermal noise $\eta(t)$, the complex envelope takes the form

$$c_v(t) = \gamma_{C_0}(t) + c_\eta(t), \qquad 0 < t \leq T. \tag{4.55}$$

where $c_\eta(t)$ is the complex envelope of the thermal noise $\eta(t)$.

Now we develop the photon counting in the two cases of (4.54) and (4.55).

4.7.4 Photon Counting in the Absence of Thermal Noise

In optical communications the detection of a digital symbol is often based on the counting of photons at the end of the symbol period, say $(0, T]$. In this section we develop the theory of photon counting in the absence of thermal noise, which will be used in Chap. 7.

From the complex envelope given by (4.54)

$$c_v(t) = \gamma_{C_0}(t), \qquad 0 < t \leq T$$

we obtain the intensity of the underlying point Poisson process as

$$\lambda(t) = \frac{1}{h\nu} P(t) = \frac{1}{h\nu} |\gamma_{C_0}(t)|^2.$$

Now the randomness of $\lambda(t)$ is not only due to the randomness of C_0 and if we fix C_0 at a given symbol of the constellation, say $C_0 = 1$, we have a nonrandom waveform (deterministic signal) and so is for the intensity $\lambda(t)$. This means that working under the condition of a given C_0 we get a nonrandom intensity and thus we can apply the statistics of simple Poisson processes. In particular (4.49) becomes

$$\bar{n}_{C_0} = \mathrm{E}[n|C_0] = \int_0^T \lambda(t)\,\mathrm{d}t = \frac{1}{h\nu} \int_0^T P(t)\,\mathrm{d}t \tag{4.56}$$

and gives the average number of arrivals with a given transmitted symbol C_0. Correspondingly, we have the conditioned Poisson distribution

$$p_n(k|C_0) := \mathrm{P}[n = k|C_0] = \mathrm{e}^{-\bar{n}_{C_0}} \frac{\bar{n}_{C_0}^k}{k!}, \qquad k = 0, 1, 2, \ldots \quad . \tag{4.57}$$

In conclusion, in the absence of thermal noise the photon counting is governed by a **Poisson distribution**. It is important to remark that in the evaluation of the probabilities (transition and correct decision) we have to work under the condition of a given symbol, so that $p_n(k|C_0)$ is just the probability distribution of interest.

4.7.5 Photon Counting in the Presence of Thermal Noise

We now develop the theory of photon counting in the presence of thermal noise, which will be used in Chap. 8. The theory is not easy as it requires a sophisticated application of doubly stochastic Poisson processes.[3]

From the complex envelope given by (4.55)

$$c_v(t) = \gamma c_0(t) + c_\eta(t), \qquad 0 < t \le T$$

we obtain the intensity of the underlying point process as

$$\lambda(t) = \frac{1}{h\nu} |\gamma c_0(t) + c_\eta(t)|^2. \tag{4.58}$$

Now we observe that the randomness of $\lambda(t)$ is not only due to the randomness of C_0 but also to the randomness of the noise $c_\eta(t)$ and to use the Poissonian statistic we have to work *under the condition of a given intensity* $\lambda(t)$. The randomness of $\lambda(t)$ can be removed by conditioning both C_0 and $c_\eta(t)$. If we know the probability density $f_\lambda(\lambda)$ of $\lambda = \lambda(t)$, we will get the unconditioned probability distribution of the arrivals as

$$p_n(k) = \int_0^\infty p_n(k|\lambda) f_\lambda(\lambda) \, d\lambda \tag{4.59}$$

with $p_n(k|\lambda)$ given by (4.50). In general the result so obtained is no more a Poisson distribution.

Things, however, are not so simple because in the final evaluation of probabilities we work "under the condition of a given symbol C_0," and therefore we want to obtain the count statistic under this condition, as in (4.57). To this end we have to replace in (4.59) the probability density $f_\lambda(\lambda)$ with the conditional density $f_\lambda(\lambda|C_0)$ and the final goal is the evaluation of the conditional probability distribution

$$p_n(k|C_0) = \int_0^\infty p_n(k|\lambda) f_\lambda(\lambda|C_0) \, d\lambda. \tag{4.60}$$

Reconsidering the structure of the intensity $\lambda(t)$ given by (4.58) we see that the term $\gamma c_0(t)$, for a given C_0, is nonrandom, while c_η is a complex random process, and the global statistical description is not given by a simple probability density, as $f_\lambda(\lambda|C_0)$. At this stage it is customary to make an approximation to complete the evaluation. Specifically, it is assumed that the waveform $\gamma c_0(t)$ and the complex noise $c_\eta(t)$ are constant (in time) in the interval $[0, T)$, so that the intensity becomes

$$\lambda = \frac{1}{h\nu} |\gamma c_0 + c_\eta|^2, \qquad 0 < t < T. \tag{4.61}$$

---

[3] In Chap. 8 we will arrive at the same conclusions in a very easy way, where the counting is obtained by a quantum measurement (see Sect. 8.5).

The average number of photons in the interval $(0, T]$ is then

$$\Lambda = \lambda T = \frac{T}{h\nu} |\gamma_{C_0} + c_\eta|^2 := |\gamma(C_0) + c|^2 \tag{4.62}$$

with

$$\gamma(C_0) := \frac{\gamma_{C_0} T}{\sqrt{h\nu}}, \qquad c := \frac{c_\eta T}{\sqrt{h\nu}}. \tag{4.63}$$

Note that $\gamma(C_0)$, under the condition of a given C_0, becomes a constant (nonrandom) and will be simply denoted by γ.

The above formulation leads to the following explicit result for the probability density of Λ:

Proposition 4.1 *Let Λ be the random variable given by $|\gamma + c|^2$, where γ is a complex constant and $c = a + ib$ is a complex Gaussian random variable with zero average, with a and b statistically independent and with the same variance σ^2. Then the probability density of Λ results in a* **Rice density**, *given by*

$$f_\Lambda(\Lambda|\gamma) = \begin{cases} \dfrac{\Lambda}{\sigma^2} \exp\left(-\dfrac{\Lambda + |\gamma|^2}{\sigma^2}\right) I_0\left(\dfrac{2|\gamma|\sqrt{\Lambda}}{\sigma}\right), & \Lambda \geq 0 \\ 0 & \Lambda < 0 \end{cases} \tag{4.64}$$

where $I_0(\cdot)$ is the Bessel function

$$I_0(x) = \frac{1}{2\pi} \int_0^{2\pi} e^{x\cos\theta} d\theta. \qquad \square$$

For the proof see the book by Gagliardi and Karp [12]. The Rice density is shown in Fig. 4.23 for $\sigma^2 = 1$ and some values of $|\gamma|^2$.

4.7.6 From Rice's Density to Laguerre's Distribution

Note that the intensity λ given by (4.61) is dimensional, being the average number of photon *per second*, while $\Lambda = \lambda T$ is adimensional, and the Rice distribution refers to Λ. The probability density of λ is obtained as $f_\lambda(\lambda) = T f_\Lambda(\lambda T)$. Applying this relation in (4.60) one gets [12]:

Proposition 4.2 *Under the assumptions of the previous proposition, the arrival number n, conditioned by the symbol C_0, has a* **Laguerre distribution** *given by*

$$p_n(k|\gamma) = \frac{\sigma^{2k}}{(\sigma^2 + 1)^{2k+1}} \exp\left(-\frac{|\gamma|^2}{\sigma^2 + 1}\right) L_k\left(-\frac{|\gamma|^2}{\sigma^2(\sigma^2 + 1)}\right) \tag{4.65}$$

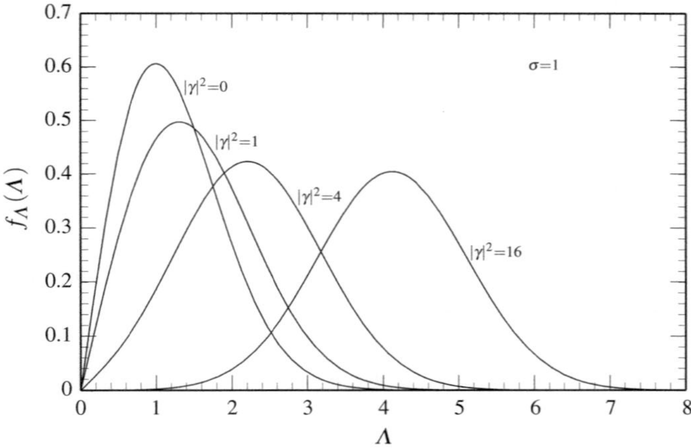

Fig. 4.23 The Rice probability density for $\sigma = 1$ and for some values of $|\gamma|^2$

where $L_k(x)$ is the ordinary Laguerre polynomial. The dependence on the symbol C_0 is obtained when the parameter γ is expressed in the form of (4.63).   □

The Laguerre distribution is illustrated in Fig. 4.24. The corresponding average and variance are given by [12]

$$\overline{n}(\gamma) = |\gamma|^2 + \sigma^2, \qquad \sigma_n^2(\gamma) = |\gamma|^2 + 2|\gamma|^2\sigma^2 + \sigma^2(\sigma^2 + 1). \tag{4.66}$$

At this point it is convenient to get an interpretation of the parameters and also to introduce an alternative notation, related to Quantum Mechanics, which will be used in Chap. 8. The reason for a double notation is to find an alignment between two theories (classical and quantum), which use consolidated different notations. The quantity $|\gamma|^2 = |\gamma(C_0)|^2 = |\gamma_{C_0}|^2 \, T/h\nu$ represents the average number of photons due to the "signal" γ_{C_0} and will be called **number of signal photons**. Analogously, the quantity $\sigma^2 = |c_\eta|^2 T/h\nu$ represents the average number of photons due to the "thermal noise" and will be called **number of thermal photons**. The alternative notations are

$$N_\gamma := |\gamma|^2 \quad \text{and} \quad \mathcal{N} := \sigma^2. \tag{4.67}$$

Then the global (average) number of photons is written as

$$\overline{n}(\gamma) := E[n|\gamma] = |\gamma|^2 + \sigma^2 = N_\gamma + \mathcal{N}. \tag{4.68}$$

Note that the variance can be written in the form

$$\sigma_n^2(\gamma) = \overline{n}(\gamma) + 2|\gamma|^2\sigma^2 + \sigma^4 = \overline{n}(\gamma) + 2N_\gamma^2\,\mathcal{N} + \mathcal{N}^2 \tag{4.69}$$

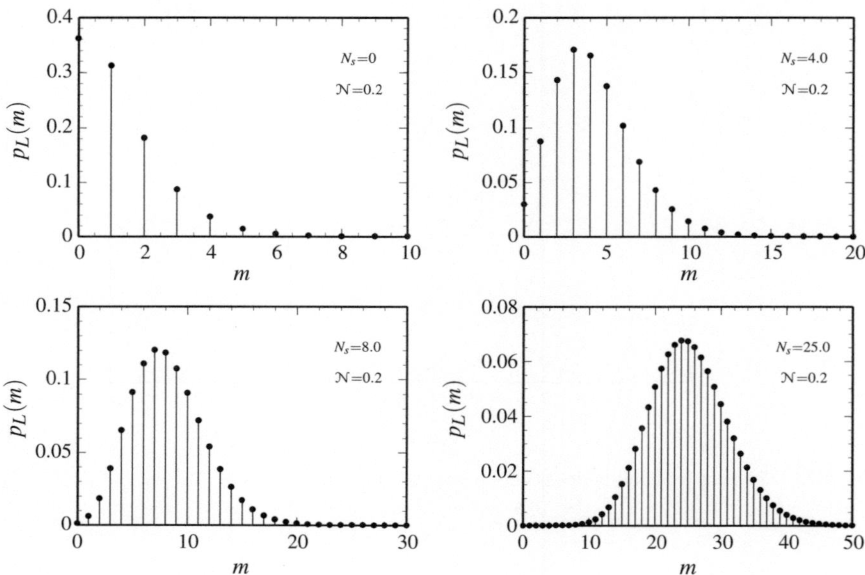

Fig. 4.24 Laguerre distribution $p_n(k)$, which describes the photon counting n in the presence of both signal and thermal noise, shown for $\mathcal{N} = 0.2$ and some values of N_γ. For $N_\gamma = 0$ (absence of signal) the Laguerre distribution degenerates to a geometric distribution

which emphasizes the difference with respect to the Poisson distribution, where average and variance coincide. Another comparison is obtained with the variance in the absence of thermal noise, namely

$$\sigma^2_{\text{Laguerre}} = \sigma^2_{\text{Poisson}} + \mathcal{N}\left(1 + 2N_\gamma^2 + \mathcal{N}\right) \text{ with } \sigma^2_{\text{Poisson}} = |\gamma|^2 = N_\gamma. \tag{4.70}$$

The Laguerre distribution (4.65) can be rewritten in terms of the average photon numbers N_γ and $\mathcal{N}$ as

$$p_n(k|\gamma) = \frac{\mathcal{N}^k}{(\mathcal{N}+1)^{k+1}} \exp\left(-\frac{N_\gamma}{\mathcal{N}+1}\right) L_k\left(-\frac{N_\gamma}{\mathcal{N}(\mathcal{N}+1)}\right). \tag{4.71}$$

The distribution has two important degenerate cases. When $N_\gamma = 0$, that is, in the absence of signal, $p_n(k|\gamma)$ becomes a geometrical distribution

$$p_n(k|0) = \frac{\mathcal{N}^k}{(\mathcal{N}+1)^{k+1}} \tag{4.72}$$

describing the photon numbers of the thermal noise. When $\mathcal{N}$ approaches zero $p_n(k|\gamma)$ becomes a Poisson distribution

$$\lim_{\mathcal{N}\to 0} p_n(k|\gamma) = \exp(-N_\gamma)\frac{N_\gamma^k}{k!} \tag{4.73}$$

describing the photon number in the absence of thermal noise.

4.8 Simplified Theory of Photon Counting and Implementation

In this section we consider the photon counting from a simpler viewpoint, and we will arrive at a possible electronic implementation of the counter. The formulation is different from that of Sect. 4.7.5, where signal and noise are jointly considered in a filtered Poisson process, leading to a Laguerre distribution for counting. Here, only the signal is modeled as a filtered Poisson process, while the noise is added as a Gaussian process.

Remarkable in this counting is the quick transition from the Poisson to the Gaussian regime with increasing Gaussian noise amount.

4.8.1 Counting with Uncorrupted Signal

First we consider an ideal photon counter, which receives at the input an instantaneous optical power $p_R(t)$ of the form

$$p_R(t) = \sum_i (h\nu)\,\delta(t - t_i)$$

and gives at the output the number of photons counted in a symbol period. Figure 4.25 shows the counting in the period $(0, T]$, where the counting starts at $t = 0$, it counts $n(t) := n(0, t]$ and terminates at $t = T$; then it is reset for the detection of a new symbol. Substantially, the counter performs the integration

$$n := n(0, T] = \frac{1}{h\nu}\int_0^T p_R(t)\mathrm{d}t.$$

From Theorem 4.1 one deduces that the probability distribution of the counting n is given by a Poisson distribution with mean

$$N_R = \int_0^T \lambda(t)\,\mathrm{d}t = \frac{1}{h\nu}\int_0^T P_R(t)\,\mathrm{d}t = \frac{E_T}{h\nu}$$

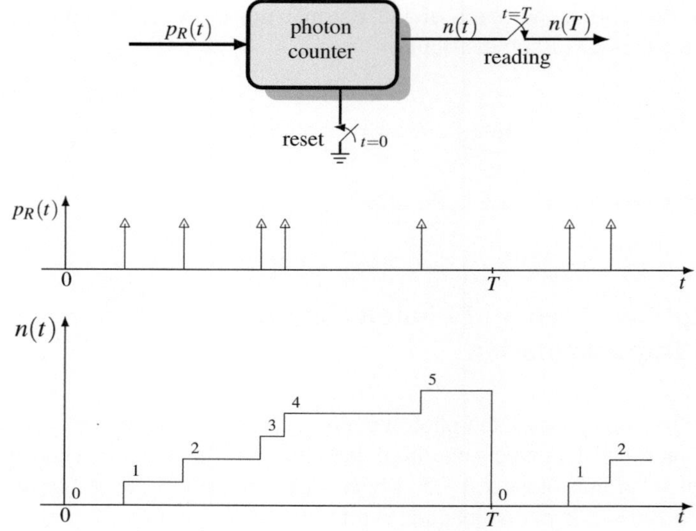

Fig. 4.25 Photon counter with input the instantaneous optical power $p_R(t)$ consisting of quanta $h\nu$ at random times t_i. The counting starts at $t = 0$ and gives the number of quanta $n(t)$ in $(0, t]$. At $t = T$ the counting gives the number of photons in the symbol period $[0, T)$ and is reset for the detection of the next symbol

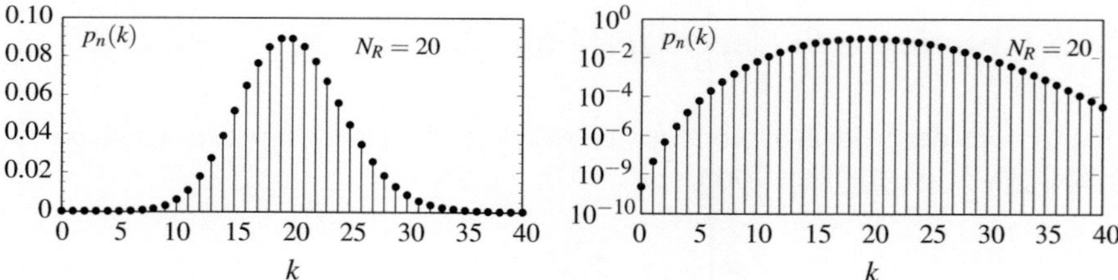

Fig. 4.26 The Poisson probability distribution with mean $N_R = 20$, in linear scale and logarithmic scale

where E_T is the optical energy in the interval $(0, T]$. Then the distribution is explicitly given by

$$p_n(k) = \mathrm{P}[n = k] = \mathrm{e}^{-N_R} \frac{(N_R)^k}{k!}, \qquad k = 0, 1, 2, \ldots \qquad (4.74)$$

and is illustrated in Fig. 4.26 both in linear and in logarithmic scale for $N_R = 20$ photons. Note in particular that the probability of $n = 0$ is $0.5 \, 10^{-10}$.

The counting n can be decomposed into its mean N_R and the deviation $\Delta n = n - N_R$, from N_R, which represents the *shot noise*. Considering that n is a Poisson variable, the variance of the shot noise is given by the mean, that is, $\sigma^2_{\Delta n} = N_R$.

Note the meaning of N_R as *number of signal photons* ($N_R = N_\gamma$).

4.8.2 Presence of a Gaussian Noise

Now we suppose that the counter output n is corrupted by a Gaussian noise η with a given variance σ^2_η, and we investigate the statistics of the random variable

$$u = n + \eta.$$

While n is a nonnegative integer, the presence of the Gaussian noise makes u a continuous random variable. Assuming n and η independent, the probability density of u is given by the convolution

$$f_u(a) = \int_{-\infty}^{+\infty} f_n(a - b)\, f_\eta(b)\, \mathrm{d}b.$$

Now the probability densities of n and η are, respectively,

$$f_n(a) = \sum_{m=0}^{\infty} p_n(m)\, \delta(a - m)$$

$$f_\eta(a) = \frac{1}{\sqrt{2\pi}\,\sigma_\eta}\, e^{-\frac{a^2}{2\sigma_\eta^2}} = \frac{1}{\sigma_\eta} \varphi\left(\frac{a}{\sigma_\eta}\right).$$

where $p_n(k)$ is given by (4.74) and $\phi(x) = (1/\sqrt{2\pi})\exp(-x^2/2)$ is the normalized Gaussian density function. Then the convolution gives

$$\begin{aligned}
f_u(a) &= \sum_{k=0}^{\infty} p_n(k)\, \frac{1}{\sigma_\eta} \varphi\left(\frac{a - k}{\sigma_\eta}\right) \\
&= \sum_{k=0}^{\infty} e^{-N_R}\, \frac{N_R^k}{k!}\, \frac{1}{\sigma_\eta} \varphi\left(\frac{a - k}{\sigma_\eta}\right),
\end{aligned} \tag{4.75}$$

which provides the *exact* statistical description of the global variable $u = n + \eta$.

The parameters in (4.75) are: the mean N_R of the Poissonian component n and the standard deviation σ_η of the Gaussian component. We expect to find: for $\sigma_\eta << N_R$ the dominance of the Poissonian component, and for $\sigma_\eta >> N_R$ the dominance of the Gaussian component. For a comparison it is convenient to consider the *Gaussian approximation* of $u = n + \eta$, where also η is considered to be Gaussian, so that u

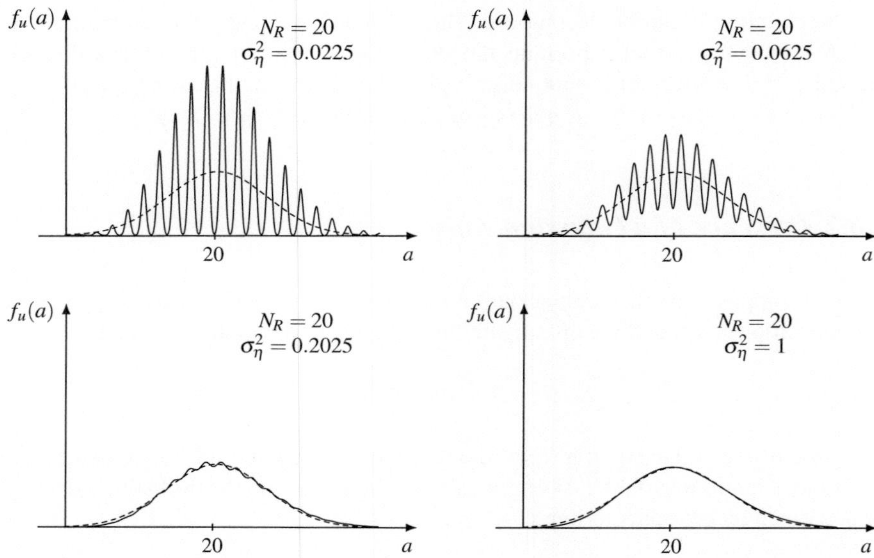

Fig. 4.27 Probability density of the random variable $u = n + \eta$ obtained with $N_R = N_\gamma = 20$ signal photons and four values of the number of thermal photons $\sigma_\eta^2 = \mathcal{N}$. The density is compared with its Gaussian approximation (*dashed line*)

has a Gaussian density with mean N_R and variance $\sigma_u^2 = N_R + \sigma_\eta^2$, that is,

$$f_u(a)_{\text{Gauss}} = \frac{1}{\sqrt{N_R + \sigma_\eta^2}}\varphi\left(\frac{a - N_R}{\sqrt{N_R + \sigma_\eta^2}}\right).$$

The comparison between the exact density and the Gaussian approximation is made in Fig. 4.27 for the fixed value of $N_R = 20$ and a few values of σ_η. Thus we realize that the transition from the Poisson regime to the Gaussian regime happens very quickly and with σ_η as small as $\sigma_\eta = 1$ the global variable is practically Gaussian.

Note that with the symbolism of the previous section:

$$N_R = N_\gamma = \text{number of signal photons}, \qquad \sigma_\eta^2 = \mathcal{N} = \text{number of thermal photons}.$$

Note also that the global output $u = n + \eta$ is not a counting variable, because it is continuous. To get an estimate of the counting statistics we have to round u to an integer. See Problem 4.11, where it is shown that u approximates a Laguerre variable.

4.8.3 Electronic Implementation of a Photon Counter

An implementation of a photon counter requires the transition from the optical domain to the electronic domain. This is provided by a photodiode which converts the optical power $p_R(t)$ to an electrical current $i(t)$, as shown in Fig. 4.28, where the photodiode is represented by its equivalent circuit. An incoming pulse of optical power $(h\nu)\,\delta(t - t_k)$ produces a pulse of electrical current $i_0(t - t_k)$ at the photodiode output; the pulse has an area equal to the electronic charge e. Then an amplifier–integrator converts the electrical current to a voltage $v(t)$. If the time constant $R_s\,C_d$ of the equivalent circuit is small with respect to the symbol period T, the current pulses have a small duration and the integration in the period $[0, T)$ gives

$$v(T) = K_I \int_0^T i(t)\,\mathrm{d}t = K_I \int_0^T \sum_k i_0(t - t_k)\,\mathrm{d}t = K_I\,e\,n = v_1 n \qquad (4.76)$$

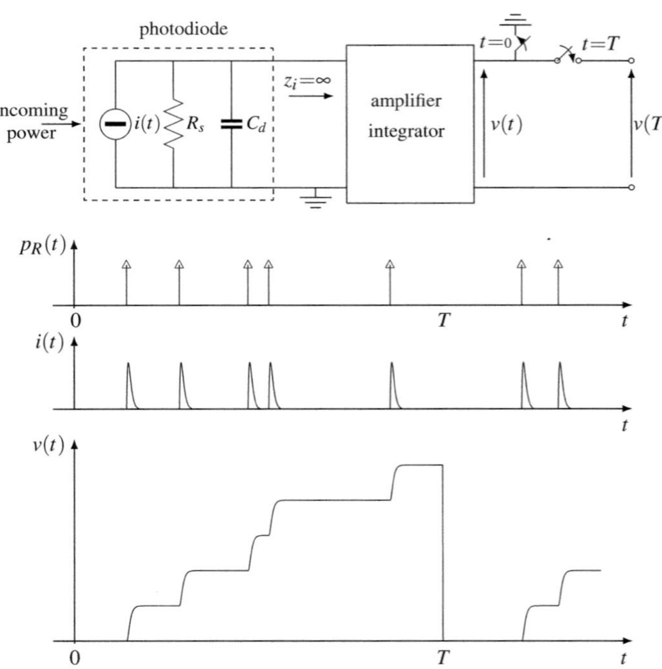

Fig. 4.28 Implementation of photon counter through a photodiode and an integrate-and-dump circuit. The photon counting is made through an electron counting. The photodiode is represented by its equivalent circuit

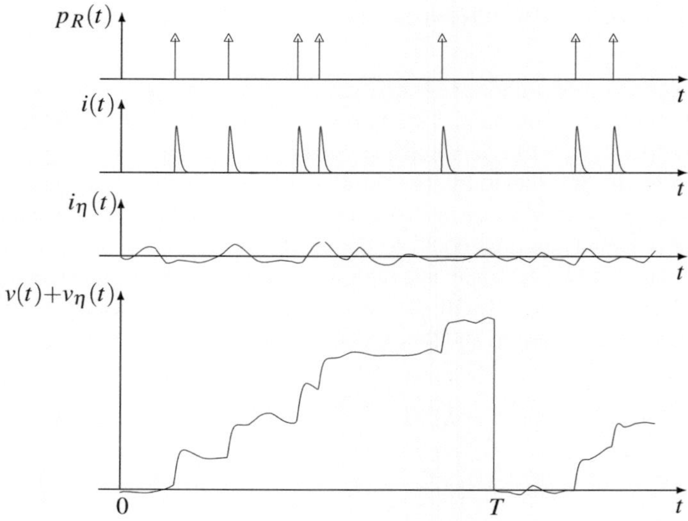

Fig. 4.29 Photon counter through the electron counting in the presence of thermal noise. The thermal noise current $i_\eta(t)$ is produced by the passive part of the photon circuitry

where n is the number of current pulses generated in the interval $(0, T]$. In (4.76) K_I is the integrator constant, e is the electron charge, and $v_1 = K_I e$ gives the contribution of each pulse.

In conclusion, in this implementation the photons are counted through the number of electrons they produce. For simplicity we have assumed that the photodiode has a unitary efficiency, so that we have the conversion $1h\nu \rightarrow 1e$, without any loss.

But to complete the analysis of the implementation we have to take into account the *thermal noise* produced by the passive parts of the circuit, and we have to add to the current $i(t)$, produced by the optical power, a contribution $i_\eta(t)$ due to thermal noise, as shown in Fig. 4.29. Thus at the integrator output at the end of the symbol period we find the voltage

$$v(T) + v_\eta(T) = v_1(n + \eta), \tag{4.77}$$

where $v_\eta(T)$ is the contribution of the thermal noise given by $v_\eta(T) = K_I \int_0^T i_\eta(t) \, \mathrm{d}t$ which is surely Gaussian and zero-mean. After the normalization we have the form $u = n + \eta$ of a Poisson variable corrupted by a Gaussian noise, as discussed above.

Problem 4.11 $\star$ Consider the counting of the random variable $u = n + \eta$, where n is Poissonian with mean N_R and η is Gaussian with zero mean and variance σ_η^2, with u and η independent. Since u is continuous, for the counting we have to introduce a rounding. Find the probability distribution of $v = \mathrm{round}(u)$.

References

1. B.M. Oliver, J.R. Pierce, C.E. Shannon, The philosophy of PCM. Proc. IRE **36**(11), 1324–1331 (1948)
2. C.E. Shannon, A mathematical theory of communication. Bell Syst. Tech. J. **27**(3), 379–423 (1948)
3. A.J. Viterbi, J.K. Omura, *Principles of Digital Communication and Coding. Dover Books on Electrical Engineering* (Dover Publications, New York, 2009)
4. A.S. Holevo, V. Giovannetti, Quantum channels and their entropic characteristics. Rep. Prog. Phys. **75**(4), 046001 (2012)
5. E. Parzen, *Stochastic Processes* (Holden Day, San Francisco, 1962)
6. G. Cariolaro, P. Franco, M. Midrio, G. Pierobon, A probabilstic model of traveling wave optical amplification. Opt. Commun. **95**(4–6), 311–318 (1993)
7. G. Cariolaro, R. Corvaja, G. Pierobon, Exact performance evaluation of lightwave systems with optical preamplifier. Eur. Trans. Telecommun. **5**(6), 757–766 (1994)
8. G. Cariolaro, P. Franco, M. Midrio, G. Pierobon, Complete statistical characterization of signal and noise in optically amplified fiber channels. IEEE J. Quantum Electron. **31**(6), 1114–1122 (1995)
9. R. McIntyre, The distribution of gains in uniformly multiplying avalanche photodiodes: theory. IEEE Trans. Electron Devices **19**(6), 703–713 (1972)
10. R.J. Glauber, The quantum theory of optical coherence. Phys. Rev. **130**, 2529–2539 (1963)
11. J. Shapiro, Quantum noise and excess noise in optical homodyne and heterodyne receivers. IEEE J. Quantum Electron. **21**(3), 237–250 (1985)
12. R. Gagliardi, S. Karp, *Optical Communications* (Wiley, New York, 1995)
13. D.J. Sakrison, *Communication Theory: Transmission of Waweforms and Digital Information* (Wiley, New York, 1968)

Chapter 5
Quantum Decision Theory: Analysis and Optimization

5.1 Introduction

We consider **the transmission of classical information through a quantum channel**, where the information carrier is given by quantum states. A system that achieves this target is called *Quantum Communications system*. Like in classical communications, in quantum communications the usual configuration applies: transmitter, channel, and receiver. Analog quantum transmission systems have been considered too [1], but, as seen in the previous chapter, according to the current trend, we limit ourselves exclusively to digital systems. So Fig. 5.1 illustrates a quantum digital system, emphasizing its essential components.

In this chapter we will develop the theory of decision applied to the combination of the quantum measure and the decision element, without any specification on the nature of the quantum states. In the following chapters the quantum decision theory will be applied to the systems in which the states are physically produced by a coherent monochromatic radiation (coherent or Glauber states).

5.1.1 General Description of a Digital Transmission System

We consider the transmission of a single[1] classical symbol $A \in \mathcal{A}$. Thus, a classical source emits a symbol among K possible symbols, $A \in \mathcal{A} = \{0, 1, \ldots, K-1\}$, with assigned *a priori* probabilities

$$q_i := \mathrm{P}[A = i], \qquad i \in \mathcal{A}. \tag{5.1}$$

---

[1] In Sect. 4.2 we justified the advantage of dealing with a single symbol instead of a sequence of symbols.

© Springer International Publishing Switzerland 2015
G. Cariolaro, *Quantum Communications*, Signals and Communication Technology,
DOI 10.1007/978-3-319-15600-2_5

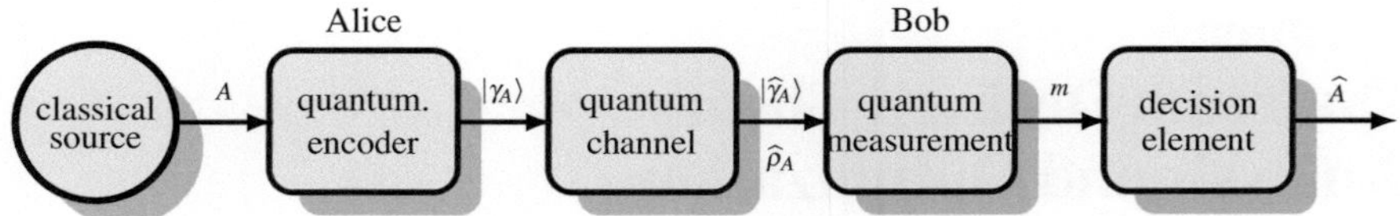

Fig. 5.1 Quantum communication system for the transmission of *classical information* through a *quantum* channel. The transmission of a single digital symbol $A \in \mathcal{A}$ is considered. In transmission a pure state $|\gamma_A\rangle$ is assumed, while in reception the quantum state may be a pure state, $|\widehat{\gamma}_A\rangle$, or a mixed state, $\widehat{\rho}_A$

The transmitter (Alice) encodes the symbol A into a quantum state $|\gamma_A\rangle$ of a Hilbert space $\mathcal{H}_T$, thus realizing the classical-to-quantum mapping $A \rightarrow |\gamma_A\rangle$. This implies that Alice is capable of preparing the quantum system $\mathcal{H}_T$ in K distinct quantum states

$$|\gamma_0\rangle, |\gamma_1\rangle, \ldots, |\gamma_{K-1}\rangle \tag{5.2}$$

which must be considered as *pure*, since they are known to Alice, because she prepares the specific state $|\gamma_i\rangle$ when the source emits the symbol $A = i$. The pure state (ket) prepared by Alice is alternatively described by the density operator $\rho_i = |\gamma_i\rangle\langle\gamma_i|$.

The channel, be it an optical fiber or the free space, modifies the density operators, introducing noise and distortion, so that the received state is in general a mixed state described by a density operator $\widehat{\rho}_A$. Then the channel performs the quantum-to-quantum mapping $\rho_A \rightarrow \widehat{\rho}_A$. As we shall see in Chap. 12, a quite general model to represent explicitly this mapping is given by the the Kraus representation [2]

$$\widehat{\rho}_A = \sum_k V_k^* \, \rho_A \, V_k \tag{5.3}$$

where $\{V_k\}$ is a class of operators.

The receiver (Bob) performs a quantum measurement on the received state $\widehat{\rho}_A$, and, to this end, he must choose a system of measurement operators $\{P_k, k \in \mathcal{M}\}$, which in general are POVM, and, in particular, projectors (see Sect. 3.8). The outcome of the measurement m is a new discrete random variable with alphabet $\mathcal{M}$, which can be seen as the *received signal*, or better, in the language of telecommunications, the *signal at the decision point*. Finally, according to the outcome m of the measurement, a decision must be made, based on a *decision criterion*, to select the symbol $\widehat{A} \in \mathcal{A}$ that was presumably transmitted. Globally, the quantum measurement combined with the decision element provides the quantum-to-classical mapping $\widehat{\rho}_A \rightarrow \widehat{A}$.

Note on symbolism. The alphabet $\mathcal{A}$ of the symbols is indicated in the form

$$\mathcal{A} = \{0, 1, \ldots, K-1\}$$

but it can take other forms (also with complex symbols) related to the modulation format. The alphabet of the measurements $\mathcal{M}$ can be different, even in cardinality, from the alphabet of the source, but as will be seen in the next section, it is not restrictive to assume that the two alphabets coincide, then, in the analysis of specific systems, we will assume $\mathcal{M} = \mathcal{A}$.

Guidelines and Preview of the Chapter

As we want to adopt a very general and complete approach, the chapter may appear long and complex. We encourage the reader to tackle it gradually, restricting the first reading to the concepts related to decision with pure states, and skipping decision based on mixed states. So the study of Chap. 7 is quite feasible, as, for its comprehension, the decision theory based on pure states is sufficient. Later on, the study can be resumed and completed, going through the decision with mixed states, a subject necessary for a full understanding of Chap. 8. Another suggestion is to read this chapter again after viewing the applications of quantum decision to quantum communications systems, developed in Chaps. 7 and 8.

We now detail the line followed in this chapter for the Quantum Decision Theory, but before we remark that this theory, here presented in the language of Telecommunications, is an important and autonomous field of Quantum Mechanics, which could be presented independently of quantum communications systems (and, in fact, in Quantum Mechanics the quantum communications systems are often ignored).

The chapter is organized in four topics.

Analysis of Quantum Decision (Sects. 5.2 to 5.7)

We begin with the *Analysis* of a general quantum communications system, where the target is the evaluation of the system's performance in terms of probabilities. Then, we deal with a specific case to let the reader become familiar with the main concepts introduced: the optimization of a binary system following *Helstrom's theory*.

Optimization of Quantum Decision (Sects. 5.8 to 5.11)

We give a general formulation of *Quantum Optimization*, which has the target of finding the measurement operators that ensure the "best performance," that is, the *maximum correct decision probability*. Optimization may be viewed in the framework of *convex linear programming* and appears to be a formidable problem because the unknowns are the measurement operators, which have severe constraints. Two main results are Holevo's and Kennedy's theorems, which provide conditions that the measurement operators must meet to be optimal.

Geometrically Uniform Symmetry (GUS) (Sects. 5.13 and 5.14)

The GUS is verified in several quantum communications systems and facilitates, in general, analysis and performance evaluation, in particular, optimization and suboptimization. We first consider the GUS for pure states and then for mixed states.

State Compression in Quantum Detection (Sect. 5.15)

In general, quantum states and measurement operators are "redundant," but it is possible and convenient to perform a compression onto a "compressed" space, where redundancy is removed. Quantum detection can be reformulated in the "compressed" space, getting properties simpler than in the original Hilbert space.

5.2 Analysis of a Quantum Communications System

In the *Analysis* of a general quantum communications system, it is assumed that both the transmitter (Alice) and the receiver (Bob) are assigned and the target consists in finding the statistical description of the system's behavior **in terms of probabilities**, exactly as in a classical communications system.

In a quantum system, probabilities come into play in two ways, and, in fact we have two sources of randomness. One is related to the source of information, which emits a symbol $A = i \in \mathcal{A}$ with a given probability $q_i = \mathrm{P}[A = i]$, which is called *a priori probabilities*. Therefore, we have a probability distribution $q_i, i \in \mathcal{A}$ of the random variable A. The other form of randomness is related to the quantum measurement, which produces another random variable $m \in \mathcal{M}$, whose statistical description is provided by Postulate 3 of Quantum Mechanics seen in Sect. 3.5. Then the Analysis of the system will be necessarily based on Probability Theory.

Next, we have to study the viewpoint of Bob, who receives a "signal" and performs the measurement. About this we can make two different hypotheses:

(1) The signal has not been contaminated, so that Bob receives the state $|\gamma_A\rangle$ that Alice associated to the symbol A.
(2) The signal has been contaminated by the channel and by *thermal noise* (also called *background noise*), and therefore Bob does not see the pure state $|\gamma_A\rangle$ any more, but instead a mixed state represented by a density operator ρ_A.

The two cases are illustrated in Fig. 5.2.

Case (1) corresponds to a transmission with an ideal noiseless channel, whereas case (2) accounts for the fact that the channel can fail to be ideal and noiseless. It is important to observe that also in case (1) Bob will not be able to make with certainty a correct decision, because it would be based on quantum measurements, which, as already seen, do not give error-free results; in the classical case, the randomness of the measurement with pure states corresponds to *shot noise*.

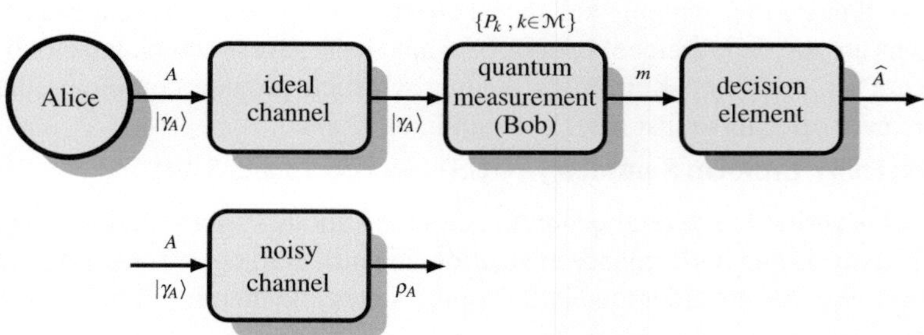

Fig. 5.2 Transmission of a classical symbol A through a quantum channel. At reception Bob performs the measurement in a quantum system in a pure state $|\gamma_A\rangle$ (ideal channel) or in a quantum system in a mixed state ρ_A (noisy channel)

The two cases can be unified considering that also in case (1), to the pure state $|\gamma_A\rangle$ one can associate the degenerate density operator $\rho_A = |\gamma_A\rangle\langle\gamma_A|$.

5.2.1 Quantum Measurement

To perform a quantum measurement, Bob chooses a *measurement operator system*

$$\{P_k, k \in \mathcal{M}\}.$$

From Postulate 3, if we know that the system under measurement is in the state $|\gamma_A\rangle$, the probability that the result of the measurement be $m = k$, is given by (see (3.26) and (3.50))

$$P[m = k|\ \gamma_A] = \langle\gamma_A|P_k|\gamma_A\rangle, \qquad k \in \mathcal{M}. \tag{5.4}$$

Clearly, this result holds if the state $|\gamma_A\rangle$ is known with certainty (pure state). If, instead, the system state is only statistically known through the density operator ρ_A (mixed state), the probability that the result of the measurement be $m = k$ is calculated according to (see (3.32) and (3.51))

$$P[m = k|\ \rho_A] = \mathrm{Tr}[\rho_A P_k], \qquad k \in \mathcal{M}. \tag{5.5}$$

Relation (5.5) includes relation (5.4), because it holds even when the system state is known, thus $\rho_A = |\gamma_A\rangle\langle\gamma_A|$ and then it suffices to recall the identity on the trace (2.37), to obtain (5.4) from (5.5).

In quantum communications systems, we must apply (5.4) when we neglect *thermal noise*, and (5.5) when we take it into account.

5.2.2 The Digital Channel from the Source to the Measurement

The steps that go from the transmitted symbol $A \in \mathcal{A}$ to the outcome of the measurement $m \in \mathcal{M}$ identify a *digital channel*, as shown in Fig. 5.3. The alphabet at the input of this channel is that of the possible symbols of the source

$$\mathcal{A} = \{0, 1, \ldots, K - 1\}, \tag{5.6}$$

whereas at the output we have the alphabet $\mathcal{M}$, which gives the possible outcomes of the measurements and can be indicated in the form

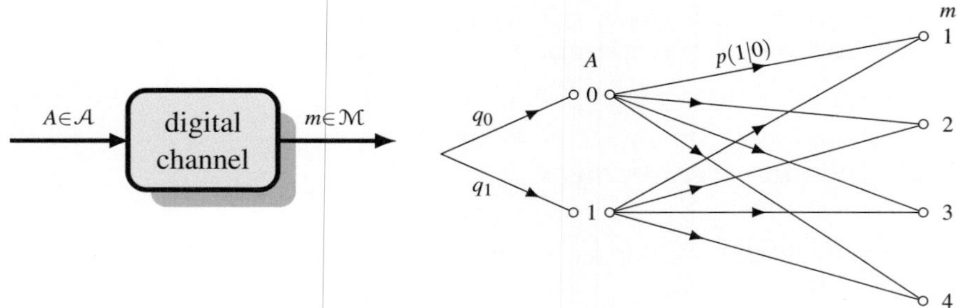

Fig. 5.3 Source-to-measurement digital channel with transition probabilities $p(k|i) = \mathrm{P}[m = k|A = i]$. In the graph, the source alphabet is $\mathcal{A} = \{0, 1\}$ and the measurement alphabet is $\mathcal{M} = \{1, 2, 3, 4\}$

$$\mathcal{M} = \{1, 2, \ldots, K'\} \tag{5.7}$$

where the cardinality K' may be different from K.

The transition probabilities of this channel are given by (5.4) or by (5.5). In fact, in the former case, thinking in terms of Probability Theory, the *event* $\{|\gamma_A\rangle = |\gamma_i\rangle\}$ coincides with the *event* $\{A = i\}$, because Alice has "prepared" the quantum system in the state $|\gamma_i\rangle$, having observed that $A = i$. Therefore, $\mathrm{P}[m = k|\,A = i] = \mathrm{P}[m = k|\,|\gamma\rangle = |\gamma_i\rangle)]$ and the transition probabilities of the channel become

$$p(k|i) := \mathrm{P}[m = k|\,A = i] = \langle\gamma_i|P_k|\gamma_i\rangle, \qquad k \in \mathcal{M}, i \in \mathcal{A}. \tag{5.8a}$$

Even in the latter case, Alice has prepared the system in the state $|\gamma_A\rangle$. However, because of the presence of noise, the state is not pure any more, but it is described by the density operator ρ_A. However, at the level of events, we still have that to $\{A = i\}$ it uniquely corresponds $\{\rho_A = \rho_i\}$, thus

$$p(k|i) := \mathrm{P}[m = k|\,A = i] = \mathrm{Tr}[\rho_i P_k], \qquad k \in \mathcal{M}, i \in \mathcal{A}. \tag{5.8b}$$

As usual, (5.8b) represents the general case, as it yields (5.8a) assuming $\rho_i = |\gamma_i\rangle\langle\gamma_i|$.

It remains to observe that, for the sake of generality, we have chosen a measurement alphabet $\mathcal{M}$, in general different from the source alphabet $\mathcal{A}$ of the symbols. For example, in Fig. 5.3 we have $\mathcal{A} = \{0, 1\}$ and $\mathcal{M} = \{1, 2, 3, 4\}$. The important constraint is that the cardinality of $\mathcal{M}$ must not be smaller than that of $\mathcal{A}$

$$|\mathcal{M}| \geq |\mathcal{A}| \qquad \rightarrow \qquad K' \geq K.$$

As we will see, the two alphabets are often chosen coincident.

5.2.3 Post-measurement Decision. Correct Decision Probability

Remaining in the general case, the decision criterion after the measurement must be expressed by partitioning the measurement alphabet in correspondence with the symbol alphabet, i.e., by finding a *partition* of $\mathcal{M}$ of the type

$$\mathcal{M}_0, \mathcal{M}_1, \ldots, \mathcal{M}_{K-1}. \tag{5.9}$$

Then the *decision criterion* becomes

$$m \in \mathcal{M}_i \quad \Longleftrightarrow \quad \widehat{A} = i. \tag{5.10}$$

For example, in Fig. 5.4, where $\mathcal{A} = \{0, 1\}$ and $\mathcal{M} = \{1, 2, 3, 4\}$, we have chosen the partitions $\mathcal{M}_0 = \{1, 2\}$ and $\mathcal{M}_1 = \{3, 4\}$.

Once chosen the decision criterion, we complete the global digital channel of the quantum system, whose input is the symbol $A \in \mathcal{A}$, and output the symbol $\widehat{A} \in \mathcal{A}$ obtained after the decision (Fig. 5.4). The transition probabilities of this global channel become

$$p_c(j|i) = P[\widehat{A} = j| A = i] = P[m \in \mathcal{M}_j| A = i]$$
$$= \sum_{k \in \mathcal{M}_j} P[m = k| A = i]. \tag{5.11}$$

Therefore, using (5.8b), we have

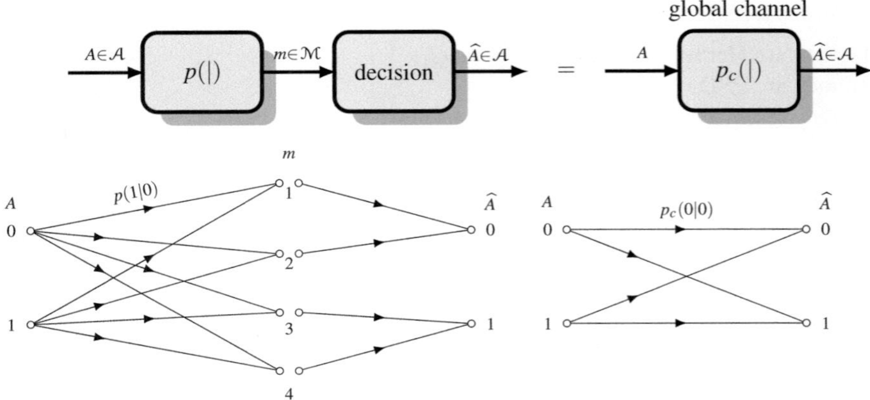

Fig. 5.4 Global digital channel with transition probability $p_c(j|i) = P[\widehat{A} = j|A = i]$

$$p_c(j|i) = \sum_{k \in \mathcal{M}_j} \mathrm{Tr}[\rho_i P_k], \qquad i, j \in \mathcal{A}. \qquad\qquad (5.12)$$

From the global transition probabilities, being also known the a priori probabilities $q_i = \mathrm{P}[A = i]$, we can calculate the *correct decision probability* as

$$P_c = \mathrm{P}[\widehat{A} = A] = \sum_{i \in \mathcal{A}} q_i \, p_c(i|i)$$

$$= \sum_{i \in \mathcal{A}} \sum_{k \in \mathcal{M}_i} q_i \, \mathrm{Tr}[\rho_i P_k] \qquad\qquad (5.13)$$

from which we obtain the *error probability*[2] as $P_e = 1 - P_c$.

5.2.4 Combination of Measurement and Post-measurement Decision

To the purpose of optimization, the decision criterion can be combined with the system of the measurement operators.

Then, given the system of the measurement operators $\{P_k, k \in \mathcal{M}\}$, and the decision criterion determined by the partition (5.9), a set of new operators is defined as follows:

$$Q_i = \sum_{k \in \mathcal{M}_i} P_k, \qquad i \in \mathcal{A}. \qquad\qquad (5.14)$$

The set of the operators $\{Q_i, i \in \mathcal{A}\}$ forms a system of POVMs, that is, with the properties (see Sect. 3.7):

(1) they are Hermitian operators, $Q_i^* = Q_i$,
(2) they are PSD, $Q_i \geq 0$,
(3) they resolve the identity, $\sum_{i \in \mathcal{A}} Q_i = I_{\mathcal{H}}$.

The proof of these properties is based on the fact that the initial operators P_k also have such properties; in particular, (3) is obtained according to

$$\sum_{i \in \mathcal{A}} Q_i = \sum_{i \in \mathcal{A}} \sum_{k \in \mathcal{M}_i} P_k = \sum_{k \in \mathcal{M}} P_k = I_{\mathcal{H}}.$$

Substituting the new operators (5.14) in (5.12) for the transition probabilities, we obtain simply

---

[2] In practice, the performance of a telecommunication system (classical or quantum) is often expressed in terms of the *error probability*, but in theoretical formulation it is more convenient to refer to the *correct decision probability*.

$$p_c(j \mid i) = \mathrm{Tr}[\rho_i Q_j], \qquad j, i \in \mathcal{A}. \tag{5.15}$$

Analogously, the *correct decision probability* becomes

$$P_c = \sum_{i \in \mathcal{A}} q_i \mathrm{Tr}[\rho_i Q_i]. \tag{5.16}$$

In particular, when the system in reception is in a pure state (absence of thermal noise), letting $\rho_i = |\gamma_i\rangle\langle\gamma_i|$ we obtain

$$P_c = \sum_{i \in \mathcal{A}} q_i \langle\gamma_i|Q_i|\gamma_i\rangle. \tag{5.16a}$$

At this point, conceptually, the quantum measurement can be performed directly with the new measurement operators Q_i (global measurement operators), and we obtain directly, as its result, the decided symbol $\widehat{A}$, as illustrated in Fig. 5.5.

In conclusion, we have seen that in principle, in reception, we perform a quantum measurement, followed by a decision, but **it is not restrictive** to include in the measurement also the final post-measurement decision, therefore the choice to make for a good performance affects only the *global measurement operators*.

We finally remark the following statement:

Proposition 5.1 *If the measurement operators $\{P_k, k \in \mathcal{M}\}$ form a projector system, also the global operators $\{Q_i, i \in \mathcal{A}\}$ form a projector system.*

Problem 5.1 ★★ Prove Proposition 5.1. *Hint*: see Sect. 3.6.4.

Problem 5.2 ★★ *Optimization of decision element.* In a post-measurement decision the decision element is a mapping: $\mathcal{M} \rightarrow \mathcal{A}$, where $|\mathcal{M}| \geq |\mathcal{A}|$, in which every point $k \in \mathcal{M}$ must be associated to a symbol $a \in \mathcal{A}$, thus creating a partition of $\mathcal{M}$ into K sets $\mathcal{M}_a, a \in \mathcal{A}$. For given a priori probabilities $\{q_i\}$ and transition probabilities $\{p_c(j|i)\}$, one can optimize the decision element with the criterion to

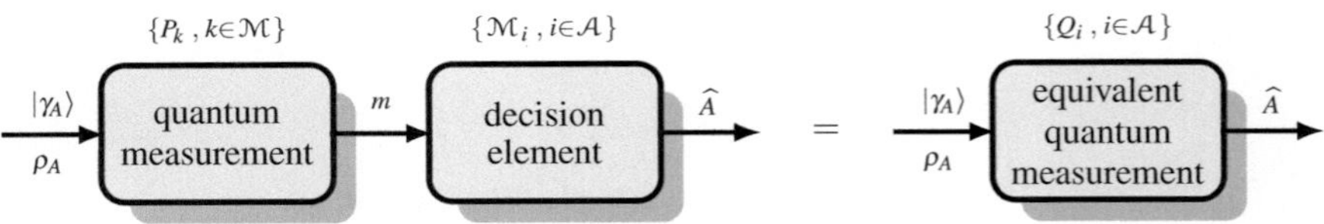

Fig. 5.5 The quantum measurement with the system of measurement operators $\{P_k, k \in \mathcal{M}\}$, followed by the decision element, is equivalent to the measurement with the system of global measurement operators $\{Q_i, i \in \mathcal{A}\}$

get the maximum correct decision probability. Prove the following statement: *Define the K decision functions as*

$$f_a(k) := q_a \, p_c(k|a), \qquad a \in \mathcal{A}, k \in \mathcal{M}.$$

Then, for each $k \in \mathcal{M}$, find the decision function $f_a(k)$ such that

$$f_a(k) \geq f_b(k), \qquad \forall b \neq a. \tag{5.17}$$

The value of a that verifies (5.17) is placed in $\mathcal{M}_a$. This defines the sets $\mathcal{M}_a$ that determine the optimum decision element.

Problem 5.3 ⋆⋆ In a binary system $\{0, 1\}$, where the a priori probabilities are $q(0) = 1/3$ and $q(1) = 2/3$, the quantum measurement, obtained with a photon counting, gives two Poisson variables with averages $\Lambda_0 = \mathrm{E}[m|A = 0] = 5$ and $\Lambda_1 = \mathrm{E}[m|A = 1] = 20$.

Apply the statement of the previous problem to find the optimum decision element.

Problem 5.4 ⋆   As in the previous problem but with $\Lambda_0 = 0$ and $\Lambda_1 = 20$ and equally likely symbols.

5.3 Analysis and Optimization of Quantum Binary Systems

To become familiar with the problem, before proceeding with the general theory, it seems useful to develop explicitly the decision theory in a binary quantum communications system, following the well-known *Helstrom theory* [1]. This theory represents one of the few cases in which explicit closed-form results are obtained.

In a quantum binary system with symbols $A \in \{0, 1\}$ the modulator (Alice) puts the system in one of the two states $|\gamma_0\rangle$ and $|\gamma_1\rangle$. We assume that the measurement alphabet $\mathcal{M}$ is still binary and coincident with the source alphabet, $\mathcal{A} = \mathcal{M} = \{0, 1\}$, and therefore we omit the post-measurement decision element. Then, for the measurement, we need two measurement operators (Hermitian and PSD) Q_0 and Q_1 that maximize the correct decision probability (optimal decision). Given that $Q_0 + Q_1 = I$, we can restrict our search to a single operator, for example, to Q_1.

5.3.1 Optimization with Mixed States (General Case)

We now proceed with the case in which the system is specified by two density operators ρ_0 and ρ_1. To calculate the probability of correct decision we use (5.16), which, as $Q_0 = I - Q_1$, yields

$$P_c = q_0 \, \mathrm{Tr}[\rho_0 Q_0] + q_1 \, \mathrm{Tr}[\rho_1 Q_1]$$
$$= q_0 \, \mathrm{Tr}[\rho_0 \, I] + \mathrm{Tr}[(q_1\rho_1 - q_0\rho_0)Q_1] \tag{5.18}$$
$$= q_0 + \mathrm{Tr}[(q_1\rho_1 - q_0\rho_0)Q_1]$$

where we have taken into account the fact that the trace of a density operator is always unitary (see Sect. 3.3.2). The correct decision probability becomes

$$P_c = q_0 + \mathrm{Tr}[D \, Q_1]$$

where

$$D := q_1\rho_1 - q_0\rho_0 = \widehat{\rho}_1 - \widehat{\rho}_0 \tag{5.19}$$

is called for convenience *decision operator* ($\widehat{\rho}_i = q_i\rho_i$ are *weighted* density operators).

Then, to maximize the correct decision probability, we must find the measurement operator Q_1 such that

$$\max_{Q_1} \mathrm{Tr}[(q_1\rho_1 - q_0\rho_0)Q_1] = \max_{Q_1} \mathrm{Tr}[D \, Q_1] \qquad q_0 + q_1 = 1.$$

To this end, let us consider the eigendecomposition (EID) of the decision operator

$$D = q_1\rho_1 - q_0\rho_0 = \sum_k \eta_k |\eta_k\rangle \langle \eta_k| \tag{5.20}$$

where η_k is the generic eigenvalue, and $|\eta_k\rangle$ the corresponding eigenvector (the η_k are assumed as distinct, so the corresponding vectors $|\eta_k\rangle$ are orthonormal). Note that D is Hermitian but not PSD, so that the η_k are real, but may be either positive or negative. We then have

$$\mathrm{Tr}[D \, Q_1] = \sum_k \eta_k \mathrm{Tr}[|\eta_k\rangle \langle \eta_k|Q_1]$$
$$= \sum_k \eta_k \langle \eta_k|Q_1|\eta_k\rangle, \tag{5.21}$$

where we have used the notable identity (2.37).

Now the crucial point for optimization is to observe that the quantity

$$\varepsilon_k := \langle \eta_k|Q_1|\eta_k\rangle$$

represents the **probability of a measurement obtained through the measurement operator** Q_1 when the system is in the state $|\eta_k\rangle$, and therefore $0 \le \varepsilon_k \le 1$. Then the maximum of the expression (5.21) is obtained by choosing, if possible, the terms with $\eta_k > 0$ and $\varepsilon_k = 1$. This choice is actually possible if we define the measurement operator Q_1 in the following way

$$Q_1 = \sum_{\eta_k > 0} |\eta_k\rangle \langle \eta_k|. \tag{5.22}$$

In fact, with this operator we obtain $\varepsilon_k = \langle \eta_k | Q_1 | \eta_k \rangle = 1$ and the required maximum is

$$\mathrm{Tr}[(q_1 \rho_1 - q_0 \rho_0) Q_1] = \sum_{\eta_k > 0} \eta_k,$$

i.e., it is given by the sum of the positive eigenvalues. With this choice, the maximum correct decision probability becomes

$$P_c = q_0 + \sum_{\eta_k > 0} \eta_k. \tag{5.23}$$

It remains to verify that the two operators obtained through the optimization

$$Q_1 = \sum_{\eta_k > 0} |\eta_k\rangle \langle \eta_k|, \qquad Q_0 = I - Q_1 = \sum_{\eta_k < 0} |\eta_k\rangle \langle \eta_k| \tag{5.24}$$

really form a *measurement operator system*. What is more, it can be shown that Q_1 and Q_0 form a *projector system* (see Problem 5.5).

In conclusion, to obtain the maximum correct decision probability in a binary system, we must perform a *projective* measurement with projectors given by (5.24).

Summary of the Optimization Procedure

We summarize the steps required to find the optimal measurement operators in a quantum binary system:

(1) we start from the EID (5.20) of the decision operator

$$D = q_1 \rho_1 - q_0 \rho_0 = \sum_{k} \eta_k |\eta_k\rangle \langle \eta_k|; \tag{5.25}$$

(2) the optimal measurement operators (projectors) Q_0 and Q_1 are calculated from (5.24);
(3) the maximum probability of a correct decision is simply given by q_0 plus the sum of the positive eigenvalues of the operator D.

It is important to observe that this result is *totally general*, in the sense that no hypothesis has been made on the density operators ρ_0 and ρ_1, which can describe even mixed states. This general result will be applied in Chap. 8 to binary quantum communications systems in the presence of thermal noise.

Problem 5.5 ⋆⋆   Prove that the operators Q_1 and Q_0, defined by (5.24), form a projector system.

Problem 5.6 ⋆⋆   Consider the following density operators:

$$\rho_0 = \frac{1}{208} \begin{bmatrix} 46 & 13-37i & -16 & 13+37i \\ 13+37i & 58 & 13-37i & -32 \\ -16 & 13+37i & 46 & 13-37i \\ 13-37i & -32 & 13+37i & 58 \end{bmatrix}$$

$$\rho_1 = \frac{1}{208} \begin{bmatrix} 58 & 29-29i & 8 & 21+29i \\ 29+29i & 58 & 29-21i & -8 \\ 8 & 29+21i & 46 & 21-21i \\ 21-29i & -8 & 21+21i & 46 \end{bmatrix}$$

First verify that they are "true" density operators. Then, assuming that they are the states in a binary transmission with a priori probabilities $q_0 = 1/5$ and $q_1 = 4/5$, find the correct decision probability P_c.

5.4 Binary Optimization with Pure States

The general theory of the previous section is now applied to a binary quantum system prepared in one of the two pure states $|\gamma_0\rangle$ and $|\gamma_1\rangle$, therefore described by the density operators

$$\rho_0 = |\gamma_0\rangle\langle\gamma_0| \qquad \rho_1 = |\gamma_1\rangle\langle\gamma_1|. \tag{5.26}$$

We will find explicit and very important results, which be applied in Chap. 7 to quantum binary communications systems in the absence of thermal noise.

5.4.1 Helstrom's Bound

To find the optimal measurement operators, we must evaluate the EID of the decision operator, which with pure state is given by

$$D = q_1\rho_1 - q_0\rho_0 = q_1|\gamma_1\rangle\langle\gamma_1| - q_0|\gamma_0\rangle\langle\gamma_0|. \tag{5.27}$$

To comprehend the nature of this operator, consider its image

$$\mathcal{D} = \operatorname{im} D = D\,\mathcal{H}$$

which is a subspace generated by the linear combination of the two kets $|\gamma_0\rangle$ and $|\gamma_1\rangle$, assumed (geometrically) independent, and whose dimension is dim $\mathcal{D} = 2$. Then the EID of D is limited to two terms (only two eigenvalues are different from zero) and the two eigenvectors $|\eta_)\rangle$ and $|\eta_1\rangle$ of D must belong to the subspace $\mathcal{D}$ and therefore are linear combinations of two states[3]

$$|\eta_0\rangle = a_{00}|\gamma_0\rangle + a_{01}|\gamma_1\rangle, \qquad |\eta_1\rangle = a_{10}|\gamma_0\rangle + a_{11}|\gamma_1\rangle. \qquad (5.28)$$

Now, the coefficients a_{ij} are obtained by applying the definition of eigenvector, that is,

$$D\,|\eta_0\rangle = \eta_0\,|\eta_0\rangle, \qquad D\,|\eta_1\rangle = \eta_1\,|\eta_1\rangle \qquad (5.29)$$

where η_0 and η_1 are the eigenvalues. Substituting (5.27) and (5.28) in (5.29), recalling that $\langle\gamma_1|\gamma_1\rangle = \langle\gamma_0|\gamma_0\rangle = 1$ and letting $X = \langle\gamma_0|\gamma_1\rangle$, we obtain

$$q_1(a_{0i}X + a_{1i})|\gamma_1\rangle - q_0(a_{0i} + a_{1i}X^*)|\gamma_0\rangle = \eta_{0i}(a_{0i}|\gamma_0\rangle + a_{1i}|\gamma_1\rangle)), \qquad i = 0, 1. \qquad (5.30)$$

But, because of the assumed *independence*, in (5.30) the coefficients of $|\gamma_1\rangle$ and $|\gamma_0\rangle$ must be equal to zeo. Hence

$$q_1(a_{i\,0}X^* + a_{i\,1}) = \eta_i\,a_{i\,1}, \qquad -q_0(a_{i\,0} + a_{i\,1}X) = \eta_i\,a_{i\,0}, \qquad i = 0, 1\,. \qquad (5.31)$$

Solving with respect to η_i we get the equation

$$\eta_i^2 - \eta_i(q_1 - q_0) - q_0q_1(1 - |X|^2) = 0$$

from which

$$\eta_{0,1} = \tfrac{1}{2}\,(q_1 - q_0 \mp R), \qquad R := \sqrt{1 - 4q_0q_1|X|^2} \qquad (5.32)$$

where $\eta_1 > 0$ and $\eta_0 < 0$.

We have only one positive eigenvalue, and (5.23) gives

$$\boxed{\begin{aligned} P_c &= \tfrac{1}{2}\left(1 + \sqrt{1 - 4q_0q_1|X|^2}\right) \\ P_e &= \tfrac{1}{2}\left(1 - \sqrt{1 - 4q_0q_1|X|^2}\right) \end{aligned}} \qquad (5.33)$$

where the parameter

$$|X|^2 = |\langle\gamma_0|\gamma_1\rangle|^2 \qquad (5.33a)$$

---

[3] This point will be clarified in Sect. 5.11, Proposition 5.4. The eigenvectors $|\eta_i\rangle$ are called *measurement vectors* because they form the measurement operators as $Q_i = |\eta_i\rangle\langle\eta_i|$.

represents the *(quadratic) superposition degree* between the two states. In the literature expressions (5.33) are universally known as **Helstrom's bound**.

The optimal projectors derive from (5.24) and become

$$Q_0 = |\eta_0\rangle\langle\eta_0|, \qquad Q_1 = |\eta_1\rangle\langle\eta_1| \tag{5.34}$$

and therefore they are of the *elementary* type, with **measurement vectors** given by the eigenvectors $|\eta_0\rangle$ and $|\eta_1\rangle$ of the decision operator D.

It remains to complete the computation of these two eigenvectors, identified by the linear combinations (5.28). Considering (5.31) we find

$$|\eta_0\rangle = a_{00}\left(|\gamma_0\rangle + \frac{q_1 X^*}{\eta_0 - q_1}|\gamma_1\rangle\right), \qquad |\eta_1\rangle = a_{11}\left(-\frac{q_0 X}{\eta_1 + q_0}|\gamma_0\rangle + |\gamma_1\rangle\right) \tag{5.35}$$

where a_{00} and a_{11} are calculated by imposing the normalization $\langle\eta_i|\eta_i\rangle = 1$. In the general case, the calculation of the eigenvectors is very complicated[4] and we prefer to carry out the evaluation with the geometric approach developed below.

To consolidate the ideas on quantum detection we anticipate a few definitions and properties on quantum detection and optimization. The linear combination (5.28) can be written in the matrix form[5]

$$M = \Gamma A \quad \text{with} \quad \Gamma = [|\gamma_0\rangle, |\gamma_1\rangle], \ M = [|\mu_0\rangle, |\mu_1\rangle, \ A = \begin{bmatrix} a_{00} & a_{01} \\ a_{10} & a_{11} \end{bmatrix}$$

where Γ is called *state matrix* and M is called *measurement matrix* (see Sect. 5.6). The target of optimization is to find the (optimal) measurement matrix M_{opt} that maximizes the correct decision probability. In Sect. 5.11 we shall see that the optimal measurement vectors are always orthogonal. This property can be written in the form $M_{\mathrm{opt}}^* M_{\mathrm{opt}} = I_2$, where I_2 is the 2×2 identity matrix.

Finally, we note that a quantum system with pure states, say $\mathcal{S}(q, \Gamma)$, is completely specified by the vector of the a priori probabilities q and by the state matrix Γ. The optimization is specified by the measurement matrix M_{opt}, which allows us to find the maximum correct decision probability as

$$P_{e,\max} = \sum_{i=0}^{K-1} |\langle\mu_i|\gamma_i\rangle|^2 = \sum_{i=0}^{K-1} |\langle M_{\mathrm{opt}}(i)|\Gamma(i)\rangle|^2 \tag{5.36}$$

where $|\mu_i\rangle = M_{\mathrm{opt}}(i)$ is the ith element of M_{opt}.

---

[4] To the author's knowledge the general expression of the eigenvectors (with X complex and not equally likely symbols) does not seem to be available in the literature.

[5] The measurement vectors, previously obtained as eigenvectors and denoted by $|\eta_i\rangle$, are hereafter denoted by $|\mu_i\rangle$.

5.4.2 Optimization by Geometric Method

The binary optimization with pure states can be conveniently developed by a geometric approach with several advantages. We first assume that the inner product $Y := \langle \gamma_0 | \gamma_1 \rangle$ is **real** and then we generalize the approach to the complex case.

The geometry of decision with two pure states $|\gamma_0\rangle$ and $|\gamma_1\rangle$ is developed in the subspace $\mathcal{D}$ generated by two states. In this hyperplane, the states are written in terms of an appropriate orthonormal basis $\{|u_0\rangle, |u_1\rangle\}$ as (Fig. 5.6)

$$\begin{aligned}
|\gamma_0\rangle &= \cos\theta\,|u_0\rangle + \sin\theta\,|u_1\rangle \\
|\gamma_1\rangle &= \cos\theta\,|u_0\rangle - \sin\theta\,|u_1\rangle
\end{aligned} \tag{5.37}$$

where

$$\cos 2\theta = \langle \gamma_0 | \gamma_1 \rangle = Y. \tag{5.38}$$

In (5.37) we have assumed that the basis vector $|u_0\rangle$ lies in the bisection determined by the state vectors, which does not represent a restriction. For now we assume the two measurement vectors $|\mu_0\rangle$ and $|\mu_1\rangle$ not necessarily optimal, but satisfying the conditions of being orthonormal, in addition to belonging to the Hilbert subspace $\mathcal{D}$. Then they can be written as

$$\begin{aligned}
|\mu_0\rangle &= \cos\phi\,|u_0\rangle + \sin\phi\,|u_1\rangle \\
|\mu_1\rangle &= \sin\phi\,|u_0\rangle - \cos\phi\,|u_1\rangle.
\end{aligned} \tag{5.39}$$

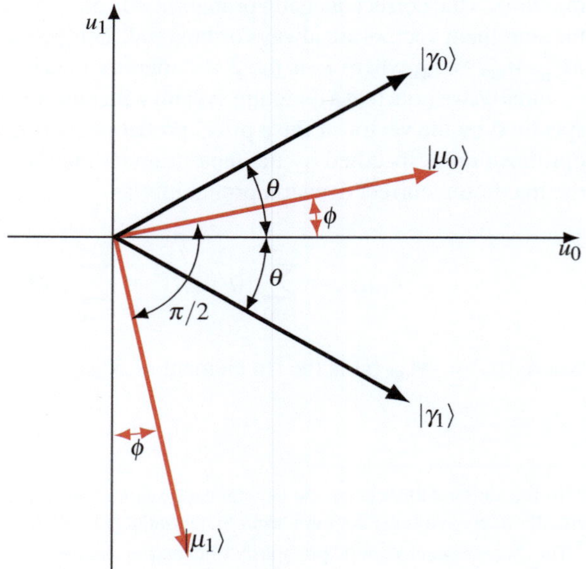

Fig. 5.6 Binary decision with generic state vectors and measurement vectors

Note that trigonometric functions take automatically into account the ket normalization and allow us to express the ket geometry of the four vectors involved by only two angles.

Considering that $\langle \mu_0 | \gamma_0 \rangle = \cos(\phi - \theta)$ and $\langle \mu_1 | \gamma_1 \rangle = \sin(\phi + \theta)$, the transition probabilities $p(j|i) := P[\hat{A}_0 = j | A_0 = i]$ are given by

$$
\begin{aligned}
p(0|0) &= \cos^2(\phi - \theta) = \tfrac{1}{2} [1 + \sin 2\theta \sin 2\phi + \cos 2\theta \cos 2\phi] \\
p(1|1) &= \sin^2(\phi + \theta) = \tfrac{1}{2} [1 + \sin 2\theta \sin 2\phi - \cos 2\theta \cos 2\phi]
\end{aligned}
\tag{5.40}
$$

and the correct detection probability turns out to be

$$
\begin{aligned}
P_c &= q_0 \cos^2(\phi - \theta) + q_1 \sin^2(\phi + \theta) \\
&= \tfrac{1}{2} [1 + (q_0 - q_1)(\cos 2\theta \cos 2\phi + \sin 2\theta \sin 2\phi)].
\end{aligned}
\tag{5.41}
$$

Here the angle θ is given through the inner product Y (see (5.38)), while the angle ϕ is unknown and is evaluated by optimization. It is immediate to see that the angle ϕ giving the maximum of P_c is given by

$$
\tan 2\phi = \frac{1}{q_0 - q_1} \tan 2\theta = \frac{1}{q_0 - q_1} \frac{\sqrt{1 - Y^2}}{Y},
\tag{5.42}
$$

which gives

$$
\sin 2\phi = \frac{1}{R} \sin 2\theta, \quad \cos 2\phi = \frac{q_0 - q_1}{R} \cos 2\theta
\tag{5.43}
$$

where $R = \sqrt{1 - 4q_0 q_1 Y^2}$. The corresponding optimal correct decision probability is

$$
P_c = \tfrac{1}{2}(1 + R) = \tfrac{1}{2} \left(1 + \sqrt{1 - 4q_0 q_1 Y^2} \right),
\tag{5.44}
$$

i.e., the Helstrom bound.

The transition probabilities (5.40), with the optimal decision, become

$$
\begin{aligned}
p(0|0) &= \tfrac{1}{2} \left[1 + (1 - Y^2 + (q_0 - q_1)Y^2)/R \right] \\
p(1|1) &= \tfrac{1}{2} \left[1 + (1 - Y^2 - (q_0 - q_1)Y^2)/R \right].
\end{aligned}
\tag{5.45}
$$

Finally, we consider the explicit evaluation of the optimal measurement vectors. The first step is finding in (5.39) the expression of the basis vectors $|u_0\rangle$ and $|u_1\rangle$ in terms of the given quantum states. For the particular choice made for these vectors we have that $|u_0\rangle$ is proportional to $|\gamma_0\rangle + |\gamma_1\rangle$ and $|u_1\rangle$ is proportional to $|\gamma_0\rangle - |\gamma_1\rangle$ (see Fig. 5.6), that is, $|u_0\rangle = H_0(|\gamma_0\rangle + |\gamma_1\rangle)$ and $|u_1\rangle = H_1(|\gamma_0\rangle - |\gamma_1\rangle)$.

The normalization gives $H_0 = 1/\sqrt{2 + 2Y}$ and $H_1 = 1/\sqrt{2 - 2Y}$. Next, in (5.39) the optimal angle is given by (5.42) and then we find the optimal measurement matrix as[6]

$$M_{\text{opt}} = \Gamma A \quad \text{with} \quad A = \frac{1}{2}\begin{bmatrix} \dfrac{\sqrt{1-L}}{\sqrt{1-Y}} + \dfrac{\sqrt{1+L}}{\sqrt{1+Y}} & \dfrac{\sqrt{1-L}}{\sqrt{1+Y}} - \dfrac{\sqrt{1+L}}{\sqrt{1-Y}} \\[3ex] \dfrac{\sqrt{1+L}}{\sqrt{1+Y}} - \dfrac{\sqrt{1-L}}{\sqrt{1-Y}} & \dfrac{\sqrt{1-L}}{\sqrt{1+Y}} + \dfrac{\sqrt{1+L}}{\sqrt{1-Y}} \end{bmatrix}$$

$$(5.46)$$

where $L = (q_0 - q_1)\, Y/R$. This completes the optimization with a real inner product.

In the general case of a **complex inner product**

$$X = |X|\, e^{i\beta}$$

we introduce the new quantum states

$$|\widetilde{\gamma}_0\rangle = |\gamma_0\rangle, \qquad |\widetilde{\gamma}_1\rangle = e^{-i\beta}|\gamma_1\rangle$$

which give the matrix relation

$$\widetilde{\Gamma} = \Gamma B, \qquad \text{with} \quad B = \begin{bmatrix} e^{-i\beta} & 0 \\ 0 & 1 \end{bmatrix}. \tag{5.47}$$

Now we have two binary systems, $\mathcal{S}(q, \Gamma)$ and $\mathcal{S}(q, \widetilde{\Gamma})$, with the same a priori probabilities, but different inner products, respectively $X = |X|\, e^{i\beta}$ and $\widetilde{X} = \langle \widetilde{\gamma}_0 | \widetilde{\gamma}_1 \rangle = e^{-i\beta}\langle \gamma_0 | \gamma_1 \rangle = |X|$. It is immediate to verify (see (5.36)) that if M_{opt} is the optimal measurement matrix for $\mathcal{S}(q, \Gamma)$, the optimal measurement matrix for $\mathcal{S}(q, \widetilde{\Gamma})$ is given by

$$\widetilde{M}_{\text{opt}} = M_{\text{opt}}\, B \quad \rightarrow \quad M_{\text{opt}} = \widetilde{M}_{\text{opt}}\, B^{-1}. \tag{5.48}$$

But the system $\mathcal{S}(q, \widetilde{\Gamma})$ has a real inner product and, with the replacement $Y \rightarrow |X|$, we can use the previous theory to find: (1) the Helstrom bound from (5.44), (2) the transition probabilities from (5.45) and (3) the optimal measurement matrix $\widetilde{M}_{\text{opt}} = \widetilde{\Gamma}\widetilde{A}$ from (5.46). Hence, from $\widetilde{M}_{\text{opt}}$ we can obtain the measurement matrix for the system $\mathcal{S}(q, \Gamma)$. In fact, by combination of (5.47) and (5.48) we find

$$M_{\text{opt}} = \Gamma A \quad \text{with} \quad A = B\widetilde{A}\, B^{-1}$$

---

[6] To express $\cos \phi$ and $\sin \phi$ from $\tan 2\phi$ we use the trigonometric identities

$$\sin \phi = 2^{-1/2}\sqrt{1 - 1/\sqrt{1 + \tan^2 2\phi}}, \qquad \cos \phi = 2^{-1/2}\sqrt{1 + 1/\sqrt{1 + \tan^2 2\phi}}$$

which hold for $0 \leq \phi \leq \pi/4$. This range of ϕ covers the cases of interest.

which gives

$$A = \begin{bmatrix} a_{00} & a_{01} \\ a_{10} & a_{11} \end{bmatrix} = \begin{bmatrix} \widetilde{a}_{00} & e^{i\beta}\widetilde{a}_{01} \\ e^{-i\beta}\widetilde{a}_{10} & \widetilde{a}_{11} \end{bmatrix}.$$

We summarize the general results as follows:

Proposition 5.2 *The optimization of the quantum decision in a binary system prepared in the pure states $|\gamma_0\rangle$ and $|\gamma_1\rangle$, having inner product $\langle\gamma_0|\gamma_1\rangle := X = |X|e^{i\beta}$ and a priori probabilities q_0 and q_1, gives the transition probabilities*

$$\begin{aligned} p(0|0) &= \tfrac{1}{2}\left[1 + (1 - |X|^2 + (q_0 - q_1)|X|^2)/R\right] \\ p(1|1) &= \tfrac{1}{2}\left[1 + (1 - |X|^2 - (q_0 - q_1)|X|^2)/R\right]. \end{aligned} \tag{5.49}$$

and the correct decision probability

$$P_c = \tfrac{1}{2}\left(1 + \sqrt{1 - 4q_0q_1|X|^2}\right). \tag{5.50}$$

The optimal measurement matrix is obtained as $M_{\mathrm{opt}} = \Gamma A$, where

$$A = \frac{1}{2}\begin{bmatrix} \dfrac{\sqrt{1-L}}{\sqrt{1-|X|}} + \dfrac{\sqrt{1+L}}{\sqrt{1+|X|}} & e^{i\beta}\left(\dfrac{\sqrt{1-L}}{\sqrt{1+|X|}} - \dfrac{\sqrt{1+L}}{\sqrt{1-|X|}}\right) \\[2em] e^{-i\beta}\left(\dfrac{\sqrt{1+L}}{\sqrt{1+|X|}} - \dfrac{\sqrt{1-L}}{\sqrt{1-|X|}}\right) & \dfrac{\sqrt{1-L}}{\sqrt{1+|X|}} + \dfrac{\sqrt{1+L}}{\sqrt{1-|X|}} \end{bmatrix}$$

with

$$R = \sqrt{1 - 4q_0q_1}, \qquad L = |(q_0 - q_1)X|/R.$$

5.4.3 Pure States with Equally Likely Symbols

With equally likely symbols ($q_0 = q_1 = \tfrac{1}{2}$) we find several simplifications. In the trigonometric approach the optimization is obtained by rotating the measurement vectors until they form the same angle with the corresponding state vectors, specifically, we have $\theta = \pi/4$, as shown in Fig. 5.7. The expressions of correct decision probabilities and of the error probability are simplified as

$$P_c = \tfrac{1}{2}\left(1 + \sqrt{1 - |X|^2}\right), \qquad P_e = \tfrac{1}{2}\left(1 - \sqrt{1 - |X|^2}\right). \tag{5.51}$$

Fig. 5.7 Optimal binary decision with equally probable symbols ($q_0 = q_1 = \frac{1}{2}$). The optimization is obtained by rotating the measurement vectors until they form the same angle with the corresponding state vectors

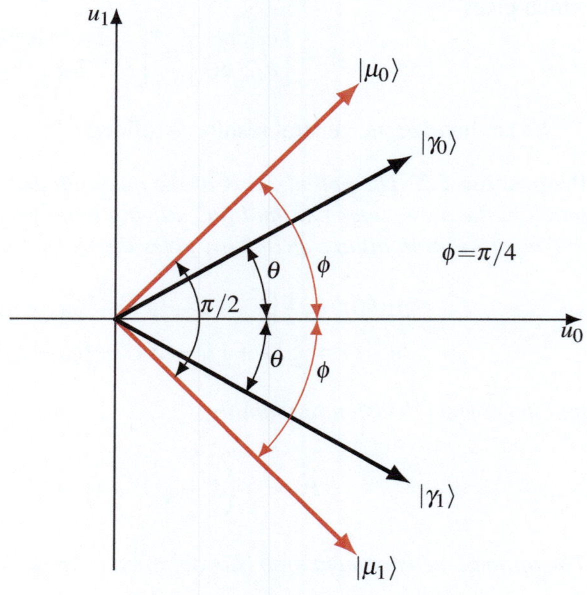

The transition probabilities become equal

$$p(0|0) = p(1|1) = \tfrac{1}{2}\left(1 + \sqrt{1 - |X|^2}\right) = P_c$$

and hence we get a *binary symmetric channel*.

The measurement vectors become

$$|\mu_0\rangle = a\,|\gamma_0\rangle + b\,e^{i\beta}\,|\gamma_1\rangle, \qquad |\mu_1\rangle = b\,e^{-i\beta}\,|\gamma_0\rangle + a\,|\gamma_1\rangle \tag{5.52}$$

where $\beta = \arg X$ and

$$a = \frac{1}{2}\left[\frac{1}{\sqrt{1 - |X|}} + \frac{1}{\sqrt{1 + |X|}}\right], \qquad b = \frac{1}{2}\left[\frac{1}{\sqrt{1 + |X|}} - \frac{1}{\sqrt{1 - |X|}}\right]. \tag{5.53}$$

Problem 5.7 ★★   Find the coefficients a_{01} and a_{11} in the expression of the measurement vectors (5.35), assuming equally likely symbols and X real.

Problem 5.8 ★★   Write the fundamental relations of the geometrical approach in matrix form, using the matrices

$$\Gamma = [|\gamma_0\rangle, |\gamma_1\rangle], \qquad U = [|u_0\rangle, |u_1\rangle], \qquad M = [|\mu_0\rangle, |\mu_1\rangle].$$

5.5 System Specification in Quantum Decision Theory

After a thorough examination of decision in a binary system, we return to the general considerations, assuming a K-ary system.

From the general analysis of Sect. 5.2 and from the choice of proceeding with global measurements, we realize that the system specification in quantum decision can be limited to the following few parameters ("players").

On the transmitter side (Alice), the players are:

(a) the a priori probabilities q_i, $i \in \mathcal{A}$,
(b) the states $|\gamma_i\rangle$, $i \in \mathcal{A}$, or the density operator ρ_i, $i \in \mathcal{A}$.

The sets $\{|\gamma_i\rangle | i \in \mathcal{A}\}$ and $\{\rho_i | i \in \mathcal{A}\}$ will be called *constellations of states*.

At the receiver side (Bob) the players are the (global) measurement operators, which must form a *measurement operator system* $\{Q_i, i \in \mathcal{A}\}$ in the sense already underlined, but worthwhile recalling:

(1) they are Hermitian operators, $Q_i^* = Q_i$,
(2) they are PSD, $Q_i \geq 0$,
(3) they give a resolution of the identity, $\sum_{i \in \mathcal{A}} Q_i = I_{\mathcal{H}}$.

There are several ways to specify the above parameters, as we shall see in the next sections, making the usual distinction between pure and mixed states.

5.5.1 Weighted States and Weighted Density Operators

In the above, the transmitter specification is composed by two players, however, the same specification can be obtained by a single player with the introduction of weighted states (already used in Sect. 3.11).

The *weighted states* are defined by

$$|\widehat{\gamma}_i\rangle = \sqrt{q_i}|\gamma_i\rangle, \qquad i \in \mathcal{A} \tag{5.54}$$

and contain the information of both the probabilities q_i and the states $|\gamma_i\rangle$. In fact, considering that the states are normalized, $\langle \gamma_i | \gamma_i \rangle = 1$, one gets

$$q_i = \langle \widehat{\gamma}_i | \widehat{\gamma}_i \rangle, \qquad |\gamma_i\rangle = (1/\sqrt{q_i}) \, |\widehat{\gamma}_i\rangle. \tag{5.55}$$

The *weighted density operators* are defined by

$$\widehat{\rho}_i = q_i \, \rho_i, \qquad i \in \mathcal{A}. \tag{5.56}$$

Then, considering that $\mathrm{Tr}[\rho_i] = 1$, one gets

$$q_i = \mathrm{Tr}[\widehat{\rho}_i], \qquad \rho_i = (1/q_i) \, \widehat{\rho}_i. \tag{5.57}$$

5.6 State and Measurement Matrices with Pure States

If the decision is taken from pure states, that is, from rank-one density operators, also the measurement operators may be chosen with rank-one, and therefore expressed by *measurement vectors* in the form $Q_i = |\mu_i\rangle\langle\mu_i|$. This was seen in Sect. 5.3 with a binary system, but it holds in general (see Kennedy's theorem in Sect. 5.11). Then, referring to an n-dimensional Hilbert space $\mathcal{H}$, the players become vectors (kets) of $\mathcal{H}$, which can be conveniently represented in the matrix form.

Now, K pure states $|\gamma_i\rangle$, interpreted as column vectors of dimension $n \times 1$, form *the state matrix*

$$\underset{n\times K}{\Gamma} = [|\gamma_0\rangle, |\gamma_1\rangle, ..., |\gamma_{K-1}\rangle]. \tag{5.58}$$

Analogously, the measurement vectors $|\mu_i\rangle$ form the *measurement matrix*

$$\underset{n\times K}{M} = [|\mu_0\rangle, |\mu_1\rangle, \ldots, |\mu_{K-1}\rangle]. \tag{5.59}$$

In particular, the measurement matrix allows us to express the resolution of the identity $\sum_{i\in\mathcal{A}} |\mu_i\rangle\langle\mu_i| = I_{\mathcal{H}}$, in the compact form

$$M M^* = I_{\mathcal{H}}. \tag{5.60}$$

The specification of the source by the state matrix Γ is sufficient in the case of equally likely symbols. With generic a priori probabilities q_i we can introduce the matrix of weighted states [3]

$$\widehat{\Gamma} = \left[|\widehat{\gamma_0}\rangle, |\widehat{\gamma_1}\rangle_1, \ldots, |\widehat{\gamma}_{K-1}\rangle\right], \tag{5.61}$$

where $|\widehat{\gamma_i}\rangle = \sqrt{q_i}|\gamma_i\rangle$.

5.7 State and Measurement Matrices with Mixed States　⇓

The state and the measurement matrices can be extended to mixed states but their introduction is less natural because the density and the measurement operators are not presented in a factorized form as in the case of pure states.

With pure states, the density operators have the factorized form $\rho_i = |\gamma_i\rangle\langle\gamma_i|$ and, with standard notation, $\rho_i = \gamma_i\gamma_i^*$, where the states $\gamma_i = |\gamma_i\rangle$ must be considered as column vectors. In the general case, the density operators do not appear as a product of two factors, but can be equally factorized in the form

$$\rho_i = \gamma_i\gamma_i^*, \qquad i = 0, 1, \ldots, K - 1 \tag{5.62}$$

where the γ_i become matrices of appropriate dimensions (and not simply column vectors). As we will see soon, if n is the dimension of the Hilbert space and h_i is the rank of ρ_i, the matrix γ_i can be chosen of dimensions $n \times h_i$. It must be observed that such factorization is not unique, and also the dimensions $n \times h_i$ are to some extent arbitrary, because h_i has the constraint $\text{rank}(\rho_i) \leq h_i \leq n$. However, the minimal choice $h_i = \text{rank}(\rho_i)$ is the most convenient (and in the following we will comply with this choice).

Similar considerations hold for the measurement operators Q_i, which, with unitary rank, have the factored form $Q_i = |\mu_i\rangle\langle\mu_i|$, but also with rank $h_i > 1$ can be factored in the form

$$Q_i = \mu_i \mu_i^* \tag{5.63}$$

where the factors μ_i are $n \times h_i$ matrices. Further on, we will realize (see Kennedy's theorem and its generalization in Sect. 5.11) that in the choice of the measurement operators it is not restrictive to assume that h_i be given by the same rank of the corresponding density operators.

By analogy with the pure states and with the measurement vectors, the factors γ_i will be called **state factors** and the factors μ_i **measurement factors** (this terminology is not standard and is introduced for the sake of simplicity). The factorization will be useful in various ways; first of all because, if the rank h_i is not full ($h_i < n$), it removes the redundancy of the operators, by gathering the information in an $n \times h_i$ rectangular matrix, instead of an $n \times n$ square matrix, and also because it often makes it possible to extend to the general case some results that are obtained with pure states.

5.7.1 How to Obtain a Factorization

The factorization of a density operator was developed in Sect. 3.11 in the context of the multiplicity of an ensemble of probabilities/states. Here the factorization is seen in a different context and, for clarity, some considerations will be repeated.

Consider a generic density operator ρ of dimensions $n \times n$ and rank h, which is always a PSD Hermitian operator. Then a factorization $\gamma\gamma^*$ can be obtained using its *reduced* EID (see Sect. 2.11 and Proposition 3.5)

$$\rho = Z_h \, D_h^2 \, Z_h^* = \sum_{i=1}^{h} d_i^2 \, |z_i\rangle\langle z_i| \tag{5.64}$$

where $D_h^2 = \text{diag}[d_1^2, \ldots, d_h^2]$ is an $h \times h$ diagonal matrix containing the h positive eigenvalues of ρ and $Z_h = [|z_1\rangle \cdots |z_h\rangle]$ is $n \times h$. Letting $D_h = \sqrt{D_h^2} = \text{diag}[d_1, \ldots, d_h]$, we see immediately that

$$\gamma = Z_h \, D_h \tag{5.65}$$

is a *factor* of ρ.

From the EID (5.64) it results that the density operator is decomposed into the sum of the elementary operators $d_i^2 \, |z_i\rangle\langle z_i|$, where d_i^2 has the meaning of the probability that the quantum system described by the operator ρ be in the state $|z_i\rangle$, exactly in the form in which the density operator has been introduced (see (3.7)). Then the factor γ turns out to be a collection of h vectors

$$\gamma = [d_1 \, |z_1\rangle, \ldots, d_h \, |z_h\rangle] \tag{5.66}$$

where the $|z_i\rangle$ are orthonormal (as taken from a unitary matrix Z of an EID).[7]

Reconsidering the theory developed in Sect. 3.11, we find that γ is a *minimum factor* of ρ and, more specifically, a *minimum orthogonal factor*.

Example 5.1 Consider the Hilbert space $\mathcal{H} = \mathbb{C}^4$, where we assume as basis

$$|b_1\rangle = \begin{bmatrix} \frac{1}{2} \\ \frac{1}{2} \\ \frac{1}{2} \\ \frac{1}{2} \end{bmatrix}, \quad |b_2\rangle = \begin{bmatrix} \frac{1}{2} \\ -\frac{i}{2} \\ -\frac{1}{2} \\ \frac{i}{2} \end{bmatrix}, \quad |b_3\rangle = \begin{bmatrix} \frac{1}{2} \\ -\frac{1}{2} \\ \frac{1}{2} \\ -\frac{1}{2} \end{bmatrix}, \quad |b_4\rangle = \begin{bmatrix} \frac{1}{2} \\ \frac{i}{2} \\ -\frac{1}{2} \\ -\frac{i}{2} \end{bmatrix}.$$

From this basis we build the density operator

$$\rho = \frac{1}{3}|b_1\rangle\langle b_1| + \frac{2}{3}|b_2\rangle\langle b_2| = \begin{bmatrix} \frac{1}{4} & \frac{1}{12}-\frac{i}{6} & -\frac{1}{12} & \frac{1}{12}+\frac{i}{6} \\ \frac{1}{12}+\frac{i}{6} & \frac{1}{4} & \frac{1}{12}-\frac{i}{6} & -\frac{1}{12} \\ -\frac{1}{12} & \frac{1}{12}+\frac{i}{6} & \frac{1}{4} & \frac{1}{12}-\frac{i}{6} \\ \frac{1}{12}-\frac{i}{6} & -\frac{1}{12} & \frac{1}{12}+\frac{i}{6} & \frac{1}{4} \end{bmatrix}$$

which has eigenvalues $\left\{\frac{2}{3}, \frac{1}{3}, 0, 0\right\}$ and therefore has rank $h = 2$. Its reduced EID $\rho = Z_h \, D_h^2 \, Z_h^*$ is specified by the matrices

$$Z_h = \begin{bmatrix} \frac{i}{2} & \frac{1}{2} \\ -\frac{1}{2} & \frac{1}{2} \\ -\frac{i}{2} & \frac{1}{2} \\ \frac{1}{2} & \frac{1}{2} \end{bmatrix} \qquad D_h^2 = \begin{bmatrix} \frac{2}{3} & 0 \\ 0 & \frac{1}{3} \end{bmatrix} \qquad Z_h^* = \begin{bmatrix} -\frac{i}{2} & -\frac{1}{2} & \frac{i}{2} & \frac{1}{2} \\ \frac{1}{2} & \frac{1}{2} & \frac{1}{2} & \frac{1}{2} \end{bmatrix}.$$

Now, to obtain a factor γ of ρ we use (5.65), which gives the 4×2 matrix

---

[7] Another way to obtain a factorization is given by Choleski's decomposition (see Sect. 2.12.5).

$$\gamma = Z_h \sqrt{D^2} = \begin{bmatrix} \frac{i}{2} & \frac{1}{2} \\ -\frac{1}{2} & \frac{1}{2} \\ -\frac{i}{2} & \frac{1}{2} \\ \frac{1}{2} & \frac{1}{2} \end{bmatrix} \begin{bmatrix} \sqrt{\frac{2}{3}} & 0 \\ 0 & \sqrt{\frac{1}{3}} \end{bmatrix} = \begin{bmatrix} \frac{i}{\sqrt{6}} & \frac{1}{2\sqrt{3}} \\ -\frac{1}{\sqrt{6}} & \frac{1}{2\sqrt{3}} \\ -\frac{i}{\sqrt{6}} & \frac{1}{2\sqrt{3}} \\ \frac{1}{\sqrt{6}} & \frac{1}{2\sqrt{3}} \end{bmatrix}.$$

As regards the factorization of the measurement operator, say $Q = \mu\mu^*$, similar considerations hold, provided that the operator Q is known. However, in quantum detection Q is not known and it should be determined by optimization. In this context the unknown may become μ, then giving Q as $\mu\mu^*$ and the factorization is no more required.

5.7.2 State and Measurement Matrices

The definition of these matrices can be extended to mixed states by expressing the density operators and the corresponding measurement operators through their factors. The state matrix Γ is obtained by juxtaposing the factors γ_i, intended as blocks of columns of dimensions $n \times h_i$

$$\underset{n \times H}{\Gamma} = \begin{bmatrix} \gamma_0, \gamma_1, \ldots, \gamma_{K-1} \end{bmatrix} \tag{5.67}$$

where the number of columns H is given by the global number of the columns of the factors γ_i

$$H = h_0 + h_1 + \cdots + h_{K-1}.$$

We can make explicit Γ bearing in mind that each factor γ_i is a collection of h_i kets (see (5.66)). For example, for $K = 2, h_0 = 2, h_1 = 3$ we have

$$\Gamma = [\gamma_0, \gamma_1] = [|\gamma_{01}\rangle, |\gamma_{02}\rangle, |\gamma_{11}\rangle, |\gamma_{12}\rangle, |\gamma_{13}\rangle] \tag{5.68}$$

where $|\gamma_{0i}\rangle$ are the kets of γ_0 and $|\gamma_{1i}\rangle$ are the kets of γ_1.

Analogously, the measurement matrix M is obtained by juxtaposing the factors μ_i, intended as blocks of columns

$$\underset{n \times H}{M} = \begin{bmatrix} \mu_0, \mu_1, \ldots, \mu_{K-1} \end{bmatrix}. \tag{5.69}$$

Even the resolution of the identity (5.60) is extended to mixed states. In fact,

$$M M^* = \sum_{i \in \mathcal{A}} \mu_i \mu_i^* = \sum_{i \in \mathcal{A}} Q_i = I_{\mathcal{H}}. \tag{5.70}$$

Clearly, these last definitions are the most general and include the previous ones when the ranks are unitary ($h_i = 1$ and $H = K$).

Also the definition of the *matrix of weighted states*, given by (5.61) for pure states, can be extended to mixed states [3], namely

$$\widehat{\Gamma} = \left[\widehat{\gamma}_0, \widehat{\gamma}_1, \ldots, \widehat{\gamma}_{K-1}\right] = \left[\sqrt{q_0}\gamma_0, \sqrt{q_1}\gamma_1, \ldots, \sqrt{q_{K-1}}\gamma_{K-1}\right], \qquad (5.71)$$

where the weighted states can be obtained as a factorization of weighted density operators, namely $\widehat{\rho}_i = q_i \rho_i = \sqrt{q_i}\gamma_i \sqrt{q_i}\gamma_i^* = \widehat{\gamma}_i \widehat{\gamma}_i^*$.

5.7.3 Probabilities Expressed Through Factors

In quantum decision, probabilities can be computed from the factors γ_i and μ_i of the density operators and of the measurement operators. Recalling the expression of the transition probabilities, given by (5.15), we obtain explicitly

$$p_c(j|i) = \text{Tr}[Q_j \rho_i] = \text{Tr}[\mu_j \mu_j^* \gamma_i \gamma_i^*]. \qquad (5.72)$$

Analogously, from (5.16) we obtain the correct decision probability

$$P_c = \sum_{i \in \mathcal{A}} q_i \, \text{Tr}[Q_i \rho_i] = \sum_{i \in \mathcal{A}} q_i \, \text{Tr}[\mu_i \mu_i^* \gamma_i \gamma_i^*]. \qquad (5.73)$$

In the evaluation of these probabilities it is convenient to introduce the *matrix of mixed products*

$$\underset{H \times H}{B} := M^* \Gamma = \begin{bmatrix} b_{0,0} & \cdots & b_{0,K-1} \\ \vdots & \ddots & \vdots \\ b_{k-1,0} & \cdots & b_{K-1,K-1} \end{bmatrix}, \qquad b_{ij} := \mu_i^* \gamma_i \qquad (5.74)$$

where $\dim b_{ij} = h_i \times h_j$. Then, using the cyclic property of the trace, we find

$$\boxed{p_c(j|i) = \text{Tr}[b_{ji}^* b_{ji}], \qquad P_c = \sum_{i \in \mathcal{A}} q_i \text{Tr}[b_{ii}^* b_{ii}].} \qquad (5.75)$$

Finally, it must be observed that state and measurement factors are not uniquely determined by the corresponding operators. In fact, if γ_i is a factor of ρ_i, also $\tilde{\gamma}_i = \gamma_i Z$, where Z is any matrix with the property $ZZ^* = I_{h_i}$, is a factor of ρ_i, as follows from $\tilde{\gamma}_i \tilde{\gamma}_i^* = \gamma_i ZZ^* \gamma_i^* = \gamma_i \gamma_i^* = \rho_i$. However, the multiplicity of the factors has no influence on the computation of the probabilities, as can be verified from the above expressions.

Problem 5.9 ⋆⋆ From the following normalized states of $\mathcal{H} = \mathbb{C}^4$

$$
|\gamma_1\rangle = \begin{bmatrix} \dfrac{2}{\sqrt{13}} \\[4pt] \dfrac{2}{\sqrt{13}} \\[4pt] \dfrac{2}{\sqrt{13}} \\[4pt] \dfrac{1}{\sqrt{13}} \end{bmatrix}
\quad
|\gamma_2\rangle = \begin{bmatrix} \dfrac{1}{2} \\[4pt] \dfrac{1}{2} \\[4pt] \dfrac{1}{2} \\[4pt] \dfrac{1}{2} \end{bmatrix}
\quad
|\gamma_3\rangle = \begin{bmatrix} \dfrac{1}{2} \\[4pt] -\dfrac{i}{2} \\[4pt] -\dfrac{1}{2} \\[4pt] \dfrac{i}{2} \end{bmatrix}
\quad
|\gamma_4\rangle = \begin{bmatrix} \dfrac{2}{\sqrt{13}} \\[4pt] -\dfrac{2i}{\sqrt{13}} \\[4pt] -\dfrac{1}{\sqrt{13}} \\[4pt] \dfrac{2i}{\sqrt{13}} \end{bmatrix}
\quad
|\gamma_5\rangle = \begin{bmatrix} \dfrac{1}{\sqrt{13}} \\[4pt] -\dfrac{2i}{\sqrt{13}} \\[4pt] -\dfrac{2}{\sqrt{13}} \\[4pt] \dfrac{2i}{\sqrt{13}} \end{bmatrix}
$$

form the density operators

$$
\rho_1 = \frac{3}{4}|\gamma_1\rangle\langle\gamma_1| + \frac{1}{4}|\gamma_2\rangle\langle\gamma_2|, \qquad
\rho_2 = \frac{3}{4}|\gamma_3\rangle\langle\gamma_3| + \frac{1}{8}|\gamma_4\rangle\langle\gamma_4| \frac{1}{8}|\gamma_5\rangle\langle\gamma_5|
$$

and find their *minimum* factors γ_1 and γ_2. Find also factorizations in which the matrices γ_1 and γ_2 have the same dimensions.

Problem 5.10 ⋆ Consider the transition probabilities given by (5.72). Prove that, if γ_i is replaced by $\gamma_i Z$, with $ZZ^* = I_h$, and μ_j by $\mu_j W$, with $WW^* = I_h$, the transition probabilities do not change.

Problem 5.11 ⋆⋆ Prove that the measurement matrix M defined by (5.59) and its generalization to mixed states (5.69), allows us to express the resolution of the identity in the form $MM^* = I_{\mathcal{H}}$.

5.8 Formulation of Optimal Quantum Decision

The viewpoint for the Optimal Quantum Decision is the following: the a priori probabilities and the constellation (of pure states or of mixed stated) are assumed as given, whereas the measurement operator system is unknown and should be determined to meet the decision criterion, given by **the maximization of the correct decision probability**.

Then, considering the general expression of the correct decision probability, given by (see (5.16))

$$
P_c = \sum_{i \in \mathcal{A}} q_i \operatorname{Tr}[\rho_i Q_i]
$$

the *optimal measurement operators* Q_i must be determined from

$$
\max_{\{Q_i\}} \sum_{i=0}^{K-1} q_i \operatorname{Tr}[\rho_i Q_i]. \tag{5.76}
$$

If the operators are expressed through their factors (see (5.62) and (5.63)), (5.76) becomes

$$\max_{\{\mu_i\}} \ \sum_{i=0}^{K-1} q_i \, \mathrm{Tr}[\gamma_i \gamma_i^* \mu_i \mu_i^*]. \tag{5.77}$$

Finally, if the states are pure, we have the simplification

$$\max_{\{|\mu_i\rangle\}} \ \sum_{i=0}^{K-1} q_i \, \mathrm{Tr}[|\gamma_i\rangle\langle\gamma_i|\mu_i\rangle\langle\mu_i|] = \max_{\{|\mu_i\rangle\}} \ \sum_{i=0}^{K-1} q_i \, |\langle\gamma_i|\mu_i\rangle|^2. \tag{5.78}$$

In the last relation we have used the identity (2.37) over the trace, $\mathrm{Tr}[A|u\rangle\langle u|] = \langle u|A|u\rangle$, with $A = |\gamma_i\rangle\langle\gamma_i|$ and $|u\rangle = |\mu_i\rangle$.

5.8.1 Optimization as Convex Semidefinite Programming (CSP)

Starting from the Hilbert space $\mathcal{H}$ on which the quantum decision is defined, it is convenient to introduce the following classes (Fig. 5.8):

- the class $\mathcal{B}$ of the Hermitian operators defined on $\mathcal{H}$,
- the subclass $\mathcal{B}_0$ of the PSD Hermitian operators,
- the class $\mathcal{M}$ of the K-tuples $\mathbf{Q} = [\,Q_0, \ldots, Q_{K-1}\,]$, $Q_i \in \mathcal{B}$ of Hermitian operators,
- the subclass $\mathcal{M}_0$ of $\mathcal{M}$ consisting of the K-tuples $\mathbf{Q}$, whose elements Q_i are PSD Hermitian, $Q_i \in \mathcal{B}_0$, and, globally, resolve the identity on $\mathcal{H}$, that is, $\sum_i Q_i = I_{\mathcal{H}}$.

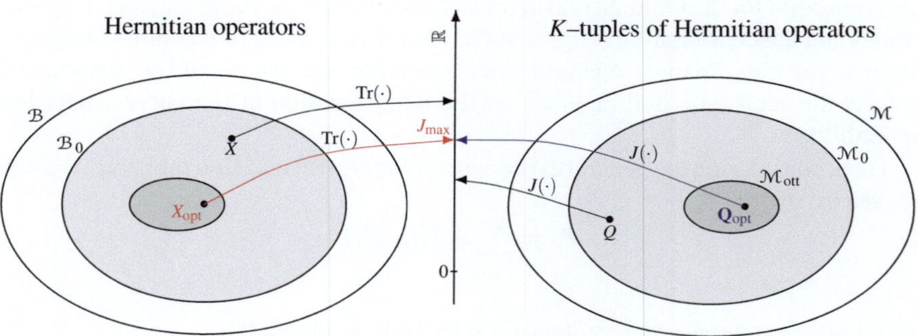

Fig. 5.8 Classes in the quantum decision for the determination of optimal measurement operators. On the *right*, the class $\mathcal{M}$ formed by the K-tuples of Hermitian operators and the subclass $\mathcal{M}_0$ constituted by systems of measurement operators $\mathbf{Q}$; in $\mathcal{M}_0$ the functional $J(\mathbf{Q})$ is defined, which has a maximum $J_{\max}$ when $\mathbf{Q}$ becomes optimal. On the *left*, the class $\mathcal{B}$ of the Hermitian operators X and the subclass $\mathcal{B}_0$ of the positive semidefinite X; in general $\mathrm{Tr}(X) \geq J_{\max}$, but for particular $X = X_{\mathrm{opt}}$ it results $\mathrm{Tr}(X_{\mathrm{opt}}) = J_{\max}$

In other words, each K-tuple $\mathbf{Q} \in \mathcal{M}_0$ identifies a valid *measurement operator system*.

With these premises, the problem of the optimal decision can be treated in the framework of *convex programming*. Starting from the data specified by the *weighted* density operators

$$\widehat{\rho}_i = q_i \rho_i, \qquad i = 0, \ldots, K - 1, \tag{5.79}$$

we must determine a measurement operator system $Q \in \mathcal{K}_0$ that maximizes the quantity

$$J(\mathbf{Q}) = \sum_{i=0}^{K-1} \operatorname{Tr}[\widehat{\rho}_i Q_i], \qquad \mathbf{Q} \in \mathcal{M}_0. \tag{5.80}$$

We are dealing with a problem of *convex semidefinite optimization* because the K-tuple $\mathbf{Q}$ must be found on a convex set: in fact, given two K-tuples $\mathbf{P}$ and $\mathbf{Q}$ of $\mathcal{M}_0$ and given any λ with $0 < \lambda < 1$, it can be easily shown that the convex linear combination $\lambda\mathbf{P} + (1 - \lambda)\mathbf{Q}$ is still formed by a K-tuple of $\mathcal{M}_0$. Therefore, by definition, $\mathcal{M}_0$ is a convex set. Within such set, it results:

Proposition 5.3 *The functional $J(\mathbf{Q})$, which gives the correct decision probability P_c, in $\mathcal{M}_0$ admits the maximum*

$$J_{\max} = \max_{\mathbf{Q}\in\mathcal{M}_0} J(\mathbf{Q}) = J(\mathbf{Q}_{\mathrm{opt}}).$$

This maximum gives the maximum of the correct decision probability, $P_{c,\max} = J_{\max} = J(\mathbf{Q}_{\mathrm{opt}})$, and $\mathbf{Q}_{\mathrm{opt}}$ is by definition an optimal system of measurement operators.

This proposition will be proved in Appendix section "Proof of Holevo's Theorem".

5.9 Holevo's Theorem

The following theorem, stated by Holevo in (1972) [4], completely characterizes the optimal solution, and is probably one of the most important results of the theory of quantum decision in the last decades.

Theorem 5.1 (Holevo's Theorem) *In a K-ary system characterized by the weighted density operators $\widehat{\rho}_i = q_i \rho_i$, the measurement operators Q_i are optimal if and only if, having defined the operator*

$$L = \sum_{i=0}^{K-1} Q_i \widehat{\rho}_i, \tag{5.81}$$

it follows that the operators $L - \widehat{\rho}_i$ are PSD, that is,

$$L - \widehat{\rho}_i \in \mathcal{B}_0 \tag{5.82}$$

and, for each $i = 0, \ldots, K - 1$,

$$(L - \widehat{\rho}_i)Q_i = 0_{\mathcal{H}}. \tag{5.83}$$

Holevo's theorem, which will be proved in Appendix section "Proof of Holevo's Theorem", determines the conditions that must be verified by an optimal system of measurement operators $\mathbf{Q}_{\mathrm{opt}}$, but does not provide any clue on how to identify it.

An equivalent form of Holevo's theorem, but, as we will see, more appropriate for numerical computation, has been proved by Yuen et al. [5] and, recently, in a detailed form, by Eldar et al. [3]. The result is obtained by transforming the original problem into a *dual problem*, according to a well-known technique of linear programming.

Theorem 5.2 (Dual theorem) *In a K-ary system characterized by the weighted density operators $\widehat{\rho}_i = q_i \rho_i$, the measurement operators Q_i are optimal if and only if there exists a PSD operator, $X \in \mathcal{B}_0$, such that $\mathrm{Tr}[X]$ is minimal,*

$$T_{\min} = \min_{X \in \mathcal{B}_0} \mathrm{Tr}[X] \tag{5.84}$$

and for every $j = 0, \ldots, K - 1$ the operators $X - \widehat{\rho}_j$ are PSD, $X - \widehat{\rho}_j \in \mathcal{B}_0$. The optimal operators Q_i satisfy the conditions

$$(X - \widehat{\rho}_i)Q_i = 0_{\mathcal{H}}. \tag{5.85}$$

The minimum obtained for $\mathrm{Tr}[X]$ coincides with the requested maximum of $J(\mathbf{Q})$

$$T_{\min} = J_{\max} = P_{c,\max}. \tag{5.86}$$

Notice that the conditions imposed on the operators for optimality $X - \widehat{\rho}_i$ are the same as those indicated in Holevo's theorem, imposed on operators $X - \widehat{\rho}_i$. To understand why the dual theorem leads to a lower computational complexity, suppose that the Hilbert space be of finite dimensions n. In Holevo's theorem we must look for a K-tuple of Hermitian operators Q_i, for a total of Kn^2 unknowns; instead, in the dual theorem we must look for the Hermitian matrix X, for a total of n^2 unknowns.

Example 5.2 We want to check that the projectors Q_0 and Q_1, evaluated in Sect. 5.3 with Helstrom's theory, satisfy the conditions of Holevo's theorem. For $K = 2$, the operator (5.81) becomes, bearing in mind the resolution constraint of the identity $Q_0 + Q_1 = I$,

$$L = Q_0\widehat{\rho}_0 + Q_1\widehat{\rho}_1 = (I - Q_1)\widehat{\rho}_0 + Q_1\widehat{\rho}_1 = \widehat{\rho}_0 + Q_1 D$$

where $D = \widehat{\rho}_1 - \widehat{\rho}_0$ is the decision operator introduced in Helstrom's theory (see (5.20)). The conditions (5.83) give

$$Q_1 D Q_0 = 0, \qquad Q_0 D Q_1 = 0$$

which are mutually equivalent. We can verify them using the expressions (5.25) and (5.24), and the orthonormality. We obtain

$$Q_1 D Q_0 = \sum_{\eta_k > 0} |\eta_k\rangle\langle\eta_k| \sum_m \eta_m |\eta_m\rangle\langle\eta_m| \sum_{\eta_h < 0} |\eta_h\rangle\langle\eta_h| = 0.$$

The conditions (5.83) become

$$L - \widehat{\rho}_0 = Q_1 D \geq 0, \qquad L - \widehat{\rho}_1 = -Q_0 D \geq 0.$$

We have

$$Q_1 D = \sum_{\eta_h > 0} |\eta_h\rangle\langle\eta_h| \sum_m \eta_m |\eta_m\rangle\langle\eta_m| = \sum_{\eta_h > 0} \eta_h |\eta_h\rangle\langle\eta_h|$$

which is PSD because $\eta_h > 0$ and $|\eta_h\rangle\langle\eta_h|$ are elementary projectors. Analogously, it can be proved that $-Q_0 D \geq 0$.

5.10 Numerical Methods for the Search for Optimal Operators

As already said, only in some particular cases the problem of the determination of the optimal measurement operators and of the maximum correct decision probability has closed-form solutions. In the other cases, we either restrict ourselves to search for near-optimal solutions, with the SRM measurements, or we must resort to numerical computation. As we are dealing with problems of convex programming, which fall under a very general class of problems, we can use existing very sophisticated software packages, like LMI (linear matrix inequalities) and CSP (convex semidefinite programming), both operating in the MatLab$^{©}$ environment [6, 7].

5.10.1 The MatLab Procedure Cvx

The use of this procedure is conceptually very simple. For the application of Holevo's theorem, in the general case, all it takes is to provide, as input data, the K weighted density operators $\widehat{\rho}_i$, with the constraints

$$Q_i \geq 0, \qquad i = 0, \ldots, K-1, \qquad \sum_{i=0}^{K-1} Q_i = I$$

and to request as output the K measurement operators Q_i that *maximize*

$$J(\mathbf{Q}) = \sum_{i=0}^{K-1} \text{Tr}[\widehat{\rho}_i Q_i].$$

Resorting to the dual theorem reduces the computational complexity. Inputting the $\widehat{\rho}_i$, with the constraints

$$X - \widehat{\rho}_i \geq 0, \qquad i = 0, \ldots, K-1$$

the user asks for the operator X of *minimal trace*. From X we obtain the optimal measurement operators as solutions of the equations $(X - \widehat{\rho}_i)Q_i = 0$. Clearly, the computation is simplified because the search is limited to the single operator X.

We write the `MatLab` procedure in the binary case, which is easily extended to an arbitrary K.

```
%%%%%%%%%%%%%%%%%%%%%%%%%%%%%%%%%%%%%%%%%%%%%%%%%%%%%%%
% cvx  procedure applied to Holevo's theorem
%%%%%%%%%%%%%%%%%%%%%%%%%%%%%%%%%%%%%%%%%%%%%%%%%%%%%%%
cvx_begin SDP
   variable Q0(dim, dim) hermitian
   variable Q1(dim, dim) hermitian
   maximize(trace(Q0*R0+Q1*R1))
   subject to
    Q0>0;
    Q1>0;
    Q0==eye(dim)-Q1;
cvx_end

Pc_Holevo=trace(Q0*R0+Q1*R1);


%%%%%%%%%%%%%%%%%%%%%%%%%%%%%%%%%%%%%%%%%%%%%%%%%%%%%%%
% cvx procedure apply to the dual problem
%%%%%%%%%%%%%%%%%%%%%%%%%%%%%%%%%%%%%%%%%%%%%%%%%%%%%%%

cvx_begin SDP
   variable Q(dim, dim) hermitian
   minimize(trace(Q))
   subject to
    Q>R0;
```

```
       Q>R1;
   cvx_end

   Pc_dual=trace(Q);
```

5.10.2 Example

We input the weighted density operators

$$\widehat{\rho}_0 = \frac{1}{2} \begin{bmatrix} 0.29327 & 0.29327 & 0.29327 & 0.17788 \\ 0.29327 & 0.29327 & 0.29327 & 0.17788 \\ 0.29327 & 0.29327 & 0.29327 & 0.17788 \\ 0.17788 & 0.17788 & 0.17788 & 0.12019 \end{bmatrix}$$

$$\widehat{\rho}_1 = \frac{1}{2} \begin{bmatrix} 0.23558 & 0.24519i & -0.22596 & -0.24519i \\ -0.24519i & 0.26442 & 0.24519i & -0.26442 \\ -0.22596 & -0.24519i & 0.23558 & 0.24519i \\ 0.24519i & -0.26442 & -0.24519i & 0.26442 \end{bmatrix}.$$

the "Holevo" procedure gives as output

$$P_e = 0.009316144.$$

$$Q_0 = \begin{bmatrix} 0.502788 & 0.259735 & 0.247032 & -0.0141425 \\ 0.259735 & 0.278855 & 0.259735 & 0.256145 \\ 0.247032 & 0.259735 & 0.502788 & -0.0141425 \\ -0.0141425 & 0.256145 & -0.0141425 & 0.715569 \end{bmatrix}$$

$$Q_1 = \begin{bmatrix} 0.497212 & -0.259735 & -0.247032 & 0.0141425 \\ -0.259735 & 0.721145 & -0.259735 & -0.256145 \\ -0.247032 & -0.259735 & 0.497212 & 0.0141425 \\ 0.0141425 & -0.256145 & 0.0141425 & 0.284431 \end{bmatrix}$$

The "dual" procedure gives as the output

$$P_e = 0.009316139$$

$$X = \begin{bmatrix} 0.263349 & 0.138595 & 0.0314089 & 0.0971245 \\ 0.138595 & 0.264951 & 0.138595 & -0.0387083 \\ 0.0314089 & 0.138595 & 0.263349 & 0.0971245 \\ 0.0971245 & -0.0387083 & 0.0971245 & 0.199034 \end{bmatrix}$$

Hence we find the same minimum error probability, as expected (the Helstrom procedure gives $P_e = 0.00936141$). The negligible differences are due to the different way of numerical computations.

5.11 Kennedy's Theorem

Holevo's theorem has general validity, because it is concerned with optimal decision in a system specified through density operators, which does not rule out the possibility that the states may be pure. Instead, Kennedy's theorem [8] is about a system in which there is a constellation of K pure states

$$|\gamma_0\rangle, |\gamma_1\rangle, \ldots, |\gamma_{K-1}\rangle. \tag{5.87}$$

Theorem 5.3 (Kennedy's theorem) *In a K-ary system specified by K **pure states** $|\gamma_0\rangle, \ldots, |\gamma_{K-1}\rangle$, the optimal projectors (which maximize the correct decision probability) are always elementary, that is, they have the form*

$$Q_i = |\mu_i\rangle\langle\mu_i|, \quad i = 0, 1, \ldots, K - 1 \tag{5.88}$$

where the measurement vectors $|\mu_i\rangle$ must be **orthonormal**.

The theorem is proved in Appendix section "Proof of Kennedy's Theorem".

5.11.1 Consequences of Kennedy's Theorem

With Kennedy's Theorem the search for the optimal decision is substantially simplified, as it is restricted to the search for K *orthonormal measurement vectors*

$$|\mu_0\rangle, |\mu_1\rangle, \ldots, |\mu_{K-1}\rangle$$

from which the optimal projectors are built, using (5.88). The simplification lies in the fact that, instead of searching for K matrices, it suffices to search for K vectors.

Example 5.3 In the binary case, we have seen that the optimal projectors are given by (5.34), where both Q_0 and Q_1 are elementary projectors. In addition, $|\mu_0\rangle$ and $|\mu_1\rangle$ are orthonormal.

From now on, the Hilbert space $\mathcal{H}$ will be assumed of finite dimension n, even though, in the applications to quantum communications systems, the dimensions become infinite ($n = \infty$). When the decision is made starting from K pure states, a fundamental role is played by the *subspace* generated from the states

$$\mathcal{U} = \mathrm{span}(|\gamma_0\rangle, |\gamma_1\rangle, \ldots, |\gamma_{K-1}\rangle) \subseteq \mathcal{H}. \tag{5.89}$$

The dimension of this space, $r = \dim \mathcal{U}$, is equal to K if the states $|\gamma_i\rangle$ are linearly independent (not necessarily orthonormal), and lower than K if the states are linearly independent; so, in general

$$r = \dim \mathcal{U} \leq K \leq \dim \mathcal{H} = n.$$

In any case, it is very important to observe that:

Proposition 5.4 *It is not restrictive to suppose that the measurement vectors $|\mu_i\rangle$ belong to the space generated by the states*

$$|\mu_i\rangle \in \mathcal{U} \tag{5.90}$$

because any component of the $|\mu_i\rangle$ belonging to the complementary $\mathcal{U}^{\perp}$ has no influence on the decision probabilities.

In fact, if we decompose $|\mu_j\rangle$ into the sum

$$|\mu_j\rangle = |\mu_j'\rangle + |\mu_j''\rangle, \qquad |\mu_j'\rangle \in \mathcal{U}, \qquad |\mu_j''\rangle \in \mathcal{U}^{\perp}$$

the transition probabilities become

$$p_c(j|i) = |\langle \mu_j | \gamma_i \rangle|^2 = |\langle \mu_j' | \gamma_i \rangle|^2$$

where $\langle \mu_j'' | \gamma_i \rangle = 0$ as $|\mu_j''\rangle \in \mathcal{U}^{\perp}$ is orthogonal to $|\gamma_i\rangle \in \mathcal{U}$.

Proposition 5.4 is illustrated in Fig. 5.9, where it is evidenced that the states and the measurement vectors belong to the common subspace $\mathcal{U}$. In harmony with Proposition 5.4, we have:

Proposition 5.5 *For the measurement operators, the resolution of the identity can be substituted by the resolution of the generalized identity*

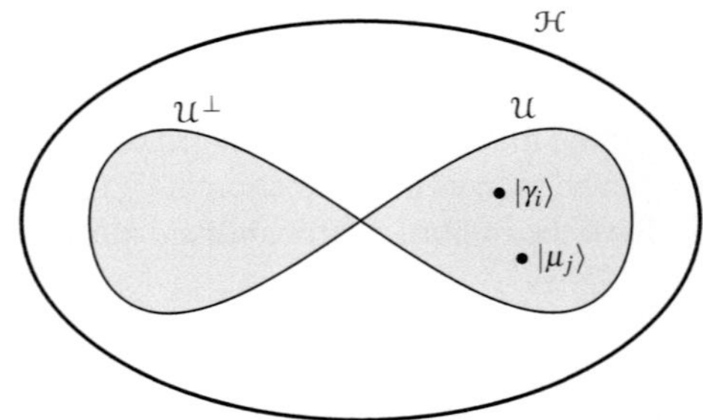

Fig. 5.9 The measurement vectors $|\mu_j\rangle$ belong to the subspace $\mathcal{U}$ generated by the constellation of the states $|\gamma_i\rangle$

$$\sum_{i=0}^{K-1} |\mu_i\rangle\langle\mu_i| = P_{\mathcal{U}} \qquad (5.91)$$

where $P_{\mathcal{U}}$ is the projector of $\mathcal{H}$ onto $\mathcal{U}$.

For the proof of this proposition see Sect. 3.7.2. A consequence of the (5.90) is the following:

Proposition 5.6 *The measurement vectors are given by a linear combination of the states*

$$|\mu_i\rangle = \sum_{j=0}^{K-1} a_{ij}|\gamma_j\rangle, \qquad (5.92)$$

where the coefficients a_{ij} are in general complex.

Proposition 5.7 *With decision from pure states, the transition probabilities become*

$$p_c(j|i) = |\langle\mu_i|\gamma_j\rangle|^2 \qquad (5.93)$$

and the correct decision probability is given by

$$P_c = \sum_{i=0}^{K-1} q_i \, |\langle\mu_i|\gamma_j\rangle|^2. \qquad (5.94)$$

5.11.2 Applications of Kennedy's Theorem to Holevo's Theorem

In a decision starting from pure states, the optimal measurement vectors must satisfy Holevo's theorem with $Q_i = |\mu_i\rangle\langle\mu_i|$ and $\widehat{\rho}_i = q_i \, |\gamma_i\rangle\langle\gamma_i|$. Then, assuming that the $|\mu_i\rangle$ belong to the same subspace $\mathcal{U}$ of the states, the geometry relative to the two vector systems is determined by the inner products

$$b_{ij} = \langle\mu_i|\gamma_j\rangle, \qquad i, j = 0, 1, \ldots, K - 1. \qquad (5.95)$$

Assuming that the $|\mu_i\rangle$ form an orthonormal basis of $\mathcal{U}$ (Fig. 5.10), the inner product b_{ij} can be seen as the projection of $|\gamma_i\rangle$ along the axis $|\mu_j\rangle$. We observe also that the b_{ij} have the important probabilistic meaning

$$p_c(j|i) = |b_{ij}|^2.$$

Using the mixed inner products b_{ij}, from Holevo's theorem we obtain:

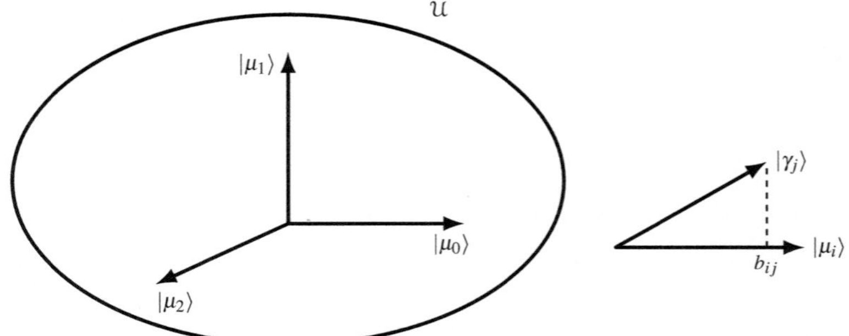

Fig. 5.10 Coordinate systems of $\mathcal{U}$ done by the measurement vectors $|\mu_i\rangle$ and meaning of the mixed inner product $b_{ij} = \langle \gamma_j | \mu_i \rangle$

Corollary 5.1 *In a K-ary system with a constellation of pure states $|\gamma_0\rangle, \ldots, |\gamma_{K-1}\rangle$, the optimal measurement vectors $|\mu_i\rangle$ must verify the conditions*

$$(q_j\, b_{ij} b_{jj}^* - q_i\, b_{ii} b_{ji}^*)|\mu_i\rangle\langle\mu_j| = 0, \qquad \forall i, \forall j \tag{5.96a}$$

$$\sum_{j=0}^{K-1} q_j b_{jj}|\mu_j\rangle\langle\gamma_j| - q_i|\gamma_i\rangle\langle\gamma_i| \geq 0, \qquad \forall i. \tag{5.96b}$$

Relation (5.96a) allows us to write the following conditions on the inner products

$$q_j\, b_{ij} b_{jj}^* - q_i\, b_{ii} b_{ji}^* = 0 \tag{5.97}$$

which can be seen as a nonlinear system of $(K-1)K/2$ equations in the K^2 unknowns b_{ij}. We can add to this other equations derived from the Fourier expansion of the states $|\gamma_i\rangle$ with basis $|\mu_j\rangle$ (see (2.51)), which assumes the form

$$|\gamma_i\rangle = \sum_{j=0}^{K-1} (\langle\mu_j|\gamma_i\rangle)|\mu_j\rangle = \sum_{j=0}^{K-1} b_{ji}|\mu_j\rangle.$$

Then, expressing the inner products $\langle\gamma_i|\gamma_j\rangle$, which we assumed as known, we obtain the relations

$$\sum_{k=0}^{K-1} b_{ki}^* b_{kj} = \langle\gamma_i|\gamma_j\rangle \tag{5.98}$$

which constitute the $(K+1)K/2$ equations.

In principle, we can try to solve this nonlinear system, which admits solutions if the states are linearly independent, and eventually we can verify whether, with these solutions, even the conditions (5.96b) are verified. However, we can see that even in the binary case the search for an exact solution turns out to be rather complicated. We could proceed in numerical form, but in this case it is more convenient to adopt the method derived from the geometric interpretation, as we are going to illustrate.

5.11.3 Geometric Interpretation of Optimization

We consider the subspace $\mathcal{U}$ generated by the states $|\gamma_i\rangle$ in which an orthogonal system of coordinate has been introduced, made of the measurement vectors $|\mu_j\rangle$. The correct decision probability can be expressed in the forms

$$P_c = \sum_{i=0}^{K-1} q_i\, p_c(i|i) = \sum_{i=0}^{K-1} q_i\, |b_{ii}|^2$$

where b_{ij} are the inner products (5.95). If such products are real numbers, we can define the angle θ_i between $|\gamma_i\rangle$ and $|\mu_i\rangle$ from

$$\sin^2\theta_i = 1 - b_{ii}^2$$

and then the error probability can be written as

$$P_e = 1 - P_c = \sum_{i=0}^{K-1} q_i\, \sin^2\theta_i.$$

The angles θ_i are illustrated in Fig. 5.11 for $K = 2$.

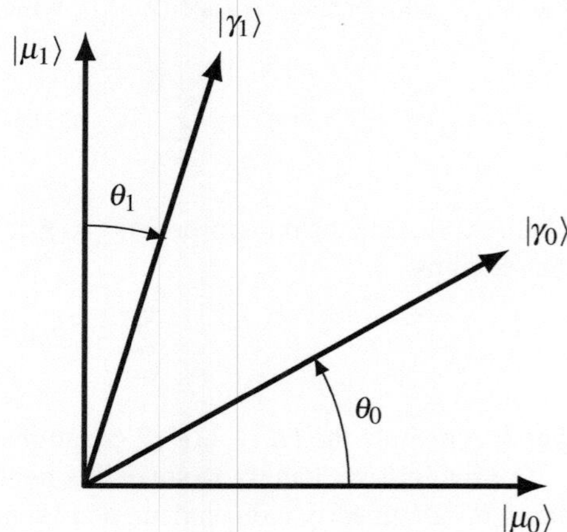

Fig. 5.11 Angles between measurement vectors and states

To minimize P_e we must rotate the constellation of the vectors $|\gamma_i\rangle$ around the respective axes $|\mu_i\rangle$ until a minimum is reached.

This optimization technique has recently been used by the scientists of JPL because it makes it possible to obtain useful results even in the presence of thermal noise [9–11].

5.11.4 Generalization of Kennedy's Theorem

Recently [3], Kennedy's theorem has been partially extended to mixed states and precisely:

Theorem 5.4 *In a system specified by K density operators $\rho_0, \ldots, \rho_{K-1}$, the optimal measurement operators Q_i (maximizing the correct decision probability) have rank not higher than that of the corresponding density operators*

$$\operatorname{rank}(Q_i) \leq \operatorname{rank}(\rho_i), \qquad i = 0, 1, \ldots, K - 1. \tag{5.99}$$

The connection with the original theorem can be understood considering the consequences on the factors of the operators. If $h_i = \operatorname{rank}(\rho_i)$, the corresponding factor γ_i is an $n \times h_i$ matrix and the measurement factor μ_i has dimensions $n \times \tilde{h}_i$, with $\tilde{h}_i \leq h_i = \operatorname{rank}(\rho_i)$, but it is not restrictive to suppose that it has the same dimensions $n \times h_i$ as γ_i (and so we will suppose in the following). In particular, if the ranks are unitary, the factors become kets, $\gamma_i = |\gamma_i\rangle$ and $\mu_i = |\mu_i\rangle$, as established by Kennedy's theorem.

Also the considerations made on the subspace $\mathcal{U}$ generated by the states (see (5.89) and Proposition 5.4) can be generalized. It must be remembered that the state factors are a collection of kets of $\mathcal{H}$ and the state matrix Γ collects these kets. Then the subspace $\mathcal{U}$ is generated according to

$$\mathcal{U} = \operatorname{span}\{\text{kets of } \Gamma\} = \operatorname{Im} \Gamma$$

and Proposition 5.4 is extended by saying that it is not restrictive to suppose that the kets of the measurement vectors $|\mu_i\rangle$ belong to the space generated by the states

$$\operatorname{Im} \mu_i \subseteq \mathcal{U}. \tag{5.100}$$

5.12 The Geometry of a Constellation of States

We continue with the study of decision, investigating the geometry generated by the states in the Hilbert space. The basic tools used herein are the eigendecomposition (EID) and the singular value decomposition (SVD). We will refer to pure states, and only at the end of this section the concepts will be extended to mixed states. Also, we refer to equal a priori probabilities, which imply that $q_i = 1/K$; to get general results the states should be replaced by weighted states.

5.12.1 State Matrix and Measurement Matrix

In Sect. 5.7.2 we introduced the state matrix Γ and the measurement matrix M, which, with pure states, result in

$$\underset{n \times K}{\Gamma} = [|\gamma_0\rangle, |\gamma_1\rangle, ..., |\gamma_{K-1}\rangle], \qquad \underset{n \times K}{M} = [|\mu_0\rangle, |\mu_1\rangle, ..., |\mu_{K-1}\rangle].$$

With these matrices, the problem of decision becomes: given the state matrix Γ, find the measurement matrix M. We have seen that the measurement vectors are given by a linear combination of the states (see (5.92)), that is,

$$|\mu_i\rangle = \sum_{j=0}^{K-1} a_{ij} |\gamma_j\rangle ; \tag{5.101}$$

this combination in matrix terms can be written as

$$\underset{n \times K}{M} = \Gamma\, A, \qquad \underset{K \times K}{A} = [a_{ij}]. \tag{5.102}$$

At this point, the problem is already simplified, because it is sufficient to search for the coefficient matrix A, which is $K \times K$ and therefore of smaller dimensions than the dimensions $n \times K$ of the measurement matrix (where n can become infinite).

It will be useful to compare the matrix expression (5.102) with the following:

$$\underset{n \times K}{M} = \underset{n \times n}{C}\ \underset{n \times K}{\Gamma}, \qquad \underset{n \times n}{C} = [c_{ij}] \tag{5.103}$$

which, differently from the linear combination (5.102), gives the relation

$$|\mu_i\rangle = C\, |\gamma_i\rangle, \tag{5.103a}$$

in which the single vector $|\gamma_i\rangle$ is transformed to the vector $|\mu_i\rangle$, with same index i.

Example 5.4 We write explicitly relations (5.102) and (5.103) in the binary case with the purpose of showing how to deal with composite matrices, whose entries are vectors instead of scalar elements.

The matrices Γ and M in an n-dimensional Hilbert space, where the kets must be considered as column vectors of size n, are

$$\Gamma = [|\gamma_1\rangle, |\gamma_2\rangle] = \begin{bmatrix} \gamma_{11} & \gamma_{12} \\ \vdots & \vdots \\ \gamma_{n1} & \gamma_{n2} \end{bmatrix}, \qquad M = [|\mu_1\rangle, |\mu_2\rangle] = \begin{bmatrix} \mu_{11} & \mu_{12} \\ \vdots & \vdots \\ \mu_{n1} & \mu_{n2} \end{bmatrix}.$$

For $K = 2$ relation (5.102) becomes

$$\underset{1\times2}{M} = \underset{1\times2}{\Gamma} \; \underset{2\times2}{A} \quad \rightarrow \quad [|\mu_1\rangle, |\mu_2\rangle] = [|\gamma_1\rangle, |\gamma_2\rangle] \begin{bmatrix} a_{11} & a_{12} \\ a_{21} & a_{22} \end{bmatrix} \tag{5.104a}$$

and more explicitly

$$\underset{n\times2}{M} = \underset{n\times2}{\Gamma} \; \underset{2\times2}{A} = \begin{bmatrix} \mu_{11} & \mu_{12} \\ \vdots & \vdots \\ \mu_{n1} & \mu_{n2} \end{bmatrix} = \begin{bmatrix} \gamma_{11} & \gamma_{12} \\ \vdots & \vdots \\ \gamma_{n1} & \gamma_{n2} \end{bmatrix} \begin{bmatrix} a_{11} & a_{12} \\ a_{21} & a_{22} \end{bmatrix}. \tag{5.104b}$$

The different dimensions, as appearing in the two writings above, are justified as follows: in (5.104a) the kets are regarded a single objects of dimensions 1×1, whereas in (5.104b) they become $1 \times n$ column vectors.

For $K = 2$ relation (5.103) becomes

$$\underset{1\times2}{M} = \underset{1\times1}{C} \; \underset{1\times2}{\Gamma} \quad \rightarrow \quad [|\mu_1\rangle, |\mu_2\rangle] = C \, [|\gamma_1\rangle, |\gamma_2\rangle] \tag{5.105a}$$

and more explicitly

$$\underset{n\times2}{M} = \underset{n\times n}{C} \; \underset{n\times2}{\Gamma} \quad \rightarrow \quad \begin{bmatrix} \mu_{11} & \mu_{12} \\ \vdots & \vdots \\ \mu_{n1} & \mu_{n2} \end{bmatrix} = \begin{bmatrix} c_{11} & \cdots & c_{1n} \\ \vdots & \ddots & \vdots \\ c_{n1} & \cdots & c_{nn} \end{bmatrix} \begin{bmatrix} \gamma_{11} & \gamma_{12} \\ \vdots & \vdots \\ \gamma_{n1} & \gamma_{n2} \end{bmatrix} \tag{5.105b}$$

Now in (5.105a) the matrix C must be regarded as a single object of dimension 1×1, and in fact, using this interpretation, it gives explicitly the relation

$$|\mu_1\rangle = C \, |\gamma_1\rangle, \qquad |\mu_2\rangle = C \, |\gamma_2\rangle$$

in agreement with (5.103a).

Problem 5.12 $\star$ Write the relations of Example 5.4 using the results of Helstrom's theory.

5.12.2 Matrices of the Inner Products and of the Outer Products

From the state matrix $\Gamma = [|\gamma_0\rangle, |\gamma_1\rangle, \ldots, |\gamma_{K-1}\rangle]$ two matrices can be formed

$$\underset{K \times K}{G} = \Gamma^* \Gamma, \qquad \underset{n \times n}{T} = \Gamma \Gamma^*. \tag{5.106}$$

The matrix G, called **Gram's matrix**, is the *matrix of inner products* with elements

$$G_{ij} = \langle \gamma_i | \gamma_j \rangle \tag{5.107}$$

while the matrix T gives the sum of the K *outer products*

$$T = \sum_{i=0}^{K-1} |\gamma_i\rangle \langle \gamma_i|. \tag{5.108}$$

These statements can be verified indicating with γ_{ri} the rth element of the column vector $|\gamma_i\rangle$, and performing the operations indicated in (5.106). As T is the sum of elementary operators in the Hilbert space $\mathcal{H}$, also T can be considered an operator of $\mathcal{H}$, which is sometimes called **Gram's operator** (see [12]).

The matrices (5.106) have the following properties:

(1) they are Hermitian semidefinite positive,
(2) both have the same rank as the matrix Γ,
(3) they have the same eigenvalues different from zero (and positive).

Let us prove (3). If λ is an eigenvalue of G, it follows that $G|v\rangle = \lambda|v\rangle$, where $|v\rangle$ is the eigenvector. Then, multiplying this relation by Γ we have

$$\Gamma G|v\rangle = \Gamma \Gamma^* \Gamma |v\rangle = T\Gamma|v\rangle = \lambda \Gamma|v\rangle$$

hence $T|u\rangle = \lambda|u\rangle$ with $|u\rangle = \Gamma|v\rangle$. Then λ is also an eigenvalue of T with eigenvector $\Gamma|v\rangle$. Analogously, we can see that if $\lambda \neq 0$ is an eigenvalue of T with eigenvector $|u\rangle$, we have that λ is also an eigenvalue of G with eigenvector $\Gamma^*|u\rangle$.

The properties (1), (2), and (3) have obvious consequences on the EID of G and of T. Indicating with r the rank and with $\sigma_1^2, \ldots, \sigma_r^2$ the positive eigenvalues, we obtain

$$T = U \Lambda_T U^* = \sum_{i=1}^{r} \sigma_i^2 |u_i\rangle \langle u_i| = U_r \Sigma_r^2 U_r^* \tag{5.109a}$$

$$G = V \Lambda_G V^* = \sum_{i=1}^{r} \sigma_i^2 |v_i\rangle \langle v_i| = V_r \Sigma_r^2 V_r^* \tag{5.109b}$$

where

- U is a $n \times n$ unitary matrix,
- $\{|u_i\rangle\}$ is an orthonormal basis of $\mathcal{H}$ formed by the columns of the matrix U,
- Λ_T is an $n \times n$ diagonal matrix whose first r diagonal elements are the positive eigenvalues σ_i^2, and the other $n - r$ diagonal elements are null,
- V is an $K \times K$ unitary matrix,
- $\{|v_i\rangle\}$ is an orthonormal basis of $\mathbb{C}^K$ formed by the columns of V,
- Λ_G is an $K \times K$ diagonal matrix whose first r diagonal elements are the positive eigenvalues σ_i^2 and the other $n - r$ diagonal elements are null,
- U_r and V_r are formed by the first r columns of U and V, respectively,
- $\Sigma_r^2 = \mathrm{diag}[\sigma_1^2, \ldots, \sigma_r^2]$.

In (5.109) appear both the *full* form and the *reduced* form of the EIDs (see Sect. 2.11).

5.12.3 Singular Value Decomposition of Γ

Combining the EIDs of Gram's operator T and of Gram's matrix G we obtain the SVD of the state matrix Γ. The result is (see [13])

$$\Gamma = U \Sigma V^* = \sum_{i=1}^{r} \sigma_i |u_i\rangle \langle v_i| = U_r \Sigma_r V_r^* \qquad (5.110)$$

where U, V, U_r, V_r, and Σ_r are the matrices that appear in the previous EIDs, Σ is an $n \times K$ diagonal matrix whose first r diagonal elements are given by the square root σ_i of the positive eigenvalues σ_i^2 of T and G and the other diagonal elements are null.

Before discussing and applying the above decompositions, let us develop a couple of examples.

Example 5.5 Consider a binary system ($K = 2$) on $\mathcal{H} = \mathbb{C}^4$, where the two states are specified by the matrix

$$\Gamma = \frac{1}{2} \begin{bmatrix} 1 & 1 \\ -1 & 1 \\ 1 & -1 \\ -1 & 1 \end{bmatrix}.$$

The matrices G and T become respectively

$$G = \Gamma^* \Gamma = \begin{bmatrix} 1 & -\frac{i}{2} \\ \frac{i}{2} & 1 \end{bmatrix} \qquad T = \Gamma \Gamma^* = \frac{1}{4} \begin{bmatrix} 2 & -1-i & 1+i & 0 \\ -1+i & 2 & -2 & 1+i \\ 1-i & -2 & 2 & -1-i \\ 0 & 1-i & -1+i & 2 \end{bmatrix}.$$

The eigenvalues of G are $\sigma_1^2 = 3/2$ and $\sigma_2^2 = 1/2$ and the corresponding EID is

$$G = V \Lambda_G V^* \quad \text{with} \quad V = \frac{1}{2}\begin{bmatrix} -1 & 1 \\ 1 & 1 \end{bmatrix}, \qquad \Lambda_G = \Sigma_r^2 = \begin{bmatrix} \frac{3}{2} & 0 \\ 0 & \frac{1}{2} \end{bmatrix}$$

which coincides with the reduced EID, because $r = K = 2$. The eigenvalues of T are $\sigma_1^2 = 3/2$, $\sigma_2^2 = 1/2$, $\sigma_3 = \sigma_4 = 0$ and the corresponding reduced EID is

$$T = U_r \Sigma_r^2 U_r^* \quad \text{with} \quad U_r = \begin{bmatrix} 0 & 1 \\ \frac{1}{\sqrt{3}} & 0 \\ -\frac{1}{\sqrt{3}} & 0 \\ \frac{1}{\sqrt{3}} & 0 \end{bmatrix}, \qquad U_r^* = \begin{bmatrix} 0 & \frac{1}{\sqrt{3}} & -\frac{1}{\sqrt{3}} & \frac{1}{\sqrt{3}} \\ 1 & 0 & 0 & 0 \end{bmatrix},$$

$$\Sigma_r^2 = \begin{bmatrix} \frac{3}{2} & 0 \\ 0 & \frac{1}{2} \end{bmatrix}.$$

The reduced SVD of Γ is: $\Gamma = U_r \Sigma_r V^*$, where the factors are specified above.

Example 5.6 Consider a constellation composed by two coherent states with real parameters $\pm\alpha$ (see Sect. 3.2.2)

$$|\gamma_1\rangle = |-\alpha\rangle, \qquad |\gamma_2\rangle = |\alpha\rangle, \qquad |\gamma_1\rangle, |\gamma_1\rangle \in \mathcal{G}, \alpha \in \mathbb{R}$$

which, as well known, must be defined on an infinite-dimensional Hilbert space. The purpose of the example is to show that, in spite of the infinite dimensions, eigenvalues and eigenvectors can be developed in finite terms (at least for the parts that are connected to the following applications).

The expressions of the two states are (see (3.4))

$$|\gamma_1\rangle = \sum_{n=0}^{\infty} e^{-\alpha^2/2}\frac{(\alpha)^n}{\sqrt{n!}}|n\rangle, \qquad |\gamma_2\rangle = \sum_{n=0}^{\infty} e^{-\alpha^2/2}\frac{(-\alpha)^n}{\sqrt{n!}}|n\rangle \qquad (5.111)$$

an so the corresponding matrix becomes

$$\Gamma = [|\gamma_1\rangle, |\gamma_2\rangle] = \sum_{n=0}^{\infty} \frac{e^{-\alpha^2/2}}{\sqrt{n!}}[\alpha^n, (-\alpha)^n]|n\rangle \qquad (5.112)$$

and has dimensions $\infty \times 2$. We can easily see that these two vectors are linearly independent and therefore the rank of Γ is $r = K = 2$.

Gram's matrix is 2×2 and becomes

$$G = \begin{bmatrix} \langle\gamma_1|\gamma_1\rangle & \langle\gamma_1|\gamma_2\rangle \\ \langle\gamma_2|\gamma_1\rangle & \langle\gamma_2|\gamma_2\rangle \end{bmatrix} = \begin{bmatrix} 1 & \gamma_{12} \\ \gamma_{12} & 1 \end{bmatrix} \qquad (5.113)$$

where (see (7.10)) $\gamma_{12} = e^{-2\alpha^2}$, whereas Gram's operator T is infinite dimensional and has a rather complicated expression, that can be obtained from (5.111) developing the outer products as follows:

$$T = |\gamma_1\rangle\langle\gamma_1| + |\gamma_2\rangle\langle\gamma_2|.$$

The eigenvalues of G are given by the solution of the equation

$$\det(G - \lambda\,I) = (1 - \lambda)^2 - \gamma_{12}^2 = 0$$

and therefore we have, with the notation of (5.113)

$$\sigma_1^2 = 1 + \gamma_{12}, \qquad \sigma_2^2 = 1 - \gamma_{12} \tag{5.114}$$

and the normalized eigenvectors are

$$|v_1\rangle = \begin{bmatrix} \frac{1}{\sqrt{2}} \\ \frac{1}{\sqrt{2}} \end{bmatrix}, \qquad |v_2\rangle = \begin{bmatrix} \frac{1}{\sqrt{2}} \\ -\frac{1}{\sqrt{2}} \end{bmatrix}.$$

In this way, we have performed the spectral decomposition of G in the form (5.109b) with

$$V = \frac{1}{\sqrt{2}} \begin{bmatrix} 1 & 1 \\ 1 & -1 \end{bmatrix}, \qquad \Lambda_G = \begin{bmatrix} \sigma_1^2 & 0 \\ 0 & \sigma_2^2 \end{bmatrix}.$$

The spectral decomposition of T, given by (5.109a), requires the computation of the eigenvectors $|u_1\rangle, |u_2\rangle$ which are of infinite dimension. In principle, such computation can be done, but it is very complicated, and so the vectors, for now, are left indicated in a nonexplicit form.

The singular value decomposition of Γ results in

$$\Gamma = \sigma_1|u_1\rangle\langle v_1| + \sigma_2|u_2\rangle\langle v_2|$$

where the singular values are $\sigma_{1,2} = \sqrt{1 \pm \gamma_{12}}$.

5.12.4 Spaces, Subspaces, Bases, and Operators

In the above decompositions several spaces and subspaces come into play. The reference environment is the Hilbert space $\mathcal{H}$, which is assumed of dimension n. We then have the subspace generated by the states

$$\mathcal{U} = \mathrm{span}(|\gamma_0\rangle, |\gamma_1\rangle, \ldots |\gamma_{K-1}\rangle)$$

of dimension r, which is also the subspace where the measurement vectors $|\mu_i\rangle$ operate (see Fig. 5.9). The unitary operator

$$\underset{n \times n}{U} = [\,|u_1\rangle, \ldots, |u_n\rangle\,] \; : \; \mathcal{H} \to \mathcal{H}$$

provides with its n columns an orthonormal basis for $\mathcal{H}$, while its first r columns, corresponding to the non-null eigenvalues σ_i^2, form a basis for the subspace $\mathcal{U}$

$$\mathcal{U} = \mathrm{span}(|u_1\rangle, \ldots, |u_r\rangle) \subseteq \mathcal{H}.$$

These r eigenvectors were collected in the matrix U_r that appears in the reduced EID of T (see (5.109a)); the remaining $n - r$ eigenvectors $|u_{r+1}\rangle, \ldots, |u_n\rangle$ generate the complementary space $\mathcal{U}^\perp$. Then the following resolutions are found

$$\sum_{k=1}^{n} |u_k\rangle\langle u_k| = U\, U^* = I_{\mathcal{H}}, \qquad \sum_{k=1}^{r} |u_k\rangle\langle u_k| = U_r\, U_r^* = P_{\mathcal{U}} \qquad (5.115)$$

where $P_{\mathcal{U}}$ is the *projector* on $\mathcal{U}$. Analogously, the unitary operator ($K \times K$ matrix)

$$V = [\,|v_1\rangle, \ldots, |v_K\rangle\,] \; : \; \mathbb{C}^K \to \mathbb{C}^K$$

provides with its K columns a basis for $\mathbb{C}^K$, while its first r columns provide a basis for an r-dimensional subspace $\mathcal{V}$ of $\mathbb{C}^K$.

$$\mathrm{span}(|v_1\rangle, \ldots, |v_r\rangle) = \mathcal{V} \subseteq \mathbb{C}^K.$$

We obtain the resolutions

$$\sum_{k=1}^{K} |v_k\rangle\langle v_k| = V\, V^* = I_K, \qquad \sum_{k=1}^{r} |v_k\rangle\langle v_k| = V_r\, V_r^* = P_{\mathcal{V}}. \qquad (5.116)$$

The state matrix defines a *linear transformation*[8]

$$\Gamma \; : \; \mathbb{C}^K \to \mathcal{H}$$

because it "accepts" at the input a ket $|v\rangle \in \mathbb{C}^K$ and produces the ket $\Gamma |v\rangle \in \mathcal{H}$. The image of Γ is

$$\mathrm{im}\, \Gamma = \mathcal{U}.$$

---

[8] The term *operator*, in practice represented by a square matrix, is reserved to *linear transformations* from one space to the same space.

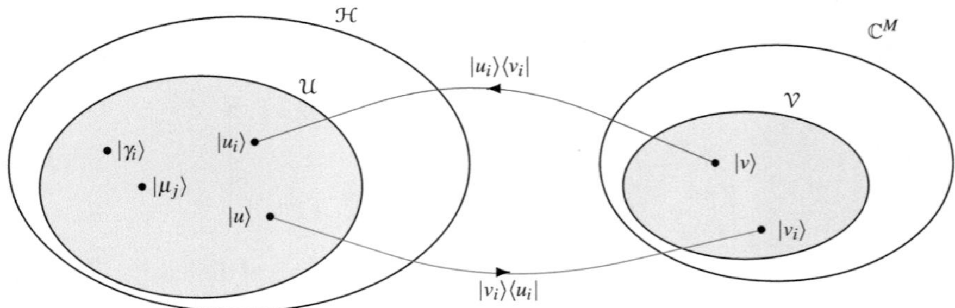

Fig. 5.12 Spaces and subspaces generated by a constellation of states. In *red* and *green* the tranjectors

Analogously, the adjoint matrix $\Gamma^* : \mathcal{H} \to \mathbb{C}^K$ operates on a ket $|u\rangle \in \mathcal{H}$ and returns the ket $\Gamma^*|u\rangle \in \mathbb{C}^K$. The image of Γ^* is: im $\Gamma^* = \mathcal{V}$. The connection between $\mathbb{C}^K$ and $\mathcal{H}$ is made by the elementary operators $|u_i\rangle\langle v_i|$ appearing in the SVD (5.110). These operators transform a ket $|v\rangle$ of $\mathbb{C}^K$ to the ket

$$|u_i\rangle\langle v_i|v\rangle = k_i|u_i\rangle \in \mathcal{H}, \quad \text{with} \quad k_i = \langle v_i|v\rangle$$

and, because they provide a transfer (from $\mathbb{C}^K$ to $\mathcal{H}$ and from $\mathcal{H}$ to $\mathbb{C}^K$), they are named "transjectors" in [14] (Fig. 5.12).

Analogously, the connection between $\mathcal{H}$ and $\mathbb{C}^K$ is done by the elementary operators $|v_i\rangle\langle u_i|$ of the SVD (5.110).

5.12.5 The Geometry with Mixed States

All the above considerations, referring to pure states, can be extended in a rather obvious way to mixed states with some dimensional changes. The starting point is the matrix of the states, which now collects the factors γ_i of the density operators ρ_i

$$\underset{n \times H}{\Gamma} = \left[\gamma_0, \gamma_1, \ldots, \gamma_{K-1}\right] \tag{5.117}$$

where the number of the columns $H = h_0 + h_1 + \cdots h_{K-1}$ is given by the total number of columns of the state factors γ_i. As we have seen in (5.68), this matrix can be considered as a collection of H kets of $\mathcal{H}$, which generate the subspace $\mathcal{U}$, whose dimension r is always given by the rank of Γ.

Gram's operator has the expressions

$$\underset{n \times n}{T} = \Gamma\Gamma^* = \sum_{i=0}^{K-1} \gamma_i \gamma_i^* = \sum_{i=0}^{K-1} \rho_i \tag{5.118}$$

and therefore can be directly evaluated from the ρ_i, without finding their factorizations. Its dimensions remain $n \times n$. Instead, Gram's matrix becomes $H \times H$ and has the structure

$$\underset{H \times H}{G} = \Gamma^* \, \Gamma = \begin{bmatrix} \gamma_0^* \, \gamma_0 & \cdots & \gamma_0^* \, \gamma_{K-1} \\ \vdots & \ddots & \vdots \\ \gamma_{K-1}^* \, \gamma_0 & \cdots & \gamma_{K-1}^* \, \gamma_{K-1} \end{bmatrix} \tag{5.119}$$

where the $\gamma_i^* \, \gamma_j$ are not ordinary inner products, but matrices of dimensions $h_i \times h_j$.

Finally, the subspace $\mathcal{V}$ becomes of dimensions $H \geq K$. This part concerning mixed states will be further developed in Chap. 8.

5.12.6 Conclusions

We have seen that a constellation of states (or of state factors) gathered in the matrix Γ, can be defined on the Hilbert space $\mathcal{H}$ and, more precisely, on its subspace $\mathcal{U}$, generating several operators.

It remains to evaluate the measurement matrix M identifying the measurement operators. To get specific results we must state the objective, which, in the context of quantum communications, is the *maximization of the correct decision probability*. An alternative objective, which brings to a suboptimal solution, is to *minimize the quadratic error between the states and the corresponding measurement vectors*. This technique, called *square root measurement* (SRM), will be seen in the next chapter.

5.13 The Geometrically Uniform Symmetry (GUS)

The set of the states (constellation) can have a symmetry that facilitates its study and its performance evaluation. The kind of symmetry that allows for these simplifications is called *geometrically uniform symmetry* (GUS) and is verified in several quantum communications systems, like the quantum systems obtained with the modulations PSK and PPM and all the binary systems.[9]

5.13.1 The Geometrically Uniform Symmetry with Pure States

A constellation of K pure states

$$\{|\gamma_0\rangle, |\gamma_1\rangle, \ldots, |\gamma_{K-1}\rangle\}$$

---

[9] The interest of the GUS is confined to the case in which the a priori probabilities are equal $(q_i = 1/K)$.

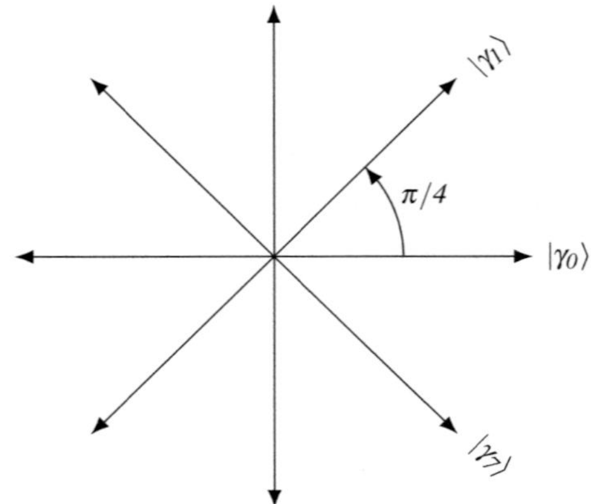

Fig. 5.13 Constellation of states with the geometrically uniform symmetry in the complex plane $\mathbb{C}$. The reference state is $|\gamma_0\rangle = 1$ (the complex number 1) and the symmetry operator is $S = e^{i\pi/4}$

has the *geometrically uniform symmetry* when the two properties are verified:

(1) the K states $|\gamma_i\rangle$ are obtained from a single reference state $|\gamma_0\rangle$ in the following way

$$|\gamma_i\rangle = S^i |\gamma_0\rangle, \qquad i = 0, 1, \dots, K - 1 \tag{5.120a}$$

where S is a unitary operator, called *symmetry operator*;

(2) the operator S is a Kth root of the identity operator in the sense that

$$S^K = I_{\mathcal{H}}. \tag{5.120b}$$

An elementary example of constellation that verifies the GUS is given by the K roots of unity in the complex plane, as shown in Fig. 5.13 for $K = 8$.

In the presence of the GUS, the specification of the constellation is limited to the reference state $|\gamma_0\rangle$ and to the symmetry operator S. In addition, it simplifies the decision, because, as we shall see for the optimal decision, we can choose the measurement vectors with the same symmetry as the states, that is,

$$|\mu_i\rangle = S^i |\mu_0\rangle, \qquad i = 0, 1, \dots, K - 1. \tag{5.121}$$

In the next chapter we will verify that the PSK and PPM systems have the GUS. Here we limit ourselves to the binary case.

5.13.2 All Binary Constellations Have the GUS

A constellation of two arbitrary states, $|\gamma_0\rangle$ and $|\gamma_1\rangle$, is always geometrically uniform, with symmetry operator S defined by [14]

$$S = I_{\mathcal{H}} - 2\frac{|w\rangle\langle w|}{\langle w|w\rangle} \tag{5.122}$$

where $|w\rangle = |\gamma_1\rangle - |\gamma_0\rangle$ if the two states have inner product $X := \langle\gamma_0|\gamma_1\rangle$ real. In this case S is a "reflector," which reflects a state with respect to the hyperplane (bisector) determined by the vectors $|\gamma_0\rangle$ and $|\gamma_1\rangle$. It can be verified from definition (5.122) that S is unitary and $S^2 = I_{\mathcal{H}}$ (see problems).

If the inner product X is complex, $X = |X|e^{i\phi}$, we modify $|\gamma_1\rangle$ as $|\tilde{\gamma}_1\rangle = e^{-i\phi}|\gamma_1\rangle$ and apply (5.122) to the states $|\gamma_0\rangle$ and $|\tilde{\gamma}_1\rangle$, which have a real inner product. This does not represent any restriction because $|\gamma_1\rangle$ and $|\tilde{\gamma}_1\rangle$ differ by a phase factor and therefore represent the same physical state.

5.13.3 The GUS with Mixed States

The definition of GUS is now extended to mixed states. A constellation of K density operators

$$\{\rho_0, \rho_1, \ldots, \rho_{K-1}\}$$

has the *geometrically uniform symmetry* when the following two properties are verified:

(1) the K operators ρ_i are obtained from a single reference operator ρ_0 as

$$\rho_i = S^i \rho_0 (S^i)^*, \qquad i = 0, 1, \ldots, K - 1 \tag{5.123}$$

where S is a unitary operator called *symmetry operator*;
(2) the operator S is a Kth root of the identity operator

$$S^K = I_{\mathcal{H}}. \tag{5.123b}$$

This extension is in harmony with the fact that with pure states the density operators become $\rho_i = |\gamma_i\rangle\langle\gamma_i|$. In addition, with the factorization of the density operators, $\rho_i = \gamma_i \gamma_i^*$, relation (5.123) gives

$$\gamma_i = S^i \gamma_0, \qquad i = 0, 1, \ldots, K - 1 \tag{5.124}$$

which generalizes (5.120a). In the context of optimal decision [3] we will prove that the same symmetry is transferred to the measurement operators, and also to the measurement factors, namely,

$$\mu_i = S^i \mu_0, \qquad i = 0, 1, \ldots, K - 1. \tag{5.125}$$

5.13.4 Generalizations of the GUS

The GUS can be generalized in two ways. We limit ourselves to introducing the two generalizations in the case of pure states. In the first generalization [3], we have L reference states $|\gamma_{01}\rangle, \ldots, |\gamma_{0L}\rangle$, instead of a single state $|\gamma_0\rangle$, and the constellation is subdivided into L subconstellations generated by a single symmetry operator S in the form $|\gamma_{ik}\rangle = S^i |\gamma_{0k}\rangle$. An example of modulation that has this kind of *composite* GUS is the Quadrature Amplitude Modulation (QAM), which will be seen in Chap. 7.

In the second type of generalization [14], we have K distinct symmetry operators S_i, made up of K unitary matrices forming a multiplicative group, and each state of the constellation is generated in the form $|\gamma_i\rangle = S_i |\gamma_0\rangle$ from a single reference state $|\gamma_0\rangle$.[10]

5.13.5 Eigendecomposition of the Symmetry Operator

The EID of the symmetry operator S plays an important role in the analysis of Communications Systems having the GUS. We give the two equivalent forms of EIDs of S (see Sects. 2.10 and 2.11)

$$S = \sum_{i=1}^{k} \lambda_i \, P_i, \qquad S = Y \, \Lambda \, Y^* = \sum_{i=0}^{n-1} \overline{\lambda}_i |y_i\rangle\langle y_i| \tag{5.126}$$

where $\{\lambda_i, i = 1, \ldots, k\}$ are the distinct eigenvalues of S, $\{P_i, i = 1, \ldots, k\}$ form a projector system, that is, with $P_i P_j = \delta_{ij} P_i$, Y is an $n \times n$ unitary matrix, and $\Lambda = \mathrm{diag}[\overline{\lambda}_1, \ldots, \overline{\lambda}_n]$ contains the nondistinct eigenvalues. In general, the distinct eigenvalues λ_i have a multiplicity $c_i \geq 1$.

Considering that S is a unitary operator, the $\overline{\lambda}_i$ have unitary amplitude and, because $S^K = I_{\mathcal{H}}$, the eigenvalues have the form

$$\lambda_i = W_K^{r_i}, \qquad 0 \leq r_i < K \tag{5.127}$$

where $W_K := \mathrm{e}^{\mathrm{i}2\pi/K}$ and r_i are integers. Now, in the second EID, collecting the elementary projectors $|y_j\rangle\langle y_j|$ with a common eigenvalue, we arrive at the form

$$S = \sum_{i=0}^{K-1} W_K^i \, Y_i \, Y_i^* \tag{5.128}$$

---

[10] In the literature [3] the set of the states that satisfy (5.120) is called *cyclic state set*, whereas the term *geometrically uniform symmetry* indicates the general case, which is obtained with a multiplicative group of unitary matrices.

where Y_i are $n \times c_i$ matrices, with c_i the multiplicity of $\lambda_i = W_K^i$. Note that the projectors are given by $P_i = Y_i Y_i^*$.

Example 5.7 In the PSK the symmetry operator is given by

$$S = \mathrm{diag}[W_K^k, k = 0, 1, \ldots, K - 1]. \tag{5.129}$$

As S is diagonal, its EID is immediately found as $S = I_n S I_n^*$, with I_n the identity matrix. For example, for $K = 3$ and $n = 6$, we have tree distinct eigenvalues

$$\Lambda = \mathrm{diag}[1, W_3, W_3^2, 1, W_3, W_3^2]$$

and the EID results in

$$S = \begin{bmatrix} 1\,0\,0\,0\,0\,0 \\ 0\,1\,0\,0\,0\,0 \\ 0\,0\,1\,0\,0\,0 \\ 0\,0\,0\,1\,0\,0 \\ 0\,0\,0\,0\,1\,0 \\ 0\,0\,0\,0\,0\,1 \end{bmatrix} \begin{bmatrix} 1 & 0 & 0 & 0 & 0 & 0 \\ 0 & W_3 & 0 & 0 & 0 & 0 \\ 0 & 0 & W_3^2 & 0 & 0 & 0 \\ 0 & 0 & 0 & 1 & 0 & 0 \\ 0 & 0 & 0 & 0 & W_3 & 0 \\ 0 & 0 & 0 & 0 & 0 & W_3^2 \end{bmatrix} \begin{bmatrix} 1\,0\,0\,0\,0\,0 \\ 0\,1\,0\,0\,0\,0 \\ 0\,0\,1\,0\,0\,0 \\ 0\,0\,0\,1\,0\,0 \\ 0\,0\,0\,0\,1\,0 \\ 0\,0\,0\,0\,0\,1 \end{bmatrix}$$

Now, to obtain the form (5.128), we must collect in the matrices Y_i the eigenvectors corresponding to the eigenvalues W_3^i. Thus

$$Y_0 = \begin{bmatrix} 1\,0 \\ 0\,0 \\ 0\,0 \\ 0\,1 \\ 0\,0 \\ 0\,0 \end{bmatrix}, \qquad Y_1 = \begin{bmatrix} 0\,0 \\ 1\,0 \\ 0\,0 \\ 0\,0 \\ 0\,1 \\ 0\,0 \end{bmatrix}, \qquad Y_2 = \begin{bmatrix} 0\,0 \\ 0\,0 \\ 1\,0 \\ 0\,0 \\ 0\,0 \\ 0\,1 \end{bmatrix}.$$

5.13.6 *Commutativity of S with T*

An important property with GUS, proved in Appendix section "Commutativity of the Operators T and S", is given by:

Proposition 5.8 *Gram's operator and the symmetry operator of the GUS commute*

$$TS = ST. \tag{5.130}$$

This leads to the simultaneous diagonalization (see Theorem 2.4) of T and S, stated by

$$T = U\,\Sigma^2\,U^*, \qquad S = U\Lambda U^*. \tag{5.131}$$

Note that, in general, the eigenvalues of the symmetry operator are multiple and then the diagonalization of S is not unique (see Theorem 2.3 of Sect. 2.11). This multiplicity will be used to find useful simultaneous decompositions, as will be seen at the end of Chap. 8.

Problem 5.13 ★★   Prove that the quantum states of $\mathcal{H} = \mathbb{C}^4$

$$|\gamma_0\rangle = \frac{1}{2}[1, -1, 1, -1]^{\mathrm{T}}, \qquad |\gamma_1\rangle = \frac{1}{2}[1, 1, -1, 1]^{\mathrm{T}}$$

verify the GUS for a binary transmission. Find the symmetry operator S, verify that S has the properties of a symmetry operator and that $|\gamma_1\rangle$ is obtained from $|\gamma_0\rangle$ as $|\gamma_1\rangle = S|\gamma_0\rangle$.

Problem 5.14 ★   Find the EID of the symmetry operator S of the previous problem.

Problem 5.15 ★★   Prove that the two quantum states of $\mathcal{H} = \mathbb{C}^4$

$$|\gamma_0\rangle = \frac{1}{2}[1, -1, 1, -1]^{\mathrm{T}}, \qquad |\gamma_1\rangle = \frac{1}{2}[1, 1, -i, 1]^{\mathrm{T}}$$

verify the GUS for a binary transmission, and find the corresponding symmetry operator S. Note that in this case the inner product $X := \langle\gamma_0|\gamma_1\rangle$ is complex.

5.14 Optimization with Geometrically Uniform Symmetry

In the general case of *weighted* density operators the geometrically uniform symmetry (GUS) is established by the condition

$$\widehat{\rho}_i = S^i \widehat{\rho}_0 (S^i)^*, \qquad i = 0, 1, \ldots, K - 1. \tag{5.132}$$

In such case, the search for the optimal measurement operators is simplified because the data are restricted to the reference operator $\widehat{\rho}_0$ and to the symmetry operator S, and in addition the search can be restricted to the measurement operator Q_0 only.

5.14.1 Symmetry of the Measurement Operators

The GUS is transferred also to the measurement operators, according to

Proposition 5.9 *If the weighted density operators have the GUS, established by (5.132), it is not restrictive to suppose that also the optimal measurement operators have the GUS, with the same symmetry operator, namely*

$$Q_i = S^i \, Q_0 \, S^{-i}, \qquad i = 0, 1, \ldots, K - 1. \tag{5.133}$$

Proof Holevo's theorem ensures that there exists a system of optimal measurement operators $\mathbf{Q} = \mathbf{Q}_{\mathrm{opt}} \in \mathcal{M}_0$ that maximizes the functional $J(\mathbf{Q})$ defined by (5.80). The point here is to prove that from this system, which does not necessarily enjoy the GUS, another system can be obtained $\tilde{\mathbf{Q}} \in \mathcal{M}_0$ that enjoys the GUS and has the same properties as the original system. To this end, we define

$$\tilde{Q}_0 = \frac{1}{K} \sum_{i=0}^{K-1} S^{-i} Q_i S^i, \qquad \tilde{Q}_i = S^i \tilde{Q}_0 S^{-i} \quad, \quad i = 1, \ldots, K - 1.$$

We soon verify that the new operators are PSD. In addition

$$\sum_{i=0}^{K-1} \tilde{Q}_i = \frac{1}{K} \sum_{i=0}^{K-1} \sum_{j=0}^{K-1} S^{i-j} Q_j S^{-(i-j)} = \frac{1}{K} \sum_{j=0}^{K-1} \sum_{k=0}^{K-1} S^k Q_j S^{-k}$$

where the periodicity of the symmetry operator S is used. Then

$$\sum_{i=0}^{K-1} \tilde{Q}_i = \frac{1}{K} \sum_{k=0}^{K-1} S^k \sum_{j=0}^{K-1} Q_j S^{-k} = \frac{1}{K} \sum_{k=0}^{K-1} S^k S^{-k} = I_{\mathcal{H}}.$$

We conclude that the new operators $\tilde{Q}_i$ are legitimate measurement operators. We have also

$$J(\tilde{\mathbf{Q}}) = \sum_{i=0}^{K-1} \mathrm{Tr}[\widehat{\rho}_i \, \tilde{Q}_i] = \sum_{i=0}^{K-1} \mathrm{Tr}[S^i \widehat{\rho}_0 \, \tilde{Q}_0 S^{-i}]$$

$$= \sum_{i=0}^{K-1} \mathrm{Tr}[\widehat{\rho}_0 \, \tilde{Q}_0] = K \, \mathrm{Tr}[\widehat{\rho}_0) \, \tilde{Q}_0]$$

$$= \mathrm{Tr}\left[\widehat{\rho}_0 \sum_{i=0}^{K-1} S^{-i} Q_i S^i \right] = \sum_{i=0}^{K-1} \mathrm{Tr}[S^i \widehat{\rho}_0 S^{-i} Q_i] = J(\mathbf{Q})$$

so that even the new measurement operators are optimal. $\square$

We must observe also that, choosing measurement operators that enjoy the GUS, for the maximum correct decision probability we simply have

$$P_{c\mathrm{max}} = J(\mathbf{Q}_{\mathrm{opt}}) = K \, \mathrm{Tr}[\widehat{\rho}_0 \, Q_{0,\mathrm{opt}}] \tag{5.134}$$

where $Q_{0,\mathrm{opt}}$ (to be found) identifies the optimal measurement operator system.

5.14.2 Holevo's Theorem with GUS

From the previous results, Holevo's theorem becomes:

Theorem 5.5 (Holevo's theorem with GUS) *In a K-ary system characterized by the weighted density operators $\widehat{\rho}_i = q_i \rho_i$, that enjoy the GUS according to (5.132), the optimal measurement operators Q_i can be chosen with the same GUS, according to (5.133). Then the reference operator Q_0 produces a system of optimal operators if and only if, having defined the operator*

$$L = \sum_{i=0}^{K-1} S^i Q_0 \widehat{\rho}_0 S^{-i}, \tag{5.135}$$

we have that the operator $L - \widehat{\rho}_0$ is PSD and verifies the condition $(L - \widehat{\rho}_0)Q_0 = 0_{\mathcal{H}}$. We also have that S commutes with L.

In fact, the operator L is obtained by (5.83) substituting the symmetry expressions (5.132) and (5.133). We can also verify that

$$L = S^i L S^{-i} \quad \text{for every integer} \quad i \tag{5.136}$$

from which we obtain, in particular, that S and L commute. From (5.136) we can prove that, if $L - \widehat{\rho}_0$ is PSD, so are $L - \widehat{\rho}_i$, and that, if $(L - \widehat{\rho}_0)Q_0 = 0_{\mathcal{H}}$, also $(L - \widehat{\rho}_i)Q_i = 0_{\mathcal{H}}$, so that all the conditions of Holevo's theorem are verified.

Even the dual theorem is simplified taking the following form [15]:

Theorem 5.6 (Dual theorem with GUS) *In a K-ary system characterized by the weighted density operators $\widehat{\rho}_i = q_i \rho_i$ that enjoy the GUS with symmetry operators S, a measurement operator system $\{Q_i\}$ that enjoy the GUS is optimal if there exists a PSD operator X with the properties: (1) $X \geq \rho_0$, (2) $X S = S X$, and (3) $\mathrm{Tr}[X]$ is minimal. The operator Q_0 that generates the optimal operators satisfies the condition $(X - \widehat{\rho}_0)Q_0 = 0_{\mathcal{H}}$ and the minimum obtained for $\mathrm{Tr}[X]$ coincides with the requested maximum of $J(\mathbf{Q})$.*

In the assumed conditions we have in fact that $X = S^i X S^{-i}$ for every i. Thus $X - \widehat{\rho}_i = S^i (X - \widehat{\rho}_0)S^{-i}$ is PSD and $(X - \widehat{\rho}_i)Q_i = S^i (X - \widehat{\rho}_0)S^{-i} = 0_{\mathcal{H}}$, in such a way that the conditions of the theorem dual to Holevo's theorem are satisfied.

Note that, in the presence of GUS, the quantum source and the optimal decision become completely specified by the symmetry operator S and by the reference operators ρ_0 and Q_0 (or by their factors γ_0 and μ_0). This has a consequence also in the simplification of convex linear programming (CSP).

5.14.3 Numerical Optimization with MatLab$^{©}$

In the presence of GUS, referring to Theorem 5.6, the input data are reduced to the weighted density $\widehat{\rho}_0$ and to the symmetry operator S. The constraints to be applied are

$$X - \widehat{\rho}_0 \geq 0, \qquad XS = SX$$

and the requested output is the operator X of minimal trace.

In MatLab the use of the `cvx` procedure seen in Sect. 5.10 becomes

```
%%%%%%%%%%%%%%%%%%%%%%%%%%%%%%%%%%%%%%%%%%%%%%%%%%%%%%%%%%
% cvx procedure applied to the dual problem with GUS
%%%%%%%%%%%%%%%%%%%%%%%%%%%%%%%%%%%%%%%%%%%%%%%%%%%%%%%%%%
 cvx_begin
variables X(dim)
minimize(trace(X))
subject to
 X>rho0;
 X*S==S*X;
  cvx_end
```

Applications of this simplified procedure to Quantum Communications systems will be seen in Chaps. 7 and 8.

5.15 State Compression in Quantum Detection

Quantum detection is formulated in an n-dimensional (possibly infinite) Hilbert space $\mathcal{H}$, but in general, the quantum states and the corresponding measurement operators span an r-dimensional subspace $\mathcal{U}$ of $\mathcal{H}$, with $r \leq n$. Quantum detection could be restricted to this subspace, but the operations involved are redundant for $r < n$, since the kets in $\mathcal{U}$ have n components, as the other kets of $\mathcal{H}$. It is possible and convenient to perform a compression from the subspace $\mathcal{U}$ onto a "compressed" space $\overline{\mathcal{U}}$, where the redundancy is removed (kets are represented by r components). We will show that in the "compressed" space the quantum detection can be perfectly reformulated without loss of information, and some properties become simpler than in the original (uncompressed) Hilbert space $\mathcal{H}$ [16].

State compression has some similarity with **quantum compression**, which will be developed in Chap. 12 in the framework of Quantum Information Theory. Both techniques have the target of representing quantum states more efficiently, but state compression does not consider the information content (entropy) of the states and is based only on geometrical properties.

Before proceeding it is convenient to recall the dimensions, which have a fundamental role in this topic:

- n: dimension of the Hilbert space $\mathcal{H}$,
- K: size of the alphabet,
- r: common rank of Γ, G, and T and dimension of the compressed space $\overline{\mathcal{H}}$.

- pure states $\quad \underset{n \times K}{\Gamma}, \quad \underset{K \times K}{G}, \quad \underset{n \times n}{T},$ $\qquad\qquad\qquad$ (5.137a)

- mixed states $\quad \underset{H \times K}{\Gamma}, \quad \underset{H \times H}{G}, \quad \underset{n \times n}{T},$ $\qquad\qquad\qquad$ (5.137b)

where $H = h_0 + \cdots h_{K-1}$ with h_i the number of columns of the factors γ_i and μ_i.

We will refer to mixed states since they represent the general case and the most interesting one with compression.

5.15.1 State Compression and Expansion

To find the compression operation (and also the expansion) we rewrite the SVD of the state matrix Γ, given by (5.110)

$$\underset{n \times H}{\Gamma} = U \, \Sigma \, V_r^* = U_r \, \Sigma_r \, V_r^* = \sum_{i=1}^{r} \sigma_i |u_i\rangle \langle v_i| \qquad (5.138)$$

where $U = [|u_1\rangle, \ldots, |u_n\rangle]$ is an $n \times n$ unitary matrix, $V_r = [|v_1\rangle, \ldots, |v_r\rangle]$ is an $r \times r$ unitary matrix, Σ is an $n \times r$ diagonal matrix whose first r diagonal entries $\sigma_1, \ldots \sigma_r$ are the (positive) singular values, and the other diagonal entries are zero, $\Sigma_r = \mathrm{diag}\{\sigma_1, \ldots \sigma_r\}$ is $r \times r$ diagonal, $U_r = [|u_1\rangle, \ldots, |u_r\rangle]$ is formed by the first r columns of U. We also recall that U_r gives the *projector* operator onto $\mathcal{U}$ as (see (5.115))

$$\sum_{i=1}^{r} |u_i\rangle \langle u_i| = U_r \, U_r^* = P_{\mathcal{U}}. \qquad (5.139)$$

In the r-dimensional subspace $\mathcal{U}$ the kets $|u\rangle$ have n components, as in the rest of $\mathcal{H}$, but it is possible to compress each $|u\rangle \in \mathcal{U}$ into a ket $|\overline{u}\rangle$, with $r \leq n$ components, without loss of information. The key remark is that for a ket $|u\rangle$ of $\mathcal{U}$ the projection coincides with the ket $|u\rangle$ itself

$$\boxed{P_{\mathcal{U}} |u\rangle = |u\rangle, \qquad \forall \, |u\rangle \in \mathcal{U}.} \qquad (5.140)$$

Considering (5.139) we can split the identity (5.140) into the pair

$$|\overline{u}\rangle = U_r^* |u\rangle, \qquad |u\rangle = U_r |\overline{u}\rangle \qquad \forall |u\rangle \in \mathcal{U}$$

where the first relation represents a **compression**, with *compressor* U_r^*, and the second an **expansion**, with *expander* U_r. The compressor U_r^* generates the r-dimensional subspace

$$\overline{\mathcal{U}} := U_r^* \, \mathcal{U} = \{|\overline{u}\rangle = U_r^* |u\rangle, \, |u\rangle \in \mathcal{U}\}$$

and the expander U_r restores the original subspace as $\mathcal{U} = U_r \overline{\mathcal{U}}$. In particular the compressed Hilbert space is given by

$$\boxed{\overline{\mathcal{H}} := \overline{\mathcal{U}} = U_r^* \, \mathcal{U}.} \tag{5.141}$$

Compression and expansion are schematically depicted in Fig. 5.14.

Now, all the detection operations in the original Hilbert space $\mathcal{H}$ can be transferred into the compressed space $\overline{\mathcal{U}}$ (here we mark compressed objects with an overline, as $\overline{\mathcal{U}}$). In the transition from $\mathcal{U}$ onto $\overline{\mathcal{U}}$ the geometry of kets is preserved (isometry). In fact, if $|u\rangle, |v\rangle \in \mathcal{U}$ and $|\overline{u}\rangle, |\overline{v}\rangle \in \overline{\mathcal{U}}$ are the corresponding compressed kets, we find for the inner products: $\langle \overline{u}|\overline{v}\rangle = \langle u| P_{\mathcal{U}}|v\rangle = \langle u|v\rangle$. In $\overline{\mathcal{U}}$ the state matrix becomes

$$\underset{r \times r}{\overline{\Gamma}} = \underset{r \times n}{U_r^*} \, \underset{n \times r}{\Gamma} \tag{5.142}$$

and collects the compressed states $\overline{\gamma}_i = U_r^* \gamma_i$. From $\overline{\Gamma}$ we can restore Γ by expansion, as $\Gamma = U_r \overline{\Gamma}$. Analogously, for the measurement matrix we find $\overline{M} = U_r^* M$

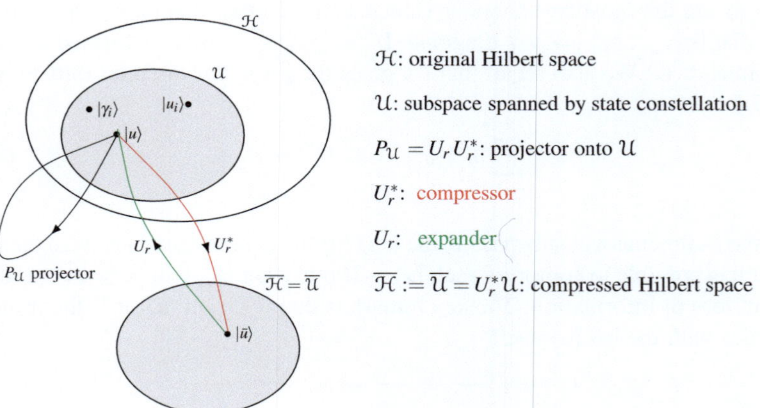

Fig. 5.14 The geometry for quantum compression: passage from the subspace $\mathcal{U}$ to the "compressed" space $\overline{\mathcal{U}} = U_r^* \mathcal{U}$, where U_r^* is the compressor. $\overline{\mathcal{U}}$ gives the compressed Hilbert space $\overline{\mathcal{H}}$

and $M = U_r \overline{M}$. For the density operators $\rho_i = \gamma_i \gamma_i^*$ and the measurement operators $\Pi_i = \mu_i \mu_i^*$ the compression/expansion give

$$\underset{r \times r}{\overline{\rho}_i} = U_r^* \, \rho_i \, U_r, \qquad \underset{n \times n}{\rho_i} = U_r \, \overline{\rho}_i \, U_r^*$$

$$\underset{r \times r}{\overline{\Pi}_i} = U_r^* \, \Pi_i \, U_r, \qquad \underset{n \times n}{\Pi_i} = U_r \, \overline{\Pi}_i \, U_r^*. \tag{5.143}$$

Note that, while $U_r \, U_r^*$ gives the projector $P_{\mathcal{U}}$, $U_r^* \, U_r$ gives the identities

$$\boxed{U_r \, U_r^* = P_{\mathcal{U}}, \qquad U_r^* \, U_r = I_r.} \tag{5.144}$$

In fact $U_r^* \, U_r = \sum_{i=1}^r |u_i\rangle\langle u_i|$, where $|u_i\rangle$ are orthonormal.

5.15.2 Properties in the Compressed Space

We review some properties in the compressed space, starting from the corresponding properties in the original Hilbert space.

Gram operator. The Gram operator $T := \Gamma \Gamma^*$ acting on the original Hilbert space $\mathcal{H}$ has dimension $n \times n$. In the compressed Hilbert space $\overline{\mathcal{H}}$ it becomes

$$\underset{r \times r}{\overline{T}} = U_r^* \, U_r \, \Sigma_r^2 \, U_r^* \, U_r = \Sigma_r^2, \tag{5.145}$$

and therefore the *compressed Gram operator is always diagonal*. On the other hand, the Gram matrix $G := \Gamma^* \, \Gamma$ does not change: $\overline{G} = G$. In fact, compression preserves inner products (see Property (1) in the next subsection).

Probabilities. The relation giving the transition probabilities is exactly preserved in the transition to the compressed space, namely (see Problem 5.16)

$$\boxed{p(j|i) = \mathrm{Tr}[\Pi_j \, \rho_i] = \mathrm{Tr}[\overline{\Pi}_j \, \overline{\rho}_i].} \tag{5.146}$$

Hence the relation for the probability of a correct detection

$$P_c = \sum_{i=0}^{K-1} q_i \mathrm{Tr}[\Pi_i \, \rho_i] = \sum_{i=0}^{K-1} q_i \mathrm{Tr}[\overline{\Pi}_i \, \overline{\rho}_i]. \tag{5.147}$$

This result is very important: it states that, once obtained the compressed operators, for the evaluation of the system performance, **it is not required to return back**

to the original uncompressed space. This conclusion is particularly important in the optimization with convex semidefinite programming (CSP), where the numerical evaluations can be completely carried out in the compressed space.

5.15.3 Compression as a Linear Mapping

Relation (5.144) defines a linear mapping connecting the subspace $\mathcal{U}$ to the compressed space $\overline{\mathcal{H}}$

$$U_r^* : \rho \in \mathcal{U} \;\to\; \overline{\rho} \in \overline{\mathcal{H}}. \tag{5.148}$$

This mapping has several interesting properties:

(1) the compressor U_r^* preserves inner products[11]: $\langle \overline{x}|\overline{y}\rangle = \langle x|y\rangle$, $|x\rangle, |y\rangle \in \mathcal{U}$,
(2) the compression preserves the PSD condition: $\rho \geq 0 \;\to\; \overline{\rho} \geq 0$,
(3) the compression is *trace preserving*: $\mathrm{Tr}[\overline{\rho}] = \mathrm{Tr}[\rho]$.
(4) the compression preserves the quantum entropy: $S(\overline{\rho}) = S(\rho)$ (see Chap. 12).

We prove statement (1). If $|x\rangle, |y\rangle \in \mathcal{U}$, we get $\langle \overline{x}|\overline{y}\rangle = \langle x|U_r U_r^*|y\rangle = \langle x|P_{\mathcal{U}}|y\rangle$, where $P_{\mathcal{U}}|y\rangle = |y\rangle$ by the fundamental property (5.140). Hence $\langle \overline{x}|\overline{y}\rangle = \langle x|y\rangle$. Similar is the proof of statement (2). The proof of (3) and (4) will be seen in Sect. 12.6.

A final comment. In the context of quantum channels, which will be seen in Sect. 12.8, a compression mapping may be classified as a *noiseless quantum channel*. This is essentially due to the fact that compression is a reversible transformation.

5.15.4 State Compression with GUS

The GUS is preserved in the compressed space (see Problem 5.17 for the proof).

Proposition 5.10 *If the states γ_i have the GUS with generating state γ_0 and symmetry operator S, then the compressed states $\overline{\gamma}_i$ have the GUS with generating state $\overline{\gamma}_0 = U_r^* \gamma_0$ and symmetry operator $\overline{S} = U_r^* S U_r$.*

The simultaneous diagonalization of T and S seen in Proposition 5.10 is also useful to establish other properties related to the GUS. In fact, by choosing the compressor U_r^* from the common eigenvector matrices U as in Eq. (5.131), we find the further properties:

Proposition 5.11 *With the simultaneous diagonalization the compressed symmetry operator becomes diagonal, with diagonal entries formed by the first r diagonal entries of the matrix Λ.*

---

[11] An operator from one space to another space is called *isometric* if it preserves norms and inner products [17].

In fact, decomposing Λ in the form diag$[\Lambda_r, \Lambda_c]$, where Λ_r is $r \times r$ and Λ_c is $(n-r) \times (n-r)$, we get

$$\overline{S} = U_r^* \, U \, \Lambda \, U^* \, U_r = [I_r \; 0] \begin{bmatrix} \Lambda_r & 0 \\ 0 & \Lambda_c \end{bmatrix} \begin{bmatrix} I_r \\ 0 \end{bmatrix} = \Lambda_r.$$

Proposition 5.12 *With the simultaneous diagonalization the compressed Gram operator is simply given by*

$$\overline{T} = \mathrm{diag}[K \, \overline{\rho}_0(i, i), i = 1, \ldots, r]$$

where $\overline{\rho}_0(i, i)$ are the diagonal entries of the compressed generating density operator $\overline{\rho}_0$.

In fact, $\overline{T} = \sum_{i=0}^{K-1} \overline{S}^i \, \overline{\rho}_0 \, \overline{S}^{-i}$, where $\overline{S}$ is diagonal. Then, the i, j entry is given by

$$\overline{T}(i, j) = \sum_{k=0}^{K-1} \overline{S}^k(i, i) \, \overline{\rho}_0(i, j) \, \overline{S}^{-k}(j, j).$$

In particular, considering that $\overline{S}$ is unitary diagonal, the diagonal entries are

$$\overline{T}(i, i) = \sum_{k=0}^{K-1} \overline{S}^k(i, i) \, \overline{\rho}_0(i, i) \, \overline{S}^{-k}(i, i) = K \, \overline{\rho}_0(i, i)$$

and the evaluation can be limited to these diagonal entries, since $\overline{T}$ is diagonal (see (5.145)) (in general $\overline{\rho}_0$ is not diagonal).

5.15.5 *Compressor Evaluation*

The leading parameter in compression is the **dimension of the compressed space** r, which is given by the rank of the state matrix Γ, but also by the rank of the Gram matrix G and of the Gram operator T. For the evaluation of the compressor we can use the reduced SVD of Γ, or the reduced EID of G and of T. In any case, for the choice, it is important to have in mind the dimensions of these matrices shown in (5.27).

With pure states, where often the dimension n of the Hilbert space is greater than the alphabet size K and the kets of Γ are linearly independent, r **is determined by the alphabet size** K and the EID of the Gram matrix, of dimension $K \times K$, becomes the natural choice.

With mixed states the choice depends on the specific application. In several cases of practical interest, n may be very large, so that the decompositions represent a very hard numerical task. But, in the presence of GUS, the computational complexity can be reduced, using the commutativity of the Gram operator T with the symmetry operator S. This will be seen in detail in the next chapters in correspondence with the specific applications (see the last two sections of Chap. 8).

Problem 5.16 $\star\star$ Prove that the evaluation of the transition probabilities in the compressed space is based on the same formula as in the uncompressed space, that is,

$$p(j|i) = \mathrm{Tr}[\Pi_j\, \rho_i] = \mathrm{Tr}[\overline{\Pi}_j\, \overline{\rho}_i].$$

Hint: Use orthonormality relationship $U_r^*\, U_r = I_r$, where I_r is the $r \times r$ identity matrix.

Problem 5.17 $\star\star\star$ Prove Proposition 5.10, which states that the GUS is preserved after a compression. *Hint:* Use orthonormality relationship $U_r^*\, U_r = I_r$, where I_r is the $r \times r$ identity matrix.

Problem 5.18 $\star\star$ Consider the state matrix of $\mathcal{H} = \mathbb{C}^4$

$$\Gamma = \frac{1}{2} \begin{bmatrix} 1 & 1 \\ -1 & 1 \\ 1 & -1 \\ -1 & 1 \end{bmatrix}$$

Find the compressor U_r^* and the compressed versions of the state matrix Γ and of the Gram operator T.

Problem 5.19 $\star\star$ Consider a binary transmission where the quantum states are specified by the state matrix of the previous problem. Apply Helstrom's theory with $q_0 = 1/3$ to find the probability of a correct decision P_c. Then apply the compression and evaluate P_c from the compressed states.

Problem 5.20 $\star\star$ Consider the binary constellation of Problem 5.13, where we determined the symmetry operator S. Find the compressor U_r^* showing, in particular, that the compressed symmetry operator $\overline{S}$ is diagonal.

Appendix

Proof of Holevo's Theorem

We refer to the classes introduced at the beginning of Sect. 5.8 and illustrated in Fig. 5.8. We start by proving Proposition 5.3. The set $\mathcal{M}$ of the K-tuples of Hermitian operators is closed with respect to addition and multiplication by a real number (the sum of two Hermitian operators and the product of a Hermitian operator by a real scalar are Hermitian operators), so it is in fact a real (Kn^2)-dimensional vector space. In such a space, the operation

$$(\mathbf{P}, \mathbf{Q}) = \sum_{i=0}^{K-1} \text{Tr}[P_i Q_i], \qquad \mathbf{P}, \mathbf{Q} \in \mathcal{M},$$

enjoys the property $(\mathbf{P}, \mathbf{Q}) = (\mathbf{Q}, \mathbf{P})$, a consequence of the cyclic property of the trace, as well as of the property $(\mathbf{P}, \mathbf{P}) \geq 0$ with $(\mathbf{P}, \mathbf{P}) = 0$, only if $\mathbf{P}$ is formed by null operators. Therefore, we are dealing with an operation of **inner product** and so $\mathcal{M}$ is a Hilbert space. In this space, the subset $\mathcal{M}_0$, formed by the K-tuples of PSD operators and resolving the identity, is closed and bounded, and therefore compact. From the classical Weierstrass theorem, in such a set, the continuous functional $J(\mathbf{Q})$ admits a maximum.

We now move on to Holevo's theorem, proving that the conditions indicated are sufficient conditions for maximization. Let $\mathbf{Q} = [Q_0, \ldots, Q_{K-1}] \in \mathcal{M}_0$, where the Q_i satisfy the conditions (5.83) and (5.82), and let $\mathbf{P} = [P_0, \ldots, P_{K-1}]$ be an arbitrary K-tuple of $\mathcal{M}_0$. Then, recalling the definition of L given by (5.81)

$$\sum_{i=0}^{K-1} \text{Tr}[P_i \widehat{\rho}_i] = \text{Tr}[L] + \sum_{i=0}^{K-1} \text{Tr}[P_i (\widehat{\rho}_i - L)]$$

$$= \sum_{i=0}^{K-1} \text{Tr}[Q_i \widehat{\rho}_i] - \sum_{i=0}^{K-1} \text{Tr}[P_i (L - \widehat{\rho}_i)].$$

On the other hand, because the trace of the product of PSD operators is nonnegative, for every i we have $\text{Tr}[P_i (L - \widehat{\rho}_i)] \geq 0$ and $J(\mathbf{P}) \leq J(\mathbf{Q})$. Therefore, the system $\mathbf{Q}$ is optimal and the sufficiency of the hypothesis of Holevo's theorem is proved.

We can also prove that the definition of L and the condition (5.82) imply the condition (5.83). In fact, we can write

$$0 = \text{Tr}[L] - \sum_{i=0}^{K-1} \text{Tr}[Q_i \widehat{\rho}_i] = \sum_{i=0}^{K-1} \text{Tr}[Q_i (L - \widehat{\rho}_i)].$$

As all the terms of the last sum are nonnegative, it must be $\text{Tr}[(L - \widehat{\rho}_i) Q_i] = 0$ for every i, then $(L - \widehat{\rho}_i) Q_i = 0_{\mathcal{H}}$.

The necessity of the conditions of Holevo's theorem is based on continuity considerations. Let $\mathbf{Q} \in \mathcal{M}_0$ be an optimal system and let U_{jk}, $j, k = 0, \ldots, K - 1$ be operators such that

$$\sum_{j=0}^{K-1} U_{jm}^* U_{jn} = \delta_{mn} I_{\mathcal{H}}.$$

Then, having defined the operators $P_j = S_j^* S_j$, with

$$S_j = \sum_{k=0}^{K-1} U_{jk} Q_k^{1/2},$$

it is easy to verify that $\mathbf{P} = [P_0, \ldots, P_{K-1}] \in \mathcal{M}_0$ and, from the optimality of $\mathbf{Q}$, it must be $J(\mathbf{P}) \leq J(\mathbf{Q})$.

We now appropriately particularize the operators U_{jk}, $j, k = 0, \ldots, K-1$, imposing that $U_{jj} = I_{\mathcal{H}}$ for $j = 2, \ldots, K-1$, and

$$\begin{bmatrix} U_{00} & U_{01} \\ U_{10} & U_{11} \end{bmatrix} = \exp\left(\varepsilon \begin{bmatrix} 0_{\mathcal{H}} & -A^* \\ A & 0_{\mathcal{H}} \end{bmatrix} \right)$$

with $\varepsilon > 0$ arbitrarily small and A arbitrary linear operator. Finally, we suppose that all the other operators U_{jk} be null. We then verify that

$$\begin{bmatrix} U_{00} & U_{01} \\ U_{10} & U_{11} \end{bmatrix}^* \begin{bmatrix} U_{00} & U_{01} \\ U_{10} & U_{11} \end{bmatrix} = \begin{bmatrix} I_{\mathcal{H}} & 0_{\mathcal{H}} \\ 0_{\mathcal{H}} & I_{\mathcal{H}} \end{bmatrix}$$

so that the operators U_{jk} satisfy the above conditions. The operators P_j, $j = 2, \ldots, K-1$ coincide with the operators Q_j, while, neglecting the infinitesimals ε^2 and those of higher order, we obtain

$$U_{00} = U_{11} = I_{\mathcal{H}}, \qquad U_{01} = -\varepsilon A^*, \qquad U_{10} = \varepsilon A$$

$$S_0 = Q_0^{1/2} - \varepsilon A^* Q_1^{1/2}, \qquad S_1 = Q_1^{1/2} + \varepsilon A Q_0^{1/2}$$

and eventually

$$P_0 = Q_0 - \varepsilon(Q_1^{1/2} A Q_0^{1/2} + Q_0^{1/2} A^* Q_1^{1/2})$$
$$P_1 = Q_1 + \varepsilon(Q_0^{1/2} A^* Q_1^{1/2} + Q_1^{1/2} A Q_0^{1/2}).$$

It follows that

$$\begin{aligned}
J(\mathbf{P}) - J(\mathbf{Q}) &= \sum_{j=0}^{K-1} \operatorname{Tr}[\widehat{\rho}_j (P_j - Q_j)] \\
&= \operatorname{Tr}[\widehat{\rho}_0 (P_0 - Q_0)] + \operatorname{Tr}[\widehat{\rho}_1 (P_1 - Q_1)] \\
&= \varepsilon \operatorname{Tr}[(\widehat{\rho}_1 - \widehat{\rho}_0)(Q_1^{1/2} A Q_0^{1/2} + Q_0^{1/2} A^* Q_1^{1/2})] \\
&= \varepsilon \operatorname{Tr}[Q_0^{1/2}(\widehat{\rho}_1 - \widehat{\rho}_0) Q_1^{1/2} A + Q_1^{1/2}(\widehat{\rho}_1 - \widehat{\rho}_0) Q_0^{1/2} A^*].
\end{aligned}$$

As the coefficient of ε must be null to ensure that the difference be non positive for every value of the arbitrary operator A, it must be $Q_0^{1/2}(\widehat{\rho}_1 - \widehat{\rho}_0)Q_1^{1/2} = 0_{\mathcal{H}}$ or, equivalently, $Q_0(\widehat{\rho}_1 - \widehat{\rho}_0)Q_1 = 0_{\mathcal{H}}$. As the reasoning can be repeated for every couple of indexes i and j, it follows that it must be $Q_i(\widehat{\rho}_j - \widehat{\rho}_i)Q_j = 0_{\mathcal{H}}$, that is, $Q_i\widehat{\rho}_j Q_j = Q_i\widehat{\rho}_i Q_j$. Summing both sides with respect to i, we obtain for every j, $\widehat{\rho}_j Q_j = LQ_j$, coinciding with (5.83). At this point, it should be proved that the operators $L - \widehat{\rho}_j$ are PSD. For a rigorous (and very technical) proof of the result, please refer to [3].

Proof of Kennedy's Theorem

Kennedy's theorem (Theorem 5.3) can be derived in a generalized form from Holevo's theorem. We recall that this requires in the first place that the operators $L - \widehat{\rho}_i$ be PSD for every i. If we assume that the eigenvalues of the operators $\widehat{\rho}_i$ span over the entire Hilbert space $\mathcal{H}$, the operator L is positive definite and has rank n. From the optimality conditions of Holevo's theorem

$$(L - \widehat{\rho}_i)Q_i = 0_{\mathcal{H}},$$

we have first of all that, if $|y\rangle$ belongs to the image of the operator Q_i, that is, if there exists $|x\rangle \in \mathcal{H}$ such that $|y\rangle = Q_i|x\rangle$, then $(L - \widehat{\rho}_i)|y\rangle = 0$, and $|y\rangle$ belongs to the null space of the operator $L - \widehat{\rho}_i$. We then have that the image of Q_i is a subspace contained in the null space $\mathcal{N}(L - \widehat{\rho}_i)$ of $L - \widehat{\rho}_i$, therefore its dimension, coinciding with the rank of Q_i, is not greater than the dimension of the null space $\mathcal{N}(L - \widehat{\rho}_i)$ and this yields the inequality

$$\text{rank}(Q_i) \le \dim(\mathcal{N}(L - \widehat{\rho}_i)) = n - \text{rank}(L - \widehat{\rho}_i).$$

From the subadditivity of the rank, i.e., from $\text{rank}(A + B) \le \text{rank}(A) + \text{rank}(B)$, letting $A = L - \widehat{\rho}_i$ and $B = \widehat{\rho}_i$, we obtain $n = \text{rank}(L) \le \text{rank}(L - \widehat{\rho}_i) + \text{rank}(\widehat{\rho}_i)$, which, substituted in the above inequality, yields $\text{rank}(Q_i) \le \text{rank}(\widehat{\rho}_i)$.

Let us now consider the special case in which we have n pure states $|\gamma_i\rangle$, linearly independent, generating the n-dimensional space $\mathcal{H}$, so that the operators $\widehat{\rho}_i$ have rank 1. Then the optimal measurement operators Q_i must have rank not greater than 1, and therefore either be null, or have the form $Q_i = |\mu_i\rangle\langle\mu_i|$. As it must be

$$\sum_{i=0}^{n-1} Q_i = \sum_{i=0}^{n-1} |\mu_i\rangle\langle\mu_i| = I_{\mathcal{H}},$$

the vectors $|\mu_i\rangle$ cannot be null and must be linearly independent. Furthermore, as, for every j,

$$|\mu_j\rangle = \sum_{i=0}^{n-1} |\mu_i\rangle\langle\mu_i|\mu_j\rangle$$

from the comparison of the two sides, we obtain $\langle\mu_i|\mu_j\rangle = \delta_{ij}$ and the measurement vectors are orthonormal.

Commutativity of the Operators T and S

Let us prove Proposition 5.10. Using (5.120a) in the definition of Gram's operator (5.108) and remembering that S is a unitary operator, so that $S^* = S^{-1}$, we obtain

$$T = \sum_{i=0}^{K-1} |\gamma_i\rangle\langle\gamma_i| = \sum_{i=0}^{K-1} S^i|\gamma_0\rangle\langle\gamma_0|S^{-i}$$

hence

$$T S = \sum_{i=0}^{K-1} S^i|\gamma_0\rangle\langle\gamma_0|S^{-i+1} = SS^{-1}\sum_{i=0}^{K-1} S^i|\gamma_0\rangle\langle\gamma_0|S^{-i+1}$$

$$= S\sum_{i=0}^{K-1} S^{i-1}|\gamma_0\rangle\langle\gamma_0|S^{-i+1} = S\sum_{k=0}^{K-1} S^k|\gamma_0\rangle\langle\gamma_0|S^{-k} = S T$$

where in the last step we exploited the periodicity of S^i with respect to i.

References

1. C.W. Helstrom, J.W.S. Liu, J.P. Gordon, Quantum-mechanical communication theory. Proc. IEEE **58**(10), 1578–1598 (1970)
2. K. Kraus, *States, Effect and Operations: Fundamental Notions of Quantum Theory*, Lecture Notes in Physics, vol. 190 (Springer, New York, 1983)
3. Y.C. Eldar, A. Megretski, G.C. Verghese, Optimal detection of symmetric mixed quantum states. IEEE Trans. Inf. Theory **50**(6), 1198–1207 (2004)
4. A.S. Holevo, Statistical decision theory for quantum systems. J. Multivar. Anal. **3**(4), 337–394 (1973)
5. H.P. Yuen, R. Kennedy, M. Lax, Optimum testing of multiple hypotheses in quantum detection theory. IEEE Trans. Inf. Theory **21**(2), 125–134 (1975)
6. M. Grant, S. Boyd, CVX: Matlab software for disciplined convex programming, version 2.1. March 2014, http://cvxr.com/cvx
7. M. Grant, S. Boyd, Graph implementations for nonsmooth convex programs, in: *Recent Advances in Learning and Control*, ed. by V. Blondel, S. Boyd, H. Kimura, (eds). Lecture Notes in Control and Information Sciences. (Springer, 2008), pp. 95–110, http://stanford.edu/~boyd/graph_dcp.html

8. R.S. Kennedy, A near-optimum receiver for the binary coherent state quantum channel. Massachusetts Institute of Technology, Cambridge (MA), Technical Report, January 1973. MIT Research Laboratory of Electronics Quarterly Progress Report 108
9. V. Vilnrotter and C.W. Lau, Quantum detection theory for the free-space channel. NASA, Technical Report, August 2001. Interplanetary Network Progress (IPN) Progress Report 42–146
10. V. Vilnrotter, C.W. Lau, Quantum detection and channel capacity using state-space optimization. NASA, Technical Report, February 2002. Interplanetary Network Progress (IPN) Progress Report 42–148
11. V. Vilnrotter, C.W. Lau, Binary quantum receiver concept demonstration. NASA, Technical Report, Interplanetary Network Progress (IPN) Progress Report 42–165, May 2006
12. K. Kato, M. Osaki, M. Sasaki, O. Hirota, Quantum detection and mutual information for QAM and PSK signals. IEEE Trans. Commun. **47**(2), 248–254 (1999)
13. R.A. Horn, C.R. Johnson, *Matrix Analysis* (Cambridge University Press, Cambridge, 1998)
14. Y.C. Eldar, G.D. Forney, On quantum detection and the square-root measurement. IEEE Trans. Inf. Theory **47**(3), 858–872 (2001)
15. A. Assalini, G. Cariolaro, G. Pierobon, Efficient optimal minimum error discrimination of symmetric quantum states. Phys. Rev. A **81**, 012315 (2010)
16. G. Cariolaro, R. Corvaja, G. Pierobon, Compression of pure and mixed states in quantum detection. in Global Telecommunications Conference, vol. 2011 (GLOBECOM, IEEE, 2011), pp. 1–5
17. A.S. Holevo, V. Giovannetti, Quantum channels and their entropic characteristics. Rep. Prog. Phys. **75**(4), 046001 (2012)

Chapter 6
Quantum Decision Theory: Suboptimization

6.1 Introduction

In the previous chapter we have seen that **optimization** is very difficult, and exact solutions are only known in few cases (binary systems and systems where the state constellation have the *geometrically uniform symmetry*, GUS). To overcome the difficulty, **suboptimization** is considered.

In quantum communications the most important suboptimal criterion is based on the minimization of the quadratic error between the states and the measurement vectors, known by the acronym LSM (least squares measurements), and also SRM (square root measurements), because its solution is based on the square root of an operator.

From a historical point of view, we must start from quantum SRM (square root measurement), introduced by Hausladen and other authors in 1996 [1], who proposed as measurement matrix $M = T^{-1/2}\Gamma$, where T is Gram's operator and $T^{-1/2}$ is its inverse square root. With this choice, the quantum decision is not in general optimal, but it gives a good approximation of the performance ("pretty good" is the judgment given by the authors and very often echoed in the literature).

The quantum least squares measurements (LSM) were subsequently developed by Eldar and Forney, in two articles [2, 3], deserving particular attention, because they formalize the whole problem in a very clear and general way, establishing a connection between the LSM and other types of measurements. In particular, they proved that the LSM technique produces the same results as the SRM technique, and precisely that the optimal measurement matrix (which minimizes the quadratic error) can be obtained both from Gram's operator and from Gram's matrix in the following way.

$$M_0 = T^{-\frac{1}{2}}\Gamma = \Gamma G^{-\frac{1}{2}}. \tag{6.1}$$

An important result is concerned with the SRM in the presence of GUS, which gives the **optimal decision for pure states**, allowing the exact evaluation of the error probability. Recently [4, 5], the SRM technique has been systematically applied to

© Springer International Publishing Switzerland 2015

G. Cariolaro, *Quantum Communications*, Signals and Communication Technology,
DOI 10.1007/978-3-319-15600-2_6

the performance evaluation of most popular quantum communications systems. From a computational viewpoint, the SRM can be improved with the technique of quantum compression [6], which has been introduced at the end of the previous chapter.

Organization of the Chapter

The SRM technique is mainly based on the SVD of the state matrix Γ and on the EID of the Gram matrix G and of the Gram operator T, developed in Sect. 5.12 of the previous chapter. For this reason these decompositions are recalled before developing the SRM.

The SRM for pure states is developed in Sects. 6.2 and 6.3 and extended to mixed states in Sect. 6.4. In Sects. 6.5 and 6.6 the SRM is developed assuming that the state constellations have the GUS.

In Sect. 6.7 the SRM technique is combined with the compression technique, showing the advantage of working in a compressed space, mainly in the presence of GUS. Finally, in Sect. 6.8 the quantum Chernoff bound is introduced as a further technique of suboptimization in quantum detection.

Recall from the Previous Chapter

For convenience we reconsider the main matrices and the related decompositions seen in Sect. 5.12 of the previous chapter, which are useful to SRM.

- *State and measurement matrices*

$$\text{pure states} \quad \underset{n \times K}{\Gamma} = [|\gamma_0\rangle, |\gamma_1\rangle, \ldots, |\gamma_{K-1}\rangle], \quad \underset{n \times K}{M} = [|\mu_0\rangle, |\mu_1\rangle, \ldots, |\mu_{K-1}\rangle].$$

$$\text{mixed states} \quad \underset{n \times H}{\Gamma} = [\gamma_0, \gamma_1, \ldots, \gamma_{K-1}], \quad \underset{n \times H}{M} = [\mu_0, \mu_1, \ldots, \mu_{K-1}].$$

Relations

$$M = \Gamma A, \quad M = C \Gamma. \tag{6.2}$$

In particular, the second relation gives

$$|\mu_i\rangle = C |\gamma_i\rangle \quad \text{or} \quad \mu_i = C \gamma_i. \tag{6.2a}$$

Singular value decomposition of Γ

$$\Gamma = U \Sigma V^* = \sum_{i=1}^{r} \sigma_i |u_i\rangle \langle v_i| = U_r \Sigma_r V_r^* \tag{6.3}$$

- *Gram's matrix and Gram's operator*

$$\text{pure states} \qquad \underset{K \times K}{G} = \Gamma^* \Gamma , \qquad \underset{n \times n}{T} = \Gamma \Gamma^* \qquad (6.4a)$$

$$\text{mixed states} \qquad \underset{H \times H}{G} = \Gamma^* \Gamma , \qquad \underset{n \times n}{T} = \Gamma \Gamma^* \qquad (6.4b)$$

Relations

$$\text{pure states} \qquad G_{ij} = \langle \gamma_i | \gamma_j \rangle , \qquad T = \sum_{i=0}^{K-1} |\gamma_i\rangle\langle\gamma_i| \qquad (6.5a)$$

$$\text{mixed states} \qquad G_{ij} = \gamma_i^* \gamma_j , \qquad T = \sum_{i=0}^{K-1} \rho_i . \qquad (6.5b)$$

Eigendecompositions

$$T = U \Lambda_T U^* = \sum_{i=1}^{r} \sigma_i^2 |u_i\rangle\langle u_i| = U_r \Sigma_r^2 U_r^* \qquad (6.6a)$$

$$G = V \Lambda_G V^* = \sum_{i=1}^{r} \sigma_i^2 |v_i\rangle\langle v_i| = V_r \Sigma_r^2 V_r^* . \qquad (6.6b)$$

6.2 Square Root Measurements (SRM)

6.2.1 Formulation

Considering the equivalence between LSM and SRM, we find it convenient to introduce the topic in the sense of LSM, but we will use the more consolidated acronym SRM.

In the case of pure states, the measurement vectors $|\mu_i\rangle$ are chosen with the criterion of making the differences between the states and the measurement vectors, $|e_i\rangle = |\gamma_i\rangle - |\mu_i\rangle$, as "small" as possible (Fig. 6.1), and more specifically we look

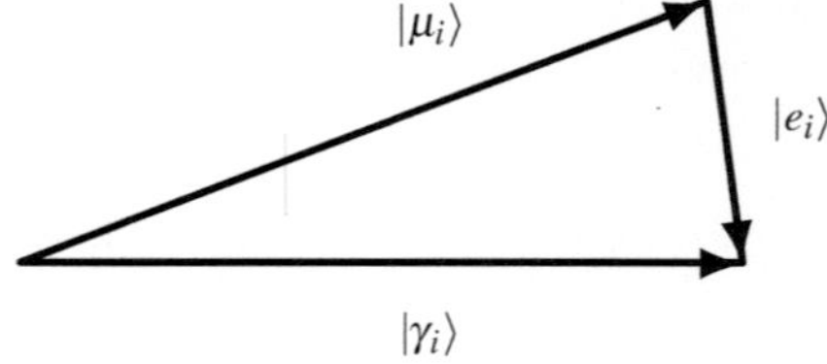

Fig. 6.1 In the LSM method the quadratic average of the "errors" $|e_i\rangle = |\gamma_i\rangle - |\mu_i\rangle$ is minimized

for **the measurement vectors** $|\mu_i\rangle$ **that minimize the quadratic error**

$$\mathcal{E} = \sum_{i=0}^{K-1} \langle e_i | e_i \rangle = \sum_{i=0}^{K-1} ((\langle \gamma_i | - \langle \mu_i |)(|\gamma_i \rangle - \mu_i))$$

with the constraint of resolution of the identity

$$MM^* = \sum_{i=0}^{K-1} |\mu_i\rangle\langle\mu_i| = I_{\mathcal{H}} \quad \rightarrow \quad P_{\mathcal{U}} \tag{6.7}$$

where $I_{\mathcal{H}}$ can be replaced by $P_{\mathcal{U}}$ (see Proposition 5.5).

Introducing the difference between the state matrix and the measurement matrix: $E = [|e_0\rangle, |e_1\rangle, \ldots, |e_{K-1}\rangle] = \Gamma - M$, the quadratic error can be written in the form

$$\mathcal{E} = \mathrm{Tr}[E^* E] = \mathrm{Tr}[E E^*]\,. \tag{6.8}$$

We observe that if the vectors $|\gamma_i\rangle$ were orthonormal, the minimum of $\mathcal{E}$, satisfying the constraint (6.7), would be trivially $|\mu_i\rangle = |\gamma_i\rangle$, $1 \leq i \leq K$, which yields $\mathcal{E} = 0$.

The above can be extended to mixed states, for which the error is considered between the state factors and the measurement factors, $e_i = \gamma_i - \mu_i$, and the quadratic error is still given by (6.8). In any case we assume **equiprobable symbols**, that is, with equal a priori probabilities, $q_i = 1/K$, but the SRM method can be extended to generic a priori probabilities q_i substituting the states $|\gamma_i\rangle$ with the weighted states $\sqrt{q_i}|\gamma_i\rangle$ (see [2]).

As we will see, the SRM method always leads to explicit results and, in general, provides a good overestimation of the error probability.

6.2.2 Computation of the Optimal Measurement Matrix

Now we search for the optimal measurement matrix, $M = M_0$, that minimizes the quadratic error $\mathcal{E}$. Even though in quantum communications the states are always independent and so the rank of the state matrix is $r = K$, for greater generality (and for the interest that the general case will have with the extension of the method to mixed states), we suppose that Γ has a generic rank r. We obtain:

Theorem 6.1 *The measurement matrix M that minimizes the quadratic error $\mathcal{E}$ with the constraint (6.7), is given by*

$$M_0 = \sum_{i=1}^{r} |u_i\rangle\langle v_i| = U_r V_r^*\,, \tag{6.9}$$

that is, by the sum of the r transjectors $|u_i\rangle\langle v_i|$ seen in the (reduced) SVD of the state matrix Γ. The minimum quadratic error is

$$\mathcal{E}_{\min} = \sum_{i=0}^{K-1} (1 - \sigma_i)^2 \,,$$

where σ_i are the square roots of the eigenvalues (i.e., the singular values) of the Gram operator and of the Gram matrix (see (6.6)).

Note that here we indicate as *optimum* the measurement matrix giving the minimum square error (representing the "best" solution in this context). In general this does not provide the optimum decision, which minimizes the error probability.

Proof We follow the demonstration by [2] with some simplification. In the expression (6.8) of the quadratic error we take explicitly the trace with respect to the orthonormal basis $|u_i\rangle$, seen in the EID of Gram's operator in the previous chapter (see (5.109a)). In this way we find

$$\mathcal{E} = \mathrm{Tr}[EE^*] = \sum_{i=1}^{n} \langle u_i | EE^* | u_i \rangle = \sum_{i=1}^{n} \langle d_i | d_i \rangle \qquad (6.10)$$

where

$$|d_i\rangle := E^*|u_i\rangle = (\Gamma - M)^*|u_i\rangle \,. \qquad (6.10a)$$

Let us now consider the reduced SVD of Γ^* (see (6.3))

$$\Gamma^* = V_r \Sigma_r U_r^* = \sum_{i=1}^{r} \sigma_i |v_i\rangle\langle u_i|$$

which gives $\Gamma^*|u_i\rangle = \sigma_i|v_i\rangle$. Now, letting

$$|a_i\rangle = M^*|u_i\rangle \,, \qquad i = 1, \ldots, r \qquad (6.11)$$

(6.10a) becomes $|d_i\rangle = (\Gamma - M)^*|u_i\rangle = \sigma_i|v_i\rangle - |a_i\rangle$ and the ith component of the quadratic error results in

$$\begin{aligned}
\mathcal{E}_i = \langle d_i | d_i \rangle &= \sigma_i^2 \langle v_i | v_i \rangle + \langle a_i | a_i \rangle - \sigma_i \langle v_i | a_i \rangle - \sigma_i \langle a_i | v_i \rangle \\
&= \sigma_i^2 + 1 - \sigma_i \langle v_i | a_i \rangle - \sigma_i \langle a_i | v_i \rangle \,.
\end{aligned}$$

The minimum of $\mathcal{E}_i$ is reached when the quantity $\sigma_i \langle v_i | a_i \rangle + \sigma_i \langle a_i | v_i \rangle$ is maximum. Because of the constraint $|\langle v_i | a_i \rangle| \le 1$, this quantity is maximum when $|a_i\rangle = |v_i\rangle$, that is, when (6.11) becomes $|v_i\rangle = M^*|u_i\rangle$ for an appropriate $M = M_0$. From this relation, (6.9) follows.

6.2.3 Consequences of the Result

From the expression (6.9) of the optimal measurement matrix, it can be soon verified:

Corollary 6.1 *If the states $|\gamma_i\rangle$ are linearly independent, that is, if the rank of Γ is $r = K$, the optimal measurement vectors result orthonormal, $\langle\mu_i|\mu_j\rangle = \delta_{ij}$, and therefore the corresponding measurement operators $Q_i = |\mu_i\rangle\langle\mu_i|$ constitute a projector system.*

In fact, let us consider the (optimal) Gram's matrix of the measurement vectors

$$M_0^* M_0 = \sum_{i=0}^{K-1} |v_i\rangle\langle u_i| \sum_{j=0}^{K-1} |u_j\rangle\langle v_j| = \sum_{i=0}^{K-1} |v_i\rangle\langle v_i| = I_K$$

where (5.116) has been used. To conclude, it suffices to observe that *Gram's matrix of the measurement vectors M* has as elements the inner products $\langle\mu_i|\mu_i\rangle$ (see (5.107)).

In addition, we find what was anticipated by (6.1):

Corollary 6.2 *The optimal measurement matrix $M_0 = U_r V_r^*$ can be calculated also from the expressions*

$$M_0 = \Gamma(\Gamma^*\Gamma)^{-1/2} = \Gamma G^{-1/2} \tag{6.12a}$$

$$M_0 = (\Gamma\Gamma^*)^{-1/2}\Gamma = T^{-1/2}\Gamma \tag{6.12b}$$

where $G^{-1/2}$ and $T^{-1/2}$ are the inverse square roots of G and T that are obtained from the corresponding reduced EIDs in the following way

$$G^{-1/2} = V_r\,\Sigma_r^{-1}\,V_r^*, \qquad T^{-1/2} = U_r\,\Sigma_r^{-1}\,U_r^*. \tag{6.13}$$

For example, the proof of (6.12a) is carried out using the expression $G^{-1/2}$ introduced above, and the reduced SVD of Γ. We obtain

$$\Gamma G^{-1/2} = U_r\Sigma_r V_r^* V_r \Sigma_r^{-1} V^* = U_r V_r^* = M_0$$

where we took into account that $V_r^* V_r = I_r$ and that $\Sigma\Sigma_r^{-1} = I_r$.

On the inverse square roots. In (6.13) we have formally introduced the inverse square roots $G^{-1/2}$ and $T^{-1/2}$ of G and of T. To obtain these roots we start from the corresponding reduced EIDs and we operate on the common diagonal matrix $\Sigma_r^2 = \mathrm{diag}[\sigma_1^2, \dots, \sigma_r^2]$, taking its inverse square root $\Sigma_r^{-1} = \mathrm{diag}[1/\sigma_1, \dots, 1/\sigma_r]$, where the σ_i^2 are all positive, and therefore there are no indeterminacy problems. In general, $G^{-1/2}$ and $T^{-1/2}$ should be intended as *pseudoinverses* (according to Moore–Penrose formula [7, 8]). The Moore–Penrose pseudoinverse is based on the full EID (not reduced) $U\Lambda U^*$, by taking the reciprocal of each nonzero element on the diagonal, leaving the zeros in place. Here we prefer to use the **reduced EID**, where the diagonal matrix is regular and has no problem for its inversion. In any case note that the inversion can bring about some surprising result. For example, it can be verified that the relation $T^{-1/2}T^{1/2} = U_r U_r^*$ does not yield, in general, the identity $I_{\mathcal{H}}$, but it gives the projector $P_{\mathcal{U}} = U_r U_r^*$ and only if $r = K$ one actually produces the identity $I_{\mathcal{H}}$.

The path followed so far to introduce the inverse square roots, based on the *reduced* EIDs, is slightly unusual; in fact, the EIDs are normally considered full, and this entails the complication of having to introduce several diagonal matrices [2].

Problem 6.1 Prove that $T^{-1/2}T^{1/2}$ does not yield, in general, the identity $I_{\mathcal{H}}$, but the projector $P_{\mathcal{U}} = U_r U_r^*$. Only if $r = K$ one actually produces the identity $I_{\mathcal{H}}$.

Problem 6.2 $\star\star$ Consider the following state matrix of $\mathcal{H} = \mathbb{C}^4$

$$
\Gamma = \begin{bmatrix} \frac{1}{2} & \frac{1}{2} \\ -\frac{1}{2} & \frac{1}{2} \\ \frac{1}{2} & -\frac{1}{2} \\ -\frac{1}{2} & \frac{1}{2} \end{bmatrix}
$$

Find the inverse square root $G^{-1/2}$ and $T^{-1/2}$ based on the two approaches: (1) the Moore–Penrose pseudoinverse and (2) the reduced EID.

6.3 Performance Evaluation with the SRM Decision

With the SRM method, we have seen that the optimal measurement matrix has three distinct expressions

$$
M_0 = U_r V_r^* = T^{-1/2}\Gamma = \Gamma G^{-1/2} . \tag{6.14}
$$

The first expression is bound to the reduced SVD of the measurement matrix Γ, while the other two are obtained from the reduced EIDs of T and G, respectively.

From the measurement matrix, which collects the measurement vectors $|\mu_i\rangle$, the measurement operators can be computed as $Q_i = |\mu_i\rangle\langle\mu_i|$ and from these the performance of the quantum system. The transition probabilities result in

$$
p_c(j|\,i) = \mathrm{Tr}[\rho_i Q_j] = |\langle\mu_j|\gamma_i\rangle|^2 \tag{6.15}
$$

and the correct decision probability (with equiprobable symbols)

$$
P_c = \frac{1}{K} \sum_{i=0}^{K-1} |\langle\mu_i|\gamma_i\rangle|^2. \tag{6.16}
$$

We now discuss the three possible methods, expressing them in a form suitable for computation.

6.3.1 Method Based on the SVD of the State Matrix

The optimal measurement matrix, evaluated according to the expression

$$M_0 = U_r\, V_r^*\,, \tag{6.17}$$

can be obtained directly from the reduced SVD of the state matrix, which has the form (see (5.110)): $\Gamma = U_r\, \Sigma_r\, V_r^*$. Therefore, in this expression it suffices to suppress the diagonal matrix to obtain the optimal measurement matrix. This is the most direct method as it does not require to calculate the inverse root square of a matrix.

6.3.2 Method Based on Gram's Operator

Let us start from *Gram's operator*,

$$T = \Gamma\, \Gamma^* = \sum_{i=0}^{K-1} |\gamma_i\rangle\langle\gamma_i|\,, \tag{6.18}$$

which is a positive semidefinite Hermitian operator (see Sect. 2.10.4). Then it is possible to define its square root, using the EID (5.109a), which is, in the reduced form, $T = U_r\, \Sigma_r^2\, U_r^*$ and gives

$$T^{-\frac{1}{2}} = U_r\, \Sigma_r^{-1}\, U_r^* \tag{6.19}$$

from which we obtain the optimal measurement matrix as

$$M_0 = T^{-\frac{1}{2}}\Gamma. \tag{6.20}$$

At this point we observe that (6.20) falls into the form (5.103), that is, $M = C\,\Gamma$, and then the measurement vectors are simply obtained according to

$$\boxed{|\mu_i\rangle = T^{-\frac{1}{2}}|\gamma_i\rangle} \tag{6.21}$$

from which we get the elementary measurement operators as

$$Q_i = |\mu_i\rangle\langle\mu_i| = T^{-\frac{1}{2}}|\gamma_i\rangle\langle\gamma_i|T^{-\frac{1}{2}}. \tag{6.22}$$

6.3.3 Method Based on Gram's Matrix

We start from *Gram's matrix*

$$\underset{K\times K}{G} = \Gamma^*\,\Gamma = \begin{bmatrix} \langle\gamma_0|\gamma_0\rangle & \cdots & \langle\gamma_0|\gamma_{K-1}\rangle \\ \vdots & \ddots & \vdots \\ \langle\gamma_{K-1}|\gamma_0\rangle & \cdots & \langle\gamma_{K-1}|\gamma_{K-1}\rangle \end{bmatrix} \tag{6.23}$$

which is obtained by computing the inner products $\langle\gamma_i|\gamma_j\rangle$. We then evaluate the reduced EID, which has the form

$$G = V_r\,\Sigma_r^2\,V_r^* \tag{6.24}$$

where the diagonal matrix Σ_r^2 is the same as the one appearing in the previous EID. From this we compute the inverse square root

$$G^{-\frac{1}{2}} = V_r\,\Sigma_r^{-1}\,V_r^* \tag{6.25}$$

and then we obtain the optimal measurement matrix as

$$M_0 = \Gamma\,G^{-\frac{1}{2}}. \tag{6.26}$$

This form is of the type (5.102), that is, $M = \Gamma\,A$, which expresses in a compact form the fact that the (optimal) measurement vectors are given by a linear combination of the states, that is,

$$|\mu_i\rangle = \sum_{j=0}^{K-1} a_{ij}|\gamma_j\rangle.$$

Now, as $A = G^{-1/2}$ and therefore $a_{ij} = (G^{-1/2})_{ij}$, the measurement vectors result explicitly in

$$|\mu_i\rangle = \sum_{j=0}^{K-1} (G^{-1/2})_{ij}|\gamma_j\rangle. \tag{6.27}$$

The transition probabilities are computed from the *mixed inner products* $b_{ij} = \langle\mu_i|\gamma_j\rangle$, which define the $K \times K$ matrix $B = M^*\Gamma$ (see (5.74)). Now, from (6.24) and (6.25) we have

$$B := M^* \Gamma = G^{-1/2} \Gamma^* \Gamma = G^{-1/2} G = G^{1/2} \, . \tag{6.28}$$

Then the matrix of the mixed inner products becomes simply $G^{1/2}$ and since $p_c(j|i) = |b_{ij}|^2$ we have

$$\boxed{p_c(j|\,i) = \left| (G^{\frac{1}{2}})_{ij} \right|^2} \tag{6.29}$$

from which we obtain the correct decision probability with equiprobable symbols

$$\boxed{P_c = \frac{1}{K} \sum_{i=0}^{K-1} \left| (G^{\frac{1}{2}})_{ii} \right|^2 \, .} \tag{6.30}$$

Example 6.1 Consider a binary system ($K = 2$) on $\mathcal{H} = \mathbb{C}^4$, in which the two states are specified by the matrix

$$\Gamma = \big[|\gamma_1\rangle, \, |\gamma_2\rangle \big] = \frac{1}{2\sqrt{13}} \begin{bmatrix} 5 & 1 \\ 3 - 2\,\mathrm{i} & 3 + 2\,\mathrm{i} \\ 1 & 5 \\ 3 + 2\,\mathrm{i} & 3 - 2\,\mathrm{i} \end{bmatrix} \tag{6.31}$$

which has rank $r = K = 2$. The reduced SVD of Γ becomes: $\Gamma = U_r \Sigma V_r^*$, where

$$U_r = \frac{1}{2} \begin{bmatrix} 1 & 1 \\ 1 & -\mathrm{i} \\ 1 & -1 \\ 1 & \mathrm{i} \end{bmatrix}, \qquad \Sigma_r = \begin{bmatrix} \sqrt{\frac{18}{13}} & 0 \\ 0 & \sqrt{\frac{8}{13}} \end{bmatrix}, \qquad V = V_r = \frac{1}{\sqrt{2}} \begin{bmatrix} 1 & 1 \\ 1 & -1 \end{bmatrix} \, .$$

Then, from (6.17) we get the optimal measurement matrix

$$M_0 = U_r V^* = \frac{1}{2} \begin{bmatrix} 1 & 1 \\ 1 & -\mathrm{i} \\ 1 & -1 \\ 1 & \mathrm{i} \end{bmatrix} \frac{1}{\sqrt{2}} \begin{bmatrix} 1 & 1 \\ 1 & -1 \end{bmatrix} = \frac{1}{2\sqrt{2}} \begin{bmatrix} 2 & 0 \\ 1 - \mathrm{i} & 1 + \mathrm{i} \\ 0 & 2 \\ 1 + \mathrm{i} & 1 - \mathrm{i} \end{bmatrix} \, . \tag{6.32}$$

The measurement vectors become then

$$|\mu_1\rangle = \frac{1}{2\sqrt{2}} \begin{bmatrix} 2 \\ 1 - \mathrm{i} \\ 0 \\ 1 + \mathrm{i} \end{bmatrix}, \qquad |\mu_2\rangle = \frac{1}{2\sqrt{2}} \begin{bmatrix} 0 \\ 1 + \mathrm{i} \\ 2 \\ 1 - \mathrm{i} \end{bmatrix}$$

and their orthonormality can be verified, in particular $\langle\mu_1|\mu_2\rangle = 0$, in agreement with Corollary 6.1. We can now compute the transition probabilities from (6.15), that is, $p_c(j|\,i) = |\langle\mu_j|\gamma_i\rangle|^2$. We obtain the matrix

$$p_c = \begin{bmatrix} \frac{25}{26} & \frac{1}{26} \\ \frac{1}{26} & \frac{25}{26} \end{bmatrix} \tag{6.33}$$

from which we have that the error probability with equiprobable symbols results in $P_e = 1 - P_c = \frac{1}{26}$.

We leave it to the reader to verify that the other two performance evaluation methods, based on the reduced EIDs of G and of T, lead to the same results found with the SVD of Γ.

Problem 6.3 ⋆⋆ Consider the state matrix Γ given by (6.31) of Example 6.1. Check that the methods based on the EIDs of G and T give the same transition probabilities as obtained with the SVD of Γ.

Problem 6.4 ⋆⋆ With the data of the previous problem, find the relations

$$\mu_1 = C\,\gamma_1\,, \qquad \mu_2 = C\,\gamma_2.$$

These relations are somewhat intriguing since they lead to think that μ_1 depends only on γ_1 and not on γ_2 and μ_2 only on γ_2. Explain why not.

6.3.4 Properties of the SRM

We have seen that the operators of the SRM can always be calculated in a rather simple manner for any constellation of states $|\gamma_i\rangle$ and therefore for any quantum communications system. This is already a first advantage. It remains to understand whether the SRM are optimal or close to optimal. It has been proved by Holevo in 1979 [9] that the SRM are *asymptotically optimal*, in the sense that they become optimal in practice when the average number of photons is large enough. Furthermore, these measurements become optimal when the constellation of the states enjoys the geometrically uniform symmetry (GUS) (see below).

Another advantage of the SRM regards their practical implementation. In fact, receivers based on the SRM have already been implemented (in 1999), using QED cavities [10].

6.4 SRM with Mixed States

The SRM method is normally used for decision in the presence of pure states, but recently [11] this method was extended to systems described by density operators (mixed states), allowing for the evaluation of the performance of quantum communications systems even *in the presence of thermal noise*.

The technique behind this generalization, consisting in passing from pure states to density operators, is the usual factorization of the density operators

$$\rho_i = \gamma_i \gamma_i^*$$

which allows us to proceed in basically the same way as seen with pure states, using the following correspondence

$$\boxed{\begin{array}{rcl} \text{state } |\gamma_i\rangle & \rightarrow & \text{state factor } \gamma_i \\ \text{measurement vector } |\mu_i\rangle & \rightarrow & \text{measurement factor } \mu_i \end{array}}$$

whose consequences are summarized in Table 6.1. As done with pure states, we keep the hypothesis of equiprobable symbols.

Table 6.1 The SRM method with pure states and with mixed states

Operation	Pure states	Mixed states
Density operators		$\rho_0, \ldots, \rho_{K-1}$
States/factor states	$|\gamma_0\rangle, \ldots, |\gamma_{K-1}\rangle$	$\gamma_0, \ldots, \gamma_{K-1}$
State matrix	$\Gamma = [|\gamma_0\rangle, \ldots, |\gamma_{K-1}\rangle]$	$\Gamma = [\gamma_0, \ldots, \gamma_{K-1}]$
Gram's matrix	$G = \Gamma^* \Gamma = \big[\langle \gamma_i | \gamma_j \rangle\big]$	$G = \Gamma^* \Gamma = \big[\gamma_i^* \gamma_j\big]$
Gram's operator	$T = \Gamma \Gamma^* = \sum_{i=0}^{K-1} |\gamma_i\rangle\langle\gamma_i|$	$T = \Gamma \Gamma^* = \sum_{i=0}^{K-1} \gamma_i \gamma_i^*$
Measurement vectors/factors	$|\mu_i\rangle = T^{-\frac{1}{2}} |\gamma_i\rangle$	$\mu_i = T^{-\frac{1}{2}} \gamma_i$
Measurement matrix	$M = T^{-\frac{1}{2}} \Gamma = \Gamma G^{-\frac{1}{2}}$	$M = T^{-\frac{1}{2}} \Gamma = \Gamma G^{-\frac{1}{2}}$
Mixed product matrix	$B = M^* \Gamma = \big[\langle \mu_i | \gamma_j \rangle\big] = G^{1/2}$	$B = M^* \Gamma = \big[\mu_i^* \gamma_j\big] = G^{1/2}$
Measurement operators	$Q_i = |\mu_i\rangle\langle\mu_i|$	$Q_i = \mu_i \mu_i^*$
Transition probabilities $p(j|i)$	$|\langle \mu_j | \gamma_i \rangle|^2 = |b_{ji}|^2$	$\mathrm{Tr}[\mu_j \mu_j^* \gamma_i \gamma_i^*] = \mathrm{Tr}[b_{ji}^* b_{ji}]$
Correct decision probability P_c	$\dfrac{1}{K} \sum_{i=0}^{K-1} |b_{ii}|^2$	$\dfrac{1}{K} \sum_{i=0}^{K-1} \mathrm{Tr}[b_{ii}^* b_{ii}]$

6.4.1 Discussion of the Method

Having obtained the factors γ_i of the density operators ρ_i, we form the state matrix

$$\underset{n \times H}{\Gamma} = \left[\gamma_0, \gamma_1, \ldots, \gamma_{K-1}\right] \tag{6.34}$$

where H is the total number of columns, and from this we obtain Gram's operator (see (5.118)) and Gram's matrix (see (5.119))

$$\underset{n \times n}{T} = \Gamma \Gamma^*, \qquad \underset{H \times H}{G} = \Gamma^* \Gamma.$$

Theorem 6.1 and the subsequent corollaries still hold, so the optimal measurement matrix can be calculated from three distinct expressions

$$M_0 = U_r V_r^* = T^{-1/2}\Gamma = \Gamma G^{-1/2}. \tag{6.35}$$

The first expression is bound to the reduced SVD of the measurement matrix Γ, while the other two are obtained from the reduced EIDs of T and G.

From $M_0 = \left[\mu_0, \mu_1, \ldots, \mu_{K-1}\right]$ we get the measurement factors μ_i and, from these, the measurement operators $Q_i = \mu_i \mu_i^*$. The relation giving the mixed product matrix $B = [b_{ij}] = [\mu_i^* \gamma_j]$ still holds

$$B = M^* \Gamma = G^{1/2}. \tag{6.36}$$

Finally, we obtain the transition probabilities and the correct decision probability from (5.75)

$$p_c(j|i) = \mathrm{Tr}[b_{ji}^* \, b_{ji}], \qquad P_c = \sum_{i \in \mathcal{A}} q_i \mathrm{Tr}[b_{ii}^* \, b_{ii}]. \tag{6.37}$$

where now b_{ij} is the i, j block of the matrix $G^{1/2}$.

Example 6.2 Let us consider Problem 5.9 of the previous chapter, where starting from two density operators ρ_0 and ρ_1 of $\mathcal{H} = \mathbb{C}^4$, we found the factors γ_0 of dimensions 4×2 and γ_1 of dimensions 4×3. From these factors, the 4×5 state matrix is formed

$$\Gamma = [\gamma_0, \gamma_1] = \begin{bmatrix} -0.54117 & -0.02018 & -0.47937 & -0.06934 & 0.03124 \\ -0.54117 & -0.02018 & i\,0.51339 & 0.0 & i\,0.02917 \\ -0.54117 & -0.02018 & 0.479370 & -0.06934 & -0.03124 \\ -0.33238 & 0.09857 & -i\,0.51339 & 0.0 & -i\,0.02917 \end{bmatrix}.$$

From Γ we obtain the 4×4 Gram's operator

$$
T = \Gamma\, \Gamma^* =
\begin{bmatrix}
0.52885 & 0.29327 + i\,0.24519 & 0.06731 & 0.17788 - i\,0.24519 \\
0.29327 - i\,0.24519 & 0.55769 & 0.29327 + i\,0.24519 & -0.08654 \\
0.06731 & 0.29327 - i\,0.24519 & 0.52885 & 0.17788 + i\,0.24519 \\
0.17788 + i\,0.24519 & -0.08654 & 0.17788 - i\,0.24519 & 0.38462
\end{bmatrix}
$$

and the 5×5 Gram's matrix

$$
G = \Gamma^*\, \Gamma =
\begin{bmatrix}
0.98906 & 0.0 & -i\,0.10719 & 0.07505 & -i\,0.00609 \\
0.0 & 0.01094 & -i\,0.06096 & 0.00280 & -i\,0.00346 \\
i\,0.10719 & i\,0.06096 & 0.98673 & 0.0 & \\
0.07505 & 0.00280 & & 0.00962 & \\
i\,0.00609 & i\,0.00346 & 0.0 & 0.0 & 0.00365
\end{bmatrix} .
$$

The four eigenvalues of T are all positive and precisely

$$
1.04854 \quad 0.941288 \quad 0.101754 \quad 0.0647906
$$

and it can be verified that G has the same positive eigenvalues (the fifth eigenvalue of G is null).

As T has full rank, its inverse square root must be intended in the ordinary sense, and results in

$$
T^{-1/2} =
\begin{bmatrix}
6.55880 & -3.71728 - i\,3.44423 & -0.73775 + i\,1.90902 & -1.85711 + i\,2.39197 \\
-3.71728 + i\,3.44423 & 8.04954 & -3.71728 - i\,3.44423 & 0.66455 \\
-0.73775 - i\,1.90902 & -3.71728 + i\,3.44423 & 6.55880 & -1.85711 - i\,2.39197 \\
-1.85711 - i\,2.39197 & 0.66455 & -1.85711 + i\,2.39197 & 6.11094
\end{bmatrix} .
$$

From $T^{-1/2}$ we obtain the measurement factors

$$
\mu_0 = T^{-1/2}\gamma_0 =
\begin{bmatrix}
0.48977 + i\,0.01653 & -0.25565 - i\,0.26963 \\
0.54077 & -0.00921 \\
0.48977 - i\,0.01653 & -0.25565 + i\,0.26963 \\
0.47493 & 0.60823
\end{bmatrix}
$$

$$
\mu_1 = T^{-1/2}\gamma_1 =
\begin{bmatrix}
-0.02278 - i\,0.49923 & -0.40600 + i\,0.18138 & 0.09726 - i\,0.40602 \\
0.50687 & 0.48486 - i\,0.07894 & -0.45743 \\
-0.02278 + i\,0.49923 & -0.44257 - i\,0.04322 & 0.09726 + i\,0.40602 \\
-0.49203 & 0.26908 - i\,0.04381 & 0.29679
\end{bmatrix} .
$$

Finally, from (6.37) we obtain the transition probabilities $p_c(j|i)$, whose matrix is

$$
p_c =
\begin{bmatrix}
0.986242 & 0.013758 \\
0.01084 & 0.013758
\end{bmatrix}
$$

hence, with equiprobable symbols, we have

$$P_c = 0.986242 \qquad P_e = 0.013758.$$

Computation of P_e from the eigenvalues (Helstrom). Having considered a binary case, we know how to compute the optimal projectors according to Helstrom's theory, which is based on the eigenvalues of the decision operator $D = \frac{1}{2}(\rho_1 - \rho_0)$. The eigenvalues of D become

$$\{-0.977483, 0.977145, 0.00422318, -0.00388438\}$$

So, applying (5.22), we obtain

$$P_c = 0.981368 \qquad P_e = 0.018632$$

and we realize that the SRM gives an *underestimate* of the error probability.

6.5 SRM with Geometrically Uniform States (GUS)

The geometrically uniform symmetry (GUS) has been introduced in Sect. 5.13. Now, it is evident that the GUS on a constellation of states leads to a symmetry also on the measurement vectors, with remarkable simplifications, but the most important consequence is that, with pure states, the SRM method in the presence of GUS provides the *optimal* decision (maximizing the correct decision probability), as will be seen toward the end of this section.

6.5.1 Symmetry of Measurement Operators Obtained with the GUS

In Proposition 5.9 we have seen that if the state constellation has the GUS, also the *optimal* measurement operators have the same symmetry. We now prove that this property also holds for the measurement operators obtained with the SRM, which are not optimal in general. [1]

---

[1] It is useful to recall that we call *optimal* the measurement operators obtained with the maximization of the correct decision probability, while the measurement operators obtained with the SRM minimize the quadratic error between the measurement vectors and the state vectors.

In Proposition 5.8 we have seen that the Gram operator T and the symmetry operator S commute and therefore they are simultaneously diagonalizable

$$T S = S T \iff T = U \Lambda_T U^*, \qquad S = U \Lambda_S U^*. \tag{6.38a}$$

Then, also the powers of T and of S are simultaneously diagonalizable and therefore commute

$$T^\alpha = U \Lambda_T^\alpha U^*, \qquad S^\beta = U \Lambda_S^\beta U^* \iff T^\alpha S^\beta = S^\beta T^\alpha. \tag{6.38b}$$

In particular $T^{-\frac{1}{2}}$ commutes with S^i for every i

$$T^{-\frac{1}{2}} S^i = S^i T^{-\frac{1}{2}}, \qquad i = 0, 1, \ldots, K - 1. \tag{6.39}$$

Then, combining (6.21) with (6.39) we obtain

$$|\mu_i\rangle = T^{-\frac{1}{2}}|\gamma_i\rangle = T^{-\frac{1}{2}} S^i |\gamma_0\rangle = S^i T^{-\frac{1}{2}}|\gamma_0\rangle .$$

The above result can be formulated as follows:

Theorem 6.2 *If a constellation of states $|\gamma_i\rangle$ has the GUS with symmetry operator S, also the measurement vectors obtained with the SRM have the GUS with the same symmetry operator, namely*

$$|\mu_i\rangle = S^i |\mu_0\rangle , \qquad i = 0, 1, \ldots, K - 1 \tag{6.40}$$

where
$$|\mu_0\rangle = T^{-\frac{1}{2}}|\gamma_0\rangle . \tag{6.40a}$$

Thus, from (6.40), all the measurement vectors can be obtained from the reference vector $|\mu_0\rangle$. This property is then transferred to the measurement operators Q_i with the usual rules.

6.5.2 Consequences of the GUS on Gram's Matrix

When the states have the GUS, Gram's matrix becomes circulant and the SRM methodology can be developed to arrive at explicit results.

We recall that a matrix $G = [G_{ij}]$ of dimensions $K \times K$ is called *circulant* if its elements depend only on the difference of the indexes, modulo K, that is, they are of the type

$$G_{ij} = r_{i-j \,(\mathrm{mod}\,K)}. \tag{6.41}$$

For example for $K = 4$ we have the structure

$$G = \begin{bmatrix} r_0 & r_1 & r_2 & r_3 \\ r_3 & r_0 & r_1 & r_2 \\ r_2 & r_3 & r_0 & r_1 \\ r_1 & r_2 & r_3 & r_0 \end{bmatrix}$$

and the elements of the rows are obtained as permutations of those of the first row. Therefore, a circulant matrix is completely specified by its first row, which for convenience we will call *circulant vector*.

Now, from (5.120) it results that the inner products

$$G_{ij} = \langle \gamma_i | \gamma_j \rangle = \langle \gamma_0 | (S^*)^i S^j | \gamma_0 \rangle$$
$$= \langle \gamma_0 | S^{j-i} | \gamma_0 \rangle = r_{i-j \,(\mathrm{mod}\,K)}$$

depend upon the difference $i - j \pmod{K}$, so ensuring that Gram's matrix is (Hermitian) circulant with circulant vector

$$[r_0, r_1, \ldots, r_{K-1}] = [1, \langle \gamma_0 | S | \gamma_0 \rangle, \ldots, \langle \gamma_0 | S^{K-1} | \gamma_0 \rangle].$$

The EID of a circulant Gram's matrix is expressed through the matrix of the DFT (Discrete Fourier Transform), given by

$$W_{[K]} = \frac{1}{\sqrt{K}} \begin{bmatrix} 1 & 1^0 & 1^{-1} & \cdots & 1^{-(K-1)} \\ 1 & W_K^{-1} & W_K^{-2} & \cdots & W_K^{-2(K-1)} \\ \vdots & & \vdots & & \\ 1 & W_K^{-(K-1)} & W_K^{-2(K-1)} & \cdots & W_K^{-(K-1)(K-1)} \end{bmatrix} \quad (6.42)$$

where $W_K = \mathrm{e}^{\mathrm{i}2\pi/K}$. From the orthonormality condition

$$\sum_{s=0}^{K-1} \frac{1}{K} W_K^{rs} = \delta_{r0} \quad (6.43)$$

it can be verified that the columns of $W_{[K]}$

$$|w_p\rangle = \frac{1}{\sqrt{K}} \left[W_K^{-p}, W_K^{-2p}, \ldots, W_K^{-p(K-1)} \right]^{\mathrm{T}}, \qquad p = 0, 1, \ldots, K - 1 \quad (6.44)$$

form an orthonormal basis of $\mathbb{C}^K$, i.e., $\langle w_p | w_q \rangle = \delta_{pq}$.

Theorem 6.3 *A circulant Gram's matrix* $G = [G_{ij}] = [r_{i-j \,(\mathrm{mod}\,K)}]$ *has the following EID*

$$G = W^* \Lambda W = \sum_{p=0}^{K-1} \lambda_p |w_p\rangle\langle w_p| \quad \text{with} \quad W := W_{[K]} \tag{6.45}$$

where the eigenvalues are given by the DFT of the circulant vector

$$\lambda_p = \sum_{q=0}^{K-1} G_{0q} W_K^{-pq} = \sum_{q=0}^{K-1} r_q W_K^{-pq} \tag{6.45a}$$

and $\Lambda = \text{diag}\,[\lambda_0, \lambda_1, \ldots, \lambda_{K-1}]$.

The theorem is proved in Appendix section "On the EID of a Circulant Matrix".

6.5.3 Performance Evaluation

From Theorem 6.3 we soon find the square roots

$$G^{\pm\frac{1}{2}} = \sum_{p=0}^{K-1} \lambda_p^{\pm\frac{1}{2}} |w_p\rangle\langle w_p| = W^* \Lambda^{\pm\frac{1}{2}} W \tag{6.46}$$

whose elements are given by

$$(G^{\pm\frac{1}{2}})_{ij} = \frac{1}{K} \sum_{p=0}^{K-1} \lambda_p^{\pm\frac{1}{2}} W_K^{-p(i-j)}. \tag{6.46a}$$

We can then evaluate the transition probabilities from (6.29), where the element ij is computed from (6.46a); thus

$$p_c(j|i) = \left| \frac{1}{K} \sum_{p=0}^{K-1} \lambda_p^{\frac{1}{2}} W_K^{-p(i-j)} \right|^2, \quad i, j = 0, 1, \ldots, K-1 \tag{6.47}$$

in particular, the diagonal transition probabilities are found to be all equal

$$p_c(i|i) = \left[\frac{1}{K} \sum_{p=0}^{K-1} \lambda_p^{\frac{1}{2}} \right]^2 \quad (\text{independent of } i) \tag{6.47a}$$

and therefore the correct decision probability (6.30) becomes explicitly

$$P_c = \left[\frac{1}{K} \sum_{p=0}^{K-1} \lambda_p^{\frac{1}{2}} \right]^2. \tag{6.48}$$

The measurement vectors $|\mu_i\rangle$ are obtained as linear combination of the states (see (6.27)), but considering that they have the GUS, their evaluation can be limited to the reference vector, given by

$$|\mu_0\rangle = \sum_{j=0}^{K-1} (G^{-1/2})_{ij} |\gamma_j\rangle. \tag{6.49}$$

In conclusion, when Gram's matrix G is circulant, to evaluate the measurement vectors and their performance, it suffices to compute their eigenvalues, given simply by the DFT of the first row of G. This methodology will be applied to PSK and PPM modulations in the next chapter.

6.5.4 Optimality of SRM Decision with Pure States Having the GUS

We now prove that the SRM decision, when the states have the GUS, realizes the minimum error probability.

Proposition 6.1 *When the constellation of* **pure states** *verifies the GUS, the SRM becomes* **optimum**, *achieving the minimum error probability.*

For the poof we use Holevo's theorem, in the version given by Corollary 5.1. With equiprobable symbols, the first conditions of Holevo's theorem result from (5.97)

$$b_{ij}\, b_{jj}^* - b_{ii}\, b_{ji}^* = 0\,, \qquad \forall i, \forall j \tag{6.50}$$

where $b_{ij} = \langle \mu_i | \gamma_j \rangle$ are the mixed products. Their matrix is given by (see (6.28))

$$B = G^{1/2} = W^* \Lambda^{1/2} W$$

and it is symmetric. Its elements b_{ij} depend only upon the difference $i - j$, as indicated also by (6.46a), and then they can be expressed in the form

$$b_{ij} = f(j - i) \quad \text{with} \quad b_{ij}^* = f(i - j)\,.$$

Therefore, from (6.50) it follows $f(j - i)f^*(0) - f(0)f(j - i) = 0$, which is verified because $f(0)$ is real.

For a proof of the second condition, we address the reader to [2].

6.5.5 Helstrom's Bound with SRM

In Sect. 5.13 we have seen that a binary constellation always satisfies the GUS, and hence the SRM gives the optimal decision. Then with the SRM approach we have to obtain the Helstrom bound.

In the binary case the state matrix is $\Gamma = [|\gamma_0\rangle, |\gamma_1\rangle]$ and Gram's matrix is

$$G = \Gamma^* \Gamma = \begin{bmatrix} 1 & X \\ X^* & 1 \end{bmatrix}, \qquad X := \langle \gamma_0 | \gamma_1 \rangle.$$

and in general is not circulant because $X^* \neq X$ and hence we cannot apply the approach based on the DFT.

We evaluate the square roots of G by hand. Assuming that $G^{1/2}$ has the form[2]

$$G^{1/2} = \begin{bmatrix} a & b \\ b^* & a \end{bmatrix}$$

we find the conditions

$$a^2 + |b|^2 = 1, \qquad 2ab = X$$

which give

$$a^2 + \frac{|X|^2}{4a^2} = 1 \quad \rightarrow \quad a^4 - \frac{1}{4}|X|^2 = 0.$$

The solution is

$$a = \frac{1}{\sqrt{2}} \sqrt{1 + \sqrt{1 - |X|^2}}$$

and

$$b = \frac{X}{2a} = \frac{X}{\sqrt{2}|X|} \sqrt{1 - \sqrt{1 - |X|^2}} = \frac{e^{i\beta}}{\sqrt{2}} \sqrt{1 - \sqrt{1 - |X|^2}}$$

where $\beta = \arg X$. As a check

$$G^{1/2} G^{1/2} = \begin{bmatrix} 1 & X \\ X^* & 1 \end{bmatrix} = G.$$

From $G^{1/2}$ we have the correct decision probability from (6.30) as

$$P_c = \frac{1}{2} \left[|(G^{1/2})_{00}|^2 + |(G^{1/2})_{00}|^2 \right] = \frac{1}{2} \left[1 + \sqrt{1 - |X|^2} \right]$$

that is, the Helstrom bound.

---

[2] The assumption that the diagonal elements are equal is in agreement with a Sasaki's et al. [12] theorem, which states that in a optimal decision the square root of the Gram matrix must have all the diagonal elements equal.

Next we evaluate the optimal measurement vectors. Considering that $\det G^{1/2} = \sqrt{1 - |X|^2}$, the inverse of $G^{1/2}$ is

$$G^{-1/2} = \frac{1}{\sqrt{1 - |X|^2}} \begin{bmatrix} a & -b \\ -b^* & a \end{bmatrix}.$$

Using the identities

$$\frac{1}{1 - |X|} \pm \frac{1}{1 + |X|} = \sqrt{2} \frac{\sqrt{1 \pm \sqrt{1 - |X|^2}}}{\sqrt{1 - |X|^2}}$$

we find (with $\beta = \arg X$)

$$G^{-1/2} = \frac{1}{2} \begin{bmatrix} \frac{1}{1-|X|} + \frac{1}{1+|X|} & e^{i\beta}(\frac{1}{1-|X|} - \frac{1}{1+|X|}) \\ e^{-i\beta}(\frac{1}{1-|X|} - \frac{1}{1+|X|}) & \frac{1}{1-|X|} + \frac{1}{1+|X|} \end{bmatrix}$$

which gives the measurement matrix as $M = \Gamma G^{-1/2}$.

When the inner product is real the Gram matrix turns out to be circulant and the approach based on the DFT can be applied to get the Helstrom bound (see Problem 6.5).

Table 6.2 The SRM method in general, and with geometrically uniform symmetry (GUS)

Operation	General case	With GUS
Constellation of states	$\lvert\gamma_0\rangle, \ldots, \lvert\gamma_{K-1}\rangle$	$\lvert\gamma_i\rangle = S^i \lvert\gamma_0\rangle$
State matrix Γ	$[\lvert\gamma_0\rangle, \ldots, \lvert\gamma_{K-1}\rangle]$	$[\lvert\gamma_0\rangle, \ldots, S^{K-1}\lvert\gamma_0\rangle]$
Gram's matrix $G = \Gamma^* \Gamma$	$[\langle\gamma_i\lvert\gamma_j\rangle]$	$[\langle\gamma_0\lvert S^{j-i}\lvert\gamma_0\rangle] = W^* \Lambda W$
Gram's operator $T = \Gamma \Gamma^*$	$\displaystyle\sum_{i=0}^{K-1} \lvert\gamma_i\rangle\langle\gamma_j\rvert$	$\displaystyle\sum_{i=0}^{K-1} S^i\lvert\gamma_0\rangle\langle\gamma_0\rvert S^{-i}$
Measurement vectors	$\lvert\mu_i\rangle = T^{-\frac{1}{2}}\lvert\gamma_i\rangle$	$\lvert\mu_0\rangle = T^{-\frac{1}{2}}\lvert\gamma_0\rangle$, $\lvert\mu_i\rangle = S^i\lvert\mu_0\rangle$
Measurement matrix M	$T^{-\frac{1}{2}}\Gamma = \Gamma G^{-\frac{1}{2}}$	$T^{-\frac{1}{2}}\Gamma = \Gamma W^* \Lambda^{-\frac{1}{2}} W$
Mixed product matrix B	$M^* \Gamma = G^{\frac{1}{2}}$	$G^{\frac{1}{2}} = W^* \Lambda^{\frac{1}{2}} W$
Transition probabilities $p_c(j\lvert i)$	$\lvert\langle\mu_j\lvert\gamma_i\rangle\rvert^2 = \lvert b_{ji}\rvert^2$	$\left\lvert \dfrac{1}{K} \displaystyle\sum_{p=0}^{K-1} \lambda_p^{\frac{1}{2}} W_K^{-p(i-j)} \right\rvert^2$
Correct decision probability	$P_c = \dfrac{1}{K} \displaystyle\sum_{i=0}^{K-1} \lvert b_{ii}\rvert^2$	$P_c = \left[\dfrac{1}{K} \displaystyle\sum_{p=0}^{K-1} \lambda_p^{\frac{1}{2}} \right]^2$

Summary of the SRM Method

Table 6.2 summarizes the computation procedure of the SRM method, on the left in the general case, and on the right in the presence of GUS.

Problem 6.5 ⋆⋆   Apply the SRM approach to find the optimal decision in a binary system with equiprobable symbols and with a real inner product X.

6.6 SRM with Mixed States Having the GUS

We have seen that the (GUS) can be extended from pure states to density operators, with the condition

$$\rho_i = S^i \rho_0 \, (S^i)^*, \qquad i = 0, 1, \ldots, K - 1. \tag{6.51}$$

This extension entails, for the factors, the relation

$$\gamma_i = S^i \, \gamma_0, \qquad i = 0, 1, \ldots, K - 1$$

and the same symmetry is transferred to the measurement operators,

$$Q_i = S^i \, Q_0 \, (S^i)^*, \qquad i = 0, 1, \ldots, K - 1 \tag{6.52}$$

as well as to the measurement factors

$$\mu_i = S^i \mu_0, \qquad i = 0, 1, \ldots, K - 1.$$

With the SRM method in the presence of GUS, the performance evaluation becomes simpler, as already seen with the pure states, but some complication arises, due to the fact that Gram's matrix is not circulant, but *block circulant* [4]. However, we can still manage to formulate the computation based on the DFT, arriving at results explicit enough.

Relation (6.35) still holds, in particular

$$M_0 = T^{-1/2} \Gamma = \Gamma G^{-1/2}$$

so that we have two possible approaches.

6.6.1 Gram Operator Approach

This approach is based on the evaluation of the inverse square root $T^{-1/2}$ of the Gram operator T, and the reference measurement operator is given by (see (6.40a))

$$Q_0 = T^{-1/2} \rho_0 \, T^{-1/2}. \tag{6.53}$$

Proposition 6.2 *The transition probabilities with mixed states having the GUS can be obtained from the reference density operator and the inverse square root of the Gram operator as*

$$p_c(j|i) = \mathrm{Tr}\left[S^{i-j} \rho_0 \, S^{-(i-j)} \, Q_0 \right] \tag{6.54}$$

with Q_0 given by (6.53). The correct decision probability is given by the synthetic formula

$$\boxed{P_c = \mathrm{Tr}\left[(\rho_0 \, T^{-1/2})^2 \right].} \tag{6.55}$$

In fact,

$$\begin{aligned}
p_c(j|i) = \mathrm{Tr}[\rho_i \, Q_j] &= \mathrm{Tr}\left[S^i \rho_0 \, S^{-i} \, S^j \, Q_0 \, S^{-j} \right] \\
&= \mathrm{Tr}\left[S^{i-j} \rho_0 \, S^{-(i-j)} \, Q_0 \right].
\end{aligned}$$

Then, using (6.53), we obtain

$$P_c = \mathrm{Tr}\left[\rho_0 T^{-1/2} \rho_0 \, T^{-1/2} \right] = \mathrm{Tr}\left[T^{-1/2} \rho_0 T^{-1/2} \rho_0 \right]$$

and (6.55) follows at once.

6.6.2 Gram Matrix Approach

With the Gram matrix it is less trivial to get useful results, because they need the EID of the symmetry operator, given by (5.128)

$$S = \sum_{i=0}^{K-1} W_k^i \, P_i \,,$$

where P_i are projectors. The Gram matrix is formed by the blocks of order h_0

$$G_{ij} = \gamma_i^* \gamma_j = \gamma_0^* S^{j-i} \gamma_0 = \sum_{k=0}^{K-1} W_K^{k(j-i)} \gamma_0^* P_k \gamma_0 = \frac{1}{K} \sum_{k=0}^{K-1} W_K^{k(j-i)} D_k \tag{6.56}$$

where

$$D_k := K \gamma_0^* P_k \, \gamma_0. \tag{6.57}$$

Then we find that the (i, j) block of G has the structure

$$G_{ij} = r_{i-j \, (\mathrm{mod}\, K)}. \tag{6.58}$$

Since G_{ij} depends only on the difference $(j - i) \bmod K$, the matrix G turns out to be *block circulant*, with blocks of the same order $h_k = h_0$. Then one can extend what seen with pure states in Sect. 6.5.3, operating on the blocks, instead of on the scalar elements, to get the explicit factorization of G, namely[3]

$$G = W_{(h_0)} \, D \, W^*_{(h_0)}$$

where $D = \mathrm{diag}[D_0, \ldots, D_{K-1}]$ and $W_{(h_0)}$ is the $K h_0 \times K h_0$ block DFT matrix

$$W_{(h_0)} = \frac{1}{\sqrt{K}} \begin{bmatrix} 1 & 1 & 1 & \cdots & 1 \\ 1 & W_K^{-1} & W_K^{-2} & \cdots & W_K^{-2(K-1)} \\ \vdots & & \vdots & & \\ 1 & W_K^{-(K-1)} & W_K^{-2(K-1)} & \cdots & W_K^{-(K-1)(K-1)} \end{bmatrix} \otimes I_{h_0}. \tag{6.59}$$

As a consequence, the diagonal blocks are given as the DFT of the first block row of G, namely

$$D_k = \sum_{s=0}^{K-1} W_K^{-ks} \, G_{0s}. \tag{6.60}$$

Now we have to find the square root of G and this can be done as seen with pure states, but acting on blocks instead of on scalars. We find

$$G^{1/2} = W_{(h_0)} \, D^{1/2} \, W^*_{(h_0)}$$

where

$$D^{1/2} = \mathrm{diag}[D_0^{1/2}, \ldots, D_{K-1}^{1/2}].$$

In particular, the (i, j) block is given by

$$\boxed{(G^{1/2})_{ij} = \frac{1}{K} \sum_{k=0}^{K-1} W_K^{k(j-i)} D_k^{1/2}.} \tag{6.61}$$

Note that we have found the alternative expressions (6.57) and (6.60) for the diagonal blocks D_k, where the first is based on the EID of the symmetry operator.

---

[3] This is not a standard EID, because the diagonal blocks D_i are not diagonal matrices.

This EID is used in the proof, but we can use the alternative expression (6.57) to avoid its evaluation (which may be difficult).

To summarize:

Proposition 6.3 *With the GUS the (i, j) block of the Gram matrix can be written in the forms*

$$G_{ij} = \gamma_i^* \gamma_j = \gamma_0^* S^{j-i} \gamma_0 = \frac{1}{K} \sum_{k=0}^{K-1} W_K^{k(j-i)} D_k \tag{6.62}$$

where the matrices D_k of order h_0 are Hermitian PSD given by

$$D_k = \sum_{i=0}^{K-1} \gamma_0^* \gamma_i W_K^{-ki}. \tag{6.63}$$

The (i, j) block of the matrix $G^{1/2}$ is given by relation (6.61).

Now, using expression (6.61), one can obtain the transition probabilities from (6.37) with $b_{ij} = (G^{1/2})_{ij}$. For the correct decision probability one finds

$$P_c = \mathrm{Tr}\left[\frac{1}{K} \sum_{k=0}^{K-1} D_k^{1/2} \right]^2. \tag{6.64}$$

The reference measurement factor μ_0 can be obtained as in (6.29), that is,

$$\mu_0 = \sum_{j=0}^{K-1} (G^{-1/2})_{ij} \gamma_j. \tag{6.65}$$

Remark on optimality. Differently from the case of pure states, the SRM method with GUS is not optimal in general with mixed states. In fact, for optimality, the further condition is required for the reference factors [11]

$$b_{00} = \mu_0^* \gamma_0 = \alpha\, I \tag{6.66}$$

where I is the identity matrix, and α a proportionality constant. Note that $b_{00} = (G^{1/2})_{00}$. As we will see, the PSK and PPM systems verify the GUS even in the presence of noise, but do not verify the further condition (6.66), hence the SRM method is not optimal.

Application to generalized GUS. The above theory of SRM for mixed states with GUS can be used for pure states having the first form of generalized GUS introduced in Sect. 5.13.4. This possibility will be applied in Sect. 7.11 to Quantum Communications systems using the QAM modulation.

Problem 6.6 ★ Write explicitly the block DFT matrix, defined by (6.59), for $K = 4$ and $h_0 = 2$ and prove that it is a unitary matrix.

Problem 6.7 ★★ Prove in general that the block DFT matrix, defined by (6.59), is a unitary matrix.

Problem 6.8 ★★★ Extend Theorem 6.3 on circulant matrices to block circulant matrices.

Problem 6.9 ★★ To check the fundamental formulas of the SRM with mixed states having the GUS, consider the following degenerate case of reference state factor in a quaternary system

$$\gamma_0 = \frac{1}{\sqrt{3}}[|\beta_0\rangle, |\beta_0\rangle, |\beta_0\rangle]$$

where $|\beta_0\rangle$ is an arbitrary pure state, and the symmetry operator S generates the other state factor in the form $\gamma_i = S^i \gamma_0$, $i = 1, 2, 3$. Find the correct decision probability P_c.

6.7 Quantum Compression with SRM

The technique of compression seen at the end of the previous chapter, for the reduction of redundancy in quantum states, can be applied to the detection based on the SRM. In practice, quantum compression is useful in numerical computations because it reduces the size of the matrices. For instance in quantum communications using the PPM format the computational complexity may become huge and the compression allows us to get results otherwise not reachable.

We recall that compression preserves the GUS and therefore we can apply the very efficient technique that combines the SRM with the GUS, *after the state compression*.

We now review the main simplifications achieved with the application of the compression to the SRM.

6.7.1 Simplification with Compression in the General Case

We first recall that all the detection probabilities can be evaluated in the compressed space exactly as in the uncompressed space, as stated by relations (5.146) and (5.147). Also, in the compressed space, the Gram operator is always diagonal (see (5.145)).

It is also convenient to recall the dimensions of the quantities involved in the compression. We refer to **mixed states**, from which we have the case of pure states as a particularization. Before compression the dimensions are

γ_i	μ_i	Γ	M	G	T
$n \times h_i$	$n \times h_i$	$n \times H$	$n \times H$	$H \times H$	$n \times n$

where $H = h_0 + \cdots + h_{K-1}$ with h_i the rank ρ_i. After compression we have

$$\underset{r \times h_i}{\overline{\gamma}_i} \qquad \underset{r \times h_i}{\overline{\mu}_i} \qquad \underset{r \times K}{\overline{\Gamma}} \qquad \underset{r \times K}{\overline{M}} \qquad \underset{H \times H}{\overline{G}} \qquad \underset{r \times r}{\overline{T}} .$$

The case of **pure states** is obtained by setting

$$h_i = 1, \qquad H = K.$$

6.7.2 *Simplification of Compression in the Presence of GUS*

The GUS is preserved in the compressed space, as stated by Proposition 5.10.

The main property in the presence of GUS is that the Gram operator T commutes with the symmetry operator S, that is, T and S becomes simultaneously diagonalizable, as stated by

$$T = U \, \Sigma^2 \, U^*, \qquad S = U \, \Lambda \, U^*. \tag{6.67}$$

This allows us to establish simple formulas for both T and S, as in Propositions 5.11 and 5.12. A sophisticated technique to find a very useful simultaneous diagonalization is descried at the end of Chap. 8.

Problem 6.10 $\star\star$ Solve Problem 6.3 introducing compression.

6.8 Quantum Chernoff Bound

We have seen that the SRM is a suboptimal method that gives an upper bound of the error probability in a quantum communications system. Another suboptimal method is given by the *quantum Chernoff bound*, which recently received a great attention, especially for Gaussian quantum states [13], as a simple mean to estimate the performance of quantum discrimination [14–16].

The Chernoff bound is usually employed in Telecommunications and Probability Theory to establish an upper bound to the error probability [17] or more in general to bound the probability that a random variable exceeds a certain quantity, based on the knowledge of the characteristic function or of the moments of the random variable. The extension of the Chernoff bound to quantum systems, leading to the *quantum Chernoff bound*, is considered in several works [13–16, 18], employing the bound as a tool to estimate the error probability in the discrimination of quantum states, both for single-mode and for multi-mode states. The Chernoff bound can be seen also as a distance measure between operators. The Chernoff distance has been investigated, for example, in [14–16] and related to other distiguishability measures, such as fidelity.

In a recent paper [19] Corvaja has compared the Chernoff bound, and other bounds, with the SRM bound in terms of both performance and complexity.

6.8.1 Formulation

The quantum Chernoff bound has the limitation that it can be applied only to binary quantum systems. For the binary case, where the states are described by the density operators ρ_0 and ρ_1, the quantum Chernoff bound states that error probability can be bounded by the expression

$$P_e \leq \frac{1}{2} \inf_{0 \leq s \leq 1} \mathrm{Tr}\left[\rho_0^s \rho_1^{1-s}\right] \tag{6.68}$$

where s is a real parameter. Therefore, the bound requires the evaluation of the fractional power of operators (in practice of a square matrix) for all the values of the minimization parameter s. This is obtained with an eigendecomposition of the kind

$$\rho_i = U_i \, \Lambda_i U_i^* \quad \longrightarrow \quad \rho_i^s = U_i \, \Lambda_i^s U_i^*. \tag{6.69}$$

Although in general the bound requires the minimization with respect to the real value s, when the Gaussian states have the same covariance matrix or the same thermal noise component and no relative displacement (see Chap. 11), the optimum is attained for $s = 1/2$. In this case the square root of the density operators must be evaluated and the bound becomes

$$P_e \leq \frac{1}{2}\mathrm{Tr}\left[\sqrt{\rho_0}\sqrt{\rho_1}\right]. \tag{6.70}$$

In the comparison reported in [19] it is shown that for mixed states the SRM solution provides a tighter bound than the Chernoff bound in the binary case, with a comparable numerical complexity. Moreover, the SRM has the advantage that it can be applied also to the general K-ary case.

Problem 6.11 ⋆⋆   Consider the binary system specified by the pure states

$$|\gamma_0\rangle = \frac{1}{\sqrt{13}}[5, 3 - 2i, 1, 3 + 2i]^{\mathrm{T}}, \qquad |\gamma_1\rangle = \frac{1}{2\sqrt{13}}[1, 3 + 2i, 5, 3 - 2i]^{\mathrm{T}}.$$

Check that: (1) Hestrom's theory gives $P_e = 1/26$, (2) the Chernoff bound gives $P_e = 25/338$.

Appendix

On the EID of a Circulant Matrix

Let us prove Theorem 6.3. To this end, consider the matrix

$$Z := W^* G \quad \text{with} \quad W = W_{[K]}. \tag{6.71}$$

From inspection of the structure of the element Z_{ij} of Z and bearing in mind the condition (6.41), we have

$$
\begin{aligned}
Z_{ij} &= \frac{1}{\sqrt{K}} \sum_{t=0}^{K-1} W_K^{it} G_{tj} = \frac{1}{\sqrt{K}} \sum_{t=0}^{K-1} W_K^{it} \, r_{j-t \,(\mathrm{mod}\, K)} \\
&= \frac{1}{\sqrt{K}} \left(\sum_{t=0}^{j} W_K^{it} \, r_{j-t} + \sum_{t=j+1}^{K-1} W_K^{it} \, r_{K+j-t} \right).
\end{aligned}
$$

Letting $k = j - t$ in the first summation, and $k = K + j - t$ in the second, we have

$$
\begin{aligned}
Z_{ij} &= \frac{1}{\sqrt{K}} \left(\sum_{k=0}^{j} W_K^{i(j-k)} r_k + \sum_{k=j+1}^{K-1} W_K^{i(K+j-k)} r_k \right) \\
&= \frac{1}{\sqrt{K}} \sum_{k=0}^{K-1} W_K^{i(j-k)} r_k = \frac{1}{\sqrt{K}} W_K^{ij} \sum_{k=0}^{K-1} W_K^{-ik} r_k \\
&= \frac{1}{\sqrt{K}} W_K^{ij} \lambda_i
\end{aligned}
$$

where (see (6.45a))

$$\lambda_i := \frac{1}{\sqrt{K}} \sum_{k=0}^{K-1} W_K^{-ik} r_k.$$

From the above result we infer that the matrix Z can be written in the form

$$Z = \Lambda W^*. \tag{6.72}$$

Then, to obtain (6.45) from (6.71) and (6.72), it suffices to recall that W is unitary, then $W^* = W^{-1}$.

References

1. P. Hausladen, R. Jozsa, B. Schumacher, M. Westmoreland, W.K. Wootters, Classical information capacity of a quantum channel. Phys. Rev. A **54**, 1869–1876 (1996)
2. Y.C. Eldar, G.D. Forney, On quantum detection and the square-root measurement. IEEE Trans. Inf. Theory **47**(3), 858–872 (2001)
3. Y.C. Eldar, G.D. Forney, Optimal tight frames and quantum measurement. IEEE Trans. Inf. Theory **48**(3), 599–610 (2002)
4. G. Cariolaro, G. Pierobon, Performance of quantum data transmission systems in the presence of thermal noise. IEEE Trans. Commun. **58**(2), 623–630 (2010)
5. G. Cariolaro, G. Pierobon, Theory of quantum pulse position modulation and related numerical problems. IEEE Trans. Commun. **58**(4), 1213–1222 (2010)
6. G. Cariolaro, R. Corvaja, G. Pierobon, Compression of pure and mixed states in quantum detection, in *2011 IEEE Global Telecommunications Conference (GLOBECOM, 2011)*, pp. 1–5 (2011)
7. R. Penrose, A generalized inverse for matrices. Math. Proc. Camb. Philos. Soc. **51**, 406–413 (1955)
8. R.A. Horn, C.R. Johnson, *Matrix Analysis* (Cambridge University Press, Cambridge, 1998)
9. A.S. Holevo, Statistical decision theory for quantum systems. J. Multivar. Anal. **3**(4), 337–394 (1973)
10. M. Sasaki, T. Sasaki-Usuda, M. Izutsu, O. Hirota, Realization of a collective decoding of code-word states. Phys. Rev. A **58**, 159–164 (1998)
11. Y.C. Eldar, A. Megretski, G.C. Verghese, Optimal detection of symmetric mixed quantum states. IEEE Trans. Inf. Theory **50**(6), 1198–1207 (2004)
12. M. Sasaki, K. Kato, M. Izutsu, O. Hirota, Quantum channels showing superadditivity in classical capacity. Phys. Rev. A **58**, 146–158 (1998)
13. C. Weedbrook, S. Pirandola, R. Garcí-Patrón, N.J. Cerf, T.C. Ralph, J.H. Shapiro, S. Lloyd, Gaussian quantum information. Rev. Mod. Phys. **84**, 621–669 (2012)
14. K.M.R. Audenaert, M. Nussbaum, A. Szkoła, F. Verstraete, Asymptotic error rates in quantum hypothesis testing. Commun. Math. Phys. **279**(1), 251–283 (2008)
15. K.M.R. Audenaert, J. Calsamiglia, R. Muñoz Tapia, E. Bagan, L. Masanes, A. Acin, F. Verstraete, Discriminating states: the quantum Chernoff bound. Phys. Rev. Lett. **98**, paper no. 160501 (2007)
16. J. Calsamiglia, R. Muñoz Tapia, L. Masanes, A. Acin, E. Bagan, Quantum chernoff bound as a measure of distinguishability between density matrices: application to qubit and Gaussian states. Phys. Rev. A **77**, 032311 (2008)
17. J.M. Wozencraft, *Principles of Communication Engineering* (Wiley, New York, 1965)
18. M. Nussbaum, A. Szkola, The Chernoff lower bound for symmetric quantum hypothesis testing. Ann. Stat. **37**(2), 1040–1057 (2009)
19. R. Corvaja, Comparison of error probability bounds in quantum state discrimination. Phys. Rev. A **87**, paper no. 042329 (2013)

Chapter 7
Quantum Communications Systems

7.1 Introduction

The quantum decision theory, developed in the previous two chapters, is now applied to quantum communications systems where the nature of the states that carry the information is specified. A *constellation of K quantum states*, to which to commit a symbol belonging to a K-ary alphabet, corresponds, in the classical version, to a K-ary modulation format. We still consider states that operate at optical frequencies (*optical* quantum systems), because at radio frequencies quantum phenomena are not appreciable. In practice, the quantum states are usually treated as *coherent states* of a coherent monochromatic radiation emitted by a laser. For these states there exists a universal model, proposed by Glauber, that will be introduced in the next section.

Also *squeezed states* as a candidate carrier for quantum communications are considered. Squeezed light is an efficient form of optical radiation, which is obtained from a laser radiation in several ways, mainly based on parametric amplifiers.

In this chapter, we shall first examine *binary* systems, presenting the quantum versions of the OOK (on–off keying) and 2PSK (phase-shift keying) modulations. Then we shall move to *multilevel* systems, and examine the quantum versions of the QAM (quadrature amplitude), PSK, and PPM (pulse position) modulations. All the above-mentioned systems will be examined *in the absence of thermal noise*, which, instead, will be considered in the next chapter. Thus, in this chapter, the scheme of Fig. 7.1 will be followed, in which the channel is ideal and the received state is directly given by the transmitted state. As already observed, neglecting thermal noise does not mean that the analysis will be done in the absence of noise; because we shall take into account the fact that quantum measurements are affected by an intrinsic randomness, corresponding, in the classical model, to *shot noise*.

Organization of the Chapter

The next two sections deal with the definition and properties of coherent states and how to provide a constellation of coherent states. Section 7.5 develops the theory of *classical* optical systems where the decision is based on photon counting.

© Springer International Publishing Switzerland 2015

G. Cariolaro, *Quantum Communications*, Signals and Communication Technology,
DOI 10.1007/978-3-319-15600-2_7

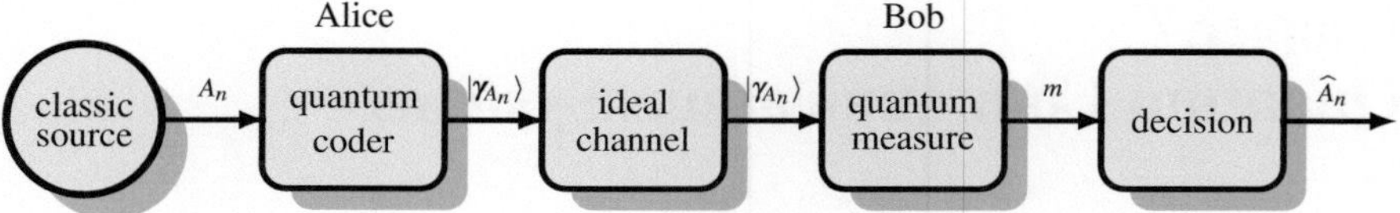

Fig. 7.1 Quantum communications system for digital transmission. $\{A_n\}$ is a sequence of classical symbols of information that Alice conveys into a sequence of quantum states $\{|\gamma_{A_n}\rangle\}$. Bob, in each symbol period, performs a quantum measurement to argue, from the result m of the measurement, which symbol was transmitted

The subsequent sections, from Sects. 7.9 to 7.13, develop the specific quantum communications system with the modulation format listed above.

In the two final sections, we will develop quantum communications with squeezed states with a comparison of the performance with that obtained with coherent states.

As explained in Chap. 4, only digital systems will be considered. For binary systems, we shall use the general theory of binary optimization, essentially Helstrom's theory, developed in Sect. 5.4. For multilevel systems, for which an explicit optimization theory is not available, we shall use the square root measurements (SRM) decision developed in Chap. 6 and, when convenient, we compare SRM results with the ones obtained with convex semidefinite programming (CSP).

7.2 Overview of Coherent States

A general model of the quantum state created by an electromagnetic field at a certain (optical) frequency is given by a coherent quantum state according to Glauber's theory. This model is now formulated in detail in a form suitable to deal with quantum communications systems, but without entering in theoretical considerations. In Chap. 11 coherent states will be fully developed in the framework of quantum information as *continuous quantum states* and also as *Gaussian quantum states*.

7.2.1 Glauber's Representation

The *coherent* radiation emitted by a laser is modeled as a *coherent state*. It has been demonstrated [1–3] that the coherent states of a *single mode* can be represented in a Hilbert space of infinite dimensions, through an orthonormal basis $\{|n\rangle, n = 0, 1, 2, \ldots\}$, where the states are called *number states*, because $|n\rangle$ *contains exactly n photons*. They are also called *number eigenstates* and *Fock states*.

To this basis, the *number operator* is associated, which is defined by

$$N = \sum_{n=0}^{\infty} n|n\rangle \langle n. \tag{7.1}$$

Then N has eigenvectors $|n\rangle$ with eigenvalues n and the spectrum of N is given by the set of naturals, $\sigma(N) = \{0, 1, 2, \ldots\}$.

In this mathematical context, a generic *coherent state* (or Glauber state) is expressed as follows:

$$|\alpha\rangle = e^{-\frac{1}{2}|\alpha|^2} \sum_{n=0}^{\infty} \frac{\alpha^n}{\sqrt{n!}} |n\rangle. \tag{7.2}$$

where α is a complex amplitude whose meaning is

$$|\alpha|^2 = \text{average number of photons in the state } |\alpha\rangle. \tag{7.3}$$

Therefore, according to (7.2), to each point α of the complex plane $\mathbb{C}$, a coherent state is associated whose physical meaning is given by (7.3). Thus, the more α moves away from the origin of $\mathbb{C}$, the higher becomes the photonic intensity associated to the state $|\alpha\rangle$.

The set of coherent states will be indicated by

$$\mathcal{G} = \{|\alpha\rangle, \alpha \in \mathbb{C}\} \quad : \quad \text{coherent states} \tag{7.4}$$

and then the notation $|\alpha\rangle \in \mathcal{G}$ will be used to distinguish one of these specific kets from the other numerous kets that we will meet. It is interesting to observe that letting $\alpha = 0$ in (7.2) we obtain

$$|\alpha\rangle_{\alpha=0} = |n\rangle_{n=0} \tag{7.5}$$

that is, with $\alpha = 0$ we obtain the state $|0\rangle$ of the Fock basis, called *ground state*.

Remark The notations of Quantum Mechanics are powerful, but sometimes subtle. In this context, it is important to distinguish the complex number $\alpha \in \mathbb{C}$ from the coherent state $|\alpha\rangle \in \mathcal{G}$, which is a ket of the infinite dimensional Hilbert space $\mathcal{H}$, generated by the basis $\{|n\rangle | n = 0, 1, 2, \ldots\}$. The fundamental relation (7.2) is a mapping $\mathbb{C} \rightarrow \mathcal{G}$, where $\mathcal{G} \subset \mathcal{H}$. For instance, $\alpha = 3 - i4 \in \mathbb{C}$ is mapped onto the coherent state $|3 - i4\rangle \in \mathcal{G}$, whose full expression is

$$|3 - i4\rangle = \exp\left[-\frac{1}{2}|3 - i4|^2\right] \sum_{n=0}^{\infty} \frac{(3 - i4)^n}{\sqrt{n}} |n\rangle.$$

7.2.2 Link with Poisson's Regime

To find the relationship between the representation of a coherent state $|\alpha\rangle \in \mathcal{G}$ and Poisson's regime, we set up a quantum measurement (Fig. 7.2) with the number

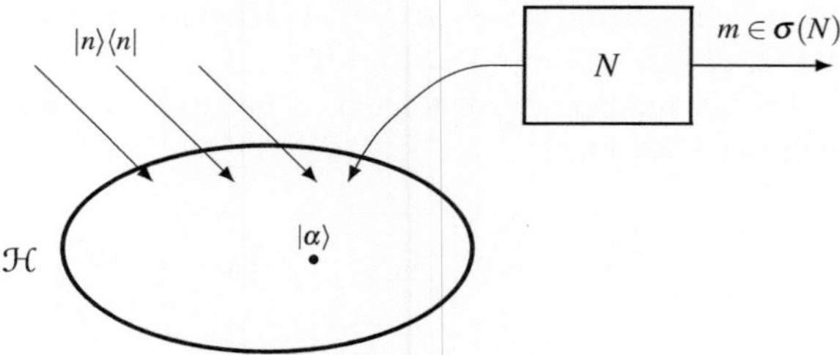

Fig. 7.2 Quantum measurement in a Hilbert space of the coherent state $|\alpha\rangle$ with an observable given by the number operator N. The elementary projectors $|n\rangle\langle n|$ are formed by the number states $|n\rangle$

operator N, interpreted as an *observable* (see Sect. 3.6). The outcome m of the measurement gives the number of photons of the quantum system in the state $|\alpha\rangle$. Then the probability that the measurement gives the outcome $m = i$ turns out to be

$$
P[m = i|\alpha] = |\langle i|\alpha\rangle|^2 = \left|\sum_{n=0}^{\infty} e^{-\frac{1}{2}|\alpha|^2} \frac{\alpha^n}{\sqrt{n!}} \langle i|n\rangle\right|^2
$$
$$
= \left|e^{-\frac{1}{2}|\alpha|^2} \frac{\alpha^i}{\sqrt{i!}}\right|^2 = e^{-|\alpha|^2} \frac{|\alpha|^{2i}}{i!}.
\tag{7.6}
$$

Therefore,

$$
P[m = i|\alpha] = e^{-N_\alpha} \frac{(N_\alpha)^i}{i!} \quad \text{with} \quad N_\alpha = |\alpha|^2.
\tag{7.7}
$$

It can also be verified that the average of m is

$$
E[m|\alpha] = \langle \alpha|N|\alpha\rangle = |\alpha|^2 = N_\alpha.
\tag{7.8}
$$

In conclusion, the outcome of the measurement m is a Poisson random variable with average $N_\alpha = |\alpha|^2$.

7.2.3 *Degree of Superposition of Coherent States*

It is important to evaluate the degree of superposition of two distinct coherent states $|\alpha\rangle$ and $|\beta\rangle$, within the geometry given by the inner product. We have

Proposition 7.1 *The inner product of two coherent states is given by*

$$
\boxed{\langle \alpha|\beta\rangle = e^{-\frac{1}{2}(|\alpha|^2 + |\beta|^2 - 2\alpha^*\beta)}.}
\tag{7.9}
$$

Hence **two distinct coherent states are never orthogonal** *(Fig. 7.3).*

Fig. 7.3 Two distinct coherent states are never orthogonal: $\langle\alpha|\beta\rangle \neq 0$

In fact, from (7.2) we have

$$\langle\alpha|\beta\rangle = e^{-\frac{1}{2}(|\alpha|^2+|\beta|^2)} \sum_{m=0}^{\infty}\sum_{n=0}^{\infty} \frac{(\alpha^*)^m \beta^n}{\sqrt{m!n!}} \langle m|n\rangle$$

$$= e^{-\frac{1}{2}(|\alpha|^2+|\beta|^2)} \sum_{m=0}^{\infty} \frac{(\alpha^*\beta)^m}{m!} = e^{-\frac{1}{2}(|\alpha|^2+|\beta|^2)} e^{\alpha^*\beta}.$$

and (7.9) follows   □

The *(quadratic)* degree of superposition of two states is expressed by

$$\boxed{|X|^2 := |\langle\alpha|\beta\rangle|^2 = e^{-|\alpha-\beta|^2}, \qquad |\alpha\rangle, |\beta\rangle \in \mathcal{G}} \tag{7.10}$$

where $X = \langle\alpha|\beta\rangle$.

7.2.4 Tensor Product of Coherent States ⇓

The tensor product of two or more coherent states will be particularly relevant to PPM modulation and in general for *vector* modulations.

Let $|\alpha\rangle$ be the tensor product of two coherent states

$$|\alpha\rangle = |\alpha_1\rangle \otimes |\alpha_2\rangle, \qquad |\alpha_1\rangle, |\alpha_2\rangle \in \mathcal{G}.$$

Then, for each of the two factors, the previous result holds: To the state $|\alpha_i\rangle$ a Poisson variable m_i can be associated, with average $E[m_i|\alpha_i] = |\alpha_i|^2$. The global number of photons m associated to the composite state $|\alpha\rangle$ is given by the sum of the two random variables $m = m_1 + m_2$, where m_1 and m_2 are statistically independent. Therefore, m is again a Poisson variable with average $E[m|\alpha] = E[m_1|\alpha_1] + E[m_2|\alpha_2] = |\alpha_1|^2 + |\alpha_2|^2$. This result can be easily generalized to the tensor product of N Glauber states

$$|\alpha\rangle = |\alpha_1\rangle \otimes |\alpha_2\rangle \otimes \cdots \otimes |\alpha_N\rangle$$

and we find, in particular, that the total number of photons $m = m_1 + m_2 + \cdots m_N$ associated to the composite state $|\alpha\rangle$ is still a Poisson variable with average given by

$$E[m\,|\,\alpha] = |\alpha_1|^2 + |\alpha_2|^2 + \cdots + |\alpha_N|^2. \tag{7.11}$$

7.2.5 Coherent States as Gaussian States ▽

In Chap. 11, in the framework of continuous quantum variables, coherent states will be defined as eigenkets of the annihilator operator, acting in an infinite dimensional bosonic Hilbert space $\mathcal{H}$. Then, from this abstract definition, the infinite dimensional representation (7.2) is obtained. An alternative representation is considered in the so-called *phase space*, where a quantum state, pure or mixed, is represented by its *Wigner function* $W(x, y)$, a real function of two real variables, having properties similar to the joint probability density of two continuous random variables. Thus we pass from an infinite dimensional Hilbert space $\mathcal{H}$ to the two-dimensional real space $\mathbb{R}^2$, with notable advantages.

The Wigner function $W(x, y)$ allows us to define *Gaussian quantum states*, as the quantum states having as Wigner function the Gaussian bivariate form

$$W(x, y) = \frac{1}{2\pi\sqrt{\det V}}\exp\left[-\frac{1}{2}\frac{V_{22}(x-\overline{q})^2 + V_{11}(y-\overline{p})^2 - 2V_{12}(x-\overline{q})(y-\overline{p})}{\det V}\right] \tag{7.12}$$

where V_{ij} are the covariances and $\overline{q}$, $\overline{p}$ are the mean values ($\det V = V_{11}V_{22} - V_{12}^2$). Hence a Gaussian state is completely specified by the mean vector and by the covariance matrix

$$\bar{X} = \begin{bmatrix} \overline{q} \\ \overline{p} \end{bmatrix}, \qquad V = \begin{bmatrix} V_{11} & V_{12} \\ V_{12} & V_{22} \end{bmatrix}. \tag{7.13}$$

To emphasize this property, a Gaussian state in general represented by a density operator is symbolized as $\rho(\bar{X}, V)$.

We shall see that a coherent state $|\alpha\rangle$ is a special case of Gaussian states with the simple specification

$$\bar{X} = \begin{bmatrix} \overline{q} \\ \overline{p} \end{bmatrix} = \begin{bmatrix} \Re\alpha \\ \Im\alpha \end{bmatrix}, \qquad V = \begin{bmatrix} 1 & 0 \\ 0 & 1 \end{bmatrix} = I_2. \tag{7.14}$$

Then the Wigner function of a coherent state results in

$$W(x, y) = \frac{1}{2\pi}\exp\left[-\frac{1}{2}\left((x-\overline{q})^2 + (y-\overline{p})^2\right)\right] \tag{7.15}$$

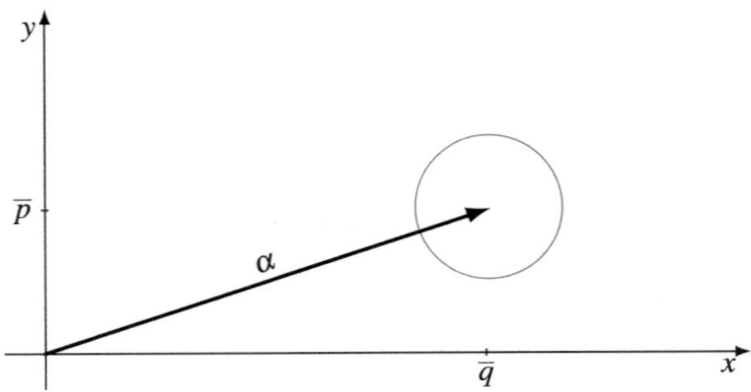

Fig. 7.4 Contour level of the Wigner function $W(x, y)$ of a pure coherent state $|\alpha\rangle$ (in *red*). The mean vector $(\bar{q}, \bar{p}) = (\Re\alpha, \Im\alpha)$ gives the center of the contour

and in the x, y plane it is often represented by a *contour level* obtained by the equation $W(x, y) = L$, where $L > 0$ is a reference level. For a coherent state, this contour is a circle centered at $(\bar{q}, \bar{p})$, as shown in Fig. 7.4.

Problem 7.1 ★   Prove that the inner product $X = \langle\alpha|\beta\rangle$ of two coherent states is real if and only if $\arg\alpha - \arg\beta = 0$ or $\arg\alpha - \arg\beta = \pm\pi$.

Problem 7.2 ★★   The map (7.2) gives for any $\alpha \in \mathbb{C}$ a coherent state $|\alpha\rangle$. Given $|\alpha\rangle$ is it possible to find the complex number α?

Problem 7.3 ★★   Examine the effect of the introduction of a phasor $z = e^{i\varphi}$ into the complex parameter α that identifies the state $|\alpha\rangle$, that is, evaluate $|e^{i\varphi}\alpha\rangle$.

Problem 7.4 ★★★   Let $|\alpha\rangle = |\alpha_1\rangle\otimes|\alpha_2\rangle$ be a two-mode coherent states. The number of photons m_i associated to each component state is a Poisson variable with mean $\Lambda_i = |\alpha_i|^2$. Considering that m_1 and m_2 are statistically independent (see Sect. 3.10), prove that the total number of photons $m = m_1 + m_2$ is a Poisson variable.

Hint: use the characteristic function given by (4.23).

7.3 Constellations of Coherent States

We recall that the target of a quantum communications system is the transmission of a sequence of classical symbols $\{A_n\}$ through a sequence of quantum states $\{|\gamma_{A_n}\rangle\}$, which in practice are often coherent states. Thus, in general, with a K-ary alphabet $\mathcal{A} = \{0, 1, \ldots, K - 1\}$, Alice must be able to prepare a **constellation** of K coherent states

$$\mathcal{S} = \{|\gamma_0\rangle, |\gamma_1\rangle, \ldots, |\gamma_{K-1}\rangle\} \tag{7.16}$$

to realize the c $\to$ q mapping

$$A_n \in \mathcal{A} \quad \to \quad |\gamma_{A_n}\rangle \in \mathcal{S},$$

which must be bijective. This operation may be called *quantum encoding*.

Now a problem to investigate is the choice of the constellation, of course, with the purpose of realizing a high-performance quantum transmission system. One way to decide about the choice, as we shall see in this section, is to get the "inspiration" from the optical transmission systems that we shall briefly call *classical systems*. This approach has also the advantage of allowing us a comparison between the performances of two kinds of systems, classical and quantum.

In Sect. 4.4, we have seen that optical communications use two kinds of modulations, incoherent and coherent; but in the present context, the right comparison is with classical coherent modulations which make use just of a coherent radiation emitted by a laser, as done in quantum communications. A classical K-ary coherent modulation, in general nonlinear, is specified by K complex waveforms

$$\gamma_0(t), \gamma_1(t), \ldots, \gamma_{K-1}(t) \tag{7.17}$$

of duration limited[1] to the signaling interval $[0, T]$, with the rule that if $A_n \in \mathcal{A}$ is the nth source symbol, the modulator forms a signal with *complex envelope*[2] [4]

$$c(t) = \gamma_{A_n}(t) \qquad 0 \le t < T.$$

With a sequence of symbols $\{A_n\}$, the complete expression of the complex envelope becomes

$$c(t) = \sum_{n=-\infty}^{+\infty} \gamma_{A_n}(t - nT), \tag{7.18}$$

from which a real modulated signal is obtained as

$$v(t) = \Re\, c(t)\, e^{i2\pi \nu t} \tag{7.19}$$

where ν is the carrier optical frequency. The comparison between a classical modulator and a quantum encoder is depicted in Fig. 7.5.

To proceed from the classical system, characterized by the waveforms $\gamma_i(t)$, $i \in \mathcal{A}$, to the quantum system with coherent state constellation $|\gamma_i\rangle$, $i \in \mathcal{A}$, we must "remove" in some way the time dependence, which is not present in coherent states. For some kinds of modulations the solution is straightforward; for others, it is less obvious.

---

[1] Some classical coherent modulations use a duration greater than one symbol period.

[2] See Sect. 4.7 for the definition of complex envelope.

Fig. 7.5 Comparison of a classical modulator (*left*) with a quantum encoder

7.3.1 State Constellations from Scalar Modulations

In some modulations, like PSK and QAM, the waveforms (7.17) are of the form

$$\gamma_i(t) = \gamma_i\, h(t)\,, \qquad i \in \mathcal{A} = \{0, 1, \ldots, K-1\}, \tag{7.20}$$

where $h(t)$ is a real pulse; for example, rectangular between 0 and T, and γ_i are complex numbers. The complex envelope $c(t)$ of the modulated signal is then produced by an encoder, mapping the symbols $i \in \mathcal{A}$ into the complex symbols γ_i, and by an interpolator with impulse response $h(t)$. The resulting complex envelope becomes

$$c(t) = \sum_{n=-\infty}^{+\infty} C_n\, h(t - nT)\,, \tag{7.21}$$

where $\{C_n\}$ is the sequence of complex symbols obtained by the mapping $A_n = i \rightarrow C_n = \gamma_i$ (Fig. 7.6).

In this way, a *constellation of complex symbols* is identified

$$\mathcal{C} = \{\gamma_0, \gamma_1, \ldots, \gamma_{K-1}\}\,, \qquad \gamma_i \in \mathbb{C} \tag{7.22}$$

from which one can form the *constellation of coherent states*

$$\mathcal{S} = \{|\gamma_0\rangle, |\gamma_1\rangle, \ldots, |\gamma_{K-1}\rangle\}\,, \qquad |\gamma_i\rangle \in \mathcal{G} \tag{7.23}$$

that are in a one-to-one correspondence with the constellation of complex symbols $\mathcal{C}$ (Fig. 7.7).

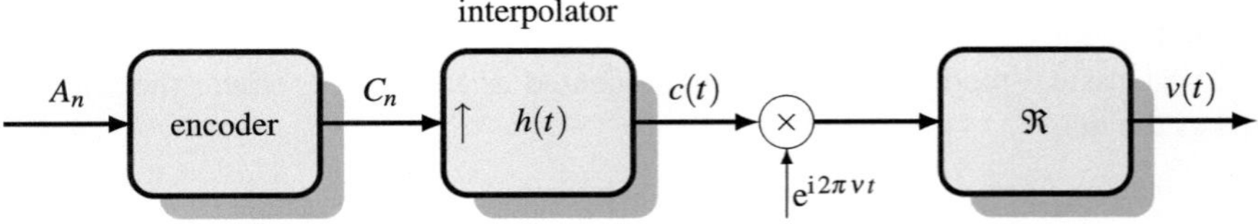

Fig. 7.6 Scheme of a classical **scalar modulator**. The encoder maps the source symbols $A_n \in \mathcal{A}$ into the complex symbols $C_n \in \mathcal{C}$. The interpolator maps the complex symbols into the complex envelope $c(t)$

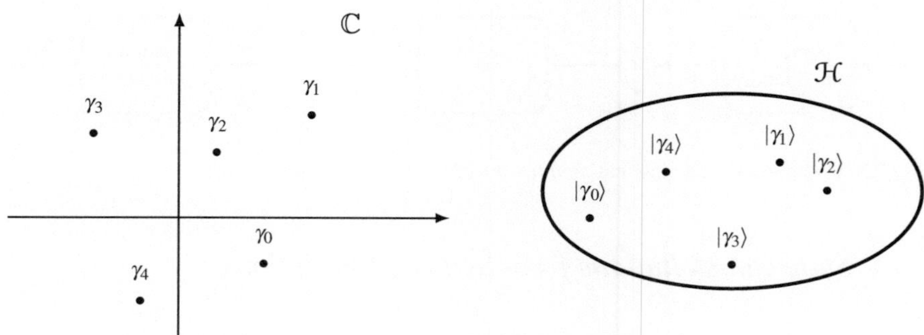

Fig. 7.7 Constellation of complex symbols $\mathcal{C}$ and corresponding constellation of coherent states $\mathcal{S}$: each complex symbol $\gamma \in \mathcal{C}$ is mapped into a coherent state $|\gamma\rangle \in \mathcal{G}$

7.3.2 State Constellations from Vector Modulations ⇓

The previous procedure, consisting in directly creating the constellation of coherent states from the constellation of symbols, is not always possible, because in general the K waveforms (7.17) cannot be expressed in the form (7.20). To remove the time dependence, we can proceed in the following way [4]. We take a basis of functions, $h_1(t), \ldots, h_N(t)$, orthonormal in the interval $[0, T)$, where, in general, $N \leq K$, and we expand the waveforms (7.17) on this basis, namely

$$\gamma_i(t) = \sum_{j=1}^{N} \gamma_{ij}\, h_j(t)\,, \qquad i = 0, 1, \ldots, K-1 \tag{7.24a}$$

where the coefficients are given by

$$\gamma_{ij} = \int_0^T \gamma_i(t)\, h_j^*(t)\, \mathrm{d}t\,, \qquad j = 1, \ldots, N. \tag{7.24b}$$

The vectors of the complex coefficients

$$\gamma_i = (\gamma_{i1}, \ldots, \gamma_{iN})\,, \qquad i = 0, 1, \ldots, K-1 \tag{7.25}$$

uniquely identify the waveform $\gamma_i(t)$.

The classical modulator can be implemented as in Fig. 7.8, where the encoder makes the map

$$A_n = i \in \mathcal{A} \quad \rightarrow \quad C_n = \gamma_i \in \mathbb{C}^N$$

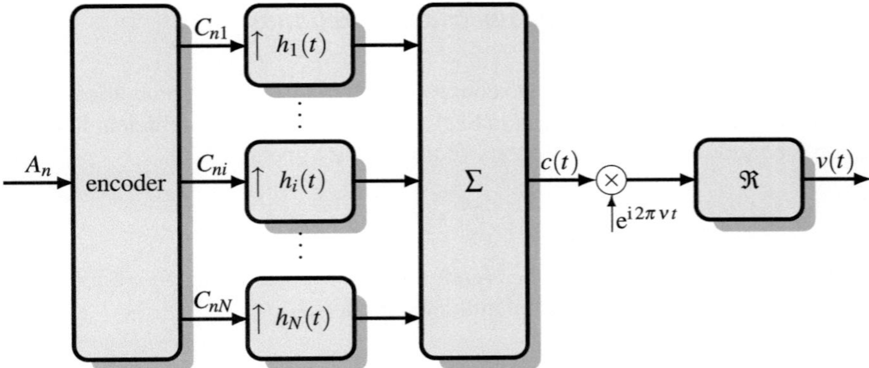

Fig. 7.8 Scheme of a classical **vector modulator**. The encoder maps the source symbols $A_n \in \mathcal{A}$ into a vector of complex symbols $C_n = [C_{n1}, \ldots, C_{ni}, \ldots, C_{nN}]^{\mathrm{T}}$. The bank of interpolators maps the vectors C_n into the complex envelope $c(t)$

with

$$C_n = [C_{n1}, \ldots, C_{nN}], \qquad \gamma_i = [\gamma_{i1}, \ldots, \gamma_{iN}].$$

Then from the vector C_n, a bank of interpolators forms the complex envelope $c(t)$ of the modulated signal, as

$$c(t) = \sum_{n=-\infty}^{\infty} \sum_{i=1}^{N} C_{ni}\, h_i(t - nT). \tag{7.26}$$

This generalizes the scalar modulation, which is obtained with $N = 1$.

The general procedure just described allows us to identify a constellation of complex vectors $\{\gamma_i, i = 0, \ldots, K - 1\}$ with $\gamma_i \in \mathbb{C}^N$. Now, to introduce the coherent states, we must consider a composite Hilbert space, given by the tensor product $\mathcal{H} = \mathcal{H}_0 \otimes \mathcal{H}_0 \otimes \cdots \otimes \mathcal{H}_0$ of N equal Hilbert spaces $\mathcal{H}_0$. In this composite space, the states become the *tensor product* of coherent states and, through (7.25), to each symbol $i \in \mathcal{A}$ the **tensor product of coherent states** is associated

$$|\gamma_i\rangle = |\gamma_{i1}\rangle \otimes |\gamma_{i2}\rangle \otimes \cdots \otimes |\gamma_{iN}\rangle \tag{7.27}$$

that, with i varying in $\mathcal{A}$, forms the desired constellation of coherent states. In the context of continuous variables of Chap. 11, the tensor product of N coherent states (7.27) is called N-**mode coherent state**.

An example in which we use this method of forming a composite constellation of coherent states will be seen in PPM modulation developed in Sect. 7.13 (see also Problem 7.5).

7.3.3 Construction of a Symmetric Constellation

An innovative way to obtain a state constellation is based on the geometrically uniform symmetry (GUS), introduced in Sect. 5.13. To this end, it is sufficient finding a unitary operator S having the property of a symmetry operator S

$$S^K = I_{\mathcal{H}} \tag{7.28}$$

that is, S must be a Kth root of the identity operator. Then, fixing an arbitrary state $|\gamma_0\rangle \in \mathcal{H}$, one gets a K-ary constellation of states as

$$|\gamma_i\rangle = S^i |\gamma_0\rangle, \qquad i = 0, 1, \ldots, K - 1. \tag{7.29}$$

More generally, one can fix an arbitrary density operator ρ_0 acting on the Hilbert space $\mathcal{H}$ to get a constellation of density operators as (see (5.123))

$$\rho_i = S^i \rho_0 (S^i)^*, \qquad i = 0, 1, \ldots, K - 1. \tag{7.30}$$

In this way, we can generate infinitely many constellations having the very useful property represented by the GUS. After the choice of S and of the reference state $|\gamma_0\rangle$ or ρ_0, one achieves "interesting practical properties" for the quantum communications system based on the corresponding constellation. A nontrivial problem is finding a unitary operator with the property (7.28), especially in the case of infinite dimensions, as is for coherent states.

Problem 7.5 ★★ Show that the PPM must be considered a vector modulation. Find explicitly the waveform $\gamma_i(t)$ and the vector γ_i of the coefficients.

Problem 7.6 ★★ The n-DFT matrix $W_{[n]}$ is unitary and has the property $W_{[n]}^n = I_n$. Then it allows for the construction of n-ary constellations in $\mathcal{H} = \mathbb{C}^n$. Find a quaternary constellation using $S = W_{[4]}$ and reference state $|\gamma_0\rangle = [1, 1, 0, 0]^{\mathrm{T}}$. Also prove that the four states are linearly independent.

7.4 Parameters in a Constellation of Coherent States

In the previous section, we have investigated how to form interesting constellations of coherent states for quantum communications systems. In this section, we want to clarify how a given constellation format can be parametrized to modify the photonic flux therein, expressed, e.g., in terms of the number of signal photons per symbol. In fact, we are interested in the evaluation of the system performance in a given range of this parameter.

Note that the constellation of coherent states $\mathcal{S}$ given by (7.23) can be structured in matrix form as

$$\Gamma = [|\gamma_0\rangle, |\gamma_1\rangle, \ldots, |\gamma_{K-1}\rangle] \tag{7.31}$$

and becomes the *state matrix*. In practical modulation formats (that will be considered further on), the states of $\mathcal{S}$ are always *independent* (in the sense of vector spaces), and then the state matrix has always full rank, i.e., $\mathrm{rank}(\Gamma) = K$. From the state matrix, we obtain the *Gram's matrix*, a $K \times K$ matrix formed by the inner products between the couples of states

$$G = \Gamma^* \Gamma = \left[\langle \gamma_i | \gamma_j \rangle \right], \qquad |\gamma_i\rangle, |\gamma_j\rangle \in \mathcal{G}$$

that can be calculated using (7.9). Also G has always full rank and, because the states are not orthogonal, all the entries of G are different from zero.

Even in the N-dimensional case, when the states are given by the tensor product of N component states (see (7.27)), to calculate the state superposition, we evaluate the inner products

$$\langle \gamma_i | \gamma_j \rangle = \langle \gamma_{i1} | \gamma_{j1} \rangle \, \langle \gamma_{i2} | \gamma_{j2} \rangle \, \cdots \, \langle \gamma_{iN} | \gamma_{jN} \rangle. \tag{7.32}$$

In this relation we have borne in mind that the inner product of states, given by a tensor product, is obtained as a product of the inner products of the component states (see relation (2.100)). Each of the inner products of the component states is evaluated from (7.9).

7.4.1 Number of Signal Photons in a Constellation

From (7.8) we have that the average number of photons associated to the coherent state $|\gamma\rangle \in \mathcal{G}$ is given by the squared norm of the complex amplitude γ

$$N_\gamma = |\gamma|^2.$$

In a constellation of coherent states, we introduce the *signal photons per symbol*. To this end, we observe that the generic symbol of the constellation, $C \in \mathcal{C}$, must be considered as a random variable with probability $\mathrm{P}[C = \gamma], \gamma \in \mathcal{C}$, and also the average number of photons N_C associated to C becomes a random variable; the statistical average of N_C,

$$N_s = \mathrm{E}[N_C] = \sum_{\gamma \in \mathcal{C}} \mathrm{P}[C = \gamma] N_\gamma = \sum_{\gamma \in \mathcal{C}} \mathrm{P}[C = \gamma] |\gamma|^2, \tag{7.33}$$

defines the *average number of photons per symbol*, briefly **number of signal photons per symbol**. Now, given the one-to-one correspondence $A = i \Leftrightarrow C = \gamma_i$, the probability of these two events turns out to be equal to the prior probability q_i. Therefore, we have

$$N_s = \sum_{i \in \mathcal{A}} q_i |\gamma_i|^2 \quad \text{(photons/symbol)}.$$

In particular, with equally likely symbols, the number of signal photons per symbol becomes

$$\boxed{N_s = \frac{1}{K} \sum_{i \in \mathcal{A}} |\gamma_i|^2 = \frac{1}{K} \sum_{\gamma \in \mathcal{C}} |\gamma|^2.} \tag{7.34}$$

Finally, remembering that, with equiprobable symbols, there are $\log_2 K$ bit/symbol, we find that the **number of signal photons per bit** is given by

$$N_R = \frac{N_s}{\log_2 K} \quad \text{(photons/bit).} \tag{7.35}$$

⇓ We have seen above that in an N-dimensional constellation $\mathcal{C}$, whose states are N-mode coherent states, $|\gamma\rangle = |\gamma_1\rangle \otimes |\gamma_2\rangle \otimes \ldots \otimes |\gamma_N\rangle$, the average number of photons associated to the composite state $|\gamma\rangle$ results in (see (7.11))

$$N_\gamma = |\gamma_1|^2 + |\gamma_2|^2 + \cdots + |\gamma_N|^2 \tag{7.36}$$

where $\gamma = [\gamma_1, \gamma_2, \ldots, \gamma_N]$. Consequently, the number of signal photons per symbol must be evaluated according to

$$N_s = \sum_{\gamma \in \mathcal{C}} \mathrm{P}[C = \gamma] N_\gamma \tag{7.37}$$

with N_γ given by (7.36), and the sum is extended to the N-dimensional constellation. Of course, with equiprobable symbols we have $\mathrm{P}[C = \gamma] = 1/K$ and the number of signal photons per bit is still given by (7.35).

Sensitivity of a receiver. In telecommunications an important parameter is the sensitivity, which is defined as the minimum value of a parameter of the receiver that guarantees a given value of the error probability P_e, typically $P_e = 10^{-9}$. In optical communications (classical or quantum), the parameter is often given by the number of photons per bit N_R. Thus we say, e.g., that a quantum receiver has the sensitivity of $N_R = 11.5$ photons/bit.

7.4.2 Scale Factor and Shape Factor of a Constellation

The constellation (7.22) of complex symbols $\mathcal{C}$, from which we can *directly* obtain the constellation (7.23) of coherent states $\mathcal{S}$, contains a scale factor linked to the photonic intensity, but modulation formats are usually specified in a normalized form. In the evaluation of a system's performance, it is worthwhile to underline this aspect by expressing the symbols γ_i in the form $\bar{\gamma}_i \Delta$, where $\bar{\gamma}_i$ are *normalized* symbols and Δ is the **scale factor**. Then it is convenient to introduce a *normalized* constellation

$$\mathcal{C}_0 = \{\bar{\gamma}_0, \bar{\gamma}_1, \ldots, \bar{\gamma}_{K-1}\}$$

from which one obtains the scaled constellation as $\mathcal{C} = \{\bar{\gamma}_0 \Delta, \bar{\gamma}_1 \Delta, \ldots, \bar{\gamma}_{K-1}\Delta\}$ and hence the constellation of coherent states as

$$\mathcal{S} = \{|\bar{\gamma}_0 \Delta\rangle, |\bar{\gamma}_1 \Delta\rangle, \ldots, |\bar{\gamma}_{K-1}\Delta\rangle\}.$$

The scale factor appears in the number of signal photons per symbol, given by (7.34), which can be written in the form

$$N_s = \frac{1}{K} \sum_{\gamma \in \mathcal{C}} |\gamma|^2 = \Delta^2 \frac{1}{K} \sum_{\bar{\gamma} \in \mathcal{C}_0} |\bar{\gamma}|^2 = \mu_K \Delta^2 \tag{7.38}$$

where

$$\mu_K := \frac{1}{K} \sum_{\bar{\gamma} \in \mathcal{C}_0} |\bar{\gamma}|^2 \tag{7.39}$$

is a characteristic parameter of the constellation, which we call **shape factor**. For example, in the PSK modulation, the normalized constellation consists of K points on the *unit* circle

$$\mathcal{C}_0 = \{e^{i2\pi m/K} \mid m = 0, 1, \ldots, K-1\}$$

whereas the scaled constellation is given by K points on the circle of radius Δ

$$\mathcal{C} = \{\Delta\, e^{i2\pi m/K} \mid m = 0, 1, \ldots, K-1\}.$$

In this case, the shape factor is $\mu_K = 1$. An example where $\mu_K \neq 1$ is given by the QAM modulation.

7.4.3 Summary of Constellation Formats

To conclude these two sections on constellations of coherent states, it is convenient
to recall that the target of a quantum communications system is the transmission
of a classical information, encoded in a *classical* symbol sequence $\{A_n\}$, through a
sequence of *quantum* states $\{|\gamma_{A_n}\rangle\}$, as illustrated in Fig. 7.1. Thus, a key operation is
the c $\rightarrow$ q mapping $A_n \rightarrow |\gamma_{A_n}\rangle$. This finally explains why we have constellations
in both classical and quantum domain.

Here, we wish to summarize the constellations introduced above which are all
useful to proceed on. Starting from a symbol alphabet, which was indicated in the
form $\mathcal{A} = \{0, 1, \ldots, K - 1\}$, we have introduced:

- a constellation of *normalized complex symbols* $\mathcal{C}_0 = \{\bar{\gamma}_0, \bar{\gamma}_1, \ldots, \bar{\gamma}_{K-1}\}$,
- a constellation of *scaled complex symbols* $\mathcal{C} = \{\gamma_0, \gamma_1, \ldots, \gamma_{K-1}\}$, with $\gamma_i = \bar{\gamma}_0 \Delta$,
  where Δ is a scale factor,
- a constellation of *coherent states* $\mathcal{S} = \{|\gamma_0\rangle, |\gamma_1\rangle, \ldots, |\gamma_{K-1}\rangle\}$, where $|\gamma_i\rangle$ is the
  coherent state uniquely determined by the scaled complex symbol γ_i, according
  to relation (7.2).

Note that $\mathcal{C}_0$ and $\mathcal{C}$ live in the field of complex numbers $\mathbb{C}$, while $\mathcal{S}$ lives in the infinite
dimensional Hilbert space $\mathcal{H}$.

Problem 7.7 $\star\star$ ∇ Find the shape factor μ_k of the 16-QAM constellation (see
Fig. 7.28).

7.5 Theory of Classical Optical Systems

We want to compare the performance of a quantum communications system with that
of the *corresponding* classical communications system, i.e., not based on quantum
measurements, but on an optical detection (see semiclassical detection in Chap. 4).

In the formulation of the transmitter and the receiver, it is convenient to introduce
two distinct schemes: One working at the level of *instantaneous optical power* and
the other one working on the *complex envelope*. In fact in the semiclassical theory of
an optical system, both the optical power and the complex envelope must be jointly
considered, as remarked in Sect. 4.7.

7.5.1 Scheme for Instantaneous Optical Powers

We recall that a monochromatic radiation at the optical frequency ν can be modeled
as an *instantaneous optical power*, which is formed by the energy quanta of size $h\nu$
and has the impulsive expression

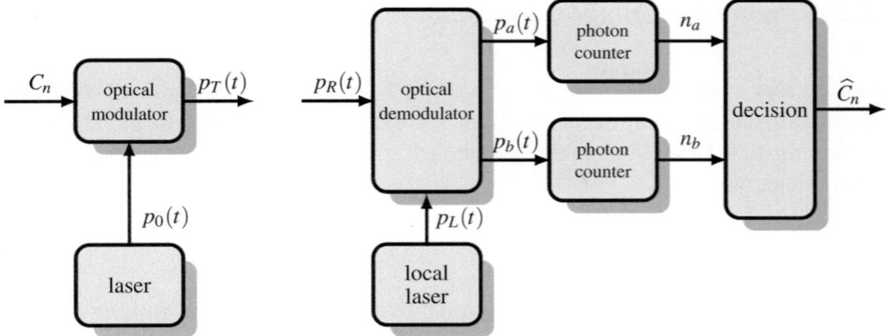

Fig. 7.9 Scheme of a classical modulator and demodulator for the instantaneous optical power. In the figure, the initial encoder $A_n \rightarrow C_n$ and the final decoder $\widehat{C}_n \rightarrow \widehat{A}_n$ are omitted

$$p(t) = \sum_k (h\nu)\, \delta(t - t_k) \tag{7.40}$$

where the arrival instants $\{t_k\}$ are represented by a *doubly stochastic Poisson process*, specified by its random intensity $\lambda(t)$.

Referring to digital systems, the information to be transmitted is first conveyed in a sequence of symbols $\{A_n\}$, $A_n \in \{0, 1, \ldots, K-1\}$ and then, for convenience, in a sequence of complex symbols $\{C_n\}$ belonging to a given (normalized) constellation $\mathcal{C}_0$. Then the first part of the transmitter is an encoder, which provides the map $A_n \rightarrow C_n$. The task of a digital modulator is to modify the laser beam in each symbol period $(nT, nT + T)$ in dependence of the symbol C_n falling in this period.[3] If there are no further processing, as we suppose, the output of the modulator gives the instantaneous transmitted power $p_T(t)$, as shown in Fig. 7.9.

In the receiver, the incoming instantaneous power $p_R(t)$, an attenuated version of $p_T(t)$, is combined with the instantaneous power $p_L(t)$ of a local laser tuned at the same frequency ν as the laser in the transmitter (homodyne detection) or at a different frequency (etherodyne detection). The task of the demodulator is the production of two distinct instantaneous powers $p_a(t)$ and $p_b(t)$ to feed two photon counters, which count the photon numbers in each symbol period as

$$n_a = \frac{1}{h\nu} \int_{nT}^{nT+T} p_a(t)\, \mathrm{d}t, \qquad n_b = \frac{1}{h\nu} \int_{nT}^{nT+T} p_b(t)\, \mathrm{d}t.$$

The reason of this double path is due to the fact that n_a and n_b are real (integer) and the receiver has to give an estimated version $\{\widehat{C}_n\}$ of the complex sequence $\{C_n\}$.

In the case of binary systems, where the symbols C_n are real, the double path is not necessary and the detection is based only on a single photon counting. In the following, for brevity, we will consider only homodyne detection.

---

[3] The practical implementation of this operation will be seen in Sect. 9.2.

7.5.2 Scheme for Complex Envelopes

In an optical system, the complex envelope $V(t)$ (denoted by $c_v(t)$ in Sect. 4.7) contains all the information useful both for the signal analysis and the statistical analysis. In fact from $V(t)$, we can obtain the *signal* $v(t)$, present in the form of electric field, as

$$v(t) = \Re V(t)\, e^{i2\pi\, v\, t}\,. \tag{7.41}$$

Also, the average power $P(t)$ is proportional to $|V(t)|^2$ and, by appropriate normalization of the electric field, it can be directly written as

$$P(t) = |V(t)|^2\,. \tag{7.42}$$

On the other hand, the average power is connected to the instantaneous power, modeled as a doubly stochastic filtered Poisson process, through Campbell's theorem according to

$$P(t) = \mathrm{E}[p(t)|\lambda] = (h\nu)\,\lambda(t)\,, \tag{7.43}$$

where $\mathrm{E}[\cdot|\lambda]$ denotes the conditional expectation "with a given $\lambda(t)$." This holds for the powers $p_T(t)$, $p_R(t)$, $p_a(t)$, and $p_b(t)$ in the scheme of Fig. 7.9. The fundamental remark is that from the complex envelope $V(t)$, we can obtain the intensity $\lambda(t)$ which gives the full statistical description of the doubly stochastic Poisson processes involved.

At the level of complex envelope, the modulator scheme is essentially the one anticipated in Fig. 7.6, where, starting from the complex sequence $\{C_n\}$, the complex envelope of the modulated signal is obtained with an interpolator according to (7.21), that is,

$$V_T(t) = \sum_{n=-\infty}^{+\infty} C_n V_0\, h(t - nT) \tag{7.44}$$

where V_0 is the amplitude of the carrier produced by the laser. This corresponds to the transmitter instantaneous power $p_T(t)$.

The scheme of the demodulator is extremely simple. To the incoming complex envelope $V_R(t)$, corresponding to the received instantaneous power $p_R(t)$, the amplitude V_L is added for the upper path and the amplitude $i\, V_L$ to the lower path to get

$$V_a(t) = V_R(t) + V_L, \quad V_b(t) = V_R(t) + i\, V_L \tag{7.45}$$

as shown in Fig. 7.10.

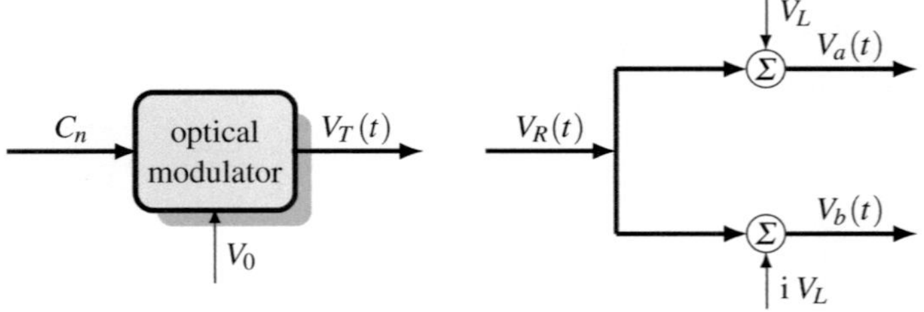

Fig. 7.10 Scheme of a classical modulator and demodulator for the complex envelopes

In the following, we make the assumption that the interpolator impulse response $h(t)$ is unitary in $(0, T)$, so that $V_T(t)$ is simplified as

$$V_T(t) = C_0 \, V_0, \quad 0 < t < T. \tag{7.46}$$

Correspondingly (7.45) become

$$V_a(t) = C_0 \, V_R + V_L, \quad V_b(t) = C_0 \, V_R + iV_L \tag{7.47}$$

where V_R is the amplitude of the received carrier.

7.5.3 Scheme for Signals. Quadrature Modulator

The scheme for the complex envelope is sufficient for the analysis of an optical system. Now we consider the scheme for the *signal*, which is more detailed and may have interest for the implementation of the system.

Signals are obtained from complex envelopes according to relation (7.41). Letting $C_n = A_n + iB_n$, from (7.44); we find that the modulated signal results in

$$v_T(t) = \Re \sum_{n=-\infty}^{+\infty} (A_n + iB_n) \, h(t - nT) \, V_0 \, e^{\,i\,2\pi\nu t}$$

$$= \sum_{n=-\infty}^{+\infty} [A_n V_0 \, h(t - nT) \, \cos 2\pi\nu t - B_n V_0 \, h(t - nT) \, \sin 2\pi\nu t]. \tag{7.48}$$

The interpretation of these relations leads to the scheme of Fig. 7.11, called *quadrature modulator*. The carrier $V_T \cos 2\pi\nu t$ is produced by a laser tuned at the frequency ν and the quadrature carrier $-V_T \sin 2\pi\nu t$ is obtained by shifting the carrier $V_T \cos 2\pi\nu t$.

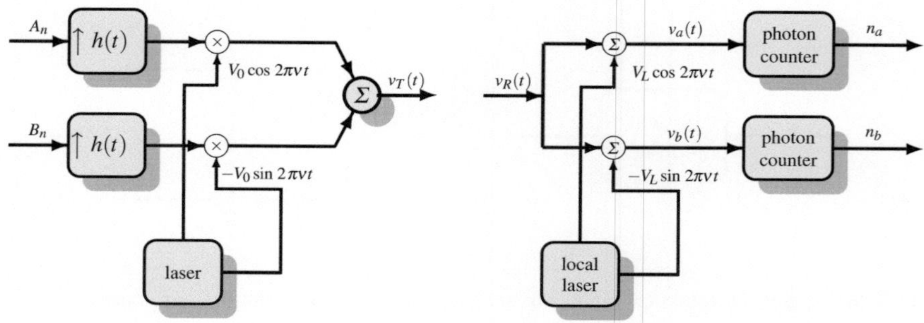

Fig. 7.11 Implementation of a coherent optical system based on a quadrature modulator. On the *left* the transmitter and on the *right* the homodyne receiver

With the simplification of (7.46), (7.48) gives

$$v_T(t) = A_0 V_0 \cos 2\pi v t - B_0 V_0 \sin 2\pi v t , \qquad 0 < t < T. \qquad (7.49)$$

At reception, the modulated signal becomes

$$v_R(t) = A_0 V_R \cos 2\pi v t - B_0 V_R \sin 2\pi v t , \qquad 0 < t < T, \qquad (7.50)$$

and the constant complex envelopes V_L and i V_L give

$$\Re \, V_L e^{i 2\pi v t} = V_L \cos 2\pi v t , \qquad \Re \, i \, V_L e^{i 2\pi v t} = -V_L \sin 2\pi v t.$$

These carriers are provided by a local laser, tuned with the transmission laser (homodyne reception). Finally, (7.47) gives for $0 < t < T$

$$v_a(t) = \Re \, [C_0 V_R + V_L] e^{i 2\pi v t} = (A_0 V_R + V_L) \cos 2\pi v t - B_0 V_R \sin 2\pi v t$$
$$v_b(t) = \Re \, [C_0 V_R + i \, V_L)] e^{i 2\pi v t} = A_0 V_R \cos 2\pi v t - (B_0 V_R + V_L) \sin 2\pi v t.$$
$$(7.51)$$

These signals feed the photon counters.

7.5.4 Photon Counting and Detection

The count in the interval $(0, T]$ yields two values, n_a and n_b, from which a decision must be taken on the transmitted symbol C_0; n_a and n_b are conditioned Poisson variables and therefore characterized by their averages $\bar{n}_a(C_0) := \mathrm{E}[n_a|C_0]$ and $\bar{n}_b(C_0) := \mathrm{E}[n_b|C_0]$, the condition being "given a transmitted symbol C_0." These averages are obtained dividing the corresponding energies in a symbol period T by the quantum $h\nu$. Considering that the complex envelopes are constant in $(0, T)$, we have $E_a = P_a T = |V_a|^2 T$ and $E_b = P_b T = |V_b|^2 T$, and then

$$\bar{n}_a(C_0) = H|C_0\, V_R + V_L|^2 = H\left[(A_0 V_R + V_L)^2 + (B_0 V_R)^2\right]$$

$$\bar{n}_b(C_0) = H|C_0\, V_R + i\, V_L|^2 = H\left[(A_0 V_R)^2 + (B_0 V_R + V_L)^2\right] \tag{7.52}$$

where $H = T/(h\nu)$.

At this point, we assume that the local carrier has an amplitude V_L much greater than V_R, which allows us to get the following approximations

$$\bar{n}(A_0) = H(2A_0 V_R V_L + V_L^2), \qquad \bar{n}(B_0) = H(2B_0 V_R V_L + V_L^2), \tag{7.53a}$$

where now the upper counting depends only on A_0 and the lower counting only on B_0. The averages can be expressed in "numbers" by letting $N_L = H V_L^2$ and $N_R = H V_R^2$ to get

$$\bar{n}(A_0) = 2\sqrt{N_L\, N_R}\, A_0 + N_L, \qquad \bar{n}(B_0) = 2\sqrt{N_L\, N_R}\, B_0 + N_L. \tag{7.53b}$$

The numbers of photons n_a and n_b can be decomposed as

$$\begin{aligned} n_a &= \bar{n}(A_0) + u_a = A_0\, U_0 + N_L + u_a \\ n_b &= \bar{n}(B_0) + u_b = B_0\, U_0 + N_L + u_b \end{aligned}, \qquad U_0 := 2\sqrt{N_L\, N_R} \tag{7.54}$$

where

- $U_0 A_0$ and $U_0 B_0$ are the *useful signals*,
- N_L is a bias,
- u_a and u_b are the *shot noises*.

We compose for convenience the two countings into a complex one to get

$$z_0 = n_a + i n_b = C_0\, U_0 + N_L + i\, N_L + u_a + i u_b \tag{7.55}$$

which is the standard form of the "signal at the decision point" in a quadrature modulator. Note that

$$\bar{n}(C_0) := \bar{n}(A_0) + i\, \bar{n}(B_0) = C_0\, U_0 + N_L(1 + i), \qquad C_0 \in \mathcal{C}_0$$

generates a constellation of "received values", with center the point $N_L(1 + i)$ of the complex plane $\mathbb{C}$. The constellation is illustrated in Fig. 7.12 in the case of 8-PSK.

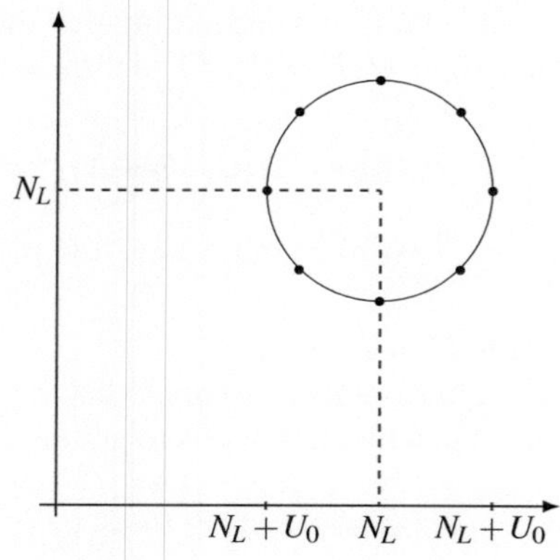

Fig. 7.12 Constellation of "received values" in the complex plane of optical 8-PSK. The points of the constellations are given by $N_L(1+\mathrm{i}) + U_0\,\mathrm{e}^{\mathrm{i}2\pi k/8}$, $k = 0, 1\ldots, 7$

7.5.5 *Correct Decision Probability*

In principle, it is possible to evaluate the correct decision probability $P_c = \mathrm{P}[\widehat{C}_0 = C_0]$ from the statistical description of the integer random variables n_a and n_b. These variables can be considered statistically independent, and therefore described by two conditioned Poisson distributions $p_{n_a}(k|A_0)$ and $p_{n_b}(k|B_0)$, which in turn are specified by their averages $\bar{n}(A_0)$ and $\bar{n}(B_0)$ given by (7.53). The preliminary step is the choice of the decision regions $\{\mathcal{R}(\gamma)|\gamma \in \mathcal{C}_0\}$, which has to form a partition of the set of integer pairs $\{(k_1, k_2)|k_1, k_2 = 0, 1, 2, \ldots\}$. Then we have the decision criterion

$$\widehat{C}_0 = \gamma \quad \text{if} \quad (n_a, n_b) \in \mathcal{R}(\gamma). \tag{7.56}$$

Correspondingly, the transition probabilities are given by

$$p(\gamma'|\gamma) := \mathrm{P}[\hat{C}_0 = \gamma'\,|\,C_0 = \gamma] \sum_{(k_1,k_2)\in\mathcal{R}(\gamma')} p_{n_a}(k_1|\Re\gamma)\,p_{n_b}(k_2|\Im\gamma) \tag{7.57}$$

and the correct decision probability, with equally likely symbols, by

$$P_c = \frac{1}{K}\sum_{\gamma\in\mathcal{C}} p(\gamma|\gamma). \tag{7.58}$$

The decision regions should be optimized to maximize P_c.

This procedure will be applied in the next chapter (Sect. 8.6) to a specific case (a BPSK system). In general, it is cumbersome and does not give readable results because only numerical evaluations are possible. The alternative is the **Gaussian approximation**, *where it is assumed that the photon numbers n_a and n_b are independent Gaussian random variables.* This allows us to simplify the analysis and to arrive at very simple results.

Note that n_a and n_b are Poisson random variables and it may appear to be strange that discrete random variables, described by (mass) probability distributions, are approximated by continuous random variables, described by probability densities. The approximation does not work in counting, but in the evaluation of the transition probabilities and of the error probability. In Sect. 8.6, we will compare the exact evaluation of probabilities, obtained with the Poisson statistics, and the approximate evaluation, obtained with the Gaussian assumption. We will see that the Gaussian approximation gives a very accurate evaluation of the exact probabilities. This conclusion holds in general in the presence of a strong photonic intensity [5] (here ensured by the assumption $V_L \gg V_0$).

With the Gaussian approximation, n_a and n_b become specified by their conditional means $\bar{n}(A_0) := \mathrm{E}[n|A_0]$ and $\bar{n}(B_0) := \mathrm{E}[n_b|B_0]$ and by their variances $\sigma^2(A_0) = \bar{n}(A_0)$ and $\sigma^2(B_0) = \bar{n}(B_0)$. For the latter, a further simplification[4] can be introduced by neglecting in (7.53) $2A_0\sqrt{N_R N_L}$ and $2B_0\sqrt{N_R N_L}$ with respect to N_L, so that they become equal, $\sigma_n^2 := \sigma^2(B_0) = \sigma^2(B_0) = N_L$, and independent of the symbols. Then their joint probability density results in

$$f_{n_a}(a|A_0)f_{n_b}(a|B_0) = \frac{1}{2\pi\sigma_n^2}\exp\left[-\frac{(a-\bar{n}(A_0))^2 + (b-\bar{n}(B_0))^2}{2\sigma_n^2}\right]$$
$$= \frac{1}{\sigma_n}\phi\left(\frac{a-\bar{n}(A_0)}{\sigma_n}\right)\frac{1}{\sigma_n}\phi\left(\frac{b-\bar{n}(B_0)}{\sigma_n}\right). \tag{7.59}$$

The decision regions $\{R(\gamma) \mid \gamma \in \mathcal{C}_0\}$ become a partition of the complex plane. Then the transition probabilities are given by

$$p_c(\gamma'|\gamma) := \mathrm{P}[\hat{C}_0 = \gamma' \mid C_0 = \gamma] = \int_{R(\gamma')} f_{n_a}(a|\Re\gamma)f_{n_b}(a|\Im\gamma)\,\mathrm{d}a\,\mathrm{d}b. \tag{7.60}$$

The correct decision probability P_c, with equally likely symbols, is still given by (7.58).

The above probabilities depend only on the SNR, which results in

$$\Lambda = \frac{U_0^2}{\sigma_n^2} = 4\,N_R \tag{7.61}$$

and is related to the number of signal photons contained in the **received power** $P_R = V_R^2$. In fact, considering that $V_R(t) = C_0\,V_R$, the received power is given by $P_R = |C_0|^2 V_R^2$, and therefore the number of signal photons associated to the symbol C_0 is $|C_0|^2 H V_R^2 = |C_0|^2 N_R$. This can be related to the **number of signal photons per symbol** N_s as (see (7.34))

$$N_s = \frac{1}{K}\sum_{\gamma \in \mathcal{C}_0} |\gamma|^2 N_R = \mu_K N_R \tag{7.62}$$

---

[4] This simplification is not possible for the means given by (7.53) because they represent the useful signal. Otherwise the information on symbols would be lost.

where μ_K is the shape factor of the constellation (see (7.39)). Then $\Lambda = 4N_s/\mu_K$. As we will see, in the cases of interest, with an optimized choice of the decision region, the error probability is a function of the SNR Λ expressed by function $Q(x)$ (see Problem 7.8).

The above theory on classical optical systems is quite long and contains a lot of relations, but the net result for the evaluation of the performance is extremely simple.

Proposition 7.2 *In a classical optical system, where the local carrier has an amplitude V_L much greater than the received carrier amplitude V_R, the shot noise may be considered Gaussian. With equally likely symbols and optimized decision regions $\{R(\gamma), \gamma \in \mathcal{C}_0\}$, the minimum error probability turns out to be a simple function of the SNR*

$$\boxed{\Lambda = \frac{4N_s}{\mu_K}} \tag{7.63}$$

expressed through the complementary normalized Gaussian distribution $Q(x)$. In (7.63) N_s is the number of signal photons per symbol and μ_K is shape factor of the constellation.

Problem 7.8 $\star$ Consider the 4-QAM (which is equivalent to 4-PSK) where the normalized constellation is $\mathcal{C}_0 = \{\gamma = \pm 1 + \pm i\}$ and the constellation of received values is given by

$$\{(\pm 1 + \pm i)U_0 + (1 + i)N_L\}.$$

Find the optimal decision regions and prove that the minimum error probability P_c is given by $P_c = 1 - \left(1 - Q(\sqrt{\Lambda})^2\right)$ with $\Lambda = 4N_R$.

7.6 Analysis of Classical Optical Binary Systems

In classical optical systems, the transmitted symbol C_0 has in general a complex format; but in the binary case, without restrictions, we can assume a real format. Then the general schemes of the previous section (Figs. 7.9, 7.10, and 7.11) are simplified because the double path is reduced to a single path.

For the sake of comparison with other schemes, it is convenient to express the system performance (error probability) in terms of the *average number of photons per bit N_R*, which is given in general by

$$N_R = q_0 N_R(0) + q_1 N_R(1), \qquad q_0 + q_1 = 1$$

where $q_i = P[A_0 = i]$ are the a priori probabilities and $N_R(i) = E[n|A_0 = i]$ the average number of photons associated to the symbol $A_0 = i$. Usually we will consider equiprobable symbols, so that the average number of photon per bit becomes

$$N_R = \frac{1}{2} N_R(0) + \frac{1}{2} N_R(1). \tag{7.64}$$

In the following analysis, we will use the notations for signals and complex envelopes:

- $v_0(t)$ V_0 optical carrier at the transmitter,
- $v_T(t)$ $V_T(t)$ transmitted optical signal,
- $v_R(t)$ $V_R(t)$ received optical signal,
- $v_L(t)$ V_L local optical carrier at the reception.

We assume that the channel is ideal, so that

$$v_R(t) = v_T(t).$$

7.6.1 Binary System with Incoherent Detection (OOK Modulation)

In the classical formulation, a monochromatic wave at frequency ν emitted by a laser can be represented by a sinusoidal signal

$$v_0(t) = V_0 \cos 2\pi \nu t \tag{7.65}$$

where the amplitude V_0 gives the optical power as (see Sect. 7.5)

$$P = V_0^2. \tag{7.66}$$

The simplest optical communications system uses amplitude modulation (OOK) and incoherent detection, as shown in Fig. 7.13. The OOK modulator is a special case of the general modulator of Fig. 7.10 with the encoding mapping the identity, $A_0 \;\rightarrow\; C_0 = A_0$, which gives the modulated signal

$$v_T(t) = \Re C_0\, V_0\, \mathrm{e}^{\mathrm{i}\,2\pi \nu t} = A_0\, V_0 \cos 2\pi \nu t, \qquad 0 < t < T.$$

In practice, in the symbol period $(0, T)$ the transmitter associates a zero field to the symbol $A_0 = 0$ and the field $V_0 \cos 2\pi \nu t$ to the symbol $A_0 = 1$.

Figure 7.14 shows a sequence of binary symbols and the corresponding modulated signal. This is obtained by amplitude modulating the laser beam of frequency ν or, more simply, by switching on and off the laser itself according to the source symbol to be transmitted. At the receiver, a photodetector transforms the incident field into an electrical current from which a photon counting can be obtained, as seen in Sect. 4.8.

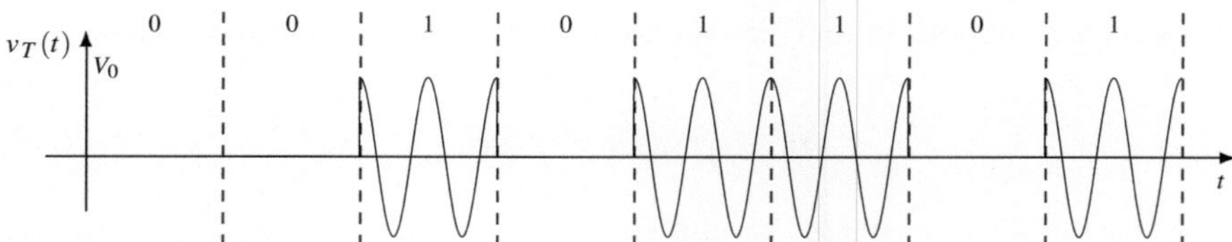

Fig. 7.13 A realization of a binary sequence and corresponding OOK signal

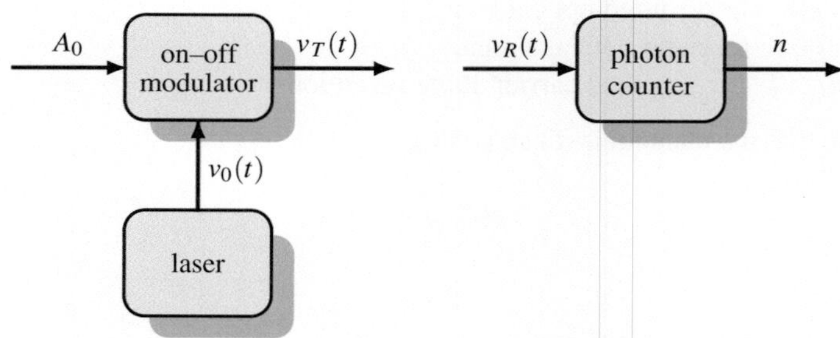

Fig. 7.14 Binary optical system with amplitude on–off modulation and uncoherent detection

Considering that with the transmission of the symbol $A_0 = 0$, the number of photons is null, $n = 0$, in (7.64) we have $N_R(0) = 0$, and therefore

$$N_R = \frac{1}{2}N_R(1).$$

At reception the photon count receiver uses the decision criterion

$$\widehat{A_0} = \begin{cases} 0 & \text{if } n = 0 \\ 1 & \text{if } n \geq 1, \end{cases} \tag{7.67}$$

where n is the number of photons counted in a symbol period. Then, with the transmission of the symbol $A_0 = 0$, we always have a correct decision

$$P_e(0) = 0. \tag{7.68a}$$

When $A_0 = 1$ the number of arrivals n is a Poisson variable with average $N_R(1)$, and therefore with (conditioned) distribution

$$p_n(k|1) = e^{-N_R(1)} \frac{N_R(1)^k}{k!}, \qquad k = 0, 1, \ldots$$

and we have an error when $n = 0$, which occurs with probability

$$P_e(1) = p_n(0|1) = e^{-N_R(1)} = e^{-2N_R}. \tag{7.68b}$$

The average error probability in the classical system is therefore

$$\boxed{P_{e,\text{classical}} = \frac{1}{2}e^{-2N_R}} \tag{7.69}$$

where equally likely symbols are assumed.

In optical communications this probability is called the **quantum limit** [6] or **shot noise limit**, and it is the optimum for any detection that does not exploit the coherence property of the optical beam. Notice, in fact, that in this classical context the decision criterion (7.67) is optimal (see Problem 5.4). The receiver scheme is called *direct detection* of the incident light pulses. The main advantage of this approach is its simplicity. In particular, phase and frequency instabilities of the laser source are well tolerated. Moreover, at the receiver direct detection is used and phase sensitive devices are avoided.

7.6.2 Quantum Interpretation of Photon Counting in OOK

The above scheme, known as on–off keying (OOK) modulation, has a simple quantum equivalent, employing the coherent states $|0\rangle$ and $|\alpha\rangle$, with $\alpha > 0$, where the photon counting can be treated as a quantum measurement with not optimal measurement operators.

The quantum measurement realized by the photon counter is obtained with the elementary projectors $|n\rangle\langle n|$, where $|n\rangle$ is the number state (see Fig. 7.2), and the outcome of the measurement is given by the number of photons n. The transition probabilities in the measurement are

$$p(i|\alpha) = \mathrm{P}[n = i|\alpha] = e^{-|\alpha|^2}\frac{|\alpha|^{2i}}{i!}, \qquad p(i|0) = \mathrm{P}[n = i|0] = \delta_{i0}.$$

The alphabet of the measurement is then $\mathcal{M} = \{0, 1, 2, \ldots\}$, and it is different from the alphabet $\mathcal{A} = \{0, 1\}$ of the source (see Sect. 5.2). To find the global performance, we must introduce a decision criterion consisting in the partitioning of $\mathcal{M}$ into two decision regions $\mathcal{M}_0$ and $\mathcal{M}_1$ to obtain two global measurement operators (see Sect. 5.2.3). The optimal partition is $\mathcal{M}_0 = \{0\}$ and $\mathcal{M}_1 = \{1, 2, \ldots\}$ (Fig. 7.15) and so we have the global projectors

$$Q_0 = |0\rangle\langle 0| \qquad Q_1 = \sum_{n=1}^{\infty}|n\rangle\langle n|. \tag{7.70}$$

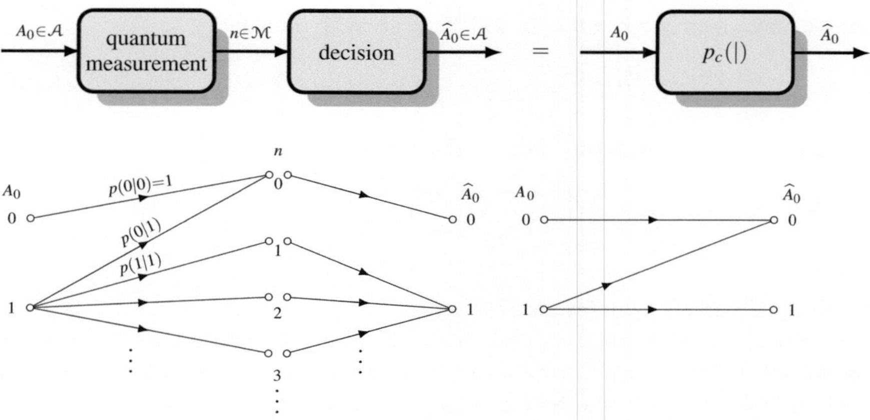

Fig. 7.15 Quantum interpretation of the decision made via a photon counter in an OOK system. The outcome of quantum measurement is given by the number of photons n present in the state $|\alpha\rangle$ or $|0\rangle$. The decision converts the measurement alphabet $\mathcal{M} = \{0, 1, 2, \ldots\}$ into the binary alphabet $\mathcal{A} = \{0, 1\}$, thus realizing a binary channel

The global transition probabilities, from (5.15), are $p_c(0|0) = \mathrm{Tr}[\rho_0 \, Q_0]$ and $p_c(0|1) = \mathrm{Tr}[\rho_1 \, Q_0]$, where $\rho_1 = |\alpha\rangle\langle\alpha|$ and $\rho_0 = |0\rangle\langle 0|$. Then

$$p_c(0|0) = \langle 0|0\rangle\langle 0|0\rangle = 1$$

$$p_c(0|1) = \langle\alpha|Q_0|\alpha\rangle = |\langle\alpha|0\rangle|^2 = \mathrm{e}^{-|\alpha|^2} = \mathrm{e}^{-2N_R}. \tag{7.71}$$

The performance is lower than that of the quantum version of the OOK, which will be seen in Sect. 7.9, because the projectors (7.70) are suboptimal. We recall, in fact, that with pure states, the optimal measurement operators must be elementary with measurement vectors arranged symmetrically with respect to the coherent states (Fig. 7.16); whereas (7.70) Q_1 has infinite rank and Q_0 is elementary with measurement vector $|\mu_0\rangle$ coinciding with the state $|0\rangle$.

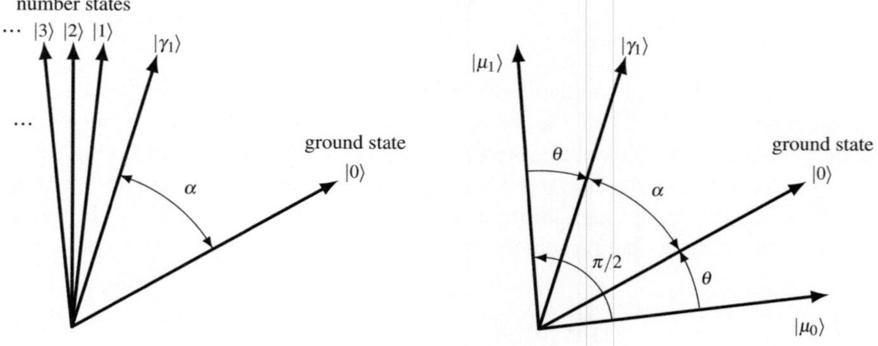

Fig. 7.16 Decision with a photon counter (*left*) and optimal decision

7.6.3 Binary System with Coherent Detection (BPSK Modulation)

A more sophisticated scheme of classical optical communications uses binary phase-shift keying (BPSK) modulation (Fig. 7.17). The BPSK modulator is a special case of the general modulator of Fig. 7.10 with the encoding mapping

$$A_0 \;\rightarrow\; C_0 = e^{iA_0\,\pi} = \begin{cases} +1 & A_0 = 0 \\ -1 & A_0 = 1 \end{cases}$$

which gives the modulated signal

$$v_T(t) = \Re\, C_0\, V_0\, e^{\,i\,2\pi\,\nu t} = V_0\cos(2\pi\,\nu t + A_0\pi)\,, \qquad 0 < t < T \qquad (7.72)$$

where V_0 is the amplitude of the carrier $v_0(t) = V_0\cos 2\pi\nu t$. Figure 7.18 shows a sequence of binary symbols and the corresponding BPSK signal, which in the interval $(nT, nT + T)$ is given by $V_0\cos(2\pi\,\nu t + A_n\,\pi)$.

BPSK with Homodyne Detection

Since the modulated signals $v_R(t) = v_T(t)$ for different symbols have the same optical energy, and hence the same photon counting, direct detection cannot discriminate between them. Then the receiver adds to the incoming field $v_R(t) = V_R\cos(2\pi\,\nu t + A_n\,\pi)$ the field $V_L\cos 2\pi\,\nu t$, generated by a "local" laser tuned at the same frequency as $v_0(t)$, to get the signal

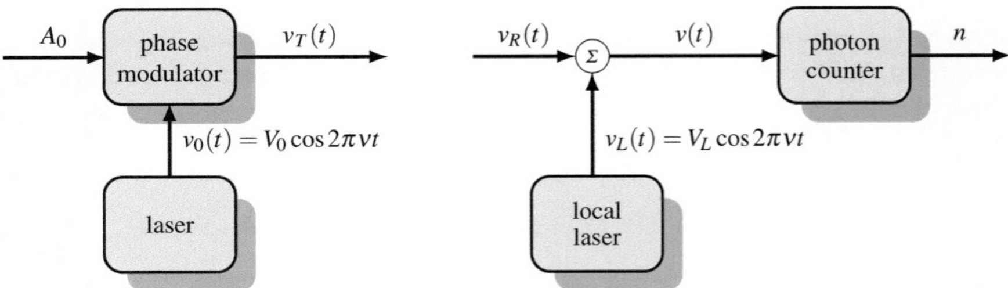

Fig. 7.17 Scheme of a binary coherent optical system with BPSK modulation. The receiver is called *homodyne* because the frequency of the local laser is the same as the frequency of modulation carrier

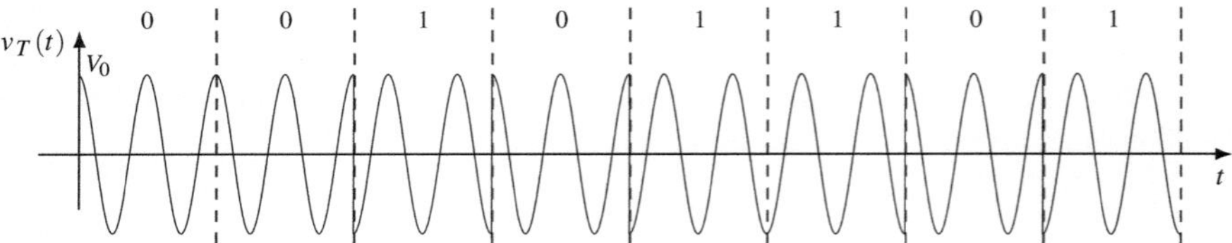

Fig. 7.18 A realization of a binary sequence and corresponding BPSK signal

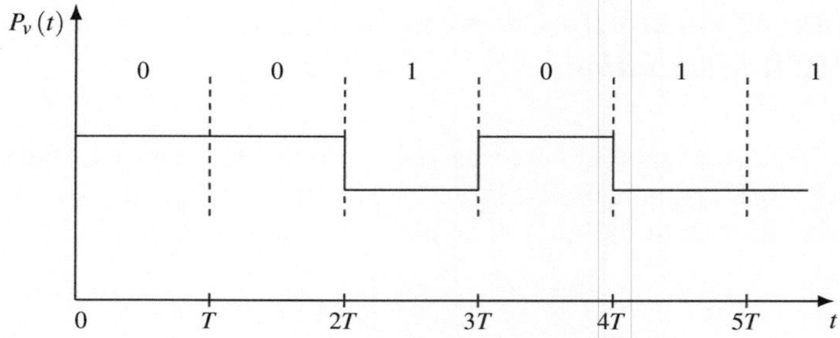

Fig. 7.19 Example of the optical power $P_R(t)$ after the introduction of the local carrier in a homodyne receiver

$$v(t) = V_R \cos\left(2\pi \nu t + A_0 \pi\right) + V_L \cos\left(2\pi \nu t\right). \tag{7.73}$$

As in Sect. 7.5.4, we assume that the local carrier has an amplitude V_L much greater than that of the received signal, $V_L \gg V_0$. Since $\cos\left(2\pi \nu t + A_0 \pi\right) = \cos A_0 \pi \, \cos 2\pi \nu t$, the power becomes

$$P_v(t) = (V_R \cos \pi A_0 + V_L)^2 = V_R^2 + V_L^2 + 2 V_R V_L \cos A_0 \pi \tag{7.74}$$

which is illustrated in Fig. 7.19 for a sequence of source symbols. Applying this power to a photon counter, we obtain a number of arrivals n in a symbol period, which can be decomposed in the form

$$n = \bar{n}(A_0) + u$$

where $\bar{n}(A_0) = \mathrm{E}[n|A_0]$ is the useful signal and the fluctuation u is the shot noise. Now, from the theory of semiclassical detection developed in the previous section, the number of signal photons is given by the photonic intensity $P_v(t)/h\nu$ integrated over $(0, T)$, and therefore it results in

$$\bar{n}(A_0) = H\left(V_L^2 + V_R^2 + 2 V_R V_L \cos \pi A_0\right) = N_L + N_R + U_0 \cos \pi A_0 \tag{7.75}$$

where $N_L + N_R = H(V_L^2 + V_R^2)$ is a bias term, $U_0 = 2\sqrt{N_L N_R}$, and $U_0 \cos \pi A_0$ is the symbol-dependent part. The variance, coinciding with the average, is

$$\sigma_n^2(A_0) = N_L + N_R + U_0 \cos \pi A_0 \cong N_L, \tag{7.76}$$

where the approximation follows from the hypothesis $V_L \gg V_0$. In conclusion, the decision on the transmitted symbol A_0 is made on the value

$$\boxed{n = N_L + N_R + U_0 \cos(\pi A_0) + u.} \tag{7.77}$$

At this point we introduce the **Gaussian approximation**, *where it is assumed that the photon number n is a Gaussian random variable* and hence specified by the mean $\bar{n}(A_0)$ and by the variance $\sigma^2(A_0) = N_L$, which is independent of the symbol A_0. As seen in the previous section, the Gaussian assumption allows us to simplify the analysis and to arrive at a very simple result.

By choosing the decision rule as

$$\widehat{A}_0 = \begin{cases} 1 & n \le N_L + N_R \\ 0 & n > N_L + N_R \end{cases} \tag{7.78}$$

we obtain the error probability

$$P_e = Q\left(\frac{U_0}{\sigma_n}\right) = Q\left(\sqrt{\Lambda}\right), \tag{7.79}$$

where the SNR is given by $\Lambda = U_0^2/\sigma_n^2 = 4\,N_R$. Note that $N_R = N_R(0) = N_R(1)$ gives the number of signal photons per bit. In conclusion, the error probability in the classical BPSK with homodyne receiver is given by

$$\boxed{P_{e,\text{classical}} = Q\left(\sqrt{4N_R}\right).} \tag{7.80}$$

This error probability is known as the **standard quantum limit**. The result is in agreement with Proposition 7.2.

Comparison with incoherent detection (OOK) shows that the performances of the homodyne detection are better, as illustrated in Fig. 7.20, where the error probability P_e is plotted versus the average number of signal photons per bit N_R. On the other hand, the implementation of an efficient homodyne scheme implies some complications, in that it requires the presence of a local laser that must be accurately tuned in frequency and phase with the source laser.

BPSK with Superhomodyne Reception

We suppose that at the reception we have available a laser (local oscillator) producing a radiation $v_L(t)$ with the same amplitude, frequency and phase of the carrier at the transmitter, that is,

$$v_L(t) = V_L \cos(2\pi v t) \quad \text{with} \quad V_L = V_0. \tag{7.81}$$

This local carrier is added to the received modulated signal, yielding (Fig. 7.21)

$$v(t) = V_0 \cos(2\pi v t + A_0 \pi) + V_0 \cos(2\pi v t)$$

$$= \begin{cases} 2V_0 \cos(2\pi v t) & A_0 = 0 \\ 0 & A_0 = 1. \end{cases} \tag{7.82}$$

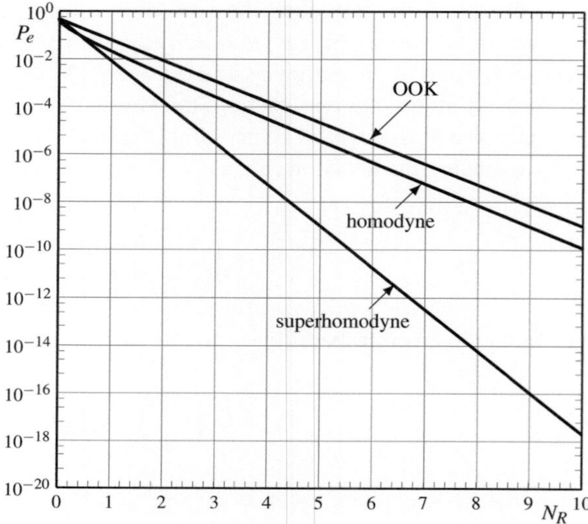

Fig. 7.20 Comparison of error probability P_e versus average number of signal photons per bit N_R in classical binary optical systems

Then the number of signal photons in a symbol period becomes

$$N_v(0) = \frac{4P_R T}{h\nu}, \qquad N_v(1) = 0. \tag{7.83}$$

Using a photon counter, the decision is based on the number of arrivals n by the rule

$$\widehat{A}_0 = \begin{cases} 1 & \text{if } n = 0 \\ 0 & \text{if } n > 0. \end{cases} \tag{7.84}$$

Then, similarly to relation (7.69), we get the error probability

$$P_{e,\text{classical}} = \frac{1}{2}P_e(0) = \frac{1}{2}e^{-N_v(0)} \tag{7.85}$$

The interesting thing is that the number of signal photons per bit at the reception, i.e., before adding the carrier, is $N_R = P_R T/(h\nu)$ and it is equal to one fourth of $N_v(0)$, so that the relation (7.85) becomes

$$\boxed{P_{e,\text{classical}} = \frac{1}{2}e^{-4N_R}} \tag{7.86}$$

which represents the *super quantum limit* [6]. So we have a great improvement over the homodyne detection, as shown in Fig. 7.20; because the power introduced by the local oscillator creates a more favorable situation for a correct decision. But the implementation of superhomodyne is very difficult in that it requires the presence of

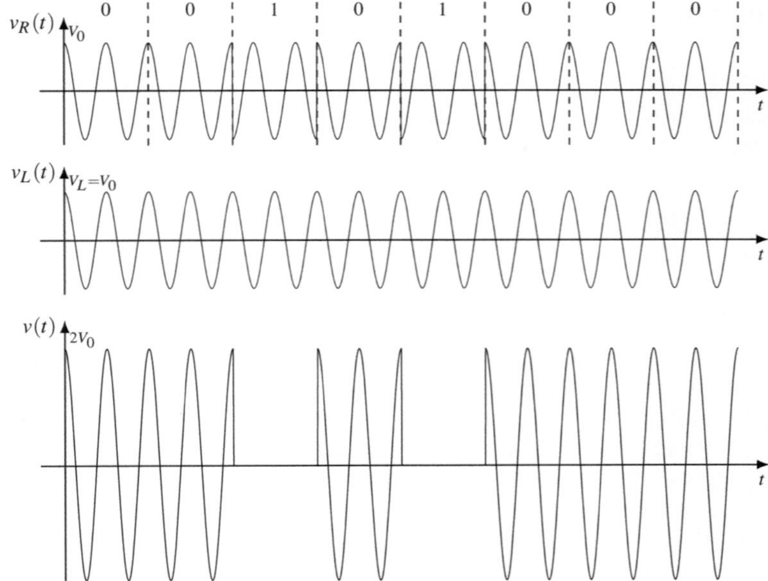

Fig. 7.21 Signals in classical BPSK superhomodyne reception

a local laser that must be accurately tuned with the source laser, not only in frequency and phase but also in amplitude.

Problem 7.9 ★★★ The error probability in classical homodyne BPSK has been evaluated assuming equiprobable symbols. When the symbols are not equiprobable the number of signal photons per bit N_R is still independent of the symbols and gives the SNR as $\Lambda = 4N_R$. The only change is in the evaluation is the decision element, given for equiprobable symbol by (7.78), as

$$
\widehat{A}_0 = \begin{cases} 1 & n \le S \\ 0 & n > S \end{cases}
$$

where S is the threshold to be optimized.

Find the optimal decision threshold and prove that the minimum error probability is given by

$$
P_e = q_1 \, Q\left(\sqrt{\Lambda} + \frac{1}{2\sqrt{\Lambda}} \log \frac{q_1}{q_0}\right) + q_0 \, Q\left(\sqrt{\Lambda} - \frac{1}{2\sqrt{\Lambda}} \log \frac{q_1}{q_0}\right). \tag{7.87}
$$

7.7 Quantum Decision with Pure States

In a K-ary quantum communications system, in the absence of thermal noise, we use the quantum decision theory developed in the previous two chapters, limited to *pure states*. We recall the main ideas and the available methods.

The source (Alice) is characterized by a constellation of coherent states $\mathcal{S} = \{|\gamma_0\rangle, |\gamma_1\rangle, ..., |\gamma_{K-1}\rangle\}$, which can be collected in the state matrix

$$\underset{n \times K}{\Gamma} = [|\gamma_0\rangle, |\gamma_1\rangle, ..., |\gamma_{K-1}\rangle] \tag{7.88}$$

where n is the dimension of the underlying Hilbert space (n may be infinite, and it really is infinite in Glauber's representation). The specification of the source requires also the definition of the prior probabilities $q_i = \mathrm{P}[A_0 = i] = \mathrm{P}[C_0 = \gamma_i]$, but usually throughout the chapter we shall assume equiprobable symbols $q_i = 1/K$.

The goal is to find an *optimal* system of measurement operators $Q_i, i \in \mathcal{A}$, that is, minimizing error probability. Kennedy's theorem (Theorem 5.3) states that, with pure states, the optimal measurement operators must be elementary, i.e., in the form $Q_i = |\mu_i\rangle\langle\mu_i|$. We can then limit our search to the measurement vectors, specified by the *measurement matrix*

$$\underset{n \times K}{M} = [|\mu_0\rangle, |\mu_1\rangle, \ldots, |\mu_{K-1}\rangle]. \tag{7.89}$$

These vectors are given as a linear combination of the states, as established by the relation

$$\boxed{M = \Gamma A} \tag{7.90}$$

where A is a $K \times K$ complex matrix.

Kennedy's theorem states also that the optimal measurement vectors $|\mu_i\rangle$ must be orthogonal, and therefore the corresponding measurement operators $Q_i = |\mu_i\rangle\langle\mu_i|, i \in \mathcal{A}$ must form a system of projectors. Unfortunately, Kennedy's theorem, as well as Holevo's theorem, do not provide explicit solutions. To compute optimal solutions, we could resort to the numeric programming methods outlined in Sect. 5.8, but, luckily enough, we can get help from the geometrically uniform symmetry (GUS), which is verified in the majority of quantum communications systems. In fact, square root measurement (SRM) decision, which is, in general, suboptimal, with pure states and in the presence of GUS becomes optimal (see Sect. 6.5.4). It is then appropriate to explicitly recall the SRM methodology, that gives good results even in the absence of GUS.

From the measurement vectors, we calculate the transition probabilities $p(j|i) = |\langle\mu_j|\gamma_i\rangle|^2$ and then the probability of correct decision.

7.7.1 Recall of SRM Approach

We summarize the main steps of the SRM theory developed in Sect. 6.3.

Starting from the constellation of K coherent states $\mathcal{C} = \{|\gamma_0\rangle, \ldots, |\gamma_{K-1}\rangle\}$, we evaluate in sequence

(1) Gram's matrix of the inner products $G = [\langle\gamma_i|\gamma_j\rangle|]$, $i, j = 0, 1, \ldots, K - 1$, calculated according to (7.9). In the cases of interest, the matrix G, which is $K \times K$, has rank K.

(2) The spectral decomposition (EID) of G

$$G = V \Lambda_G V^* = \sum_{i=0}^{K-1} \sigma_i^2 \, |v_i\rangle\langle v_i| . \tag{7.91}$$

From this EID we find the eigenvalues σ_i^2 and the orthonormal basis $\{|v_i\rangle\}$.

(3) The square roots of G

$$G^{\pm\frac{1}{2}} = V \Lambda_G^{\pm\frac{1}{2}} V^*. \tag{7.92}$$

(4) The transition probabilities according to (see (6.29))

$$p_c(i|j) = \left|\left(G^{\frac{1}{2}}\right)_{ij}\right|^2 \tag{7.93}$$

and the error probabilities (with equiprobable symbols)

$$P_e = 1 - \frac{1}{K} \sum_{i=0}^{K-1} \left|\left(G^{\frac{1}{2}}\right)_{ii}\right|^2 . \tag{7.94}$$

(5) The measurement vectors as linear combination of the states according to

$$M = \Gamma \, G^{\frac{1}{2}} \quad \rightarrow \quad |\mu_i\rangle = \sum_{j=0}^{K-1} \left(G^{-\frac{1}{2}}\right)_{ij}|\gamma_j\rangle. \tag{7.95}$$

With geometrically uniform symmetry (GUS). If the states $|\gamma_i\rangle$ have the GUS, Gram's matrix becomes *circulant* and its EID is given by

$$G = W_{[K]} \Lambda_G W_{[K]}^* = \sum_{i=0}^{K-1} \sigma_i^2 |w_i\rangle\langle w_i|, \tag{7.96}$$

where the vectors $|w_i\rangle$ are the columns of the DFT matrix $W_{[K]}$

$$|w_i\rangle = \frac{1}{\sqrt{K}} \left[W_K^{-i}, W_K^{-2i}, \ldots, W_K^{-i(K-1)} \right]^{\mathrm{T}}, \qquad i = 0, 1, \ldots, K - 1 \tag{7.97}$$

and the eigenvalues are given by the DFT of the first row $[r_0, r_1, \ldots, r_{K-1}]$ of the matrix G

$$\lambda_i = \sigma_i^2 = \sum_{k=0}^{K-1} r_k W_K^{-ki}, \qquad r_k = \langle \gamma_0 | \gamma_k \rangle. \tag{7.98}$$

The square roots of G have elements

$$\left(G^{\pm \frac{1}{2}} \right)_{ij} = \frac{1}{K} \sum_{p=0}^{K-1} \lambda_p^{\pm \frac{1}{2}} W_K^{-p(i-j)} \tag{7.99}$$

and in particular the diagonal elements are all equal. Therefore, the error probability is simply given by

$$P_e = 1 - \left| \left(G^{\frac{1}{2}} \right)_{00} \right|^2. \tag{7.100}$$

7.8 Quantum Binary Communications Systems

We develop the analysis of a quantum binary system, in which the information is carried by two coherent states. In this section, we assume that the constellation of the two states be generic; whereas, in the subsequent sections, two specific modulation formats will be developed (OOK and BPSK).

In the binary case, the optimal decision can be obtained in explicit form by Helstrom's theory and also by the geometric approach, seen in Sect. 5.4. With equiprobable symbols, we can also use the SRM method, which provides an optimal decision (see Sect. 6.5).

7.8.1 Binary Systems with Coherent States

To implement a Quantum Communications binary system, the transmitter (laser) is placed in one of two distinct coherent states $|\gamma_0\rangle$, $|\gamma_1\rangle \in \mathcal{G}$, which can be collected in the state matrix $\Gamma = [|\gamma_0\rangle, |\gamma_1\rangle]$. The geometry is completely specified by the inner product $X := \langle \gamma_0 | \gamma_1 \rangle$, which can be calculated explicitly from (7.9).

The optimal decision is based on two measurement operators Q_0 and Q_1, with $Q_0 + Q_1 = I$, which, by Kennedy's theorem (see Sect. 5.11), must be in the form $Q_0 = |\mu_0\rangle\langle\mu_0|$ e $Q_1 = |\mu_1\rangle\langle\mu_1|$, and therefore are identified by two measurement vectors $|\mu_0\rangle$ and $|\mu_1\rangle$, which form the measurement matrix

$$M_{\text{opt}} = [|\mu_0\rangle, |\mu_1\rangle].$$

Still by Kennedy's theorem, the two measurement vectors must be orthogonal, $\langle\mu_0|\mu_1\rangle = 0$, so that the quantum measurement is always projective.

7.8.2 Recall of Helstrom's Theory and of Geometric Approach

We briefly recall the theory of optimal binary decision developed in Sects. 5.3 and 5.4. The results of this theory are completely specified by the a priori probabilities q_0 and q_1 and by the (quadratic) superposition degree of the states $|X|^2 = |\langle\gamma_0|\gamma_1\rangle|^2$, which can be calculated in explicit form from (7.9), obtaining

$$|X|^2 = e^{-|\gamma_0-\gamma_1|^2}. \tag{7.101}$$

The optimal measurement matrix is related to the state matrix as $M = \Gamma A$ where the matrix A is given explicitly by (5.39).

The error probability, known as the Helstrom bound, is given by

$$\boxed{P_e = \frac{1}{2}\left(1 - \sqrt{1 - 4q_0q_1|X|^2}\right).} \tag{7.102}$$

Case of Equiprobable Symbols

When the symbols are equiprobable, which is the case of main interest, we have a few simplifications. The matrix A becomes

$$A = \frac{1}{2}\begin{bmatrix} \dfrac{1}{\sqrt{1+|X|}} + \dfrac{1}{\sqrt{1-|X|}} & \dfrac{1}{\sqrt{1+|X|}} - \dfrac{1}{\sqrt{1-|X|}} \\ \dfrac{1}{\sqrt{1+|X|}} - \dfrac{1}{\sqrt{1-|X|}} & \dfrac{1}{\sqrt{1+|X|}} + \dfrac{1}{\sqrt{1-|X|}} \end{bmatrix}. \tag{7.103}$$

The error probability is simplified as

$$\boxed{P_e = \tfrac{1}{2}\left[1 - \sqrt{1 - |X|^2}\right].} \tag{7.104}$$

Problem 7.10 $\star\star$ Prove that with the optimization the a posteriori probabilities $q(i|i) := P[A_0 = i|\hat{A}_0 = i]$ are equal and coincide with the correct decision probability P_c.

Problem 7.11 ⋆   Prove that in a binary system with equiprobable symbols, the error probability can be expressed as function of $N_R(0)$, $N_R(1)$, and of the relative phase of the complex parameters γ_0 and γ_1. that determine the coherent states.

7.9 Quantum Systems with OOK Modulation

The constellation consists of the states (Fig. 7.22)

$$|\gamma_0\rangle = |0\rangle \ , \ |\gamma_1\rangle = |\Delta\rangle \ \in \mathcal{G} \tag{7.105}$$

where $|0\rangle$ is the ground state and the state $|\Delta\rangle$ is determined by the number Δ which is not restrictive to assume real and positive. The quadratic superposition of the two states is $|\langle 0|\Delta\rangle|^2 = e^{-\Delta^2}$. The number of signal photons associated to the symbol $A_0 = 0$ is $N_R(0) = 0$, while the one associated to the symbol $A_0 = 1$ is $N_R(1) = \Delta^2$. The number of signal photons per bit is then

$$N_R = \tfrac{1}{2} N_R(0) + \tfrac{1}{2} N_R(1) = \tfrac{1}{2} N_R(1)$$

and the quadratic superposition of the two states can be written in the meaningful form

$$|X|^2 = e^{-2N_R}.$$

From (7.104), we obtain that the error probability of the OOK quantum system with equiprobable symbols becomes

$$\boxed{P_e = \tfrac{1}{2}\left[1 - \sqrt{1 - e^{-2N_R}}\right].} \tag{7.106}$$

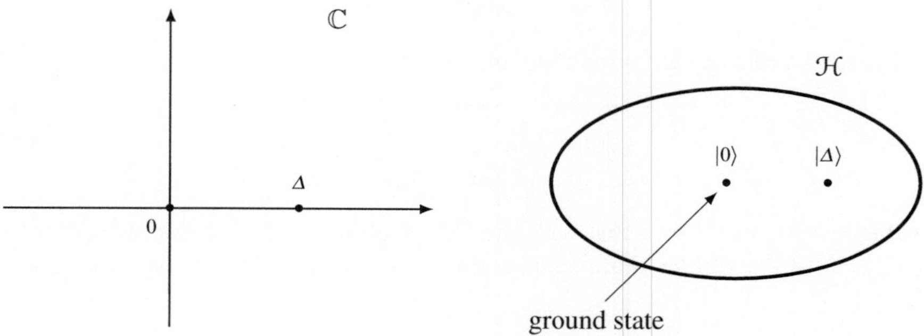

Fig. 7.22 Constellation of symbols and states in OOK modulation

The evaluation of the measurement vectors does not exhibit any specific simplification in (7.103), where now the inner product X can be expressed in the form $X = \mathrm{e}^{-N_R}$.

7.9.1 Comparison with Classical OOK Optical Systems

The *classical* OOK system was developed in Sect. 7.6.1, where we found that the error probability, with equiprobable symbol, is given by

$$P_{e,\text{classical}} = \tfrac{1}{2}\,\mathrm{e}^{-2N_R}. \tag{7.107}$$

The comparison between the $P_{e,\text{classical}}$ of the classical receiver, given by (7.107), and the P_e of the quantum receiver, given by (7.106), is shown in Fig. 7.23 as a function of the average number of photons per bit N_R. The asymptotic behavior of (7.106) becomes (by the approximation $1 - \sqrt{1-x} \simeq \tfrac{1}{2}x$ for x small)

$$P_e = \tfrac{1}{4}\mathrm{e}^{-2N_R} \qquad N_R \gg 1$$

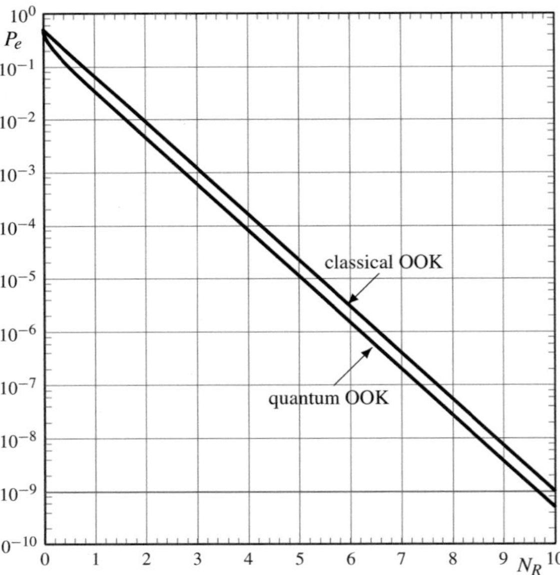

Fig. 7.23 Comparison of quantum and classical OOK

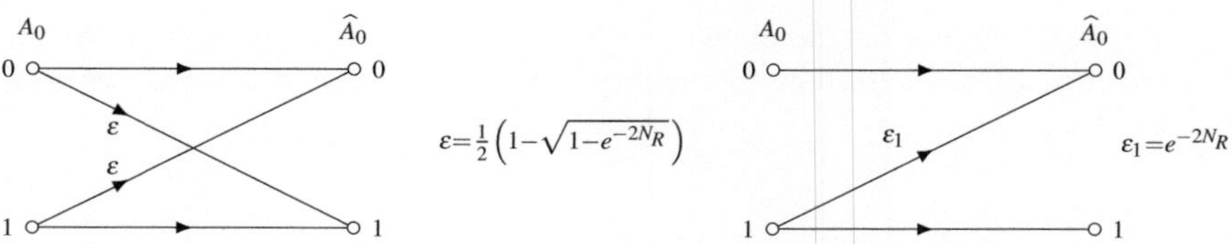

Fig. 7.24 Symmetric binary channel realized by the quantum optimal decision and asymmetric binary channel realized by photon count decision

that is, one half of the classical case. Thus we not have a great improvement in quantum OOK with respect to classical OOK (in error probability a relevant improvement is expressed in decades). The sensititvities in the two kinds of OOK are

$$N_R = 9.668 \text{ photons/bit}, \qquad N_{R,\text{classical}} = 10.015 \text{ photons/bit}.$$

Another comparison is between the channels realized by the two kinds of receiver: With the quantum receiver (optimized with equiprobable symbols) we obtain a symmetric channel, notwithstanding that the constellation is unbalanced, while, with the receiver based on photon count, the channel turns out to be very asymmetric (Fig. 7.24).

7.10 Quantum Systems with BPSK Modulation

In the BPSK quantum system, the symbol $A_0 = 0$ (phase $\varphi = 0$) is encoded into a coherent state $|\Delta\rangle$ with a given amplitude Δ and the symbol $A_0 = 1$ (phase $\varphi = \pi$) into the coherent state $|-\Delta\rangle$ (Fig. 7.25)

$$|\gamma_0\rangle = |\Delta\rangle \, , \ |\gamma_1\rangle = |-\Delta\rangle \ \in \mathcal{G}. \tag{7.108}$$

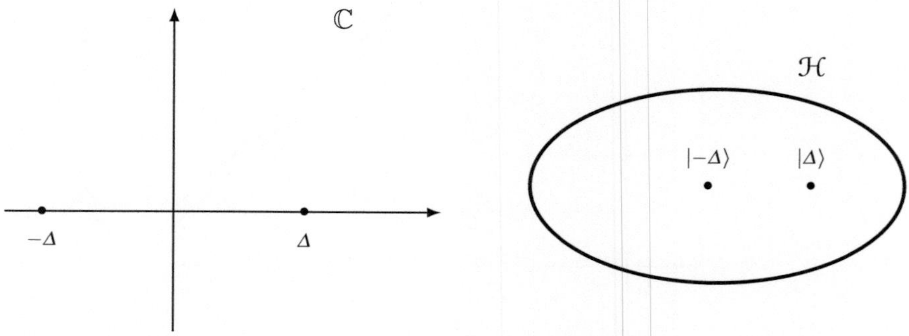

Fig. 7.25 Constellation of symbols and states in 2-PSK modulation

Obviously, the number of signal photons associated to the two states is equal

$$N_R(0) = N_R(1) = |\Delta|^2 = N_R$$

and the (quadratic) superposition degree of the two states becomes

$$|X|^2 = e^{-|\Delta - (-\Delta)|^2} = e^{-4|\Delta|^2} = e^{-4N_R}$$

which yields the error probability

$$P_e = \frac{1}{2}\left[1 - \sqrt{1 - e^{-4N_R}}\right]. \tag{7.109}$$

Compared to the quantum OOK modulation, we have an improvement, because the term at the exponent $4N_R$ in place of $2N_R$.

7.10.1 Comparison with Classical BPSK Optical System

The *classical* BPSK system was developed in Sect. 7.6.2, where we found that the error probability, with equiprobable symbols, is given by

$$P_{e,\text{classical}} = Q\left(\sqrt{4N_R}\right) \tag{7.110}$$

where $Q(x)$ is the normalized complementary Gaussian distribution. The Fig. 7.26 shows the comparison between the P_e of the classical homodyne receiver and the P_e of the quantum receiver.

In this case, the improvement is relevant. For instance for $N_R = 5$ photons per bit, we have $P_e = 0.515 \; 10^{-9}$ and $P_{e,\text{classical}} = 0.387 \; 10^{-5}$, and the improvement is of the order of four decades! The sensititvities in the two kinds of BPSK are

$$N_R = 4.837 \text{ photons/bit}, \qquad N_{R,\text{classical}} = 8.913 \text{ photons/bit}.$$

Generic a Priori Probabilities

Usually we consider equally probable symbols, but it may be interesting to see the comparison when the a priori probabilities q_0 and $q_1 = 1 - q_0$ are different.

In both quantum and classical BPSK we have $N_R(0) = N_R(1)$ and then the number of signal photons per bit $N_R = q_0 N_R(0) + q_1 N_R(1)$ is independent of q_0. In the classical BPSK, the error probability is given by (7.87) of Problem 7.9, that is,

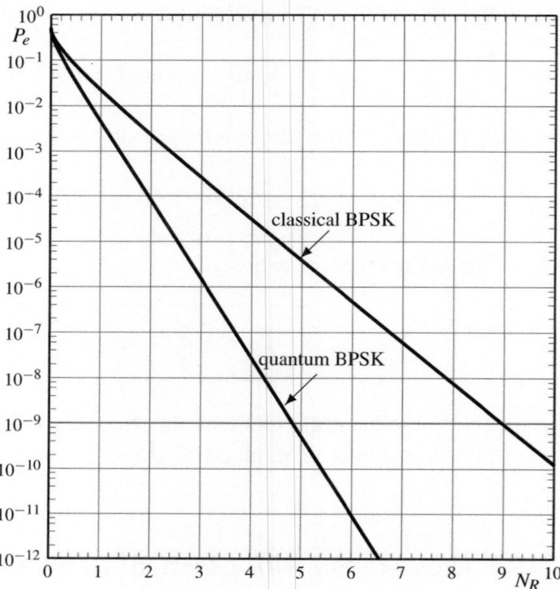

Fig. 7.26 Comparison between quantum and classical BPSK (with equal a priori probabilities)

$$P_{e,\text{classical}} = q_1\, Q\left(\sqrt{4N_R} + \frac{1}{2\sqrt{4N_R}}\log\frac{q_1}{q_0}\right) + q_0\, Q\left(\sqrt{4N_R} - \frac{1}{2\sqrt{4N_R}}\log\frac{q_1}{q_0}\right).$$
$$(7.111)$$

In the quantum BPSK (7.109) becomes

$$P_e = \frac{1}{2}\left[1 - \sqrt{1 - q_0 q_1 e^{-4N_R}}\right]. \tag{7.112}$$

The comparison is shown in Fig. 7.27. Note in particular that for $N_R = 0$ in both systems the error probability becomes $P_e = q_0$, so that it reduces with q_0 and also for $N_R > 0$ it is reduced when q_0 becomes smaller. This may lead to think that the performance of a quantum BPSK improves when the a priori probabilities are unbalanced. This is not true because the performance of a system is given not only by the error probability, but also by the entropy and by the capacity (see Chap. 12). With $q_0 = \frac{1}{2}$ the entropy H of a binary source is $H = 1$ bit per symbol, while with $q_0 = 0.01$ the entropy is reduced to $H = 0.08$ bit per symbol.

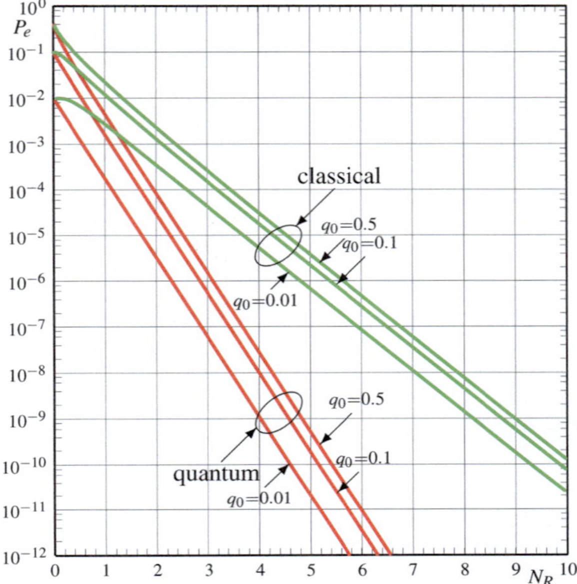

Fig. 7.27 Comparison between quantum and classical BPSK with non equiprobable symbols. The error probability P_e is reduced when q_0 becomes smaller

7.11 Quantum Systems with QAM Modulation

Quadrature amplitude modulation (QAM) is one of the most interesting format in radio frequency (RF) transmission, and can also be proposed for coherent optical modulation (classical system) and for quantum modulation.

QAM is the first example of *multilevel* modulation we are considering, with typically $K = L^2$ levels, that is, $K = 4$, $K = 9$, $K = 16$, and so on. For this format, the optimal quantum detection is not available, and suboptimal solutions must be adopted. We will apply the SRM technique which gives a good overestimate of the error probability [7], with a check by convex semidefinite programming (CSP).

7.11.1 Classical and Quantum QAM Formats

The alphabet of QAM modulation consists of a constellation of $K = L^2$ equally spaced points on a square grid of the complex plane, which can be defined starting from the L–ary balanced alphabet

$$\mathcal{A}_L = \{-(L-1) + 2(i-1)\,|\, i = 1, 2, \ldots, L\} \qquad \text{with} \quad L = 2, 3, 4, \ldots$$

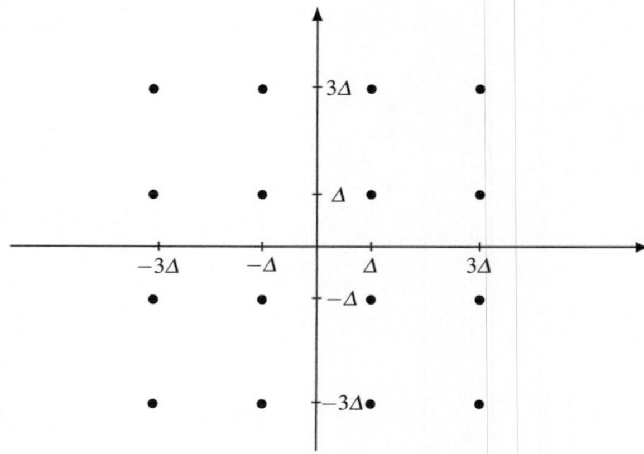

Fig. 7.28 Constellation of 16-QAM with scale factor Δ

In particular

$$
\begin{aligned}
\mathcal{A}_3 &= \{-2, 0, +2\} \\
\mathcal{A}_4 &= \{-3, -1, +1, +3\} \\
\mathcal{A}_5 &= \{-4, -2, 0, +2, +4\}.
\end{aligned}
$$

The K-ary QAM constellation is then formed by the complex numbers

$$
\mathcal{C} = \{\Delta(u + iv) \mid u, v \in \mathcal{A}_L\}
$$

where Δ is the scale factor and 2Δ gives the spacing of symbols in the constellation, with Δ real and positive. Figure 7.28 illustrates the constellation for $L = 4$ (16-QAM). Notice that the 4-QAM, obtained with $L = 2$, is equivalent to the 4-PSK, which will be developed in the next section.

To obtain the constellation of the coherent states in quantum QAM it suffices to assign to each symbol γ of the constellation $\mathcal{C}$ the corresponding coherent state $|\gamma\rangle$. Then the constellation of the coherent states becomes

$$
\mathcal{S} = \left\{ |\gamma_{uv}\rangle = |\Delta(u + iv)\rangle \mid u, v \in \mathcal{A}_L \right\}.
$$

For example, for the 16-QAM, where $L = 4$ and $\mathcal{A}_4 = \{-3, -1, +1, +3\}$, we have the following coherent states listed in *lexicographic* order (see Sect. 2.13)

$$
\begin{aligned}
u = -3 \quad & v = -3 \quad & |\gamma_0\rangle = |\gamma_{-3,-3}\rangle = |\Delta(-3 - 3i)\rangle \\
u = -3 \quad & v = -1 \quad & |\gamma_1\rangle = |\gamma_{-3,-1}\rangle = |\Delta(-3 - i)\rangle \\
u = -3 \quad & v = +1 \quad & |\gamma_2\rangle = |\gamma_{-3,+1}\rangle = |\Delta(-3 + i)\rangle
\end{aligned}
$$

$$u = -3 \qquad v = +3 \qquad |\gamma_3\rangle = |\gamma_{-3,+3}\rangle = |\Delta(-3 + 3i)\rangle$$

$$u = -1 \qquad v = -3 \qquad |\gamma_4\rangle = |\gamma_{-1,-3}\rangle = |\Delta(-1 - 3i)\rangle$$

$$\vdots \qquad\qquad \vdots$$

$$u = +3 \qquad v = -3 \qquad |\gamma_{12}\rangle = |\gamma_{+3,-3}\rangle = |\Delta(3 - 3i)\rangle$$

$$u = +3 \qquad v = -1 \qquad |\gamma_{13}\rangle = |\gamma_{+3,-1}\rangle = |\Delta(3 - i)\rangle$$

$$u = +3 \qquad v = +1 \qquad |\gamma_{14}\rangle = |\gamma_{+3,+1}\rangle = |\Delta(3 + i)\rangle$$

$$u = +3 \qquad v = +3 \qquad |\gamma_{15}\rangle = |\gamma_{+3,+3}\rangle = |\Delta(3 + 3i)\rangle$$

7.11.2 Performance of Quantum QAM Systems

We consider the decision based on the SRM method, recalled in Sect. 7.7.1. We start from the construction of Gram's matrix G, whose elements are the inner products

$$\langle \gamma_{uv} | \gamma_{u'v'} \rangle = \langle \Delta(u + iv) | \Delta(u' + iv') \rangle.$$

Remembering (7.9), we get

$$\langle \gamma_{uv} | \gamma_{u'v'} \rangle = \exp\{-\tfrac{1}{2}\Delta^2[(u' - u)^2 + (v' - v)^2 - 2i(u'v - v'u)]\}$$
$$u, v, u', v' \in \mathcal{A}_L. \tag{7.113}$$

The only problem in building the Gram matrix G is the ordering of the four-index elements in a standard (bidimensional) matrix. To this end, we can use the *lexicographic order* indicated above.

The main point of the SRM technique is the spectral decomposition of G, according to (7.91), namely,

$$G = V \Lambda_G V^* = \sum_{i=0}^{K-1} \sigma_i^2 \, |v_i\rangle\langle v_i|$$

which identifies the eigenvalues σ_i^2 and the orthonormal basis $|v_i\rangle$, $i = 1, 2, \ldots, K$, and also the square roots $G^{\pm \frac{1}{2}} = V \Lambda_G^{\pm \frac{1}{2}} V^*$. We can then compute the transition probabilities from (7.93) and the error probability from (7.94), that is,

$$p(j|i) = |(G^{\frac{1}{2}})_{ij}|^2, \qquad P_e = 1 - \frac{1}{K}\sum_{i=0}^{K-1} \left[(G^{\frac{1}{2}})_{ii}\right]^2. \tag{7.114}$$

As usual, the performance is evaluated as a function of the number of signal photons per symbol N_s, given in general by (7.34). For the QAM we find

$$N_s = \frac{1}{K} \sum_{i=0}^{K-1} |\gamma_i|^2 = \frac{1}{K} \sum_{u \in \mathcal{A}_L} \sum_{v \in \mathcal{A}_L} |\gamma_{uv}|^2$$

$$= \frac{1}{K} \Delta^2 \sum_{u \in \mathcal{A}_L} \sum_{v \in \mathcal{A}_L} (u^2 + v^2) = \frac{2L}{K} \Delta^2 \sum_{u \in \mathcal{A}_L} u^2$$

$$= \frac{2L}{K} \Delta^2 \sum_{i=1}^{L} [-(L-1) + 2(i-1)]^2.$$

The result is[5]

$$N_s = \frac{2}{3}(L^2 - 1)\Delta^2 = \frac{2}{3}(K-1)\Delta^2 \tag{7.115}$$

so that the *shape factor* (7.39) of the QAM constellation is given by

$$\mu_K = \frac{2}{3}(K-1). \tag{7.115a}$$

For example, for the 16-QAM we have $N_s = 10\Delta^2$ and $\mu_K = 10$.

Finally, from N_s, we get the number of signal photons per bit as

$$N_R = N_s / \log_2 K.$$

Remark As noted above, the 4-QAM may be viewed as a 4-PSK, for which an exact (non-numerical) evaluation of the SRM is possible. This exact evaluation can be used as a test to check the numerical accuracy of higher order QAM.

7.11.3 An Alternative Evaluation Using the Generalized GUS

The QAM modulation has not the ordinary GUS, but it verifies the first form of generalized GUS introduced in Sect. 5.13, where there are L reference states $|\gamma_0\rangle, \ldots, |\gamma_L\rangle$, instead of a single state $|\gamma_0\rangle$, and the K-ary constellation is subdivided into L subconstellations generated by a common symmetry operator S in the form $|\gamma_{ik}\rangle = S^i |\gamma_k\rangle$, $k = 1, \ldots, L$, $i = 0, 1, \ldots, K/L - 1$.

---

[5] Using the identities [8]

$$\sum_{i=1}^{n} i = \frac{1}{2}n(n+1), \qquad \sum_{i=1}^{n} i^2 = \frac{1}{6}n(n+1)(2n+1).$$

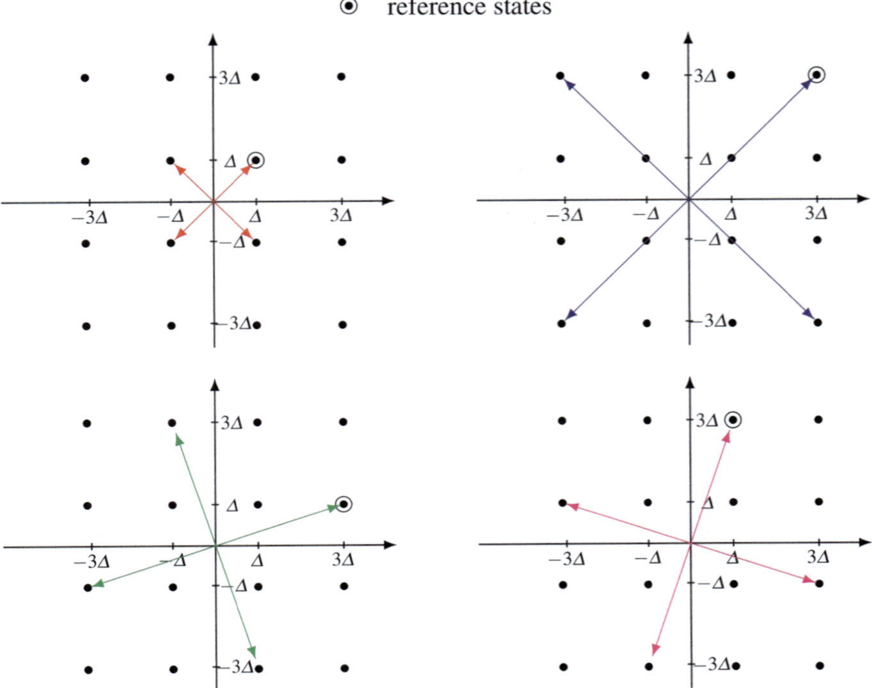

Fig. 7.29 Constellation of 16-QAM and its decomposition into four 4-PSK constellations

In particular, the 16-QAM constellation can be decomposed into 4-PSK constellations, as shown in Fig. 7.29. The reference states are the states belonging to the first quadrant of the complex plane, namely

$$|\gamma_0\rangle = |\Delta(1+i)\rangle \,, \quad |\gamma_1\rangle = |\Delta(3+3i)\rangle \,, \quad |\gamma_2\rangle = |\Delta(3+i)\rangle \,, \quad |\gamma_3\rangle = |\Delta(1+3i)\rangle. \tag{7.116}$$

The symmetry operator is the rotation operator of 4-PSK, $S = R(\pi/2)$, which allows us to generate all the other states of the 16-QAM by rotating the reference states into the other three quadrants. In fact, if we consider the state vector of the reference states $\tilde{\gamma}_0 = [|\gamma_0\rangle, |\gamma_1\rangle, |\gamma_2\rangle, |\gamma_3\rangle]$ and apply the rotation operator in the form

$$\tilde{\Gamma} = [\tilde{\gamma}_0, S\tilde{\gamma}_0, S^2\tilde{\gamma}_0, S^3\tilde{\gamma}_0]$$

we obtain a 16×16 Gram matrix $\tilde{G} = \tilde{\Gamma}^*\tilde{\Gamma}$, which contains the same inner products of the Gram matrix of the previous approach. But, for the different ordering, $\tilde{G}$ turns out to be **block circulant**.

At this point, although we are dealing with pure states, we can represent the 16 states through four density operators, ρ_0, ρ_1, ρ_2, and ρ_3, where ρ_0 collects the four reference states with a fictitious probability $1/4$, that is,

$$\rho_0 = \gamma_0\,\gamma_0^* = \tfrac{1}{4}(|\gamma_0\rangle\langle\gamma_0| + |\gamma_1\rangle\langle\gamma_1| + |\gamma_2\rangle\langle\gamma_2| + |\gamma_3\rangle\langle\gamma_3|)$$

and the other density operators are obtained by the GUS relation

$$\rho_i = S^i\,\rho_0\,S^{-i}, \qquad i - 1, 2, 3.$$

Hence we can apply the theory of the SRM with GUS for **mixed states** (see Proposition 6.3), which requires the evaluation of the matrices (where now K becomes L)

$$D_k = \sum_{i=0}^{L-1} \gamma_0^* \gamma_i W_L^{-ki} \tag{7.117}$$

and of their square roots $D_k^{1/2}$, where in the present case $L = 4$. Finally, one gets the the correct decision probability as

$$P_c = \mathrm{Tr}\left[\frac{1}{L}\sum_{k=0}^{L-1} D_k^{1/2}\right]^2.$$

This new approach gives exactly the same performance that we find with the previous SRM approach where the generalized GUS was not considered, with the advantage of a reduced computational complexity. In the specific case of 16-QAM, the evaluation is confined to the square roots of 4×4 matrices, instead of the square root of a 16×16 matrix.

The technique developed for the 16-QAM can be applied to constellations of any order. In particular, the constellation of 64-QAM can be decomposed into 16-PSK constellations, and the evaluation of the square roots is still confined to 4×4 matrices, instead of the square root of a 64×64 matrix.

7.11.4 Performance of the Classical Optical QAM System

The scheme of modulation and demodulation falls under the general classical scheme of Fig. 7.11. We have seen that the signal at the decision point is (see (7.55))

$$z = C_0\,U_0 + u_a + i\,u_b$$

where $C_0 = A_0 + i\,B_0$ is the transmitted symbol, U_0 is the amplitude, u_a and u_b are statistically independent Gaussian noises with null average and the same variance σ_u^2.

Fig. 7.30 Decision regions for the 16-QAM system constellation

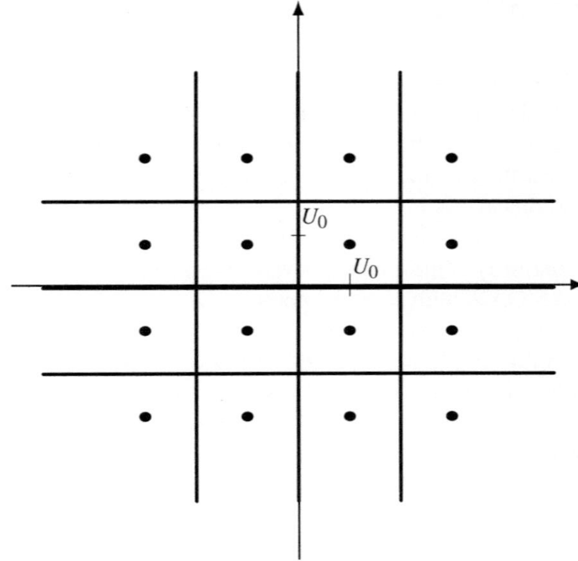

To calculate the error probability, we must choose the decision regions on the complex plane. With equiprobable symbols, the optimal decision regions are found in a straightforward way, as illustrated in Fig. 7.30 for the 16-QAM. In particular for the *inner* symbols of the constellation, the decision regions are squares with sides of length $2U_0$, centered in the corresponding symbols. Using the procedure outlined in Sect. 7.5, one obtains the following expression for the error probability [9]

$$P_{e,\text{classical}} = 1 - \left[1 - 2\left(1 - \frac{1}{L}\right) Q\left(\frac{U_0}{\sigma_u}\right) \right]^2 \tag{7.118}$$

where $Q(x)$ is the normalized complementary Gaussian distribution.

The result depends on the cardinality $K = L^2$ and on the SNR ratio $\Lambda = U_0^2/\sigma_u^2$, which can be expressed as a function of the average number of photons per symbol N_s (see (7.63)), that is,

$$\boxed{\Lambda = \frac{4N_s}{\mu_K} \quad \text{with} \quad \mu_K = \frac{2}{3}(K - 1).} \tag{7.119}$$

This result is in agreement with the conclusions of Proposition 7.2.

7.11.5 Comparison of Quantum and Classical QAM Systems

We are now able to compare the two QAM systems: The classical optical version, in which the error probability is given by (7.118) and the quantum optical version, in which P_e is evaluated numerically from (7.114) by the SRM procedure. In both cases, the parameters are the number of levels $K = L^2$ and the number of signal photons/symbol N_s.

The comparison, made in Fig. 7.31 for $K = 16$ and $K = 64$, shows the clear superiority of the quantum QAM system with respect to the classical one. For instance in 16-QAM with $N_s = 50$ photons/symbols we find $P_{e,\text{classical}} = 1.161\ 10^{-5}$, while in the quantum system $P_e = 1.546\ 10^{-9}$; in 64-QAM with $N_s = 200$ photons/symbol we find $P_{e,\text{classical}} = 2.231\ 10^{-5}$ and $P_e = 4.674\ 10^{-9}$. In both cases the improvement obtained with the quantum system is of about four decades.

In Fig. 7.32 the 16-QAM is compared to the 64-QAM as a function of the number of signal photons per bit N_R. The sensitivity at $P_e = 10^{-9}$ is $N_R = 12.783$ photons/bit in 16-QAM and $N_R = 36.035$ photons/bit in 64-QAM.

7.11.6 Comparison of CSP and SRM Evaluation

The QAM format does not have the GUS, and therefore the SRM approach does not give the minimum error probability. For this reason, we have evaluated the minimum error probability also by convex semidefinite programming, implemented in MatLab by the CVX procedure, which gives (numerically) this minimum. The results of the two approaches are shown in the following table for the 16-QAM

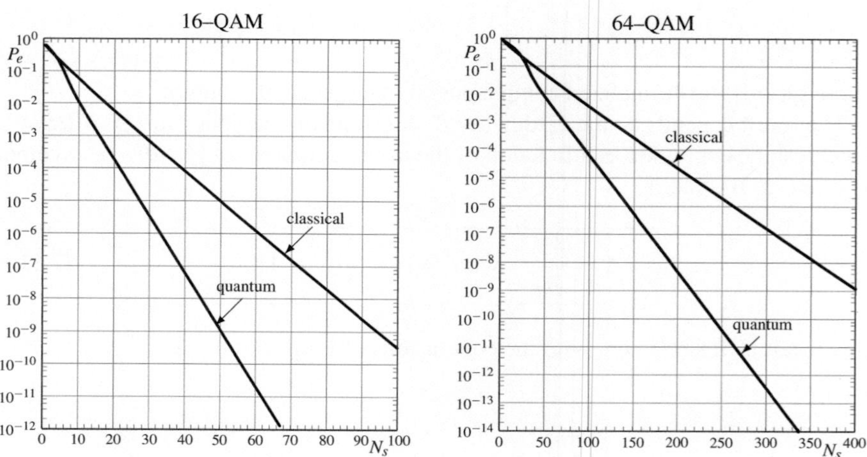

Fig. 7.31 Comparison of quantum and classical 16–QAM and 64–QAM

Fig. 7.32 Error probability of quantum QAM versus the *number of signal photons per bit N_R*

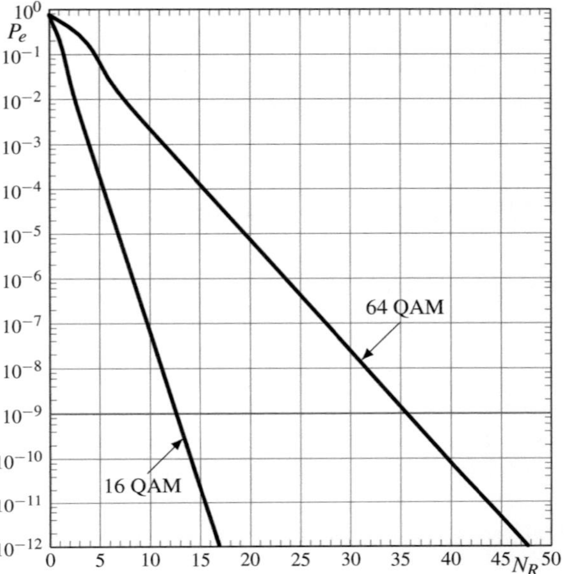

N_s	P_e with CSP	P_e with SRM
0.1	0.86919	0.884949
0.5	0.764902	0.784679
1	0.675115	0.690197
1.5	0.597074	0.608958
2.5	0.461457	0.467108
4.5	0.239407	0.240096
6.5	0.0913599	0.0910951
9.5	0.0188974	0.0188975

The two evaluations are very close, especially for large values of N_s, and cannot be distinguished in a log plot (recall that in the evaluation of P_e, decades are relevant, not decimals),

The conclusion is that the SRM approach is recommended also for the QAM (for the other formats the SRM gives the minimum of P_e).

7.12 Quantum Systems with PSK Modulation

Also PSK (phase-shift keying) modulation is one of the best known and most often used formats at radio frequency and at optical frequencies. The BPSK = 2-PSK format has been already seen in Sect. 7.10 as a special case of quantum binary systems.

Quantum K-ary PSK systems were analyzed by several authors and in particular by Kato et al. [7], using the SRM technique. In this case, the constellation of the states enjoys the *geometrically uniform symmetry* and then the SRM technique gives an optimal quantum receiver.

7.12.1 Classical and Quantum PSK Format

The constellation of the PSK modulation consists of K points uniformly distributed along a circle of the complex plane

$$\mathcal{C} = \{\Delta\, W_K^m \mid m = 0, 1, \ldots, K - 1\} \tag{7.120}$$

where the scale factor Δ is given by the radius of the circle and $W_K = \mathrm{e}^{\mathrm{i}2\pi/K}$. The constellation is illustrated in Fig. 7.33 for some values of K.

In the quantum version, the states are obtained by simply associating to every complex symbol γ of the constellation (7.120) the corresponding coherent state, which is given by

$$|\gamma_m\rangle = |\Delta\, W_K^m\rangle = \mathrm{e}^{-\frac{1}{2}\Delta^2} \sum_{n=0}^{\infty} \frac{(\Delta\, W_K^m)^n}{\sqrt{n!}}\,|n\rangle\,, \qquad m = 0, 1, \ldots, K - 1. \tag{7.121}$$

In this constellation, all the coherent states have the same number of signal photons given by

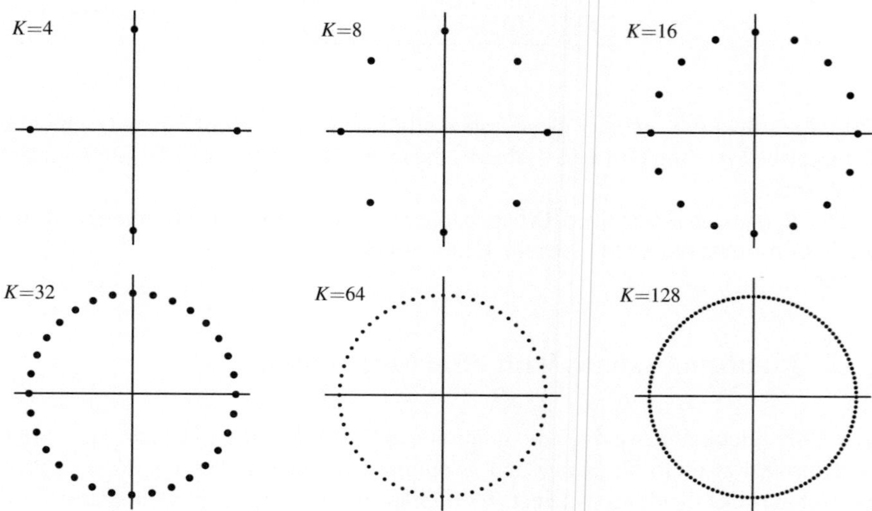

Fig. 7.33 Constellations of PSK modulation

$$N_s = \Delta^2. \tag{7.122}$$

Constellation (7.121) enjoys the GUS, that is, it satisfies the conditions (5.120a) and (5.120b) for an appropriate unitary operator S. In this case S is obtained from the *rotation operator*, which is defined as

$$R(\phi) := \exp(i\phi N)$$

where N is the number operator given by (7.1). Specifically we have

$$S = R\left(\frac{2\pi}{K}\right) = \exp\left(\frac{i2\pi}{K}N\right) = W_K^N. \tag{7.123}$$

The GUS property is verified for all constellations of Gaussian states generated by the rotation operator, as we will prove in Sect. 11.20. In Problem 7.9 we propose a specific proof for the PSK constellations where the key is that the operator $R(\phi)$ rotates a coherent state $|\alpha\rangle$ in the form (see 11.20)

$$R(\phi)\,|\alpha\rangle = |e^{i\phi}\alpha\rangle\,, \tag{7.124}$$

and the result is again a coherent state. In other words, the class of coherent states is closed under rotations.

7.12.2 *Performance of Quantum PSK Systems*

For the decision we apply the SRM, which gives an optimal result. Then, for the performance evaluation, we follow the procedure described in Sect. 7.8.3, taking into account that the PSK constellation satisfies the GUS.

The generic element p, q of Gram's matrix $G = [G_{pq}]$ is the inner product $G_{pq} = \langle\gamma_p|\gamma_q\rangle$ obtained from (7.9) with $\alpha = \Delta W_K^p$ and $\beta = \Delta W_K^q$, namely,

$$G_{pq} = \exp[-\Delta^2(1 - W_K^{q-p})]\,, \qquad p, q = 0, 1, \ldots, K - 1. \tag{7.125}$$

As predicted (by the GUS), the element p, q depends only on the difference $q - p$; and therefore Gram's matrix becomes *circulant*. The eigenvalues are obtained computing the DFT of the first row of Gram's matrix, that is,

$$\lambda_i = \sum_{k=0}^{K-1} G_{0k} W_K^{-ki} \tag{7.126}$$

and the corresponding eigenvectors are given by the columns of the DFT matrix

$$|w_i\rangle = \frac{1}{\sqrt{K}} \left[1, W_K^{-i}, W_K^{-2i}, \ldots, W_K^{-(K-1)i} \right]^{\mathrm{T}}.$$

Thus the matrices $G^{\pm\frac{1}{2}}$ are obtained from (7.99), where the element i, j is given by

$$(G^{\pm\frac{1}{2}})_{ij} = \frac{1}{K} \sum_{p=0}^{K-1} \lambda_p^{\pm\frac{1}{2}} W_K^{(j-i)p}.$$

The measurement vectors are computed as linear combination of the states according to (7.95), i.e.,

$$|\mu_i\rangle = \sum_{j=0}^{K-1} (G^{-\frac{1}{2}})_{ij} |\gamma_j\rangle.$$

Finally, the error probability with equiprobable symbols is simply

$$P_e = 1 - \frac{1}{K^2} \left(\sum_{i=0}^{K-1} \sqrt{\lambda_i} \right)^2. \tag{7.127}$$

Therefore, to calculate P_e it suffices to evaluate the eigenvalues according to (7.126) and to apply (7.127). As usual, P_e can be expressed as a function of the number of signal photons per symbol N_s, given by (7.122), and of the number of levels K. In fact, the Gram matrix G depends only on $\Delta^2 = N_s$ and K, and so is for the square root of G and subsequent relations. This conclusion is in agreement with Proposition 7.2.

7.12.3 Performance of Classical PSK Systems and Comparison

The classical optical PSK system falls into the general model of quadrature modulation (with homodyne receiver) seen in Sect. 7.5. The signal at the decision point becomes

$$z_0 = C_0 U_0 + u_a + iu_b$$

where C_0 is the transmitted symbol, $C_0 \in \mathcal{C}_0 = \{W_K^i \mid i = 1, \ldots, K\}$, u_a and u_b are independent zero-mean Gaussian components with the same variance σ_u^2. At this point, we introduce the count parameters, recalling that

$$U_0 = (2V_R V_L)H \qquad \sigma_u^2 = H V_L^2$$

and that in this case the number of signal photons contained in the received power is

$$N_R = N_s = H V_R^2.$$

So the SNR becomes

$$\Lambda = \frac{U_0^2}{\sigma_u^2} = 4N_s$$

the shape factor μ_K being unitary.

To find the error probability, we must partition the complex plane into K decision regions. Even in this case, with equiprobable symbols, such regions are easily found, as illustrated in Fig. 7.34 for the 8-PSK, and then we apply the general relation (7.60). Given the nature of the constellation, it is convenient to do a coordinate change in the probability density, from Cartesian to polar coordinates. The exact computation is only known for $K = 2$ and $K = 4$ and it yields (see (7.80))

$$K = 2 \qquad P_{e,\text{classical}} = Q(2\sqrt{N_s})$$
$$K = 4 \qquad P_{e,\text{classical}} = 1 - \left[1 - Q(\sqrt{2N_s})\right]^2 \qquad (7.128)$$

where we recall that for $K = 4$ the PSK coincides with the QAM. For $K > 4$ the exact computation is not known, and we resort to the inequality [10, Chap.10]

$$P_{e,\text{classical}} < P'_e = \exp\left(-\frac{1}{2}\frac{U_0^2}{\sigma_u^2}\sin^2\frac{\pi}{K}\right) = \exp\left(-2N_s\sin^2\frac{\pi}{K}\right). \qquad (7.129)$$

The comparison between the classical and the quantum system has been done in Fig. 7.26 for the 2-PSK. The comparison of 4-PSK and 8-PSK is done in Fig. 7.35, where the error probability is plotted as a function of the number of signal photons per symbol N_s. Even in this case we notice a striking superiority of the quantum system. For instance in 4-PSK with $N_s = 10$ photon/symbol we find $P_e = 1.030 \ 10^{-9}$ and $P_{e,\text{classical}} = 7.742 \ 10^{-6}$; in 8-PSK with $N_s = 30$ photon/symbol we find $P_e = 1.166 \ 10^{-8}$ and $P_{e,\text{classical}} = 7.742 \ 10^{-6}$. In both cases, the improvement of the quantum system is of several decades.

Fig. 7.34 Decision regions in 8-PSK

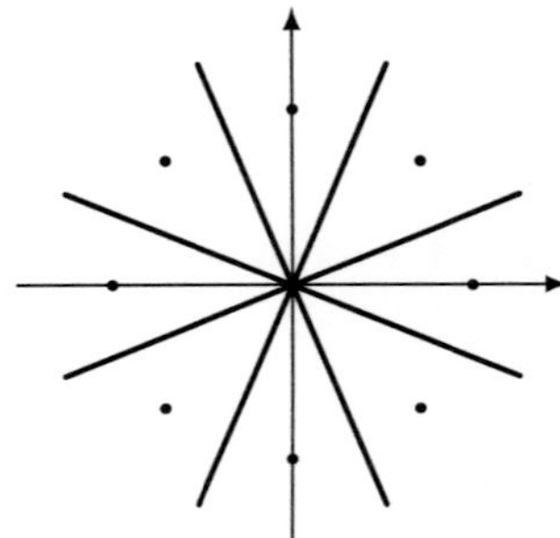

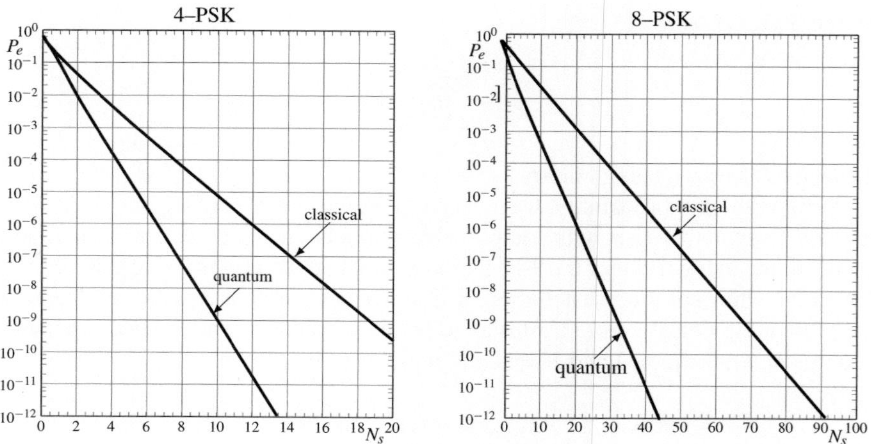

Fig. 7.35 Comparison of quantum and classical 4-PSK and 8-PSK

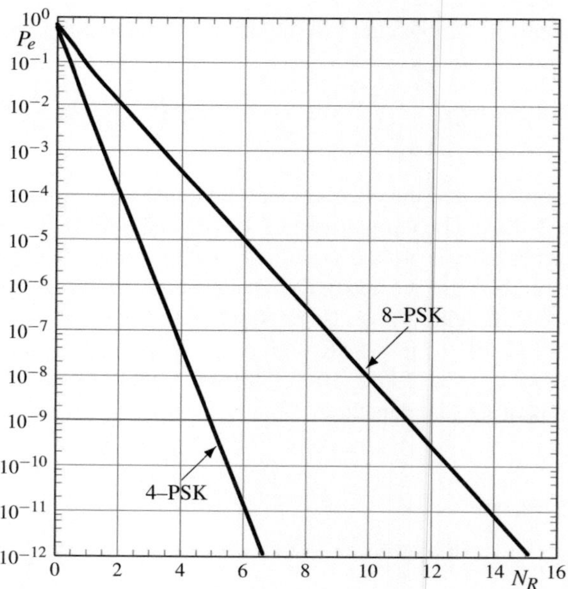

Fig. 7.36 Error probability of quantum PSK versus the *number of signal photons per bit N_R*

Figure 7.36 compares quantum 4-PSK and quantum 8-PSK as a function of the number of signal photons per bit N_R. The sensitivity at $P_e = 10^{-9}$ is $N_R = 5.001$ photons/bit in 4-PSK and $N_R = 11.402$ photons/bit in 8-PSK.

Problem 7.12 ⋆⋆⋆   Prove that the operator S defined by (7.123) is the symmetry operator of the K-PSK modulation.

Problem 7.13 ⋆   Find explicitly the formula for the error probability P_e of quantum 4-PSK system, with the target to show that P_e depends only on $\Delta^2 = N_s$.

7.13 Quantum Systems with PPM Modulation

Pulse position modulation (PPM) is widely adopted in free space optical transmission, and is a candidate for deep-space transmission, also in quantum form [11, 12].

The analysis of a quantum PPM system has been done in a famous article [4] by Yuen, Kennedy and Lax, who found the optimal elementary projectors using an algebraic method developed "ad hoc" for this kind of modulation. Here we shall propose an original method based on the SRM and on the property of quantum PPM of verifying the GUS [13]. It seems odd that such property has not been remarked by other authors; because, on one hand, it is very intuitive, and, on the other hand, it makes it possible to directly achieve the same optimal result.

7.13.1 Classical PPM Format

In the classical version, the symbol period T is subdivided into K parts with spacing $T_0 = T/K$, obtaining K "positions." Then, to the symbol $i \in \mathcal{A} = \{0, 1, \ldots, K-1\}$ the waveform is associated consisting of a rectangle in the ith position iT_0 of the symbol period

$$\gamma_i(t) = \begin{cases} \Delta & iT_0 < t < (i+1)T_0 \\ 0 & \text{elsewhere} \end{cases} \qquad i = 0, 1, \ldots, K-1 \qquad (7.130)$$

where $\Delta > 0$ is the scale factor (see Fig. 4.12). But, we will adopt the specular format

$$\gamma_i(t) = \begin{cases} \Delta & (K-1-i)T_0 < t < (K-i)T_0 \\ 0 & \text{elsewhere} \end{cases} \qquad i = 0, 1, \ldots, K-1 \quad (7.131)$$

where the ith position becomes $(K-1-i)T_0$ instead of iT_0, as illustrated in Fig. 7.37 for $K = 4$. The reason of this choice is due to the fact that it simplifies the formulation of the symmetry operator in the quantum version.

To waveforms (7.131), K binary words can be associated of length K

$$\gamma_i = [\gamma_{i,K-1}, \ldots, \gamma_{i,1}, \gamma_{i,0}], \qquad i = 0, 1, \ldots, K-1$$

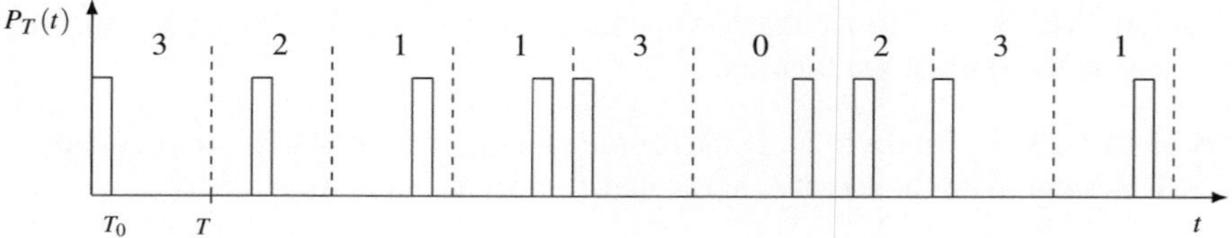

Fig. 7.37 Realization of transmitted symbols and corresponding optical power in classic 4-PPM modulation

where $\gamma_{ij} = \Delta\,\delta_{ij}$. For example, for $K = 4$ the words are

$$\gamma_0 = \begin{bmatrix} 0 & 0 & 0 & \Delta \end{bmatrix} \qquad \gamma_1 = \begin{bmatrix} 0 & 0 & \Delta & 0 \end{bmatrix}$$
$$\gamma_2 = \begin{bmatrix} 0 & \Delta & 0 & 0 \end{bmatrix} \qquad \gamma_3 = \begin{bmatrix} \Delta & 0 & 0 & 0 \end{bmatrix}.$$

As outlined in Problem 7.5, it can be verified that PPM modulation is a special case of *vector* modulation seen in Sect. 7.3.2.

7.13.2 Quantum PPM Format

We have seen in Sect. 7.3.2 that the quantum formulation of a vector modulation must be done over a composite Hilbert space, given by the tensor product $\mathcal{H} = \mathcal{H}_0 \otimes \mathcal{H}_0 \otimes \cdots \otimes \mathcal{H}_0$ of K equal Hilbert spaces $\mathcal{H}_0$, into each of which Glauber's representation must have been introduced, and the states are given by the *tensor product* of K coherent states and become **K–mode Gaussian states**.

In the specific case of PPM, the states become

$$|\gamma_i\rangle = |\gamma_{i,K-1}\rangle \otimes \cdots \otimes |\gamma_{i,1}\rangle \otimes |\gamma_{i,0}\rangle, \qquad i = 0, 1, \ldots K - 1 \tag{7.132}$$

with

$$|\gamma_{ij}\rangle = \begin{cases} |\Delta\rangle & i = j \\ |0\rangle & i \neq j \end{cases} \tag{7.132a}$$

where $|\Delta\rangle$ is a coherent state with parameter Δ, and $|0\rangle$ is the "ground state". For example, for $K = 4$ we have the four states

$$|\gamma_0\rangle = |0\rangle \otimes |0\rangle \otimes |0\rangle \otimes |\Delta\rangle \qquad |\gamma_1\rangle = |0\rangle| \otimes 0\rangle \otimes |\Delta\rangle \otimes |0\rangle$$
$$|\gamma_2\rangle = |0\rangle \otimes |\Delta\rangle \otimes |0\rangle \otimes |0\rangle \qquad |\gamma_3\rangle = |\Delta\rangle \otimes |0\rangle \otimes |0\rangle \otimes |0\rangle. \tag{7.133}$$

7.13.3 Geometrically Uniform Symmetry (GUS) of PPM

As we shall see in Sect. 11.20, the constellations of Gaussian states generated by the rotation operator $R(\phi)$ verify the GUS, also in the K–mode. This is the case of K-ary PPM. But the parameter ϕ becomes a $K \times K$ Hermitian matrix and the exponential defining $R(\phi)$ becomes difficult to handle. Here we prefer finding directly the symmetry operator S, and in Chap. 11 we will prove that S can be expressed by the K–mode rotation operator.

The symmetry operator S of the quantum PPM format (7.132) can be defined as follows: S is an operator of the composite Hilbert space $\mathcal{H}$ that causes a shift to the left by one position (modulo K) of the factors of the tensor product of the states, moving the first factor to second position, the second to third , and the Kth factor to first. For example, for $K = 4$, the action of S is as follows

$$S|\gamma_{i3}\rangle \otimes |\gamma_{i2}\rangle \otimes |\gamma_{i1}\rangle \otimes |\gamma_{i0}\rangle = |\gamma_{i2}\rangle \otimes |\gamma_{i1}\rangle \otimes |\gamma_{i0}\rangle \otimes |\gamma_{i3}\rangle. \tag{7.134}$$

Then, going on with $K = 4$, with the states of (7.113) we can see that, starting from the state $|\gamma_0\rangle = |0\rangle \otimes |0\rangle \otimes |0\rangle \otimes |\Delta\rangle$, the other states can be obtained in the following way:

$$S\,|0\rangle \otimes |0\rangle \otimes |0\rangle \otimes |\Delta\rangle = |0\rangle \otimes |0\rangle \otimes |\Delta\rangle \otimes |0\rangle = |\gamma_1\rangle$$
$$S^2|0\rangle \otimes |0\rangle \otimes |0\rangle \otimes |\Delta\rangle = |0\rangle \otimes |\Delta\rangle \otimes |0\rangle \otimes |0\rangle = |\gamma_2\rangle$$
$$S^3|0\rangle \otimes |0\rangle \otimes |0\rangle \otimes |\Delta\rangle = |\Delta\rangle \otimes |0\rangle \otimes |0\rangle \otimes |0\rangle = |\gamma_3\rangle$$

while the application of S^4 brings back to the initial state, and therefore $S^4 = I_{\mathcal{H}}$.

The considerations we just made "in words" can be translated to "formulas," but this is not so simple, because the symmetry operator S is not separable, but it operates between the K factors of the composite Hilbert space $\mathcal{H} = \mathcal{H}_0^{\otimes K}$. For the symmetry operator, the following result applies, recently demonstrated in [13].

Proposition 7.3 *Let n be the dimension of the component Hilbert spaces $\mathcal{H}_0$, and therefore $N = n^K$ is the dimension of $\mathcal{H}_0^{\otimes K}$. Then the symmetry operator of the K-ary PPM has the following expression*

$$S = \sum_{k=0}^{n-1} w_n(k) \otimes I_L \otimes w_n^*(k), \tag{7.135}$$

where $\otimes$ is Kronecker's product, $w_n(k)$ is a column vector of length n, with null elements except for one unitary element at position k, and I_L is the identity matrix of order $L = n^{K-1}$. $\square$

For example, for $n = 2$ and $K = 3$ (3-PPM) we have

$$w_2(0) = \begin{bmatrix} 1 \\ 0 \end{bmatrix}, \qquad w_2(1) = \begin{bmatrix} 0 \\ 1 \end{bmatrix}, \qquad I_4 = \begin{bmatrix} 1 & 0 & 0 & 0 \\ 0 & 1 & 0 & 0 \\ 0 & 0 & 1 & 0 \\ 0 & 0 & 0 & 1 \end{bmatrix}$$

and therefore from (7.135)

$$S = w_2(0) \otimes I_4 \otimes w_2^*(0) + w_2(1) \otimes I_4 \otimes w_2^*(1)$$

$$= \begin{bmatrix} 1 \\ 0 \end{bmatrix} \otimes \begin{bmatrix} 1 & 0 & 0 & 0 \\ 0 & 1 & 0 & 0 \\ 0 & 0 & 1 & 0 \\ 0 & 0 & 0 & 1 \end{bmatrix} \otimes [1\ 0] + \begin{bmatrix} 0 \\ 1 \end{bmatrix} \otimes \begin{bmatrix} 1 & 0 & 0 & 0 \\ 0 & 1 & 0 & 0 \\ 0 & 0 & 1 & 0 \\ 0 & 0 & 0 & 1 \end{bmatrix} \otimes [0\ 1]$$

and, developing the products, we obtain the 16×16 matrix (remember that the tensor product for matrices becomes Kronecker's product, see Sect. 2.13)

$$S = \begin{bmatrix}
1 & 0 & 0 & 0 & 0 & 0 & 0 & 0 & 0 & 0 & 0 & 0 & 0 & 0 & 0 & 0 \\
0 & 0 & 1 & 0 & 0 & 0 & 0 & 0 & 0 & 0 & 0 & 0 & 0 & 0 & 0 & 0 \\
0 & 0 & 0 & 0 & 1 & 0 & 0 & 0 & 0 & 0 & 0 & 0 & 0 & 0 & 0 & 0 \\
0 & 0 & 0 & 0 & 0 & 0 & 1 & 0 & 0 & 0 & 0 & 0 & 0 & 0 & 0 & 0 \\
0 & 0 & 0 & 0 & 0 & 0 & 0 & 0 & 1 & 0 & 0 & 0 & 0 & 0 & 0 & 0 \\
0 & 0 & 0 & 0 & 0 & 0 & 0 & 0 & 0 & 0 & 1 & 0 & 0 & 0 & 0 & 0 \\
0 & 0 & 0 & 0 & 0 & 0 & 0 & 0 & 0 & 0 & 0 & 0 & 1 & 0 & 0 & 0 \\
0 & 0 & 0 & 0 & 0 & 0 & 0 & 0 & 0 & 0 & 0 & 0 & 0 & 0 & 1 & 0 \\
0 & 1 & 0 & 0 & 0 & 0 & 0 & 0 & 0 & 0 & 0 & 0 & 0 & 0 & 0 & 0 \\
0 & 0 & 0 & 1 & 0 & 0 & 0 & 0 & 0 & 0 & 0 & 0 & 0 & 0 & 0 & 0 \\
0 & 0 & 0 & 0 & 0 & 1 & 0 & 0 & 0 & 0 & 0 & 0 & 0 & 0 & 0 & 0 \\
0 & 0 & 0 & 0 & 0 & 0 & 0 & 1 & 0 & 0 & 0 & 0 & 0 & 0 & 0 & 0 \\
0 & 0 & 0 & 0 & 0 & 0 & 0 & 0 & 0 & 1 & 0 & 0 & 0 & 0 & 0 & 0 \\
0 & 0 & 0 & 0 & 0 & 0 & 0 & 0 & 0 & 0 & 0 & 1 & 0 & 0 & 0 & 0 \\
0 & 0 & 0 & 0 & 0 & 0 & 0 & 0 & 0 & 0 & 0 & 0 & 0 & 1 & 0 & 0 \\
0 & 0 & 0 & 0 & 0 & 0 & 0 & 0 & 0 & 0 & 0 & 0 & 0 & 0 & 0 & 1
\end{bmatrix} . \qquad (7.136)$$

We leave it to the reader to check that, using the properties of Kronecker's product of Sect. 2.13, from (7.135) we obtain that the matrix S has dimensions $n^K \times n^K$, it is unitary, and has the property $S^K = I_{n^K}$.

For later use it is important to evaluate explicitly the EID of the symmetry operator S in the form (5.128)

$$S = \sum_{i=0}^{K-1} W_K^i \, Y_i \, Y_i^* \qquad (7.137)$$

where the columns of the matrices Y_k are formed by the eigenvectors corresponding to the eigenvalues $\lambda_i = W_K^i$. The explicit evaluation of such eigenvectors is long and cumbersome, and is developed in [13], using the fact that for the PPM the operator S is a *permutation matrix*. The important thing is that this evaluation can be done "analytically" for every n and K, without resorting to numeric evaluation, which could be prohibitive for high values of $N = n^K$. For example, for $K = 4$, $n = 2$, the eigenvalues 1, W_4, W_4^2, W_4^3 have multiplicities respectively 6, 3, 4, and 3, and the corresponding eigenvectors form the matrices

$$
\begin{array}{cccc}
Y_0 & Y_1 & Y_2 & Y_3
\end{array}
$$

$$
\begin{bmatrix}
1 & 0 & 0 & 0 & 0 & 0 \\
0 & 0 & 0 & \frac{1}{2} & 0 & 0 \\
0 & 0 & 0 & \frac{1}{2} & 0 & 0 \\
0 & 0 & 0 & 0 & \frac{1}{2} & 0 \\
0 & 0 & 0 & \frac{1}{2} & 0 & 0 \\
0 & 0 & \frac{1}{\sqrt{2}} & 0 & 0 & 0 \\
0 & 0 & 0 & 0 & \frac{1}{2} & 0 \\
0 & 0 & 0 & 0 & 0 & \frac{1}{2} \\
0 & 0 & 0 & \frac{1}{2} & 0 & 0 \\
0 & 0 & 0 & 0 & \frac{1}{2} & 0 \\
0 & 0 & \frac{1}{\sqrt{2}} & 0 & 0 & 0 \\
0 & 0 & 0 & 0 & 0 & \frac{1}{2} \\
0 & 0 & 0 & 0 & \frac{1}{2} & 0 \\
0 & 0 & 0 & 0 & 0 & \frac{1}{2} \\
0 & 0 & 0 & 0 & 0 & \frac{1}{2} \\
0 & 1 & 0 & 0 & 0 & 0
\end{bmatrix}
\begin{bmatrix}
0 & 0 & 0 \\
-\frac{i}{2} & 0 & 0 \\
-\frac{1}{2} & 0 & 0 \\
0 & -\frac{i}{2} & 0 \\
\frac{i}{2} & 0 & 0 \\
0 & 0 & 0 \\
0 & -\frac{1}{2} & 0 \\
0 & 0 & -\frac{i}{2} \\
\frac{1}{2} & 0 & 0 \\
0 & \frac{1}{2} & 0 \\
0 & 0 & 0 \\
0 & 0 & \frac{1}{2} \\
0 & \frac{i}{2} & 0 \\
0 & 0 & \frac{i}{2} \\
0 & 0 & -\frac{1}{2} \\
0 & 0 & 0
\end{bmatrix}
\begin{bmatrix}
0 & 0 & 0 & 0 \\
0 & -\frac{1}{2} & 0 & 0 \\
0 & \frac{1}{2} & 0 & 0 \\
0 & 0 & -\frac{1}{2} & 0 \\
0 & -\frac{1}{2} & 0 & 0 \\
-\frac{1}{\sqrt{2}} & 0 & 0 & 0 \\
0 & 0 & \frac{1}{2} & 0 \\
0 & 0 & 0 & -\frac{1}{2} \\
0 & \frac{1}{2} & 0 & 0 \\
0 & 0 & \frac{1}{2} & 0 \\
\frac{1}{\sqrt{2}} & 0 & 0 & 0 \\
0 & 0 & 0 & \frac{1}{2} \\
0 & 0 & -\frac{1}{2} & 0 \\
0 & 0 & 0 & -\frac{1}{2} \\
0 & 0 & 0 & \frac{1}{2} \\
0 & 0 & 0 & 0
\end{bmatrix}
\begin{bmatrix}
0 & 0 & 0 \\
\frac{i}{2} & 0 & 0 \\
-\frac{1}{2} & 0 & 0 \\
0 & \frac{i}{2} & 0 \\
-\frac{i}{2} & 0 & 0 \\
0 & 0 & 0 \\
0 & -\frac{1}{2} & 0 \\
0 & 0 & \frac{i}{2} \\
\frac{1}{2} & 0 & 0 \\
0 & \frac{1}{2} & 0 \\
0 & 0 & 0 \\
0 & 0 & \frac{1}{2} \\
0 & -\frac{i}{2} & 0 \\
0 & 0 & -\frac{i}{2} \\
0 & 0 & -\frac{1}{2} \\
0 & 0 & 0
\end{bmatrix} .
$$

7.13.4 Performance of Quantum PPM Systems

Considering that the PPM states have the GUS, the SRM gives an optimal detection, then, in performance evaluation, the same results will have to be found as those of Yuen et al. [4] with a different methodology.

Applying the method summarized in Sect. 7.7.1, about the SRM detection in the presence of GUS, the performance evaluation is articulated as follows. Gram's matrix G has as element i, j the inner product $G_{ij} = \langle \gamma_i | \gamma_j \rangle$, where now the states $|\gamma_i\rangle$ are composite. We recall that the inner product of two states, each generated by the tensor product of K component states, is given by the product of the K inner products of the component states, that is,

$$
\langle \gamma_i | \gamma_j \rangle = \langle \gamma_{i0} | \gamma_{j0} \rangle \langle \gamma_{i1} | \gamma_{j1} \rangle \cdots \langle \gamma_{iK-1} | \gamma_{jK-1} \rangle. \tag{7.138}
$$

Then, from (7.9), we get

$$\langle \gamma_i | \gamma_j \rangle = \begin{cases} 1 & i = j \\ e^{-|\Delta|^2} & i \neq j. \end{cases} \tag{7.139}$$

For example, in the case $K = 4$ the inner product $\langle \gamma_0 | \gamma_2 \rangle$ results in

$$\langle \gamma_0 | \gamma_2 \rangle = \langle \Delta | 0 \rangle \, \langle 0 | 0 \rangle \, \langle 0 | \Delta \rangle \, \langle 0 | 0 \rangle$$

$$= e^{-\frac{1}{2}|\Delta|^2} \, 1 \quad e^{-\frac{1}{2}|\Delta|^2} \, 1 = e^{-|\Delta|^2}.$$

We observe that the same energy E is associated to all symbols, and, according to (7.132), to each composite state the same signal photons are associated, given by

$$N_s = |\Delta|^2 = \text{number of signal photons/symbol}. \tag{7.140}$$

Therefore Gram's matrix becomes

$$G = \begin{bmatrix} 1 & |X|^2 & \dots & |X|^2 \\ |X|^2 & 1 & \dots & |X|^2 \\ \vdots & & \ddots & \\ |X|^2 & |X|^2 & \dots & 1 \end{bmatrix}$$

where $|X|^2$ is the quadratic superposition degree of the component states ($X = \langle \gamma_i \gamma_j \rangle$, $i \neq j$), given by

$$|X|^2 = |\langle \Delta 0 |^2 = e^{-|\Delta|^2} = e^{-N_s}).$$

Notice that G is a *circulant* matrix, as a consequence of the GUS.

Considering the GUS, the eigenvalues of G are given by the DFT of the first row $[1, |X|^2, \dots, |X|^2]$, and therefore

$$\lambda_i = \sum_{k=0}^{K-1} G_{0k} W_K^{-ki} = 1 + |X|^2 \sum_{k=1}^{K-1} W_K^{-ki}.$$

Recalling the orthogonality condition

$$\sum_{k=0}^{K-1} W_K^{-ki} = \begin{cases} K & i = 0 \\ 0 & i \neq 0 \end{cases} \tag{7.141}$$

we have

$$\lambda_i = \begin{cases} 1 + (K - 1)|X|^2 & i = 0 \\ 1 - |X|^2 & i = 1, \dots, K - 1. \end{cases}$$

The square roots of G become

$$G^{\pm\frac{1}{2}} = \sum_{i=0}^{K-1} \lambda_i^{\pm\frac{1}{2}} |w_i\rangle\langle w_i|.$$

The transition probabilities are computed from (7.93) and result in

$$p_c(i|j) = \frac{1}{K^2}\left|\sum_{p=0}^{K-1} \lambda_p^{1/2} W_K^{-p(i-j)}\right|^2 = \frac{1}{K^2}\left|\lambda_0^{1/2} + \lambda_1^{1/2}\sum_{p=1}^{K-1} W_K^{-p(i-j)}\right|^2$$

and, from (7.141),

$$p_c(j|i) = \begin{cases} K^{-2}\left(\lambda_0^{1/2} - \lambda_1^{1/2}\right)^2 & i \neq j \\ K^{-2}\left(\lambda_0^{1/2} + (K-1)\lambda_1^{1/2}\right)^2 & i = j \end{cases} \tag{7.142}$$

where

$$\lambda_0 = 1 + (K-1)|X|^2, \qquad \lambda_1 = 1 - |X|^2, \qquad \text{with} \quad |X|^2 = e^{-N_s}.$$

The error probability is computed from (7.94) and becomes

$$\boxed{P_e = 1 - \frac{1}{K^2}\left(\sqrt{1 + (K-1)|X|^2} + (K-1)\sqrt{1 - |X|^2}\right)^2} \tag{7.143}$$

in perfect agreement with the results of [4].

Finally, the measurement vectors are obtained from (7.95), namely,

$$|\mu_i\rangle = \sum_{j=0}^{K-1} a_{ij}|\gamma_j\rangle \quad , \qquad a_{ij} = \begin{cases} K^{-2}\left(\lambda_0^{-1/2} - \lambda_1^{-1/2}\right)^2 & i \neq j \\ K^{-2}\left(\lambda_0^{-1/2} + (K-1)\lambda_1^{-1/2}\right)^2 & i = j. \end{cases}$$

7.13.5 *Performance of Classical PPM Systems*

We use the notations

- A_0 transmitted word,
- B_0 received word,
- $\widehat{A}_0$ decided word.

Let us consider, for simplicity, the case $K = 4$ with the standard (non specular) format and with a unitary scale factor ($\Delta = 1$), in which the possible transmitted words A_0 are

$$\gamma_0 = \begin{bmatrix} 1 & 0 & 0 & 0 \end{bmatrix}, \qquad \gamma_1 = \begin{bmatrix} 0 & 1 & 0 & 0 \end{bmatrix}$$
$$\gamma_2 = \begin{bmatrix} 0 & 0 & 1 & 0 \end{bmatrix}, \qquad \gamma_3 = \begin{bmatrix} 0 & 0 & 0 & 1 \end{bmatrix}. \tag{7.144}$$

With a photon counter, the symbol 0 is always received correctly, whereas the symbol 1 may be received as 0, with an error probability e^{-N_s}. Then we have five possible received words: the four correct words (7.144) and the wrong word $\begin{bmatrix} 0 & 0 & 0 & 0 \end{bmatrix}$, and we have to decide to which correct word the wrong word should be associated. The optimum criterion (with equiprobable symbols) is to associate the wrong word $\begin{bmatrix} 0 & 0 & 0 & 0 \end{bmatrix}$ to whatever correct word, for example to γ_0 (Fig. 7.38). Then the decision criterion becomes

$$\widehat{A}_0 = \gamma_0 \quad \text{if} \quad B_0 = \begin{bmatrix} 1 & 0 & 0 & 0 \end{bmatrix} \quad \text{or} \quad B_0 = \begin{bmatrix} 0 & 0 & 0 & 0 \end{bmatrix}$$
$$\widehat{A}_0 = \gamma_1 \quad \text{if} \quad B_0 = \begin{bmatrix} 0 & 1 & 0 & 0 \end{bmatrix}$$
$$\widehat{A}_0 = \gamma_2 \quad \text{if} \quad B_0 = \begin{bmatrix} 0 & 0 & 1 & 0 \end{bmatrix}$$
$$\widehat{A}_0 = \gamma_3 \quad \text{if} \quad B_0 = \begin{bmatrix} 0 & 0 & 0 & 1 \end{bmatrix}$$

and we can get an error only in the last three cases, each with probability e^{-N_s}. Thus,

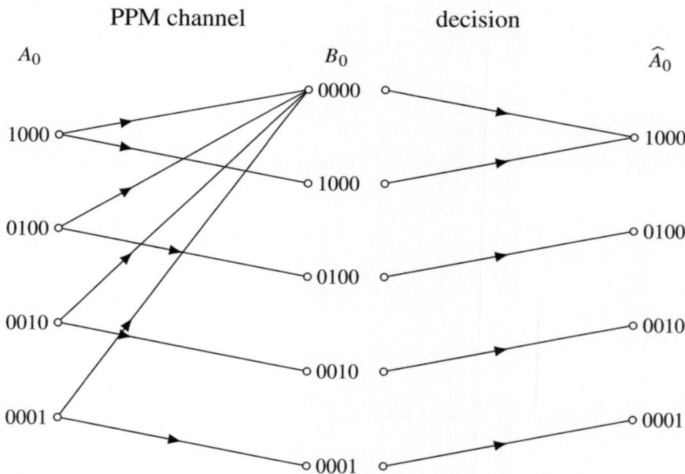

Fig. 7.38 Channel and decision criterion of a classical 4-PPM. A_0 is the transmitted word, B_0 the received word, and $\widehat{A}_0$ the decided word

$$P_{e,\text{classical}} = \frac{1}{4}[P_e(\gamma_1) + P_e(\gamma_2) + P_e(\gamma_3) + P_e(\gamma_4)]$$

$$= \frac{1}{4}[0 + e^{-N_s} + e^{-N_s} + e^{-N_s}] = \frac{3}{4}e^{-N_s}.$$

The general result is

$$P_{e,\text{classical}} = \frac{K-1}{K}e^{-N_s}. \tag{7.145}$$

7.13.6 Comparison in the Binary Case

For binary quantum PPM, from (7.143) we get

$$P_c = \frac{1}{4}\left[\sqrt{1 - |X|^2} + \sqrt{1 + |X|^2}\right]^2 = \frac{1}{2}\left[1 + \sqrt{1 - |X|^4}\right]$$

where $|X|^4 = e^{-2N_s} = e^{-2N_R}$, with $N_R = N_s$ the number of signal photons per bit. The error probability is therefore

$$P_e = \frac{1}{2}\left[1 - \sqrt{1 - e^{-2N_R}}\right], \tag{7.146}$$

the same result found for the OOK format (see (7.106)).

Instead, in the classical case, from (7.145) we have $P_{e,\text{classical}} = \frac{1}{2}e^{-N_R}$. The comparison of these results is shown in Fig. 7.39.

7.13.7 Comparison in the K-ary Case

In the quantum case, the error probability is given by (7.143), which can be rewritten in the form

$$P_e = \frac{K-1}{K^2}\left[K - (K-2)(1 - |X|^2) + 2\sqrt{(1 - |X|^2)(1 + (K-1)|X|^2)}\right]$$

where the superposition degree $|X|^2$ can be expressed as a function of the number of signal photons per symbol N_s, or of the number of signal photons per bit N_R

$$|X|^2 = e^{-N_s} = e^{-N_R \log_2 K}.$$

In the classical case the error probability is given by (7.145).

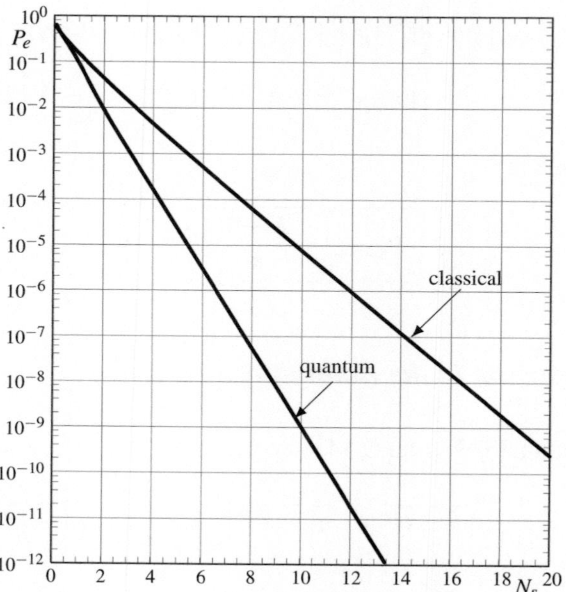

Fig. 7.39 Comparison of quantum and classical 2-PPM

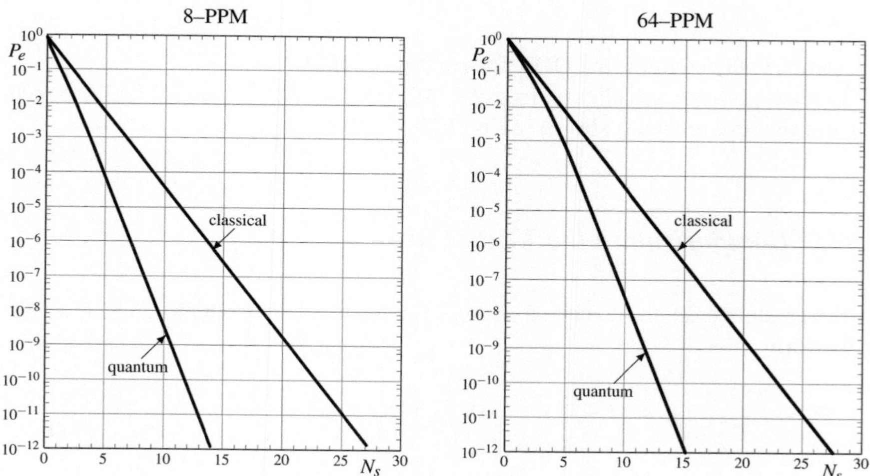

Fig. 7.40 Comparison between quantum and classical 8-PPM and 64-PPM in terms of number of signal photons per symbol N_s

The comparison between the two systems is illustrated for the 8-PPM and 64-PPM in Fig. 7.40 as a function of the number of signal photons per symbol N_s. Even in this case we notice a striking superiority of the quantum system.

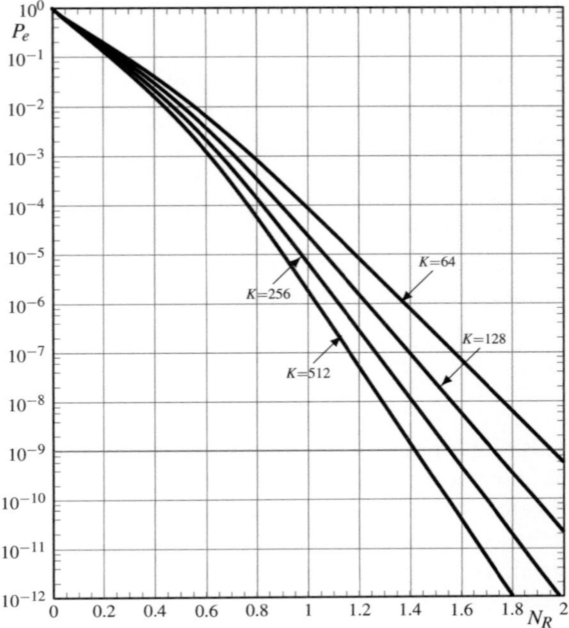

Fig. 7.41 Error probability of quantum PPM as a function of the *number of signal photons per bit* N_R

For instance in 8-PPM with $N_s = 10$ photon/symbol we find $P_e = 3.660 \ 10^{-9}$ and $P_{e,\text{classical}} = 3.972 \ 10^{-5}$; in 64–PPM with $N_s = 10$ photon/symbol we find $P_e = 3.421 \ 10^{-9}$ and $P_{e,\text{classical}} = 4.469 \ 10^{-5}$. In both cases the improvement of the quantum system is of several decades.

In Fig. 7.41 the error probability of the quantum PPM is plotted as a function of the number of signal photons per bit N_R for four values of K. We realize that quantum PPM receivers have an extraordinary sensitivity, specifically

2-PPM	$N_R = 9.66849$	photons/bit
4-PPM	$N_R = 5.10889$	photons/bit
8-PPM	$N_R = 3.54713$	photons/bit
16-PPM	$N_R = 2.75561$	photons/bit
32-PPM	$N_R = 2.27708$	photons/bit
64-PPM	$N_R = 1.95665$	photons/bit
128-PPM	$N_R = 1.72722$	photons/bit
256-PPM	$N_R = 1.55486$	photons/bit
512-PPM	$N_R = 1.42072$	photons/bit
1024-PPM	$N_R = 1.31332$	photons/bit

7.14 Overview of Squeezed States

Up to now we have considered Quantum Communications based on coherent states. In these last two sections, we consider the promising possibility to use squeezed states.

We have seen that a coherent state $|\alpha\rangle$ is completely determined by a complex parameter α. A squeezed state, and more precisely, a squeezed-displaced state, say $|z, \alpha\rangle$ may be seen as a generalization of a coherent state, because of the dependence on two complex parameters, the *displacement* α and the *squeeze factor* $z = re^{i\theta}$. In particular, setting z to zero, the squeezed-displaced state gives back the coherent state

$$|0, \alpha\rangle = |\alpha\rangle \in \mathcal{G}. \tag{7.147}$$

This simple property allows us to say that, using squeezed states in quantum communications with appropriate parameters, the performance cannot be worse than with coherent states. As we shall see, the squeeze factor z allows us to control the photon statistic in such a way that, by choosing z in an appropriate range, we get a considerable improvement of the system performance. This opportunity has been long recognized [14].

The theory of **squeezed-displaced states** will be formulated in Sect. 11.15 in the context of continuous variables, where it is shown that they represent the **most general form of Gaussian states**. In this section, we give the essential properties of squeezed states that are needed for Quantum Communications.

We shall use the following notations

- $|z, \alpha\rangle$: squeezed-displaced state
- $|0, \alpha\rangle = |\alpha\rangle$: coherent state
- $|z, 0\rangle$: squeezed state or squeezed vacuum state.

7.14.1 Definition and Properties of Squeezed-Displaced States

Squeezed states live in the same Hilbert space as coherent states, that is an infinite dimensional Hilbert space where the Fock basis has been introduced. Squeezed-displaced states are the result of two distinct operations applied to the vacuum state: A squeezing and a displacement. They may be specified by the Fock expansion, whose Fourier coefficients result in

$$|z, \alpha\rangle_n = \frac{\sqrt{n!}}{\mu} \left(\frac{\beta}{\mu}\right)^n \mathcal{H}_n\left(\frac{\mu\nu}{\beta^2}\right) \exp\left(-\frac{1}{2}|\beta|^2 - \beta^2\frac{\nu^*}{2\mu}\right). \tag{7.148}$$

where $\mathcal{H}_n(x)$ are the polynomials (of degree $\lfloor n/2 \rfloor$)

$$\mathcal{H}_n(x) := \sum_{j=0}^{\lfloor n/2 \rfloor} \frac{1}{(n-2j)!j!}\, x^j. \tag{7.149}$$

and

$$\mu = \cosh r\,, \qquad \nu = \sinh r \exp(\mathrm{i}\,\theta)\,, \qquad \beta = \mu\alpha - \nu\alpha^*. \tag{7.149a}$$

The deduction of (7.148) from the operations of squeezing and displaciment is made in Sect. 11.15.[6]

The class of squeezed-displaced states has two special subclasses, which are obtained for $z = 0$ (absence of squeezing) and for $\alpha = 0$ (absence of displacement). In the first case we have **coherent states** with coefficients (see (7.2))

$$|0, \alpha\rangle_n = \mathrm{e}^{-\frac{1}{2}|\alpha|^2}\, \frac{\alpha^n}{\sqrt{n!}} \tag{7.150}$$

In the second case we have **squeezed vacuum states** with coefficients

$$|z, 0\rangle_n = \sqrt{\mathrm{sech}\, r}\, \sum_{n=0}^{\infty} \frac{\sqrt{(2n)!}}{2^n n!}\lambda^n |2n\rangle \qquad \lambda = \tanh r\, \mathrm{e}^{\mathrm{i}\theta} \tag{7.151}$$

that is, the state $|z, 0\rangle$ is given by a linear combination of *even photon number states*, which means that the probability that the state contains an odd number of photons is zero.

7.14.2 Statistics of Squeezed-Displaced States

The probability distribution of the number of photons in a squeezed state is obtained by squaring the Fourier coefficients (7.148), that is,

$$p_n(i) := \mathrm{P}[n = i] = ||z, \alpha\rangle_i|^2 \tag{7.152}$$

This distribution is illustrated in Fig. 7.42 for $\alpha = 3$ and four values of r with $\theta = 0$. Note that for $r = 0$, absence of squeezing, $p_n(i)$ becomes a Poisson distribution, but in general it may be far form the Poisson shape, and sometimes this is classified as *sub-Poissonian statistic*. Under certain conditions, this statistics may be controlled acting on the squeeze factor [14].

The mean photon number in a squeezed-displaced state is given by [4]

---

[6] The Fock expansion of squeezed-displaced states was first established by [4], who expressed the Fourier coefficients in terms of Hermite polynomials $H_n(x)$. The equivalent formulation in terms of the polynomials $\mathcal{H}_n(x)$ appear to be more direct.

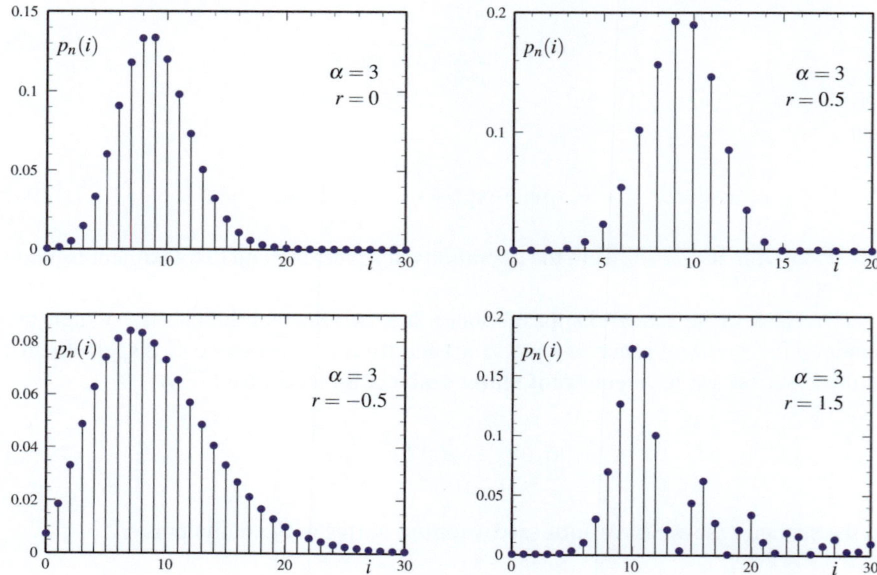

Fig. 7.42 The probability distribution of photon number in squeezed-displaced states, $p_n(i) = \mathrm{P}[n = i \mid |(\alpha, r)\rangle]$, for $\alpha = 3$ and different values of r

$$\bar{n}_{|z,\alpha\rangle} = |\alpha|^2 + \sinh^2 r \tag{7.153}$$

and the variance of the photon number is given by [15]

$$\sigma^2_{n_{|z,\alpha\rangle}} = |\alpha|^2 \left[\mathrm{e}^{-2r} \cos^2 \theta + \mathrm{e}^{2r} \sin^2 \theta \right] + \frac{1}{2} \sinh^2 2r. \tag{7.154}$$

Clearly these parameters, illustrated in Fig. 7.43, confirm the non-Poissonian statistic, because the mean and the variance are different.

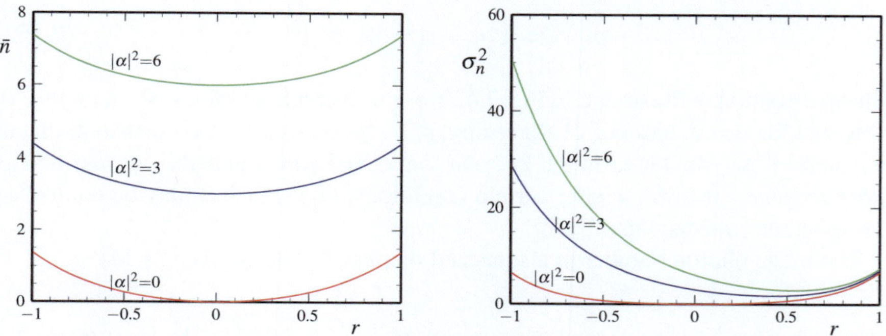

Fig. 7.43 Mean photon number $\bar{n}$ and variance σ_n^2 in squeezed-dispaced states versus the squeeze factor r for three values of $|\alpha|^2$

7.14.3 Degree of Superposition of Squeezed-Displaced States

The most important parameter for Quantum Communications is given by the inner product, which must be studied in detail.

Proposition 7.4 *The inner product of two squeezed-displaced states was evaluated by Yuen [4, Eq. 3.25] and reads*

$$\langle z_1, \alpha_1 | z_0, \alpha_0 \rangle = A^{-\frac{1}{2}} \exp\left[-\frac{A\left(|\beta_1|^2 + |\beta_0|^2\right) - 2\beta_1\beta_0^* + B\,\beta_1^{*2} - B^*\beta_0^2}{2A} \right]$$

(7.155)

where

$$
\begin{aligned}
z_i &= r_i e^{i\theta_i}, \qquad i = 0, 1 \\
\mu_i &= \cosh(r_i), \qquad \nu_i = \sinh(r_i)e^{i\theta_i} \\
\beta_i &= \mu_i\alpha_i - \nu_i\alpha_i^* \\
A &= \mu_0\mu_1^* - \nu_0\nu_1^*, \qquad B = \nu_0\mu_1 - \mu_0\nu_1
\end{aligned}
$$

(7.155a)

To study this complicate expression, we begin with remarking the dependence on the displacements α_i and on the squeeze factors z_i: The parameters β_0 and β_1 depend on both, while *all the other parameters depend only on the squeeze factors*. We can write (7.155) as

$$\langle z_1, \alpha_1 | z_0, \alpha_0 \rangle = A^{-\frac{1}{2}} \exp\left[-\sum_{i=0}^{1}\sum_{j=0}^{1}\left(a_{ij}\,\alpha_i\alpha_j + b_{ij}\,\alpha_i\alpha_j^* + d_{ij}\,\alpha_i^*\alpha_j^* \right) \right]$$

having at the exponent a bi-quadratic structure in α_0, α_1, α_0^*, and α_1^*, whose coefficients a_{ij}, b_{ij}, d_{ij} depend *only on the squeeze factors*.

For $\alpha_0 = \alpha_1 = 0$ (absence of displacement), (7.155) gives

$$
\begin{aligned}
\langle z_1, 0 | z_0, 0 \rangle &= A^{-\frac{1}{2}} = (\mu_0\mu_1^* - \nu_0\nu_1^*)^{-\frac{1}{2}} \\
&= (\cosh r_0 \cosh r_1 - \sinh r_0 \sinh r_1 e^{i(\theta_0 - \theta_1)})^{-\frac{1}{2}}
\end{aligned}
$$

which is in agreement with the expression obtained in [16] for the inner product of two squeezed vacuum states

$$\langle z_1, 0 | z_0, 0 \rangle = \sqrt{\operatorname{sech} r \,\operatorname{sech} r_0} \,\Big/\, \sqrt{1 - e^{i(\theta_0 - \theta)}\tanh r \tanh r_0}.$$

(7.156)

For $z_0 = z_1 = 0$ (absence of squeezing) we get $A = 1$, $B = 0$, $\beta_i = \alpha_i$, and then

$$\langle 0, \alpha_1 | 0, \alpha_0 \rangle = \exp\left[-\frac{1}{2}\left(|\alpha_1|^2 + |\alpha_0|^2 - 2\alpha_1\alpha_0^* \right) \right]$$

which is the formula we got for the inner product of two coherent states (see (7.9)).

7.14.4 Squeezed-Displaced States as Gaussian States

As said above, squeezed-displaced states are the most general Gaussian states. The state $|z, \alpha\rangle$ depends on two complex parameters

$$z = r\,\mathrm{e}^{\mathrm{i}\theta}, \qquad \alpha = \Delta\,\mathrm{e}^{\mathrm{i}\varepsilon}. \tag{7.157}$$

We examine in detail the Wigner function $W(x, y)$ of the state (7.157), which is completely determined by the mean value and by the covariance matrix (see Sect. 7.2.5). Now, in $|z, \alpha\rangle$ the squeezing part does not give any contribution to the mean value, so we have

$$\begin{bmatrix} \overline{q} \\ \overline{p} \end{bmatrix} = \begin{bmatrix} \Re\alpha \\ \Im\alpha \end{bmatrix} = \begin{bmatrix} \Delta\cos\varepsilon \\ \Delta\sin\varepsilon \end{bmatrix}. \tag{7.158}$$

On the other hand the covariance matrix of the displacement component is the identity, so that the covariance matrix depends only on the squeeze factor as

$$\begin{aligned} V &= \begin{bmatrix} V_{11} & V_{12} \\ V_{12} & V_{22} \end{bmatrix} \\ &= \begin{bmatrix} \cosh^2 r + \sinh^2 r + \cos\theta\,\sinh 2r & \sin\theta\,\sinh 2r \\ \sin\theta\,\sinh 2r & \cosh^2 r + \sinh^2 r - \cos\theta\,\sinh 2r \end{bmatrix}. \end{aligned} \tag{7.159}$$

Considering that $\det V = 1$, the Wigner function results in

$$W(x, y) = \frac{1}{2\pi}\exp\left\{ -\frac{1}{2}\left[V_{22}(x - \overline{q})^2 + V_{11}(y - \overline{p})^2 - 2V_{12}(x - \overline{q})(y - \overline{p}) \right] \right\}. \tag{7.160}$$

A convenient representation of $W(x, y)$ in the x, y plane is given by a *contour level*, which represents the curve given by the relation $W(x, y) = L$, with $L > 0$ real. In general, these curves are *tilted* ellipses as shown in Fig. 7.44. The ellipses have the common center given by the displacement α, and the main axis is tilted by the angle $\frac{1}{2}\theta$. The lengths of the main axis and of the minor axis are proportional to e^{2r} and to e^{-2r}, respectively, and so they are independent of the squeeze phase θ.

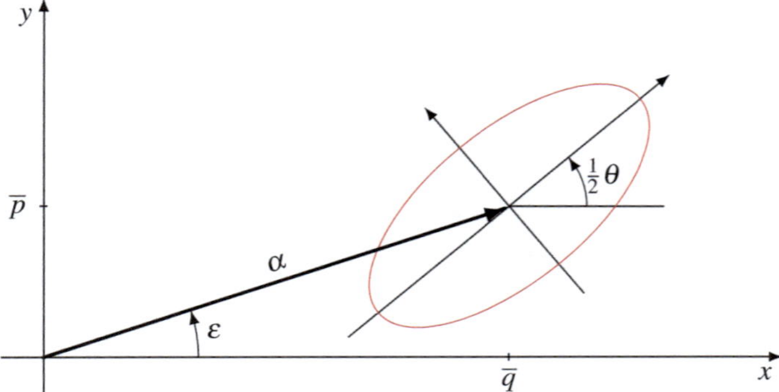

Fig. 7.44 Contour level of the Wigner function W(x,y) (in *red*) of a squeezed-displaced state $|z, \alpha\rangle$ with $z = r\,e^{i\theta}$ and $\alpha = \Delta\,e^{i\varepsilon}$. The mean vector $(\bar{q}, \bar{p}) = (\Delta\cos\varepsilon, \Delta\sin\varepsilon)$ gives the displacement amount and determines the center of the elliptic countour. The main axis of the *ellipse* is tilted with respect to the x axis of the angle $\frac{1}{2}\theta_0$

7.14.5 *Constellations of Squeezed-Displaced States with GUS*

In Sect. 11.20, we will prove that the application of the rotation operator $R(\phi)$ to a squeezed-displaced state $|z, \alpha\rangle$, with $z = r\,e^{i\theta}$ and $\alpha = \Delta e^{i\varepsilon}$, gives back the new squeezed-displaced state

$$R(\phi)|z, \alpha\rangle = |ze^{i\,2\phi}, \alpha e^{i\,\phi}\rangle = |re^{i\,(2\phi+\theta)}, \Delta e^{i\,(\phi+\varepsilon)}\rangle$$

that is, with the modification of squeeze factor $z \to ze^{i\,2\phi}$ and of the displacement $\alpha \to \alpha e^{i\,\phi}$. In other words, the class of squeezed-displaced states is closed under rotations.

The above properties allow us to construct K-ary PSK constellations having the GUS, using as symmetry operator $S = R(2\pi/K)$. If $|z_0, \alpha_0\rangle$ is a reference squeezed-displaced state, the constellation has the form

$$\mathcal{S} = \{S^k\,|z_0, \alpha_0\rangle = |z_0 e^{i\,2k\,2\pi/K}\,,\ \alpha_0 e^{i\,k\,2\pi/K}\rangle\,,\ k = 0, 1, \ldots, K - 1\}. \quad (7.161)$$

Figure 7.45 shows two 8-PSK constellations having the GUS with coherent states and squeezed-displaced states. The circles and the ellipses around the states represent the contour levels of the Wigner function of each state; the eccentricity of the ellipse depends only the squeeze factor r.

As seen for coherent states the GUS will allow us to find an optimal detection (minimum error probability) with the SRM approach.

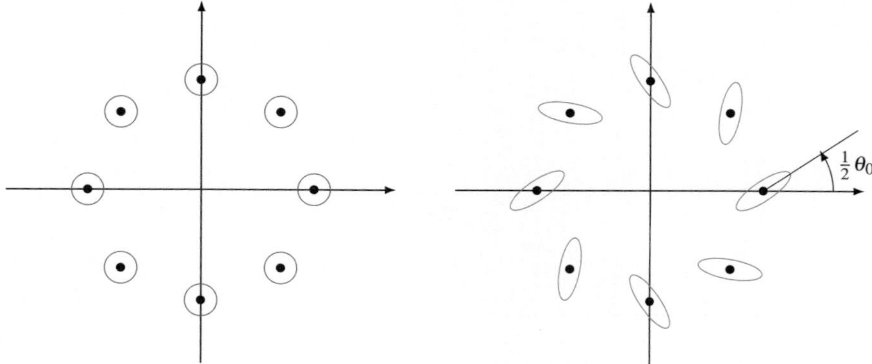

Fig. 7.45 8-PSK constellations of coherent states (*left*) and of squeezed-displaced states (*right*) with the GUS. In both constellations the reference state has a real and positive displacement ($\alpha = \Delta$, $\varepsilon = 0$). The squeeze phase of the reference state is θ_0 and the corresponding *ellipse* is tilted of $\frac{1}{2}\theta_0$

7.15 Quantum Communications with Squeezed States

In this section, we evaluate the performance (error probability) in quantum communications systems where the information carrier is given by squeezed-displaced states instead of coherent states. We consider only PSK communications systems[7]; and therefore it is natural to choose constellations having the GUS. Then, for a given modulation order K, we have to choose a reference state of the constellation, $|z_0, \alpha_0\rangle$, because the other states are generated through the rotation operator, as indicated in (7.161).

In the choice of the reference state $|z_0, \alpha_0\rangle$, where $z_0 = r_0\, e^{i\theta_0}$, without restriction we can assume α_0 real and positive. This parameter, together with r_0, determines the average number of photons contained in the state, which is given by

$$\overline{n}_{|z_0,\alpha_0\rangle} = |\alpha_0|^2 + \sinh^2 r_0. \tag{7.162}$$

This number is very important because, in a PSK constellation with equally likely symbols, it also gives the average number of signal photons per symbol N_s. Now, choosing r_0 as a parameter that quantify the squeezing amount, it remains to choose the squeeze phase θ_0 and this will be done by taking θ_0 that minimizes the error probability.

Considering that the performance of PSK essentially depends on the quadratic superposition between the states of the constellation

$$|X|^2 = |\langle r_0 e^{i\theta_0}, \alpha_0 | r_1 e^{i\theta_1}, \alpha_1\rangle|^2$$

---

[7] Recently also the PPM with squeezed-displaced states has been considered [17].

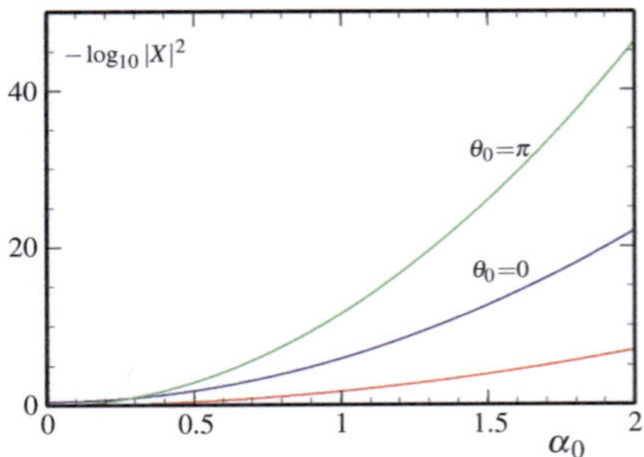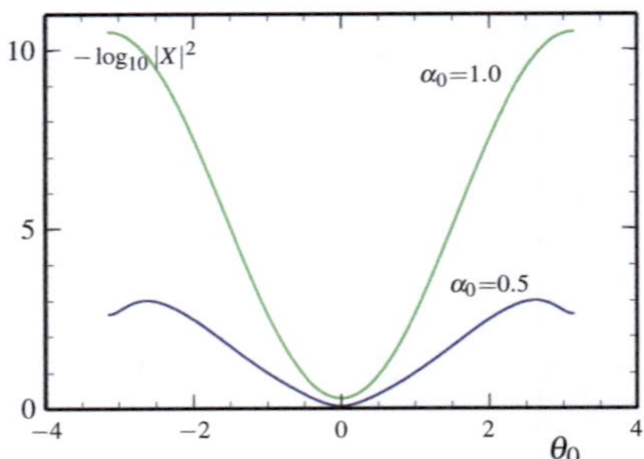

Fig. 7.46 Square of the inner product $X = \langle \alpha_0, r_0 e^{i\theta_0} | \alpha_1, r_1 e^{i\theta_1} \rangle$, represented by $-\log_{10} |X|^2$, for a BPSK constellation. On the *left*, the plot is versus the amplitude α_0 of the displacement for coherent states (*red*) and for squeezed-displaced states with $r_0 = 0.9$ and two values of θ_0. On the *right*, the plot is versus θ_0 with $r_0 = 0.9$ and two values of α_0

it is important to learn how $|X|^2$ depends on squeeze and displacement parameters. This is considered in Fig. 7.46 for the BPSK, where $r_1 e^{i\theta_1} = r_0 e^{i(\theta_0 + 2\pi)}$ and $\alpha_1 = \alpha_0 e^{i\pi}$. On the left of the figure $|X|^2$ is plotted versus α_0 for coherent states (red curve) and also for squeezed-displaced states for a fixed value of r_0 and two values of θ_0. It is remarkable the great improvement obtained with squeezed states, especially at the increase of the displacement amount α_0. The right of the figure shows the strong dependence of $|X|^2$ on the squeeze phase θ_0 and hence the importance of an appropriate choice of this parameter.

7.15.1 BPSK with Squeezed States

The BPSK constellation of squeezed-displaced states is obtained from (7.161) with $K = 2$, namely

$$\mathcal{S} = \{ |r_0 e^{i\theta_0}, \alpha_0\rangle, |r_0 e^{i\theta_0} e^{i2\pi}, \alpha_0 e^{i\pi}\rangle \}. \tag{7.163}$$

For the evaluation of the error probability we can apply Helstrom's theory (see (7.102)), which gives, with equally likely symbols,

$$P_e = \frac{1}{2}\left(1 - \sqrt{1 - |X|^2}\right). \tag{7.164}$$

Thus, the only parameter needed is the quadratic superposition $|X|^2$ between the two states of the constellation. We have seen that $|X|^2$ is a function of r_0, α_0, and θ_0. Now, fixing r_0 and α_0, we have the number of signal photons per symbol N_s as

$$N_s = |\alpha_0|^2 + \sinh^2 r_0. \tag{7.165}$$

and we choose the squeeze phase θ_0 that achieves the minimum error probability.

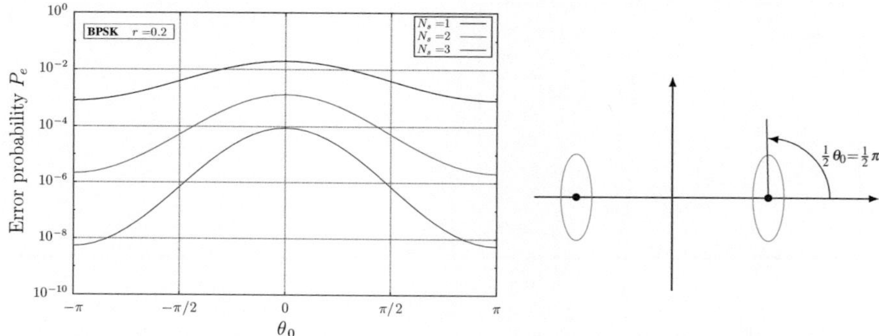

Fig. 7.47 BPSK system with squeezed-displaced states. On the *left*, the error probability P_e versus the squeeze phase θ_0 for three values of the number of signal photons per symbol N_s. All the curves have a minimum for $\theta_0 = \pi$. On the *right*, the optimal BPSK constellation where the ellipse of the reference state is tilted of $\theta_0 = \pi/2$ because the optimal squeeze phase is $\theta_0 = \pi$

In Fig. 7.47 the error probability is plotted versus θ_0 for three values of N_s. Clearly, the minimum of P_e is obtained for $\theta_0 = \pi$, which means that in optimal BPSK constellation the ellipses appear to be vertically tilted, as shown at the right of the figure.

Finally, in Fig. 7.48 we compare the error probability P_e versus N_s obtained with coherent states and squeezed-displaced states. It is remarkable that the performance of the BPSK is highly improved with the presence of squeezing.

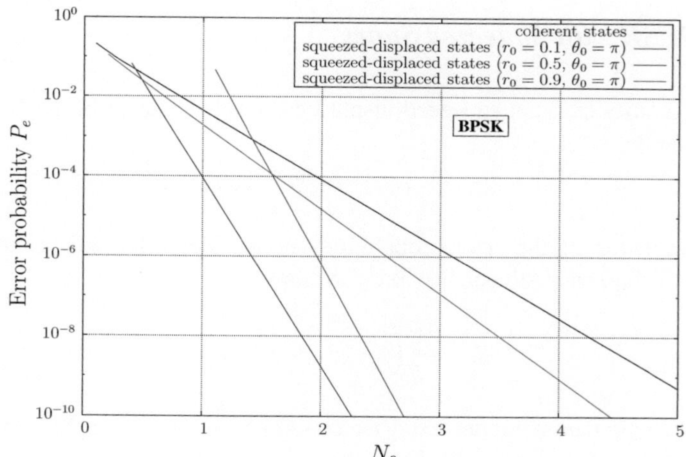

Fig. 7.48 Error probability P_e versus the number of signal photons per symbol N_s in 4-PSK. The *dark curve* refers to coherent states, while the *colored curves* refer to squeezed-displaced states with different values of the squeeze factor r_0 and optimal squeeze phase $\theta_0 = -\pi/2$. The curve do not start at $N_s = 0$, because $N_s = |\alpha_0|^2 + \sinh^2 r_0$ and for $N_s < \sinh^2 r_0$ there is no room for the displacement α_0

7.15.2 4-PSK with Squeezed States

The 4-PSK constellation of squeezed-displaced states is obtained from (7.161) with $K = 4$, namely $\mathcal{S} = \{|z_i, \alpha_i\rangle\}$, where

$$z_i = z_0 e^{i\,2k\,2\pi/4}, \qquad \alpha_i = \alpha_0 e^{i\,k\,2\pi/2}, \qquad i = 0, 1, 2, 3. \tag{7.166}$$

Considering that the constellation has the GUS, for the evaluation of the error probability we apply the SRM approach, which turns out to be optimum. From Sect. 7.7 we recall that the evaluation of P_e using the SRM is obtained as follows:

(1) Evaluation of the inner products

$$G_{pq} = \langle z_p, \alpha_p | z_q, \alpha_q \rangle, \qquad p, q = 0, 1, 2, 3.$$

(2) Evaluation of the eigenvalues $\lambda_i = \sum_{k=0}^{3} G_{0k}\, W_4^{-ki}$, and finally

$$P_e = 1 - \left(\frac{1}{4} \sum_{i=0}^{3} \sqrt{\lambda_i} \right)^2. \tag{7.167}$$

Also in this case P_e is a function of $r_0, \alpha_0,$ and θ_0 and the number of signal photons per symbol N_s is still given by (7.165). Again, we choose r_0 and α_0 as parameters and we search for the squeeze phase θ_0 that gives the minimum error probability.

In Fig. 7.49 the error probability is plotted versus θ_0 for three values of N_s. Clearly, the minimum of P_e is obtained for $\theta_0 = -\pi/2$, which means that in the optimal 4-PSK constellation the reference ellipse is tilted of $\frac{1}{2}\theta_0 = -\pi/4$, as shown on the right of the figure (see Fig. 7.45).

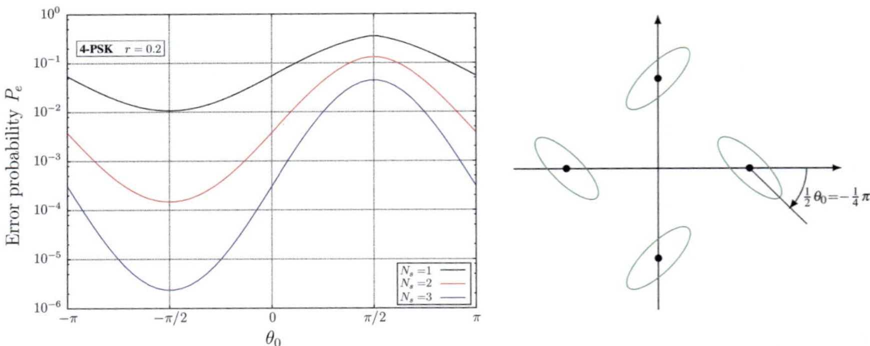

Fig. 7.49 4-PSK system with squeezed-displaced states. On the *left*, the error probability P_e versus the squeeze phase θ_0 for three values of the number of signal photons per symbol N_s All the curves have a minimum for $\theta_0 = -\pi/2$. On the *right* the optimal 4-PSK constellation; the ellipse of the reference state is tilted by $\frac{1}{2}\theta_0 = -\pi/4$ because the optimal squeeze phase is $\theta_0 = -\pi/2$

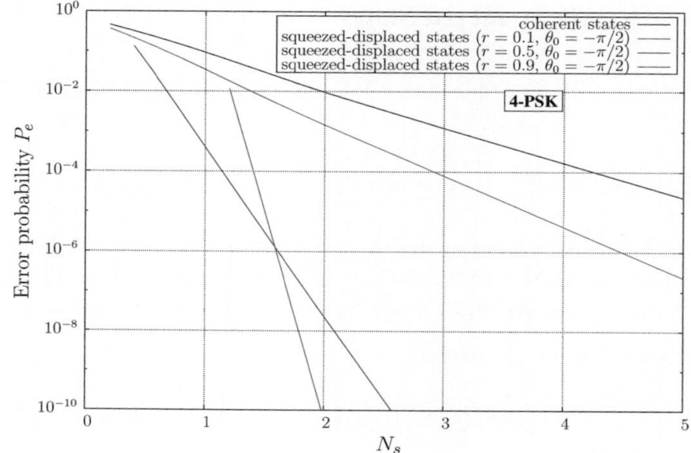

Fig. 7.50 Error probability P_e versus the number of signal photons per symbol N_s in 4-PSK. The *dark curve* refers to coherent states, while the *colored curves* refer to squeezed-displaced states with different values of the squeeze factor r_0 and optimal squeeze phase $\theta_0 = -\pi/2$. The curve do not start at $N_s = 0$, because $N_s = |\alpha_0|^2 + \sinh^2 r_0$ and for $N_s < \sinh^2 r_0$ there is no room for the displacement α_0

Finally, in Fig. 7.50 we compare the error probability P_e versus. N_s obtained with coherent states and with squeezed-displaced states. We realize that also the 4-PSK is highly improved with the presence of squeezing.

7.15.3 Conclusions

In this chapter, we have considered coherent states as the standard carrier for data transmission in quantum communications systems. On the other hand, in this last section, we have seen the possibility of a huge improvement using squeezed light, but we have limited the analysis only to the systems 2PSK and 4PSK for the reason that squeeze technique is not promising for the immediate future because of losses and excess noise present in this technique and because of the complexity and power required. However, quantum optics is making rapid progress so that quantum communications with squeezed states merits a special attention.

References

1. R.J. Glauber, The quantum theory of optical coherence. Phys. Rev. **130**, 2529–2539 (1963)
2. R.J. Glauber, Coherent and incoherent states of the radiation field. Phys. Rev. **131**, 2766–2788 (1963)

3. C.W. Helstrom, J.W.S. Liu, J.P. Gordon, Quantum-mechanical communication theory. Proc. IEEE **58**(10), 1578–1598 (1970)
4. H.P. Yuen, R. Kennedy, M. Lax, Optimum testing of multiple hypotheses in quantum detection theory. IEEE Trans. Inf. Theory **21**(2), 125–134 (1975)
5. J. Shapiro, Quantum noise and excess noise in optical homodyne and heterodyne receivers. IEEE J. Quantum Electron. **21**(3), 237–250 (1985)
6. J.E. Mazo, J. Salz, On optical data communication via direct detection of light pulses. Bell Syst. Tech. J. **55**, 347–360 (1976)
7. K. Kato, M. Osaki, M. Sasaki, O. Hirota, Quantum detection and mutual information for QAM and PSK signals. IEEE Trans. Commun. **47**(2), 248–254 (1999)
8. I.S. Gradshteyn, I.M. Ryzhik, *Tables of Integrals, Series, and Products*, 7th edn. (Elsevier, Amsterdam, 2007)
9. J.G. Proakis, *Digital Communications* (McGraw-Hill, New York, 2001)
10. G. Cariolaro, *Modulazione analogica, discreta e numerica* (Edizioni Progetto, Padova, 1996)
11. A. Waseda, M. Sasaki, M. Takeoka, M. Fujiwara, M. Toyoshima, A. Assalini, Numerical evaluation of PPM for deep-space links. IEEE/OSA J. Opt. Commun. Netw. **3**(6), 514–521 (2011)
12. M. Sasaki, A. Waseda, M. Takeoka, M. Fujuwara, H. Tanaka, Quantum information technology for power minimum info-communications, in *Toward Green ICT*, ed. by R. Prasad, S. Ohmori, D. Simunic (River Publishers, Denmark, 2010). ch. 15
13. G. Cariolaro, G. Pierobon, Theory of quantum pulse position modulation and related numerical problems. IEEE Trans. Commun. **58**(4), 1213–1222 (2010)
14. R.E. Slusher, B. Yurke, Squeezed light for coherent communications. J. Lightwave Technol. **8**(3), 466–477 (1990)
15. M.S. Kim, F.A.M. de Oliveira, P.L. Knight, Properties of squeezed number states and squeezed thermal states. Phys. Rev. A **40**, 2494–2503 (1989)
16. X.B. Wang, T. Hiroshima, A. Tomita, M. Hayashi, Quantum information with Gaussian states. Phys. Rep. **448**(1–4), 1–111 (2007)
17. G. Cariolaro, R. Corvaja, G. Pierobon, Gaussian states and geometrically uniform symmetry. Phys. Rev. A **90**(4), 042309 (2014)

Chapter 8
Quantum Communications Systems with Thermal Noise

8.1 Introduction

The analysis of Quantum Communications systems developed in the previous chapter ignored thermal noise, sometimes called *background noise*. In this chapter we will consider such noise, which in practical Quantum Communications is always present, although it is neglected by most researchers, at least nowadays.

The general scheme of quantum data transmission seen in Sect. 5.2 is reconsidered in Fig. 8.1, with the purpose of evidencing the parameters that apply in the presence of thermal noise.

At transmission Alice "prepares" the quantum system $\mathcal{H}$ in one of the coherent states $|\gamma_i\rangle$, $i \in \mathcal{A}$, as in the previous chapter. These states are pure ("certain"), but the thermal noise, which originates in the receiver and may be conventionally ascribed to the quantum channel, removes the "certainty" of the states $|\gamma_i\rangle$, and therefore they must be described by density operators. Then, if the transmitted state is $|\gamma_i\rangle$, at reception Bob finds the "noisy" density operator $\rho(\gamma_i)$, with nominal state $|\gamma_i\rangle$, whose expression will be seen in the next sections.

Unfortunately, the analysis and especially the optimization in the presence of thermal noise becomes very difficult, and the reason is due to the representation through density operators, whose mathematical structure is *intrinsically nonlinear*, while in the classical case thermal noise is simply represented as an *additive* Gaussian noise.

The chapter begins with the quantum representation of thermal noise, where the density operators are expressed as a continuum of coherent states. This representation is formulated in a Hilbert space with infinite dimensions, but to get a numerical evaluation the density operators are approximated by matrices of finite dimensions, so that the quantum decision theory seen in Chaps. 5 and 6 can be applied.

For the optimization of the measurement operators, explicit results are available only for binary systems and are provided by Helstrom's theory, which holds also in the presence of noise. For multilevel Quantum Communications systems we can apply the numerical optimization, and especially the square root measurement (SRM), which is suboptimal but gives a good approximation of the system performance.

© Springer International Publishing Switzerland 2015

G. Cariolaro, *Quantum Communications*, Signals and Communication Technology,

DOI 10.1007/978-3-319-15600-2_8

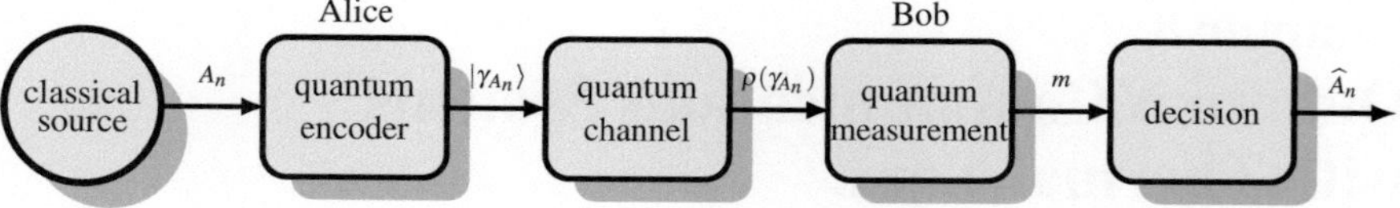

Fig. 8.1 Quantum Communications systems in the presence thermal noise: Alice produces pure states $|\gamma_{A_n}\rangle$ and Bob receives *noisy* states $\rho(\gamma_{A_n})$ and performs a quantum measurement to argue which symbol was transmitted

We will also develop the theory of classical (or semiclassic) optical systems to provide a comparison with the corresponding Classical Communications systems. As already seen in the previous chapter in the absence of thermal noise, also in the presence of thermal noise, Quantum Communications systems perform better than classical counterparts.

Some of the topics developed in this chapter are original and represent an advanced research. Among them, the SRM method applied to mixed states and the analysis in the presence of noise of QAM, PSK, and PPM quantum systems [1, 2]. In some numerical computations we will find it convenient to use the compression technique introduced at the end of Chap. 5, and also this approach is new.

Organization of the Chapter

Sections 8.2–8.4 deal with the representation of noisy coherent states, according to Glauber's theory [3, 4], whose environment is given by an infinite dimensional Hilbert state. The approximation to finite dimension, needed for numerical computations, is considered with great detail to ensure an acceptable evaluation of the system performance, which in practice is always given by the error probability.

Section 8.5 formulates the theory of classical optical communications in the presence of thermal noise starting from the theory developed in the absence of thermal noise in Chap. 7. This theory is used in the subsequent sections to compare classical and quantum systems.

Section 8.6 checks the validity of the Gaussian approximation, usually adopted in classical detection theory.

Section 8.7 develops the general theory of Quantum Communication systems in the presence of thermal noise, which is applied from Sect. 8.8 to the end of the chapter to the specific communications systems (OOK, QAM, PSK, and PPM) considered in the previous chapter in the absence of thermal noise.

A special attention is paid to PPM systems, where the numerical methods encounter serious computation problems because their complexity increases exponentially with the order of modulation. In Sect. 8.12 the performance is evaluated with the methods used for the other systems, while in Sect. 8.13 we apply the state compression, which allows us to reach ranges of performance not possible with the other methods.

8.2 Representation of Thermal Noise

The environment in which thermal noise is represented is an infinite dimensional Hilbert space, where the Fock basis $\{|n\rangle, n = 0, 1, 2, \ldots\}$ of the *number states* is introduced, the same considered in Sect. 7.2 for the representation of coherent states. It is convenient to recall the expression of these states, namely

$$|\alpha\rangle = e^{-\frac{1}{2}|\alpha|^2} \sum_{n=0}^{\infty} \frac{\alpha^n}{\sqrt{n}} |n\rangle \tag{8.1}$$

where α is a complex parameter that specifies the coherent state $|\alpha\rangle$. This representation is just the tool used to specify the density operator which accounts for the presence of thermal noise.

8.2.1 Glauber's Theory on Thermal Noise

The environment is a resonant cavity in thermal equilibrium at a given absolute temperature T_0, where the description of the electromagnetic field in the cavity, for a given fixed *mode*, is given by the density operator [3, 4]

$$\rho_{\text{th}} = \frac{1}{\pi \mathcal{N}} \int_{\mathbb{C}} \exp\left(-\frac{|\alpha|^2}{\mathcal{N}}\right) |\alpha\rangle\langle\alpha| \, d\alpha \tag{8.2}$$

where $|\alpha\rangle$ is the coherent state defined in (8.1) and the integration is done with respect to the complex variable α.[1] The integrand has a bidimensional Gaussian profile (Fig. 8.2) with a dispersion determined by the number of thermal photons $\mathcal{N}$. The parameter $\mathcal{N}$, already introduced in Chap. 4, is defined by

$$\mathcal{N} = \frac{1}{\exp\left(h\nu/kT_0\right) - 1} \tag{8.3}$$

with h Plank's constant, k Boltzmann's constant, ν the frequency of the specific mode, and T_0 the absolute temperature of the cavity. The interpretation of $\mathcal{N}$ is *average number of thermal photons* associated to the mode; as done in Sect. 4.7.3, $\mathcal{N}$ will be called **number of thermal photons**. For instance, if the cavity is at the

---

[1] More specifically, letting $\alpha = x + iy$, the integration must be interpreted in the form

$$\int_{\mathbb{C}} f(\alpha) \, d\alpha = \int_{-\infty}^{+\infty} \int_{-\infty}^{+\infty} f(x + iy) \, dx \, dy.$$

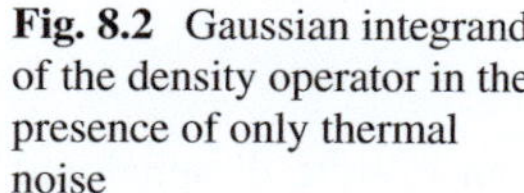

Fig. 8.2 Gaussian integrand of the density operator in the presence of only thermal noise

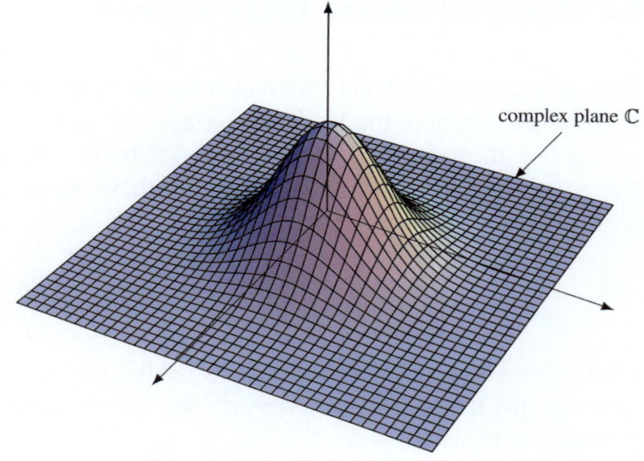

ambient temperature ($T_0 = 290$ K) and the system works at optical frequencies, say at $\nu = 2\ 10^{14}$ Hz, corresponding to $\lambda = 1.5\ \mu$m, one gets

$$h\nu = 1.3\ 10^{-19} \text{J} \qquad kT_0 = 4\ 10^{-21} \text{J} \qquad \mathcal{N} = 7.8\ 10^{-15} \ .$$

The state described by the density operator ρ_{th} is called **thermal state** and represents a fundamental notion in the theory of continuous variables (see Sect. 11.9). According to (8.2), ρ_{th} is given by a linear combination of a *continuum* of coherent states $|\alpha\rangle$. Considering the expression (8.1) of the coherent state $|\alpha\rangle$, one can develop ρ_{th} as a function of the eigenstates (Fock states) $|n\rangle$ of the number operator N, giving

$$\rho_{\text{th}} = \sum_{n=0}^{\infty} \frac{\mathcal{N}^n}{(\mathcal{N}+1)^n} |n\rangle \langle n| \ . \tag{8.4}$$

We briefly describe how the expansion (8.4) is obtained. Substituting the coherent state (8.1) in (8.2) we get a double summation, say in m and n, where the mixed terms ($m \neq n$) have the same expression as the odd moments of a Gaussian random variable with zero mean, which are zero. Then, continuing with a single summation we arrive at (8.4) (see Problem 8.1).

One can easily check, with the measurement setup seen in Sect. 7.2 and illustrated in Fig. 7.2, that the average of the outcome m of such measurement is given by (see (3.45) and Problem 8.2)

$$\mathrm{E}[m|\rho_{\text{th}}] = \mathrm{Tr}(\rho_{\text{th}}N) = \mathcal{N} \ . \tag{8.5}$$

Hence, the parameter $\mathcal{N}$ has actually the meaning of average number of photons due to the thermal noise. One can also verify, considering the standard quantum measurement of counting, that the probability that the outcome of the measurement m be the integer k, is given by the *geometrical distribution* (see Problem 8.1)

$$p_m(k|\rho_{\text{th}}) = \frac{\mathcal{N}^k}{(\mathcal{N}+1)^k}, \qquad k = 0, 1, 2, \ldots \tag{8.6}$$

From (8.4) it follows that ρ_{th} has the following (diagonal) matrix representation in the Fock basis

$$\rho_{\text{th}} \quad \to \quad R = [R_{mn}] \quad \text{with} \quad R_{mn} = \delta_{mn} \frac{\mathcal{N}^n}{(\mathcal{N}+1)^n} \, . \tag{8.7}$$

8.2.2 Signal in the Presence of Thermal Noise

If, in addition to the thermal noise, a coherent signal with complex envelope γ is present, the global statistical description is again given by a density operator with the Gaussian structure (8.2), but modified in the form

$$\rho(\gamma) = \frac{1}{\pi \mathcal{N}} \int_{\mathbb{C}} \exp\left(-\frac{|\alpha - \gamma|^2}{\mathcal{N}}\right) |\alpha\rangle\langle\alpha| \, d\alpha \tag{8.8}$$

where the center of the Gaussian profile, which in the absence of signal is given by the origin of the complex plane, is displaced by the quantity γ (Fig. 8.3). The meaning of this parameter is

$$N_\gamma := |\gamma|^2 \;=\; \text{number of signal photons} \tag{8.9}$$

while $\mathcal{N}$ preserves the meaning of number of thermal photons.

As in the case of pure noise, the operator $\rho(\gamma)$ can be expressed in terms of the eigenstates $|n\rangle$ of the number operator. The expansion is given by [5]

Fig. 8.3 Gaussian integrand of the density operator in the presence of both signal and thermal noise

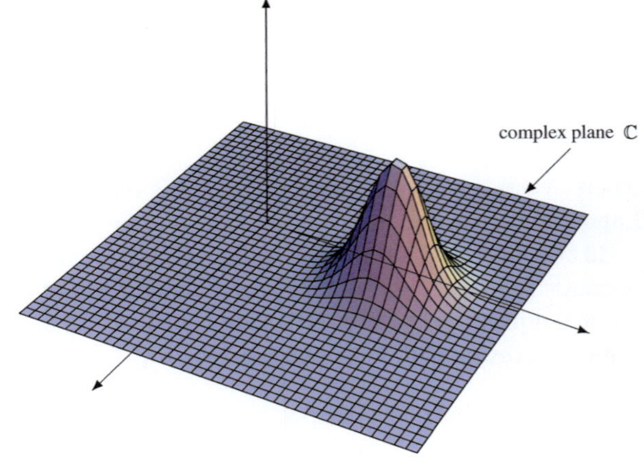

$$\rho(\gamma) = \sum_{m=0}^{\infty} \sum_{n=0}^{\infty} R_{mn}(\gamma)\, |m\rangle\langle n| \tag{8.10}$$

where the coefficients have the expression: for $m \leq n$

$$R_{mn}(\gamma) = \frac{\mathcal{N}^n}{(\mathcal{N}+1)^n}\sqrt{\frac{m!}{n!}}\left(\frac{\gamma^*}{\mathcal{N}}\right)^{n-m}\exp\left(-\frac{|\gamma|^2}{\mathcal{N}+1}\right)L_m^{(n-m)}\left(1-\frac{|\gamma|^2}{\mathcal{N}}\right) \tag{8.10a}$$

while for $m > n$ one uses the Hermitian symmetry $R_{nm}(\gamma) = R_{mn}^*(\gamma)$. In (8.10a), $L_m^{(n-m)}(x)$ is the generalized Laguerre polynomial of degree m and of parameter $n-m$, which is given by

$$L_m^{(n-m)}(x) = \sum_{k=0}^{m}(-1)^k \frac{(2m-n)!}{(n-m-k)!(n-k)!}\, x^k\,. \tag{8.10b}$$

As we can see, differently from the case of pure noise, now the density operator $\rho(\gamma)$ is no more diagonal.

One can check, with the quantum measurement recalled above, that the average of the outcome of the measurement m is

$$\mathrm{E}[m|\rho(\gamma)] = |\gamma|^2 + \mathcal{N} = N_\gamma + \mathcal{N}, \tag{8.11}$$

with the clear meaning of *global average number of photons*. The variance results in (see Sect. 8.5)

$$\sigma_n^2(\rho(\gamma)) = N_\gamma + 2N_\gamma\,\mathcal{N} + \mathcal{N}(\mathcal{N}+1)\,. \tag{8.12}$$

Moreover, we find that the probability that the measurement's outcome m be the integer k is given by the *Laguerre distribution*

$$p_n(k|\rho(\gamma)) := \mathrm{P}[m = k|\rho(\gamma)] = R_{kk}(\gamma) = \langle k|\rho(\gamma)|k\rangle$$

$$= \frac{\mathcal{N}^k}{(\mathcal{N}+1)^{k+1}}\exp\left(-\frac{N_\gamma}{\mathcal{N}+1}\right)L_k\left(-\frac{N_\gamma}{\mathcal{N}(\mathcal{N}+1)}\right) \tag{8.13}$$

where $N_\gamma = |\gamma|^2$ and $L_k(x) = L_k^{(0)}(x)$ is the ordinary Laguerre polynomial. The Laguerre distribution has been illustrated in Fig. 4.24.

In conclusion, the photon counting in a quantum system described by the density operator $\rho(\gamma)$ is given by a Laguerre random variable, and only in the absence of thermal noise ($\gamma = 0$) it does become a Poisson random variable. One can reach the same conclusion with a classical formulation (see Sects. 4.7 and 8.5).

Remark Note that in the perspective of quantum channels, developed in Chap. 12, the parameter γ may be regarded as the *signal* bearing the information, which becomes the *signal state* $|\gamma\rangle$ after the c $\to$ q mapping $\gamma \to |\gamma\rangle$, and finally is transformed into the *noisy signal state* $\rho(\gamma)$ by the q $\to$ q mapping $|\gamma\rangle \to \rho(\gamma)$.

Nomenclature. The operator $\rho(\gamma)$ given by (8.8) will be called *Glauber density operator*, corresponding to the *nominal* coherent state $|\gamma\rangle$, which in turn is specified by the *complex symbol* γ. One can check that the operator ρ_{th}, given by (8.4), is the Glauber density operator, corresponding to the ground state, that is, $\rho_{th} = \rho(0)$. We always say "in the presence of thermal noise", not abbreviated as "in the presence of noise", to recall that also in the absence of thermal noise, the noise is always present in the form of shot noise, because of the uncertainty of the outcomes in quantum measurements.

Problem 8.1 $\star\star\star$ Starting from the integral representation (8.2) of the density operator ρ_{th}, find the Fock representation (8.4). *Hint*: use polar coordinates.

Problem 8.2 $\star$ Organize a quantum measurement with the system in the state ρ_{th} defined by (8.2). The outcome m should have the geometrical distribution $p_m(k|\rho_{th})$ given by (8.6).

Problem 8.3 $\star$ Prove (8.5), that is, $\mathrm{E}[m|\rho_{th}] = \mathrm{Tr}(\rho_{th} N) = \mathcal{N}$, where ρ_{th} is the density operator of thermal noise given by (8.4) and N is the number operator.

Problem 8.4 $\star\star$ Representations (8.8) and (8.10) on thermal noise hold for $\gamma \neq 0$ and $\mathcal{N} > 0$. Find and discuss the representations in the degenerate cases $\gamma = 0$ (absence of signal) and $\mathcal{N} = 0$ (absence of noise).

8.3 Noisy Coherent States as Gaussian States ∇

In Sect. 7.2.5 we have seen that a pure coherent state $|\gamma\rangle$ is a Gaussian state with mean value and covariance matrix given by

$$\bar{X} = \begin{bmatrix} \Re\gamma \\ \Im\gamma \end{bmatrix}, \qquad V = \begin{bmatrix} 1 & 0 \\ 0 & 1 \end{bmatrix} = I_2 . \tag{8.14}$$

Also a noisy coherent state $\rho(\gamma)$ turns out to be Gaussian with the same mean vector, but with the covariance matrix modified as

$$V = (1 + 2\mathcal{N}) \begin{bmatrix} 1 & 0 \\ 0 & 1 \end{bmatrix} \tag{8.15}$$

where $\mathcal{N}$ is the number of thermal photons. In Fig. 8.4 it is compared the contour level (see Sect. 7.2) of the Wigner function of a noisy coherent state $\rho(\gamma)$ and of the corresponding pure state $|\gamma\rangle$.

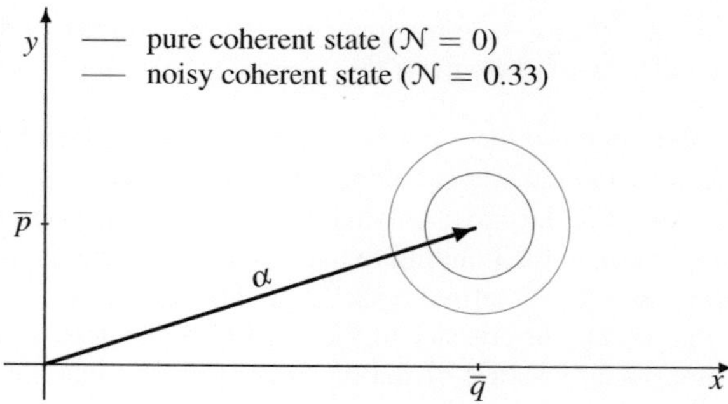

Fig. 8.4 Contour level of the Wigner function $W(x, y)$ of a pure coherent state $|\alpha\rangle$ (in *red*) and of noisy coherent state $\rho(|\alpha\rangle)$ (in *green*). The mean vector $(\bar{q}, \bar{p}) = (\Re\alpha, \Im\alpha)$ gives the center of the countour. The variances $\sigma^2 = 1$ and $\sigma^2 = 1 + 2\mathcal{N}$, respectively, are proportional to the radius of the circles

8.3.1 The Channel as Additive–Noise Gaussian Channel

In Quantum Information a channel is called *Gaussian* if it maps a Gaussian state into a Gaussian state. This is the case of the channel that we are considering in Quantum Communications systems, where at the input we have the noiseless coherent state $|\gamma\rangle$ and at the output the noisy coherent state $\rho(\gamma)$, both being Gaussian.

We now see how this specific Gaussian channel should be formulated in the family of Gaussian channels (Fig. 8.5). In Sect. 12.8 we shall see that, in the single bosonic mode, a Gaussian channel is specified by a triplet (S, B, d), where S and B are real 2×2 matrices and d is a vector in $\mathbb{R}^2$. The mean value $\bar{X}$ and the covariance matrix V of the input state $\rho(\bar{X}, V)$ are transformed by the Gaussian channel, as follows

$$\bar{X} \ \rightarrow \ S\bar{X} + d, \quad V \ \rightarrow \ SVS^{\mathrm{T}} + B \ . \tag{8.16}$$

The matrix B is the noise parameter. For $B = 0$ the Gaussian channel is noiseless, while for $B \neq 0$ it becomes noisy. In particular, in an **additive-noise channel**, S and B have the simple forms

$$S = \begin{bmatrix} 1 & 0 \\ 0 & 1 \end{bmatrix} = I_2, \quad B = \mathcal{N}_0 \begin{bmatrix} 1 & 0 \\ 0 & 1 \end{bmatrix}, \quad \mathcal{N}_0 > 0 \tag{8.17}$$

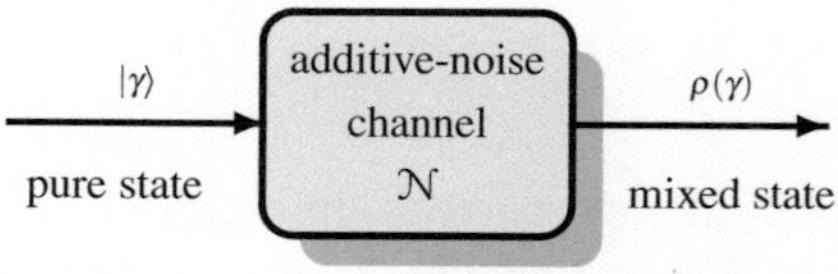

Fig. 8.5 A pure Gaussian state $|\gamma\rangle$ is sent through an additive noise channel specified by the number of thermal photons $\mathcal{N}$. At the output the noisy state is still Gaussian, but becomes mixed and described by a density operator $\rho(\gamma)$

while $d = 0$. This channel does not modify the mean value, but only and the covariance matrix as

$$V \ \rightarrow \ V + B = (1 + \mathcal{N}_0)\, I_2 \ . \tag{8.18}$$

Thus we find (8.15), with $\mathcal{N}_0 = 2\mathcal{N}$, where $\mathcal{N}$ is the number of thermal photons in $\rho(\gamma)$.

In conclusion, the channel we are considering in Quantum Communications systems in the presence of thermal noise is an additive noise Gaussian channel, specified by the matrices (8.17).

8.4 Discretization of Density Operators

In a quantum transmission with coherent states in the presence of thermal noise the constellation

$$\rho_i = \rho(\gamma_i), \qquad i = 0, 1, \ldots, K - 1 \tag{8.19}$$

is formed by Glauber density operators defined by (8.8). To proceed, one must approximate this expression, which gives each ρ_i as a linear combination of a *continuum* of coherent states, with a *finite* expression, given by the matrix representation (8.10), but limited to a finite number n of terms, that is,

$$\rho(\gamma) \simeq \sum_{h=0}^{n-1} \sum_{k=0}^{n-1} R_{hk}(\gamma)\, |h\rangle\langle k| := R(\gamma) \ . \tag{8.20}$$

In such a way, the infinite dimensional density operator is approximated by a square matrix $R(\gamma) = [R_{hk}(\gamma)]$ of finite dimension $n \times n$.

8.4.1 Spectral Decomposition (EID) and Factorization

The *finite* representation (8.20) can be elaborated through the EID. Considering that ρ is Hermitian and positive semidefinite, and that these properties also hold for its approximation, the EID of R can be expressed in the forms

$$R = Z\, \Lambda_\rho\, Z^* = \sum_{i=1}^{h} d_i^2 |z_i\rangle\langle z_i| = Z_h D_h Z_h^* \tag{8.21}$$

where

- $h \leq n$ is the rank of R,
- Z is an $n \times n$ unitary matrix, which forms with its columns the orthonormal basis $\{|z_i\rangle\}$,
- Λ_ρ is an $n \times n$ diagonal matrix whose first h diagonal elements are the positive eigenvalues d_i^2 and the other $n - h$ diagonal elements are zero,
- Z_h is a $h \times n$ matrix, which collects the first h columns of Z,
- D_h is a $h \times h$ diagonal matrix with diagonal elements d_i^2.

The meaning of the eigenvalues is obtained from the definition of density operator, specifically, d_i^2 gives the probability that the quantum system be in the state $|u_i\rangle$.

Letting[2]

$$\beta = Z_h \sqrt{D_h} \tag{8.22a}$$

where $D_h = \mathrm{diag}[d_1, \ldots, d_h]$, one gets

$$\rho \simeq R = \beta \, \beta^* \tag{8.22b}$$

which gives the *factorization of the density operator* (see Sect. 5.7), approximated by the matrix R. In such a way, from the K density operators ρ_i one gets so many *state factors* β_i.

Example 8.1 We want to find the EID and the factorization of a Glauber density operator $\rho(\gamma)$ making appropriate checks on the accuracy. We consider the following data

$$\gamma = 1.41421 \;\rightarrow\; N_\gamma = \gamma^2 = 2.0 \quad \mathcal{N} = 0.2.$$

The real value of γ implies that the matrix R is real. With $n = 11$ we find the 11×11 matrix

$$R = \begin{bmatrix}
0.15740 & 0.18549 & 0.15458 & 0.10518 & 0.06198 & 0.03266 & 0.01572 & 0.00700 & 0.00292 & 0.00115 & 0.00043 \\
0.18549 & 0.24484 & 0.22589 & 0.16857 & 0.10810 & 0.06159 & 0.03186 & 0.01518 & 0.00674 & 0.00281 & 0.00111 \\
0.15458 & 0.22589 & 0.22905 & 0.18659 & 0.12982 & 0.07981 & 0.04433 & 0.02258 & 0.01067 & 0.00472 & 0.00197 \\
0.10518 & 0.16857 & 0.18659 & 0.16513 & 0.12424 & 0.08224 & 0.04897 & 0.02665 & 0.01341 & 0.00629 & 0.00278 \\
0.06198 & 0.10810 & 0.12982 & 0.12424 & 0.10073 & 0.07161 & 0.04564 & 0.02650 & 0.01418 & 0.00706 & 0.00330 \\
0.03266 & 0.06159 & 0.07981 & 0.08224 & 0.07161 & 0.05453 & 0.03713 & 0.02297 & 0.01306 & 0.00689 & 0.00340 \\
0.01572 & 0.03186 & 0.04433 & 0.04897 & 0.04564 & 0.03713 & 0.02695 & 0.01774 & 0.01070 & 0.00598 & 0.00312 \\
0.00700 & 0.01518 & 0.02258 & 0.02665 & 0.02650 & 0.02297 & 0.01774 & 0.01239 & 0.00793 & 0.00469 & 0.00258 \\
0.00292 & 0.00674 & 0.01067 & 0.01341 & 0.01418 & 0.01306 & 0.01070 & 0.00793 & 0.00537 & 0.00335 & 0.00195 \\
0.00115 & 0.00281 & 0.00472 & 0.00629 & 0.00706 & 0.00689 & 0.00598 & 0.00469 & 0.00335 & 0.00221 & 0.00136 \\
0.00043 & 0.00111 & 0.00197 & 0.00278 & 0.00330 & 0.00340 & 0.00312 & 0.00258 & 0.00195 & 0.00136 & 0.00087
\end{bmatrix}$$

---

[2] In Sect. 5.7 the factor of a density operator was denoted by γ, but here this symbol denotes the complex number determining the density operator $\rho(\gamma)$. Thus, in this chapter the factor is denoted by β.

The rank of this matrix is surely 11, as follows from the eigenvalues evaluated with a great accuracy

$$
\begin{bmatrix}
1 & 0.8333264138373294 \\
2 & 0.13884015974224564 \\
3 & 0.023018143481200053 \\
4 & 0.003692051210944246 \\
5 & 0.0005318568489146765 \\
6 & 0.00006324572935641604 \\
7 & 5.936804006779007\ 10^{-6} \\
8 & 4.292297763228036\ 10^{-7} \\
9 & 2.2892724563994347\ 10^{-8} \\
10 & 8.158936327309927\ 10^{-10} \\
11 & 1.4844074185916688\ 10^{-11}
\end{bmatrix}
$$

The trace of R, given by the sum of the eigenvalues, results in 0.99947826060, which is very close to 1, which confirms a very good approximation (we recall that the trace of the original density operator is exactly unitary).

At this stage we fix an "accuracy" to neglect the very small eigenvalues; for instance the accuracy 0.001 leads to take only the four eigenvalues:

$$
\begin{bmatrix} 0.83333 & 0.13884 & 0.02302 & 0.00369 \end{bmatrix}.
$$

This implies assigning to the matrix a virtual rank $h = 4$. With this choice of h we get the second form of spectral decomposition indicated in (8.21). The matrix Z_h is formed by the first four columns of the matrix Z and results in

$$
Z_h =
\begin{bmatrix}
0.36788 & 0.52044 & -0.52202 & -0.43077 \\
0.52027 & 0.36793 & 0.00115 & 0.31205 \\
0.52026 & -0.00011 & 0.36969 & 0.30091 \\
0.42479 & -0.30054 & 0.30048 & -0.13492 \\
0.30037 & -0.42492 & -0.00173 & -0.35518 \\
0.18997 & -0.40303 & -0.27097 & -0.21395 \\
0.10968 & -0.31018 & -0.38948 & 0.07896 \\
0.05862 & -0.20719 & -0.37374 & 0.30469 \\
0.02931 & -0.12426 & -0.28995 & 0.38433 \\
0.01382 & -0.06831 & -0.19473 & 0.34854 \\
0.00618 & -0.03489 & -0.11720 & 0.26102
\end{bmatrix}
$$

while the diagonal matrix D_h becomes

$$
D_h =
\begin{bmatrix}
0.83333 & 0.00000 & 0.00000 & 0.00000 \\
0.00000 & 0.13884 & 0.00000 & 0.00000 \\
0.00000 & 0.00000 & 0.02302 & 0.00000 \\
0.00000 & 0.00000 & 0.00000 & 0.00369
\end{bmatrix}.
$$

Finally, from Z_h and D_h we get the factor β of dimensions 11×4

$$\beta = Z_h \sqrt{D_h} = \begin{bmatrix} 0.33583 & 0.19392 & -0.07920 & -0.02617 \\ 0.47493 & 0.13710 & 0.00017 & 0.01896 \\ 0.47493 & -0.00004 & 0.05609 & 0.01828 \\ 0.38778 & -0.11199 & 0.04559 & -0.00820 \\ 0.27420 & -0.15833 & -0.00026 & -0.02158 \\ 0.17342 & -0.15017 & -0.04111 & -0.01300 \\ 0.10012 & -0.11558 & -0.05909 & 0.00480 \\ 0.05351 & -0.07720 & -0.05670 & 0.01851 \\ 0.02676 & -0.04630 & -0.04399 & 0.02335 \\ 0.01261 & -0.02545 & -0.02954 & 0.02118 \\ 0.00564 & -0.01300 & -0.01778 & 0.01586 \end{bmatrix} .$$

8.4.2 Approximation Criteria

In the EID and in the previous factorizations we have two approximations:

- the approximation R of the density operator ρ by a finite number n of terms,
- the approximation of the true rank n of the matrix R by the *virtual* rank h, which allows us to find the approximate factor $\widetilde{\beta}$ of ρ of dimensions $n \times h$.

For both approximations we need to establish a criterion that guarantees a given accuracy.

For the first approximation the *trace criterion* seems to be more appropriate. We recall that a density operator has always unitary trace and that $\mathrm{Tr}[\rho] = 1$ represents also the normalization condition of the Laguerre distribution $p_n(k|\gamma)$, which gives the meaning of the diagonal elements (see (8.13)). Hence, only if $\mathrm{Tr}[R]$ gives a value close to one, the approximation is satisfactory, and, having fixed an accuracy ε, the number of terms $n = n_\varepsilon$ must be calculated according to the condition

$$\mathrm{Tr}[R] = \sum_{k=0}^{n_\varepsilon-1} R_{kk} = \sum_{k=0}^{n_\varepsilon-1} p_n(k|\gamma) \geq 1 - \varepsilon.$$

Of course, n_ε depends on N_γ and on $\mathcal{N}$, and also on ε. For instance, with $N_\gamma = 10$ and $\mathcal{N} = 0.1$ we find that in order to achieve the accuracy $\varepsilon = 0.00001$, $n_\varepsilon = 32$ terms are needed.

For the second approximation one can use the *reconstruction criterion*, based on the square error between the matrix R and its "reconstruction" $\widetilde{\beta}\widetilde{\beta}^*$, where both R and $\widetilde{\beta}\widetilde{\beta}^*$ are $n \times n$ matrices, with n previously evaluated. Hence, having fixed the accuracy v, we evaluate the virtual rank $n_v = h$ such that

$$\mathrm{mse}(R - \widetilde{\beta}\widetilde{\beta}^*) \leq v$$

where mse denotes the square error, given by the square moduli of the elements, divided by the number of elements n^2. For instance, with $N_\gamma = 10$ and $\mathcal{N} = 0.1$ we

Table 8.1 Values of n_ε and n_ν as a function of N_γ and $\mathcal{N}$ that ensure the accuracy $\varepsilon = \nu = 10^{-5}$

$N_\gamma \rightarrow$	0.5		1.0		5		10		15		25	
	n_ε	n_ν	n_ε	n_ν	n_ε	n_ν	n_ε	n_ν	n_ε	n_ν	n_ε	n_ν
$\mathcal{N} = 0.001$	7	2	10	2	21	2	31	2	40	2	57	2
$\mathcal{N} = 0.01$	7	3	9	3	20	2	30	2	39	2	55	2
$\mathcal{N} = 0.1$	9	4	11	4	22	4	32	4	41	4	57	4
$\mathcal{N} = 1.0$	21	12	24	12	38	11	51	11	62	10	81	10
$\mathcal{N} = 2.0$	33	18	36	18	52	17	66	17	78	16	99	16
$\mathcal{N} = 3.0$	45	24	49	24	67	23	83	22	97	21	121	20

find that, to achieve the accuracy $\nu = 0.00001$, $n_\nu = 4$ is needed, so that the factor $\widetilde{\beta}$ has dimensions $n_\varepsilon \times n_\nu = 32 \times 4$.

Table 8.1 gives the values of n_ε and n_ν required to achieve the accuracy $\varepsilon = \nu = 10^{-5}$ for several values of N_γ and of $\mathcal{N}$.

The accuracy is related to the range of error probability P_e we want to explore, e.g., to evaluate $P_e \sim 10^{-2}$ an accuracy of 10^{-3} is sufficient, but to evaluate $P_e \sim 10^{-8}$ the accuracy should become 10^{-9}. Thus we get the *rule of thumb* in the choice of the parameters ν and ε

$$\nu = \varepsilon = \frac{1}{10} P_e. \tag{8.23}$$

Alternative discretization. The discretization seen above is based on the matrix representation of the density operator. An alternative discretization, considered at JPL of NASA [6], is based on the subdivision of the integration of the density operator given by (8.8) into a finite number of regions of the complex plane. This discretization procedure, which gives results very close to the previous ones, is outlined in Appendix section "Alternative Discretization".

8.5 Theory of Classical Optical Systems with Thermal Noise

To compare the performance of a Quantum Communications system with that of the corresponding Classical Communications system we reconsider the *theory of optical detection* developed in Sect. 7.5 in the absence of thermal noise. The presence of thermal noise complicates such a theory, but the final result is quite simple. Essentially, the thermal noise modifies the arrival distribution, which changes from the Poisson form to a Laguerre form.

8.5.1 Classical Optical Decision with Thermal Noise

We follow closely the theory of optical decision in the absence of thermal noise developed in Sect. 7.5 by introducing the modifications due to the presence of such

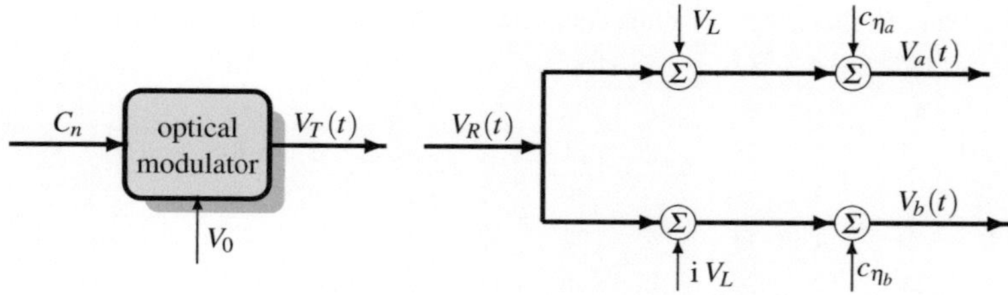

Fig. 8.6 Scheme of a classical modulator and demodulator for the complex envelopes with thermal noise at receiver

noise. Then we reconsider the scheme for the complex envelope of Fig. 7.10, which becomes as in Fig. 8.6, where $\{C_n\}$ is the sequence of complex symbols to be transmitted, V_0 is the amplitude of the carrier produced by the transmitter laser, and V_L is the amplitude of the carrier of the receiver laser. Now, at the receiver, the outputs of the two paths are given by (see (7.47))

$$V_a(t) = C_0\, V_R + V_L + c_{\eta_a}. \quad V_b(t) = C_0\, V_R + i\, V_L + c_{\eta_b}. \tag{8.24}$$

where c_{η_a} and c_{η_b} are the complex envelopes of the thermal noises. Relations (8.24) hold in the symbol period $(0, T)$ for the detection of the zeroth symbol C_0; $V_a(t)$ and $V_b(t)$ are constant in this period according to the simplifications made in the previous chapter.

The presence of thermal noise modifies the statistics of photon arrivals, as shown in the theory of semiclassical detection of Sect. 4.7. The numbers of arrivals n_a and n_b in a symbol period become Laguerre random variables and therefore governed by the Laguerre distribution

$$p_n(k|\gamma) = \frac{\mathcal{N}^k}{(\mathcal{N}+1)^{k+1}}\, \exp\left(-\frac{N_\gamma}{\mathcal{N}+1}\right)\, L_k\left(-\frac{N_\gamma}{\mathcal{N}(\mathcal{N}+1)}\right). \tag{8.25}$$

where $N_\gamma = |\gamma|^2$ is the number of signal photons and $\mathcal{N}$ is the number of thermal photons. We recall that the conditional mean and variance of a Laguerre random variable are given by

$$\bar{n}(\gamma) = N_\gamma + \mathcal{N}, \qquad \sigma_n^2(\gamma) = \bar{n}(\gamma) + 2N_\gamma\mathcal{N} + \mathcal{N}^2. \tag{8.26}$$

In particular, the structure of the variance emphasizes the difference with respect to the Poisson distribution, where average and variance coincide. Both relations (8.25) and (8.26) hold for the arrival numbers n_a and n_b. Also, in the presence of thermal noise, we can use the **Gaussian approximation**, which holds better than in the absence of thermal noise (see Sect. 8.6). Then the performance is completely determined by the signal-to-noise ratio (SNR), which in the absence of thermal noise is given by (see Sect. 7.5.5)

$$\Lambda = \frac{U_0^2}{\sigma_n^2} == \frac{4\,N_s}{\mu_K}$$

where U_0 is the amplitude of the useful signal, σ_n^2 is the common variance of n_a and n_b, N_s is the number of signal photons per symbol, and μ_K is the shape factor of the constellation. Now, the SNR must be modified according to the Laguerre statistics. From (8.26) we see that the average is increased by the term $\mathcal{N}$, but this is only a bias and does not modify the amplitude of the useful signal U_0 with respect to the Poissonian case. The only relevant modification comes from the variance, which is increased. With the homodyne detection, where the amplitude of the local carrier is much larger than the one of the received signals ($V_L \gg V_R$)), one gets $N_\gamma \gg \mathcal{N}$ and hence

$$\sigma_{\text{Laguerre}}^2 \simeq \sigma_{\text{Poisson}}^2 \,(1 + 2\mathcal{N}) \quad \text{with} \quad \sigma_{\text{Poisson}}^2 = N_\gamma. \tag{8.27}$$

In conclusion, the performance of a classical optical system in the presence of thermal noise is evaluated in the same way as in the absence of such noise, by simply increasing the variances by the factor $(1 + 2\mathcal{N})$. Consequently, we have a deterioration of the SNR.

Proposition 8.1 *The performance of a Classical Communication system in the presence of thermal noise can be obtained from the performance established in the absence of thermal noise by decreasing the SNR in the form*

$$\boxed{\Lambda_{\text{th}} = \frac{4N_s}{\mu_K}\,\frac{1}{1 + 2\mathcal{N}} = \Lambda\,\frac{1}{1 + 2\mathcal{N}}} \tag{8.28}$$

where Λ is the SNR in the absence of thermal noise and $\mathcal{N}$ is the number of thermal photons.

We recall that in the cases of interest, with equally likely symbols and optimized decision regions, the minimum error probability turns out to be a simple function of the SNR expressed through the complementary normalized Gaussian distribution $Q(x)$.

8.5.2 Alternative Deduction from Glauber's Representation

The *semiclassical theory* just recalled gives the same results as the *quantum theory*, as we shall see now with the photon counting.

Let us consider a quantum system specified by a Glauber density operator $\rho(\gamma)$ with nominal state $|\gamma\rangle$ and number of thermal photons $\mathcal{N}$. Now, setting up a quantum measurement with the elementary operators $|k\rangle\langle k|$, where $|k\rangle$ are the number states

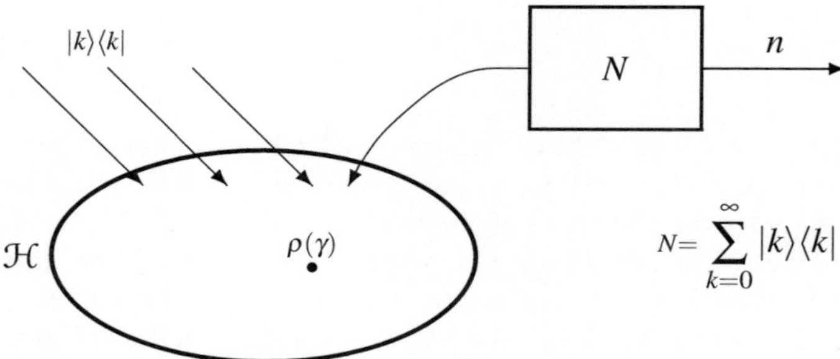

Fig. 8.7 Quantum measurement in a Hilbert space, prepared in the noisy state $\rho(\gamma)$, with the number operator N as observable. The elementary projectors $|k\rangle \langle k|$ are formed by the number states $|k\rangle$. The outcome of the measurement is the photon number n

(Fig. 8.7), one gets as result an integer random variable n, having distribution (see (8.25) and (8.10))

$$p_n(k|\rho(\gamma)) := \mathrm{P}[n = k|\rho(\gamma)] = \langle k|\rho(\gamma)|k\rangle = R_{kk} = p_n(k|\gamma) \, . \qquad (8.29)$$

Hence, we find the same result as in the semiclassical theory.

This alternative deduction of Laguerre distribution deserves a comment. In Sect. 4.7 we have seen the conceptual and mathematical difficulties to prove that the photon counting in the presence of thermal noise has a Laguerre distribution. Here, with the tools of Quantum Mechanics, the deduction is immediate; in particular, the powerful synthesis of the Glauber representation should be recognized.

8.6 Check of Gaussianity in Classical Optical Detection

We have seen that in the presence of thermal noise the number of photons arriving in a symbol period has a Laguerre distribution, which depends on the number of signal photons N_γ and on the number of thermal photons $\mathcal{N}$. In the absence of thermal noise, $\mathcal{N} = 0$, the distribution degenerates to a Poisson distribution. In the literature, see, e.g., [7], it is customary, "in the presence of a strong photonic intensity", to make the **Gaussian approximation**, which allows us to simplify the analysis and to get very simple results. The tendency to Gaussianity was already realized in the implementation of a photon counter at the end of Chap. 4.

We have adopted the Gaussian approximation in the previous chapter in the evaluation of performance of classical optical systems and we apply this approximation also in the present chapter. However, it remains to establish, in a quantitative form, the meaning of "a strong photonic intensity". In this section, we give an answer to this problem and we will arrive at the conclusion that the Gaussian approximation works very well also with a "moderate photonic intensity". To obtain a simple and clear formulation, we consider the homodyne detection in a BPSK system in the presence of thermal noise, but including in the formulation also the case of absence of thermal noise.

8.6.1 *Exact Evaluation of Probabilities*

We consider the classical BPSK with a homodyne receiver and a final photon counting in the presence of thermal noise. Referring to the semiclassical theory the number of photons $n = n(A_0)$ has a Laguerre distribution with mean and variance, respectively,

$$\bar{n}(A_0) = N_\gamma(A_0) + \mathcal{N}, \qquad \sigma_n^2(A_0) = N_\gamma(A_0) + 2N_\gamma(A_0)\mathcal{N} + \mathcal{N}(\mathcal{N}+1) \quad (8.30)$$

where A_0 is a binary symbol. On the other hand, the homodyne receiver gives the signal power $V_0 + V_L^2 + 2V_0 V_L \cos \pi A_0$, which can be converted to "numbers" by multiplying by $H = T/h\nu$. Thus we have

$$N_\gamma(A_0) = N_R + N_L + 2\sqrt{N_R N_L} \cos \pi A_0 \qquad (8.31)$$

where

- $N_R = HV_0^2$ is the (average) number of signal photons/bit,
- $N_L = HV_L^2$ is the (average) number of photons introduced by the local carrier,
- $N_\gamma(A_0)$ is the global average due to the received power and to the local carrier (the thermal noise gives the contribution $\mathcal{N}$ to the average number of photons).

In the binary case we have two distinct Laguerre distributions, $p_n(k|1)$ and $p_n(k|0)$, which are illustrated in Fig. 8.8 for $\mathcal{N} = 0.3$, $N_\gamma(1) = 10$, and $N_\gamma(0) = 25$.

The first step is the evaluation of the optimal decision regions, which have the forms $\mathcal{R}(1) = \{0, 1, \ldots, S_o - 1\}$ and $\mathcal{R}(0) = \{S_o, S_o + 1, S_o + 2, \ldots\}$. The optimal threshold S_o is an integer determined by the conditions $p_n(k|1) > p_n(k|0)$ for $k \in \mathcal{R}(1)$ and $p_n(k|0) \geq p_n(k|1)$ for $k \in \mathcal{R}(0)$. In the case of Fig. 8.8 the threshold is $S_o = 12$. Then we have the cross-transition probabilities as

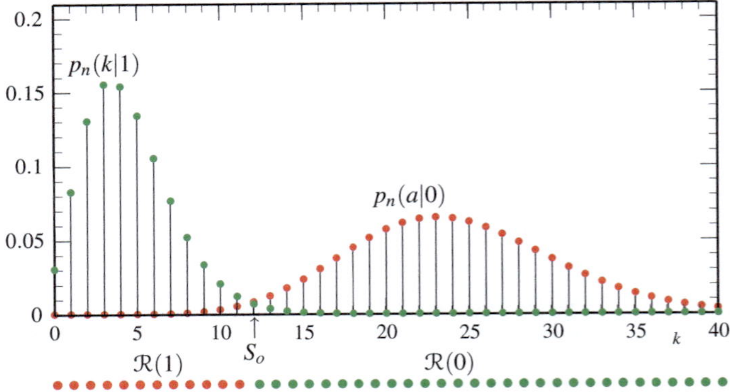

Fig. 8.8 Laguerre distributions $p_n(k|1)$ and $p_n(k|0)$ of photon counting in BPSK and corresponding optimal decision regions $\mathcal{R}(1)$ and $\mathcal{R}(0)$

$$p(0|1) = P[n \geq S_o | A_0 = 1] = \sum_{k=S_o-1}^{\infty} p_n(k|1)$$

$$p(1|0) = P[n < S_o | A_0 = 0] = \sum_{k=0}^{S_o} p_n(k|0).$$

Note that $p(0|1) = 1 - p(1|1) = 1 - \sum_{k=0}^{S_o-1} p_n(k|1)$ so that in any case we have an evaluation in finite terms. The error probability (with equally likely symbols) is then given by

$$P_e = \tfrac{1}{2} p(0|1) + \tfrac{1}{2} p(1|0) . \tag{8.32}$$

8.6.2 Gaussian Approximation

In the Gaussian approximation the probability distributions $p_n(k|A_0)$ are replaced by the probability densities

$$f_n(a|A_0) = \frac{1}{\sigma_n(A_0)} \phi\left(\frac{a - \overline{n}(A_0)}{\sigma_n(A_0)}\right), \qquad A_0 = 0, 1$$

which are illustrated in Fig. 8.9.

Considering that $\overline{n}(0) > \overline{n}(1)$ we denote the threshold in the form

$$S = \overline{n}(1) + \beta\,[\overline{n}(0) - \overline{n}(1)] \quad \text{with} \quad 0 < \beta < 1 .$$

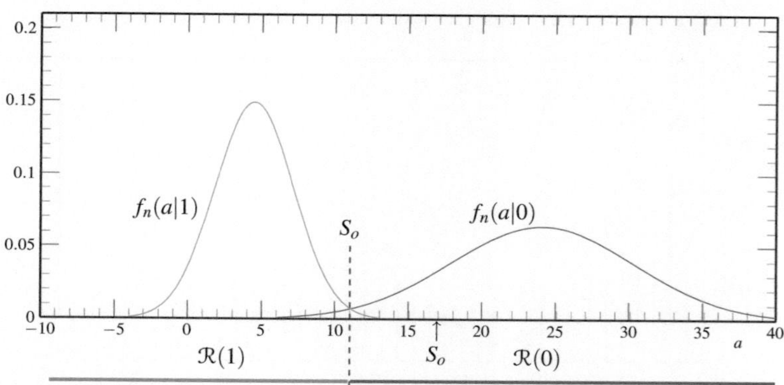

Fig. 8.9 Gaussian probability densities $f_n(a|1)$ and $f_n(a|1)$ $A_0 = 0, 1, 2$ and corresponding optimal decision regions $\mathcal{R}(1)$ and $\mathcal{R}(0)$. The optimal threshold is determined by the condition $f_n(S_o|1) = f_n(S_o|1)$

Then the cross-transition probabilities are given by

$$p_{\text{Gauss}}(1|0) = Q\left(\beta\,\Lambda_0^{1/2}\right), \qquad p_{\text{Gauss}}(0|1) = Q\left((1-\beta)\,\Lambda_1^{1/2}\right) \tag{8.33}$$

where Λ_0 and Λ_1 are the SNRs

$$\Lambda_0 = \frac{[\bar{n}(0) - \bar{n}(1)]^2}{\sigma_n^2(0)}, \qquad \Lambda_1 = \frac{[\bar{n}(0) - \bar{n}(1)]^2}{\sigma_n^2(1)} . \tag{8.34}$$

The error probability becomes

$$P_{e,\text{Gauss}} = \tfrac{1}{2} Q\left(\beta\,\Lambda_0^{1/2}\right) + \tfrac{1}{2} Q\left((1-\beta)\,\Lambda_1^{1/2}\right) \tag{8.35}$$

and it depends on the threshold S through the parameter β. The optimal threshold S_o is determined by the condition $f_n(S_o|1) = f_n(S_o|0)$ and its evaluation is cumbersome because the variances are different. We find that S_o is determined by the following value of β

$$\beta_o = \frac{1}{c-1}\left\{\left[c + \frac{c-1}{2\Lambda_0}\log c\right]^{1/2} - 1\right\}, \qquad c := \frac{\sigma_n^2(0)}{\sigma_n^2(1)} = \frac{\Lambda_1}{\Lambda_0} . \tag{8.36}$$

8.6.3 Asymptotic Behavior of the Gaussian Approximation

We now introduce the condition $N_L \gg N_R, \mathcal{N}$ in the above relations. The numbers $N_\gamma(A_0)$ become

$$N_\gamma(0) = N_L(1 + N_R/N_L + 2\sqrt{N_R/N_L}) \;\to\; N_L$$
$$N_\gamma(1) = N_L(1 + N_R/N_L - 2\sqrt{N_R/N_L}) \;\to\; N_L .$$

Analogously, we find

$$\sigma_n^2(A_0) \;\to\; N_L(1 + 2\mathcal{N}), \qquad c \to 1$$

and also

$$\lim_{c\to 1}\beta_o = \lim_{c\to 1}\frac{1}{c-1}\left\{\left[c + \frac{c-1}{2\Lambda_0}\log c\right]^{1/2} - 1\right\} = \frac{1}{2} .$$

In the SNR we have $\bar{n}(0) - \bar{n}(1) = 4\sqrt{N_R N_L}$ and

$$\Lambda_0 = \frac{16 N_R N_L}{\sigma_n^2(0)} \;\to\; \frac{16 N_R}{1 + 2\mathcal{N}}, \qquad \Lambda_1 = \frac{16 N_R N_L}{\sigma_n^2(1)} \;\to\; \frac{16 N_R}{1 + 2\mathcal{N}} .$$

In conclusion, the asymptotic error probability is

$$P_{e,\text{as.}} = Q(\tfrac{1}{2}\Lambda_0^{1/2}) = Q(\Lambda^{1/2}) \quad \text{with} \quad \Lambda = \frac{4N_R}{1+2\mathcal{N}} \tag{8.37}$$

which is the expression used both in the absence (Chap. 7) and in the presence of thermal noise (present chapter).

Note that the asymptotic cross-transition probabilities become equal and given by the error probability $p_{\text{as}}(0|1) = p_{\text{as}}(1|0) = P_{e,\text{as}}$.

8.6.4 Numerical Evaluation

We have seen the exact probabilities $p(0|1)$, $p(1|0)$, and P_e, their Gaussian approximations $p_{\text{Gauss}}(0|1)$, $p_{\text{Gauss}}(1|0)$, and $P_{e,\text{Gauss}}$, and the asymptotic form of the latter. In all the three cases the probabilities can be evaluated from the parameters

$$N_R, \qquad N_L, \qquad \mathcal{N}.$$

We have systematically evaluated the exact error probability versus N_L for fixed values of N_R and $\mathcal{N}$ and compared it with the Gaussian approximation and the asymptotic Gaussian approximation. An example of evaluation is shown in Fig. 8.10 for $N_R = 3.0$ and $\mathcal{N} = 0.3$. Note that the exact error probability has a jumping behavior due to the discrete choice of threshold S_o. The asymptotic value is reached at about $N_L = 25$, but considering that a precise value as $P_e = 3.123\ 10^{-3}$ has not a practical interest and $P_e = 3\ 10^{-3}$ is sufficient for the applications, we see that the value obtained with $N_L = 10$ is a good approximation.

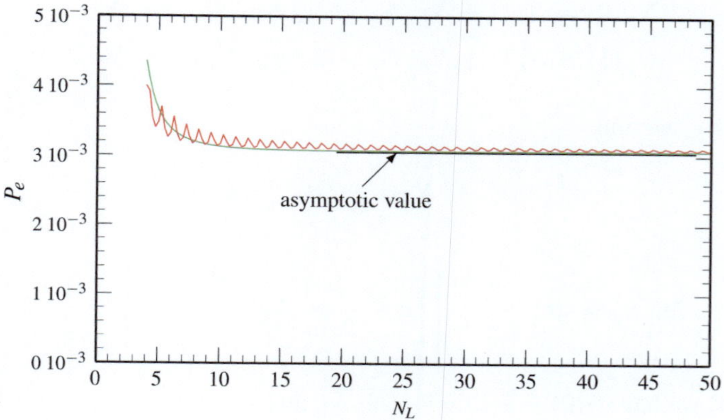

Fig. 8.10 Error probability in BPSK versus N_L for $N_R = 3.0$ and $\mathcal{N} = 0.3$. In *red* the exact value and in *green* the Gaussian approximation

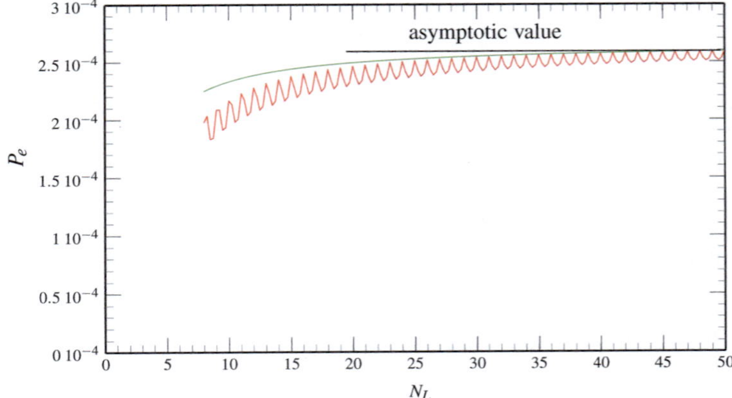

Fig. 8.11 Error probability in BPSK versus N_L for $N_R = 3.0$ and $\mathcal{N} = 0$ (absence of thermal noise). In *red* the exact value and in *green* the Gaussian approximation

We have also made evaluations **in the absence of thermal noise**, that is, with $\mathcal{N} = 0$, where the Laguerre distribution becomes a Poisson distribution. An example is shown in Fig. 8.11. In this case, the jumping behavior of the exact error probability is more evident and the asymptotic value is reached with about $N_L = 50$. Also in this case the conclusion is that the asymptotic Gaussian approximation given by (8.37) represents a very good estimation of the error probability in Classical Communications systems.

A Numerical Example

N_L	P_e	$P_{e,\text{Gauss}}$	$P_{e,\text{as.}}$
10	0.000792231	0.0812565	0.0126737
30	0.0813477	0.0400633	0.0126737
50	0.0400787	0.0298845	0.0126737
150	0.0184381	0.0187132	0.0126737
170	0.0173542	0.0180197	0.0126737
190	0.0164959	0.0174692	0.0126737

8.7 Quantum Communications Systems with Thermal Noise

We reconsider the general scheme of a Quantum Communications system in the presence of thermal noise shown in Fig. 8.1, where Alice "prepares" the quantum system $\mathcal{H}$ in one of the coherent states $|\gamma_i\rangle$, $i \in \mathcal{A}$, as in the previous chapter, but now Bob receives the "noisy" Glauber density operator $\rho(\gamma_i)$, with nominal state $|\gamma_i\rangle$. The expression of $\rho(\gamma_i)$ is obtained from (8.8) with $\gamma = \gamma_i$

$$\rho_i = \rho(\gamma_i) = \frac{1}{\pi \mathcal{N}} \int_{\mathbb{C}} \exp\left(-\frac{|\alpha - \gamma_i|^2}{\mathcal{N}}\right) |\alpha\rangle \langle\alpha| \mathrm{d}\alpha . \tag{8.38}$$

As seen above, these operators have infinite dimensions, but they are approximated by square matrices R_i of finite dimensions $n \times n$. With this approximation, we operate in an n-dimensional Hilbert space (isomorphic to $\mathbb{C}^n$). To simplify the notation, the finite-dimensional approximating matrices will be denoted by the same symbol as the density operators, that is, by ρ_i.

Before proceeding it is convenient to recall the several constellations involved in the analysis, summarized in Sect. 7.4.3:

- constellation of *normalized complex symbols* $\mathcal{C}_0 = \{\bar{\gamma}_0, \bar{\gamma}_1, \dots, \bar{\gamma}_{K-1}\}$,
- constellation of Δ–*scaled complex symbols* $\mathcal{C} = \{\gamma_0, \gamma_1, \dots, \gamma_{K-1}\}$, with $\gamma_i = \bar{\gamma}_0 \Delta$,
- constellation of *coherent states* $\mathcal{S} = \{|\gamma_0\rangle, |\gamma_1\rangle, \dots, |\gamma_{K-1}\rangle\}$.
  To the above we now add the
- constellation of density operators

$$\mathcal{S}_\rho = \{\rho_0, \rho_1, \dots, \rho_{K-1}\} \quad \text{with} \quad \rho_i = \rho(\gamma_i) \tag{8.39}$$

which collects the possible noisy states seen by Bob.

8.7.1 *Quantum Decision in the Presence of Thermal Noise*

Now we assume as known the constellation of the density operators $\mathcal{S}_\rho$ and apply the decision theory developed in Chaps. 5 and 6, more specifically, the part concerning *mixed states*, which is now briefly recalled.

For each of the $n \times n$ matrix ρ_i, we evaluate the factor β_i of dimensions $n \times h_i$, where h_i is the rank of ρ_i. From these factors, regarded as blocks of h_i columns, we form the state matrix

$$\underset{n \times H}{\Gamma} = \left[\beta_0, \beta_1, \dots, \beta_{K-1}\right] \tag{8.40}$$

where the number of columns H is given by the global number of the factor columns $H = h_0 + h_1 + \cdots + h_{K-1}$. In such a way, the states seen by Bob are described by the constellation of density operators (8.39), as well as by the state matrix (8.40). The source specification is completed by the prior probabilities $q_i = \mathrm{P}[A_n = i] = \mathrm{P}[C_n = \gamma_i]$; but for simplicity in this chapter, we suppose equally likely symbols, that is, $q_i = 1/K$.

The target is to find the *optimal* measurement operators $Q_i, i \in \mathcal{A}$ (which minimize the error probability). The generalized Kennedy's theorem (see Sect. 5.11) states that, in the case of mixed states, the optimal measurement operators can be factored in the form $Q_i = \mu_i \mu_i^*$, where the *measurement factors* μ_i have the same dimensions $N \times h_i$ as the corresponding state factors β_i. Hence, one constructs the measurement matrix

$$\underset{n \times H}{M} = [\mu_0, \mu_1, \ldots, \mu_{K-1}] \tag{8.41}$$

which is constrained by the completeness condition (5.70). For the evaluation of probabilities we calculate the mixed product matrix

$$B = [b_{ij}] = M^* \, \Gamma = [\mu_i^* \beta_j] \,.$$

Then one gets the transition probabilities and the correct decision probability as (see (5.75))

$$p_c(j|\,i) = \mathrm{Tr}\left[b_{ji}^* \, b_{ji}\right], \qquad P_c = \frac{1}{K} \sum_{i=0}^{K-1} \mathrm{Tr}\left[b_{ii}^* \, b_{ii}\right] \,. \tag{8.42}$$

We recall that the explicit optimal solution is only known for binary systems (Helstrom's theory), while for multilevel systems ($K \geq 3$) to get optimal solutions one has to use numeric programming techniques, convex semidefinite programming (CSP), or the square root measurement (SRM); the latter gives suboptimal results but with a fair approximation of the minimum error probability.

8.7.2 CSP Optimization with Mixed States

In Sect. 5.10 we have seen the numerical optimization based on convex semidefinite programming (CSP) for pure states. This optimization can also be used for a constellation of mixed states. Here we give the MatLab implementation (CVX) for a constellation of $K = 16$ mixed states:

```
cvx_begin SDP
 variable X(dim, dim) hermitian
 minimize(trace(X))
   subject to
     X>rho0;     X>rho1;   ... ; X>rho15;
cvx_end
copt=cvx_optval
t=(copt);t=trace(X)
Pe=1.0-t
```

This procedure will be applied in Sect. 8.9 for QAM and, with a specific version, for PPM in the final sections.

8.7.3 SRM Decision with Mixed States

We summarize the fundamental steps to evaluate the Quantum Communications performance with the SRM method, developed in Sect. 6.3.

From the state matrix (8.40) we evaluate the Gram operator

$$\underset{n \times n}{T} = \Gamma \, \Gamma^* = \sum_{i=0}^{K-1} \beta_i \beta_i^* = \sum_{i=0}^{K-1} \rho_i \tag{8.43}$$

and the Gram matrix

$$\underset{H \times H}{G} = \Gamma^* \, \Gamma = [\beta_i^* \, \beta_j] \, . \tag{8.44}$$

From Theorem 6.1 and the subsequent corollaries, the measurement matrix that minimizes the square error has three distinct expressions

$$M_0 = U_r \, V_r^* = T^{-1/2} \Gamma = \Gamma G^{-1/2} \, . \tag{8.45}$$

The first one is related to the reduced SVD of the state matrix Γ, while the other two are obtained from the reduced EID of T and of G, respectively.

At this point it is important to remark the differences with respect to the decision with pure states. With mixed states the dimension of the Gram operator is still $n \times n$, where n is the dimension of the Hilbert space, theoretically infinite, but in practice determined by the accuracy we assume to approximate the density operators. The dimension $H \times H$ of the Gram matrix G is determined by the "virtual" rank of the density operators. Without restriction we may suppose that the ranks are equal, $h_i = h_0$, and hence the size of the matrix G, consisting of $K \times K$ blocks of dimensions $h_0 \times h_0$, results in $H = K \, h_0$.

If $n > H$ we use the Gram operator approach, otherwise, the Gram matrix approach.

Gram Matrix Approach

(1) Evaluate the reduced EID $G = V_r \Sigma_r^2 V_r^*$.
(2) Find the inverse square root $G^{-1/2} = V_r \Sigma_r^{-1} V_r^*$.
(3) Subdivide the matrices $G^{\pm 1/2}$ into blocks $\left(G^{\pm \frac{1}{2}}\right)_{ij}$ of dimensions $h_i \times h_j$.
(4) Evaluate the measurement matrix as $M = \Gamma \, G^{-1/2}$.
(5) Considering that $B = G^{-1/2}$, evaluate the transition probabilities from (8.42), with b_{ij} given by $(G^{-1/2})_{ij}$.

The computational complexity of the whole procedure is concentrated in the EID of the $H \times H$ Gram matrix.

Gram Operator Approach

(1) Evaluate the reduced EID $T = U_r \Sigma_r^2 U_r^*$.
(2) Find the inverse square root $T^{-1/2} = U_r \Sigma_r^{-1} U_r^*$.

(3) Find the measurement vectors as $\mu_i = T^{-1/2}\beta_i$, where β_i are the factors of the density operators.
(4) Evaluate the transition probabilities and the correct decision probability from (8.42), where $b_{ij} = \mu_j^* \beta_i$.

The computational complexity of the whole procedure is concentrated in the EID of the $n \times n$ Gram operator.

8.7.4 SRM in the Presence of GUS

We recall from Sects. 5.13 and 6.6 that the *geometric uniform symmetry* (GUS) with mixed stated implies the existence of a unitary operator (symmetry operator) S with the property $S^K = I_{\mathcal{H}}$ and that the constellation $\mathcal{S}_\rho$ of K density operators can be obtained by the reference operator ρ_0 according to

$$\rho_i = S^i \rho_0 S^{-i}, \qquad i = 0, 1, \ldots, K - 1 . \tag{8.46}$$

The factors verify the symmetry conditions in the form

$$\beta_i = S^i \beta_0, \qquad \mu_i = S^i \mu_0, \qquad i = 0, 1, \ldots, K - 1 . \tag{8.47}$$

With the GUS the SRM method is simplified, as in the case of pure states, with some complications because the Gram matrix G, instead of circulant, is *block circulant*.

Given the reference state vector β_0 (or the reference density operator ρ_0), and the symmetry operator S, the correct decision probability P_c is evaluated as follows.

Gram Operator Approach (see Proposition 6.2):

(1) Evaluate the inverse square root $T^{-1/2}$ of T.
(2) Evaluate the reference measurement operator as $Q_0 = T^{-1/2} \rho_0 T^{-1/2}$.
(3) Evaluate the transition probabilities as $p_c(j|i) = \text{Tr}\left[S^{i-j} \rho_0 S^{-(i-j)} Q_0 \right]$.
(4) The correct decision probability is obtained as

$$P_c = \text{Tr}\left[\rho_0 T^{-1/2} \right]^2 . \tag{8.48}$$

The computational complexity is confined to the evaluation of $T^{-1/2}$.

Gram Matrix Approach (see Proposition 6.3):

(1) Evaluate the matrices of dimension $h_0 \times h_0$

$$D_k = \sum_{i=0}^{K-1} \beta_0^* \beta_i W_K^{-ki}, \qquad \beta_i = S^i \beta_0 . \tag{8.49}$$

(2) Evaluate by EID the square roots $D_k^{1/2}$.

(3) Evaluate the (i, j) block of $G^{1/2}$ as

$$(G^{1/2})_{ij} = \frac{1}{\sqrt{K}} \sum_{k=0}^{K-1} W_K^{k(j-i)} D_k^{1/2} \,.$$

(4) Evaluate the transition probabilities as $p_c(j|i) = \mathrm{Tr}\left[(G^{1/2})_{ji}^{*} (G^{1/2})_{ji}\right]$.

(5) The correct decision probability is given by

$$P_c = \mathrm{Tr}\left(\frac{1}{K} \sum_{k=0}^{K-1} D_k^{1/2} \right)^2 . \tag{8.50}$$

The computational complexity is confined to the evaluation of the square roots $D_k^{1/2}$.

Remark With mixed states, also in the presence of GUS, the decision obtained with the SRM in general is not optimal.

8.8 Binary Systems in the Presence of Thermal Noise

In the quantum decision theory developed in Sect. 5.3 we have seen the optimization of a quantum binary system (Helstrom's theory), which is valid also for mixed states. We briefly recall this theory assuming equally likely symbols ($q_0 = q_1 = \frac{1}{2}$):

(1) We start from the EID of the decision operator

$$D = \frac{1}{2}\left[\rho(\beta_1) - \rho(\beta_0)\right] = \sum_k \eta_k |\eta_k\rangle \langle \eta_k| \tag{8.51}$$

where η_k are the eigenvalues and $|\eta_k\rangle$ the corresponding eigenvectors,

(2) The two optimal measurement operators are given by the sum of the elementary projectors $|\eta_k\rangle\langle\eta_k|$, and specifically

$$Q_0 = \sum_{\eta_k<0} |\eta_k\rangle\langle\eta_k|, \qquad Q_1 = \sum_{\eta_k>0} |\eta_k\rangle\langle\eta_k| \tag{8.52}$$

(3) The maximum correct decision probability is obtained as the sum of the positive eigenvalues according to

$$P_c = \frac{1}{2} + \sum_{\eta_k>0} \eta_k \quad \rightarrow \quad P_e = \frac{1}{2} - \sum_{\eta_k>0} \eta_k. \tag{8.53}$$

Using the expansion of $\rho(\gamma)$ given by (8.10), for the decision operator D one gets the following matrix representation $[D_{mn}]$ of D

$$D_{mn} = \frac{1}{2}[R_{mn}(\gamma_1) - R_{mn}(\gamma_0)], \qquad m, n = 0, 1, 2, \ldots$$

To proceed we need to approximate this matrix to finite dimensions $n \times n$ according to the criterion developed in Sect. 8.4.

8.8.1 Application to BPSK Modulation

The constellation of scaled symbols results in

$$\gamma_0 = -\alpha \qquad \gamma_1 = \alpha.$$

To get the error probability we evaluate the matrix representation of the decision operator D limited to a finite number n of terms.

Let us consider an explicit example with the following data

$$q_0 = q_1 = \tfrac{1}{2}, \qquad \mathcal{N} = 0.2, \qquad \alpha = 1 \;\rightarrow\; N_s = 1.$$

The evaluation of the matrix D is limited to 30×30 terms and here we write only the upper left portion

$$D = \begin{bmatrix}
0.000 & 0.302 & 0.000 & 0.086 & 0.000 & 0.013 & 0.000 & 0.001 & 0.000 & 0.000 \\
0.302 & 0.000 & 0.219 & 0.000 & 0.058 & 0.000 & 0.009 & 0.000 & 0.001 & 0.000 \\
0.000 & 0.219 & 0.000 & 0.117 & 0.000 & 0.030 & 0.000 & 0.005 & 0.000 & 0.001 \\
0.086 & 0.000 & 0.117 & 0.000 & 0.053 & 0.000 & 0.013 & 0.000 & 0.002 & 0.000 \\
0.000 & 0.058 & 0.000 & 0.053 & 0.000 & 0.021 & 0.000 & 0.005 & 0.000 & 0.001 \\
0.013 & 0.000 & 0.030 & 0.000 & 0.021 & 0.000 & 0.008 & 0.000 & 0.002 & 0.000 \\
0.000 & 0.009 & 0.000 & 0.013 & 0.000 & 0.008 & 0.000 & 0.003 & 0.000 & 0.001 \\
0.001 & 0.000 & 0.005 & 0.000 & 0.005 & 0.000 & 0.003 & 0.000 & 0.001 & 0.000 \\
0.000 & 0.001 & 0.000 & 0.002 & 0.000 & 0.002 & 0.000 & 0.001 & 0.000 & 0.000 \\
0.000 & 0.000 & 0.001 & 0.000 & 0.001 & 0.000 & 0.001 & 0.000 & 0.000 & 0.000 \\
& & & & \vdots & & & & & \ddots
\end{bmatrix}$$

The corresponding eigenvalues result in

$$[\lambda_1, \lambda_2, \lambda_3, \ldots] = [-0.4063, 0.4063, -0.05621, 0.05621, -0.00581, 0.00581, \ldots]$$

with an alternation of positive and negative values. The error probability P_e is found by summing the positive eigenvalues, according to (8.53)

$$P_e = \frac{1}{2} - \sum_{\eta_k > 0} \eta_k \simeq 0.031118.$$

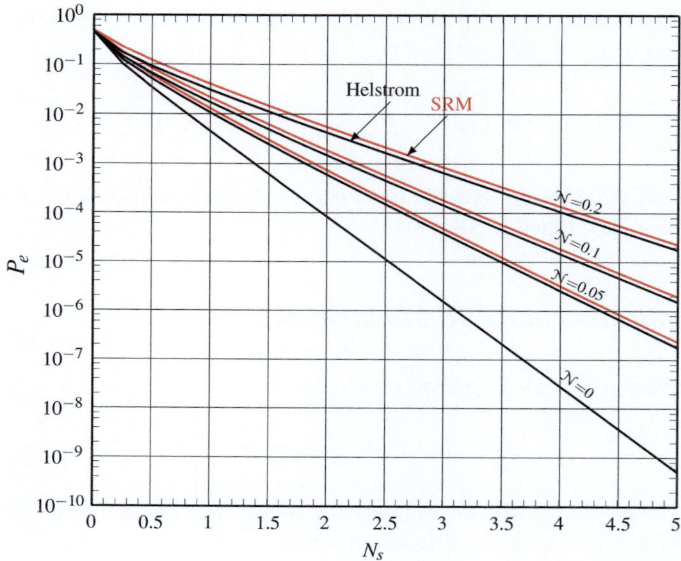

Fig. 8.12 Error probability in BPSK system as a function of N_s for some values of $\mathcal{N}$. Comparison between the SRM method and Helstrom optimal decision. For $\mathcal{N}=0$ the values coincide because SRM method becomes optimal

In the absence of thermal noise ($\mathcal{N}=0$) the error probability would be $P_e = 0.00460$, and we see that the presence of thermal noise with $\mathcal{N}=0.2$ leads to a worsening in P_e by one decade.

The error probability is shown in Fig. 8.12 as a function of the number of signal photons per symbol N_s for some values of the number of thermal photons $\mathcal{N}$. Note that for $\mathcal{N}=0$ the density operators degenerate to the form $\rho(\pm\alpha) = |\pm\alpha\rangle\langle\pm\alpha|$ and the evaluation can be done in exact form according to (7.109), that is,

$$P_e = \tfrac{1}{2}\left[1 - \sqrt{1 - e^{-4N_s}}\right].\tag{8.54}$$

Figure 8.12 also shows the error probability obtained with the SRM method, which is not optimal with mixed states, but it gives a good approximation.

Classical BPSK System and Comparison

The performance of a classical optical BPSK system, in the absence of thermal noise, has been evaluated in Sect. 7.10. With a homodyne receiver we found that the signal at the decision point has the form given by (7.77), that is,

$$V_0 + U_0 \cos(\pi A_0) + u$$

where V_0 is a bias term, U_0 is the useful amplitude, $\cos(\pi A_0) = \pm 1$, and u is the shot noise, which, with a strong local carrier, may be considered Gaussian. In the same section we also evaluated the signal-to-noise ratio $\Lambda = 4N_R = 4N_s$, which gave the expression $Q(\sqrt{4N_s})$ for the error probability.

According to the theory of Sect. 8.5, for the evaluation of the error probability in the presence of thermal noise it is sufficient to decrease the SNR according to (8.28) of Proposition 8.1, where the decreasing is due to the passage from the Poisson regime to the Laguerre regime. In such a way one gets

$$P_{e,\text{classic}} = Q\left(\Lambda\frac{1}{1+2N}\right) = Q\left(\frac{4N_s}{1+2N}\right). \tag{8.55}$$

This result gives the error probability as a function of the parameters N_s and $\mathcal{N}$ considered for the quantum BPSK system, and thus it allows for a comparison, which is shown in Fig. 8.13. One can realize the clear superiority of quantum systems also in the presence of thermal noise, mainly for low values of the number of thermal photons $\mathcal{N}$.

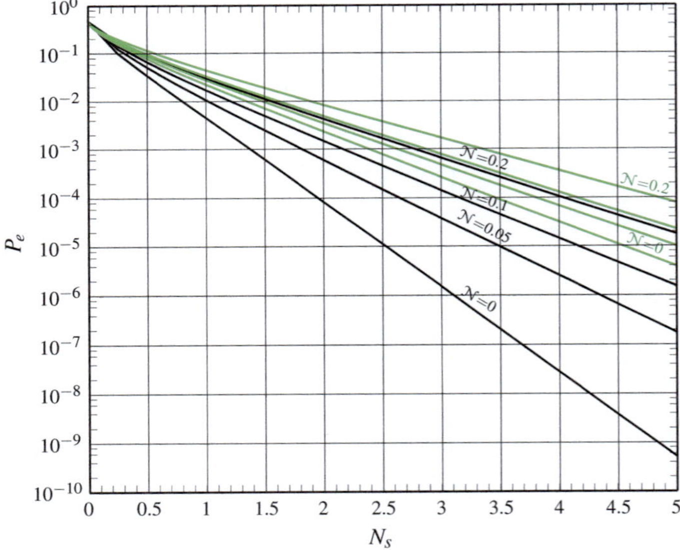

Fig. 8.13 Error probability in BPSK system as a function of N_s for some values of $\mathcal{N}$. *Black lines* refer to the quantum system, *green lines* to the classical homodyne system

8.8.2 *Application of OOK Modulation*

The constellation of scaled symbols is

$$\gamma_0 = 0 \qquad \gamma_1 = \alpha$$

where the coherent state $|\gamma_0\rangle = |0\rangle$ is the ground state and the corresponding density operator is obtained from (8.2); we find a diagonal matrix representation (see (8.4)) with elements

$$R_{mn}(0) = (1 - v)v^n \delta_{mn}, \qquad v = \mathcal{N}/(\mathcal{N} + 1).$$

For the state $|\alpha\rangle$ with $\alpha \neq 0$ the matrix is not diagonal and must be calculated from (8.10a). Apart from the difference due to the presence of the ground state, the error probability is calculated as in the previous case.

For instance, with

$$q_0 = q_1 = \tfrac{1}{2}, \qquad \mathcal{N} = 0.2, \qquad \alpha = \sqrt{2} \;\rightarrow\; N_s = \tfrac{1}{2}N_\alpha = 1$$

the error probability results in $P_e = \simeq 0.1046$, while in the absence of thermal noise ($\mathcal{N} = 0$) we would have $P_e = 0.03506$.

Figure 8.14 shows the error probability P_e as a function of the number of signal photons per symbol N_s for some values of $\mathcal{N}$. Note that also in this case with $\mathcal{N} = 0$

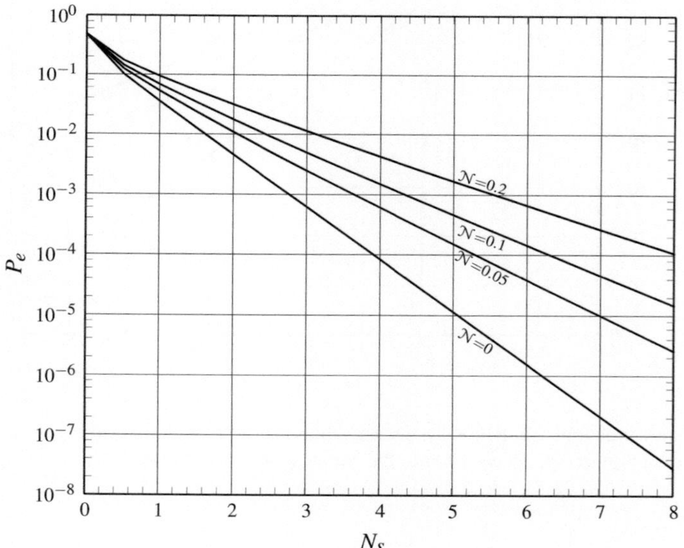

Fig. 8.14 Error probability in OOK system as a function of N_s for some values of $\mathcal{N}$

the evaluation can be done in a closed form from (7.2) of the previous chapter, that is,

$$P_e = \frac{1}{2}\left[1 - \sqrt{1 - e^{-2N_s}}\right] . \tag{8.56}$$

One can realize that also for the OOK the quantum system is superior to the corresponding classical system, but in a less relevant form with respect to the BPSK system.

Comment. As said above, apart from the binary case, no general solutions are available for the optimization of Quantum Communications systems, for which we will develop suboptimal techniques. The results obtained with the binary case are useful to test the approximations achieved with the suboptimal techniques.

8.9 QAM Systems in the Presence of Thermal Noise

The quantum QAM (quadrature amplitude modulation) systems have been analyzed in Sect. 7.11 with pure states, that is, in the absence of thermal noise, whereas now they are analyzed with density operators, that is, in the presence of background noise. To evaluate the performance we use the SRM method extended to the density operators. As the QAM constellation does not enjoy the uniform geometry, we cannot use the simplifications that such symmetry implies.

From Sect. 7.11 we recall that the constellation of the QAM modulation is constituted by $K = L^2$ points equally spaced on a square grid of the complex plane, defined through the L-ary alphabet $\mathcal{A}_L = \{-(L-1)+2(i-1)|\ i = 1, 2, \ldots, L\}$ with $L = 2, 3, 4, \ldots$. The K-ary QAM constellation of scaled symbols is then formed by the complex numbers

$$\mathcal{C} = \{\gamma_{uv} = \Delta(u + iv)|\ u, v \in \mathcal{A}_L\}$$

where Δ is a scale factor. The constellation of the 16-QAM system that will be analyzed here in the presence of noise, is illustrated in Fig. 8.15.

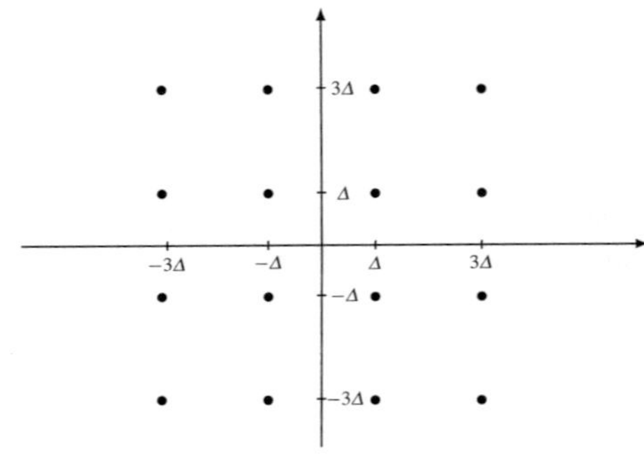

Fig. 8.15 Constellation of the 16-QAM with scale factor Δ

To evaluate the performance **in the absence of thermal noise**, we started from the Gram matrix G, whose elements are the inner products $\langle \gamma_{uv} | \gamma_{u'v'} \rangle$, $u, v, u', v' \in \mathcal{A}_L$ (see (7.113)). Having computed the eigenvalues and eigenvectors of G, we calculated the square root $G^{\frac{1}{2}}$, and hence the error probability according to

$$P_{e,0} = 1 - \frac{1}{K} \sum_{i=0}^{K-1} \left[(G^{\frac{1}{2}})_{ii} \right]^2 . \tag{8.57}$$

The performance was evaluated as a function of the number of signal photons per symbol, which for the QAM with equiprobable symbols results in

$$N_s = \frac{2}{3}(L^2 - 1)\Delta^2 = \frac{2}{3}(K - 1)\Delta^2 \tag{8.58}$$

with a shape factor $\mu_K = \frac{2}{3}(K - 1)$.

In the presence of thermal noise we have to consider a constellation of $K = L^2$ Glauber density operators

$$\rho_{uv} = \rho(\Delta(u + \mathrm{i}\,v)), \qquad u, v \in \mathcal{A}_L$$

which does not verify the GUS. We apply the SRM method recalled in Sect. 8.7. For the evaluation of the error probability one can use both the Gram operator and the Gram matrix, choosing the one requiring a lower computational complexity. In the numerical evaluation, the main problem is to handle the approximations in an appropriate form. In this modulation format, the number of signal photons associated to the symbol $\gamma = \gamma_{uv}$, given by $N_\gamma = |(u + \mathrm{i}v)\Delta|^2$, is not uniform and varies from $N_\gamma = 2\Delta^2$ for the inner symbols to $N_\gamma = 2(L - 1)^2\Delta^2$ for the corner symbols. Then the reduced dimensions of the Hilbert space $n = n_\varepsilon$ must be chosen based on the maximum

$$N_{\gamma,\max} = 2(L - 1)^2\Delta^2$$

and, assuming N_s as the fundamental parameter to express the final result (recall that Δ^2 can be expressed in terms of N_s using (7.115)). For the choice of n_ε we must bear in mind that

$$N_{\gamma,\max} = 3[(L - 1)^2/(L^2 - 1)] N_s = 3(L - 1)/(L + 1) N_s .$$

In particular, in the 16-QAM we have $N_{\gamma,\max} = 1.8N_s$.

We give a numerical example to illustrate the dimensions of the various matrices. With $N_s = 4$ and $\mathcal{N} = 0.1$ we obtain $N_{\gamma,\max} = 7.2$ and, choosing $n_\varepsilon = 40$, an accuracy $\varepsilon = 10^{-7}$ is ensured. The matrices $\rho_i \simeq R_i$ are 40×40, and are factored in the form $\beta_i \beta_i^*$, where the factors β_i are 40×8. The dimensions are

$$\underset{40 \times 118}{\beta_i} \qquad \underset{40 \times 40}{T} \qquad \underset{118 \times 118}{G} \quad .$$

Therefore, it is more convenient to calculate $T^{-1/2}$ instead of $G^{\pm 1/2}$. With these choices we find the following diagonal transition probabilities

- inner states : 0.87575 • side states : 0.91650 • corner states : 0.94777

and the error probability results in $P_e = 0.08587$.

We have applied systematically the SRM approach to evaluate the error probability P_e in the 16-QAM, following the procedure indicated above. The results are illustrated in Fig. 8.16, where P_e is shown as a function of the number of signal photons per symbol N_s for some values of the number of thermal photons $\mathcal{N}$. In particular, we find that the shape of the function for $\mathcal{N} = 0$ (absence of noise) is in perfect agreement with the results of the previous chapter with pure states. To ensure an accuracy of $\varepsilon = 10^{-7}$, the Hilbert space has been chosen with a dimension of $n = n_\varepsilon = 130$.

The same figure shows a comparison of the performance of the quantum receiver with the classical homodyne receiver, for which the evaluation has been done considering the degradation of the signal-to-noise ratio according to (8.28). The advantage of the quantum receiver, which in absence of noise is about 3 dB, quickly decreases with increasing noise quantified by the parameter $\mathcal{N}$.

8.9.1 Comparison of CSP and SRM Evaluation

The SRM approach used above does not give the minimum error probability. For this reason with have evaluated the minimum error probability by the convex semi-definite programming (CSD), implemented in MatLab by the CVX procedure (see Sect. 8.7.2), which gives (numerically) this minimum. The results of the two evaluation are compared in Fig. 8.17 for the 16-QAM, where the error probability P_e is plotted as a function of the number of signal photons per symbol N_s for a few values of the number of thermal photons $\mathcal{N}$. We realize that the SRM gives an overestimation of P_e, which may be acceptable in practice.

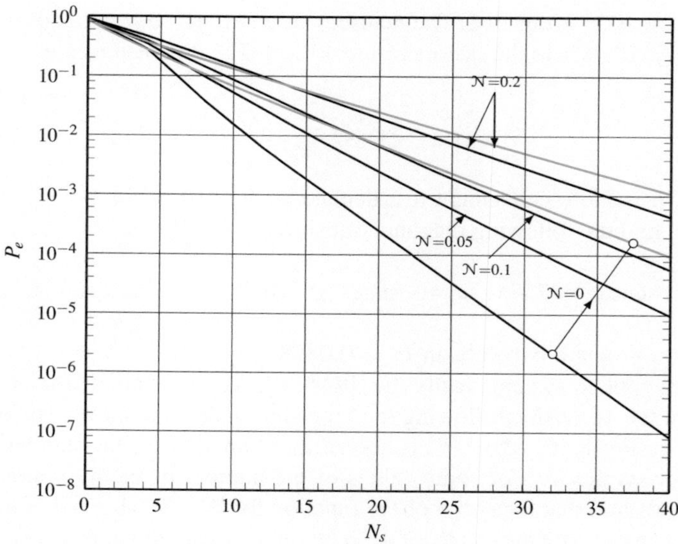

Fig. 8.16 Error probability in 16-QAM versus N_s for some values of $\mathcal{N}$. *Black lines* refer to quantum detection and *green lines* to classical homodyne detection

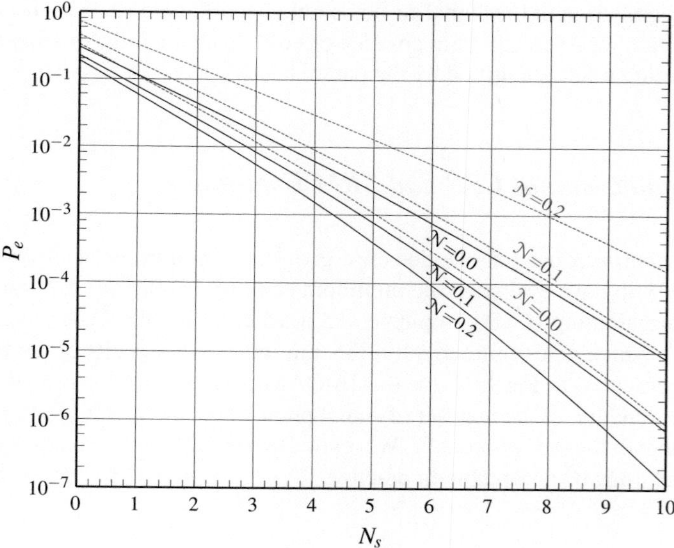

Fig. 8.17 Comparison of error probability in 16-QAM versus N_s for some values of $\mathcal{N}$. *Black lines* refer to CVX evaluation and *red lines* to SRM evaluation

8.10 PSK Systems in the Presence of Thermal Noise

The PSK (phase shift keying) quantum systems have been studied in Sect. 7.12 with pure states. Now they will be analyzed with density operators, which account for the presence of thermal noise.

From Sect. 7.12 we recall that the scaled constellation of the PSK modulation consists of K points uniformly distributed on a circle of the complex plane

$$\mathcal{C} = \{\Delta\, W_K^m \mid m = 0, 1, \ldots, K - 1\} \tag{8.59}$$

where the scale factor Δ is given by the radius of the circle and $W_K = e^{i2\pi/K}$. We have also seen that the constellation satisfies the *geometrically uniform symmetry* (GUS) with symmetry operator

$$S = \exp\left(\frac{i2\pi}{K} N\right) = \exp(W_K N) \tag{8.60}$$

where N is the number operator. This allowed us to establish that the SRM method with pure states (**absence of thermal noise**) is optimal and then the minimal error probability can be easily obtained using the Gram matrix approach. Specifically, (see (7.126) and (7.127)) we calculate the eigenvalues as the DFT of the first line of the Gram matrix, from which we obtained the minimal error probability.

In the presence of thermal noise we have the constellation of Glauber density operators
$$\rho_m = \rho(\Delta\, W_K^m), \qquad m = 0, 1, \ldots, K - 1$$

which verifies the GUS with symmetry operator given by (8.60).

The BPSK modulation in the presence of noise has been considered in Sect. 8.8, where the performance obtained by Helstrom's method, which is optimal (the only approximation regards the truncation of the density operators with a matrix of finite dimensions), has been compared with that of the SRM method (see Fig. 8.12). For $K \geq 3$ no optimal measurement operators are exactly known, therefore we apply the SRM method generalized to include the density operators. In a PSK quantum system, differently from the QAM, the treatment of approximations in the numerical evaluation is simpler because the number of signal photons associated to the symbols γ is the same over the whole constellation, that is, $N_\gamma = \Delta^2 = N_s$. In addition, the PSK constellation verifies the GUS, then we can use the simplifications implied by such symmetry, seen in Sect. 8.7.3.

Based on the required accuracy, we choose the dimension $n \times h_0$ of the reference factor γ_0. Then we have

$$\underset{n \times h_0}{\gamma_0} \qquad \underset{n \times n}{T} \qquad \underset{K h_0 \times K h_0}{G} \qquad \underset{h_0 \times h_0}{D_k} \quad .$$

where D_k are the matrices defined by (8.49). We shall see that it is convenient to adopt the Gram matrix approach, which requires doing the EID of the D_k to compute their square root, and eventually we compute the transition probabilities and the error probability.

8.10.1 4-PSK Modulation

We illustrate a specific example of evaluation, and we then discuss the performance in general and make a comparison with the classical 4-PSK system.

Example 8.2 We assume the following data

$$N_s = 4.0 \quad \Delta = \sqrt{N_s} = 2.0 \quad \mathcal{N} = 0.2 \,.$$

With the accuracy $\varepsilon = 10^{-7}$ we must approximate the dimensions for the density operator ρ_0 with $n = 60$. With this accuracy we have a virtual rank of $h_0 = 9$ and we obtain a factor β_0 of dimensions $n \times h_0 = 60 \times 9$. Therefore, T has dimensions $n \times n = 60 \times 60$ while G has dimensions $H \times H = 36 \times 36$. But the EIDs of these matrices are not required and it is sufficient to take the EID of the matrices D_k which have dimensions 9×9.

The transition probabilities result in

$$p_c = \begin{bmatrix} 0.98758 & 0.00602 & 0.00039 & 0.00602 \\ 0.00602 & 0.98758 & 0.00602 & 0.00039 \\ 0.00039 & 0.00602 & 0.98758 & 0.00602 \\ 0.00602 & 0.00039 & 0.00602 & 0.98758 \end{bmatrix}$$

and so

$$P_c = 0.98758 \qquad P_e = 0.01242 \,.$$

As one can see, the diagonal probabilities are equal, in agreement with the fact that the PSK verifies the GUS. However, the "further" condition (6.66) is not verified for the optimality (see Problem 8.5).

Following the lines indicated in the examples, we have systematically calculated the error probability as a function of the number of signal photons per symbol $N_s = \Delta^2 = 2N_R$ for some values of the noise parameter $\mathcal{N}$ (Fig. 8.18). We have verified that for a very small value of $\mathcal{N}$ as 0.00002, the results practically coincide with those obtained in the absence of noise.

Comparison with the Classical 4-PSK System

Figure 8.18 shows the shape of the error probability in the classical 4-PSK system, obtained with the same thermal noise parameters. To this end, we applied the theory developed in Sect. 8.5.1, relative to the quadrature modulation. We recall from the previous chapter (see Sect. 7.12), that in the absence of thermal noise the error

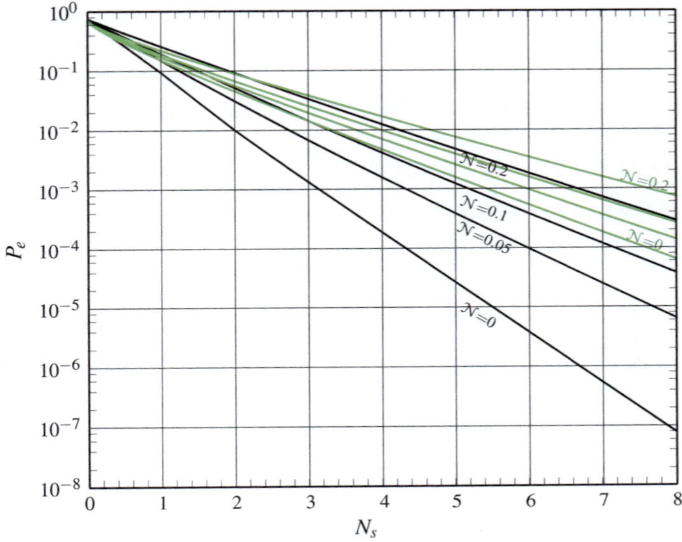

Fig. 8.18 Error probability in 4-PSK as a function of the number of signal photons per symbol N_s for some values of the number of thermal photons N. *Black lines* refer to the quantum system and *green lines* to the classic homodyne system

probability in the 4-PSK system results in $P_e = Q(\sqrt{\Lambda})$, where the signal-to-noise ratio is $\Lambda = 4N_s$. To obtain the error probability in the presence of thermal noise it is sufficient to change such ratio according to (8.28). We then have

$$P_{e,\text{classic}} = Q\left(\sqrt{\frac{4N_s}{1 + 2N}}\right). \tag{8.61}$$

The comparison in the figure shows the superiority of the quantum system with respect to the classical one, especially for low levels of thermal noise; with increasing N this superiority weakens.

8.10.2   8-PSK Modulation

Let us start out with an example.

Example 8.3 We assume the following data

$$N_s = 9.0 \quad \Delta = \sqrt{N_s} = 3.0 \quad N = 0.2 .$$

To guarantee the accuracy of the previous example we must choose $n = 64$ (multiple of 8), and $h_0 = 9$, which involves a factor β_0 of dimensions $n \times h_0 = 64 \times 9$ and then the dimensions of β become: $n \times H = 64 \times 72$. Therefore, T has dimensions 64×64, whereas G has dimensions 72×72. The matrices D_k have dimensions 9×9, and so their EID is not a problem.

Developing the computation, we find that the transition probabilities result in

$$
p_c =
\begin{bmatrix}
0.95698 & 0.02148 & 0.00002 & 0.00000 & 0.00000 & 0.00000 & 0.00002 & 0.02148 \\
0.02148 & 0.95698 & 0.02148 & 0.00002 & 0.00000 & 0.00000 & 0.00000 & 0.00002 \\
0.00002 & 0.02148 & 0.95698 & 0.02148 & 0.00002 & 0.00000 & 0.00000 & 0.00000 \\
0.00000 & 0.00002 & 0.02148 & 0.95698 & 0.02148 & 0.00002 & 0.00000 & 0.00000 \\
0.00000 & 0.00000 & 0.00002 & 0.02148 & 0.95698 & 0.02148 & 0.00002 & 0.00000 \\
0.00000 & 0.00000 & 0.00000 & 0.00002 & 0.02148 & 0.95698 & 0.02148 & 0.00002 \\
0.00002 & 0.00000 & 0.00000 & 0.00000 & 0.00002 & 0.02148 & 0.95698 & 0.02148 \\
0.02148 & 0.00002 & 0.00000 & 0.00000 & 0.00000 & 0.00002 & 0.02148 & 0.95698
\end{bmatrix}
$$

and then

$$
P_c = 0.95698 \quad P_e = 0.04302 .
$$

Even in this case the diagonal probabilities are equal, but the "further" optimality condition (6.66) is not verified, therefore, the SRM method gives an overestimate of the minimal error probability.

Along the lines of the example, the error probability has been computed as a function of the number of signal photons per symbol $N_s = \Delta^2 = 3\,N_R$ for some values of the number of thermal photons $\mathcal{N}$ (Fig. 8.19). We have verified that for a very small value of $\mathcal{N}$, as 0.00002, the results practically coincide with those obtained in the absence of noise, according to (7.127).

Comparison with the Classical 8-PSK

Figure 8.19 shows also the shape of the error probability of the classical 8-PSK system. This is calculated bearing in mind (7.129) of the previous chapter, that is, $P_e < P'_e = \exp\left(-2N_s \sin^2 \pi/K\right)$, holding in the absence of thermal noise, and modifying the signal-to-noise ratio in the usual way, i.e.,

$$
P_{e,\mathrm{classic}} < P'_e = \exp\left(-\frac{2N_s}{1 + 2\mathcal{N}}\,\sin^2\frac{\pi}{K}\right)
\tag{8.62}
$$

which with $K = 8$ gives a good approximation.

As in the previous case, the comparison shows the superiority of the quantum system with respect to the classical one, especially for low levels of the thermal noise.

Problem 8.5 $\star\star$ Check that the further condition (6.66), that is, $\mu_0^* \beta_0 = \alpha I$, is not verified in 4-PSK with the data of Example 8.1.

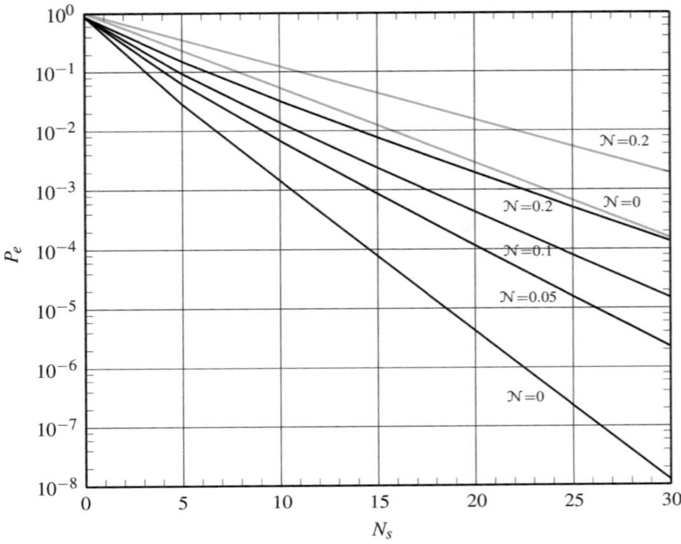

Fig. 8.19 Error probability in 8-PSK as a function of the number of signal photons per symbol N_s for some values of the number of thermal photons $\mathcal{N}$. *Continuous lines* refer to the quantum system and *dashed lines* to the classic homodyne system

8.11 PPM Systems in the Presence of Thermal Noise

We now examine PPM (pulse position modulation) systems in the presence of thermal noise. We point out that this subject is currently being investigated for the potential application of this modulation format to deep-space transmission [6, 8]. The present treatment is the result of a recent research [2] and is very detailed.

Symbolism and Dimensions

The formalization required for the quantum PPM, in the framework of a composite Hilbert space, requires a thorough attention to symbolism and dimensions of the various entities, which we list below for a general K-ary system:

n dimension of the component Hilbert spaces $\mathcal{H}_0$,
$N = n^K$ dimensions of the composite Hilbert space $\mathcal{H} = \mathcal{H}_0^{\otimes K}$,
h rank of the component density operators ρ^0 and ρ^1,
$H = h^K$ rank of the composite density operators ρ_i,
$r = KH = Kh^K$ rank of the state matrix Γ.

8.11.1 Formulation

From Sect. 7.13 of the previous chapter we recall that a K-ary quantum PPM system must be defined on a Hilbert space $\mathcal{H}$ given by the tensor product $\mathcal{H}_0^{\otimes K}$ of K equal Hilbert spaces $\mathcal{H}_0$, where the coherent state associated to the symbol i has the composite form

$$|\beta_i\rangle = |\beta_{i,K-1}\rangle \otimes \cdots \otimes |\beta_{i,1}\rangle \otimes |\beta_{i,0}\rangle, \qquad i = 0, 1, \ldots K - 1 \tag{8.63}$$

with $|\beta_{ij}\rangle = |\Delta\rangle$ for $i = j$ and $|\beta_{ij}\rangle = |0\rangle$ for $i \neq j$, where $|\Delta\rangle$ is the coherent state with *real* parameter Δ, and $|0\rangle$ is the "ground state". For example, for $K = 4$ we have the four states listed in (7.133). We recall that all the states have the same number of signal photons per symbol, given by

$$N_s = \Delta^2 .$$

In the presence of thermal noise the representation must be done in terms of density operators, so that to the symbol i we associate the operator

$$\rho_i = \rho_{i,K-1} \otimes \cdots \otimes \rho_{i,1} \otimes \rho_{i,0}, \qquad i = 0, 1, \ldots K - 1 \tag{8.64}$$

with

$$\rho_{ij} = \begin{cases} \rho(\Delta) & i = j \\ \rho(0) & i \neq j \end{cases} \tag{8.64a}$$

where $\rho(\Delta)$ is the density operator given by (8.1), and $\rho(0)$ is the density operator corresponding to the "ground state", given by (8.2). For example, for $K = 4$ we obtain the four composite density operators

$$\rho_0 = \rho(0) \otimes \rho(0) \otimes \rho(0) \otimes \rho(\Delta)$$
$$\rho_1 = \rho(0) \otimes \rho(0) \otimes \rho(\Delta) \otimes \rho(0)$$
$$\rho_2 = \rho(0) \otimes \rho(\Delta) \otimes \rho(0) \otimes \rho(0)$$
$$\rho_3 = \rho(\Delta) \otimes \rho(0) \otimes \rho(0) \otimes \rho(0) .$$

We now consider the factorization of the composite density operators ρ_i, required to apply the various decision methods. We have the following important result (see Appendix section "Proof of Proposition 8.2 on Factorization").

Proposition 8.2 *Let*

$$\rho^0 = \beta^0 \, (\beta^0)^*, \qquad \rho^1 = \beta^1 \, (\beta^1)^*$$

be the factorizations of the component density operators. Then the factorizations of the composite operators, $\rho_i = \beta_i \, \beta_i^*$, turn out to be given by the factors

$$\beta_i = \beta_{i,K-1} \otimes \cdots \otimes \beta_{i,1} \otimes \beta_{i,0}, \qquad i = 0, 1, \ldots K - 1 . \quad \square \tag{8.65}$$

Notice the perfect symmetry of expressions (8.63), (8.64) and (8.65).

8.11.2 Computation of the Dimensions and the Matrices

The numeric evaluation of the performance requires moving from infinite-dimensional operators to finite-dimensional matrices, and this is not so obvious in PPM because the operators are given by tensor products. It is then appropriate to observe that:

Proposition 8.3 *If A is an operator with (matrix) representation $A_{\mathcal{A}}$ obtained with a basis $\mathcal{A}$ and B is an operator with representation $B_{\mathcal{B}}$ obtained with a basis $\mathcal{B}$, then the representation of the operator obtained with the tensor product, $A \otimes B$, is given by Kronecker's product (see Sect. 2.13.3) of the two matrices, $A_{\mathcal{A}} \otimes B_{\mathcal{B}}$. We remind also that, if $A_{\mathcal{A}}$ and $B_{\mathcal{B}}$ have dimensions, respectively, $m_A \times n_A$ and $m_B \times n_B$, the dimensions of $A_{\mathcal{A}} \otimes B_{\mathcal{B}}$ become $m_A\, m_B \times n_A\, n_B$.*

The extension of these properties to an arbitrary number of factors is straightforward. $\square$

Then the representation of the composite density operators (8.63) can be obtained as follows. The component operators $\rho(\Delta)$ and $\rho(0)$, are approximated by $n \times n$ matrices of appropriate dimensions

$$\rho(\Delta) \simeq \underset{n \times n}{R(\Delta)}, \qquad \rho(0) \simeq \underset{n \times n}{R(0)} \tag{8.66}$$

where $R(\Delta) = [R_{mn}(\Delta)]$ is given by (8.10) and $R(0) = [R_{mn}(0)]$ by (8.7); $R(0)$ is a diagonal matrix. In other words, **with the approximation** (8.66) **we choose the dimension** n **of the component Hilbert spaces** $\mathcal{H}_0$. From Proposition 8.3 the matrix representation of the ρ_i results in

$$R_i = R_{i,K-1} \otimes \cdots \otimes R_{i,1} \otimes R_{i,0}, \qquad i = 0, 1, \ldots K - 1 \tag{8.67}$$

with

$$R_{ij} = \begin{cases} R(\Delta) & \text{for } i = j \\ R(0) & \text{for } i \neq j . \end{cases}$$

In (8.67) we have the Kronecker product of K matrices and, with the approximation (8.66), we have dimensions $\dim(R_i) = n^K \times n^K$, where $N = n^K$ gives the dimension of the composite space $\mathcal{H} = \mathcal{H}_0^{\otimes K}$.

Table 8.2 Dimensions and ranks in quantum K-PPM $(N = n^K, H = h^K)$

Parameter	Symbol	Dimensions	Rank
Elementary factors	β^0, β^1	$n \times h$	h
Elementary density operators	ρ^0, ρ^1	$n \times n$	h
Composite factors	β_i	$N \times H = n^K \times h^K$	$H = h^K$
Composite density operators	ρ_i	$N \times N = n^K \times n^K$	$H = h^K$
State matrix	Γ	$N \times KH = n^K \times Kh^K$	$r = KH = Kh^K$
Gram operator	T	$N \times N = n^K \times n^K$	$KH = Kh^K$
Gram matrix	G	$KH \times KH = Kh^K \times Kh^K$	$KH = Kh^K$

For the factorization we apply Proposition 8.2, which states that it suffices to find the factorization of the component matrices

$$R(\Delta) = \beta^1 \beta^{1*}, \qquad R(0) = \beta^0 \beta^{0*}$$

where β^1 is computed through the EID, while for β^0 we have directly $\beta^0 = \sqrt{R(0)}$ because $R(0)$ is diagonal. In general, the dimensions of the elementary factors β^0 and β^1 are $n \times n$ and $n \times h_0$, where h_0 is the rank of $R(\Delta)$, but, profiting from the fact that these dimensions have a degree of freedom (see Sect. 8.4), it is expedient to adopt the same dimensions $n \times h$ for both elementary factors. In practice, n and h are chosen simultaneously with the criterion of the trace based on a predetermined accuracy. With this choice we obtain the dimensions and the (virtual) ranks summarized in Table 8.2.

8.11.3 Computational Complexity and Method Comparison

The above listed dimensions exhibit an exponential increase with the order K of the PPM modulation, and this causes a very serious problem of computational complexity. In Glauber's representation, it is mandatory to make an accurate choice of the finite dimension n of the component Hilbert spaces $\mathcal{H}_0$ to ensure an adequate approximation for the elementary density operators. Such value of n depends on the number of signal photons $N_s = \Delta^2$, which in turn depends on the error probability range P_e that we want to examine for the evaluation. For the sake of clarity, consider the following tables, regarding the 4-PPM system with a number of thermal photons equal to $\mathcal{N} = 0.05$.

- $P_e \simeq 10^{-2}$ $\quad \rightarrow \quad$ $N_s \simeq 3$ $\quad \rightarrow \quad$ $n \simeq 10 \;\; N = n^K \simeq 10^4$
- $P_e \simeq 10^{-3}$ $\quad \rightarrow \quad$ $N_s \simeq 4.5$ $\quad \rightarrow \quad$ $n \simeq 15 \;\; N = n^K \simeq 5 \, 10^4$
- $P_e \simeq 10^{-4}$ $\quad \rightarrow \quad$ $N_s \simeq 6.5$ $\quad \rightarrow \quad$ $n \simeq 20 \;\; N = n^K \simeq 16 \, 10^4$
- $P_e \simeq 10^{-5}$ $\quad \rightarrow \quad$ $N_s \simeq 8$ $\quad \rightarrow \quad$ $n \simeq 30 \;\; N = n^K \simeq 81 \, 10^4$

From this we deduce that, while the dimension n of the component spaces stays within relatively small values, the dimension $N = n^K$ of the composite space grows very quickly up to huge values (tens of thousands) for the matrices that must be stored and processed. At this point, to reach interesting ranges of P_e, we must compare the various methods available to identify the method ensuring the minimum computational complexity. To this end, an important role is played by the GUS, which, as we have seen, is verified by the PPM.

To understand the role of dimensions N and r, it is essential to have an idea of the practical limits in the main numerical tools, i.e., EID and CSP. In our experience with standard personal computers, we have found that EID can be efficiently implemented up to the order of three thousand. On the other hand, the available CSP implementations, as the LMI (Linear Matrix Inequality) Toolbox and the CVX, both implemented in MatLab, require the introduction of "data" (complex matrices) and "conditions" that, in our experience can be limited to matrices of order up to three hundred. Thus, as an indication for a discussion, we write

$$N_{\mathrm{EID}} = 3\,000 \qquad N_{\mathrm{CSP}} = 300\,. \tag{8.68}$$

Optimal Approach

This approach is available for:

- pure states ($\mathcal{N} = 0$), for any order K, already used in the previous chapter,
- mixed states for $K = 2$ according to Helstrom's theory, which requires an $N \times N$ EID,
- CSP, considering the GUS of PPM, with at the input the reference density operator ρ_0 of dimensions $N \times N$ and the symmetry operator S (see below).

SRM Approach

This approach is always applicable, for each order K, both with pure states and with mixed states. Two possible modes are given:

- through Gram's operator, which requires one EID of an $N \times N$ matrix,
- through Gram's matrix, which requires K EIDs of $H \times H$ matrices.

As $H \ll N$, we should use the second option, with a limit of $H \simeq 3000$.

Comparison with the Classical Optical PPM

Decision in a classical PPM system is based on photon counting. Such counting, in the presence of thermal noise, has a Laguerre distribution (see Sect. 8.5.2). The corresponding error probability has been computed by Helstrom et al. [4] (chap. 6) and results in

$$P_{e,\mathrm{classic}} = \frac{1}{K} \sum_{i=2}^{K} (-1)^i \binom{K}{i} \exp\left[-\frac{(1-\mathcal{N})^{i-1} - \mathcal{N}^{i-1}}{(1-\mathcal{N})^i - \mathcal{N}^i} N_s \right]\,. \tag{8.69}$$

Notice that in the absence of thermal noise ($\mathcal{N} = 0$), (8.69) degenerates to

$$P_{e,\text{classic}} = \frac{K-1}{K} \mathrm{e}^{-N_s} .\tag{8.70}$$

found in the previous chapter (see (7.145)).

8.11.4 Application to SRM Decision in the Presence of GUS

In the previous chapter (Sect. 7.13) we have seen that the PPM format enjoys the Geometrically Uniform Symmetry (GUS) with symmetry operator S defined by (Sect. 7.135), that is,

$$S = \sum_{k=0}^{n-1} w_n(k) \otimes I_{n^{K-1}} \otimes w_n^*(k) ,\tag{8.71}$$

where $w_n(k)$ is a column vector of length n with ith element $\delta_{ik}, i = 0, \dots, K-1$.

Considering the presence of the GUS we can apply the two procedures outlined in Sect. 8.7.3: the Gram operator approach, where the correct decision probability is given by

$$P_c = \mathrm{Tr}\,[\rho_0\, T^{-1/2}]^2 .$$

and the Gram matrix approach, where P_c is given

$$P_c = \frac{1}{K} \mathrm{Tr}\Big[\Big\{ \sum_{k=0}^{K-1} D_k^{1/2} \Big\}^2 \Big].$$

8.12 PPM Performance Evaluation (Without Compression)

In this section, we evaluate the PPM performance using the more efficient available methods *without resorting to the state compression*. Due to the huge computational complexity, in 3-PPM we are able to apply the SRM approach and partially the CSP, and in 4-PPM only the SRM approach.

8.12.1 Performance of 2-PPM

In 2-PPM in the presence of noise no numerical problems are encountered because the dimensions are relatively small. The optimal performance can be evaluated with

Helstrom's method, while the SRM method turns out to be suboptimal. In Helstrom's method, the decision operator must be built

$$D = \rho_1 - \rho_0$$

where $\rho_0 = \rho(\Delta) \otimes \rho(0)$ and $\rho_1 = \rho(0) \otimes \rho(\Delta)$, and its eigenvalues must be computed.

Example 8.4 We now want to compare Helstrom's and SRM methods for the values of $N_s = 2.0$, 4.0 and 6.0 and the thermal noise parameter $\mathcal{N}$. To ensure an accuracy of 10^{-8} in the approximation of the elementary density operators, $n = 20$ and $h = 5$ must be chosen. Then the composite operators ρ_0 and ρ_1 have dimensions $N \times N$ with $N = 20^2 = 400$ and the virtual rank results in $H = 5^2 = 25$. The numeric evaluation yields the following results for the P_e

	Helstrom	SRM
$N_s = 2.0$	0.01746593348035197	0.02176673057747613
$N_s = 4.0$	0006038940756354361	0.000739329477730144
$N_s = 6.0$	0.0000382325373916581060	0.00004720917345124587

Following the formulation of the example, the two methods have been systematically compared, and the results are as shown in Fig. 8.20, where the error probability P_e is plotted as a function of the number of signal photons per symbol N_s for some values of $\mathcal{N}$. As can be seen, the SRM method provides a slight overestimate of the

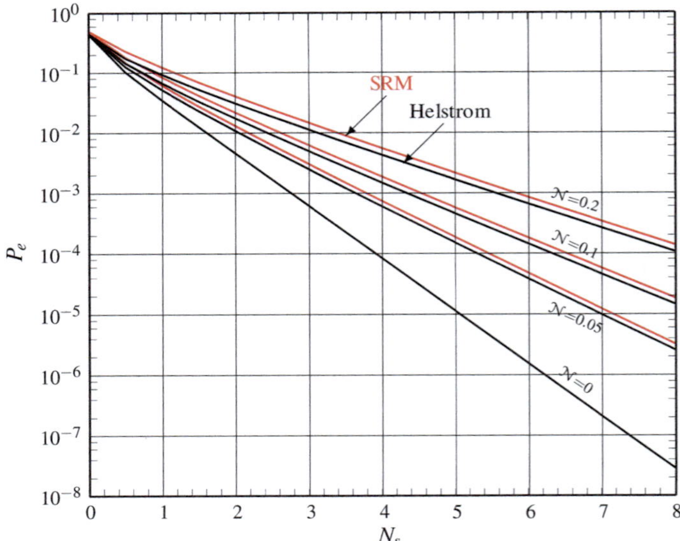

Fig. 8.20 Error probability in 2-PPM as a function of the number of photons per symbol N_s for some values of the number of thermal photons $\mathcal{N}$. Comparison between Helstrom's method (optimal) and SRM method (suboptimal)

error probability; the overestimate factor is only in the order of 1.3–1.4, and therefore quite acceptable for an error probability.

8.12.2 Performance of Quantum 3-PPM and Comparison with Classical System

With $K = 3$ only the SRM method is available and partially the CSP (see below). For an efficient evaluation it is convenient to take into account the fact that the PPM format verifies the GUS, and the SRM evaluation is done using Gram's matrix following the method of Sect. 8.11.4. To obtain an error probability in the order of 10^{-5} we must choose as parameters (see Table 8.2):

- $n = 40, \quad h = 8$,

and therefore $N = n^3 = 40^3 = 64\,000$, $H = 8^3 = 512$, values that take to the limit the evaluation power of a very good PC.

The results are shown in Fig. 8.21, which gives the error probability in 3-PPM as a function of N_s, for some values of $\mathcal{N}$. The same figure shows also the error probability of a classical 3-PPM system using a photon counter; this probability is calculated using Helstrom's formula (8.69). We can observe the clear advantage of the quantum system over the classical one, especially for low values of the thermal noise

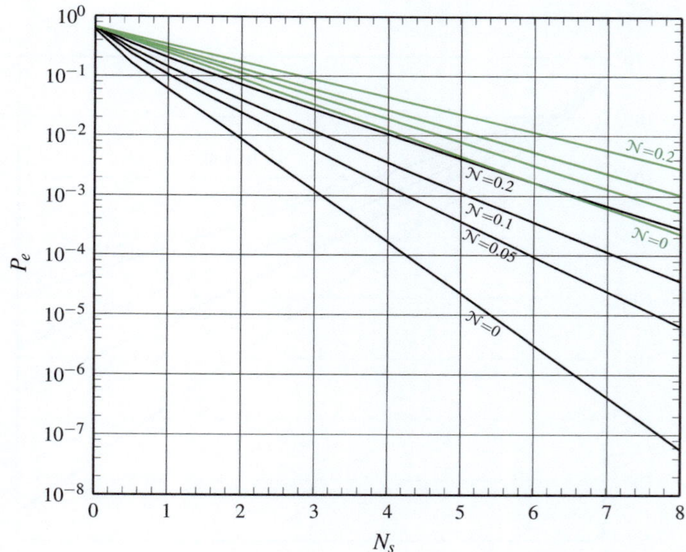

Fig. 8.21 Error probability in 3-PPM as a function of the number of signal photons per symbol N_s for some values of the number of thermal photons $\mathcal{N}$. *Black lines* refer to the quantum system and *green lines* to the classic homodyne system

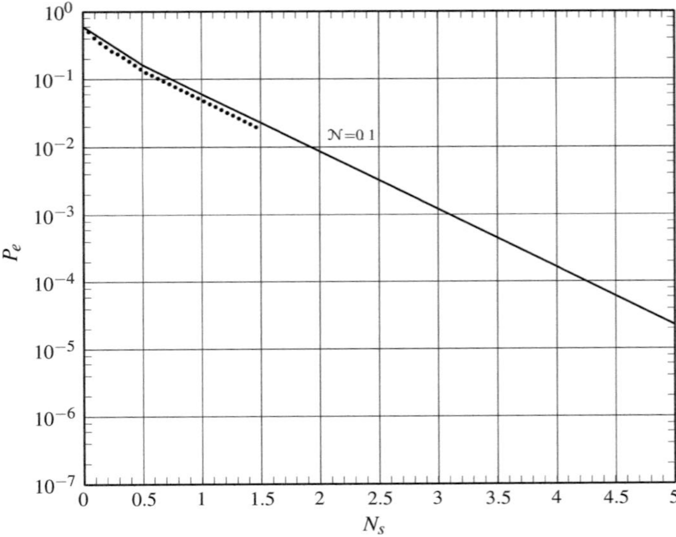

Fig. 8.22 Error probability in 3-PPM as a function of the number of signal photons per symbol N_s for $\mathcal{N} = 0.1$. The *continuous line* refers to the SRM method (suboptimal) and the partial *dashed lines* to the optimal numerical method based on CSP

parameter $\mathcal{N}$. For example, with $N_s = 6$ and $\mathcal{N} = 0.05$ with the quantum 3-PPM we obtain $P_e = 10^{-4}$, whereas with the classical 3-PPM, we get $P_e = 4.3\,10^{-3}$.

With 3-PPM it is possible to use (partially) the evaluation based on CSP, which provides an *optimal* evaluation, and then make a comparison with the SRM method, which is *suboptimal*. The comparison is shown in Fig. 8.22, which gives the error probability in 3-PPM as a function of N_s for $\mathcal{N} = 0.1$, and confirms the "pretty good" approximation of the SRM method. The comparison is only partial (up to about $N_s = 1.5$) and is obtained with density matrices 150×150, using the LMI toolbox of MatLab; this optimal evaluation requires processing times of a few hours for each point.

8.12.3 Performance in 4-PPM

With $K = 4$, for the direct evaluation only the SRM method is available, and it is not possible to implement the optimal evaluation based on CSP. The parameters required by the SRM method have already been anticipated above and give the results illustrated in Fig. 8.23, which shows the shape of the error probability as function of N_s for some values of $\mathcal{N}$. The same figure also shows the error probability of a 4-PPM system obtained using a photon counter (classical decision); for the evaluation of this probability one can use (8.69). Even for 4-PPM the superiority of the quantum system over the classical one is quite evident.

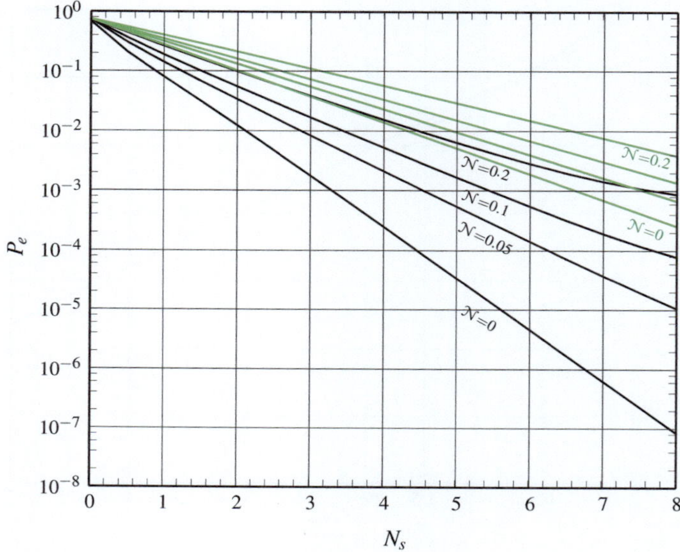

Fig. 8.23 Error probability in 4-PPM versus N_s for some values of $\mathcal{N}$. *Black lines* refer to quantum detection and *green lines* to classical photon-counter detection

8.13 PPM Performance Evaluation Using State Compression

In the previous section we have seen that numeric evaluation of the performance in the presence of thermal noise is limited to low orders ($K \leq 4$), that is, to 3-PPM and to 4-PPM. In particular the CSP method is only partially applicable to 3-PPM.

In this section, we will see how the numerical evaluation can be improved by the state compression, allowing us to explore a range of the performance not possible otherwise.

8.13.1 Recall: Methodology and Benefits of State Compression

State compression was introduced at the end of Chap. 5, Sect. 5.5, and reconsidered in Sect. 6.7 in the context of SRM. The philosophy of compression is the possibility to remove the redundancy of quantum states passing from the original Hilbert space $\mathcal{H}$ to a "compressed" Hilbert space $\overline{\mathcal{H}}$ of smaller dimension. In the K-ary PPM we have $\dim \mathcal{H} = N = n^K$ and $\dim \overline{\mathcal{H}} = r$. In the compressed space the quantum detection is achieved exactly in the same way as in the original Hilbert space. Particularly attractive is the compression when the constellation of states has the GUS, and this is the case of PPM.

For the moment, suppose to know the compressor U_r^*, which allows for the passage from the original Hilbert space to the compressed space as $\overline{\mathcal{H}} = U_r^* \mathcal{H}$. Then all operators and matrices can be transferred into the compressed space and used therein for the quantum detection. We recall in particular

- the factors β_i and μ_i are compressed as

$$\overline{\beta}_i = U_r^* \beta_i, \qquad \overline{\mu}_i = U_r^* \mu_i$$

- the operators ρ_i and Q_i are compressed as

$$\overline{\rho}_i = U_r^* \rho_i U_r, \qquad \overline{Q}_i = U_r^* Q_i U_r$$

- the Gram operator T and the symmetry operator S are compressed as

$$\overline{T} = U_r^* T U_r, \qquad \overline{S} = U_r^* S U_r$$

  and both become diagonal.
- the correct decision probability can be evaluated in the compressed space as

$$P_c = \sum_{i=0}^{K-1} q_i \operatorname{Tr} \overline{Q}_i \overline{\rho}_i, \qquad q_i = 1/K \ .$$

Now we consider the problem of the compressor evaluation. In general, the natural procedure for the evaluation of the compressor U_r^* is the SVD of the state matrix Γ, which here has the dimension $N \times H = n^K \times h^K$, so that it is exceedingly large and the SVD represents a very hard numerical task. But, in the presence of GUS, we have an alternative derived from the commutativity of the Gram operator T with the symmetry operator S. This property in used in Appendix section "Simultaneous Diagonalization of S and T" to get the procedure outlined below for the compression evaluation.

We have seen that in PPM the EID of $S = U_S \Lambda U_S^*$ is known in closed form. The eigenvalues in Λ are multiple so that the EID is not unique and we use the commutativity $T S = S T$ to find from the consequent simultaneous diagonalization

$$T = U \Sigma^2 U^*, \qquad S = U \Lambda U^* \tag{8.72}$$

the convenient unitary matrix U and then the compressor U_r^*. The procedure is articulated in the following steps [9]:

(1) We assume that, possibly after a reordering, the multiple eigenvalues in Λ occur contiguously on the diagonal. Then Λ has the block diagonal form

$$\Lambda = \operatorname{diag}[\lambda_1 I_1, \ldots, \lambda_k I_k] \tag{8.73}$$

where λ_i are the k distinct eigenvalues of S and I_i are identity matrices of size given by the multiplicity of λ_i (recall from Sect. 5.13 that λ_i has the form $W_k^{u_i}$, where u_i are integers).

(2) The matrix $V := U_S^* T U_S$ is block diagonal with diagonal blocks V_i of the same order as I_i in (8.73), which are given by

$$V_i = \sum_{k=0}^{K-1} \lambda_k^i I_i (U_S^* \rho_0 U_S)_{ii} \lambda_k^{-i} I_i = K \,(U_S^* \rho_0 U_S)_{ii} \tag{8.74}$$

(3) Find the EID $V_i = X_i \Sigma_i^2 X_i^*$ of the blocks V_i to get the factorization $V = X \Sigma^2 X^*$ with $\Sigma^2 = \mathrm{diag}[\Sigma_1^2 \ldots, \Sigma_k^2]$ and $X = \mathrm{diag}[X_1, \ldots, X_k]$.
(4) From X we get the diagonalization (8.72) of T with $U = U_S X$.
(5) The compressor U_r^* is given by the first r rows of $U^* = X^* U_S$.

In this procedure the computational complexity is confined to the EIDs of the blocks V_i of size $\simeq N/K$.

8.13.2 Application of State Compression to SRM

In the SRM detection the measurement matrix M is given by the two equivalent expressions $M = T^{-1/2} \Gamma$ and $M = \Gamma G^{-1/2}$.

This gives two alternative approaches which are important to explore an efficient computation.

Gram operator approach. In the case of GUS the reference measurement operator is given by

$$Q_0 = T^{-1/2} \rho_0 T^{-1/2}$$

and the correct detection probability by

$$P_c = \mathrm{Tr}\left[(\rho_0 T^{-1/2})^2\right] \xrightarrow{\text{compression}} P_c = \mathrm{Tr}\left[(\overline{\rho}_0 \overline{T}^{-1/2})^2\right]$$

where $\overline{T}$ is diagonal and therefore no EID is required for the evaluation of $\overline{T}^{-1/2}$.

Gram matrix approach. The matrix G is block circulant and its decomposition is related to the discrete Fourier transform (DFT). We subdivide the Gram matrix into blocks $G_{ij} = \beta_i^* \beta_j$ of size $H \times H$ and we evaluate the matrices

$$D_j = \sum_{i=0}^{K-1} G_{0i} W_K^{-ji} \,.$$

Then, the correct detection probability is given by

$$P_c = \frac{1}{K} \text{Tr}\left[\left\{\sum_{j=0}^{K-1} D_j^{1/2}\right\}^2\right].$$

The evaluation of P_c requires finding the square roots $D_j^{1/2}$ of the matrices D_j of dimension $H \times H$ with $H = h^K$.

8.13.3 Application of State Compression to CSP

We consider the *dual problem with GUS*, stated by Theorem 5.6. which can be summarized as

Proposition 8.4 *Let ρ_0 be the reference density operator and let S be the symmetry operator. Then, the optimization requires finding a PSD Hermitian operator X satisfying the conditions: (1) $X \geq \rho_0$, (2) $XS = SX$, and (3) $\text{Tr}[X]$ is minimum.*

The "dual problem" can be transferred to the compressed space, with a dimension reduction from N to r, where the minimum trace $\text{Tr}[\overline{X}]$ gives the maximum correct decision probability. In the compressed space we have the further improvement: the symmetry operator $\overline{S}$ is diagonal, with the form $\overline{S} = \text{diag}\,[\lambda_1 I_1, \ldots, \lambda_s I_s]$, where λ_i are the distinct eigenvalues of $\overline{S}$ and the size of the identity matrix I_i is given by the multiplicity of λ_i. This implies that $\overline{X}$ becomes block diagonal with blocks X_i of dimensions as I_i (see [10]).

8.13.4 State Compression in PPM: Numerical Problems

We have seen that the compression enables us to replace the matrices of order N concerned with quantum detection (density and measurement operators, Gram operator, optimization operator and symmetry operator) with compressed matrices of order r.

To understand the role of dimensions N and r, it is essential to have an idea of the practical limits in the main numerical tools, i.e., EID and CSP. As done in the previous section, as an indication for a discussion, we assume the limits

$$N_{\text{EID}} = 3\,000 \qquad N_{\text{CSP}} = 300.$$

Compressor Evaluation

This seems to be the bottleneck of the compression theory, but the limit $N = N_{\text{EID}} = 3\,000$ is substantially high. Moreover, as seen above, it can be improved in the presence of GUS, where the simultaneous diagonalization allows for the compressor evaluation through EIDs of size N/K. So, the limit is increased according to $N/K = N_{\text{EID}}$ and we find: $N = 6\,000$ for $K = 2$, $N = 12\,000$ for $K = 4$, etc.

SRM Approach

We have two possibilities, via Gram matrix and via Gram operator, and in both cases we have to perform the EID to find the square root.

We have seen that the Gram matrix does not change with the compression, $\overline{G} = G$, and the size is $r = KH$. So, in general we have the limit $KH = N_{\text{EID}}$, which gives for the density range $H = 1\,500$ for $K = 2$, $H = 750$ for $H = 4$, etc. But, with the GUS the EIDs (of the matrices E_j) have size H. Hence, the limit becomes $H = N_{\text{EID}}$ for any K.

The direct EID of the Gram operator has the limit $N = N_{\text{EID}}$, but it can be compressed to the size $KH = r$ and becomes diagonal. Thus, no EID is required after the compression and the limit is given by compressor evaluation.

CSP Approach

In the presence of GUS the data are reduced to a single density operator (ρ_0, see Proposition 8.4) and the limit becomes $N = N_{\text{CSP}} = 300$ for all K. In the presence of both GUS and compression the data is the compressed density operator $\overline{\rho}_0$, which has size $r = KH$. The limit becomes $r = KH = N_{\text{CSP}} = 300$. Hence: $H = 150$ for $K = 2$, $H = 75$ for $K = 4$, etc. Also the condition that the compressed optimization operator is block diagonal reduces the computational complexity. In conclusion, it is convenient to use the latter opportunity.

8.13.5 Performance Evaluation

For given accuracies ε and v, the sizes n and H increase with the number of signal photons/symbol N_s, which in turn depends on the range of the error probability P_e we want to explore. To fix the ideas we give the following table, where we have chosen as number of thermal photons $\mathcal{N} = 0.05$ and the accuracies with the rule $\varepsilon = v = P_e/10$.

3-PPM

P_e	N_s	n	$N = n^K$	h	$H = h^k$	$r = KH$
10^{-1}	3	5	125	2	8	24
10^{-2}	5	10	10^3	3	27	81
10^{-3}	7	15	1.210^3	3	27	81
10^{-4}	8	20	810^3	4	64	192

4-PPM

P_e	N_s	n	$N = n^K$	h	$H = h^k$	$r = K H$
10^{-1}	3	5	625	2	16	64
10^{-2}	5	10	10^4	3	81	273
10^{-3}	7	15	$5\,10^4$	3	81	273
10^{-4}	8	20	$16\,10^4$	4	256	1024

We realize that, while the dimension n of the basic space can be confined to moderately small values, the dimension N of the composite space is considerably huge. On the other hand, the dimension of the compressed space $r = K h^K$ is confined to smaller values.

Now, a fundamental remark to reduce the computational complexity is that the PPM verifies the GUS, as shown in [2], where the expression of the symmetry operator S and its EID are obtained. The presence of GUS improves the limit of the dimension N of the compressor evaluation according to $N = N_{\text{EID}}\, K$, that is $N = 9\,000$ for the 3-PPM and $N = 12\,000$ for the 4-PPM.

Following the above tables we see that we can use the SRM-Gram matrix approach to evaluate P_e down to 10^{-5} for the 3-PPM and down to 10^{-4} for the 4-PPM. On the other hand, the direct evaluation with CSP is limited to a small range: $P_e \geq 10^{-2}$ for the 3-PPM and $P_e \geq 10^{-1}$ for the 4-PPM. But with the compression these limit are considerably improved: $P_e \geq 10^{-4}$ for the 3-PPM and $P_e \geq 10^{-3}$ for the 4-PPM.

Practical Software Implementation

The results of the numerical evaluation of the error probability P_e for 3-PPM are depicted in Fig. 8.24. In the absence of thermal noise ($\mathcal{N} = 0$) a closed-form expression of the minimum P_e is known from Helstrom's theory. In the presence of thermal noise ($\mathcal{N} = 0.05$ and $\mathcal{N} = 0.1$) the solid lines represent the suboptimal P_e obtained by SRM and evaluated as in the previous section. The evaluation of the optimal error probability leads to results confined to a limited range of N_s (up to 1.5), as evidenced by the stars ($\star$). This is due to the aforementioned numerical limitations. The evaluation has been extended to a more significant range (denoted by full dots) using the GUS of PPM. But a dramatic simplification of numerical complexity, is obtained with the compression techniques, which reduces the dimensions of the involved matrices and enables one to perform the optimization with block diagonal matrices of reduced size. Provided that data, i.e., the compressed matrix $\bar{\rho}_0$, is loaded, few line of self-explanatory code are sufficient, specifically

```
cvx_begin
  variables X(dim)
  minimize(trace(X))
  subject to
    X>rho0;
    X*S==S*X;
cvx_end
```

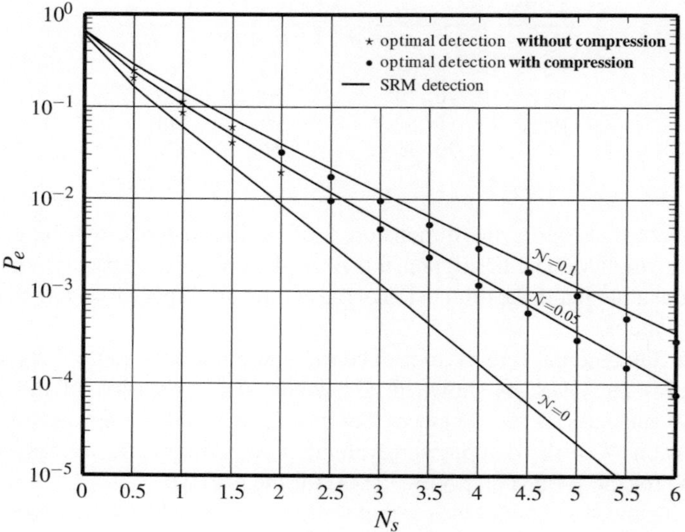

Fig. 8.24 Error probability in 3-PPM versus the number of signal photons per symbol N_s for 3 values of the number of thermal photons $\mathcal{N}$. *Solid lines* refer to SRM detection, *stars* ($\star$) and *bullets* ($\bullet$) to optimal detection. For $\mathcal{N}=0$ SRM detection coincides with optimal detection

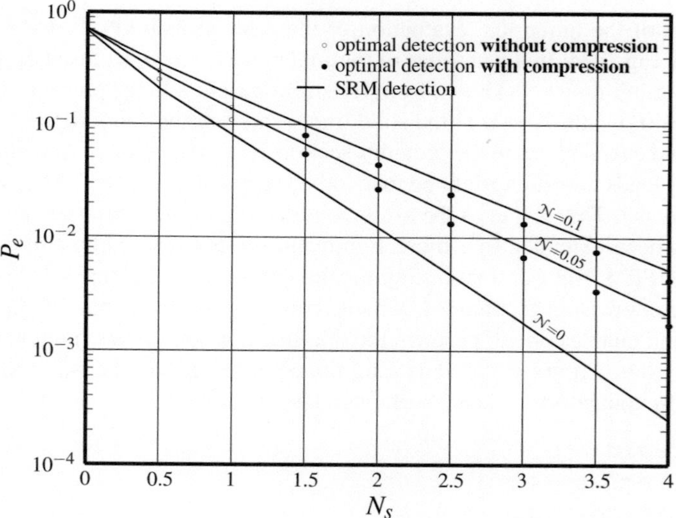

Fig. 8.25 Error probability in 4-PPM versus the number of signal photons per symbol N_s for 3 values of the number of thermal photons $\mathcal{N}$. *Solid lines* refer to SRM detection and points to optimal detection. For $\mathcal{N}=0$ (absence of thermal noise) SRM detection coincides with optimal detection

Similar remarks hold for the error probability of 4-PPM depicted in Fig. 8.25, where the direct application of the optimization is limited to a very small range and considerably extended with the compression.

Inspection on the two figures confirms that the SRM gives a "pretty good" evaluation of the quantum PPM performance.

8.14 Conclusions

We have seen the possibility to transfer quantum detection operations into a compressed space, where the redundancy of quantum states and measurement operators is completely removed. The quantum system performance, e.g., the probability of correct detection, can be evaluated in the compressed space without returning back to the original Hilbert space. In the compressed space most of the properties are improved and, in particular, some operators become diagonal. In the presence of a strong redundancy, the numerical evaluations are facilitated, in such a way that it is possible to explore ranges of the performance evaluation not possible without compression. Perhaps, also a physical realization of the quantum detection may be improved with the compression technique, provided that an optical compressor is realizable.

Appendix

Alternative Discretization

An alternative discretization of a Glauber density operator is based on the subdivision of the integration region in (8.8) into a finite number K of regions of the complex plane, namely,

$$A_k \quad \text{with} \quad \bigcup_{k=1}^{K} A_k = \mathbb{C}. \tag{8.75}$$

In such a way, one gets the discrete approximation of the density operator

$$\rho(\gamma) \simeq \sum_{k=1}^{K} h_k \, |\alpha_k\rangle\langle\alpha_k| \tag{8.76}$$

where

$$h_k = \frac{1}{\pi \mathcal{N}} \int_{A_k} \exp\left(-\frac{|\alpha - \gamma|^2}{\mathcal{N}}\right) d\alpha$$

and $|\alpha_k\rangle$ are the Glauber coherent states with complex parameter $\alpha_k \in A_k$. This integral gives the probability of the state $|\alpha_k\rangle$

$$P[s = |\alpha_k\rangle] = h_k \ . \tag{8.77}$$

The normalization of the probability is verified because the volume determined by the Gaussian profile is unitary

$$\frac{1}{\pi \mathcal{N}} \int_{\mathbb{C}} \exp\left(-\frac{|\alpha - \gamma|^2}{\mathcal{N}}\right) \, d\alpha = 1 \ . \tag{8.78}$$

To arrive at a finite form, the Glauber states in (8.76) are approximated by a finite number N of terms, that is,

$$|\alpha_k\rangle = e^{-\frac{1}{2}|\alpha_k|^2} \sum_{n=0}^{N-1} \frac{\alpha_k^n}{\sqrt{n!}} \, |n\rangle \ .$$

Partition strategy. The partition of the complex plane $\mathbb{C}$ into the regions A_k can be done in several ways. The strategy that will be considered here is:

(1) partition $\mathbb{C}$ into a finite number L circular rings centered on the nominal state $|\gamma\rangle$

$$D_i = \left\{ r_{i-1}^2 \le |\alpha - \gamma|^2 < r_i^2 \right\}, \qquad i = 1, 2, \ldots, L$$

with $r_0 = 0$ and $r_L = \infty$.
(2) subdivide the ith ring into k_i equal parts

$$A_{ip} = \left\{ \frac{2\pi}{k_i}(p-1) \le \arg(\alpha - \gamma) < \frac{2\pi}{k_i} p \right\}, \qquad p = 1, 2, \ldots, k_i$$

where for convenience we use the double subscript ip instead of k,
(3) choose the state $|\alpha_{ip}\rangle$, with α_{ip} given by the *barycenter* of the region A_{ip},
(4) assume the condition that the states are equally likely.

With such a procedure the number of states is $K = k_1 + \cdots k_L$ and the volume determined by the ring D_i results in

$$\begin{aligned}
v_i &= \frac{1}{\pi \mathcal{N}} \int_{D_i} \exp\left(-\frac{|\alpha - \gamma|^2}{\mathcal{N}}\right) \, d\alpha \\
&= \frac{1}{\pi \mathcal{N}} \int_0^{2\pi} d\phi \int_{r_{i-1}}^{r_i} dr \, r \, e^{-r^2/\mathcal{N}} \\
&= \frac{1}{2}\left(e^{-r_{i-1}^2/\mathcal{N}} - e^{-r_i^2/\mathcal{N}}\right)
\end{aligned} \tag{8.79}$$

which is subdivided into k_i equal parts, so that the equal probability condition is

$$\frac{v_i}{k_i} = \frac{1}{K} \quad \text{with} \quad K = k_1 + k_2 + \cdots + k_L .$$

By imposing such a condition, the radii of the circular rings can be evaluated by the recurrence

$$r_i = \sqrt{-\mathcal{N} \log \left(e^{-r_{i-1}^2/\mathcal{N}} - k_i/\mathcal{N} \right)}, \qquad i = 1, \ldots, L$$

letting $r_0 = 0$ and $r_L = \infty$.

Finally, the barycentric radius of the ring D_i results in

$$r_i^b = \frac{\sqrt{\mathcal{N}} \left(e^{-r_{i-1}^2/\mathcal{N}} - e^{-r_i^2 \mathcal{N}} \right)}{\sqrt{\pi} \left(\text{erf} \left(r_i/\sqrt{\mathcal{N}} \right) - \text{erf} \left(r_{i-1}/\sqrt{\mathcal{N}} \right) \right)}$$

where erf is the error function. In such a way, the coherent state $|\alpha_{ip}\rangle$ is identified by the complex number

$$\alpha_{ip} = \gamma + r_i^b \, e^{\phi_{ip}}, \qquad \phi_{ip} = \frac{2\pi}{k_i} p .$$

The number of states in this approximation of the density operator is given by the total number of subdivisions $K = k_1 + \cdots + k_L$.

Example 8.5 We consider two examples of subdivision shown in Fig. 8.26. In both cases the subdivision is into $L = 4$ rings and the inner ring is not subdivided and

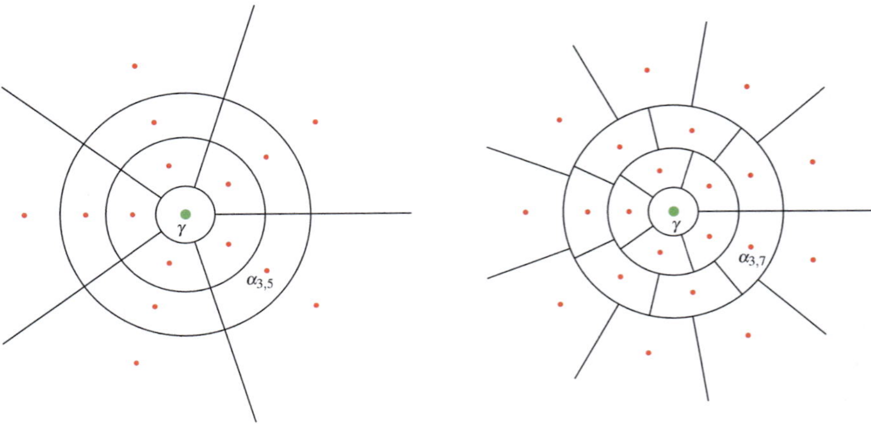

Fig. 8.26 Examples of subdivision in circular rings for the discretization of a Glauber density operator

is associated to the nominal state $|\gamma\rangle$. In the first example, the 3 outer rings are subdivided into 5 parts, and we get $K = 16$ global subdivisions, while in the second example they are respectively subdivided into 5 , 7 and 9 parts with $K = 22$ global subdivisions.

$$\text{number of rings: } L = 4 \qquad \text{number of rings: } L = 4$$
$$\text{subdivisions of rings: } 1, 5, 5, 5 \qquad \text{subdivisions of rings: } 1, 5, 7, 9$$
$$\text{global subdivisions: } K = 16 \qquad \text{global subdivisions: } K = 22.$$

We complete the evaluation in the first subdivision, assuming

$$\mathcal{N} = 0.2 \quad N_s = 2.0 .$$

With $N = 11$ we find the matrix

$$\tilde{R} = \begin{bmatrix} 0.15696 & 0.18636 & 0.15597 & 0.10627 & 0.06254 & 0.03284 & 0.01572 & 0.00696 & 0.00288 & 0.00113 & 0.00042 \\ 0.18636 & 0.24576 & 0.22704 & 0.16979 & 0.10909 & 0.06222 & 0.03217 & 0.01530 & 0.00676 & 0.00280 & 0.00109 \\ 0.15597 & 0.22704 & 0.22990 & 0.18741 & 0.13065 & 0.08054 & 0.04485 & 0.02290 & 0.01084 & 0.00479 & 0.00199 \\ 0.10627 & 0.16979 & 0.18741 & 0.16562 & 0.12463 & 0.08265 & 0.04937 & 0.02697 & 0.01363 & 0.00643 & 0.00285 \\ 0.06254 & 0.10909 & 0.13065 & 0.12463 & 0.10082 & 0.07162 & 0.04570 & 0.02661 & 0.01431 & 0.00716 & 0.00337 \\ 0.03284 & 0.06222 & 0.08054 & 0.08265 & 0.07162 & 0.05431 & 0.03688 & 0.02280 & 0.01298 & 0.00688 & 0.00341 \\ 0.01572 & 0.03217 & 0.04485 & 0.04937 & 0.04570 & 0.03688 & 0.02657 & 0.01737 & 0.01044 & 0.00583 & 0.00304 \\ 0.00696 & 0.01530 & 0.02290 & 0.02697 & 0.02661 & 0.02280 & 0.01737 & 0.01198 & 0.00758 & 0.00444 & 0.00243 \\ 0.00288 & 0.00676 & 0.01084 & 0.01363 & 0.01431 & 0.01298 & 0.01044 & 0.00758 & 0.00502 & 0.00308 & 0.00176 \\ 0.00113 & 0.00280 & 0.00479 & 0.00643 & 0.00716 & 0.00688 & 0.00583 & 0.00444 & 0.00308 & 0.00197 & 0.00117 \\ 0.00042 & 0.00109 & 0.00199 & 0.00285 & 0.00337 & 0.00341 & 0.00304 & 0.00243 & 0.00176 & 0.00117 & 0.00072 \end{bmatrix}$$

As done in Example 8.19, we perform the factorization choosing the accuracy 0.001 to neglect the very small eigenvalues. In such a way we get the 11×4 factor

$$\gamma = Z_h \, D_h = \begin{bmatrix} 0.33675 & 0.19320 & 0.07563 & 0.02214 \\ 0.47623 & 0.13659 & -0.00015 & -0.01598 \\ 0.47623 & -0.00004 & -0.05355 & -0.01550 \\ 0.38884 & -0.11157 & -0.04355 & 0.00685 \\ 0.27495 & -0.15774 & 0.00022 & 0.01825 \\ 0.17389 & -0.14962 & 0.03924 & 0.01109 \\ 0.10039 & -0.11515 & 0.05644 & -0.00395 \\ 0.05366 & -0.07692 & 0.05419 & -0.01563 \\ 0.02683 & -0.04614 & 0.04207 & -0.01983 \\ 0.01265 & -0.02537 & 0.02828 & -0.01807 \\ 0.00565 & -0.01296 & 0.01704AP20 & -0.01361 \end{bmatrix}$$

The comparison with the results of Example 8.19 shows that the numerical values are very close (the difference is of the same order of magnitude as the chosen accuracy).

Proof of Proposition 8.2 on Factorization

We exploit the rule of mixed products (ordinary products and Kronecker's products) seen in Sect. 2.13, Eq. (2.104). For example, for $K = 4$ and $i = 2$ this rule yields

$$\rho_2 = \rho(0) \otimes \rho(\alpha) \otimes \rho(0) \otimes \rho(0)$$
$$= (\gamma^0 \gamma^{0*}) \otimes (\gamma^1 \gamma^{1*}) \otimes (\gamma^0 \gamma^{0*}) \otimes (\gamma^0 \gamma^{0*})$$
$$= (\gamma^0 \otimes \gamma^1 \otimes \gamma^0 \otimes \gamma^0)(\gamma^0 \otimes \gamma^1 \otimes \gamma^0 \otimes \gamma^0)^*$$

from which

$$\rho_2 = \gamma_2 \gamma_2^* \quad \text{with} \quad \gamma_2 = \gamma^0 \otimes \gamma^1 \otimes \gamma^0 \otimes \gamma^0$$

and it is evident how the general result (8.65) can be obtained.

Simultaneous Diagonalization of S and T

We outline a procedure for finding the simultaneous diagonalization (8.72) starting from an arbitrary EID $S = U_S \Lambda U_S^*$ of the symmetry operator and using the commutativity $TS = ST$.

(1) We assume that, possibly after a reordering, the multiple eigenvalues in Λ occur contiguously on the main diagonal. Then Λ has the block diagonal form

$$\Lambda = \mathrm{diag}[\lambda_1 I_1, \ldots, \lambda_k I_k] \tag{8.80}$$

where λ_i are the distinct eigenvalues of S and I_i are identity matrices of size given by the multiplicity of λ_i.
(2) Since S commutes with T, it follows $U_S \Lambda U_S^* T = T U_S \Lambda U_S^*$ and $\Lambda U_S^* T U_S = U_S^* T U_S \Lambda$, so that Λ commutes with $V = U_S^* T U_S$.
(3) Partition the matrix $V = [V_{ij}]$ according to the partition (8.80) of Λ. Then, from $\Lambda V = V \Lambda$ we get $\lambda_i V_{ij} = \lambda_j V_{ij}$ and V_{ij} does not vanish only if $i = j$. One concludes that V is block diagonal, $V = \mathrm{diag}[V_1, \ldots, V_k]$, with blocks V_i of the same order as I_i in (8.80).
(4) Since V is PSD, each block V_i is PSD. Then, we perform the EID of the blocks $V_i = X_i \Sigma_i^2 X_i^*$ and we get the diagonalization $V = X \Sigma^2 X^*$ with $\Sigma^2 = \mathrm{diag}[\Sigma_1^2 \ldots, \Sigma_k^2]$ and $X = \mathrm{diag}[X_1, \ldots, X_k]$.
(5) By reversing the previous unitary transformations we get the diagonalization $T = U_S V U_S^* = U_S X \Sigma^2 X^* U_S^*$ and, remembering that $X_i X_i^* = I_i$, the simultaneous diagonalization $S = U_S \Lambda U_S^* = U_S X \Lambda X^* U_S^*$. Thus, we get (8.72) with $U = U_S X$.

Note that $SU_S = U_S\Lambda$ and $U_S^*S = \Lambda U_S^*$ (see point 1). Then

$$V = U_S^*TU_S = \sum_{k=0}^{m-1} U_S^*S^k\rho_0(S^*)^kU_S = \sum_{k=0}^{m-1} \Lambda^kU_S^*\rho_0U_S\Lambda^{-k}\ .$$

Since V and Λ are block–diagonal

$$V_i = \sum_{k=0}^{m-1} \lambda_k^i I_i (U_S^*\rho_0U_S)_{ii}\lambda_k^{-i} I_i = (U_S^*\rho_0U_S)_{ii} \tag{8.81}$$

and the block V_i coincides with the ith diagonal block of $U_S^*\rho_0U_S$.

References

1. G. Cariolaro, G. Pierobon, Performance of quantum data transmission systems in the presence of thermal noise. IEEE Trans. Commun. **58**(2), 623–630 (2010)
2. G. Cariolaro, G. Pierobon, Theory of quantum pulse position modulation and related numerical problems. IEEE Trans. Commun. **58**(4), 1213–1222 (2010)
3. R.J. Glauber, The quantum theory of optical coherence. Phys. Rev. **130**, 2529–2539 (1963)
4. C.W. Helstrom, J.W.S. Liu, J.P. Gordon, Quantum-mechanical communication theory. Proc. IEEE **58**(10), 1578–1598 (1970)
5. C.W. Helstrom, Quantum Detection and Estimation Theory. Mathematics in Science and Engineering, vol. 123. Academic Press, New York (1976)
6. V. Vilnrotter and C. W. Lau, Quantum detection of binary and ternary signals in the presence of thermal noise fields. NASA, Technical Report, February (2003). Interplanetary Network Progress (IPN) Progress Report 42–152
7. J. Shapiro, Quantum noise and excess noise in optical homodyne and heterodyne receivers. Quantum Electron. IEEE J. **21**(3), 237–250 (1985)
8. S. J. Dolinar, J. Hamkins, B. Moiston, and V. Vilnrotter, Optical modulation and coding, in H. Hemmati, ed. *Deep Space Optical Communications*, Wiley (2006)
9. G. Cariolaro, R. Corvaja, G. Pierobon, Compression of pure and mixed states in quantum detection, in Global Telecommunications Conference (GLOBECOM, (2011) IEEE. December **2011**, 1–5 (2011)
10. A. Assalini, G. Cariolaro, G. Pierobon, Efficient optimal minimum error discrimination of symmetric quantum states. Phys. Rev. A **81**, 012315 (2010)

Chapter 9
Implementation of QTLC Systems

9.1 Introduction

Optical communications appear as a natural evolution of digital radio frequency (RF) communications. This evolution implied a change of the frequency range of the carrier from about 10^9 Hz for radio and microwave communications to about 10^{15} Hz for optical communications (see also Sect. 4.4). The major advantage in using optical frequencies is related to the possibility of utilizing the enormous bandwidths available in the optical spectrum. Of course, owing to the very small wavelengths involved, a completely different technology was needed for the development of optical communications.

A fundamental role in optical technology has been played by the invention of laser around 1960. This component is a high-powered, almost monochromatic and very directive source, whose advent suggested the possibility of its use in long-distance optical transmissions. Indeed, the high directivity of this new source allows huge antenna gains. On the other hand, the hope related to the advent of the laser appeared somewhat cooled down by the presence of the atmospheric turbulence, a phenomenon caused by interaction of the light with the atoms. The development of optical fibers enabled the experimenters to use a waveguided channel, practically insensitive to interferences with the surrounding environment. This fact gave new impetus to optical communications, with a wide range of practical applications in the field of terrestrial communications.

Another fundamental difference between radio and optical communications arises from the fact that at optical frequencies it is not possible the use of antennas extracting an electrical signal from the electromagnetic field and only detectors sensitive to the field intensity, as photodetectors, are available. Fortunately, the development of photodetection devices as the avalanche photodetectors (APD) combined with the high energy of laser beam made possible the combination of the simplest modulation (intensity modulation) with the direct detection of the signal energy. This approach, not possible at radio frequency, has been largely utilized in terrestrial fiber optics. Moreover, usual means of the radio frequency communications as phase modulation

© Springer International Publishing Switzerland 2015

G. Cariolaro, *Quantum Communications*, Signals and Communication Technology,
DOI 10.1007/978-3-319-15600-2_9

and coherent detection can be employed in optical communications to improve the performances of the direct detection.

9.1.1 The System Model

The first optical communication systems mimicked the well-known radio frequency techniques. This gave rise to classical optical systems as incoherent detection and homodyne detection. These systems are perfectly adequate to the needs of the optical communications as long as the received field is so large that the quantum effects are negligible. On the other hand, in applications as deep space communications, the received field may be so weak that quantum effects dominate and a clear advantage is obtained using quantum detection approaches. Moreover, in other applications as quantum key distribution, the use of weak fields with quantum detection becomes essential to guarantee the security of the transmission.

As we have seen in the previous chapters, the block diagram of a point-to-point optical communications system is by no means different from the standard model and is given by the cascade of a transmitter, a physical channel, and a receiver.

The **optical transmitter** is composed of the cascade of an optical source, a modulator, and a coupling device adapting the beam to the optical transmission medium. The source (a laser) generates an electromagnetic field in the optical range. The modulator, on the basis of the digital information to be transmitted, modifies a parameter (usually the amplitude or the phase) of the electromagnetic field.

The **physical transmission channel** may be the free space or an optical fiber. On the choice of the medium is based a rough classification into **guided** and **unguided** optical transmission systems.

Finally, the **optical receiver** is the cascade of a coupling device, a demodulator, and a photodetector. The demodulating device, if any, combines the received optical field with a locally generated field. The photodetector converts the optical signal into an electric signal for the postdetection processing.

The main difference between free space and guided transmission systems resides in the coupling approach of the laser beam to the optical medium. In free space systems (Fig. 9.1) at the transmitter side an optical antenna focuses the field into a narrow beam. At the receiver side, another optical antenna refocuses the electro-magnetic beam, possibly spread by the medium, into the detection surface (see also Sect. 4.4.5). In optical fiber transmissions (Fig. 9.2) the couplings laser–fiber and fiber–detector are realized via fiber connections, adapters, or optical lenses.

Fig. 9.1 Scheme of quantum optical system in the free space

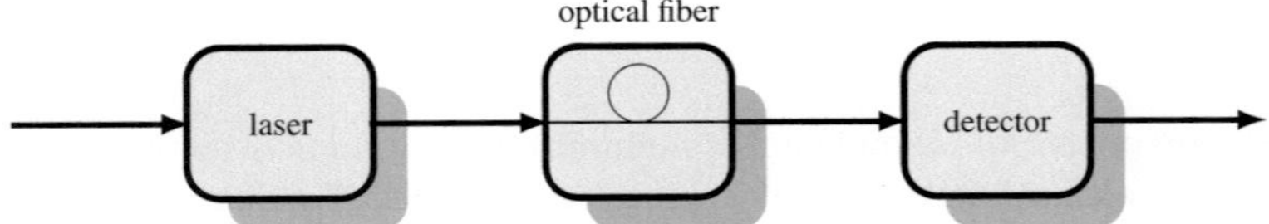

Fig. 9.2 Scheme of quantum optical guided system

9.1.2 Outline of the Chapter

In Sect. 9.2 we present the basic components of optical communication systems, namely, laser, modulators, beam splitters, and photodetectors. In particular, we give their models both from the classical and the quantum point of view. In Sect. 9.3, the major classical optical communication schemes (direct detection and homodyne detection) are presented with their quantum equivalent models. The limits of classical optical communications (shot noise limit and standard quantum limit) are introduced. In Sect. 9.4 the most popular binary quantum communication schemes are presented, starting from the Kennedy and the Sasaki–Hirota receivers. Particular attention is devoted to the analysis and interpretation of the Dolinar receiver, that attains the Helstrom bound. In Sect. 9.5 recent evolutions toward suboptimal K-ary systems are presented. In particular, multidisplacement receivers for K-PSK and K-QAM are outlined. Finally, some results on possible implementations of PPM receivers are presented.

Advice to the reader. Some topics of this chapter imply the knowledge of the continuous variables, whose fundamentals are developed in Chap. 11. Then, we strongly recommend the reader to revisit the present chapter after an adequate comprehension of some topics of Chap. 11, as bosonic operators and displacement and rotation transformations.

9.2 Components for Quantum Communications Systems

As noted in the introduction, the main components of the optical communication systems are the laser and the photodetector. Other components, as modulators, lenses, and mirrors, are used in order to improve the communication performances through modulation and demodulation techniques.

In this section, a summary description of the main components of the transmitter and the receiver of quantum communications systems is given.

9.2.1 Laser

The key component of all quantum communications systems is the laser, that provides the physical carrier for the information transmission. For a detailed analysis of principles and applications the reader may see for instance [1].

From a physical point of view, the laser is a narrow band optical amplifier with amplification provided by an active medium excited by an external source of energy (the **pump** in the technical jargon). As in many electrical oscillators, optical oscillation arises as a combined effect of the spontaneous photon emission of the active medium and of the feedback provided by an optical cavity. In order that the oscillation may start, the pump power must be above a threshold assuring that the gain of the active medium is greater than the loss. Moreover, the length of the cavity must be matched to the natural laser wavelength.

From a classical point of view, the radiation produced by a laser can be modeled as an electromagnetic wave with electric field

$$\mathbf{E}(r, t) = E_0[\, \alpha(r, t)\mathrm{e}^{\mathrm{i}2\pi \nu t} + \alpha^*(r, t)\mathrm{e}^{-\mathrm{i}2\pi \nu t}\,]\mathbf{p}(r, t) \tag{9.1}$$

originating as a single mode solution of the wave equation into the cavity and propagating in the external space along some direction z, with optical frequency ν. The vector $\mathbf{p}(r, t)$ takes into account the field polarization. The complex amplitude $\alpha(r, t)$ can be written as

$$\alpha(r, t) = \alpha_0(r)\mathrm{e}^{\mathrm{i}\phi(r,t)}, \tag{9.2}$$

where $\phi(r, t)$ is the time and space-dependent phase. In the simplest case, known as Gaussian beam, the amplitude $\alpha_0(r)$ has circular symmetry in the plane orthogonal to the propagation axis z and is given by

$$\alpha_0(r) = \frac{\alpha_0}{z + \mathrm{i}z_0} \exp\left(-\frac{\mathrm{i}(x^2 + y^2)}{2(z + \mathrm{i}z_0)}\right), \qquad r = (x, y, z), \tag{9.3}$$

where α_0 and z_0 are real constants. The corresponding intensity is

$$I(r) = \frac{\alpha_0^2}{z^2 + z_0^2} \exp\left(-\frac{z_0^2(x^2 + y^2)}{z^2 + z_0^2}\right). \tag{9.4}$$

This beam has the nice property that its shape remains unchanged in reflection and diffraction.

The field (9.1) may be put in the form

$$E(r, t) = [\, x_1(r, t)\cos(2\pi \nu t) + x_2(r, t)\sin(2\pi \nu t)\,]\,\mathbf{p}(r, t), \tag{9.5}$$

where the quadrature components x_1 and x_2 are in evidence. It can be proved that x_1 and x_2 have the same properties of the position q and the momentum p of a mechanical harmonic oscillator (see Sect. 11.3). Then, from a quantum point of view, after a suitable normalization, the components x_1 and x_2 are substituted by two quantum observables q and p satisfying the canonical correlation condition $[q, p] = 2i$. Equivalently, $\alpha(r, t)$ and its conjugate are substituted by the bosonic operators a and a^* satisfying the correlation condition $[a, a^*] = 1$ and operating in the Fock space. The formalization of these ideas will be seen in the context of continuous variables (Chap. 11), where the radiation emitted by a laser is modeled as a coherent state.

9.2.2 Beam Splitter

The beam splitter is a partially transmitting mirror (Fig. 9.3) which combines two optical beams impinging orthogonally on the mirror surface. In the case of fiber links, the device with the same role is called *fiber combiner* or *fiber coupler*, and it is usually obtained by fusing together the core of two fiber patches.

The classical model of the beam splitter, known from the nineteenth century, is as follows. We assume that the input beams have the same frequency and amplitudes α and β. Then the output beams α' and β' are related to α and β by the relations

$$\alpha' = \sqrt{1 - \tau}\,\alpha + \sqrt{\tau}\,\beta$$
$$\beta' = \sqrt{\tau}\,\alpha - \sqrt{1 - \tau}\,\beta \tag{9.6}$$

where phases have been neglected for simplicity. Since

$$|\alpha'|^2 + |\beta'|^2 = |\alpha|^2 + \beta|^2, \tag{9.7}$$

the device is lossless. The meaning of the parameter τ is apparent. If $\beta = 0$, one gets $|\alpha'|^2 = (1 - \tau)|\alpha|^2$ and $|\beta'|^2 = \tau|\alpha|^2$. Then τ is the fraction of the power transmitted

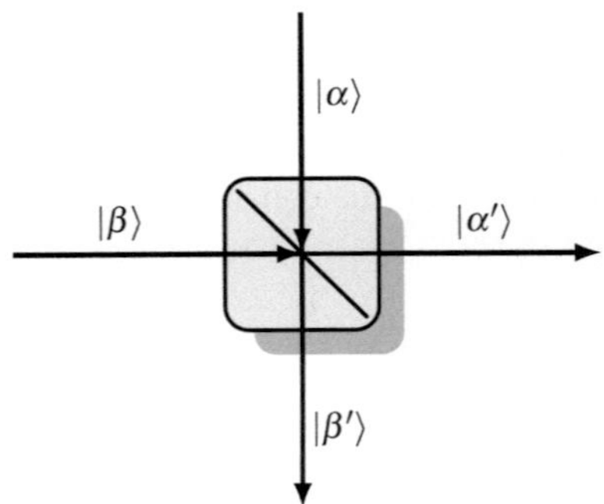

Fig. 9.3 Beam splitter with the input kets $|\alpha\rangle$ and $|\beta\rangle$ and the output kets $|\alpha'\rangle$ and $|\beta'\rangle$

through the mirror and is called the **transmissivity** of the beam splitter. If the device introduces losses due to absorption and scattering, the previous equation becomes

$$|\alpha'|^2 + |\beta'|^2 = (1 - L)(|\alpha|^2 + \beta|^2). \tag{9.8}$$

On the other hand values of L below 10^{-4} have been achieved, so that in a first approximation losses may be neglected.

For a detailed analysis of the quantum model of the beam splitter in the context of continuous variables the reader is referred to Sect. 11.17.5. Here we confine us to observe that the Heisenberg representation of the input–output relations of the beam splitter is obtained by the classical model by substituting the field amplitudes with the annihilation operators corresponding to the beams, namely,

$$a' = \sqrt{1 - \tau}\,a + \sqrt{\tau}\,b$$
$$b' = \sqrt{\tau}\,a - \sqrt{1 - \tau}\,b \tag{9.9}$$

where a and b are the annihilation operators of the input beams and a' and b' are the annihilation operators of the output beams.

Note that in the quantum model the presence of both beams is mandatory to correctly describe such a two-input two-output device. Indeed, if we ignore the annihilator b, we would have for instance $[a', a'^*] = (1 - \tau)[a, a^*] = (1 - \tau)$ in contradiction with the bosonic commutation rule. Taking into account b one gets

$$[a', a'^*] = (1 - \tau)[a, a^*] + \tau[b, b^*] = 1 \tag{9.10}$$

and the commutation relation is satisfied.

In the Schrödinger representation corresponding to the Heisenberg representation (9.9) (see Sect. 3.4) in the case of coherent states, the beam splitter transforms the input joint state $|\alpha\rangle \otimes |\beta\rangle$ to the output joint state $|\alpha'\rangle \otimes |\beta'\rangle$ with

$$|\alpha'\rangle = |\sqrt{1 - \tau}\,\alpha + \sqrt{\tau}\,\beta\rangle$$
$$|\beta'\rangle = |\sqrt{\tau}\,\alpha - \sqrt{1 - \tau}\,\beta\rangle \tag{9.11}$$

in perfect analogy with the classical interpretation.

One of the most important application of the beam splitter in optical technique is the **approximate realization of the quantum displacement**. If we apply to the second input the coherent state $|\gamma\rangle = |\beta/\sqrt{\tau}\rangle$, the coherent state at the first output becomes

$$|\sqrt{1 - \tau}\,\alpha + \gamma\rangle \tag{9.12}$$

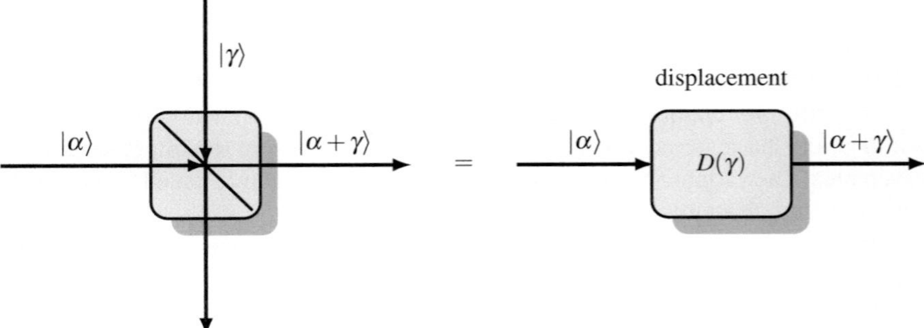

Fig. 9.4 Approximate realization of a *displacement* through a *beam splitter* (the second output of the beam splitter is not used)

approximating $|\alpha + \gamma\rangle = D(\gamma)|\alpha\rangle$ as $\tau \to 0$ (Fig. 9.4). Then, a displacement of amplitude γ may be obtained, at least approximately, with a beam splitter of low transmissivity τ driven at the second input by a high level coherent state $|\beta/\sqrt{\tau}\rangle$.

The theory of the beam splitter, formulated as a two-mode unitary operator, will be seen in Sect. 11.17.

9.2.3 Modulators

Essentially, we have *phase* modulators and *amplitude* modulators, which provide the relations

$$|\psi\rangle \;\; \to \;\; |e^{i\phi}\,\psi\rangle, \qquad |\psi\rangle \;\; \to \;\; |A\,\psi\rangle. \tag{9.13}$$

The corresponding graphical representation is illustrated in Fig. 9.5. The amplitude modulators are obtained with attenuation ($A < 1$).

In quantum transmission systems, intensity and phase modulation may be obtained by exploiting electro-optical properties of particular crystals, in which the refractive index depends on the intensity of the electric field applied to the material. Then, the

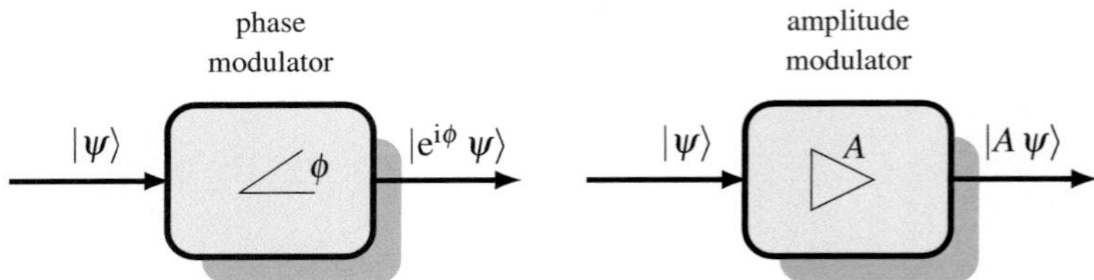

Fig. 9.5 Graphical representation of phase and amplitude modulators

phase of the output beam may be modulated by varying an electric voltage applied to the device.

Different mechanisms are used to modify the refractive index of the crystal. Electro-optic modulators exploit the so-called *Pockels cells*, waveguide made of nonlinear crystal material which can be considered equivalent to voltage-controlled waveplates. The variable electric voltage drives the phase delay induced to the optical beam traveling through the modulator.

Depending on the direction of the applied electric field, the type and the orientation of the nonlinear crystal, the phase delay may be different in the two direction of the polarization axes. The result is a polarization modulation. With the addition of polarizers at the input and output of the modulator, the change in the polarization leads to a variation of the amplitude of the output beam.

An alternative configuration, very common in fiber modulators and in integrated devices, employs this mechanism in a Mach–Zehnder interferometer. An input waveguide is split into two paths, i.e. the two arms of the interferometer, and then recombined into an output waveguide. The variable electric voltage is applied on one of these paths, resulting in an optical index modulation of one arm. The interference at the output waveguide builds the phase or intensity modulation of the beam.

Simplifying the model, the relation between the input and output beam power of an intensity modulator can be expressed as

$$I_{out} = \tau I_{in} \left[1 + \cos \left(\frac{\pi}{V_\pi} V + \Phi \right) \right] \tag{9.14}$$

where I_{in} is the input intensity, I_{out} is the output intensity, V_π is the half-wave voltage, that is the voltage required for inducing a phase change of π, and V is the modulation voltage. The coefficients τ and Φ describe a transmissivity and a phase term which take into account for losses and a mismatch between the two interferometer arms.

In the case of phase modulators, the phase variation obtained at the output is given by the affine equation

$$\phi = \frac{\pi}{V_\pi} V + \Phi \tag{9.15}$$

which involves the half-wave voltage V_π and the correction coefficient Φ.

Other type of modulators use analogous acousto-optical effects. Exploiting a piezoelectric transducer attached to the crystal, a sound wave is generated to provide a periodic refractive index grating. The traveling optical beam undergoes Bragg diffraction and propagates in a slightly different direction, enabling the possibility to build intensity (on–off) switching.

9.2.4 Photodetectors

While the other components of the optical transmission systems (laser, modulator and demodulator, transmitting and receiving antennas, and channel) have direct counterparts in a radiofrequency system, the detection in optical communications is performed almost exclusively by a photodetector, which is a very peculiar component of the optical technology, exploiting the photoelectric effect explained in quantum terms by Einstein in 1905.

The photodetection is the result of an interaction process between light and matter. Roughly speaking, a single photon in the optical beam releases an electron in the photosensitive material, which generates a pulse of electric current, converting optical power into an electric quantity. Before conversion, the electrons released by the photoelectric effect may be subjected to a multiplication procedure in which each electron generates a random number of secondary electrons.

From the classical point of view, the model may be the following. At the input, we get the instantaneous power

$$p(t) = \sum_k (h\nu)\, \delta(t - t_k)$$

where the instants t_k are the arrivals of a *doubly stochastic* Poisson process with intensity $\lambda(t)$ (see Fig. 4.28). The current produced by the photodetection can be modeled as a *filtered and marked Poisson process*, as discussed in Sects. 4.6 and 4.7, namely

$$i(t) = \sum_k g_k\, i_0(t - t_k), \tag{9.16}$$

where $i_0(t)$ is the current pulse generated by a single electrons satisfying the condition

$$\int_0^\infty i_0(t)\mathrm{d}t = e \tag{9.17}$$

with e electric charge of the electron. The coefficients g_k are independent and identically distributed random variables giving the number of electrons generated by the photon arrived at time t_k. They may take into account the random gain (if any) of the photomultiplication or the loss of photons in the material, caused by reflection and spreading, or both. In the first case the random variables g_i have mean $G > 1$, called the photomultiplication gain. In the second case g_k are binary random variable, whose mean η gives the photodetection efficiency.

The intensity $\lambda(t)$ of the Poisson process is proportional to the area A of the photodetector and to the intensity $J(t) = |\alpha(t)|^2$ of the electric field. In digital communications applications, $\lambda(t)$ turns out to be a random process, depending on the transmitted symbol. Then, the model of the photon arrivals is a doubly stochastic Poisson process.

Among practical impairments to the ideal behavior of the photodetectors, a role is played by the so-called dark current due to spontaneous emission of electrons in the photosensitive material. This is taken into account by a constant term λ_0 added to the useful intensity $\lambda(t)$.

In the decision process, the current (9.16) is integrated on the interval, say $(0, T]$, corresponding to a symbol slot, giving the quantity under decision

$$\mathcal{Q} = \int_0^T i(t)\mathrm{d}t = \sum_k^n g_k \int_0^T i_0(t - t_k)\mathrm{d}t = e\,n_T, \tag{9.18}$$

where n_T gives the electrons counting, i.e., the random number of electrons emitted in the slot symbol by the photodetector. The general statistics of n_T for a filtered and marked doubly stochastic Poisson process has been discussed in Chap. 4 (see also Fig. 4.28 for the detail of counting starting from the instantaneous power and current). On the value of the detected charge $\mathcal{Q}$, depending on the particular symbol transmitted, is based the decision process of the digital transmission scheme.

In the quantum communications applications the aim of the photodetector is limited in general to detect the presence of a positive number of photons in the optical beam, formulated as a quantum state $|\psi\rangle$. Thus a photodetector plays the role of counting the photons present in a given state $|\psi\rangle$ (Fig. 9.6). As discussed in Sect. 7.9.3, from a quantum point of view it must discriminate the vacuum state $|0\rangle$ from a coherent state

$$|\alpha\rangle = \mathrm{e}^{-|\alpha|^2/2} \sum_{n=0}^{\infty} \frac{\alpha^n}{\sqrt{n!}} |n\rangle. \tag{9.19}$$

The ideal quantum model of the detector reduces to a simple von Neumann measure with measurement operators (see (7.70))

$$Q_1 = \sum_{n=1}^{\infty} |n\rangle\langle n|, \qquad Q_0 = |0\rangle\langle 0| = I - Q_1 \tag{9.20}$$

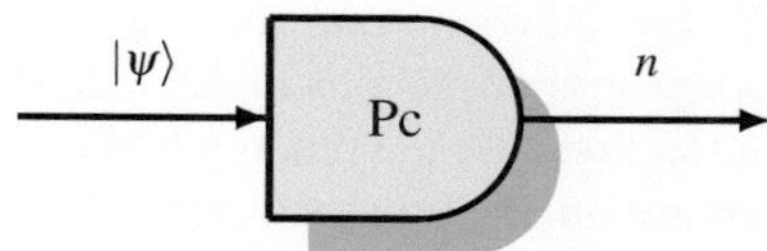

Fig. 9.6 Graphical symbol of a photon counter. The output n gives the number of photons present in the state $|\psi\rangle$

detecting 0 if and only the input state is $|0\rangle$. The resulting conditional probabilities are given by

$$p(0|0) = \langle 0|Q_0|0\rangle = 1 , \qquad p(1|1) = \langle \alpha|Q_1|\alpha\rangle = 1 - e^{-|\alpha|^2/2}. \tag{9.21}$$

If reduced efficiency and dark current are taken into account, the measurement operators becomes (see Problem 9.1)

$$Q_1 = e^{-\mu} \sum_{n=0}^{\infty} (1 - \eta)^n |n\rangle \langle n| \quad , \qquad Q_0 = I - Q_1 \tag{9.22}$$

where $\mu = \lambda_0 T$ is the average number of dark current electrons and η is the detection efficiency. The quantum state $|0\rangle$ is guessed if no dark electrons are present (with probability $e^{-\mu}$) and, for any n , n photons are missed (with probability $(1 - \eta)^n$). Note that in this case the measurement operators are POVM and not von Neumann projectors.

It must be noted that in any case the measurement performed by the photodetector is destructive in that, after the detection, the field is completely absorbed.

Problem 9.1 ⋆⋆   Consider the model of a photon counter where the dark current and the nonunitary efficiency are taken into account. Prove that the measurement operators are given by (9.22).

9.3 Classical Optical Communications Systems

9.3.1 Incoherent Detection

The simplest optical communication system uses amplitude modulation and incoherent detection (see Sect. 7.9). The transmitter associates a zero field to the binary symbol 0 and the field

$$v(t) = V_0 \cos 2\pi \nu t , \qquad 0 < t < T \tag{9.23}$$

to the binary symbol 1, where T is the symbol period. This is obtained by amplitude modulating the laser beam of frequency ν or, more simply, by switching on and off the laser itself according to the source symbol to be transmitted. At the receiver, a photodetector transforms the incident field into an electrical current, as discussed in Sect. 9.2.4.

In the absence of thermal noise, if the transmitted symbol is 0, the number of photons detected is zero; otherwise, it is a Poisson random variable n with mean value $\bar{n}$ proportional to V_0^2. An error may happen if and only if the symbol 1 is transmitted and the number of detected photons is 0. Then, the error probability turns out to be

$$P_e = \frac{1}{2}\mathrm{e}^{-\overline{n}} \tag{9.24}$$

where equally likely symbols are assumed. For the sake of comparison with other schemes it is convenient to express the error probability in terms of the *average number of photons per bit* N_R, given by $N_R = \frac{1}{2}\overline{n}$. Then

$$P_e = \frac{1}{2}\mathrm{e}^{-2N_R}. \tag{9.25}$$

The received scheme is called *direct detection* of the incident light pulses. The main advantage of this approach is its simplicity. In particular, phase and frequency instability of the laser source is well tolerated. Moreover, at the receiver direct detection is used and phase-sensitive devices are avoided.

This scheme, known as on–off keying (OOK) modulation, has a simple quantum equivalent, employing the coherent states $|0\rangle$ and $|\alpha\rangle$. As shown in Sect. 9.2.4, the photodetector can be modeled by a von Neumann measurement with projectors $Q_0 = |0\rangle\langle 0|$ e $Q_1 = I - |0\rangle\langle 0|$. The cross transition probabilities are

$$p(1|0) = \mathrm{Tr}[|0\rangle\langle 0| Q_1] = 0, \qquad p(0|1) = \mathrm{Tr}[|\alpha\rangle\langle\alpha|Q_0] = \mathrm{e}^{-|\alpha|^2} \tag{9.26}$$

so that the error probability becomes $P_e = \frac{1}{2}\mathrm{e}^{-|\alpha|^2}$. In terms of average number of photons per bit N_R, with equiprobable symbols one gets $N_R = |\alpha|^2/2$, so that we find again

$$P_e = \frac{1}{2}\mathrm{e}^{-2N_R}. \tag{9.27}$$

This result is known as the **quantum limit** (or **shot noise limit**) and is the optimum for any detection that does not exploit the coherence property of the optical beam.

9.3.2 Coherent Homodyne Detection

A more sophisticated scheme of classical optical communication uses binary phase shift keying (BPSK) modulation (see Sect. 7.10.2). The laser beam is applied to a π-phase modulator driven by the binary symbol source. As a consequence, the optical field at the receiver assumes one of the values

$$v(t) = V_0 \cos(2\pi\nu t + A_0\pi) \tag{9.28}$$

depending on the source symbol $A_0 \in \mathcal{A} = \{0, 1\}$.

Since the signals for different symbols have the same optical energy, direct detection cannot discriminate between them. In the coherent homodyne detection scheme the receiver sums to the field a high level field $V_L \cos 2\pi\nu t$ with the same frequency as $v(t)$ but with larger amplitude $(V_L \gg V_0)$ generated by a local laser. The global field

$$V_0 \cos(2\pi \nu t + A_0 \pi) + V_L \cos 2\pi \nu t = (V_0 \cos A_0\pi + V_L)\cos 2\pi \nu t \qquad (9.29)$$

applied to a photodetector produces a number of electrons which is a Poisson random variable with mean and variance proportional to

$$V_L^2 + V_0^2 + 2 \cos A_0\pi \, V_0 V_L = V_L^2 + V_0^2 + 2 B_0 \, V_0 V_L \qquad (9.30)$$

where

$$B_0 = \cos A_0\pi = \begin{cases} +1 & A_0 = 0 \\ -1 & A_0 = 1. \end{cases} \qquad (9.30\text{a})$$

Then, having subtracted the bias term $V_L^2 + V_0^2$ independent of the symbol, one obtains the useful signal proportional to $2 B_0 \, V_0 V_L$. As the amplitude V_L of the local laser field increases, an approximate Gaussian characterization of signal and noise may be adopted, so that the receiver must discriminate between two signal proportional to $2 B_0 \, V_0 V_L$ with Gaussian noise having variance

$$\sqrt{V_L^2 + V_0^2 + 2 B_0 \, V_0 V_L} \approx V_L. \qquad (9.31)$$

This can be obtained by a threshold decision device [2], that is, a device which sets a threshold and estimates the received symbol depending on whether the measured signal is above or below such a threshold. With equiprobable symbols the optimal threshold is 0 and the error probability becomes (see homodyne receiver in Sect. 7.10.2)

$$P_e = Q(2V_0) = Q(\sqrt{4N_R}), \qquad (9.32)$$

where $Q(x)$ is the Gaussian complementary distribution and N_R is the average number of photons per bit. This error probability is known as the **standard quantum limit**.

Comparison with incoherent detection shows that the performances of the homodyne detection are largely better. On the other hand the implementation of an efficient homodyne scheme implies some complications, in that it requires the presence of a local laser that must be accurately tuned in frequency and phase with the source laser.

9.4 Binary Quantum Communications Systems

The simplest quantum communication systems use binary schemes in which Alice associates to the symbol A_0 of a classical binary source, $A_0 \in \{0, 1\}$, with prior probabilities q_0 and q_1, two coherent quantum states $|\gamma_0\rangle$ and $|\gamma_1\rangle$ and Bob performs a measurement on the system by using two measurement operators Q_0 and Q_1. The most common choices are $|\gamma_0\rangle = |0\rangle$ and $|\gamma_1\rangle = |\beta\rangle$ for the On–Off Keying (OOK) scheme (see Fig. 7.22) and $|\gamma_0\rangle = |-\beta\rangle$ and $|\gamma_1\rangle = |\beta\rangle$ for the Binary Phase Shift Keying (BPSK) scheme (see Fig. 7.25).

For the sake of comparison with practical systems we begin by reviewing the ideal detection approach leading to the Helstrom bound. Next we consider in detail the OOK with direct detection and the BPSK with Kennedy's detection. Particular attention will be given to the Dolinar's receiver which promises to achieve the optimum performance, i.e., the Helstrom bound.

9.4.1 Recall of Helstrom's Theory

We reconsider the general theory of binary detection developed in Sect. 5.4.2 according to the geometric approach. The state vectors are written in terms of an appropriate orthonormal basis $\{|u_0\rangle, |u_1\rangle\}$ as

$$|\gamma_0\rangle = \cos\theta\,|u_0\rangle + \sin\theta\,|u_1\rangle, \qquad |\gamma_1\rangle = \cos\theta\,|u_0\rangle - \sin\theta\,|u_1\rangle \tag{9.33}$$

where $\cos 2\theta = \langle\gamma_0|\gamma_1\rangle = X$ is the superposition coefficient assumed to be real. The orthonormal measurement vectors are written as

$$|\mu_0\rangle = \cos\phi\,|u_0\rangle + \sin\phi\,|u_1\rangle, \qquad |\mu_1\rangle = -\sin\phi\,|u_0\rangle + \cos\phi\,|u_1\rangle. \tag{9.34}$$

Then the transition probabilities $p(j|i) := P[\hat{A}_0 = j | A_0 = i]$ are given by

$$p(0|0) = \cos^2(\phi - \theta), \qquad p(1|1) = \sin^2(\phi + \theta) \tag{9.35}$$

and the correct detection probability turns out to be

$$P_c = q_0 |\langle\mu_0|\gamma_0\rangle|^2 + q_1 |\langle\mu_1|\gamma_1\rangle|^2 = q_0 \cos^2(\phi - \theta) + q_1 \sin^2(\phi + \theta). \tag{9.36}$$

Here the angle θ is given through the superposition coefficient X, while the angle ϕ is unknown and is evaluated by optimization. We have seen that the angle ϕ giving the maximum of P_c satisfies the conditions

$$\sin 2\phi = \frac{1}{R}\sin 2\phi \quad, \qquad \cos 2\phi = \frac{q_0 - q_1}{R}\cos 2\theta \tag{9.37}$$

where $R = \sqrt{1 - 4q_0 q_1 X^2}$. The corresponding optimal correct decision probability is

$$P_c = \frac{1}{2}(1 + R) = \frac{1}{2}\left(1 + \sqrt{1 - 4q_0 q_1 X^2}\right), \tag{9.38}$$

i.e., the Helstrom bound.

We have also seen (see Problem 7.10) that the **a posteriori probabilities** $q(i|j) := \mathrm{P}[A_0 = i|\hat{A}_0 = j]$ corresponding to the optimal decision are related to the correct decision probability by

$$q(0|0) = q(1|1) = P_c. \tag{9.39}$$

In other words, the measurement modifies the a priori probabilities in the sense that, independently of the measurement result, the symbol guessed acquires a posteriori probability coinciding with the probability of correct decision.

Finally, we note that the measurement vectors are entangled linear combinations of the state vectors. Unfortunately, since in practice only photodetectors and phase-sensitive devices are available, the optimal measurement vectors are very hard to be implemented experimentally. So, for a long time suboptimal approaches have been investigated and experimented and only recently experiments demonstrating the feasibility of the optimal measurement have been accomplished.

9.4.2 Kennedy's Receiver

In 1973 Kennedy [3] proposed a very simple quantum receiver for the Binary Phase Shift Keying (BPSK). The received quantum state ($|\beta\rangle$ or $|-\beta\rangle$)) is applied to one of the inputs of a beam splitter with high transmissivity τ. To the other input of the beam splitter a quantum state $|\beta\rangle$ is applied, produced by a local laser tuned in frequency and phase with the laser of the transmitter. The corresponding displacement $D(\beta)$ changes the possible input states into $|\gamma_0\rangle = |0\rangle$ and $|\gamma_1\rangle = |2\beta\rangle$, according to an approach called **nulling technique**. Then, as in the OOK receiver, one applies to the displaced state the photodetection with measurement projectors $P_0 = |0\rangle\langle 0|$ and $P_1 = 1 - P_0$. The resulting error probability turns out to be

$$P_e = q_1 \mathrm{Tr}(|\gamma_1\rangle\langle\gamma_1| P_0) = q_1 e^{-4|\beta|^2}, \tag{9.40}$$

or, with equally likely symbols and in terms of the average number of photons per bit

$$P_e = \frac{1}{2} e^{-4N_R}. \tag{9.41}$$

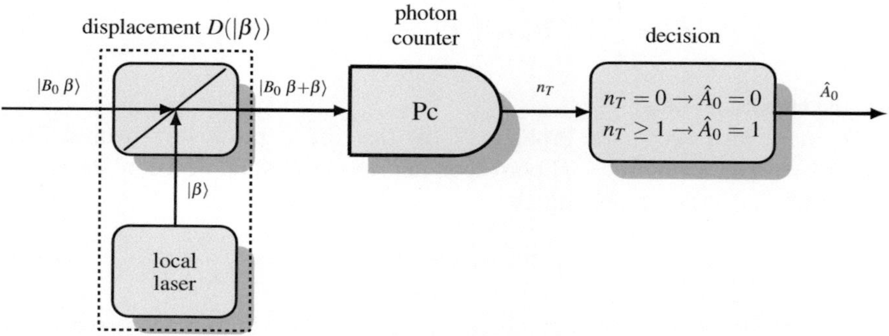

Fig. 9.7 Scheme of Kennedy's receiver. The displacement is obtained with a beam splitter feeded by the received state $|B_0\,\beta\rangle$ and the state $|\beta\rangle$ produced by a local laser. The received state takes one of the two values $|\pm\beta\rangle$ and, after the displacement, the values $|0\rangle$ and $|2\beta\rangle$, respectively. B_0 is the binary symbol $B_0 = \cos A_0\pi = \begin{cases} +1 & A_0 = 0 \\ -1 & A_0 = 1 \end{cases}$

The transmitter uses a π phase modulator driven by the input symbol. The receiver uses a local laser generating the coherent state $|\beta\rangle$ to be added to the input coherent state by a beam splitter realizing the displacement $D(\alpha)$. The scheme of the system is depicted in Fig. 9.7.

The feasibility of the Kennedy receiver has been demonstrated (see f.i. [4]). The main difficulties in the implementation are related to the presence of two lasers, the source laser and the local one, whose frequencies, phases, and levels must be accurately tuned. As a matter of fact, most practical demonstrations use a single laser source from which both the optical beam and the local beam simulating the useful carrier are derived through a beam splitter.

The performance of the Kennedy's receiver is presented in Fig. 9.8 in comparison with the performance of the OOK scheme and Helstrom's bound. The relations used are

$$P_{e,\text{OOK}} = \tfrac{1}{2}\,e^{-N_R}$$
$$P_{e,\text{Kennedy}} = \tfrac{1}{2}\,e^{-4N_R}$$
$$P_{e,\text{Helstrom}} = \tfrac{1}{2}\left[1 - \sqrt{1 - e^{-4N_R}}\right]$$

where N_R is the average number of photons per bit. The error probability plotted versus N_R shows that the Kennedy's receiver outperforms the OOK direct receiver, but is overperformed by the Helstrom's bound.

On the other hand, the Kennedy receiver does not outperforms the standard quantum limit of the homodyne detection for week signals ($N_R < 0.4$). But also for greater values of the number N_R of the received photons the performance of Kennedy's

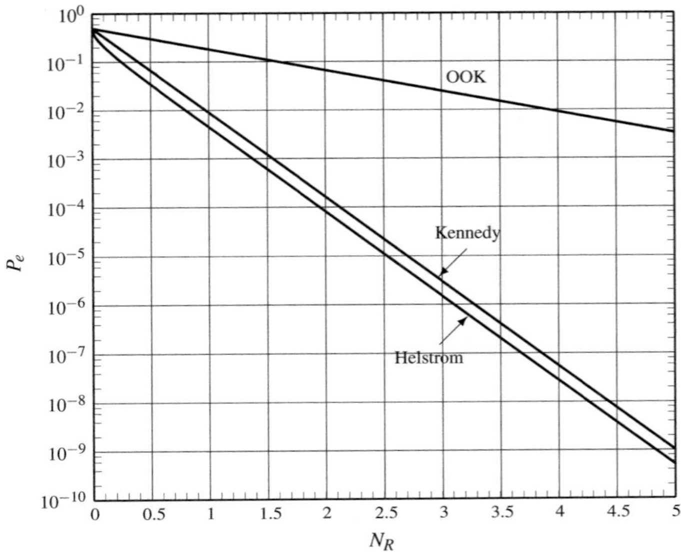

Fig. 9.8 Comparison of OOK, Kennedy's receiver, and Helstrom's bound in terms of error probability versus the average number of photons N_R

receiver in practical experiments is inferior to the standard quantum limits. Indeed, impairments of the photodetector, as reduced quantum efficiency and dark current, has relevant negative effects on the error probability [5].

9.4.3 Improved Kennedy's Receiver

Improvements to Kennedy's receiver have been suggested in recent years by Takeoka and Sasaki [5]. The basic idea is to apply to the input state a displacement $|\varepsilon\rangle$ with ε chosen in such a way that the error probability is minimized. The quantum states $|\beta\rangle$ and $|-\beta\rangle$ are displaced into the states $|\gamma_1\rangle = |\varepsilon + \beta\rangle$ and $|\gamma_0\rangle = |\varepsilon - \beta\rangle$. The error probability becomes

$$P_e = q_1 \text{Tr}(|\gamma_1\rangle \langle \gamma_1| P_0) + q_0 \text{Tr}(|\gamma_0\rangle \langle \gamma_0| P_1)$$

$$= q_1 e^{-(\varepsilon+\beta)^2} + q_0(1 - e^{-(\varepsilon-\beta)^2}). \tag{9.42}$$

(For the sake of simplicity we assume that β and ε are real). By nulling the derivative with respect to ε, we find that the displacement ε_0 minimizing P_e satisfies the transcendental equation

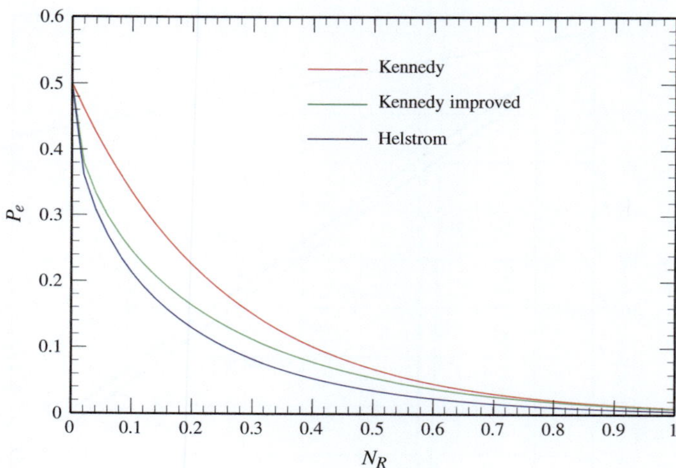

Fig. 9.9 Comparison of Kennedy's receiver, improved Kennedy's receiver, and Helstrom's bound in terms of error probability versus the average number of photons N_R

$$\frac{q_1}{q_0} = \frac{\varepsilon - \beta}{\varepsilon + \beta} e^{4\beta\varepsilon}. \tag{9.43}$$

In Fig. 9.9 the performance of the Kennedy receiver and of the improved Kennedy receiver are compared with the Helstrom bound. The relations used are

$$P_{e,\text{Kennedy}} = \tfrac{1}{2}\, e^{-4N_R}$$

$$P_{e,\text{Kennedy improved}} = \tfrac{1}{2}\left[1 + e^{-(\varepsilon_0 + \sqrt{N_R})} - e^{-(\varepsilon_0 - \sqrt{N_R})}\right]$$

$$P_{e,\text{Helstrom}} = \tfrac{1}{2}\left[1 - \sqrt{1 - e^{-4N_R}}\right]$$

where $N_R = \beta^2$ is the average number of photons per bit. For large values of β, the improvement obtained by optimizing the displacement ε appears to be negligible. On the other hand, as β goes to 0, the improved Kennedy's receiver approximates the Helstrom's bound very well and outperforms the standard quantum limit also for weak signals. This has an important consequence in the interpretation of the optimum Dolinar's receiver. The feasibility of the improved Kennedy's receiver has been recently demonstrated by Wittmann et al. [6].

Further light improvements [5] can be obtained if the input state is subjected to a displacement $D(\varepsilon)$ and to a squeezing $Z(r)$ (to be jointly optimized) before the photodetection.

9.4.4 Dolinar's Receiver

In 1973 Dolinar [7] proposed an adaptive measurement scheme, based on a combination of photon counting and feedback control, that precisely achieves the Helstrom bound. However, since the scheme requires a very precise control of an optical–electrical loop, only in 2007 Dolinar's idea has obtained a satisfactory practical implementation [4, 8].

In order to give some insight on Dolinar's approach, we consider the problem of discriminating between the states given by multiple copies

$$|\alpha_0\rangle = |\alpha\rangle \otimes \cdots \otimes |\alpha\rangle \quad , \quad |\alpha_1\rangle = |-\alpha\rangle \otimes \cdots \otimes |-\alpha\rangle \qquad (9.44)$$

in the tensorial product Hilbert space $\mathcal{H}^{\otimes n}$ where $\mathcal{H}$ is the Hilbert space spanned by the single copies $|\alpha\rangle$ and $|-\alpha\rangle$. Of course, Helstrom's theory assures that the optimum receiver gives the Helstrom's bound

$$P_c^{(n)} = \frac{1}{2}\left(1 + \sqrt{1 - 4q_0q_1 X^{2n}}\right), \qquad (9.45)$$

with $X = |\langle\alpha|-\alpha\rangle|$, so that $|\langle\alpha_0|\alpha_1\rangle| = X^n$. On the other hand, the optimal measurement vectors in $\mathcal{H}^{\otimes n}$ derived according to the Helstrom's theory are entangled vectors difficult to be realized experimentally. However, Acin et al. [9] have shown that the optimum can be achieved by adaptive local measurements on the single copies, each one taking into account the results of the previous measurements (for greater details see [10]).

Confining ourselves to the case $n = 2$, assume that the optimum measurement has been performed on the first state with correct decision probability $P_c^{(1)}$ given by (9.45) with $n = 1$. Moreover, assume that as a consequence of the measurement state $|\alpha\rangle$ has been guessed. Then, as discussed above, the a posteriori probabilities of $|\alpha\rangle$ and $|-\alpha\rangle$ become $q_0' = P_c^{(1)}$ and $q_1' = 1 - P_c^{(1)}$. If we perform an optimum measurement on the second state on the basis of the probabilities q_1' and q_0' and with corresponding new measurement vector, after the measurement we get the correct result with probability

$$P_c^{(2)} = \frac{1}{2}\left(1 + \sqrt{1 - 4(1 - P_c^{(1)})P_c^{(1)} X^2}\right) = \frac{1}{2}\left(1 + \sqrt{1 - 4q_0q_1 X^4}\right). \quad (9.46)$$

The same result is obtained if the state guessed after the first measurement is $|-\alpha\rangle$. By iterating the reasoning, the result can be generalized to n-copies states. The process can be considered as a feedback-assisted detection, in that the measurement on each copy is chosen on the basis of the result of the previous measurements.

These considerations can be applied to BPSK coherent states $|\beta\rangle$ and $|-\beta\rangle$ when they correspond to wavepackets having temporal extent of duration T. In this case, the mode can be thought as a sequence of shorter and weaker modes of duration T/n, namely,

$$|\beta\rangle = \left|\frac{\beta}{\sqrt{n}}\right\rangle \otimes \cdots \otimes \left|\frac{\beta}{\sqrt{n}}\right\rangle \tag{9.47}$$

with an analogous decomposition for $|-\beta\rangle$. Moreover, as n increases and the average number of photons per copy goes to zero, as shown above, the optimal Helstrom measurement on each copy may be conveniently approximated by an improved Kennedy receiver, i.e., a displacement followed by a photon detection.

The multiple-copy approach discussed above is mimicked by the Dolinar's receiver. Let be

$$\psi(t) = \pm\psi e^{i2\pi\nu t}, \qquad 0 \le t \le T \tag{9.48}$$

the input fields corresponding to the coherent states $|\pm\beta\rangle$. At the detector from the input field a time-varying field generated by a local laser is subtracted. The envelope of this local field is chosen between either $u_0(t)$ or $u_1(t)$, accordingly to the value of $z(t)$, a binary signal with possible values 0 and 1, giving the provisional decision at time t. Then, depending on the value of $z(t)$, the optical signal at the photon counter has enveloped either $\pm\psi - u_0(t)$ or $\pm\psi - u_1(t)$. The decision signal $z(t)$ is assumed changing at any photon arrival at the counter.

The mathematical problem is to choose the functions $u_0(t)$ and $u_1(t)$ that maximize the correct detection probability $P[z(T) = a]$, where a is the source symbol and $z(T)$ is the final decision. The problem has been solved by Geremia [11] on the basis of the dynamic programming optimality principle.

A simpler proof based on a semiclassical analysis given by Assalini et al. [10] is sketched here, under the preliminary assumption that the subtracted envelopes are opposite, namely $u_1(t) = -u_0(t)$. Provided that the transmitted symbol is $a = 0$ and consequently the received envelope is β, the process $z(t)$ can be interpreted as a telegraph process [12] alternately driven by non-homogeneous Poisson processes with rates

$$\lambda(t) = |\beta - u_0(t)|^2 \quad , \qquad \mu(t) = |\beta + u_0(t)|^2. \tag{9.49}$$

Defined the conditional probability $p_0(t) = P[z(t) = 0|a = 0]$ and $N_{t,\Delta t}$ the number of arrivals in the interval $[t, t + \Delta t)$

$$\begin{aligned}
p_0(t + \Delta) &= P[z(t) = 0, N_{t,\Delta t} = 0|a = 0] + P[z(t) = 1, N_{t,\Delta t} = 1|a = 0] + o(\Delta t) \\
&= P[N_{t,\Delta t} = 0|z(t) = 0]p_0(t) + P[N_{t,\Delta t} = 1|z(t) = 1](1 - p_0(t)) + o(\Delta t) \\
&= [1 - \lambda(t)\Delta t]p_0(t) + \mu(t)\Delta t(1 - p_0(0)) + o(\Delta t).
\end{aligned}$$

Hence the differential equation

$$p_0'(t) = \mu(t) - [\lambda(t) + \mu(t)]p_0(t) \tag{9.50}$$

follows. In a similar way can be shown that $p_1(t) = P[z(t) = 1|a = 1]$ satisfies the same differential equation, so that the probability of correct decision satisfies the differential equation

$$P'_c(t) = q_0 p'_0(t) + q_1 p'_1(t) = \mu(t) - [\lambda(t) + \mu(t)]P_c(t) \qquad (9.51)$$

independent of the symbol probabilities q_0 and q_1. If we impose

$$P_c(t) = \frac{1}{2}\left[1 + \sqrt{1 - 4q_0 q_1 e^{-4\beta^2 t}}\right] \qquad (9.52)$$

coinciding with Helstrom's bound, a simple algebra shows that the differential equation is satisfied by setting

$$u_0(t) = \frac{\beta}{\sqrt{1 - 4q_0 q_1 e^{-4\beta^2 t}}} \quad , \qquad 0 < t < T. \qquad (9.53)$$

This gives the control optical signal achieving Dolinar's bound.

A conceptual scheme of Dolinar's receiver is depicted in Fig. 9.10. An amplitude-modulated local laser produces the optical beam with complex envelope $u_0(t)$ to be added or subtracted to the input beam. The choice between $\pm u_0(t)$ is performed by

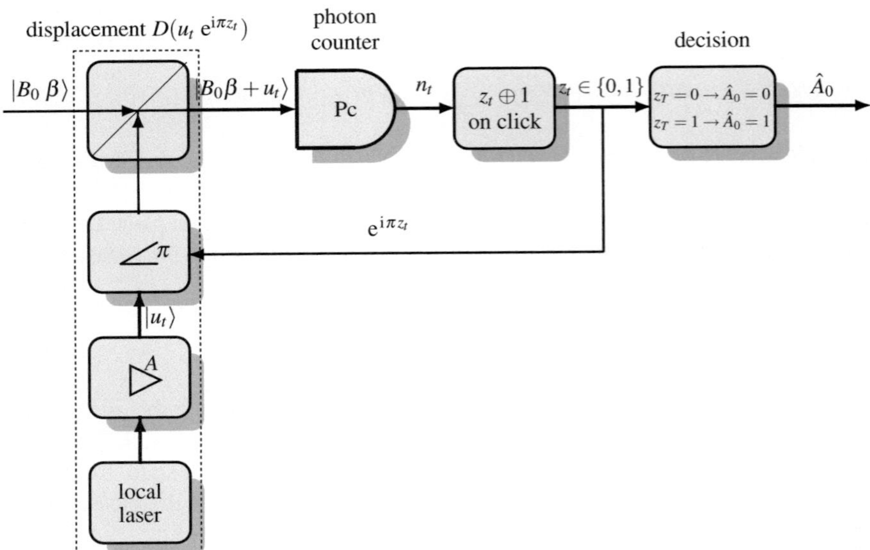

Fig. 9.10 Scheme of Dolinar's receiver: note that $z_t \oplus 1$ represents a change of 0 and 1 at every click. z_t represents the provisional symbol estimation. At the end of the symbol period, the final decision z_T is taken

a π phase modulator driven through a feedback control by the photon arrivals at the photodetector.

Of course the problems of tuning in frequency and phase the lasers encountered in Kennedy's receiver stay on. Another difficulty arises in amplitude modulating the local laser in such a way that the local envelope (9.53) is obtained. Finally, the unavoidable delays introduced by the optical–electrical feedback control may greatly reduce the performance of the system. As a consequence, the implementation of the optimal receiver out of the laboratories appears to be at present a very difficult task.

9.4.5 The Sasaki–Hirota Receiver

Sasaki and Hirota have shown [13] that in principle it is possible to achieve the Helstrom bound by considering the problem in the two-dimensional Hilbert space spanned by the states $|-\alpha\rangle$ and $|\alpha\rangle$.

Since to the input state it may be applied a displacement $|\alpha\rangle$ as in the Kennedy receiver, we may consider as input states $|0\rangle$ and $|2\alpha\rangle$. It may be easily verified that the states

$$|\eta_1\rangle = |0\rangle \quad , \qquad |\eta_2\rangle = \frac{1}{\sqrt{1-X^2}}(|2\alpha\rangle - X|0\rangle) \tag{9.54}$$

with $X = \langle 0|2\alpha\rangle$ (α is assumed to be real) form an orthonormal basis of the Hilbert space $\mathcal{H}_0$ spanned by $|0\rangle$ and $|2\alpha\rangle$. Then, consider the operator

$$U(\theta) = \cos\theta(|\eta_1\rangle\langle\eta_1| + |\eta_2\rangle\langle\eta_2|) + \sin\theta(|\eta_1\rangle\langle\eta_2| - |\eta_2\rangle\langle\eta_1|). \tag{9.55}$$

A simple algebra shows that $U(\theta)U^*(\theta) = |\eta_1\rangle\langle\eta_1| + |\eta_2\rangle\langle\eta_2|$ coincides with the identity operator in $\mathcal{H}_0$, so that $U(\theta)$ is a unitary operator in $\mathcal{H}_0$. The Sasaki–Hirota approach assumes that the unitary operator $U(\theta)$ is applied to the displaced state ($|0\rangle$ or $|2\alpha\rangle$) and that the transformed state is subjected to a von Neumann measurement with projectors

$$Q_1 = |\eta_1\rangle\langle\eta_1| , \qquad Q_2 = |\eta_2\rangle\langle\eta_2| . \tag{9.56}$$

If the error probability is computed and optimized with respect to the angle θ, the Helstrom bound is achieved. For greater mathematical details see [11, 13].

Note that, while the measurement state $|\eta_1\rangle = |0\rangle$ is a coherent state, $|\eta_2\rangle$ is not. On the other hand, since $\langle 0|\eta_2\rangle = 0$, it is innocuously substitutes the measurement operator $|\eta_2\rangle\langle\eta_2|$ with

$$\sum_{n=1}^{\infty} |n\rangle\langle n| \tag{9.57}$$

and the detection may be realized in the Glauber space with an ideal photodetector. On the contrary, the unitary operator $U(\theta)$ is not Gaussian (see Chap. 11), so that it is not realizable with usual linear optics and requires nonlinear unpractical components [13].

9.5 Multilevel Quantum Communications Systems

In spite of the fact that K-ary quantum systems can in principle achieve greater capacity than binary systems, only recently attention has been paid to implementable receivers for K-ary quantum communications with $K > 2$. Indeed, simple modulations as K-PSQ, QAM, and PPM have been frequently considered from a theoretical point of view but practical receivers are very difficult to be realized. In this section, we present some recent ideas concerning possible suboptimal receivers, in particular for K-PSK and PPM quantum systems.

9.5.1 Multiple PSK and QAM Quantum Systems

The coherent states of a K-PSK constellation are

$$|\alpha_k\rangle = |\alpha_0 e^{i2\pi k/K}\rangle, \qquad k = 0, \ldots, K - 1 \tag{9.58}$$

and may be written as $|\alpha_k\rangle = S^k|\alpha_0\rangle$ with $S = e^{i2\pi N/K}$, where N is the number operator (see Sect. 7.12.1). As shown above (see also [14]), this constellation enjoys geometrical uniform symmetry so that the SRM derived by the Gram matrix is optimal. However, the optimal measurement vectors turn to be entangled and difficult to realize. Similar considerations hold for QAM systems.

In [15], a suboptimal receiver for K-ary quantum systems is suggested based on suitable combinations of beam splitters, displacements, and photodetectors and following the ideas lying behind the Kennedy and Kennedy-improved detectors. Since the extensions to $K > 3$ appear intuitive, we confine ourselves to the case $K = 3$ (with possible states $|\alpha_0\rangle$, $|\alpha_1\rangle$ and $|\alpha_2\rangle$) and follow the scheme of Fig. 9.11. The input state is applied to a beam splitter with transmissivity τ. The first output of the beam splitter is displaced by a displacement $D(-\sqrt{1-\tau}\alpha_0)$, the second one by a displacement $D(-\sqrt{\tau}\alpha_1)$. The displaced states enter two photodetectors. Provided that the input state is $|\alpha_i\rangle$, the outputs of the beam splitter are given by $|b_1\rangle = |\sqrt{1-\tau}\alpha_i\rangle$ and $|b_2\rangle = |\sqrt{\tau}\alpha_i\rangle$ and the output of the displacements are $|c_1\rangle = |\sqrt{1-\tau}(\alpha_i - \alpha_0)\rangle$ and $|c_2\rangle = |\sqrt{\tau}(\alpha_i - \alpha_1)\rangle$. In conclusion, we have the following states in correspondence with the possible input state

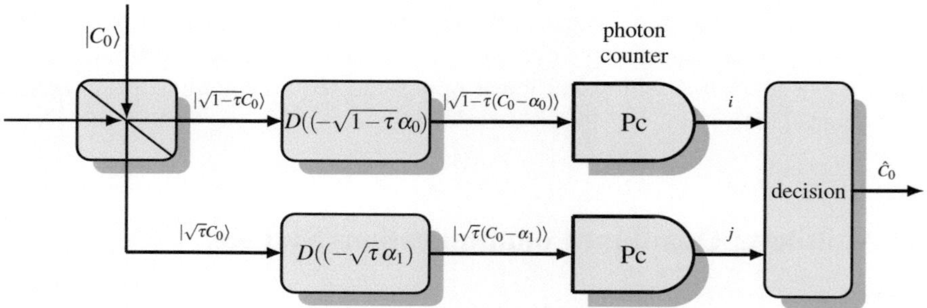

Fig. 9.11 Scheme of a ternary quantum system. C_0 is the transmitted complex (random) symbol and $|C_0\rangle$ the corresponding quantum state

$$
\begin{aligned}
|\alpha_0\rangle &\rightarrow |0\rangle \otimes |\sqrt{\tau}(\alpha_0 - \alpha_1)\rangle \\
|\alpha_1\rangle &\rightarrow |\sqrt{1-\tau}(\alpha_1 - \alpha_0)\rangle \otimes |0\rangle \\
|\alpha_2\rangle &\rightarrow |\sqrt{1-\tau}(\alpha_2 - \alpha_0)\rangle \otimes |\sqrt{\tau}(\alpha_2 - \alpha_1)\rangle.
\end{aligned}
\tag{9.59}
$$

Denoting by (i, j), $i, j \in \{0, 1\}$ the output of the photodetectors, the transition probabilities $p(i, j|\alpha_k)$ can be computed. For instance, if the input state is $|\alpha_1\rangle$, the probability that the first detector does not detect photons is $e^{-(1-\tau)|\alpha_1 - \alpha_0|^2}$, while the probability that the second detectors does not detect photons is 1, so that

$$
p(0, 0|\alpha_1) = e^{-(1-\tau)|\alpha_1 - \alpha_0|^2}.
$$

On the basis of the transition probabilities, one computes the optimum decision rule minimizing the error probability. Of course, this error probability depends on the transmissivity τ. Then a second optimization with respect to τ may be performed. The resulting error probability outperforms the standard quantum limit. Better performances can be achieved if, as in the improved Kennedy receiver, the nulling displacements $|-\alpha_0\rangle$ and $|-\alpha_1\rangle$ are substituted by optimized displacements [15].

Further improvements are possible by suitably squeezing the signals after the displacements [16].

9.5.2 Pulse Position Modulation Systems

A quantum modulation scheme that enjoys large popularity owing to its simplicity is the pulse position modulation (PPM) scheme. In this scheme, a K-ary classical symbol with alphabet $\{0, \ldots, K-1\}$ is encoded into the position of a coherent state $|\alpha\rangle$ in a sequence of $K-1$ null states. The natural environment for such modulation is the tensor Hilbert space $\mathcal{H}_0^{\otimes K}$, where $\mathcal{H}_0$ is the Fock space. The possible states are the tensor states

$$|\alpha_0\rangle = |\alpha\rangle \otimes |0\rangle \otimes \cdots \otimes |0\rangle \qquad \cdots \qquad |\alpha_{K-1}\rangle = |0\rangle \otimes |0\rangle \otimes \cdots \otimes |\alpha\rangle.$$

The corresponding modulation technique is very simple and requires only the switching of the source laser.

The symmetry of the constellation is clear, even though it is by no means trivial to evaluate the symmetry operator in the Hilbert space $\mathcal{H}_0^{\otimes K}$ [17]. The optimal error probability (see Sect. 7.12.4) turns out to be

$$P_e = \frac{K-1}{K^2} \left(\sqrt{1+(K-1)p} + \sqrt{1-p} \right)^2 \tag{9.60}$$

where $p = e^{-|\alpha|^2}$ is the probability that the state $|\alpha\rangle$ is not detected. The optimal measurement, coinciding with the SRM, enjoys the same symmetry of the states but is strongly entangled and appears very difficult to implement.

Note that in the K-PPM scheme each of the K symbols is carried by a state with average number of photons given by $|\alpha|^2$. The number of photons per bit is given bt $N_R = |\alpha|^2/\log_2 K$ and the error probability in terms of number of photons per bit is given by (9.60) with

$$p = e^{-N_R \log_2 K}.$$

Several suboptimal measurements have been proposed. The simplest idea is to measure the single pulses with direct detection. In the absence of impairments in the photodetector, the error happens only when the single nonzero state is not detected, so that, guessing at random the symbol, the error probability is

$$P_e = \frac{K-1}{K}p.$$

A more sophisticated approach [18], known as *conditionally nulling receiver*, uses the following adaptive decision strategy. During the first signaling slot, a nulling state $|-\alpha\rangle$ is added. If the photodetector does not detect a photon, one provisionally decides for $|\alpha_0\rangle$, then the photodetection continues without nulling and the decision is maintained unless some photon is detected in the subsequent slots. If some photon is detected in the first interval, the hypothesis $|\alpha_0\rangle$ is discarded and the procedure is iterated. The error probability is computed recursively. For $K = 2$, one gets $P_e^{(2)} = p^2/2$ because error occurs if and only if the state is $|\alpha_1\rangle$ and two pulses are undetected. For $K > 2$, no error occurs if the state is $|\alpha_0\rangle$, whereas in any other case (with probability $K/(K-1)$) error may occur if the nulling pulse in the first slot and the subsequent pulse are missed or if the nulling pulse is detected and an error happens in the remaining $K-1$ slots. In conclusion, the recursive relation

$$P_e^{(K)} = \frac{K-1}{K}[\, p^2 + (1-p)P_e^{(K-1)}]$$

follows. In closed form one gets [19]

$$P_e^{(K)} = \frac{1}{K}[(1-p)^K - 1 + Kp].$$

Slight performance improvements are obtained applying a nonexact (and optimized) nulling pulse as in the improved Kennedy's receiver [20]. The algorithm of the conditionally nulling receiver is mimicked applying a constant displacement $D(\varepsilon)$ with $\varepsilon \neq -\alpha$ in place of the nulling operation. A numerical optimization of the value of the displacement shows an improvements in the performance, as demonstrated in the experimental test reported in [21].

Further performance enhancements can be obtained by considering different displacement ε_i in place of the nulling operations in the slots $i = 0, \ldots, M-1$, which in general may depend on all the outcomes in the previous measurements, rather than only the last o ne.

The general structure of such a receiver is an *adaptive* scheme with local measurement in each Fock space $\mathcal{H}_0$ optimized upon all the previous outcome. Each local measurement implements a binary discrimination between the ground state $|0\rangle$ and the coherent state $|\alpha\rangle$, which may be performed with direct detection, Kennedy or Dolinar schemes depending on the design limitations or constraints.

Due to the binary outcome of each local measurement, the overall receiver strategy can be described with a binary tree, where each node corresponds to a measurement and each edge to an outcome. The binary tree is covered from the root node to the final one following the path dictated by the outcomes, performing the measurement defined in the node that come across.

Since the total number of the measurement employed grows exponentially in the cardinality K of the alphabet, a global optimization of the measurement parameter may be really demanding. However, the required numerical optimization may be lightened using a dynamic programming approach [22].

The adaptive receiver shows an improvement in the performance due to the greater flexibility of the binary discrimination scheme employed and the more general measurement sequencing, which may depend upon all the previous partial outcomes. The improvement is seen for all the alphabet cardinalities K, and in particular for $K = 2$ this receiver precisely reaches the Helstrom bound of the error probability [22].

References

1. A.E. Siegman, *Lasers* (University Science Book, Sausalito, 1986)
2. C.W. Helstrom, Quantum detection and estimation theory. *Mathematics in Science and Engineering* (Academic Press, New York, 1976)
3. R.S. Kennedy, A near-optimum receiver for the binary coherent state quantum channel. Massachusetts Institute of Technology, Cambridge (MA), Technical Report January 1973, MIT Research Laboratory of Electronics Quarterly Progress Report 108

4. C.W. Lau, V.A. Vilnrotter, S. Dolinar, J. Geremia, H. Mabuchi, Binary quantum receiver concept demonstration. NASA, Technical Report, 2006, Interplanetary Network Progress (IPN) Progress Report 42–146
5. M. Takeoka, M. Sasaki, Discrimination of the binary coherent signal: Gaussian-operation limit and simple non-Gaussian near-optimal receivers. Phys. Rev. A **78**, paper no. 022320 (2008)
6. C. Wittmann, M. Takeoka, K.N. Cassemiro, M. Sasaki, G. Leuchs, U.L. Andersen, Demonstration of near-optimal discrimination of optical coherent states. Phys. Rev. Lett. **101**, paper no. 210501 (2008)
7. S.J. Dolinar, An optimum receiver for the binary coherent state quantum channel, Massachusetts Institute of Technology, Cambridge (MA), Technical Report, October 1973, MIT Research Laboratory of Electronics, Quarterly Progress Report 111
8. R.L. Cook, P.J. Martin, J.M. Geremia, Optical coherent state discrimination using a closed-loop quantum measurement. Nature **446**(7137), 774–777 (2007)
9. A. Acín, E. Bagan, M. Baig, L. Masanes, R. Muñoz Tapia, Multiple-copy two-state discrimination with individual measurements. Phys. Rev. A, **71**, paper no. 032338 (2005)
10. A. Assalini, N. Dolla Pozza, G. Pierobon, Revisiting the Dolinar receiver through multiple-copy state discrimination theory. Phys. Rev. A **84**, paper no. 022342 (2011)
11. J. Geremia, Distinguishing between optical coherent states with imperfect detection. Phys. Rev. A **70**, paper no. 062303 (2004)
12. E. Parzen, *Stochastic Processes* (Holden Day, San Francisco, 1962)
13. M. Sasaki, O. Hirota, Optimum decision scheme with a unitary control process for binary quantum-state signals. Phys. Rev. A **54**, 2728–2736 (1996)
14. K. Kato, M. Osaki, M. Sasaki, O. Hirota, Quantum detection and mutual information for QAM and PSK signals. IEEE Trans. Commun. **47**(2), 248–254 (1999)
15. S. Izumi, M. Takeoka, M. Fujiwara, N.D. Pozza, A. Assalini, K. Ema, M. Sasaki, Displacement receiver for phase-shift-keyed coherent states. Phys. Rev. A **86**, paper no. 042328 (2012)
16. S. Izumi, M. Takeoka, K. Ema, M. Sasaki, Quantum receivers with squeezing and photon-number-resolving detectors for M-ary coherent state discrimination. Phys. Rev. A **87**, paper no. 042328 (2013)
17. G. Cariolaro, G. Pierobon, Theory of quantum pulse position modulation and related numerical problems. IEEE Trans. Commun. **58**(4), 1213–1222 (2010)
18. S.J. Dolinar, A near-optimum receiver structure for the detection of M-ary optical PPM signals. NASA, Technical Report 42–72, 1982
19. S.J. Dolinar, A class of optical receivers using optical feedback. Ph.D. dissertation, Department of Electrical Engineering and Computer Science, MIT, Cambridge (MA), 1976
20. S. Guha, J.L. Habif, M. Takeoka, Approaching Helstrom limits to optical pulse-position demodulation using single photon detection and optical feedback. J. Mod. Opt. **58**(3–4), 257–265 (2011)
21. J. Chen, J.L. Habif, Z. Dutton, R. Lazarus, S. Guha, Optical codeword demodulation with error rates below the standard quantum limit using a conditional nulling receiver. Nat. Photon. **6**(6), 374–379 (2012)
22. N. Dalla Pozza, N. Laurenti, Adaptive discrimination scheme for quantum pulse-position-modulation signals. Phys. Rev. A **89**, paper no. 012339 (2014)

Part III
Quantum Information

Chapter 10
Introduction to Quantum Information

10.1 Introduction

The theory of quantum information arose in the 1970s and enjoyed an increment of the activity in the last two decades (see history below). Its subject is the information processing with quantum states. What is interesting is that in several cases, the quantum information processing can have a great advantage with respect to classical information processing and its features often find no correspondence in the classical counterparts. The main examples of quantum information processing are the **quantum computer**, which can factorize a large number exponentially more efficiently than the classical computer, **quantum communications**, which allows for the improvement of performance of optical communications (as widely seen in Part II of this book), **quantum key distribution (QKD)**, which makes personal communications to be secure under whatever eavesdropping, and **quantum teleportation**, which can transfer quantum states to a remote party without an actual transfer of physical particles.

Quantum Information exhibits two forms, discrete and continuous; so that we have *discrete quantum information*, based on **discrete variables**, and *continuous quantum information*, based on **continuous variables**. The best known example of a discrete variable is the quantum bit or *qubit*, which has been introduced and discussed in Chap. 3. The best known example of continuous variables is provided by the quantized harmonic oscillator, which represents the fundamental tool in quantum optics and is the basis for the introduction of coherent states and more generally of Gaussian states. For this reason, the fundamentals of continuous variables and Gaussian states, not developed before, will be developed in detail in the next chapter.

An important remark is that most operations in quantum information processing can be carried out both with discrete and with continuous variables (this last possibility is a quite recent discovery). The comparison of these two possibilities should be made upon practical considerations and, more specifically, on how robustly can we manipulate quantum states. The common environment is ultimately given by light. As known, the manipulation of qubits and their combination is technically

© Springer International Publishing Switzerland 2015
G. Cariolaro, *Quantum Communications*, Signals and Communication Technology,
DOI 10.1007/978-3-319-15600-2_10

difficult because it is based on the single-photon technique with the problem of the presence of undesired multiple photons. On the other hand, the generation and the manipulation of coherent and Gaussian states are more robust and in fact most of the experiments of quantum information processing are done with Gaussian light [1]. This explains why quantum information with continuous variables has been a very hot topic in the recent years. Note that in this context "Gaussian states" may be considered a synonym of "continuous variables" because in quantum information processing continuous variables are almost always represented by Gaussian states.

Organization of Part III

First of all, we have to explain why, in the organization of this book, we have developed quantum communications before quantum information. In principle, quantum information comes before quantum communications since quantum communications may be viewed as an application of the principles of quantum information Theory. One reason is historical. In fact, it is well known that classical communications came several years before the classical Information theory and also quantum communications were developed before quantum information theory. But the true reason is considering that quantum mechanics is a very difficult discipline (Nobel laureate Feynman said); our specific choice in the organization of this book is the study of quantum communications with a minimum knowledge of quantum mechanics. But at this point, we ask the reader to improve this knowledge to achieve an adequate comprehension of quantum information. In other words, the level of difficulty in this part is higher than in Part II. We give two examples to explain the difference between the two parts.

In the fundamentals of Chap. 3, to represent mixed states, we have introduced the *density operator* and its decomposition into elementary projectors weighted by a probability distribution. This minimal notion was sufficient to develop adequately quantum optical communications, also in the presence of thermal noise. Now, for the theory of continuous variables, a more sophisticated representation for the density operator ρ is required, given by specific functions in the phase space (characteristic and Wigner functions). These functions allow for the important classification of quantum states: By definition, a quantum state ρ is Gaussian if its characteristic and Wigner functions have a Gaussian multivariate form. The representation of quantum states in the phase space will be the main topic of Chap. 11.

A second example is given by the phenomenon of *entanglement*, which was formally introduced at the end of Chap. 3. Entanglement is one of the most important properties of Quantum Mechanics and in particular of quantum information, showing an extraordinary departure from classical mechanics. It is a fundamental resource in quantum information processing for the manipulation of both quantum and classical information. Therefore, entanglement should be developed in great detail. This will be done at the end of this chapter.

The specific organization of this part (Part III) is the following. In Chap. 11, we develop the **fundamentals on continuous variables** while the fundamentals of discrete variables are not necessary because they are elementary and already developed. The primary tools for analyzing continuous-variable quantum information processing are **Gaussian states** and **Gaussian transformations**. As said above, Gaussian states are continuous-variable states that have a representation (characteristic and Wigner functions) in terms of Gaussian functions, and Gaussian transformations are those that take Gaussian states to Gaussian states. In addition to offering an easy description in terms of Gaussian functions, *Gaussian states and transformations are of great practical relevance.*

In Chap. 12 we will develop quantum information theory, which is another topic preliminary to quantum information processing. Classical information theory is a mathematical discipline, born in the field of telecommunication in 1948 thanks to Shannon [2]. Its purpose is mainly: (1) To define *information* mathematically and quantitatively, (2) to represent information in an efficient way (data compression) for storage and transmission, and (3) to ensure information protection (encoding) in the presence of noise and other impairments. Since information is essentially encoded in a physical system, and quantum mechanics is the most accurate representation of the physical world, it is natural to ask what are the limits set by Quantum Mechanics to information processing. This is developed by **Quantum Information Theory**, which is intrinsically richer and challenging than classical information theory, because of its intriguing resources, such as entanglement.

Having acquired these foundations, in the final chapter (Chap. 13), we will be ready to develop the main applications on quantum information both with discrete and continuous variables.

As a guide, we suggest the reader to revisit quantum communications (Part II) with the new perspective gained in this part.

Organization of this chapter. The organization of the present chapter is the following. In the next section, we will give the chronological history of quantum communications with the main exciting discoveries of the last 50 years. This allow us to appreciate the extraordinary activity of the research community around the world, arriving at today's hot topics. The rest of the chapter is concerned with the development of entanglement and other preliminary topics, not sufficiently considered before.

A Few Milestones in Quantum Information

1963: Seminal papers on coherent states (Glauber, Nobel Prize 2005)

1970: Quantum money (Wiesner)

1970: (mid 70s) Public key criptography (Difiie and Hellman, and Merkle)

1973: Holevo bound, one bit per qubit.

1973: Kennedy receiver

1973: Dolinar receiver

1976: Quantum detection and quantum estimation (Helstrom)

1977: RSA public key cryptosystem (Rivest, Shamir and Adleman)

1982: Computing with quantum mechanical systems (Feynman, Nobel Prize 1965)

1982: No-cloning theorem (Wotters and Zurek)

1984: BB84 protocol (Bennett, Brassard)

1990: Unified theory of Gaussian states (Ma and Rhodes)

1991: Entangled-based QKD (Ekert)

1992: B1992 protocol (Bennett)

1992: Superdense coding (Bennett and Wiesner)

1993: Teleportation (Bennett et al.)

1994: Quantum algorithms for factorization in polynomial time (Shor)

1995: Qubit (Schumacher)

1996: Quantum error correcting codes (Shor)

1996: Sasaki–Hirota receiver

1997: HSW theorem (Schumacher and Westmoreland, Holevo)

1998: Transmission through a noisy quantum channel (Barnum, Nielsen, Schumacher)

1999: Entanglement-assisted classical capacity (Bennett, Shor, Smolin, Thapliyal)

2004: Private capacity (Cai and Yeung, Devetak)

2008: Quantum Internet (Winter)

10.2 Partial Trace and Reduced Density Operators

The partial trace is a fundamental operation to handle composite quantum systems since it allows one to extract operators of component systems. Let A and B be two quantum systems forming the composite system $A\,B$, described by the Hilbert spaces $\mathcal{H} = \mathcal{H}_A \otimes \mathcal{H}_B$. Such a system is often called **bipartite system** and is pictorially illustrated in Fig.10.1.

The partial trace *over the system B* can be defined by two items [3]:

(1) For elementary operators as

$$\boxed{\mathrm{Tr}_B[|a_1\rangle\langle a_2| \otimes |b_1\rangle\langle b_2|] = |a_1\rangle\langle a_2|\, \mathrm{Tr}[|b_1\rangle\langle b_2|]} \qquad (10.1)$$

where $|a_1\rangle, |a_2\rangle$ are arbitrary kets of $\mathcal{H}_A$ and $|b_1\rangle$, $|b_2\rangle$ are arbitrary kets of $\mathcal{H}_B$. The operation on the right hand side of (10.1) is the ordinary trace operation, giving

$$\mathrm{Tr}[|b_1\rangle\langle b_2|] = \langle b_1|b_2\rangle \,. \qquad (10.1\mathrm{a})$$

(2) It is extended to arbitrary operators by **linearity**, that is,

$$\mathrm{Tr}_B[c_1\,O_1 + c_2\,O_2] = c_1\,\mathrm{Tr}_B[O_1] + c_2\,\mathrm{Tr}_B[O_1], \qquad c_1, c_2 \in \mathbb{C} \qquad (10.1\mathrm{b})$$

where O_1 and O_1 are arbitrary operators of $\mathcal{H}$.

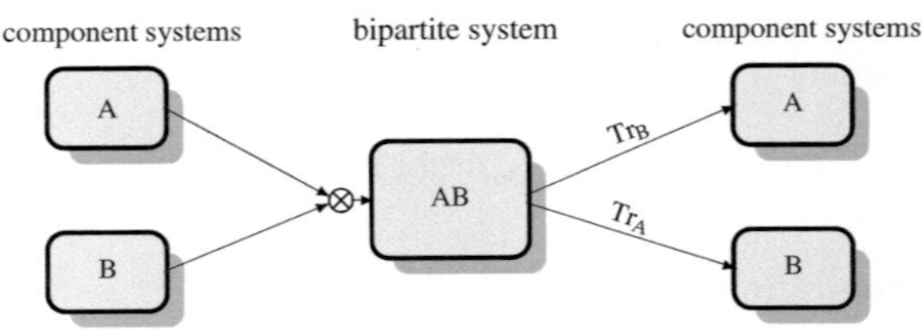

Fig. 10.1 Construction of a bipartite system AB by the *tensor product* $\otimes$ and separation of component systems A and B by the *partial trace*

The operation defined by (1) and (2) is called the partial trace *over the system B*. Analogous is the definition of the partial trace *over the system A*. Note that, while the application of the ordinary trace gives a scalar quantity, the application of the partial trace *over B* gives an operator of A and the application of the partial trace *over A* gives an operator of B.

The partial trace is usually applied to density operators. If ρ is a density operator of $\mathcal{H}$, one can "extract," or "trace out," from ρ the two operators

$$\rho_A = \mathrm{Tr}_B[\rho], \qquad \rho_B = \mathrm{Tr}_A[\rho] \tag{10.2}$$

which are called *reduced density operators*.

To see how the definition works in the evaluation of the reduced operators (10.2) we use matrix representations to get:

Proposition 10.1 *Let ρ be a density operator of the bipartite system $\mathcal{H} = \mathcal{H}_A \otimes \mathcal{H}_B$ and let $\rho(a_1, a_2, b_1, b_2)$ be a matrix representation of ρ, where a_1, a_2 refers to a basis of $\mathcal{H}_A$ and b_1, b_2 to a basis of $\mathcal{H}_B$. Then the matrix representations of the reduced operators are given by*

$$\rho_A(a_1, a_2) = \sum_b \rho(a_1, a_2, b, b), \qquad \rho_B(b_1, b_2) = \sum_a \rho(a, a, b_1, b_2). \tag{10.3}$$

Proof We express ρ in terms of its matrix representation

$$\rho = \sum_{a_1,a_2} \sum_{b_1,b_2} |a_1\rangle \otimes |b_1\rangle \rho(a_1, a_2, b_1, b_2) \langle a_2| \otimes \langle b_2|$$

$$= \sum_{a_1,a_2} \sum_{b_1,b_2} \rho(a_1, a_2, b_1, b_2) |a_1\rangle \langle a_2| \otimes |b_1\rangle \langle b_2|$$

where $\{|a\rangle\}$ is a basis of $\mathcal{H}_A$ and $\{|b\rangle\}$ is a basis of $\mathcal{H}_B$, and we have applied the mixed-product law (see (2.104)) $(A \otimes B)(C \otimes D) = (AC) \otimes (BD)$ to $(|a_1\rangle \otimes |b_1\rangle)(\langle a_2| \otimes \langle b_2|)$. Then application of (10.1) gives

$$\mathrm{Tr}_B[\rho] = \sum_{a_1,a_2} \sum_{b_1,b_2} \rho(a_1, a_2, b_1, b_2) |a_1\rangle \langle a_2| \mathrm{Tr}[|b_1\rangle \langle b_2|)]$$

where $\mathrm{Tr}[|b_1\rangle\langle b_2|] = \delta_{b_1 b_2}$. Thus we find that the matrix representation of the reduced operator $\rho_A = \mathrm{Tr}_B[\rho]$ reads as in (10.3). $\qquad\qquad\square$

∇ For the pure states, using the Schmidt decomposition (see Proposition 10.3), we find:

Proposition 10.2 *If $|\psi\rangle$ is a pure state of $\mathcal{H} = \mathcal{H}_A \otimes \mathcal{H}_B$, the reduced density operators over B is given by*

$$\rho_A = \mathrm{Tr}_B[|\psi\rangle\langle\psi|] = \sum_i d_i^2 |e_i^A\rangle\langle e_i^A| \tag{10.4}$$

where d_i are the Schmidt coefficients and $\{|e_i^A\rangle\}$ are orthonormal kets of $\mathcal{H}_A$. The corresponding state of $\mathcal{H}_A$ is pure if and only if the state $|\psi\rangle$ is separable. If the state $|\psi\rangle$ is entangled, the "reduced" state is always mixed.

10.2.1 Bell States

These important states are defined in a two-qubit system, where $\mathcal{H} = \mathcal{H}_A \otimes \mathcal{H}_B$ with $\mathcal{H}_A = \mathcal{H}_B = \mathbb{C}^2$. Let $\{|0\rangle_A, |1\rangle_A\}$ and $\{|0\rangle_B, |1\rangle_B\}$ be the bases of A and B, respectively. Then the four Bell states are defined by

$$\begin{aligned}
|\Phi^+\rangle &= \tfrac{1}{\sqrt{2}}(|0\rangle_A \otimes |0\rangle_B + |1\rangle_A \otimes |1\rangle_B) = \tfrac{1}{\sqrt{2}}(|00\rangle + |11\rangle) \\
|\Phi^-\rangle &= \tfrac{1}{\sqrt{2}}(|0\rangle_A \otimes |0\rangle_B - |1\rangle_A \otimes |1\rangle_B) = \tfrac{1}{\sqrt{2}}(|00\rangle - |11\rangle) \\
|\Psi^+\rangle &= \tfrac{1}{\sqrt{2}}(|0\rangle_A \otimes |1\rangle_B + |1\rangle_A \otimes |0\rangle_B) = \tfrac{1}{\sqrt{2}}(|01\rangle + |10\rangle) \\
|\Psi^-\rangle &= \tfrac{1}{\sqrt{2}}(|0\rangle_A \otimes |1\rangle_B - |1\rangle_A \otimes |0\rangle_B) = \tfrac{1}{\sqrt{2}}(|01\rangle - |10\rangle)
\end{aligned} \tag{10.5}$$

where we have written also the abbreviated form.

Now we consider the density operator, e.g., of the first Bell state. We find

$$\begin{aligned}
\rho = |\Phi^+\rangle\langle\Phi^+| &= \tfrac{1}{2}\big[|0\rangle_A \otimes |0\rangle_B + |1\rangle_A \otimes |1\rangle_B\big]\big[{}_A\langle 0| \otimes {}_B\langle 0| + {}_A\langle 1| \otimes {}_B\langle 1|\big] \\
&= \tfrac{1}{2}\big[|0\rangle_A \otimes |0\rangle_B\, {}_A\langle 0| \otimes {}_B\langle 0| + |0\rangle_A \otimes |0\rangle_B\, {}_A\langle 1| \otimes {}_B\langle 1| \\
&\quad + |1\rangle_A \otimes |1\rangle_B\, {}_A\langle 0| \otimes {}_B\langle 0| + |1\rangle_A \otimes |1\rangle_B\, {}_A\langle 1| \otimes {}_B\langle 1|\big]
\end{aligned}$$

where the mixed-product law gives

$$\begin{aligned}
\rho = \tfrac{1}{2}\big[&|0\rangle_A\, {}_A\langle 0| \otimes |0\rangle_B\, {}_B\langle 0| + |0\rangle_A\, {}_A\langle 1| \otimes |0\rangle_B\, {}_B\langle 1| \\
+ &|1\rangle_A\, {}_A\langle 0| \otimes |1\rangle_B\, {}_B\langle 0| + |1\rangle_A\, {}_A\langle 1| \otimes |1\rangle_B\, {}_B\langle 1|\big].
\end{aligned}$$

Now, tracing out the second qubit and considering the orthonormality of the second basis, we find

$$\rho_A = \mathrm{Tr}_B[\rho] = \tfrac{1}{2}\big[|0\rangle_A \,_A\langle 0| + |1\rangle_A \,_A\langle 1|\big] = \tfrac{1}{2} I_{\mathcal{H}_A}$$

where we have used the completeness of the first basis. Note that this reduced density operator corresponds to a mixed state because $\mathrm{Tr}[\rho_A^2] = \tfrac{1}{2} < 1$.

Note that the Bell state is given in the Schmidt form and the evaluation of the reduced operator would be immediate after Proposition 10.2, considering that the Bell state is maximally entangled with Schmidt coefficients $d_1 = d_2 = 1/\sqrt{2}$.

10.3 Overview of Entanglement

Entanglement is one of the most important properties of quantum mechanics, and in particular of quantum information. As seen in Chap. 3, it occurs in composite systems as a consequence of the superposition principle and of the fact that the reference space is given by the tensor product of the Hilbert space of the component systems. If the the subsystems are spatially separated, the so-called **nonlocality properties** are verified, showing an extraordinary departure from the classical mechanics.

There are several problems related to entanglement. The first problem lies in recognizing the absence or the presence of this intriguing phenomenon, that is, recognizing whether a given state of the composite system is *separable* or *entangled*. This is a challenging question still open. Let us consider the case of a quantum system composed by two subsystems A and B, usually called *bipartite system*, described by the Hilbert space $\mathcal{H} = \mathcal{H}_A \otimes \mathcal{H}_B$. A state ρ of $\mathcal{H}$ is said **separable** if it is given as a convex combination of product states, namely

$$\rho = \sum_k p_k \, \rho_k^A \otimes \rho_k^B \tag{10.6}$$

where $p_k \geq 0$, $\sum_k p_k = 1$, and ρ_k^A, ρ_k^B are states of $\mathcal{H}_A$ and $\mathcal{H}_B$ respectively, Otherwise the state is said **entangled**. A second problem is the measure of the entanglement amount, that is, a quantitative criterion to evaluate how much a given state is entangled, running from zero (separable state) to a maximum (maximally entangled state). Other problems are concerned with the practical entanglement generation and its preservation in quantum operations. All these problems are challenging and must be treated differently in dependence of the nature of the quantum state: continuous or discrete, finite or infinite dimensional, and pure or mixed.

Here we give a brief overview of the topic, suggesting the reader to consult the book by Nielsen and Chuang [3] for the entanglement with discrete variables, the review article by Ferraro et al. [4] for entanglement with continuous variables (Gaussian states), the paper by Schumacher for entanglement in noisy channels [5].

10.3.1 Bipartite Pure States

The presence of entanglement can be easily established in the case of pure states. Consider for simplicity a finite-dimensional composite Hilbert space $\mathcal{H} = \mathcal{H}_A \otimes \mathcal{H}_B$, where $\mathcal{H}_A$ and $\mathcal{H}_B$ have dimensions m and n, respectively, and let $\{|b_r^A\rangle\}$ and $\{|b_s^B\rangle\}$ be orthonormal bases of the component systems. Then, a basis of $\mathcal{H}$ is given by $\{|b_r^A\rangle \otimes |b_s^B\rangle , \ r = 1, \ldots, m; s = 1, \ldots, n\}$ and a state of $\mathcal{H}$ has the following Fourier expansion

$$|\psi\rangle = \sum_{r=1}^{m} \sum_{s=1}^{n} c_{rs} |b_r^A\rangle \otimes |b_s^B\rangle , \qquad c_{rs} \in \mathbb{C} \tag{10.7}$$

with c_{rs} the Fourier coefficients.

The presence of entanglement in the state (10.7) is established by the following decomposition

Proposition 10.3 (Schmidt decomposition) *A general bipartite pure state can be expressed in the form*

$$|\psi\rangle = \sum_{k=1}^{K} d_k |e_k^A\rangle \otimes |e_k^B\rangle \tag{10.8}$$

where $\sum_k d_k^2 = 1$, *and* $\{|e_k^A\rangle\}$ *and* $\{|e_k^B\rangle\}$ *are orthonormal kets of* $\mathcal{H}_A$ *and* $\mathcal{H}_B$, *respectively.*

The coefficients d_i are called the *Schmidt coefficients* and the number of the d_i different from zero is called the *Schmidt rank* of $|\psi\rangle$. From Schmidt's decomposition (which is unique), one can see that the bipartite state $|\psi\rangle$ is separable if and only if one of the Schmidt coefficients d_k is unitary and all the others are zero. In fact, if this is the case, say for $k = k_0$, one has

$$|\psi\rangle = |e_{k_0}^A\rangle \otimes |e_{k_0}^B\rangle$$

so that the bipartite state is given by the tensor product of two states. Otherwise, the state is entangled and it is *maximally entangled* when all the Schmidt coefficients d_k are equal.

Proof The coefficients c_{rs} forms an $m \times n$ matrix C, to which we can apply the singular-value decomposition (see Sect. 2.12, Theorem 2.8) to get $C = U D V$, where U and V are unitary and D is a diagonal matrix containing the singular values $\{d_1, \ldots, d_K\}$. Then

$$c_{rs} = \sum_{k} u_{rk} d_k v_{ks}$$

and

$$|\psi\rangle = \sum_{r,s}\sum_{k} u_{rk}\, d_k\, v_{ks}|b_r^A\rangle \otimes |b_r^B\rangle = \sum_{k} d_k|e_k^A\rangle \otimes |e_k^B\rangle$$

where

$$|e_k^A\rangle = \sum_{r} u_{rk}\,|b_r^A\rangle\,, \qquad |e_k^B\rangle = \sum_{s} v_{ks}\,|b_s^B\rangle\,.$$

Note that, because of normalization and orthonormality the coefficients are normalized as $\sum_{r,s}|c_{rs}|^2 = 1$ and $\sum_{k} d_k^2 = 1$.

Example 10.1 (*Bell states*). Bell states are maximally entangled states. To see why they are entangled, consider in particular the state

$$|\Phi^+\rangle = \tfrac{1}{\sqrt{2}}(|0\rangle_A \otimes |0\rangle_B + |1\rangle_A \otimes |1\rangle_B) = \tfrac{1}{\sqrt{2}}(|00\rangle + |11\rangle)$$

and note that it can be written in the form (10.7) with the matrix C given by

$$C = \frac{1}{\sqrt{2}}\begin{bmatrix} 1 & 0 \\ 0 & 1 \end{bmatrix}.$$

Then it is already given in the form of Schmidt's decomposition

$$|\Phi^+\rangle = \sum_{k=0}^{1} d_k|k\rangle_A \otimes |k\rangle_B$$

with $d_0 = d_1 = 1/\sqrt{2}$, so that the state $|\Phi^+\rangle$ is maximally entangled. The same conclusion holds for the other Bell states.

10.3.2 An Entropic Separability Criterion for Pure State ∇

We have seen that a pure bipartite state is separable if and only if its Schmidt rank is unitary and, on the opposite, it is maximally entangled if its Schmidt coefficients are all equal (to $1/\sqrt{K}$). The Schmidt decomposition has an interesting and useful interpretation in terms of quantum entropy. To see this, we consider the density operator of the given pure state, expressed in terms of the Schmidt decomposition (10.8), specifically

$$\rho = |\psi\rangle\langle\psi| = \sum_{k=1}^{K} d_k^2\, |e_k^A\rangle \otimes |e_k^B\rangle\langle e_k^B| \otimes \langle e_k^A|\,.$$

The partial traces extract the density operators

$$\rho_A = \sum_{k=1}^{K} d_k^2 \, |e_k^A\rangle \otimes \langle e_k^A| \,, \qquad \rho_B = \sum_{k=1}^{K} d_k^2 \, |e_k^B\rangle \otimes \langle e_k^B| \,. \tag{10.9}$$

Next, consider the *quantum entropies* of the two reduced density operators. These entropies are equal and given by[1]

$$S(\rho_A) = S(\rho_B) = -\sum_{k=1}^{K} d_k^2 \log_2 d_K^2 \,.$$

Considering that $0 \le S(\rho_A) \le \log_2 K$ with the minimum 0 when the pure state is separable and the maximum when the state is maximally entangled, the entropy $S(\rho_A)$ can be considered as a measure of entanglement. This measure has the advantage to be used also for infinite dimensional state, as Gaussian states (see Sect. 11.19).

10.3.3 Separability Criterion Based on Fourier Expansion

We reconsider the Fourier expansion (10.7) of a bipartite pure state of $\mathcal{H} = \mathcal{H}_A \otimes \mathcal{H}_B$

$$|\psi\rangle = \sum_r \sum_s c_{rs} |b_r^A\rangle \otimes |b_s^B\rangle \,, \qquad c_{rs} \in \mathbb{C} \tag{10.10}$$

where now r, s may range also to infinite. If the Fourier coefficients c_{rs} can be factored as

$$c_{rs} = c_r^A \, c_s^B \quad \forall r, s \tag{10.11}$$

the bipartite state is separable. Thus we have the simple criterion of separability

Proposition 10.4 *If the Fourier expansion (10.10) of a pure bipartite state verifies condition (10.11), the state is separable as $|\psi\rangle = |\psi_A\rangle \otimes |\psi_B\rangle$, where $|\psi_A\rangle = \sum_r c_r^A \, |b_r^A\rangle$ and $|\psi_B\rangle = \sum_s c_s^B \, |b_s^B\rangle$.*

The main advantage of this criterion, besides simplicity, is that it can be applied also with infinite dimensions and in particular to multimode Gaussian states (see Sect. 11.19).

---

[1] The quantum entropy of a state ρ will be introduced in Sect. 12.4 and defined as $S(\rho) = -\,\mathrm{Tr}[\rho \log_2 \rho]$. It can be calculated from the eigenvalues λ_k of ρ as $S(\rho) = -\sum_k \lambda_k \log_2 \lambda_k$ and it is constrained as $0 \le S(\rho) \le \log_2 K$. Note that in (10.9) d_k^2 are the eigenvalues of both ρ_A and ρ_B.

10.3.4 Bipartite Mixed States

We have seen that the problem of the separability of pure states, having finite dimensions, is completely dominated by Schmidt's decomposition. Despite the efforts, a general solution in the case of mixed states has not be found yet. Most of the criteria proposed so far are generally only necessary for separability. For an overview of these criteria, we suggest reference [4], where the authors arrive at the conclusion that for the quantification of entanglement, no satisfactory measure is known at present for arbitrary mixed states.

Problem 10.1 $\star\star\star$ To check Schmidt's decomposition consider a finite-dimensional bipartite system with $\mathcal{H}_A = \mathbb{C}^2$ and $\mathcal{H}_B = \mathbb{C}^4$, where the coefficient matrix C is 2×4. Suppose that the matrix has the form

$$C = \begin{bmatrix} c_{11} & c_{12} & c_{13} & c_{14} \\ c_{21} & c_{22} & c_{23} & c_{24} \end{bmatrix} = \begin{bmatrix} \frac{1}{4} & \frac{1}{4} & \frac{1}{4} & \beta \\ \frac{1}{4} & \frac{1}{4} & \frac{1}{4} & \sqrt{\frac{5}{8} - \beta^2} \end{bmatrix}$$

where β is a parameter. Find the values of β, if any, which correspond to a separable state and to a maximally entangled state.

10.4 Purification of Mixed States

A mixed state described by a density operator can be represented as a pure state provided that the quantum environment is appropriately enlarged. The procedure to get the new representation is called *purification* and plays a fundamental role in Quantum Information.

We start from a density operator ρ^A of a quantum system A. The enlargement is obtained by introducing a copy R of the system A to form the joint system AR, where the pure state $|\psi_{AR}\rangle$ is defined, such that

$$\rho^A = \mathrm{Tr}_R[|\psi_{AR}\rangle\langle\psi_{AR}|] \,. \tag{10.12}$$

In words, the pure state $|\psi_{AR}\rangle$ reduces to ρ^A, when we look at the system A alone.

Proposition 10.5 *Let ρ^A be a density operator of A and let*

$$\rho^A = \sum_i \sigma_i^2 |a_i\rangle\langle a_i| \tag{10.13}$$

*be its EID. Then a **purification** of ρ is given by the composite state*

$$|\psi_{AR}\rangle = \sum_i \sigma_i |a_i\rangle \otimes |r_i\rangle \tag{10.14}$$

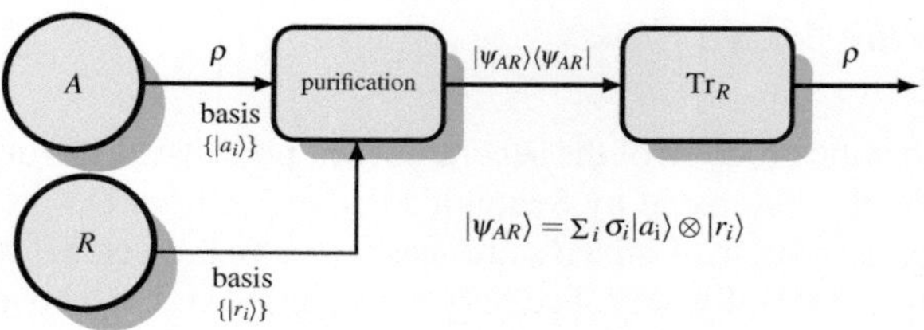

Fig. 10.2 A mixed state represented by the density operator ρ of the system A is purified with the help of a companion system R, which gives an orthonormal basis. The purification gives a pure state $|\psi_{AR}\rangle\langle\psi_{AR}| \in \mathcal{H}_A \otimes \mathcal{H}_A$. The partial trace with respect to R gives the original state ρ

where $\{r_i\}$ is a basis of R. Then the original density operator is related to the composite state as in (10.12).

Note that in general, by construction, the purified state $|\psi_{AR}\rangle$ is entangled (see Schmidt decomposition). It is not entangled if the original state ρ is pure.

The proof of the proposition is immediate

$$\mathrm{Tr}_R[|\psi_{AR}\rangle\langle\psi_{AR}|] = \sum_i \sum_j \sigma_i \sigma_j \, \mathrm{Tr}_R[|a_i\rangle \otimes |r_i\rangle\langle a_j| \otimes \langle r_j|]$$

$$= \sum_i \sum_j \sigma_i \sigma_j \, \mathrm{Tr}_R[|a_i\rangle\langle a_j| \otimes |r_i\rangle\langle r_j|]$$

$$= \sum_i \sum_j \sigma_i \sigma_j |a_i\rangle\langle a_j| \otimes \mathrm{Tr}[|r_i\rangle\langle r_j|]$$

$$= \sum_i \sum_j \sigma_i \sigma_j |a_i\rangle\langle a_j|\delta_{ij}$$

$$= \sum_i \sigma_i^2 |a_i\rangle\langle a_j| = \rho^A \; .$$

The procedure to get the pure state $|\psi_{AR}\rangle \in \mathcal{H}_A \otimes \mathcal{H}_R$ from the density operator ρ and recovery of ρ from $|\psi_{AR}\rangle$ is illustrated in Fig. 10.2.

References

1. X.B. Wang, T. Hiroshima, A. Tomita, M. Hayashi, Quantum information with Gaussian states. Phys. Rep. **448**(1–4), 1–111 (2007)
2. C.E. Shannon, A mathematical theory of communication. Bell Syst. Tech. J. **27**(3), 379–423 (1948)
3. M.A. Nielsen, I.L. Chuang, *Quantum Computation and Quantum Information* (Cambridge University Press, Cambridge, 2000)
4. A. Ferraro, S. Olivares, M. Paris, Gaussian states in continuous variable quantum information. Napoli Series on Physics and Astrophysics (ed. Bibliopolis, Napoli, 2005)
5. B.W. Schumacher, Sending entanglement through noisy quantum channels. Phys. Rev. A **54**, 2614–2628 (1996)

Chapter 11
Fundamentals of Continuous Variables

Symbols and Terminology

In the topic of this chapter the literature presents a plethora of terms, notations, and normalizations. To avoid confusion we follow closely a recent paper by Weedbrook et al. [1]

$:=$	equal by definition
$I_{\mathcal{H}}$	identity operator of $\mathcal{H}$
I_n	identity matrix of size n
A^*	adjoint of operator A
	or conjugate transpose of matrix A
A^{T}	transpose of matrix A
$\mathrm{E}[m]$	expectation of the random variable m
$\mathrm{E}[m\|\psi]$	conditional expectation with condition given by the state $|\psi\rangle$

Single Mode

$\mathcal{H}$	bosonic Hilbert space
q and p	position and momentum operators (*quadrature operators*)
a and a^*	annihilator and creation operators (*bosonic operators*)
$N = a^* a$	number operator
$|n\rangle , n = 0, 1, 2, \ldots$	number states or Fock states
$\mathcal{B}_F := \{|0\rangle, |1\rangle, |2\rangle, \ldots\}$	Fock basis

N-Mode

$\mathcal{H}_N = \mathcal{H}^{\otimes N}$	bosonic Hilbert space
$A_B = [A_1, B_1, \ldots, A_N, B_N]$	interlace of row vectors A and B
q_i and p_i	quadrature operators of the ith mode
$q = [q_1, \ldots, q_N]^{\mathrm{T}}$	vector of position operators
$p = [p_1, \ldots, p_N]^{\mathrm{T}}$	vector of momentum operators
$q_p = [q_1, p_1, \ldots, q_N, p_N]^{\mathrm{T}}$	vector of quadrature operators

© Springer International Publishing Switzerland 2015

G. Cariolaro, *Quantum Communications*, Signals and Communication Technology,

DOI 10.1007/978-3-319-15600-2_11

$Y = [q_1, \ldots, q_N, p_1, \ldots, p_N]^{\mathrm{T}}$	vector of quadrature operators
a_i and a_i^*	annihilator and creation operators of ith mode
$a = [a_1, a_2, \ldots, a_N]^{\mathrm{T}}$	(column) vector of annihilator operators
$a_\star = [a_1^*, a_2^*, \ldots, a_N^*]^{\mathrm{T}}$	(column) vector of creation operators
$|n\rangle_i, n = 0, 1, 2, \ldots$	number states or Fock states of ith mode
$\mathcal{B}_F(N)$	Fock basis (see (11.68))

$\Omega = \mathrm{diag}\,[\Omega_1, \ldots, \Omega_N], \qquad \Omega_i = \begin{bmatrix} 0 & 1 \\ -1 & 0 \end{bmatrix}$

Functions and Operators

$\mathcal{F}$	*ordinary* Fourier transform
$\mathcal{F}_s$	*symplectic* Fourier transform
$\mathcal{F}_c$	*complex* Fourier transform
$D(\xi), \xi \in \mathbb{C}^N, D(u, v), u, v \in \mathbb{R}^N$	Weyl operator
$\mathcal{D}(\lambda), \lambda \in \mathbb{C}^N, \mathcal{D}(x, y), x, y \in \mathbb{R}^N$	Fourier transform of Weyl operator
$\chi(\xi), \xi \in \mathbb{C}^N, \chi(u, v), u, v \in \mathbb{R}^N$	Wigner characteristic function
$W(\lambda), \lambda \in \mathbb{C}^N, W(x, y), x, y \in \mathbb{R}^N$	Wigner function
$w(\lambda), \lambda \in \mathbb{C}^N, w(x, y), x, y \in \mathbb{R}^N$	normalized Wigner function
$D_N(\alpha), \alpha \in \mathbb{C}^N$	N-mode displacement operator
$R_N(\phi), \phi$ Hermitian matrix	N-mode rotation operator
$Z_N(z), z$ symmetric matrix	N-mode squeeze operator
$S \quad 2N \times 2N$ real matrix	symplectic matrix
$S \quad$ operator of $\mathcal{H}^{\otimes N}$	symmetry operator

The special symbol $a_\star$ is introduced to denote the column vector of the N creation operators. The reason is that, in our conventions, a^* is the conjugate transpose of the column vector a and therefore it denotes a row vector.

11.1 Introduction

The fundamentals introduced in Chap. 3 were essentially the four postulates of Quantum Mechanics and the main tools were provided by the algebra of operators, as the eigendecomposition (EID) and sometimes the singular value decomposition (SVD). However all was limited to *discrete variables*, because we assumed the bases consisting of finite or enumerable sets of vectors, the operator EID having a finite or enumerable spectrum, and quantum measurements having a finite set (alphabet) of possible outcomes. This formulation was sufficient considering that in the subsequent chapters we developed *digital* Quantum Communications, but Quantum Information makes use of both discrete and *continuous variables*. In this chapter we extend the above fundamentals to the continuous case, where the sets become a continuum, typically given by the set of real numbers $\mathbb{R}$. The extension is by no means trivial and requires an audacious and often presumptuous use of mathematics. In this extension we follow Dirac [2], who proceeds in a "parallel form" considering, for each topic, first the discrete and then the continuous case.

The physical environment for the development of continuous variables is provided by the harmonic oscillator, which represents a very simple and general model, both in Classical and in Quantum Mechanics. It provides the basis for the theory of the electromagnetic field, in which the electromagnetic radiation is represented as a combination of harmonic oscillators. Also, it is the basis to the description of many physical systems and in particular subatomic particles. The theory of harmonic oscillators is formulated in terms of two continuous variables, called *canonical variables*, the position and the momentum of a particle. This holds for both the classical harmonic oscillator and the quantum harmonic oscillator, but in the latter the variables become Hermitian operators. Alternative continuous variables are given by the annihilator and the creation operators, called *bosonic variables*. In the theory of the harmonic oscillator, we also encounter the coherent states, the states we have extensively used in Quantum Communication systems of Part II, without knowing their definition but with the expedient of using their properties. Coherent states are simply defined as the eigenkets of the annihilator operator and in this chapter they are fully developed.

Canonical and bosonic variables are modeled in an infinite dimensional Hilbert space and their study encounters several difficulties. A simplification is obtained by transferring the representation of canonical and bosonic variables into a simpler environment, the *phase space*, where continuous-variable quantum states are fully represented by the Wigner function or by the equivalent characteristic function, both complex functions of two real variables. The passage from an infinite dimensional Hilbert space to a two-dimensional real space is quite remarkable. The Wigner and the characteristic functions allow for the introduction of *Gaussian states* and *Gaussian transformations*. Gaussian states are continuous-variable states that have a representation in terms of Gaussian functions, and Gaussian transformations are those that send Gaussian states to Gaussian states. In addition to offering an easy description in terms of Gaussian functions, Gaussian states and transformations are of great practical relevance and represent the main tool of Quantum Information processing based on continuous variables, with applications to quantum computation, quantum cryptography, and quantum communications. Coherent states are notable examples of Gaussian states, but in this chapter we will see several other examples of Gaussian states.

So far we have considered continuous variables *in the single mode*, but the most interesting applications, in particular the ones based on the entanglement, are concerned with continuous variables *in the multimode*, where the Hilbert space is obtained by N replicas (tensor product) of the space of the single mode. Thus canonical and bosonic variables become N-tuples of canonical and bosonic variables and the phase space becomes $2N$-dimensional. The extension of the theory of continuous variables to the multimode represents the main complication of the chapter.

Quantum Information with continuous variables is a hot topic and in the literature one may find several recent contributions. Perhaps the first place to start for an overview of continuous-variable quantum information is the recent review article by Weedbrook et al. (2012) [1]. This article will be the main reference of the chapter. Another review article is by Braunstein and van Loock (2005) [3]. Furthermore, there is also the recent review by Andersen et al. (2010) [4], and the books by Braunstein and Pati (2003) [5] and by Cerf et al. (2007) [6]. On Gaussian quantum information specifically there is the review article by Wang et al. (2007) [7], the lecture notes by Ferraro et al. (2005) [8], and the overview by Olivares (2011) [9]. An overview of Gaussian entanglement is presented in the review of Eisert and Plenio (2003) [10].

11.1.1 Organization of the Chapter

In Sect. 11.2 we will develop the mathematical fundamentals (bases, eigende-composition, matrix representations) of continuous variables and we will review Postulate 3 on quantum measurements letting the possible outcomes be a continuum. Section 11.3 deals with the theory of the harmonic oscillator with the introduction of canonical and bosonic variables.

In the following chapters we proceed more abstractly, sometimes we find it con-venient to introduce the concepts starting from the single mode, but more often we give directly definitions and properties for the N mode. Section 11.5 introduces the quadrature and the bosonic operator, globally called *field operators*.

Sections 11.6 and 11.7 introduce the main definitions (Weyl operator, characteris-tic and Wigner functions) in the general N-mode, including the definition of Gaussian states. Sections 11.8 and 11.9 develop in detail the general definitions in the single mode.

Section 11.10 introduces Gaussian transformations in the N-mode, which are then developed in Sects. 11.15 and 11.16 in the single mode.

Sections 11.17–11.19 are dedicated to Gaussian transformations and Gaussian states in the two-mode.

The final section deals with the geometrically uniform symmetry (GUS) inside the class of Gaussian states. This symmetry was applied in Part II in the performance evaluation of Quantum Communications systems.

The main difficulty of the chapter is due to the fact that the single mode is not sufficient to develop Quantum Information with continuous variables, and in fact the most important quantum phenomenon, the entanglement, requires at least the two-mode. Therefore the multimode is developed.

11.2 From Discrete to Continuous in Quantum Mechanics

In Quantum Mechanics formulation of Chaps. 2 and 3 we have considered some fundamentals, as bases, eigendecompositions, measurements and operators, in the *discrete* case. In this chapter, for a full development of Quantum Information, we extend the above fundamentals to the continuous case, where the sets become a con-tinuum, typically given by the set of real numbers $\mathbb{R}$. As remarked in the introduction, the extension is by no means simple and requires a non trivial mental alignment.

Formally the passage from discrete to continuous is based on two simple *replacement rules*, namely:

(1) Replace summations by integrals

$$\sum_{b \in B} f(b) \quad \rightarrow \quad \int_B db \, f(b). \tag{11.1}$$

(2) Replace Kronecker's δ_{ab} with Dirac's delta function $\delta(a - b)$.[1]

These rules must be applied everywhere. In particular, for the product of two matrices, say $A = BC$, which is defined, as usual, by the entry relation, the replacement takes the form

$$a_{rs} = \sum_{t \in T} b_{rt} \, c_{ts} \quad \rightarrow \quad a_{rs} = \int_T dt \, b_{rt} \, c_{ts} \tag{11.3}$$

where T is the range set of the index t. In the trace evaluation, using a matrix representation $[a_{rt}]$ of an operator A, the replacement is

$$\mathrm{Tr}[A] = \sum_{t \in B} a_{tt} = \quad \rightarrow \quad \mathrm{Tr}[A] = \int_T dt \, a_{tt}. \tag{11.4}$$

11.2.1 Discrete and Continuous Bases in Hilbert Space

Discrete bases in Hilbert space $\mathcal{H}$ were introduced in Sect. 2.7. We now rewrite the definition and properties in a form that is more convenient for the extension to the continuous case. A discrete basis is a set of kets of $\mathcal{H}$, $\mathcal{B} = \{|b\rangle, b \in B\}$, where the range B is a discrete set. A continuous basis is a set of kets of $\mathcal{H}$, $\mathcal{B} = \{|b\rangle, b \in B\}$, where the range set B is a continuum. The corresponding bras $\langle b|$ are obtained as the adjoint of the kets $|b\rangle$.

The properties of a discrete or of a continuous basis can be unified through the replacement rule as follows:

(1) $\mathcal{B}$ consists of orthonormal kets

$$\langle b|b'\rangle = \delta_{bb'} \quad \rightarrow \quad \langle b|b'\rangle = \delta(b - b'), \qquad b, b' \in B \tag{11.5a}$$

---

[1] We suppose that the reader be familiar with this generalized function, introduced by Dirac [2] just in this context. Here we briefly summarize the fundamental properties. The Dirac δ function is introduced by the punctual property $\delta(x - x_0) = 0$ if $x \neq x_0$ and by the integral property (*sifting property*)

$$\int_{-\infty}^{+\infty} f(x)\delta(x - x_0) \, dx = f(x_0) \tag{11.2}$$

where $f(x)$ is an arbitrary continuous function. In particular, when $f(x) = 1$, the sifting property gives $\int_{-\infty}^{+\infty} \delta(x - x_0) \, dx = 1$, which states that $\delta(x - x_0)$ has unitary area.

where $\delta_{bb'}$ is the Kronecker delta and $\delta(b-b')$ is the delta function. This property states also that the kets are normalized.

(2) $\mathcal{B}$ provides a resolution of the identity (completeness)

$$\sum_{b \in B} |b\rangle\langle b| = I_{\mathcal{H}} \quad \rightarrow \quad \int_B db\, |b\rangle\langle b| = I_{\mathcal{H}}. \tag{11.5b}$$

To see how these properties work in the continuous case, we prove that every ket $|x\rangle$ of $\mathcal{H}$ can be expanded in the form

$$|x\rangle = \int_B db\, x(b)\, |b\rangle \quad \text{with} \quad x(b) = \langle b|x\rangle. \tag{11.6}$$

In fact, by left multiplying (11.6) by $\langle b'|$ one gets

$$\langle b'|x\rangle = \int_B db\, x(b)\, \langle b'|b\rangle = \int_B db\, x(b)\, \delta(b' - b) = x(b') \tag{11.7}$$

where the sifting property (11.2) of the delta function has been used.

11.2.2 Eigensystem of Hermitian Operators. Observables

We recall the concepts of eigenvalues and eigenvectors (eigenkets) introduced in Sect. 2.6. An *eigenvalue* a of a given operator A is a complex number such that a vector $|a\rangle \in \mathcal{H}$ exists, different from zero, satisfying the following relation (eigenvalue equation)[2]

$$\boxed{A|a\rangle = a|a\rangle \quad |a\rangle \neq 0.} \tag{11.8}$$

The ket $|a\rangle$ is called *eigenket corresponding to the eigenvalue a*. The set of all the eigenvalues is called *spectrum of the operator A* and denoted $\sigma(A)$. The interpretation of the eigenvalue Eq. (11.8) is illustrated in Fig. 11.1.

Here we are only interested in Hermitian operators and we recall from Sects. 2.8 and 2.10 the properties:

(a) All the eigenvalues of a Hermitian operator are real.
(b) The eigenkets corresponding to distinct eigenvalues are orthogonal.

---

[2] We use the convention to indicate with a single letter the operator (A), the eigenvalue (a), and the eigenket $(|a\rangle)$. Another economic convention, used by Dirac [2], is a for the operator, a' for the eigenvalue and $|a'\rangle$ for the eigenket.

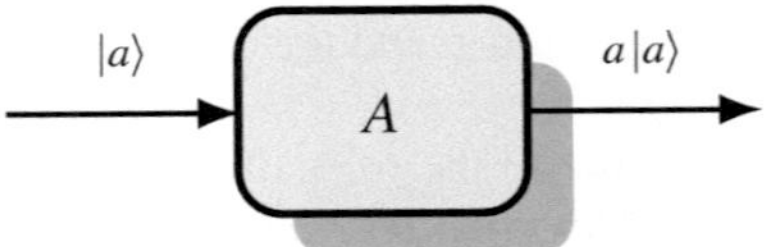

Fig. 11.1 Interpretation of an eigenvalue a and of an eigenket $|a\rangle$ of a linear operator A. The definitions are the same in the discrete and in the continuous cases

(c) If the eigenvalues are distinct (without multiplicity), the set of the corresponding eigenkets allow for the construction of an orthonormal basis $\mathcal{A} = \{|a\rangle, a \in \sigma(A)\}$.

The critical point is given by item (c), where the spectrum $\sigma(A)$ may be an enumerable set (discrete) and also a continuum. In the previous chapters we have considered only the discrete case and mainly the case of a finite spectrum and we have developed important applications, such as the Spectral Decomposition Theorem in Sect. 2.10 and the quantum measurements in Sect. 3.6. But some important Hermitian operators, related to dynamical variables, have a continuous spectrum. With a continuous spectrum properties (a) and (b) can be easily proved, but the proof of completeness, required for the construction of an orthonormal basis, is in general an impossible task [11].[3] To overcome this difficulty the expedient (not a solution) is the introduction of a definition: the Hermitian operators having a complete set of eigenvectors are called **observables**.

In conclusion, we will proceed with observables, so that the existence of a complete basis is ensured. Of course the basis must be handled in two distinct forms, depending on the nature (discrete or continuous) of the spectrum.

Example of an Observable with a Continuous Spectrum

The *reference mechanical system*, introduced in Sect. 3.4, is given by a particle of mass m constrained to move in one direction in a potential $V(q)$, where q is the coordinate of the particle. According to Classical Mechanics, the total energy of this system is given by the sum of kinetic and potential energies

$$H = \frac{1}{2}m v^2 + V(q) = \frac{1}{2}\frac{p^2}{m} + V(q) \tag{11.9}$$

where v is the velocity, $p = m v$ is the momentum, and $V(q)$ is a real function of the position coordinate q. In particular, in the *harmonic oscillator*, which will be seen in Sect. 11.3, the potential has the form $V(q) = \frac{1}{2}m\omega^2 q^2$, where ω is the (angular) frequency of the oscillator. In any case, the energy is expressed in terms of two

---

[3] Dirac, in his celebrated book [2], claimed that there is no available mathematics for solving this problem.

dynamical variables: the momentum p and the coordinate (position) q, which both may be any real number, $p, q \in \mathbb{R}$.

To treat this system according to the rules of Quantum Mechanics, **we have to replace the dynamical variables with Hermitian operators** (observables), say p and q. Then the Hamiltonian is given by

$$H = \frac{1}{2}\frac{p^2}{m} + V(q). \tag{11.10}$$

Note that also H is a Hermitian operator (observable), as can be easily verified in the general case (11.10), and in particular in the case of the harmonic oscillator, where, with $m = 1$, the Hamiltonian reads $H = (p^2 + \omega^2 q^2)/2$.

The specification of the operators describing the physical system, as p, q, and H, is not sufficient in Quantum Mechanics, but it also needs the specification of the algebra the operators must obey. In this specific case the algebra is noncommutative, as stated by the following commutation relation of the operators p and q:

$$[q, p] = i\,\hbar\, I_{\mathcal{H}} \tag{11.11}$$

where $[q, p] = qp - pq$ is the commutator and $\hbar$ is the reduced Plank constant. As usual in Quantum Mechanics, in this kind of relations the identity is usually omitted, that is, $[q, p] = i\hbar$. This noncommutativity will lead to the quantization of the energy and represents a remarkable difference with respect to Classical Mechanics, where p and q commute.

The evaluation of the eigensystem of this quantum system starts just from the commutation relation (11.11) and gives the following result:

Proposition 11.1 *In the reference mechanical system the position and the momentum operators, q and p, are observables obeying the commutation relation (11.11). They have a continuous spectrum, $\sigma(q) = \sigma(p) = \mathbb{R}$ and the eigenkets are generated from their ground states, $|0\rangle_q$ and $|0\rangle_p$, as*

$$|x\rangle = e^{-i\,x\,p/\hbar}\,|0\rangle_q\,, \qquad |y\rangle = e^{i\,yq/\hbar}\,|0\rangle_p\,. \tag{11.12}$$

For the proof see [11], where it is shown that $\sigma(q) = \sigma(p) = \mathbb{R}$.

11.2.3 Quantum Measurements with Observables

In Chap. 3 quantum measurements have been formulated by Postulate 3, where it was assumed that the possible results belong to a finite set $\mathcal{M}$, called *alphabet*. This is the case of interest for Quantum Communications, where nowadays only *digital* transmissions are considered. But, in the framework of Quantum Information, we have to extend quantum measurements to the continuous case.

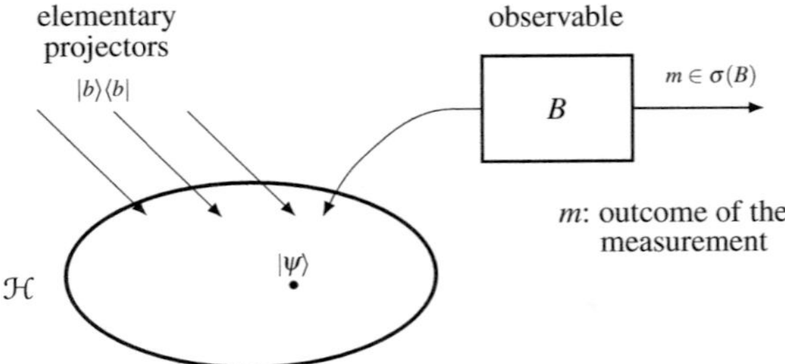

Fig. 11.2 Quantum measurement with an observable B when the system is in the state $|\psi\rangle$ of the Hilbert space $\mathcal{H}$. The elementary projectors $|b\rangle\langle b|$ are obtained from the eigenkets $|b\rangle$ of B. The range of the possible outcomes is given by the spectrum $\sigma(B)$ of the observable B

For the extension we follow the *viewpoint of observables*, as done by Dirac [2] in his original formulation of Quantum Mechanics (see also [11]). As a matter of fact, an observable provides, not only the representations of kets and operators, but also the formulation of quantum measurements, as shown in Fig. 11.2.

In Sect. 3.6 we saw that in a measurement with an observable B, having a **discrete spectrum** $\sigma(B)$, when the system is in the state $|\psi\rangle$, the probability that the measurement yield the value $m = b$ is given by (see Sect. 3.29)

$$P[m = b|\psi] = |\langle\psi|b\rangle|^2, \qquad b \in \sigma(B). \tag{11.13}$$

If the outcome is $m = b$, after the measurement the system falls into the eigenstate (see Sect. 3.31) $|\psi_{\text{post}}\rangle = |b\rangle$. The complex function $\langle\psi|b\rangle$ is called *probability amplitude*.

In conclusion, the outcome of the measurement is a *discrete random variable m*, conditioned by the state $|\psi\rangle$, with alphabet $\mathcal{M} = \sigma(B)$. The statistical description of m is given by the **mass distribution function** $p_m(b|\psi) = |\langle\psi|b\rangle|^2$, $b \in \sigma(B)$. This is really a mass distribution function, because $p_m(b|\psi) \geq 0$ and the normalization is ensured by the basis property (11.5b). In fact

$$\sum_{b\in\sigma(B)} p_m(b|\psi) = \sum_{b\in\sigma(B)} |\langle\psi|b\rangle|^2 = \sum_{b\in\sigma(B)} \langle\psi|b\rangle\langle b|\psi\rangle$$

$$= \langle\psi|I_{\mathcal{H}}|\psi\rangle = \langle\psi|\psi\rangle = 1.$$

When the observable has a **continuous spectrum**, $\sigma(B) = \mathbb{R}$, the outcome of the measurement becomes a continuous random variable $m \in \sigma(B)$. If the system before the measurement is in the state $|\psi\rangle$, the probability that the measurement yield a value m in the infinitesimal interval $(b, b + db)$ is given by

$$P[m \in (b, b + db)|\psi] = |\langle\psi|b\rangle|^2 \, db. \tag{11.14}$$

If the outcome is $m = b$, after the measurement the system falls into the eigenstate $|\psi_{\text{post}}\rangle = |b\rangle$. The complex function $\langle\psi|b\rangle$ is called *probability amplitude*.

In conclusion, the continuous random variable m is described by the **probability density function**

$$\mathrm{p}_m(b|\psi) = |\langle\psi|b\rangle|^2, \qquad b \in \sigma(B). \tag{11.15}$$

Note that the function defined by (11.15) really behaves as a probability density function, being non negative and normalized to unity. The normalization follows from the basis property (11.5b). In fact

$$\int_{\mathbb{R}} \mathrm{d}b\, \mathrm{p}_m(b|\psi) = \int_{\mathbb{R}} \mathrm{d}b\, |\langle\psi|b\rangle|^2 = \int_{\mathbb{R}} \mathrm{d}b\, \langle\psi|b\rangle\langle b|\psi\rangle$$
$$= \langle\psi|I_{\mathcal{H}}|\psi\rangle = \langle\psi|\psi\rangle = 1.$$

In the discrete case the statistical average (quantum expectation) of the random variable m can be easily obtained from the observable as (see 3.43)

$$\boxed{\mathrm{E}[m|\psi] = \langle\psi|B|\psi\rangle := \langle B\rangle.} \tag{11.16}$$

This relation holds also in the continuous case. In fact, the expectation of a continuous random variable with probability density function $\mathrm{p}_m(b|\psi) = |\langle\psi|b\rangle|^2$ is given by [12]

$$\mathrm{E}[m|\psi] = \int_{-\infty}^{+\infty} \mathrm{d}b\, b\, \mathrm{p}_m(b|\psi) = \int_{-\infty}^{+\infty} \mathrm{d}b\, b\, |\langle\psi|b\rangle|^2$$

where we can write $|\langle\psi|b\rangle|^2 = \langle\psi|b\rangle\langle b|\psi\rangle$. Then, considering the eigenvalue equation $b|b\rangle = B|b\rangle$ and the completeness (11.5b), one gets

$$\mathrm{E}[m|\psi] = \langle\psi|B\left\{\int_{-\infty}^{+\infty} \mathrm{d}b\, |b\rangle\langle b|\right\}|\psi\rangle = \langle\psi|B|\psi\rangle.$$

Relation (11.16) can be easily generalized to an arbitrary function $f(m)$ of the random variable m and in particular to the moments.

Example 11.1 To remark the difference between the discrete and the continuous case we give two examples of quantum measurements in the same system, the harmonic oscillator (see Sect. 11.3). In the first one the observable B is the *number operator* N, which has the discrete spectrum $\sigma(N) = \{0, 1, 2, \ldots\}$. Assuming that the system is in a coherent state $|\psi\rangle = |\alpha\rangle$, the mass distribution is given by (see (7.7))

$$p_m(b|\alpha) = \mathrm{e}^{-N_\alpha}\frac{N_\alpha^b}{b!}, \qquad b = 0, 1, 2, \ldots \tag{11.17}$$

that is, a Poisson distribution with average $\mathrm{E}[m|\alpha] = N_\alpha = |\alpha|^2$.

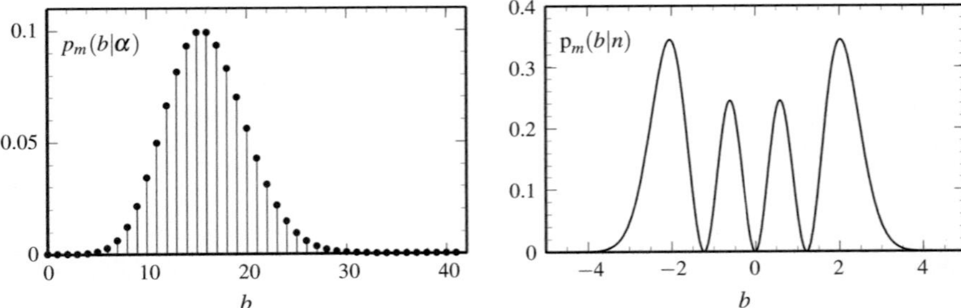

Fig. 11.3 Examples of statistical descriptions in quantum measurements. On the *left*, the outcome is a *discrete* random variable m obtained with the observable N, the number operator, when the quantum system is in a coherent state $|\alpha\rangle$; the description is given by Poisson *mass probability distribution* $p_m(b|\alpha)$. On the *right*, the outcome is a continuous random variable m obtained with the observable q, the position operator, when the quantum system is in a Fock state $|n\rangle$; the description is given by the *probability density* $\mathrm{p}_m(b|n)$ given by (11.18). The figures are obtained with $\alpha = 4.0$ and $n = 3$

In the second example the observable B is the position operator q of the harmonic oscillator, whose spectrum is $\sigma(q) = \mathbb{R}$ (see Sect. 11.3). Assuming that the system is in the Fock state $|n\rangle$, the probability density is given by [11]

$$\mathrm{p}_m(b|n) = \frac{K}{\sqrt{\pi}\, 2^n\, n!}\, H_n^2(Kb)\, \mathrm{e}^{-K^2 b^2}, \qquad b \in \mathbb{R} \tag{11.18}$$

where K is a constant of the oscillator and $H_n(x)$ are the Hermite polynomials.

The mass probability distribution (11.17) and the probability density function (11.18) are illustrated in Fig. 11.3.

11.3 The Harmonic Oscillator

The harmonic oscillator represents a very simple and general model, both in classical and in quantum mechanics. It is on the basis of the theory of the electromagnetic field, in which the electromagnetic radiation is represented as a combination of harmonic oscillators. Also, it is the basis to the description of many physical systems and in particular subatomic particles. To further stress the importance of this topic we observe that the main fundamentals of Quantum Information with continuous variables developed in this chapter are based upon the quantum harmonic oscillator.

11.3.1 The Classical Model

We reconsider the *reference model* introduced in Sect. 3.4, given by a particle of mass m constrained to move in one direction in a potential $V(q)$, where q is the coordinate of the particle. The energy/Hamiltonian of this system is given by

$$H = \frac{1}{2}\frac{p^2}{m} + V(q) \tag{11.19}$$

where p is the momentum of the particle. In the harmonic oscillator the energy is explicitly given by

$$H = \frac{1}{2}(p^2 + \omega^2 q^2) \tag{11.20}$$

where ω is the angular speed, related to the restoring force of the particle, and the mass is assumed as unitary.

The equations of motion are obtained from the Hamiltonian as [11]

$$\frac{dq}{dt} = \frac{\partial H}{\partial p} = p\,, \qquad \frac{dp}{dt} = -\frac{\partial H}{\partial q} = -\omega^2 q\,. \tag{11.21}$$

Then, by combination of (11.21) one finds

$$\frac{d^2 q}{d^2 t} = -\omega^2 q$$

whose solution is

$$\begin{aligned}
q(t) &= \quad q(0)\,\cos\omega t + [p(0)/\omega]\sin\omega t \\
p(t) &= -\omega q(0)\,\sin\omega t + p(0)\cos\omega t
\end{aligned} \tag{11.22}$$

where $q(0)$ and $p(0)$ are the values of the dynamic variables at time $t = 0$. An example of time evolution of $q(t)$ and $p(t)$ is shown in Fig. 11.4.

As we have seen, the evolution calculation is very simple, but it can be further simplified by the introduction of the new conjugate variables

$$a = \frac{1}{\sqrt{2\omega}}(\omega q + \mathrm{i}p)\,, \qquad a^* = \frac{1}{\sqrt{2\omega}}(\omega q - \mathrm{i}p) \tag{11.23}$$

which allow us to decouple the original equations (11.21). In fact, we get

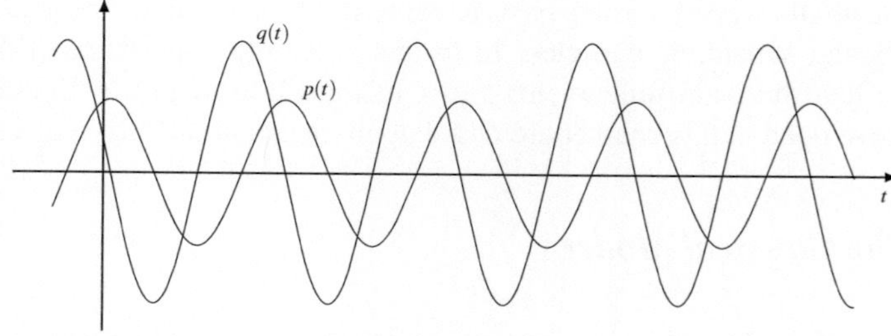

Fig. 11.4 Coordinate and momentum in a harmonic oscillator with $q(0) = 1$ and $p(0) = 1.8$

$$\frac{\mathrm{d}a}{\mathrm{d}t} = -\mathrm{i}\,\omega\,a\,, \qquad \frac{\mathrm{d}a^*}{\mathrm{d}t} = \mathrm{i}\,\omega\,a^* \tag{11.24}$$

which in practice is a single equation. Now the solution is immediate and given by

$$a(t) = a(0)\,\mathrm{e}^{-\mathrm{i}\omega t}. \tag{11.25}$$

It is easy to verify that with the new variables, the energy (11.20) becomes

$$H = \omega^2\,a\,a^*. \tag{11.26}$$

The introduction of the variables a and a^* seems a simple trick, but has an important consequence in the quantum version.

11.3.2 The Quantum Mechanical Model

As seen in general in Sect. 11.2, the passage from the classical to the quantum mechanical model, is obtained by replacing the scalar variables p and q with Hermitian operators (observables) p and q, which we indicate with the same symbol[4] used for the classical variables. Thus the Hamiltonian becomes

$$H = \frac{1}{2}(p^2 + \omega^2 q^2) \tag{11.27}$$

which is itself a Hermitian operator. But we have to add the commutation relation

$$\boxed{[q, p] = \mathrm{i}\,\hbar\,I_{\mathcal{H}}} \tag{11.28}$$

which represents the relevant difference with respect to the classical model, where q and p commute.

All these operators are independent of time, according to the Schrödinger picture. For the evaluation of the dynamic evolution we apply Postulate 2 of Quantum Mechanics, formulated in Sect. 3.4, giving the Schrödinger equation

$$\mathrm{i}\,\hbar\,\frac{\partial}{\partial t}|\psi(t)\rangle = H(t)\,|\psi(t)\rangle \tag{11.29}$$

where $|\psi(t)\rangle$ is the state (wave function) of the oscillator at time t. Considering that the Hamiltonian H is time-independent, the solution of the Schrödinger equation is given by (3.15)

---

[4] In general we denote the operators with uppercase letters, but for the operator p and q the lowercase is used almost everywhere in the literature. The same is for annihilator and creator operators a^* and a.

$$|\psi(t)\rangle = \exp\left[-\mathrm{i}\,\frac{H}{\hbar}(t-t_0)\right]|\psi(t_0)\rangle$$

$$= \exp\left[-\mathrm{i}\,\frac{p^2+\omega^2 q^2}{2\hbar}(t-t_0)\right]|\psi(t_0)\rangle \tag{11.30}$$

and the temporal evolution results in

$$U(t-t_0) = \exp\left[-\mathrm{i}\frac{p^2+\omega^2 q^2}{\hbar}(t-t_0)\right], \qquad t > t_0 . \tag{11.31}$$

The equations of motion are identical in form to the classical equations in Hamiltonian form, given by (11.21), that is,

$$\frac{\mathrm{d}q}{\mathrm{d}t} = \frac{\partial H}{\partial p} = p, \qquad \frac{\mathrm{d}p}{\mathrm{d}t} = -\frac{\partial H}{\partial q} = -\omega^2 q \tag{11.32}$$

and the solution is, by (11.22)

$$\begin{aligned} q(t) &= q\,\cos\omega t + [p/\omega]\sin\omega t \\ p(t) &= -\omega q\,\sin\omega t + p\,\cos\omega t \end{aligned} \tag{11.33}$$

where q and p are the operators in the Schrödinger picture.

11.3.3 Annihilation and Creation Operators

The decoupling variables a and a^* of the classical oscillator become the operators

$$a = \frac{1}{\sqrt{2\omega}}(\omega q + \mathrm{i}p), \qquad a^* = \frac{1}{\sqrt{2\omega}}(\omega q - \mathrm{i}p) \tag{11.34}$$

from which one gets the position and momentum operators as

$$q = \sqrt{\frac{\hbar}{2\omega}}(a^*+a), \qquad p = \mathrm{i}\sqrt{\frac{\hbar\omega}{2}}(a^*-a). \tag{11.35}$$

For the reason we shall see below, a is called **annihilation operator** and a^* **creation operator**. They are not Hermitian and like p and q, do not commute. Their commutation relation can be obtained from (11.28) and is simply given by

$$[a, a^*] = I_{\mathcal{H}}. \tag{11.36}$$

The Hamiltonian can be expressed in terms of a and a^* as

$$H = \tfrac{1}{2}\hbar\omega(aa^* + a^*a) = \hbar\omega(a^*a + \tfrac{1}{2}I_{\mathcal{H}}) \tag{11.37}$$

where (11.36) has been used.

The equations of motion (11.24) seen for the variables a and a^* hold also for the operators a and a^* [11].

11.3.4 Number Operator and Number States. Fock Basis

The **number operator** is defined as

$$\boxed{N = a^*a} \tag{11.38}$$

and it clearly is a Hermitian operator (observable). The corresponding eigenvalue equation is

$$\boxed{N\,|n\rangle = n|n\rangle} \tag{11.39}$$

where the eigenket $|n\rangle$ are called **number states** and also **Fock states**.

From the above statements we find:

Proposition 11.2 *The number operator N is an observable with the following eigensystem: the eigenvalues n are all the nonnegative integers and the eigenvectors (number states) verify the recurrences*

$$a\ |0\rangle = 0 \tag{11.40a}$$
$$a\ |n\rangle = \sqrt{n}|n-1\rangle, \qquad n \geq 1 \tag{11.40b}$$
$$a^*\ |n\rangle = \sqrt{n+1}|n+1\rangle, \qquad n \geq 0 . \tag{11.40c}$$

All the number states can be generated from the zero state (ground state) as

$$|n\rangle = \frac{(a^*)^n}{\sqrt{n!}}\ |0\rangle . \qquad \square \tag{11.41}$$

Proof For the proof we follow Dirac [2]. Starting from the commutation condition (11.36), we find $[a, a^*a] = a$ and $[a^*, a^*a] = -a^*$, that is, $Na = a(N - I_{\mathcal{H}})$ and $Na^* = a(N + I_{\mathcal{H}})$. Then, considering the eigenvalue equation (11.39), one gets the recurrences

$$Na\,|n\rangle = (n-1)\,|n\rangle , \qquad Na^*\,|n\rangle = (n+1)\,|n\rangle . \tag{11.42}$$

We recall that N is PSD and therefore its eigenvalues must be real and nonnegative, $n \geq 0$. On the other hand, Eq. (11.42) show that if n is an eigenvalue of N, also $n - 1$ and $n + 1$ are. If we assume that an eigenvalue n may be not integer, repeated iterations of the above procedure would lead to negative eigenvalues with contradiction on the PSD of N. In conclusion, the n's are all the nonnegative integers.

Now, provided that $|n\rangle$ is a normalized eigenvector of N with eigenvalue n, the first of (11.42) states that $a|n\rangle$ is an eigenvector (not unnecessarily normalized) of N with eigenvalue of $n - 1$. Then, denoting by $|n - 1\rangle$ the normalized eigenvector with eigenvalue $n - 1$, we can write $a|n\rangle = c_n|n - 1\rangle$, where c_n follows from the normalization, that is,

$$|c_n|^2 = |c_n|^2 \langle n - 1|n - 1\rangle = \langle n|a^*a|n\rangle = \langle n|N|n\rangle = n$$

so that, neglecting an irrelevant phasor, $c_n = \sqrt{n}$. Then we get in particular $a|0\rangle = 0$. Thus (11.40a) and (11.40b) are proved. In a similar way we can prove the recursion (11.40c).

The number operator N is an observable and therefore it generates a complete and orthonormal basis $\{|n\rangle, n = 0, 1, 2, \ldots\}$, formed by the number states. This basis, usually called *Fock basis*, will play a fundamental role in this chapter in the representation of continuous variables.

11.3.5 Energy Quantization. Interpretation of N, a, and a*

From (11.37) the Hamiltonian H is expressed in terms of the number operator N as

$$H = \hbar\omega \left(N + \tfrac{1}{2}I_{\mathcal{H}}\right). \tag{11.43}$$

Hence the eigenvalue equation of the Hamiltonian operator, $H|e\rangle = e|e\rangle$, can be solved using the eigensystem of N, given by Proposition 11.2. In particular, the energy eigenvalues are given by

$$e_n = \hbar\omega \left(n + \tfrac{1}{2}\right) = h\nu \left(n + \tfrac{1}{2}\right), \qquad n = 0, 1, 2, \ldots \tag{11.44}$$

where $\nu = \omega/(2\pi)$ is the oscillation frequency, and $h = 6.262 \, 10^{-34}$ J s is Planck's constant (not reduced). Notice that the fundamental eigenstate has energy $e_0 = h\nu/2 > 0$ and that the eigenstates are equally spaced by $h\nu$, the same energy as that found for light quanta. The value of $h\nu$ is very small at optical frequencies, therefore the phenomenon of energy *quantization* is not perceived at a macroscopic level. The energy quantization, which is a consequence of the commutation condition $[a, a^*] = I_{\mathcal{H}}$, becomes fundamental at subatomic level, where it allows for the unification of the particle and wave properties of light. In this case the quanta are called **photons**. But this theory applies more generally to the family of particles called **bosons**.[5]

According to Dirac's interpretation, a harmonic oscillator at the eigenstate $|n\rangle$ describes a system of n identical independent quanta, each of which has energy hv. In particular the state $|0\rangle$ has no quanta and therefore is called *vacuum state* and also *ground state*, although it has a positive energy $e_0 = \frac{1}{2}hv$.

Proposition 11.2 justifies the term **number operator** for N because a measurement with this observable yields the eigenvalues $0, 1, 2, \ldots$, which represent the number of quanta in the wave. Also, a^* is a **creation operator** because, if the oscillator is in the state $|n\rangle$ with n quanta, its application generates the state $|n+1\rangle$ with $n+1$ quanta. Analogously, the **annihilation operator** gives the state $|n-1\rangle$ with $n-1$ quanta.

11.4 Coherent States

In Part II we have seen the fundamental role played by the class of coherent states, $\mathcal{G} = \{|\alpha\rangle \,|\, \alpha \in \mathbb{C}\}$, in Quantum Communications. However, till now, we have applied coherent states using their properties, but we have not investigated their definition and formulation. Now we are ready to do this, after the development of the theory of harmonic oscillator and in particular the introduction of the bosonic operators. In fact, the definition is: a coherent state $|\alpha\rangle$ is **an eigenstate of the annihilator operator** a **with eigenvalue** α

$$a\,|\alpha\rangle = \alpha|\alpha\rangle. \tag{11.45}$$

In Sect. 11.9 we shall see that coherent states allow an alternative definition as **the transformation of the vacuum state through the Weyl operator**.

11.4.1 Fock Representation

We prove that the Fock representation of a coherent state is given by

$$|\alpha\rangle = \sum_{n=0}^{\infty} e^{-\frac{1}{2}|\alpha|^2} \frac{\alpha^n}{\sqrt{n!}} \, |n\rangle \tag{11.46}$$

---

[5] By definition *bosons* are particles which obey Bose–Einstein statistics, in contrast with *fermions* which obey Fermi–Dirac statistics. Bosons may be elementary, like photons, or composite, like mesons. A celebrated boson is Higg's boson, recently (2012) discovered at CERN in Geneva.

where $\{|n\rangle\,,\ n = 0, 1, 2, \ldots\}$ is the complete orthonormal basis formed by the number states (Fock basis). For the proof we consider the series expansion of $|\alpha\rangle$ in the Fock basis

$$|\alpha\rangle = \sum_{n=0}^{\infty} f_n\ |n\rangle\,, \qquad f_n = \langle n\,|\alpha\rangle$$

where f_n are the Fourier coefficients. To get these coefficients we use the conjugate of relation (11.40c) of Proposition 11.2, which gives the recursive formula $\sqrt{n+1}\,\langle n+1|\alpha\rangle = \langle n|a|\alpha\rangle = \alpha\,\langle n|\alpha\rangle$. This allows us to express $f_n = \langle n|\alpha\rangle$ in terms of the ground state $|0\rangle$ as

$$f_n = \langle n|\alpha\rangle = \frac{\alpha^n}{\sqrt{n!}}\langle 0|\alpha\rangle. \tag{11.47}$$

Finally, the normalization of coherent states gives

$$1 = \langle\alpha|\alpha\rangle = \sum_{n=0}^{\infty} |f_n|^2 = |\langle 0|\alpha\rangle|^2 \sum_{n=0}^{\infty} \frac{|\alpha|^{2n}}{n!} = |\langle 0|\alpha\rangle|^2 e^{|\alpha|^2}$$

so that, choosing a null phase, we get $\langle 0|\alpha\rangle = e^{-\frac{1}{2}|\alpha|^2}$, and (11.46) follows at once.

11.4.2 Coherent States as Complete Nonorthogonal Basis

The coherent states are not orthogonal. In fact, in Proposition 7.1 starting from Fock expansion (11.45) we have seen that the inner product of two coherent states is given by

$$\langle\alpha|\beta\rangle = e^{-\frac{1}{2}(|\alpha|^2+|\beta|^2-2\alpha^*\beta)} \quad\rightarrow\quad |\langle\alpha|\beta\rangle|^2 = e^{-|\alpha-\beta|^2} \tag{11.48}$$

which shows that $\langle\alpha|\beta\rangle \neq 0$ also for $\alpha \neq \beta$. However, they form a complete basis, that is, with the property

$$\frac{1}{\pi}\int_{\mathbb{C}} d\alpha\ |\alpha\rangle\,\langle\alpha| = I_{\mathcal{H}} \tag{11.49}$$

where the integration is over the complex plane $\mathbb{C}$ and $d\alpha$ must be intended as

$$d\Re\alpha\ d\Im\alpha\ . \tag{11.49a}$$

The proof of completeness is based on the identity

$$\int_{\mathbb{C}} d\alpha\ (\alpha^*)^n \alpha^m = \pi n!\delta_{nm} \tag{11.50}$$

which can be proved by letting $\alpha = |\alpha| e^{i\theta}$ and evaluating the integral. Then, using (11.50) and (11.45), we find

$$\int_{\mathbb{C}} d\alpha \, |\alpha\rangle \langle\alpha| = \pi \sum_{n=0}^{\infty} |n\rangle \langle n| = \pi \, I_{\mathcal{H}}$$

where the last passage is a consequence of the completeness of Fock basis.

The use of coherent states as a complete basis is extremely useful and find several applications [13, 14]. In particular they allow for the expansion of quantum states and operators in the sense seen in Sect. 11.2 with continuous bases. Also, they provide useful representations of density operators, as we shall see in Sect. 11.7.

11.5 Abstract Formulation of Continuous Quantum Variables

The theory of continuous quantum variables is usually developed in the framework of a tensor product of N identical Hilbert spaces, $\mathcal{H}_N = \mathcal{H}^{\otimes N}$, which is the environment of N quantized radiation modes of the electromagnetic field, corresponding to N harmonic oscillators [1, 2]. We shall follow this line in the general definitions, but for clarity we often introduce the concepts for a single mode with a subsequent extension to the N-mode.

Normalization. With reference to the theory of harmonic oscillator, hereafter we introduce the following normalization:

particle mass $m = 1$, angular speed $\omega = 1$, reduced Planck constant $\hbar = 2$.

In these normalizations we follow Weedbrook et al. [1], but in the literature we frequently find other normalizations, e.g., in Wang et al. [7] the reduced Planck constant is set to $\hbar = 1$ and in Braustein and Van Lock [3] to $\hbar = \frac{1}{2}$. The choice of normalization has a consequence in several relations and in particular in commutation relations.

The identity operator $I_{\mathcal{H}}$ in often set to 1, especially in commutation relations. Hereafter we will follow this convention.

11.5.1 Single-Mode Hilbert Space

Continuous quantum variables may be viewed as an abstract version and a generalization of the states seen in the harmonic oscillator. The environment is an infinite dimensional Hilbert space $\mathcal{H}$, where the coordinate (or position) operator q and the momentum operator p (*quadrature variables*) are introduced. These operators verify the commutation relation (11.28), which, after the normalization $\hbar = 2$ and the omission of the identity, reads as

$$[q, p] = 2\,\mathrm{i}\,, \qquad [p, q] = -2\,\mathrm{i}. \tag{11.51}$$

The Hamiltonian (11.27) becomes

$$H = \tfrac{1}{2}(q^2 + p^2). \tag{11.52}$$

From the quadrature operators the *annihilation* and the *creation* operators (*bosonic operators*) are introduced as (see (11.34))

$$a = \frac{1}{2}(q + \mathrm{i}\,p)\,, \qquad a^* = \frac{1}{2}(q - \mathrm{i}\,p) \tag{11.53}$$

whose commutation relation is still given by (11.36), that is,

$$[a, a^*] = 1\,, \qquad [a^*, a] = -1. \tag{11.54}$$

The inverse relations read

$$q = a + a^*\,, \qquad p = \mathrm{i}(a^* - a). \tag{11.55}$$

From a and a^* one gets the *number operator* as

$$N = a^* a \tag{11.56}$$

which is an observable whose eigenkets $|n\rangle$, $n = 0, 1, 2, \ldots$ are called *number states* or *Fock states*. We recall that the Fock states provide a countable orthonormal basis, $\mathcal{B}_F = \{|0\rangle, |1\rangle, |2\rangle, \ldots\}$, which has a fundamental importance in the representation of continuous quantum variables. The properties of Fock states are established in Proposition 11.2 and here they are summarized:

$$a|0\rangle = 0\,, \qquad a|n\rangle = \sqrt{n}\,|n - 1\rangle \quad \text{for } n \geq 1, \tag{11.57a}$$

$$a^*|n\rangle = \sqrt{n + 1}\,|n + 1\rangle \qquad \text{for } n \geq 0, \tag{11.57b}$$

with the important relation

$$|n\rangle = \frac{(a^*)^n}{\sqrt{n!}}\,|0\rangle \tag{11.58}$$

which establishes that all number states can be generated from the *vacuum state* $|0\rangle$.

It is worth noting how the commutation relations are usually written in the algebra of continuous quantum variables. Letting

$$B = \begin{bmatrix} B_1 \\ B_2 \end{bmatrix} = \begin{bmatrix} a \\ a^* \end{bmatrix}, \qquad X = \begin{bmatrix} X_1 \\ X_2 \end{bmatrix} = \begin{bmatrix} q \\ p \end{bmatrix} \tag{11.59}$$

and, considering that $[a, a] = 0$, $[q, q] = 0$ and $[p, p] = 0$, relations (11.54) and (11.51) are respectively written in the *symplectic form*

$$[B_i, B_j] = \Omega_{ij}, \qquad [X_i, X_j] = 2\mathrm{i}\,\Omega_{ij}, \qquad i = 1, 2 \tag{11.60}$$

where

$$\begin{bmatrix} \Omega_{11} & \Omega_{12} \\ \Omega_{21} & \Omega_{22} \end{bmatrix} = \begin{bmatrix} 0 & 1 \\ -1 & 0 \end{bmatrix}. \tag{11.61}$$

11.5.2 N-Mode Hilbert Space

In the N-mode bosonic space the environment is given by the tensor product of N identical infinite-dimensional Hilbert spaces, $\mathcal{H}_N := \mathcal{H}^{\otimes N}$, where a pair of quadrature operators (q_k, p_k) are introduced for each mode. The kth mode annihilation and creation operator are then defined as

$$a_k = \tfrac{1}{2}(q_k + \mathrm{i}\,p_k), \qquad a_k^* = \tfrac{1}{2}(q_k - \mathrm{i}\,p_k). \tag{11.62}$$

The inverse relations are

$$q_k = a_k + a_k^*, \qquad p_k = \mathrm{i}(a_k^* - a_k). \tag{11.63}$$

The rules are the same seen for the single mode, e.g., $a_k^* |n\rangle_k = \sqrt{n+1}|n+1\rangle_k$, where $|n\rangle_k$ is the number state in the kth mode. The operators of different modes commute and therefore

$$\begin{aligned} [q_i, q_j] = [p_i, p_j] = 0, \qquad & [q_i, p_j] = 2\mathrm{i}\,\delta_{ij} \\ [a_i, a_j] = [a_i^*, a_j^*] = 0, \qquad & [a_i, a_j^*] = \delta_{ij}. \end{aligned} \tag{11.64}$$

The commutation relations can be written in the symplectic form (11.60), namely

$$\boxed{[B_i, B_j] = \Omega_{ij}, \qquad [X_i, X_j] = 2\mathrm{i}\,\Omega_{ij}, \qquad i, j = 1, \ldots, 2N} \tag{11.65}$$

with the notations

$$B = [B_1, B_2, \ldots, B_{2N-1}, B_{2N}]^{\mathrm{T}} = [a_1, a_1^*, \ldots, a_N, a_N^*]^{\mathrm{T}}$$
$$X = [X_1, X_2, \ldots, X_{2N-1}, X_{2N}]^{\mathrm{T}} = [q_1, p_1, \ldots, q_N, p_N]^{\mathrm{T}} := q_p{}^{\mathrm{T}} \qquad (11.66)$$

and

$$\Omega := \mathrm{diag}\,[\Omega_1, \ldots, \Omega_N], \qquad \Omega_i = \begin{bmatrix} 0 & 1 \\ -1 & 0 \end{bmatrix}. \qquad (11.67)$$

Globally the four variables for each mode are redundant and the development of the theory could proceed with only bosonic variables or with only quadrature variables. However, both will be useful for several reasons (manipulation, interpretation and so on).

In the bosonic space $\mathcal{H}_N$ the Fock basis becomes N-mode and can be written in the form

$$\mathcal{B}_F(N) = \{|n_1\rangle_1 |n_2\rangle_2 \cdots |n_N\rangle_N, \quad n_1, n_2, \ldots, n_N = 0, 1, 2, \ldots\} \qquad (11.68)$$

where $|n_i\rangle_i$ are the Fock states in the ith mode.

Problem 11.1 $\star$ Prove relation (11.58), which states that all the number states $|n\rangle$ can be obtained from the ground state $|0\rangle$.

11.6 Phase Space Representation: Preliminaries

In a bosonic Hilbert space a state is represented in general by a density operator ρ; in particular when ρ is a projector ($\rho^2 = \rho$) the state becomes pure and can be written in the form $\rho = |\psi\rangle\langle\psi|$, with $|\psi\rangle$ a point of the Hilbert space $\mathcal{H}_N$. In any case, a density operator acting in $\mathcal{H}_N$ can be conveniently represented by a quasi-probability density, called Wigner function. This is a real function defined in the real space $\mathbb{R}^{2N}$, which is called *phase space*.[6] It represents a powerful tool for its capability of representing a density operator, which may be infinite dimensional, by a simple real function of a finite number of variables.

In the formulation of phase space representations the first step is the introduction of the Weyl operator, from which one gets the (Wigner) characteristic function and finally the Wigner function. All these quantities are defined in terms of the *exponential operator*, which therefore plays a fundamental role and is now examined in detail. Also the Fourier transform (FT) enters in these definitions and it will be convenient to investigate which form of the FT to adopt for the reason that a convenient form may simplify results and improve their interpretation.

---

[6] A more abstract definition of phase space is given in terms of the symplectic group Sp $(2N, \mathbb{R})$, which is related to the Lie groups [15], but in our formulation we will not make use of this sophisticated concept.

11.6.1 Exponential Operator Identities

The exponential of an arbitrary operator A has the same definition as in the scalar case

$$e^A := \sum_{n=0}^{\infty} \frac{1}{n!} A^n. \tag{11.69}$$

The main difference with respect to the scalar case arises when the exponent consists of two or more noncommuting operators. Here we give two important identities, proved in Appendix Section "Proof of Baker–Campbell–Hausdorff Identity".

Proposition 11.3 (Baker–Hausdorff formula) *For any two operators H and K such that their commutator $[H, K] := H K - K H$ commutes with both of them, the following identity holds:*

$$e^{H+K} = e^K \, e^H \, e^{\frac{1}{2}[H,K]} \quad \text{if} \quad [[H, K], H] = [[H, K], K] = 0. \tag{11.70}$$

Proposition 11.4 (Baker–Campbell–Hausdorff formula) *For two arbitrary operators H and K the following identity holds:*

$$\boxed{e^{xH} \, K \, e^{-xH} = \sum_{n=0}^{\infty} \frac{x^n}{n!} D_n} \tag{11.71}$$

with

$$D_0 = K, \qquad D_n = [H, D_{n-1}] \quad \text{for} \quad n \geq 1. \tag{11.71a}$$

Identity (11.71) is explicitly written as

$$e^{xH} \, K \, e^{-xH} = K + x[H, K] + \frac{x^2}{2!}[H, [H, K]] + \frac{x^3}{3!}[H, [H, [H, K]]] + \cdots \tag{11.71b}$$

where we note the presence of nested commutators.

11.6.2 Exponential of a Matrix and Related Functions

The general definition (11.69) plays a central role for the definition of other functions of an operator and of a square matrix, as trigonometric and hyperbolic functions. The specific expansions are

$$\cos A = \sum_{n=0}^{\infty} (-1)^n \frac{1}{(2n)!} A^{2n}, \qquad \cosh A = \sum_{n=0}^{\infty} \frac{1}{(2n)!} A^{2n}$$

$$\sin(A) = \sum_{n=0}^{\infty} (-1)^n \frac{1}{(2n+1)!} A^{2n+1}, \qquad \sinh A = \sum_{n=0}^{\infty} \frac{1}{(2n+1)!} A^{2n+1} \quad (11.72)$$

and the relations are

$$e^{iA} = \cos A + i \sin A, \qquad e^{A} = \cosh A + \sinh A$$

$$\cosh A = \tfrac{1}{2}\left(e^{A} + e^{-A}\right), \qquad \sinh A = \tfrac{1}{2}\left(e^{A} - e^{-A}\right) \quad (11.73)$$

$$\cosh A = \cos(iA), \qquad i \sinh A = \sin(iA).$$

All these definitions and relations for operators and matrices are exactly the same as for the corresponding scalar functions. In particular for a finite-dimensional square matrix A it is possible to find the expression of a function of the matrix in terms of the matrix elements. To this end the general solution is based on the eigendecomposition approach seen in Chap. 2. The general expressions are somewhat complicated as emphasized by the simple case of a 2×2 matrix (see Problem 11.2)

$$\exp \begin{bmatrix} a & b \\ c & d \end{bmatrix} = \frac{e^{\frac{1}{2}(a+d-\Delta)}}{2\Delta} \begin{bmatrix} d - a + \Delta + (a - d + \Delta)e^{\Delta} & 2b\left(e^{\Delta} - 1\right) \\ 2c\left(e^{\Delta} - 1\right) & a - d + \Delta(d - a + \Delta)e^{\Delta} \end{bmatrix}$$

where $\Delta = \sqrt{a^2 - 2da + d^2 + 4bc}$. Of course, we have simplifications in special cases, in particular when A is diagonal or antidiagonal. Here we give a few simple cases which will be used in the phase space representation:

$$\exp \begin{bmatrix} a & 0 \\ 0 & d \end{bmatrix} = \begin{bmatrix} e^{a} & 0 \\ 0 & e^{d} \end{bmatrix}$$

$$\exp \begin{bmatrix} 0 & b \\ b & 0 \end{bmatrix} = \begin{bmatrix} \cosh b & \sinh b \\ \sinh b & \cosh b \end{bmatrix}$$

$$\exp \begin{bmatrix} 0 & b \\ -b & 0 \end{bmatrix} = \begin{bmatrix} \cos b & \sin b \\ -\sin b & \cos b \end{bmatrix}$$

$$\exp \begin{bmatrix} 0 & ib \\ ib & 0 \end{bmatrix} = \begin{bmatrix} \cos b & i\sin b \\ i\sin b & \cos b \end{bmatrix} \qquad (11.74)$$

$$\cosh \begin{bmatrix} 0 & b \\ b & 0 \end{bmatrix} = \begin{bmatrix} \cosh b & 0 \\ 0 & \cosh b \end{bmatrix}$$

$$\sinh \begin{bmatrix} 0 & b \\ b & 0 \end{bmatrix} = \begin{bmatrix} 0 & \sinh b \\ \sinh b & 0 \end{bmatrix}.$$

11.6.3 Review of the Fourier Transform

We review in detail the different forms of defining the FT. Finally, we will make a specific choice that will be used to introduce the forms of FTs (symplectic and complex) for the single and for the multimode. This specific choice is uncommon, but it allows us to obtain simplifications for several statements.

The One-Dimensional FT

The one-dimensional FT of a complex function of a real variable $f(t), t \in \mathbb{R}$ (often called "signal") can be defined in several equivalent forms in dependence of two real parameters (α, β). The general form is (see [16], [17, Chap. 5])

$$F(x) = \sqrt{\frac{|\beta|}{(2\pi)^{1-\alpha}}} \int_{\mathbb{R}} dt f(t) \, e^{i\beta x t} \tag{11.75a}$$

where the integration is over the real set $\mathbb{R}$, $f(t)$ is the "signal" and $F(x)$ is the FT; x is a real variable as t. The inverse Fourier transform is

$$f(t) = \sqrt{\frac{|\beta|}{(2\pi)^{1+\alpha}}} \int_{\mathbb{R}} dx \, F(x) \, e^{-i\beta x t}. \tag{11.75b}$$

Hence the FT can be defined in infinitely many ways by choosing different pairs (α, β).[7] In this chapter we will make the choice $\alpha = 0$ and $\beta = 1$, which gives the symmetric forms

$$\mathcal{F} \quad F(x) = \frac{1}{\sqrt{2\pi}} \int_{\mathbb{R}} dt \, f(t) \, e^{ixt} \tag{11.76a}$$

$$\mathcal{F}^{-1} \quad f(t) = \frac{1}{\sqrt{2\pi}} \int_{\mathbb{R}} dx \, F(x) \, e^{-ixt}. \tag{11.76b}$$

By definition a signal $f_0(t)$ is an **eigenfunction of the FT with eigenvalue** λ if

$$f_0(t) \xrightarrow{\quad \mathcal{F} \quad} F_0(x) = \lambda \, f_0(x).$$

---

[7] According to Wolfram [18] the common choices for $\{\alpha, \beta\}$ are: $\{0, 1\}$ as default of `Mathematica` and in modern physics, $\{1, -1\}$ in pure mathematics and systems engineering, $\{-1, 1\}$ in classical physics, and $\{0, -2\pi\}$ in signal processing.

The choice in Weedbrook et al. is $\alpha = -1, \beta = -1$. The choice of Cahill and Glauber [14] for the introduction of the complex FT is $\{0, -2\}$; this choice has the advantage that the complex FT coincides with its inverse.

The possible eigenvalues are given by the fourth roots of unity, namely $\lambda = \{1, -1, i, -i\}$ (see [17]). Note that if $f_0(t)$ is an eigenfunction according to the choice (11.76), it is not an eigenfunction in the general case of (11.75) with $\beta \neq 1$.

11.6.4 Fourier Transform for the Single Mode

According to the choice (11.76) the **ordinary** two-dimensional FT is

$$\mathcal{F} \qquad F(x, y) = \frac{1}{2\pi} \int_{\mathbb{R}} du \int_{\mathbb{R}} dv\, f(u, v)\, e^{i(xu+yv)} \tag{11.77}$$

where the bilinear form in the exponential can be expressed as

$$x\,u + y\,v = [x, y] \begin{bmatrix} u \\ v \end{bmatrix}.$$

The **symplectic FT** is obtained with

$$[x, y]\, \Omega \begin{bmatrix} u \\ v \end{bmatrix} = [x, y] \begin{bmatrix} 0 & 1 \\ -1 & 0 \end{bmatrix} \begin{bmatrix} u \\ v \end{bmatrix} = xv - yu \tag{11.78}$$

giving

$$\mathcal{F}_s \qquad F_s(x, y) = \frac{1}{2\pi} \int_{\mathbb{R}} du \int_{\mathbb{R}} dv\, f(u, v)\, e^{i(xv-yu)}.$$

Thus the ordinary and the symplectic FTs are related as $F_s(x, y) = F(-y, x)$.

Finally, the **complex** FT is obtained from the symplectic FT by the introduction of the complex variables

$$\xi = u + iv, \qquad \lambda = x + iy$$

so that the bilinear form (11.78) becomes $xv - yu = (\lambda^*\xi - \lambda\xi^*)/2$ to get

$$\mathcal{F}_c \qquad F_c(\lambda) = \frac{1}{2\pi} \int_{\mathbb{C}} d\xi\, f(\xi)\, e^{(\lambda^*\xi - \lambda\xi^*)/2} \tag{11.79}$$

where the integration is over the complex plane $\mathbb{C}$ and $d\xi$ must be intended as $d\Re\xi\, d\Im\xi$. We may see the substantial equivalence of $\mathcal{F}_s$ and $\mathcal{F}_c$: the symplectic FT may be viewed as the real version of the complex FT.

The FTs introduced above for the single mode are summarized in Table 11.1 together with their inverses.

Table 11.1 Fourier transforms for the single mode

Type	Operator	Formula
Ordinary 2D Fourier transform	$\mathcal{F}$	$F(x, y) = \frac{1}{2\pi} \int_{\mathbb{R}} du \int_{\mathbb{R}} dv\, f(u, v)\, e^{i(xu+y,v)}$
Inverse	$\mathcal{F}^{-1}$	$f(u, v) = \frac{1}{2\pi} \int_{\mathbb{R}} dx \int_{\mathbb{R}} dy\, F(x, y)\, e^{-i(xu+yv)}$
Symplectic Fourier transform	$\mathcal{F}_s$	$F_s(x, y) = \frac{1}{2\pi} \int_{\mathbb{R}} du \int_{\mathbb{R}} dv\, f(u, v)\, e^{i(xv-yu)}$
Inverse	$\mathcal{F}_s^{-1}$	$f(u, v) = \frac{1}{2\pi} \int_{\mathbb{R}} dx \int_{\mathbb{R}} dy\, F_s(x, y)\, e^{-i(xv-yu)}$
Complex Fourier transform	$\mathcal{F}_c$	$F_c(\lambda) = \frac{1}{2\pi} \int_{\mathbb{C}} d\xi\, f(\xi)\, e^{(\lambda^*\xi - \lambda\xi^*)/2}$
Inverse	$\mathcal{F}_c^{-1}$	$f(\xi) = \frac{1}{2\pi} \int_{\mathbb{C}} d\lambda\, F_c(\lambda)\, e^{-(\lambda^*\xi - \lambda\xi^*)/2}$

11.6.5 Fourier Transforms for the N-Mode

We now extend the previous definitions to the N-mode, where the "signal" and the
FT become $2N$-dimensional, say $f(u, v), u, v \in \mathbb{R}^N$, and $F(x, y), x, y \in \mathbb{R}^N$. In
this notation the arguments are given by two vectors of length N instead of a single
vector of length $2N$ to simplify the relation between the symplectic and the complex
FT. Specifically we let

$$
\begin{aligned}
x &= [x_1, \ldots, x_N]^T \in \mathbb{R}^N, & y &= [y_1, \ldots, y_N]^T \in \mathbb{R}^N \\
u &= [u_1, \ldots, u_N]^T \in \mathbb{R}^N, & v &= [v_1, \ldots, v_N]^T \in \mathbb{R}^N \\
x_y &= [x_1, y_1, \ldots, x_N, y_N]^T \in \mathbb{R}^{2N} \\
u_v &= [u_1, v_1, \ldots, u_N, v_N]^T \in \mathbb{R}^{2N} \\
\lambda &= [\lambda_1, \ldots, \lambda_N]^T = x + i\,y \in \mathbb{C}^N \\
\xi &= [\xi_1, \ldots, \xi_N]^T = u + i v \in \mathbb{C}^N.
\end{aligned}
\tag{11.80}
$$

The relation between the real vectors and the complex vectors is provided by the
following relations, where Ω is the $2N \times 2N$ matrix defined by (11.67):

$$
\boxed{\; i x_y^T\, \Omega\, u_v = \tfrac{1}{2}[\lambda^* \xi - \xi^* \lambda]. \;}
\tag{11.81}
$$

For the proof see Problem 11.3.

The ordinary $2N$-dimensional FT is simply obtained from the two-dimensional
FT by considering that at the exponential we have to find a scalar quantity, so that
the form $i(xu + yv)$ becomes $i(xu^T + yv^T)$.

For the symplectic FT at the exponential we have to introduce the interlace of
arguments and the $2N \times 2N$ matrix Ω, specifically

Table 11.2 Fourier transforms for the N-mode

Type	Operator	Formula
Ordinary 2D Fourier transform	$\mathcal{F}$	$F(x, y) = \frac{1}{(2\pi)^N} \int_{\mathbb{R}^N} \mathrm{d}u \int_{\mathbb{R}^N} \mathrm{d}v \, f(u, v) \, \mathrm{e}^{\mathrm{i}(xu+yv)}$
Inverse	$\mathcal{F}^{-1}$	$f(u, v) = \frac{1}{(2\pi)^N} \int_{\mathbb{R}^N} \mathrm{d}x \int_{\mathbb{R}^N} \mathrm{d}y \, F(x, y) \, \mathrm{e}^{-\mathrm{i}(xu^{\mathrm{T}}+yv^{\mathrm{T}})}$
Symplectic Fourier transform	$\mathcal{F}_s$	$F_s(x, y) = \frac{1}{(2\pi)^N} \int_{\mathbb{R}^N} \mathrm{d}u \int_{\mathbb{R}^N} \mathrm{d}v \, f(u, v) \, \mathrm{e}^{\mathrm{i}(xy^{\mathrm{T}} \Omega \, uv)}$
Inverse	$\mathcal{F}_s^{-1}$	$f(u, v) = \frac{1}{(2\pi)^N} \int_{\mathbb{R}^N} \mathrm{d}x \int_{\mathbb{R}^N} \mathrm{d}y \, F_s(x, y) \, \mathrm{e}^{-\mathrm{i}(xy^{\mathrm{T}} \Omega \, uv)}$
Complex Fourier transform	$\mathcal{F}_c$	$F_c(\lambda) = \frac{1}{(2\pi)^N} \int_{\mathbb{C}^N} \mathrm{d}\xi \, f(\xi) \, \mathrm{e}^{(\lambda^*\xi-\lambda\xi^*)/2}$
Inverse	$\mathcal{F}_c^{-1}$	$f(\xi) = \frac{1}{(2\pi)^N} \int_{\mathbb{C}^N} \mathrm{d}\lambda \, F_c(\lambda) \, \mathrm{e}^{-(\lambda^*\xi-\lambda\xi^*)/2}$

$$\mathcal{F}_s \qquad F_s(x, y) = \frac{1}{(2\pi)^N} \int_{\mathbb{R}^N} \mathrm{d}u \int_{\mathbb{R}^N} \mathrm{d}v \, f(u, v) \exp\left[\mathrm{i}\, x_y{}^{\mathrm{T}} \, \Omega \, u_v\right]$$

where x_y and u_v are column vectors of length $2N$ obtained as the interlacing of vectors x, y and u, v, respectively.

With the introduction of the complex vectors (of length N) $\xi = u + \mathrm{i}v$ and $\lambda = x + \mathrm{i}y$ and use of identity (11.81), from the symplectic FT one gets the N-mode complex FT as

$$\mathcal{F}_c \qquad F_c(\lambda) = \frac{1}{(2\pi)^N} \int_{\mathbb{C}^N} \mathrm{d}\xi \, f(\xi) \, \exp[\tfrac{1}{2}(\lambda^*\xi - \lambda\xi^*)]. \tag{11.82}$$

The FTs for the N-mode are summarized in Table 11.2 together with their inverses.

Problem 11.2 ⋆⋆   Using the general definition of the exponential of a matrix, find explicitly the exponential of a 2×2 matrix.

Problem 11.3 ⋆⋆   Prove relation (11.81) linking the complex vectors and the real vectors defined by (11.80). Note that the entries of the matrix Ω can be written in the form

$$\Omega_{2(h-1)+r, 2(k-1)+s} = \delta_{hk}(\delta_{r,s-1} - \delta_{r-1,s}) = \delta_{hk}\, \varepsilon_{rs}, \qquad h, k = 1, \ldots, N \quad r, s = 1, 2$$

where

$$\varepsilon_{rs} = \begin{cases} 1 & r = 1, s = 2 \\ -1 & r = 2, s = 1 \\ 0 & \text{otherwise} . \end{cases}$$

11.7 Phase Space Representation: Definitions for the N-Mode

In this section we introduce the fundamental definitions of the phase space, given by the *characteristic function* and by the *Wigner function* in the general case of N-mode. The usefulness of these functions stems from the fact that, despite the infinite dimension of the Hilbert space modeling the harmonic oscillator, they give representations in the phase space that depend only on $2N$ real variables, or, equivalently, on N complex variables. In particular the Wigner function is the quantum analog to the classical joint probability density function of $2N$ random variables.

In the definitions we will use both the real and the complex form. In general the complex form allows us to express the formulas more compactly, while the real form may be useful in the deduction of the results.

11.7.1 The Weyl Operator

The Weyl operator (also called *displacement operator*) is defined as

$$\boxed{D(\xi) := \exp\left[\xi^{\mathrm{T}} a_\star - \xi^* a\right] , \qquad \xi \in \mathbb{C}^N}$$

(11.83)

where $a = [a_1, \ldots, a_N]^{\mathrm{T}}$ is the column vector collecting the N annihilation operators, $a_\star = [a_1^*, \ldots, a_N^*]^{\mathrm{T}}$ is the column vector collecting the N creation operators, ξ is a column vector containing N complex variables and ξ^* is a row vector. Explicitly the exponent is given by

$$\xi^{\mathrm{T}} a_\star - \xi_i^* a = \sum_{i=1}^{N} (\xi_i a_i^* - \xi_i^* a_i).$$

(11.83a)

The Weyl operator is unitary and verifies the properties

$$D(\xi) = D^*(\xi) = D^{-1}(\xi) = D(-\xi)$$

(11.84)

and in particular for $\xi = 0$ it gives the identity, $D(0) = I_{\mathcal{H}_N}$.

The real form is obtained by letting $\xi = u + \mathrm{i}\,v$ and using the relations (11.62), giving the bosonic operators in terms of quadrature operators. The expression is, after use of identity (11.81),

$$D(u, v) = \exp\left[\mathrm{i}\, q_p^{\mathrm{T}}\, \Omega\, u_v\right] , \qquad u, v \in \mathbb{R}^N$$

(11.85)

where q_p is the vector collecting the $2N$ quadrature operators in interlaced form.

11.7.2 Characteristic and Wigner Functions

The *(Wigner) characteristic function* of a density operator ρ acting in the bosonic space $\mathcal{H}_N$ is defined by the trace

$$\chi(\xi) := \text{Tr}[\rho\, D(\xi)] = \text{Tr}\left[\rho\, \exp\left(\xi^{\text{T}} a_\star - \xi^* a\right)\right], \qquad \xi \in \mathbb{C}^N. \qquad (11.86)$$

The *Wigner function* is defined as the complex Fourier transform of the characteristic function, namely[8]

$$W(\lambda) := \frac{1}{(2\pi)^N} \int_{\mathbb{C}^N} \mathrm{d}\xi\, \chi(\xi)\, \exp\left[\tfrac{1}{2}(\lambda^*\xi - \lambda\xi^*)\right], \quad \lambda \in \mathbb{C}^N. \qquad (11.87)$$

Both $\chi(\xi)$ and $W(\lambda)$ represent a density operator by functions of N complex variables. Note that they depend on ρ and a although this dependence is not indicated in the symbols.[9]

The characteristic function $\chi(\xi)$ can be straightforwardly recovered from the Wigner function through the inverse FT, namely

$$\chi(\xi) = \frac{1}{(2\pi)^N} \int_{\mathbb{C}^N} \mathrm{d}\lambda\, W(\lambda)\, \exp\left[-\frac{1}{2}(\lambda^*\xi - \lambda\xi^*)\right], \qquad \xi \in \mathbb{C}^N. \qquad (11.88)$$

From (11.86) and (11.88) we have the properties

$$\chi(0) = \text{Tr}[\rho] = 1, \qquad \frac{1}{(2\pi)^N} \int_{\mathbb{C}^N} \mathrm{d}\lambda\, W(\lambda) = 1. \qquad (11.89)$$

The density operator ρ can be recovered from the characteristic function $\chi(\xi)$ as

$$\rho = \frac{1}{\pi^N} \int_{\mathbb{C}^N} \mathrm{d}\xi\, \chi(\xi) D^*(\xi) \qquad (11.90)$$

known as *Glauber's inversion formula* [20]. It can also be recovered from the Wigner function as

$$\rho = \frac{1}{\pi^N} \int_{\mathbb{C}^N} \mathrm{d}\lambda\, W(\lambda)\, \mathcal{D}^*(\lambda) \qquad (11.91)$$

where $\mathcal{D}(\lambda)$ is the complex FT of the Weyl operator. The proof of (11.90) and (11.91) will be seen in the next section for the single mode.

The above definitions are given in terms of the bosonic operators contained in the column vectors a and $a_\star$. As seen for the Weyl operator, the characteristic and the

---

[8] This definition is quite different from the original one given by Wigner [19].

[9] Some authors, e.g., Ferraro et al. [8], use the notations $\chi[\rho](\xi)$ and $W[\rho](\lambda)$ to indicate the dependence on ρ.

Wigner functions can be expressed in terms of the quadrature operators contained in the column vectors q and p. Letting

$$\xi = u + iv, \qquad \lambda = x + iy, \qquad u, v, x, y \in \mathbb{R}^N$$

and using the relation (11.81) we obtain the real form of the two functions. The characteristic function becomes

$$\chi(u, v) = \text{Tr}[\rho\, D(u, v)] = \text{Tr}\left[\rho\, \exp\left[i\, q_p^{\text{T}}\, \Omega\, u_v\right]\right] \tag{11.92}$$

and the Wigner function is expressed as the symplectic FT of the characteristic function as

$$W(x, y) = \frac{1}{(2\pi)^N} \int_{\mathbb{R}^N} du \int_{\mathbb{R}^N} dv\, \chi(u, v) \exp\left[i\, x_y^{\text{T}}\, \Omega\, u_v\right]. \tag{11.93}$$

The inversion formulas (11.90) and (11.91), giving the density operator from the characteristic and Wigner functions, can be written in real forms in terms of $\chi(u, v)$ and $W(x, y)$.

11.7.3 Statistical Description Provided by the Wigner Function

The Wigner function $W(\lambda)$, and also its inverse Fourier transform $\chi(\xi)$, is particularly suitable for describing the effect of quantum observables that may arise from quantum mechanics and classical statistics. Note that the *normalized* Wigner function

$$w(\lambda) := \frac{1}{(2\pi)^N}\, W(\lambda) \tag{11.94}$$

has the property

$$\int_{\mathbb{C}^N} w(\lambda)\, d\lambda = 1. \tag{11.95}$$

Then $w(\lambda)$ behaves partly as a classical probability density, which allows one to calculate measurable quantities such as mean values and variances in a classical like fashion. But, in contrast to a classical probability density, the Wigner function possesses some disappointing features, due to Quantum Mechanics nature, as it can become negative. For this reason it is called *quasi-probability density*.[10]

In the N-mode bosonic space, the statistical description of each mode is concerned with the quadrature operators q_i and p_i, which are observables with a continuous nature. Now, the general statistical description would be obtained in the framework of

---

[10] The term used in Quantum Mechanics is *quasi-probability distribution*, but the reference to a probability density seems to be more appropriate.

simultaneous measurements seen in Sect. 3.8.2, in this case applied to the observables q_i and p_i, but the noncommutativity of these operators leads to the impossibility of a simultaneous quantum measurement.[11]

11.7.4 Mean Vector and Covariance Matrix

In the N-mode we have a vector of $2N$ quadrature operators

$$X := q_p = [q_1, p_1 \ldots, q_N, p_N]^{\mathrm{T}}. \tag{11.96}$$

In general the statistical description of $2N$ random variables is given by a multivariate density or a multivariate characteristic function, which in Quantum Mechanics are provided respectively by the N-mode Wigner and Wigner characteristic functions (with the disappointing features discussed above). The moments of different orders can be calculated from the characteristic function, by derivation, and from the Wigner function, by integration, as in Probability Theory. But here we shall use the trace formula:

$$\langle O \rangle = \overline{O} := \mathrm{Tr}[\rho O]$$

where O is an observable, ρ is the density operator of the quantum system and $\langle O \rangle = \overline{O}$ is the **quantum average**.[12]

Now, in preparation of Gaussian states, we deal with the means and the covariances. The **mean vector** of (11.96) is given by

$$\overline{X} = \langle X \rangle = \mathrm{Tr}[\rho X]. \tag{11.97}$$

The natural way to define the **covariance matrix** is

$$R = \left\langle \Delta X \Delta X^{\mathrm{T}} \right\rangle = \mathrm{Tr}[\rho \Delta X \Delta X^{\mathrm{T}}] \tag{11.98}$$

where $\Delta X := X - \langle X \rangle = X - \overline{X}$. As we shall see, this matrix is not symmetric, and in the context of continuous variables it is usual to define another covariance matrix V, which, by construction, turns out to be symmetric. The elements of V are defined through the *anticommutator* $\{,\}$ as

---

[11] In a recent interesting overview paper by Paris [21] it is discussed how simultaneous quantum measurements can be formulated in an enlarged Hilbert space.

[12] We recall that the terms used in Quantum Mechanics are sometimes relaxed. In a quantum measurement with an observable O, when the system is in the state ρ, the outcome is a random variable m, whose expectation (average) is given by $\mathrm{E}[m|\rho] = \mathrm{Tr}[\rho O]$. But this is often called the "average of the observable" O, which leads to think that O is random, whereas it is a nonrandom Hermitian operator, while the randomness is confined to the outcome of the measurement.

$$V_{ij} = \frac{1}{2}\langle\{\Delta X_i, \Delta X_j\}\rangle = \frac{1}{2}\langle \Delta X_i \, \Delta X_j + \Delta X_j \, \Delta X_i \rangle \qquad (11.99)$$

where the symmetry is ensured by the anticommutator, which verifies the condition $\{\Delta X_j, \Delta X_i\} = \{\Delta X_i, \Delta X_j\}$. The relation between the two covariance matrices is given by

$$R = V + i\,\Omega. \qquad (11.100)$$

This relation is developed and proved for the single mode in Problem 11.4, where, using the commutation relation (11.51), we find

$$\begin{bmatrix} V_{11} & V_{12} \\ V_{21} & V_{22} \end{bmatrix} = \begin{bmatrix} R_{11} & R_{12} \\ R_{21} & R_{22} \end{bmatrix} - i \begin{bmatrix} 0 & 1 \\ -1 & 0 \end{bmatrix}$$

which can be written in the form (11.100).

The diagonal elements of the covariance matrix give the variances of the positions and momentums and must verify the uncertainty principle, that is,

$$V_{2i-1,2i-1}\,V_{2i,2i} = R_{2i-1,2i-1}\,R_{2i,2i} \geq 1 \,, \qquad i = 1, 2, \ldots, N \qquad (11.101)$$

while the nondiagonal elements are not constrained. However, Simon et al. [15, 22] proved the more stringent condition

$$V + i\,\Omega \geq 0 \qquad (11.102)$$

where also the nondiagonal elements are constrained.

Another form of covariance matrix is defined starting from the quadrature operators ordered in the form $[q^{\mathrm{T}}, p^{\mathrm{T}}] = [q_1, \ldots, q_N, p_1, \ldots, p_N]^{\mathrm{T}}$, giving the $2N \times 2N$ matrix

$$Y = \begin{bmatrix} Y_{qq} & Y_{qp} \\ Y_{pq} & Y_{pp} \end{bmatrix} := \begin{bmatrix} \langle \Delta q \, \Delta q \rangle & \langle \Delta q \, \Delta p \rangle \\ \langle \Delta p \, \Delta q \rangle & \langle \Delta p \, \Delta p \rangle \end{bmatrix}. \qquad (11.103)$$

This matrix contains the same entries as the matrix V, but permuted. The relation is

$$V = \Pi \, Y \, \Pi^{\mathrm{T}} \qquad (11.104)$$

where Π is a permutation matrix (defined in Proposition 11.11). The form Y has the advantage to express the uncertainty in a simpler way (see Sect. 11.12.3),

11.7.5 Definition of Gaussian States

Gaussian states are defined in terms of the characteristic and Wigner functions, which should have a multivariate Gaussian form and hence they are completely specified by the mean vector and the covariance matrix. This is in perfect analogy with the classical definition of Probability Theory for Gaussian random vectors.

Definition 11.1 An N-mode bosonic state with mean $\overline{X} = \overline{q_p}$ and covariance matrix V is a Gaussian quantum state if its characteristic and Wigner functions have respectively the following real form[13]:

$$\chi(u, v) = \exp\left[-\tfrac{1}{2}{u_v}^{\mathrm{T}}\left(\Omega\, V \Omega^{\mathrm{T}}\right) u_v - \mathrm{i}(\Omega\overline{q_p})^{\mathrm{T}} u_v\right], \quad (u, v) \in \mathbb{R}^{2N} \tag{11.105a}$$

$$W(x, y) = \frac{\exp\left[-\tfrac{1}{2}(x_y - \overline{q_p})^{\mathrm{T}} V^{-1}(x_y - \overline{q_p})\right]}{(2\pi)^N \sqrt{\det\ V}}, \qquad (x, y) \in \mathbb{R}^{2N}. \tag{11.105b}$$

As we shall see, the most important states in the contest of continuous variables turn out to be Gaussian. The main property of Gaussian states lies on the extremely simple specification given by the pair $(\overline{X}, V)$. For this reason, a Gaussian state specified by a density operator is often indicated in the form $\rho(\overline{X}, V)$.

Now, we develop Definition 11.1 in the single mode, where the mean vector is $\overline{q_p} = [\overline{q}, \overline{p}]^{\mathrm{T}}$ and the covariance matrix V is 2×2. Considering that

$$-\frac{1}{2}{u_v}^{\mathrm{T}}\left(\Omega\, V \Omega^{\mathrm{T}}\right) u_v - \mathrm{i}(\Omega\overline{q_p})^{\mathrm{T}} u_v$$

$$= -\frac{1}{2}[u, v] \begin{bmatrix} 0 & 1 \\ -1 & 0 \end{bmatrix} \begin{bmatrix} V_{11} & V_{12} \\ V_{12} & V_{22} \end{bmatrix} \begin{bmatrix} 0 & -1 \\ 1 & 0 \end{bmatrix} \begin{bmatrix} u \\ v \end{bmatrix} - \mathrm{i}\ [\overline{q}\ \overline{p}] \begin{bmatrix} 0 & 1 \\ -1 & 0 \end{bmatrix} \begin{bmatrix} u \\ v \end{bmatrix}$$

$$= -\frac{1}{2}(V_{11}v^2 + V_{22}u^2 - 2V_{12}uv) - \mathrm{i}(\overline{q}v - \overline{p}\,u),$$

the characteristic function turns out to be

$$\chi(u, v) = \exp\left[-\frac{1}{2}(V_{11}v^2 + V_{22}u^2 - 2V_{12}uv) - \mathrm{i}(\overline{q}\,v - \overline{p}\,u)\right]. \tag{11.106}$$

The Wigner function $W(x, y)$, which is the symplectic FT of $\chi(u, v)$, turns out to be

$$W(x, y) = \frac{1}{2\pi\sqrt{\det V}}\exp\left[-\frac{1}{2}\frac{V_{22}(x - \overline{q})^2 + V_{11}(y - \overline{p})^2 - 2V_{12}(x - \overline{q})(y - \overline{p})}{\det V}\right] \tag{11.107}$$

where $\det V = V_{11}V_{22} - V_{12}^2$.

11.7.6 Summary of State Representations

We have seen two equivalent representations of a state ρ in the phase space: the characteristic function $\chi(\xi)$ and the Wigner function $W(\lambda)$. Each representation provides

---

[13] We do not consider the complex form because it is rather cumbersome.

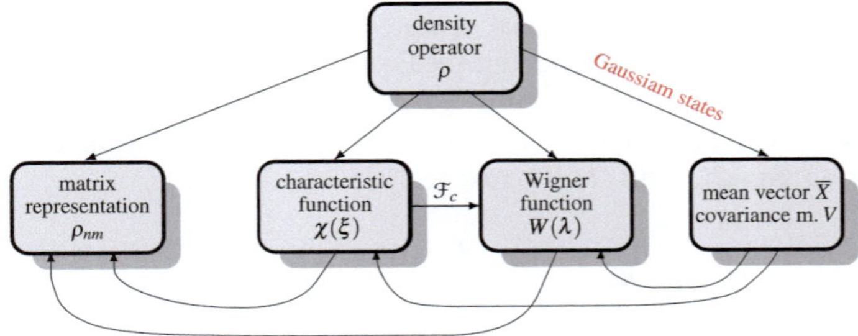

Fig. 11.5 Connections between the representations of a quantum state ρ. In principle all the connections in the graph are possible, but we have represented only the more direct ones

the full information on the given state ρ in the sense that it allows the recovery of the state. Moreover, it is possible to link each representation. We summarize these ideas, as illustrated in Fig. 11.5. The characteristic function $\chi(\xi)$ is obtained through the Weyl operator by definition (11.86) and the recovery of ρ by Glauber's inversion formula (11.90). The Wigner function $W(\lambda)$ is linked to the characteristic function $\chi(\xi)$ by the complex FT according to (11.87). From the Wigner function one can obtain directly the density operator ρ through the FT of the Weyl operator $\mathcal{D}(\lambda)$ using inversion formula (11.91).

There are several other representations of a density operator, such as the P-representation, the R-representation, and the Q-representation (for a complete review see [7]). Here we mention the P-representation, which has found widespread application in Quantum Optics. It is based on the completeness of the coherent states (see Sect. 11.4, Eq. (11.49)) and reads as

$$\rho = \int_{\mathbb{C}} d\alpha \, |\alpha\rangle \, \langle \alpha \, | P(\alpha). \tag{11.108}$$

The function $P(\alpha)$ in (11.108), called the P-representation of ρ, has properties similar to the ones of the Wigner function. Since ρ is Hermitian, $P(\alpha)$ must be real. Moreover, since $\text{Tr}\rho = 1$, it must satisfy the normalization condition $\int_{\mathbb{C}} d\alpha \, P(\alpha) = 1$, but the condition $P(\alpha) \geq 0$ is not guaranteed everywhere. The same happens for the Wigner function.

Finally, we note that for the important class of Gaussian states the representation (or specification) is simply given by the pair $(\overline{X}, V)$, from which we can calculate $\chi(\xi)$ and $W(\lambda)$, and then reconstruct the density operator.

Matrix representations. There is another important representation, not in the phase space, given by the matrix representation with the Fock basis

$$\rho_{nm} = \langle n|\rho|m\rangle, \qquad |m\rangle, |n\rangle \in \mathcal{B}_F(N). \tag{11.109}$$

From ρ_{nm} we can obtain the original operator ρ using the reconstruction formula (2.33). Note that in the general N-mode the Fock basis $\mathcal{B}_F(N)$ consists of composite Fock states, that is, $|n\rangle = |n_1\rangle_1 \cdots |n_N\rangle_N$ (see (11.68)).

The matrix representation ρ_{nm} can be obtained directly from the characteristic function $\chi(\xi)$. In fact, from (11.90) one gets

$$\rho_{nm} = \frac{1}{\pi^N} \int_{\mathbb{C}} \mathrm{d}\xi \, \chi(\xi) D^*_{mn}(\xi) \tag{11.110}$$

where $D_{mn}(\xi)$ is the matrix representation of $D(\xi)$, which is explicitly given by Proposition 11.7. Analogously from (11.91) one gets

$$\rho_{nm} = \frac{1}{\pi^N} \int_{\mathbb{C}} \mathrm{d}\lambda \, W(\lambda) \, \mathcal{D}^*_{mn}(\lambda) \tag{11.111}$$

where $\mathcal{D}^*_{mn}(\lambda)$ is explicitly given by Proposition 11.8.

11.7.7 Simplifications with Pure Gaussian States

In the case of a pure state $|\psi\rangle$ the above relations hold with $\rho = |\psi\rangle\langle\psi|$ and with a few simplifications. In particular, in the definition of the characteristic function (11.86) one can use the identity (2.37) on the trace to get

$$\chi(\xi) = \langle\psi|D(\xi)|\psi\rangle = \langle\psi|e^{a^*\xi - \xi^* a}|\psi\rangle. \tag{11.112}$$

We recall that a pure state ρ has unitary rank and $\mathrm{Tr}[\rho^2] = 1$ (see Sect. 3.3.2). The latter property leads to the following condition for the characteristic function of a pure state (see Problem 11.6):

$$\frac{1}{\pi^N} \int_{\mathbb{C}^N} \mathrm{d}\xi \, |\chi(\xi)|^2 = 1. \tag{11.113}$$

The evaluation of this integral with $\chi(\xi)$ given by (11.105a) leads to the following result (see Problem 11.7):

$$\frac{1}{\pi^N} \int_{\mathbb{C}^N} \mathrm{d}\xi \, |\chi(\xi)|^2 = (\det V)^{-1/2}. \tag{11.114}$$

Hence we find that **a Gaussian state is pure if and only if the determinant of its covariance matrix is unitary**.

Problem 11.4 ⋆⋆   Prove the relation (11.100) between the covariance matrices R and V in the single mode.

Problem 11.5 ⋆⋆   Compare conditions (11.101) and (11.102) in the single mode.

Problem 11.6 ★★★ ∇ Prove condition (11.113), which states that the characteristic function $\chi(\xi)$ refers to a pure state.
Hint Use the Fock expansion of the pure state and Proposition 11.9.

Problem 11.7 ★★★ ∇ Evaluate the integral (11.114) using Williamson's theorem (Theorem 11.2).

11.8 Phase Space Representations in the Single Mode

The general definitions of the previous section introduced for the N-mode apply in particular to the single mode. In this passage the general formulas do not change their form, but we get a few simplifications considering that the domains change as $\mathbb{C}^N \rightarrow \mathbb{C}$ and $\mathbb{R}^{2N} \rightarrow \mathbb{R}^2$. This allows us to get explicit results in an easier way. The arguments become scalar variables related by $\lambda = x + \mathrm{i}y$ and $\xi = u + \mathrm{i}v$, and relation (11.81), linking the complex forms to the real forms, becomes explicitly (see (11.78))

$$\mathrm{i}(xu - yv) = \tfrac{1}{2}(\xi\lambda^* - \lambda\xi^*)$$

so that only the real forms will get a new shape, with the light difference in the complex form due to the simplification $a_\star \rightarrow a^*$.

Thus the Weyl operator for the single mode reads as

$$\boxed{D(\xi) = e^{\xi a^* - \xi^* a}, \qquad \xi \in \mathbb{C}} \tag{11.115}$$

and its real form becomes

$$D(u, v) = e^{\mathrm{i}(vq - up)}, \qquad (u, v) \in \mathbb{R}^2. \tag{11.116}$$

The *characteristic function* becomes

$$\boxed{\chi(\xi) = \mathrm{Tr}[\rho D(\xi)] = \mathrm{Tr}\left[\rho\, e^{\xi a^* - \xi^* a}\right], \qquad \xi \in \mathbb{C}} \tag{11.117}$$

and its real form reads as

$$\chi(u, v) = \mathrm{Tr}[\rho D(u, v)] = \mathrm{Tr}\left[\rho\, e^{\mathrm{i}(vq - up)}\right], \qquad (u, v) \in \mathbb{R}^2. \tag{11.118}$$

The Wigner function becomes

$$\boxed{W(\lambda) = \frac{1}{2\pi} \int_{\mathbb{C}} \mathrm{d}\xi\, \chi(\xi)\, e^{(\xi\lambda^* - \xi^*\lambda)/2}, \qquad \lambda \in \mathbb{C}} \tag{11.119}$$

and its real form reads as

$$W(x, y) = \frac{1}{2\pi} \int_{\mathbb{R}^2} du \, dv \chi(u, v) \, e^{i(-xu + yv)} , \qquad (x, y) \in \mathbb{R}^2. \qquad (11.120)$$

11.8.1 Normal-Ordered Characteristic Function

In the single mode the Weyl operator (11.115) can also be written with the separation of the exponent as

$$D(\xi) = e^{-\frac{1}{2}|\xi|^2} e^{\xi a^*} e^{-\xi^* a}. \qquad (11.121)$$

In fact, using the commutation relation (11.54) with $H = a^* \xi$ and $K = \xi^* a$, we find $[H, K] = |\xi|^2 I_{\mathcal{H}}$. Then we can apply the exponential identity (11.70) to get (11.121).

With the form (11.121) the characteristic function reads as

$$\chi(\xi) = e^{-\frac{1}{2}|\xi|^2} \chi_N(\xi), \qquad \xi \in \mathbb{C} \qquad (11.121a)$$

where

$$\chi_N(\xi) = \text{Tr}\left[\rho e^{\xi a^*} e^{-\xi^* a}\right] \qquad (11.121b)$$

is called the *normal-ordered characteristic function*. "Normal-ordered" is a general term meaning that the operator consists of factors where the left factors involve only creation operators and the right factors involve only annihilator operators. This form will be introduced systematically in Sect. 11.13.

We recall that the trace in the Fock space is explicitly given by

$$\chi(\xi) = \sum_{n=0}^{\infty} \langle n|\rho \, e^{a^* \xi - \xi^* a} \, |n\rangle = e^{-\frac{1}{2}|\xi|^2} \sum_{n=0}^{\infty} \langle n|\rho \, \chi_N(\xi) \, |n\rangle . \qquad (11.122)$$

11.8.2 Explicit Results in the Single Mode

We now establish a few identities for the single mode, using relations (11.57a), that is,

$$a|0\rangle = 0 , \qquad a|n\rangle = \sqrt{n} \, |n-1\rangle \quad \text{for } n \geq 1.$$

We start from a specific example: the evaluation of $a^r |3\rangle$, where a is the annihilator and $|3\rangle$ is the three-photon Fock state. We find

$$a^0|3\rangle = |3\rangle\,, \qquad a|3\rangle = \sqrt{3}|2\rangle\,, \qquad a^2|3\rangle = a\sqrt{3}|2\rangle = \sqrt{3\cdot 2}|1\rangle$$

$$a^3|3\rangle = a\sqrt{3\cdot 2}|1\rangle = \sqrt{3\cdot 2\cdot 1}\,|0\rangle\,, \qquad a^4|3\rangle = a\sqrt{3\cdot 2\cdot 1}\,|0\rangle = 0.$$

Now, it is easy to prove by induction the following general result:

Proposition 11.5 *For any pair of natural integers n, r*

$$a^r\,|n\rangle \;=\; \begin{cases} \sqrt{n(n-1)\cdots(n-r+1)}\,|n-r\rangle & n \geq r \\ 0 & n < r. \end{cases} \tag{11.123}$$

Note that (11.123) can be written in the binomial form

$$a^r\,|n\rangle = \sqrt{r!\binom{n}{r}}\,|n-r\rangle\,, \qquad n, r = 0, 1, 2, \ldots \tag{11.123a}$$

without conditions because $\binom{k}{r} = 0$ for $r > k$.

Analogously, using (11.57b), that is, $a^*|n\rangle = \sqrt{n+1}\,|n+1\rangle$ for $n \geq 0$, we find

$$(a^*)^r\,|n\rangle = \sqrt{(n+1)(n+2)\cdots(n+r)}|n+r\rangle. \tag{11.124}$$

In particular, we get the important identity

$$\boxed{(a^*)^n|0\rangle = \sqrt{n!}\,|n\rangle.} \tag{11.125}$$

A third identity is related to the Weyl operator.

Proposition 11.6 *For the exponential of the annihilator operator a the following identity holds:*

$$|n, \beta\rangle := e^{\beta a}\,|n\rangle = \sum_{r=0}^{n} \mu_{nr}\,\beta^r\,|n-r\rangle \tag{11.126}$$

where $|n\rangle$ are the Fock states, $\beta \in \mathbb{C}$, and

$$\mu_{nr} = \sqrt{\frac{1}{r!}\binom{n}{r}} = \frac{1}{r!}\sqrt{\frac{n!}{(n-r)!}}\,, \qquad n, r = 0, 1, 2, \ldots\,. \tag{11.126a}$$

In particular we have

$$|0, \beta\rangle = e^{\beta a}|0\rangle = |0\rangle. \tag{11.126b}$$

Note that (11.126) defines a class of states (kets) $|n, \beta\rangle$ which depend on the complex parameter β, and gives the Fock expansion of the kets.

11.8.3 Fock Representations

The Fock representation of an operator A is given by the matrix representation obtained with the Fock basis $\mathcal{B}_F = \{|n\rangle, n = 0, 1, 0, \ldots\}$, that is, $A_{mn} = \langle m|A|n\rangle$.

The Fock representations of the Weyl operator $D(\xi)$ and of its FT $\mathcal{D}(\lambda)$, defined respectively by

$$D_{mn}(\xi) = \langle m|D(\xi)|n\rangle, \qquad \mathcal{D}_{mn}(\lambda) = \langle m|\mathcal{D}(\lambda)|n\rangle, \tag{11.127}$$

are useful to get explicit results in the phase space. They can be evaluated in terms of the *generalized* Laguerre polynomials $L_n^{(\beta)}(x)$. These polynomials have several properties [23], and in particular they form an orthonormal class, as stated by

$$\int_0^\infty e^{-x} x^\beta L_n^{(\beta)}(x) L_m^{(\beta)}(x)\, dx = \frac{\Gamma(\beta + n + 1)}{n!} \delta_{mn}. \tag{11.128}$$

Proposition 11.7 *The Fock representation of the Weyl operator (11.121), given by*

$$D_{mn}(\xi) = \langle m|D(\xi)|n\rangle = e^{-\frac{1}{2}|\xi|^2} \langle m| e^{\xi a^*} e^{-\xi^* a}|n\rangle, \tag{11.129}$$

can be expressed in the form

$$\boxed{D_{mn}(\xi) = e^{-\frac{1}{2}|\xi|^2} \sqrt{\frac{n!}{m!}}\, \xi^{m-n} L_n^{(m-n)}(|\xi|^2).} \tag{11.130}$$

where $L_n^{(m-n)}(x)$ is the generalized Laguerre polynomial.

The proof is given in Appendix Section "Proof of Fock Representation of Weyl Operator (Proposition 11.7)". The expression of $\mathcal{D}_{mn}(\lambda)$ is very close to the expression of $D_{mn}(\xi)$, as recently proved [24]:

Proposition 11.8 *The Fock representation $\mathcal{D}_{mn}(\lambda)$ of the FT of the Weyl operator is given by*

$$\mathcal{D}_{mn}(\lambda) = (-1)^{\min(m,n)} D_{mn}(\lambda). \tag{11.131}$$

In words: $\mathcal{D}_{mn}(\lambda)$ and $D_{mn}(\xi)$ have the same expression apart from a sign. Another interpretation of (11.131) is that $D_{mn}(\xi)$ **is an eigenfunction of the complex FT with eigenvalue** ± 1.

Finally, we note the following property (proved in Appendix Section "Proof of Proposition 11.9 on the Orthogonality of the D_{mn}"):

Proposition 11.9 *The functions* $D_{mn}(\xi)$, $m, n = 0, 1, 2, \ldots$, *giving the Fock representation of the Weyl operator, are orthogonal functions in* $\mathbb{C}$, *specifically*

$$\frac{1}{\pi} \int_{\mathbb{C}} d\xi \, D_{mn}^*(\xi) \, D_{rs}(\xi) = \delta_{mr}\delta_{ns}. \tag{11.132}$$

The same property holds for the functions $\mathcal{D}_{mn}(\lambda)$ (see Problem 11.9).

Problem 11.8 $\star\star\star$ Prove Glauber's inversion formula (11.90) in the single mode. *Hint*: use Fock representation and the orthogonality of the D_{mn} (see Proposition 11.9)

Problem 11.9 $\star$ Using the orthogonality of the functions $D_{mn}(\xi)$ given by (11.132), prove the orthogonality of the functions $\mathcal{D}_{mn}(\lambda)$, the Fourier transform of the $D_{mn}(\xi)$.

11.9 Examples of Continuous States in the Single Mode

In this section we evaluate the characteristic and the Wigner functions of few elementary single-mode quantum states, with the purpose of testing the Gaussianity.

11.9.1 The Vacuum State $\rho = |0\rangle \langle 0|$

The vacuum state is given by the zero Fock state $|0\rangle$ with density operator $\rho = |0\rangle \langle 0|$. We evaluate, using (11.122), the normally ordered characteristic function

$$\chi_N(\xi) = \sum_{n=0}^{\infty} \langle n|e^{\xi a^*} e^{-\xi^* a} |0\rangle \langle 0 |n\rangle = \langle 0|e^{\xi a^*} e^{-\xi^* a} |0\rangle .$$

Considering that (see (11.126)) $e^{-\xi^* a} |0\rangle = |0\rangle$ and $\langle 0|e^{\xi a^*} = \langle 0|$, we get

$$\chi_N(\xi) = 1 \quad \rightarrow \quad \chi(\xi) = e^{-\frac{1}{2}|\xi|^2} = e^{-\frac{1}{2}(u^2 + v^2)}. \tag{11.133}$$

Then, we find that *the vacuum state is Gaussian* with zero mean and covariance matrix given by the identity

$$\overline{qp} = \begin{bmatrix} \overline{q} \\ \overline{p} \end{bmatrix} = \begin{bmatrix} 0 \\ 0 \end{bmatrix}, \qquad V = I = \begin{bmatrix} 1 & 0 \\ 0 & 1 \end{bmatrix}.$$

The characteristic and Wigner functions of the vacuum state (in real form) are shown in Fig. 11.6.

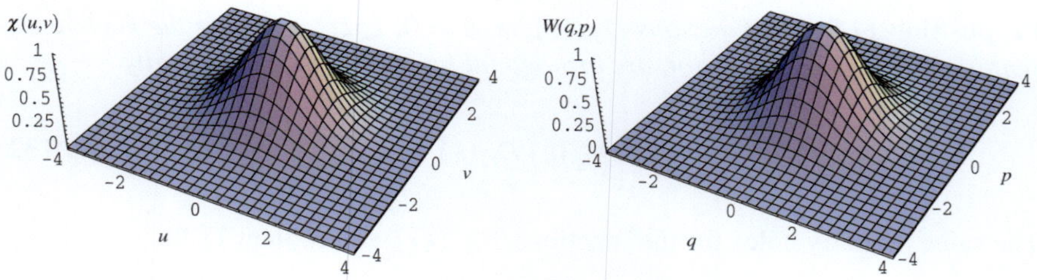

Fig. 11.6 Characteristic function and Wigner function of the vacuum state $|0\rangle$

11.9.2 Single-Photon State $\rho = |1\rangle\langle 1|$

We apply (11.122) to evaluate the normally ordered characteristic function of single-photon state $\rho = |1\rangle\langle 1|$. We find

$$\chi_N(\xi) = \sum_{n=0}^{\infty} \langle n| e^{\xi a^*} e^{-\xi^* a} |1\rangle\langle 1 | n\rangle = \langle 1| e^{\xi a^*} e^{-\xi^* a} |1\rangle. \tag{11.134}$$

Next we evaluate $e^{-\xi^* a}|1\rangle$ using Proposition 11.6 with $n = 1$ to get $e^{-\xi^* a}|1\rangle = |1\rangle - \xi^* |0\rangle$. Analogously, $\langle 1| e^{\xi a^*} = \langle 1| + \xi\langle 0|$. Hence (11.134) becomes

$$\chi_N(\xi) = (\langle 1| e^{\xi a^*})(e^{-\xi^* a}|1\rangle) = ((\langle 1| + \xi\langle 0|)(|1\rangle - \xi^* |0\rangle))$$
$$= 1 - |\xi|^2.$$

In conclusion the characteristic function is

$$\chi(\xi) = e^{-\frac{1}{2}|\xi|^2}(1 - |\xi|^2) = e^{-\frac{1}{2}(u^2+v^2)}(1 - u^2 - v^2) \tag{11.135}$$

and therefore the *one-photon state is not Gaussian*.

The characteristic and Wigner functions of the single-photon number state are shown in Fig. 11.7. Note in particular that the Wigner function in some regions of the phase space is negative.

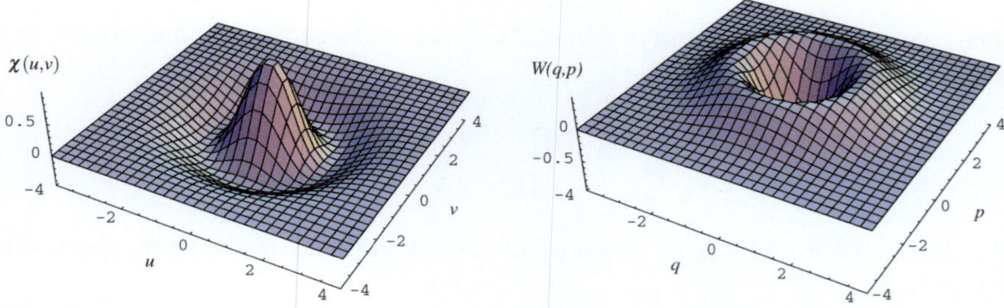

Fig. 11.7 Characteristic function and Wigner function of a single-photon number state $|1\rangle$

11.9.3 General Results for the Fock States

The characteristic function of the general Fock state $\rho = |n\rangle \langle n|$ can be calculated as above "by hand" for any n, but the evaluation becomes cumbersome as n increases. The procedure is simplified if we use Proposition 11.7, where the Fock representation of the Weyl operator is defined and evaluated. By comparison, we find immediately that the characteristic function of the pure state $|\psi\rangle = |n\rangle$ is given by (see (11.129))

$$\chi_n(\xi) = e^{-\frac{1}{2}|\xi|^2} \langle n| e^{\xi a^*} e^{-\xi^* a} |n\rangle = D_{nn}(\xi) \tag{11.136}$$

that is, by the nth diagonal element of the matrix representation of the Weyl operator. Hence from Proposition 11.7 we get

$$\chi_n(\xi) = e^{-\frac{1}{2}|\xi|^2} L_n(|\xi|^2).$$

where $L_n(x) = L_n^{(0)}(x)$ is the ordinary Laguerre polynomial.

In Sect. 11.8, Proposition 11.8, we have seen that the FT $\mathcal{D}_{mn}(\lambda)$ of $D_{mn}(\xi)$ is given by $(-1)^{\min(m,n)} D_{mn}(\lambda)$. On the other hand, the Wigner function is the FT of $\chi_n(\xi)$. Hence, from (11.136) we find $W_n(\lambda) = (-1)^n D_{nn}(\lambda)$. In other words the Wigner function $W_n(\lambda)$ of the number states is *an eigenfunction of the Fourier transform* with eigenvalues 1 or -1. To summarize

Proposition 11.10 *For the number state $|n\rangle$ the characteristic function and the Wigner function are respectively*

$$\chi_n(\xi) = e^{-\frac{1}{2}|\xi|^2} L_n(|\xi^2|)$$
$$W_n(\lambda) = (-1)^n \chi_n(\lambda) = (-1)^n e^{-\frac{1}{2}|\lambda|^2} L_n(|\lambda|^2) \tag{11.137}$$

where $L_n(x) = L_n^{(0)}(x)$ is the ordinary Laguerre polynomial of order n.

We realize that the *number states $|n\rangle$ are not Gaussian for $n > 0$.* The characteristic and Wigner functions of the five-photon number state are shown in Fig. 11.8.

11.9.4 Coherent States

The theory of coherent states was developed in Sect. 11.4, where they were defined as eigenstates of the annihilation operator a, that is, by the eigenvalue equation $a|\alpha\rangle = \alpha|\alpha\rangle$. Now we consider an alternative definition: a coherent state $|\alpha\rangle$ is a transformation of the vacuum state $|0\rangle$ obtained with the Weyl operator $D(\alpha)$ (in this context called *displacement operator*). Then, using the expression (11.121) for $D(\alpha)$, one gets

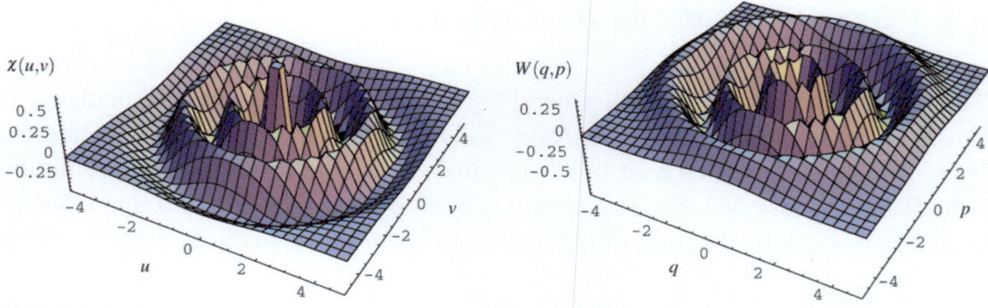

Fig. 11.8 Characteristic function and Wigner function of the five-photon number state $|5\rangle$

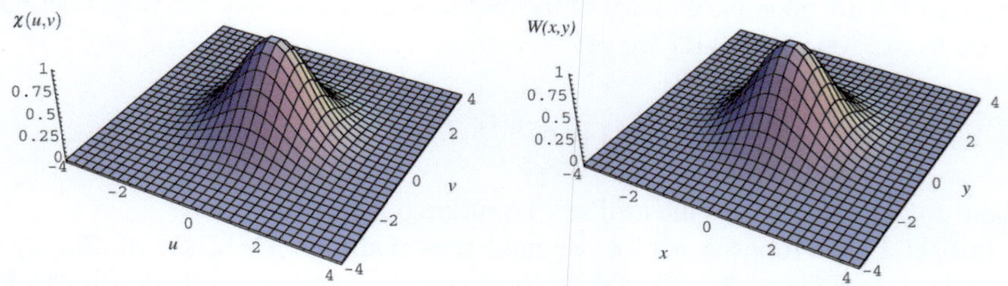

Fig. 11.9 Characteristic function and Wigner function of a coherent state $|\alpha\rangle$

$$|\alpha\rangle = D(\alpha)\,|0\rangle = e^{-\frac{1}{2}|\alpha|^2}\, e^{\alpha\, a^*}\, e^{-\alpha^*\, a}\,|0\rangle \,. \qquad (11.138)$$

To prove the equivalence it is sufficient to show that both definitions lead to the same Fock representation given by (11.46) (see Problem 11.1).

Now we consider the characteristic function. Above we have seen that the vacuum state is Gaussian with covariance matrix given by the identity and zero mean value. On the other hand a coherent state $|\alpha\rangle$ is obtained from the vacuum by transformation (11.138). In Sect. 11.10 we shall see that such transformation preserves the Gaussianity with the same covariance as the vacuum state ($V = I$) but different mean values, given by $(\overline{q}, \overline{p})^{\mathrm{T}} = (\Re\alpha, \Im\alpha)^{\mathrm{T}}$. Hence the characteristic function and the Wigner function are respectively (Fig. 11.9)

$$\chi(u, v) = e^{-\frac{1}{2}(u^2+v^2)+\mathrm{i}(\overline{q}v-\overline{p}u)}$$

$$W(x, y) = \frac{1}{2\pi} e^{-\frac{1}{2}\left[(x-\overline{q})^2+(y-\overline{p})^2\right]}. \qquad (11.139)$$

The quadrature operators have noise variance equal to one, i.e., $V_q = V_p = 1$, and, according to (11.101), this is the minimum variance reachable symmetrically by position and momentum operators. Pictorially, this symmetric behavior is represented by a contour plot consisting of circles in the x-y plane (see Fig. 11.15).

11.9.5 Thermal States

The thermal states can be defined through their characteristic function[14]

$$\chi(\xi) = e^{-\frac{1}{2}(2\mathcal{N}+1)|\xi|^2} \tag{11.140}$$

where $\mathcal{N} \geq 0$ is a parameter called *number of thermal photons* in Chap. 8. These states are very important because every Gaussian state can be decomposed into thermal states (see Proposition 11.13 of the next section). They were applied in Chap. 8 to take into account for the presence of thermal noise in Quantum Communications systems.

The density operator corresponding to the characteristic function (11.140) can be evaluated by Glauber's inversion formula (11.90) or by formula (11.110) giving the matrix representation of ρ. But, considering that the characteristic function has the form $\chi(\xi) = f(|\xi|^2)$, it is convenient to use the more direct formula (see Problem 11.11)

$$\rho_{mn} = \delta_{mn} \int_0^\infty dx\, e^{-\frac{1}{2}x} f(x)\, L_n(x) \tag{11.141}$$

with $f(x) = \exp(-\frac{1}{2}(2\mathcal{N}+1)x)$, so that

$$\rho_{mn} = \delta_{mn} \int_0^\infty dx\, e^{-(\mathcal{N}+1)x}\, L_n(x)$$

where we can use the integral [25, p. 809] $\int_0^\infty e^{-bx} L_n(x)\, dx = (b-1)^n/b^{n+1}$ to get

$$\rho_{mn} = \delta_{mn} \frac{\mathcal{N}^n}{(\mathcal{N}+1)^{n+1}}.$$

Hence, the Fock expansion of the density operator reads as

$$\rho_{\text{th}} = \sum_{n=0}^\infty \frac{\mathcal{N}^n}{(\mathcal{N}+1)^{n+1}}\, |n\rangle\langle n| \tag{11.142}$$

which establishes that thermal states have a *geometrical distribution* with mean $\mathcal{N}$ (see (8.4)).

In conclusion, thermal states are Gaussian states with zero mean and covariance matrix $V = (2\mathcal{N}+1)I_2$.

Problem 11.10 ★★★ ▽ Thermal states are defined as the bosonic states that maximize the von Neumann entropy for a fixed energy. Prove this statement using Lagrange multipliers.

---

[14] Thermal states are defined as the bosonic states that maximize the von Neumann entropy for a fixed energy, as pointed out by Weedbrook et al. [1].

Problem 11.11 ⋆⋆   Prove that, if the characteristic function $\chi(\xi)$ depends only on $|\xi|^2$, say $\chi(\xi) = f(|\xi|^2)$, the reconstruction formula (11.110) of Proposition 11.7) is simplified as

$$\rho_{nm} = \delta_{mn} \int_0^{\infty} \mathrm{d}x \; \mathrm{e}^{-\frac{1}{2}x} f(x) \, L_n(x) \tag{11.143}$$

where $L_n(x)$ is the ordinary Laguerre polynomial.

Problem 11.12 ⋆⋆   Consider the alternative definition of a coherent state given by (11.138). Show that the Fock representation of $|\alpha\rangle$ is still given by (11.46).

11.10 Gaussian Transformations and Gaussian Unitaries

In this section we restart the theory on continuous variables in the general N-mode and we will consider *quantum transformations* or *quantum operations*, which map the state of the system ρ into a new state $\tilde{\rho}$

$$\rho \;\rightarrow\; \tilde{\rho} = \Phi(\rho). \tag{11.144}$$

As we shall see in the next chapter (Sect. 12.8), a quantum transformation defines a *quantum channel*, which refers in general to an open quantum system. In closed quantum systems the map (11.144) is provided by a unitary transformation according to

$$\rho \;\rightarrow\; \tilde{\rho} = U \rho \, U^* \tag{11.145}$$

in agreement with Postulate 3 of Quantum Mechanics.

Definition 11.2   A quantum transformation is **Gaussian** when *it transforms Gaussian states into Gaussian states*.[15]   When the Gaussian operation is performed according to the unitary map (11.145) it is called **Gaussian unitary**.

It can be shown [1] that Gaussian unitaries are generated in the form $U = \exp(-\mathrm{i}H/2)$, where H is a Hamiltonian, which are second-order polynomials in the field operators q_p or in the bosonic operators a and $a_\star$. The application of such unitaries to the annihilators $a = [a_1, \dots, a_N]^{\mathrm{T}}$ leads to a transformation called **Bogoliubov transformation**, and, in terms of quadrature operators q_p, to a **symplectic transformation**.

Below we shall see that the two kinds of transformations are equivalent and we shall find their relation. Also, we shall prove that they are really Gaussian transformations, which modify the mean vector and the covariance in the form

---

[15] Note the strong analogy with the theory of stochastic processes in linear systems, where the Gaussianity is preserved.

$$\overline{X} \to S\,\overline{X} + d, \qquad V \to S\,V\,S^{\mathrm{T}} \tag{11.146}$$

where $\overline{X} = \overline{q_p}$, S is a $2N \times 2N$ real matrix, and $d \in \mathbb{R}^{2N}$. This is the key result because it allows us to specify a Gaussian transformation in terms of the parameters (S, d), which "live" in the phase space $\mathbb{R}^{2N}$.

11.10.1 Definition of Bogoliubov and Symplectic Transformations

Definition 11.3 An N-mode **Bogoliubov transformation** has the form[16]

$$a \to U^* a\, U = E\,a + F\,a_\star + y \tag{11.147}$$

where E, F are $N \times N$ complex matrices and $y \in \mathbb{C}^N$. The conditions of preserving the commutation relations (11.65) are

$$E E^* - F F^* = I_N, \qquad E F^{\mathrm{T}} = F E^{\mathrm{T}}. \tag{11.148}$$

Definition 11.4 An N-mode **symplectic transformation** has the form

$$q_p \to S\,q_p + d \tag{11.149}$$

where S is a $2N \times 2N$ real matrix and $d \in \mathbb{R}^{2N}$. The condition of preserving the commutation relations (11.65) is

$$S\,\Omega\,S^{\mathrm{T}} = \Omega. \tag{11.150}$$

A matrix S that verifies this condition is called *symplectic*.

The conditions (11.148) and (11.150) will be developed below in the single mode. For the general case they will be proved in Appendix section "About Symplectic

---

[16] It is important to remark that, for convenience, some shorthand notations are used in the algebra of matrices of operators. In fact, a, $a_\star$ and z are $N \times 1$, E and F are $N \times N$, so that $E\,a + F\,a_\star + y$ is $N \times 1$. Then, $U^* a\, U$ must be $N \times 1$, and its correct interpretation is given by the column vector [26]

$$U^* a\, U = \begin{bmatrix} U^* a_1\, U \\ \vdots \\ U^* a_N\, U \end{bmatrix}.$$

Also, the identity operator is often omitted; in particular, y should be completed as $y \otimes I_{\mathcal{H}} = [y_1 I_{\mathcal{H}}, \ldots, y_N I_{\mathcal{H}}]^{\mathrm{T}}$.

and Bogoliubov Transformations". In the same appendix we also find the relations between the parameters of the two transformations.

Proposition 11.11 *The triplet* (E, F, y) *of a Bogoliubov transformation and the pair* (S, d) *of a symplectic transformation are related as*

$$
S = \Pi \begin{bmatrix} \Re(E + F) & \Im(-E + F) \\ \Im(E + F) & \Re(E - F) \end{bmatrix} \Pi^{\mathrm{T}}, \qquad d = \Pi \begin{bmatrix} \Re y \\ \Im y \end{bmatrix} \tag{11.151}
$$

where Π *is the* $2N \times 2N$ *permutation matrix whose entries are* $\Pi_{ij} = 1$ *for* $(i, j) = (2k - 1, k)$ *and for* $(i, j) = (2k, N + k)$ *for* $k = 1, \ldots, N$, *and* $\Pi_{ij} = 0$ *otherwise. Relations (11.151) are easily inverted considering that* $\Pi \Pi^{\mathrm{T}} = I_{2N}$.

We give explicitly the permutation matrices for the first two orders:

$$
N = 1 \qquad \Pi = \begin{bmatrix} 1 & 0 \\ 0 & 1 \end{bmatrix}, \qquad \Pi^{\mathrm{T}} = \begin{bmatrix} 1 & 0 \\ 0 & 1 \end{bmatrix},
$$

$$
N = 2 \qquad \Pi = \begin{bmatrix} 1 & 0 & 0 & 0 \\ 0 & 0 & 1 & 0 \\ 0 & 1 & 0 & 0 \\ 0 & 0 & 0 & 1 \end{bmatrix}, \qquad \Pi^{\mathrm{T}} = \begin{bmatrix} 1 & 0 & 0 & 0 \\ 0 & 0 & 1 & 0 \\ 0 & 1 & 0 & 0 \\ 0 & 0 & 0 & 1 \end{bmatrix}.
$$

A symplectic transformation, $q_p \to S q_p + d$, acting on the phase space $\mathbb{R}^{2N}$, corresponds to a unitary operator, say $U_{S,d}$.

Proposition 11.12 *The unitary operator* $U_{S,d}$, *related to the symplectic transformation* $q_p \to S q_p + d$, *can be written in the form*

$$
U_{S,d} = D(\alpha) U_S, \qquad d = \Pi[\Re\alpha, \Im\alpha]^{\mathrm{T}} \tag{11.152}
$$

where the unitary operator U_S *provides the map* $q_p \to S q_p$ *and the Weyl operator* $D(\alpha) = \exp(\alpha^{\mathrm{T}} a_\star - \alpha^* a)$ *provides the map* $q_p \to q_p + d$. *The displacement* $\alpha \in \mathbb{C}^N$ *is related to the displacement* $d \in \mathbb{R}^{2N}$ *by* $d = \Pi[\Re\alpha, \Im\alpha]^{\mathrm{T}}$.

In fact, we prove that the action of the Weyl operator gives

$$
D^*(\alpha) \, a \, D(\alpha) = a + \alpha \, I_{\mathcal{H}}. \tag{11.153}
$$

To this end we use the exponential identity (11.71) with $x = 1$, $H = a^* \alpha - \alpha^* a$, and $K = a$, and consider that $[H, K] = \alpha[a, a^*] = \alpha \, I_{\mathcal{H}}$. Therefore the nested commutators D_n in (11.71) are zero, and thus we obtain $D^*(\alpha) \, a \, D(\alpha) = a + \alpha$, where, as usual, the identity is omitted. The relation between α and d is given by (11.151).

In Appendix section "Proof of the Gaussianity Preservation Theorem (Theorem 11.1)" we shall prove the following fundamental statement:

Theorem 11.1 (Gaussianity preservation theorem) *Both Bogoliubov and symplectic transformations are Gaussian. They modify the mean vector $\overline{q_p} := \overline{X}$ and the covariance matrix according to*

$$\overline{q_p} \;\to\; S\,\overline{q_p} + d\,, \qquad V \;\to\; S\,V\,S^{\mathsf{T}}. \tag{11.154}$$

11.10.2 Williamson's Theorem and Thermal Decomposition

Multimode Gaussian states can be studied with a unified approach using Williamson's Theorem, which provides a powerful tool for the interpretation and also the generation of Gaussian states. The theorem is concerned with a very interesting decomposition of the covariance matrix, which, together with the mean value, completely specified a Gaussian state.

Theorem 11.2 (Williamson) *An N-mode covariance matrix V can be decomposed in the form*

$$\boxed{\,V = S_w\,V^{\oplus}\,S_w^{\mathsf{T}}\,, \qquad V^{\oplus} = \mathrm{diag}\,[\sigma_1^2, \sigma_1^2, \ldots, \sigma_N^2, \sigma_N^2]\,} \tag{11.155}$$

where S_w is a $2N \times 2N$ symplectic matrix and the σ_i^2 are positive real values, called the **symplectic eigenvalues** *of V. The decomposition is unique up to a permutation of the element of $V^{\oplus}$.*

The evaluation of the symplectic spectrum $\{\sigma_1^2, \ldots, \sigma_N^2\}$ can be carried out from the standard eigenspectrum of the matrix $\mathrm{i}\,\Omega\,V$. In fact, this matrix is Hermitian and its eigenspectrum has the form $\{\pm\sigma_1^2, \ldots, \pm\sigma_N^2\}$.

The symplectic decomposition (11.155) provides a powerful way to handle the properties of Gaussian states. In particular the uncertainty principle can be rewritten in the simple form [1]

$$V > 0\,, \qquad V^{\oplus} \geq I_{2N} \quad \to \quad \sigma_i^2 \geq 1. \tag{11.156}$$

From Williamson's theorem we find also:

Proposition 11.13 *An arbitrary N-mode Gaussian state can be generated by the* **tensor product of N single-mode thermal states**, *with covariance matrix*

$$V_k = \begin{bmatrix} \sigma_k^2 & 0 \\ 0 & \sigma_k^2 \end{bmatrix} = \sigma_k^2\, I_2$$

and number of thermal photons $N_k = \tfrac{1}{2}(\sigma_k^2 - 1)$.

In fact, we have seen in Sect. 11.9 that a single-mode thermal state is zero mean and specified by a covariance matrix $\sigma^2 I_2$, which is related to the average photon number by $\mathcal{N} = \frac{1}{2}(\sigma^2 - 1)$. Now, the N-mode Gaussian states with the diagonal covariance matrix $V^{\oplus}$, given by (11.155), can be written in the form[17]

$$\rho_{\text{th}}(0, V^{\oplus}) = \rho(0, \sigma_1^2 I_2) \otimes \cdots \otimes \rho(0, \sigma_N^2 I_2). \tag{11.157}$$

The Gaussian state with covariance matrix V is generated from $\rho_{\text{th}}(0, V^{\oplus})$ in the form

$$\rho(0, V) = U_S \, \rho_{\text{th}}(0, V^{\oplus}) \, U_S^* \tag{11.158}$$

where U_S is the unitary operator corresponding to the symplectic transformation provided by the matrix S_w of decomposition (11.155). A non zero-mean Gaussian state is generated by introducing in (11.158) an appropriate displacement operator.

From the above relations the determinant of the covariance matrix can be expressed in terms of the symplectic eigenvalues as

$$\det(V) = \sigma_1^4 \cdots \sigma_N^4. \tag{11.159}$$

In fact, recalling that the product of the (ordinary) eigenvalues of a matrix is given be the determinant (see (2.46)), we have $\det(i\,\Omega\,V) = (-1)^N \sigma_1^4 \cdots \sigma_N^4$, where $\det(i\,\Omega\,V) = \det(i\,\Omega)\det(V) = (-1)^N \det(V)$.

A single-mode thermal state with number of thermal photons $\mathcal{N} = \frac{1}{2}(\sigma^2 - 1)$ degenerates to the vacuum state $|0\rangle$ when $\mathcal{N} = 0 \;\rightarrow\; \sigma^2 = 1$. Then Proposition 11.13 gives:

Proposition 11.14 *An N-mode Gaussian state becomes pure when $\det(V) = 1$, that is, when all its symplectic eigenvalues are unitary. It can be generated from the tensor product of the ground states of each mode.*

This statement is in agreement with the conclusion of Sect. 11.7.7.

11.11 Gaussian Transformations in the N-Mode

We recall the notations: ^{T} for transpose, * for complex conjugate of a number, transpose conjugate of a vector and of a matrix, and for the adjoint of an operator, $\star$ for the conjugate of a vector. Then for a column vector of length N we have

---

[17] In general, if a covariance matrix has the block diagonal form $V = \text{diag}\,[V_1, V_2]$, the state $\rho(0, V)$ is factored in the form $\rho(0, V_1 \otimes V_2) = \rho(0, V_1) \otimes \rho(0, V_2)$. This identity can be proved, e.g., using reconstruction formula (11.90).

$$x = \begin{bmatrix} x_1 \\ \vdots \\ x_N \end{bmatrix}, \qquad x^{\mathrm{T}} = [x_1, \ldots, x_N], \qquad x^* = [x_1^*, \ldots, x_N^*], \qquad x_\star = \begin{bmatrix} x_1^* \\ \vdots \\ x_N^* \end{bmatrix} = (x^*)^{\mathrm{T}}.$$

In the literature we find a plethora of forms for the Gaussian unitaries in single-mode, in two-mode and in the general N-mode. Here we follow the unified form developed by Ma and Rhodes [26] for the N-mode. This form, using appropriate matrix notations, turns out to have extremely similar algebraic properties as that of the single mode.

There are only three fundamental Gaussian unitaries, which are specified by the following unitary operators[18]:

(1) N-mode displacement operator

$$D_N(\alpha) := e^{\alpha^{\mathrm{T}} a_\star - \alpha^* a}, \qquad \alpha = [\alpha_1, \ldots, a_N]^{\mathrm{T}} \in \mathbb{C}^N \tag{11.160}$$

which is the same as the Weyl operator (11.83).

(2) N-mode rotation operator

$$R_N(\phi) := e^{i a^* \phi a}, \qquad \phi \quad N \times N \text{ Hermitian matrix}. \tag{11.161}$$

(3) N-mode squeeze operator[19]

$$Z_N(z) := e^{\frac{1}{2}[(a^* z a_\star - a^{\mathrm{T}} z^* a)]}, \qquad z \quad N \times N \text{ symmetric matrix}. \tag{11.162}$$

Combination of these operators allows us to get all Gaussian transformations. The corresponding Gaussian states are typically generated by applying these transformations to replicas of vacuum states and/or of coherent states.

We now list the most popular operators encountered in the literature, obtained as special cases of the fundamental unitary operators:

---

[18] The notations used at the exponentials ensure a "scalar" combination of the bosonic operators. For instance in the displacement operator we have explicitly

$$\alpha^{\mathrm{T}} a_\star - \alpha^* a = \sum_i (\alpha_i a_i^* - \alpha^* a_i).$$

[19] The squeeze operator is usually denoted by the letter S, but this is in conflict with the notation used for the symplectic matrix encountered in symplectic transformations.

- **single-mode** and **two-mode displacement operators**

$$D(\alpha) = e^{\alpha a^* - \alpha^* a} := D(a, \alpha)$$
$$D_2(\alpha_1, \alpha_2) = e^{\alpha_1 a_1^* + \alpha_2 a_2^* - \alpha_1^* a_1 - \alpha_2^* a_2} = D(a_1, \alpha_1)\, D(a_2, \alpha_2). \tag{11.163}$$

- **single-mode rotation operator**, given by (11.161) with ϕ a real number

$$R(\phi) = e^{i\phi a^* a}. \tag{11.164}$$

- N-**mode rotation operator**, $R_N(\phi)$ with the matrix ϕ having the property $R_N(N\phi) = I_{\mathcal{H}}$; it is used to express the symmetry operator of the GUS (see Sect. 11.20.2).
- **beam splitter**, given by a two-mode rotation operator (11.161) with ϕ the Hermitian matrix

$$\phi = \begin{bmatrix} 0 & -i\beta \\ i\beta & 0 \end{bmatrix} \quad \rightarrow \quad R_2(\phi) = e^{\beta(a_1^* a_2 - a_2 a_1^*)} \tag{11.165}$$

in agreement with definition of [1, 7].
- **single-mode squeeze operator**, given by (11.161) with $z = re^{i\theta}$ a complex number

$$Z(z) = e^{\frac{1}{2}[z\, a^* a^* - z^* a\, a]}. \tag{11.166}$$

- **two-mode squeeze operator**, given by (11.161) with the symmetric matrix[20]

$$z = \begin{bmatrix} 0 & z_{12} \\ z_{12} & 0 \end{bmatrix} \quad \rightarrow \quad Z_2(z) = e^{\frac{1}{2}(z_{12} a_1^* a_2^* - z_{12}^* a_1 a_2)}. \tag{11.167}$$

Other forms of the 2×2 squeeze matrix z will be seen in Sect. 11.18 with the beam splitter.

11.11.1 Evaluation of Bogoliubov and Symplectic Transformations

The development of Gaussian transformations obtained with the fundamental operators is left to the problems. The convenient technique is to develop first the Bogoliubov

---

[20] In the literature we find slightly different definitions. For instance, for the squeeze operator in the single mode: in [7] $Z(z) = \exp[-z^* a_1 a_2 + z\, a_1^* a_2^*]$ with $z = re^{i\theta}$, in [1] $Z(z) = \exp[\frac{1}{2} r(a_1 a_2 - a_1^* a_2^*)]$ wit $z = r$, and in [27] $Z(z) = \exp[z^* a_1 a_2 - z\, a_1^* a_2^*]$ with $z = re^{i\theta}$. In the two-mode: in [28] $Z_2(z)$ is given by (11.167), with $z_{12} = -2re^{i2\theta}$.

transformation using the BCH identity and then the symplectic transformation using Proposition 11.11. Note that the BCH identity must be used in the form of N "scalar" relations

$$\mathrm{e}^{xH}\, a_k\, \mathrm{e}^{-xH} = \sum_{n=0}^{\infty} \frac{x^n}{n!} D_n(k)\,, \qquad k = 1, \dots, N \tag{11.168}$$

where $D_0(k) = a_k$ and $D_n(k) = [H, D_{n-1}(k)]$ for $n \geq 1$. The results are the following.

The displacement operator gives

$$D_N^*(\alpha)\, a\, D_N(\alpha) = a + \alpha.$$

Then, the symplectic matrix is the identity $S = I_{2N}$ and the displacement vector $d = \Pi\, [\Re\alpha, \Im\alpha]^{\mathrm{T}}$.

The rotation operator gives

$$R_N^*(\phi)\, a\, R_N(\phi) = \mathrm{e}^{\mathrm{i}\phi}\, a. \tag{11.169}$$

The symplectic matrix S is expressed by trigonometric functions of the matrix ϕ. Specifically,

$$S_{\mathrm{rot}}(\phi) = \Pi \begin{bmatrix} \cos\phi & -\sin\phi \\ \sin\phi & \cos\phi \end{bmatrix} \Pi^{\mathrm{T}}, \qquad d_{\mathrm{rot}} = \begin{bmatrix} 0 \\ 0 \end{bmatrix}. \tag{11.170}$$

To find the Bogoliubov transformation corresponding to the squeeze operator, a preliminary step is the *polar decomposition* $z = r\mathrm{e}^{\mathrm{i}\theta}$ of the squeeze matrix (see Theorem 2.7), where the matrix r is obtained as the square root of the PSD matrix $z\,z^*$ and the matrix $\mathrm{e}^{\mathrm{i}\theta}$ is obtained as $r^{-1}z$. Then the Bogoliubov transformation is given by[21]

$$Z^*(z)\, a\, Z(z) = \cosh r\, a + \sinh r\ \mathrm{e}^{\mathrm{i}\theta}\, a_\star \tag{11.171}$$

which is a special case of (11.147) with $E = \cosh r$, $F = \sinh r\mathrm{e}^{\mathrm{i}\theta}$, and $y = 0$. Hence the symplectic matrix is

$$S_{\mathrm{sq}}(r\mathrm{e}^{\mathrm{i}\theta}) = \Pi \begin{bmatrix} \cosh r + \sinh r \cos\theta & \sinh r \sin\theta \\ \sinh r \sin\theta & \cosh r - \sinh r \cos\theta \end{bmatrix} \Pi^{\mathrm{T}} \tag{11.172}$$

while $d_{\mathrm{sq}} = 0$. In particular for $\theta = 0$ the symplectic matrix becomes

$$S_{\mathrm{sq}}(r) = \Pi \begin{bmatrix} \mathrm{e}^r & 0 \\ 0 & \mathrm{e}^{-r} \end{bmatrix} \Pi^{\mathrm{T}}. \tag{11.173}$$

---

[21] Notations as $\sinh r\mathrm{e}^{\mathrm{i}\theta}$ should be intended as $(\sinh r)\,(\mathrm{e}^{\mathrm{i}\theta})$.

11.11.2 Combination of the Fundamental Unitaries

The three fundamental unitaries can be combined in several ways to get new Gaussian unitaries.

Given two Gaussian unitaries U_1 and U_2, the cascade of U_1 followed by U_2 gives the Gaussian unitary $U_{12} = U_2\,U_1$. For the Bogoliubov transformation of the cascade, using (11.147) twice, we find: if (E_i, F_i, y_i) is the triplet corresponding to U_i, the triplet (E_{12}, F_{12}, y_{12}) corresponding to U_{12} is given by

$$E_{12} = E_2 E_1 + F_2 F_1^*, \quad F_{12} = E_2 F_1 + F_2 E_1^*, \quad y_{12} = E_2 y_1 + F_2 y_1^* + y_2. \tag{11.174}$$

Analogously, for the pair (S_{12}, d_{12}) of the symplectic transformation, using the relation $X \to SX + d$ (see (11.149)), we find:

$$S_{12} = S_2\,S_1\,, \qquad d_{12} = S_1\,d_1 + d_2. \tag{11.175}$$

Note that in a cascade combination one can switch the order of operators with appropriate change in the parameters, as illustrated in Fig. 11.10

$$D_N(\alpha)\,Z_N(z) = Z_N(z)\,D_N(\beta) \quad \beta = \cosh r\,\alpha - \sinh r e^{i\theta}\,\alpha^* \tag{11.176a}$$

$$Z_N(z)\,R_N(\phi) = R_N(\phi)\,Z_N(z_0) \quad z_0 = e^{-i\phi} z\,e^{-i\phi^{\mathrm{T}}} \tag{11.176b}$$

$$D_N(\alpha)\,R_N(\phi) = R_N(\phi)\,D_N(\beta) \quad \beta = e^{-i\phi}\,\alpha. \tag{11.176c}$$

The proof can be done by operating directly with the Gaussian unitaries, by imposing the condition of the form $\tilde{U}_1\tilde{U}_2 = U_1 U_2$. In alternative one can use (11.174) or (11.175) (see Problem 11.19).

For the rotation operator we find

$$R_N(\phi_1)\,R_N(\phi_2) = R_N(\phi_3). \tag{11.177}$$

where the "phase" ϕ_3 is uniquely determined by the relation $e^{i\phi_3} = e^{i\phi_1}\,e^{i\phi_2}$. Note that in general $R_N(\phi_1)\,R_N(\phi_2) \neq R_N(\phi_1 + \phi_2)$, while the equality holds if and only if

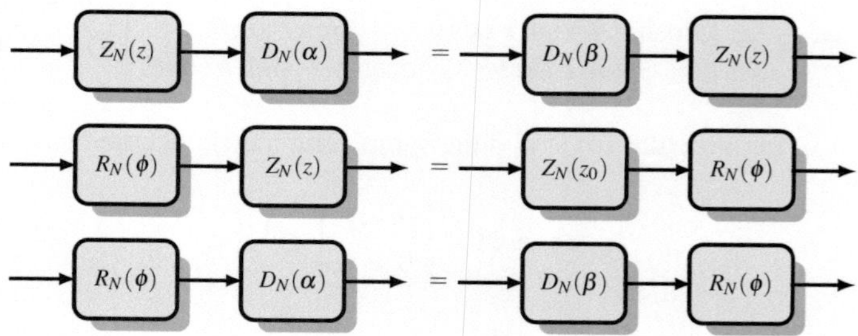

Fig. 11.10 The order of fundamental Gaussian unitaries can be inverted with appropriate modification of the parameters, as indicated by (11.176a), (11.176b) and (11.176c)

the Hermitian matrices ϕ_1 and ϕ_2 commute. Note the relevant case $R_N(\phi_0)\,R_N(\phi_0) = R_N(2\phi_0)$. For the proof the key is given by Theorem 2.5, which states that a unitary operator U can always be written in the form $U = e^{i\phi}$, with ϕ a Hermitian operator.

11.11.3 The Most General Gaussian Unitary

The importance of the fundamental unitaries lies in the following:

Theorem 11.3 *The most general Gaussian unitary is given by the combination of the three fundamental Gaussian unitaries $D_N(\alpha)$, $Z_N(z)$, and $R_N(\phi)$, cascaded in any arbitrary order, that is,*

$$Z_N(z)\,D_N(\alpha)\,R_N(\phi)\,, \qquad R_N(\phi)\,D_N(\alpha)\,Z_N(z), \quad \text{etc.}$$

The proof of this statement is easily found for the single mode by applying the singular-value decomposition to the 2×2 symplectic (real) matrix S (see [1] and Appendix section "The Most General Gaussian Unitary in the Single Mode"). For the general multimode the proof can be obtained using the Lie algebra. Specifically, Ma and Rhodes [26], generalizing a previous result obtained for the single mode [29, 30], proved that a unitary operator $e^{-iH/2}$, where H is a general N-mode quadratic Hamiltonian, can be written in the form

$$U = e^{i\gamma_N}\,Z_N(z)\,D_N(\alpha)\,R_N(\phi) \tag{11.178}$$

where $e^{i\gamma_N}$ is a phase factor (which is irrelevant for the state generation). On the other hand we can apply the switching rules (11.176) to change the order of the fundamental unitaries in (11.178), with appropriate modifications of the parameters. In other words, (11.178) can be written in six distinct orders.

It is important to remark that a Gaussian unitary is equivalent to the symplectic map (11.149) acting on the phase space $(\mathbb{R}^{2N}, \Omega)$, specified by the pair (S, d). As stated in Proposition 11.12, the Gaussian unitary can always be written in the form $U_{S,d} = D_N(\alpha)U_S$, where U_S corresponds to the map $q_p \to Sq_p$ and the displacement operator $D_N(\alpha)$ in the phase space provides the translation $q_p \to q_p + d$. By Theorem 11.3 U_S can be written as the cascade combination

$$U_S = Z_N(z)\,R_N(\phi) \quad \text{or} \quad U_S = R_N(\phi)\,Z_N(z).$$

11.11.4 Summary of Gaussian Transformations

The specification of a Gaussian transformation in the bosonic N-mode Hilbert space is provided by a unitary operator U, having the property (Gaussian unitary) that, if $\rho = \rho(\overline{X}, V)$ is a Gaussian state, also $U\,\rho\,U^*$ is a Gaussian state.

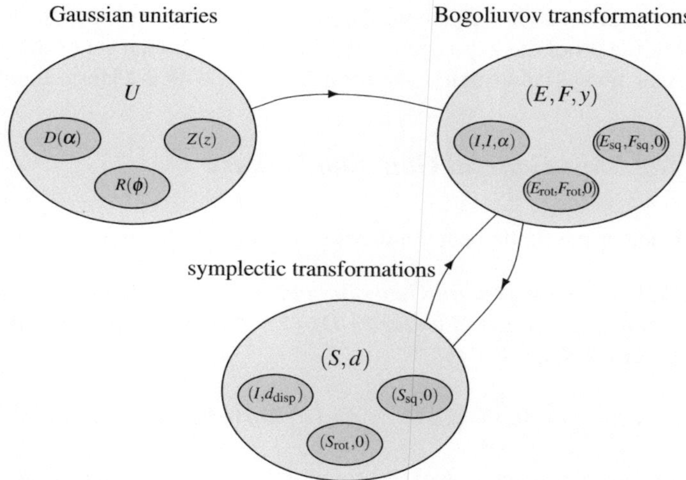

Fig. 11.11 Classes of Gaussian unitaries and related classes of Bogoliuvov and symplectic transformations. The *small ellipses* indicate the three basic Gaussian transformations

The application of the annihilation operator a to the Gaussian unitary in the form $U^* a U$ gives the Bogoliubov transformation (11.147), specified by the triplet (E, F, y) (Fig. 11.11). From the Bogoliubov transformations one gets the corresponding symplectic transformation (11.149), specified by the pair (S, d). The two kinds of transformations are equivalent to each other and in fact there is a one-to-one correspondence between the triplet (E, F, y) and the pair (S, d), as stated by Proposition 11.11. Theorem 11.1 ensures that both Bogoliubov and symplectic transformations are Gaussian and gives a simple relation between the parameters of the input and output Gaussian states.

In Theorem 11.3 we have seen that the three fundamental Gaussian unitaries (displacement, rotation, and squeezing) cover the whole class of Gaussian unitaries in the sense that their combination, in an arbitrary order, gives the most general Gaussian unitaries.

Problem 11.13 ⋆   Prove that the Bogoliubov transformation generated by the N-mode displacement operator is given by

$$D_N^*(\alpha)\, a\, D_N(\alpha) = a + \alpha.$$

Then evaluate the corresponding symplectic matrix.

Problem 11.14 ⋆⋆⋆   Prove that the N-mode rotation operator (11.161) produces the Bogoliubov transformation

$$R_N(\phi)\, a\, R_N(\phi) = \mathrm{e}^{\mathrm{i}\phi}\, a.$$

Problem 11.15 ★★   Write explicitly the symplectic matrix of a two-mode rotation operator in the two cases of matrix ϕ

$$\phi = \begin{bmatrix} \phi_0 & 0 \\ 0 & \phi_0 \end{bmatrix}, \qquad \phi = \begin{bmatrix} 0 & \phi_0 \\ \phi_0 & 0 \end{bmatrix} \qquad (\phi_0 \ \text{real}).$$

Hint: use identities (11.74) for the exponential and the expressions of Π and Π^{T} given after Proposition 11.11.

Problem 11.16 ★★★   Prove that the N-mode squeeze operator produces the Bogoliubov transformation

$$Z_N^*(z)\, a\, Z_N(z) = \cosh r\, a + \sinh r\, \mathrm{e}^{\mathrm{i}\theta}\, a_\star, \tag{11.179}$$

where the symmetric matrix z is written in the polar form $z = r\, \mathrm{e}^{\mathrm{i}\theta}$.

11.12 N-Mode Gaussian States

Williamson's theorem states that all Gaussian states can be obtained from thermal states through Gaussian unitaries. On the other hand, we have seen that all Gaussian unitaries can be obtained as cascade combination of the three fundamental unitaries (displacement, rotation, and squeezing). By elaboration of these ideas we shall see that the most general Gaussian states can be generated from thermal noise by application of the three fundamental Gaussian unitaries. For pure states the result is much simpler in that the rotation can be dropped.

11.12.1   The Most General Mixed Gaussian State

The proof of Theorem 11.3 on the most general Gaussian unitary is obtained using the Lie algebra related the bosonic Hilbert space $\mathcal{H}^{\otimes N}$. The search for the most general Gaussian state requires to work both in the Hilbert space and in the phase space.

We have seen that the specification of a Gaussian state is given by the pair $(\overline{X}, V)$, where the mean vector $\overline{X}$ and the covariance matrix V can be handled separately. According to Proposition 11.13, the application of Williamson's theorem gives the so called *thermal decomposition* of a Gaussian state, as

$$\rho(0, V) = U_S\, \rho_{\mathrm{th}}(0, V^{\oplus})\, U_S^* \tag{11.180}$$

where U_S is the unitary operator corresponding to the symplectic matrix S_w of decomposition (11.155) and

$$\rho_{\text{th}}\,(0,\,V^{\oplus}) = \rho(0,\,\sigma_1^2\,I_2) \otimes \cdots \otimes \rho(0,\,\sigma_N^2\,I_2). \tag{11.181}$$

A non zero-mean Gaussian state is generated by introducing in (11.180) an appropriate displacement operator. In conclusion, by combination of the previous statements:

Theorem 11.4 *The most general N-mode Gaussian state is generated from thermal state (11.181) by application of the three fundamental unitaries as*

$$\rho(d,\,V) = D_N(\alpha)\,R_N(\phi)\,Z_N(z)\,\rho_{\text{th}}\,(0,\,V^{\oplus})\,Z_N^*(z)\,R_N^*(\phi)\,D_N^*(\alpha) \tag{11.182}$$

where the order $D_N(\alpha)\,R_N(\phi)\,Z_N(z)$ can be arbitrarily changed with suitable modification of the arguments.

In (11.182) the vector $d \in \mathbb{R}^{2N}$ is related to the vector $\alpha \in \mathbb{C}^N$ as stated by Proposition 11.12, that is, $d = \Pi[\Re\alpha,\,\Im\alpha]^{\text{T}}$.

11.12.2 The Most General Pure Gaussian State

Theorem 11.4 holds in general for mixed states. For pure states, where all the symplectic eigenvalues are unitary, the thermal state degenerates to the product of N replicas of the vacuum state $|0\rangle$, say $|0_N\rangle$, and then $\rho(d,\,V) = |\psi(d,\,V)\rangle\langle\psi(d,\,V)|$, with

$$|\psi(d,\,V)\rangle = D_N(\alpha)\,R_N(\phi)\,Z_N(z)\,|0_N\rangle.$$

But we can invert the order of the squeezing and the rotation with the change $z \rightarrow e^{i\phi}\,z\,e^{i\phi^{\text{T}}}$, where, after the change, $R_N(\phi)|0_N\rangle = |0_N\rangle$. In conclusion:

Corollary 11.1 *The most general N-mode pure Gaussian state is obtained from the N replica $|0_N\rangle$ of the vacuum as*

$$|\psi(d,\,V)\rangle = D_N(\alpha)\,Z_N(z)\,|0_N\rangle. \tag{11.183}$$

In words, the most general N-mode Gaussian pure state is a squeezed–displaced state (or a displaced–squeezed state). This result is important for Quantum Communications, where the information carrier is provided by Gaussian states, so that the most general analysis of a Quantum Communications system can be limited to squeezed–displaced states.

To emphasize the generation of pure Gaussian states we introduce the following notations:

- $|\alpha\rangle = D_N(\alpha)\,|0_N\rangle$:   displaced (or coherent) state
- $|z\rangle = Z_N(z)\,|0_N\rangle$:   squeezed state
- $|\alpha,\,\phi\rangle = R_N(\phi)\,D_N(\alpha)\,|0_N\rangle$:   displaced–rotated state

- $|z, \alpha\rangle = D_N(\alpha)\, Z_N(z)\, |0_N\rangle$: squeezed–displaced state
- $|z, \alpha, \phi\rangle = R_N(\phi)\, D_N(\alpha)\, Z_N(z)\, |0_N\rangle$: squeezed–displaced–rotated state,

where the order of the terms refers to the generation, e.g., "squeezed–displaced" means that first the squeeze operator is applied and then the displacement operator.

Using these terms we can say that **the class of squeezed–displaced states coincides with the class of pure Gaussian states**, as established by Corollary 11.1. In particular, it is easy to see that a squeezed–displaced–rotated $|z, \alpha, \phi\rangle$ is a squeezed–displaced state with the following modification of the arguments (see Theorem 11.8):

$$|z, \alpha, \phi\rangle = \left| e^{i\phi} z\, e^{i\phi^{\mathrm{T}}}, \alpha\, e^{i\phi} \right\rangle.$$

In words, the rotation is absorbed by the squeezing and displacement.

11.12.3 Covariance Matrix of an N-Mode Gaussian State

A general procedure to evaluate the covariance matrix of the Gaussian state $|z, \alpha\rangle = D_N(\alpha)\, Z(z)|0_N\rangle$ is based on Gaussian preservation theorem (Theorem 11.1), which states that a symplectic transformation with matrix S modifies the covariance matrix as $V \;\to\; S\,V\,S^{\mathrm{T}}$. Considering that the covariance matrix of the vacuum state is the identity I_{2N}, we find

$$V = S_{\mathrm{disp}}\, S_{\mathrm{sq}}\, I_{2N}\, S_{\mathrm{sq}}^{\mathrm{T}}\, S_{\mathrm{disp}}^{\mathrm{T}} = S_{\mathrm{sq}}\, S_{\mathrm{sq}}^{\mathrm{T}} \tag{11.184}$$

where the displacement $D_N(\alpha)$ does not give any contribution because $S_{\mathrm{disp}} = I_{2N}$. The symplectic matrix S_{sq} is given by (11.172) and the covariance matrix can be written in the form (see (11.104) and [26])

$$V = \Pi \begin{bmatrix} Y_{qq} & Y_{qp} \\ Y_{qp}^{\mathrm{T}} & Y_{pp} \end{bmatrix} \Pi^{\mathrm{T}} \tag{11.185}$$

where

$$\begin{aligned}
Y_{qq} &= \frac{1}{2}\left[\cosh 2r + \sinh 2r\, e^{i\theta} + \cosh 2r^{\mathrm{T}} + \sinh 2r^{\mathrm{T}}\, e^{-i\theta^{\mathrm{T}}}\right] \\
Y_{pp} &= \frac{1}{2}\left[\cosh 2r - \sinh 2r\, e^{i\theta} + \cosh 2r^{\mathrm{T}} - \sinh 2r^{\mathrm{T}}\, e^{-i\theta^{\mathrm{T}}}\right] \\
Y_{qp} &= \frac{1}{2i}\left[-\cosh 2r + \sinh 2r\, e^{i\theta} + \cosh 2r^{\mathrm{T}} - \sinh 2r^{\mathrm{T}}\, e^{-i\theta^{\mathrm{T}}}\right]
\end{aligned} \tag{11.185a}$$

satisfying

$$Y_{qq}Y_{pp} = I_N + Y_{qp}^2 \ . \tag{11.186}$$

When the squeeze matrix is real and symmetric: $z = z^* \geq 0$, that is, $e^{i\theta} = e^{-i\theta}$ and $r = r^{\mathrm{T}} = (z^2)^{1/2}$, we have

$$Y_{qp} = 0 \, , \qquad Y_{qq}Y_{pp} = I_N \tag{11.187}$$

and the N-mode Gaussian state is a minimum uncertainty state in the sense of (11.187).

11.13 Normal Ordering of Gaussian Unitaries ⇓

In Quantum Mechanics a product of operators is usually said to be normal-ordered (also called Wick-ordered) when all creation operators are to the left of all annihilation operators in the product. The process of putting a product into normal order is called *normal ordering*. In Sect. 11.8 we saw an example of normal ordering related to the single-mode Weyl operator. In this section we extend this ordering to N-mode Gaussian unitaries in a complete general form, following Ma and Rhodes [26]. These authors used a very sophisticated approach based on Lie algebra for the deduction of the general result. An alternative approach is developed in the book by Louisell [11].

The usefulness of the normal ordering is mainly related to the generation of quantum states from the vacuum state $|0_N\rangle$ and it is based on the property that, if U is an exponential operator of the form $U = \exp[f(a, a_\star)\,a]$, where $f(a, a_\star)$ is an arbitrary function of the bosonic operators a and $a_\star$, then

$$\boxed{\exp[f(a, a_\star)\,a]\,|0_N\rangle = |0_N\rangle.} \tag{11.188}$$

In words, when U acts on the vacuum state, it can be replaced by the identity. For the proof of (11.188) it is sufficient to expand the exponential and then apply the identity $a_i^k|0\rangle = 0$ to each mode (see (11.57)).

Notation. The normally ordered form of a product $f(a, a^*)$ of annihilation and creation operators is denoted by $: f(a, a^*) :$. It denotes an operator coinciding with $f(a, a^*)$ but ordered in such a way that the creation operators appears at the left of the annihilation operators. The normally ordered form is obtained by repeated applications of the identity $a\,a^* = 1 + a^*a$. For instance: $a^*aa^* = a^*(1 + a^*a) = a^* + a^*a^*a$. Hence $: a^*aa^* := a^* + a^*a^*a$.

11.13.1 *Ordering of the Most General Gaussian Unitary*

We have seen that the most general Gaussian unitaries is given by the cascade combination of the three fundamental unitaries. The normal-ordered form of a rotated–displaced–squeezed unitary reads as

$$Z_N(z)\, D_N(\alpha)\, R_N(\phi) = K_0\, B(\alpha, a)\, C(a)\, F(\alpha, a) \tag{11.189}$$

where
$$K_0 := |\det S|^{1/2} \exp\left[-\tfrac{1}{2}(\alpha^*\alpha + \alpha^{\mathrm{T}} T^*\alpha)\right] \in \mathbb{C} \tag{11.189a}$$

and

$$B(\alpha, a) := \exp\left[\alpha^{\mathrm{T}} S^{\mathrm{T}} a_\star + \tfrac{1}{2} a^* T\, a_\star\right]$$

$$C(a) := \sum_{n=0}^{\infty} \frac{:[a^*(S\,\mathrm{e}^{\mathrm{i}\phi} - I_N)\,a]^n:}{n!} \qquad :: \text{ over all ordered terms} \tag{11.189b}$$

$$F(\alpha, a) := \exp\left[-(\alpha^{\mathrm{T}} T^* + \alpha^*)\,\mathrm{e}^{\mathrm{i}\phi}\, a - \tfrac{1}{2} a^{\mathrm{T}} \mathrm{e}^{\mathrm{i}\phi^{\mathrm{T}}} T^* \mathrm{e}^{\mathrm{i}\phi}\, a\right]$$

with z, S, and T the $N \times N$ matrices

$$z = r\mathrm{e}^{\mathrm{i}\theta}, \qquad S = \operatorname{sech} r, \qquad T = \tanh r\, \mathrm{e}^{\mathrm{i}\theta}. \tag{11.189c}$$

When we are interested in other permutations of the unitaries we can apply the switching rules (11.176) to get the corresponding ordered form. Here we develop the squeezed–displaced unitary $D_N(\alpha)\, Z_N(z)$, which is obtained from (11.189) by dropping the rotation operator and switching the order of squeeze and displacement. Then we get

$$D_N(\alpha)\, Z_N(z) = K_0\, B(\beta, a)\, C(a)\, F(\beta, a) \tag{11.190}$$

where
$$\beta := \cosh r\, \alpha - \sinh r\mathrm{e}^{\mathrm{i}\theta}\, \alpha^*$$
$$K_0 := |\det S|^{1/2} \exp\left[-\tfrac{1}{2}(\beta^*\beta + \beta^{\mathrm{T}} T^*\beta)\right] \tag{11.190a}$$

and
$$B(\beta, a) := \exp\left[\beta^{\mathrm{T}} S^{\mathrm{T}} a_\star + \tfrac{1}{2} a^* T\, a_\star\right]$$
$$C(a) := \exp[a^* \log(S)\, a] \tag{11.190b}$$
$$F(\beta, a) := \exp\left[-(\beta^{\mathrm{T}} T^* + \beta^*)\, a - \tfrac{1}{2} a^{\mathrm{T}} T^*\, a\right].$$

Note that property (11.188) holds for both $C(a)$ and $F(\beta, a)$.

11.13.2 Fock Expansion of the Most General Pure Gaussian State

The most general pure Gaussian state is given by

$$\boxed{|z, \alpha\rangle = D_N(\alpha)\, Z_N(z)\, |0_N\rangle} \tag{11.191}$$

where $|0_N\rangle$ is the N-mode vacuum state. The Fock expansion reads as

$$|z, \alpha\rangle = \sum_{n=0}^{\infty} |z, \alpha\rangle_n |n\rangle \tag{11.192}$$

where the Fourier coefficients are given by

$$|z, \alpha\rangle_n := \langle n | D_N(\alpha)\, Z_N(z)\, |0_N\rangle. \tag{11.193}$$

In (11.192) the summation is over all the N-tuples $n = (n_1, \ldots, n_N)$ of nonnegative integers and $|n\rangle = |n_1\rangle_1 \otimes \cdots \otimes |n_N\rangle_N := |n_1\rangle_1 \cdots |n_N\rangle_N$ are the Fock numbers in the N-mode.

The Fock expansion allows us to get an explicit form of the quantum state and the Fourier coefficients are seen as probability amplitudes. Specifically

$$p_n(k) := \mathrm{P}[n_1 = k_1, \ldots, n_N = k_N] = \big|\, |z, \alpha\rangle_k \,\big|^2 \tag{11.194}$$

gives the joint probability of the presence of k_1 photons in the first mode, of k_2 photons in the second mode, etc.

For the evaluation of the Fock expansion it is convenient to adopt the normal-ordered form (11.190) together with property (11.188). In fact, in (11.190b) both $C(a)$ and $F(\beta, a)$ are normal-ordered so that (11.191) gives

$$|z, \alpha\rangle = K_0\, B(\beta, a)\, |0_N\rangle = K_0 \exp\left[\alpha^{\mathrm{T}} S^{\mathrm{T}} a_\star + \tfrac{1}{2} a^* T a_\star\right] |0_N\rangle \tag{11.195}$$

where K_0 is a complex constant. Thus we only need to develop the exponential operator $B(\beta, a)$. The expansion of this exponential leads to a multivariate form of the type

$$B(\beta, a) = \sum_{n=0}^{\infty} b(n_1, \ldots, n_N)(a_1^*)^{n_1} \cdots (a_N^*)^{n_N} \tag{11.196}$$

where, to each power of the creator operators we can apply the identity (11.125), that is, $(a^*)^n |0\rangle = \sqrt{n!}\,|n\rangle$, to get from (11.191)

$$|z, \alpha\rangle = K_0 \sum_{n=0}^{\infty} b(n_1, \ldots, n_N)\sqrt{n_1! \cdots n_N!}\, |n_1\rangle \cdots |n_N\rangle \qquad (11.197)$$

which gives the Fock expansion of a general N-mode Gaussian state.

It remains to find the coefficients in (11.196). A general solution may be the following: in the expression of $B(\beta, a)$ we replace the creators by complex variables, say $a_i^* \rightarrow u_i$, so that $B(\beta, u)$ becomes a *generating function*. Hence we can apply the Taylor expansion to $B(\beta, u)$ to capture the coefficients as

$$b(n_1, \ldots, n_N) = \frac{1}{n_1! \cdots n_N!} \frac{\partial^{n_1 + \cdots + n_N} B(\beta, u)}{\partial u_1^{n_1} \cdots \partial u_N^{n_N}}\Bigg|_{u_1 = 0, \ldots, u_N = 0}.$$

This holds in the general case. As we shall see, in the specific cases, the generating function can be related to well known generating functions, as the ones of Laguerre and Hermite polynomials, thus achieving synthetic results.

We have presented the application of the normal-ordered form of Gaussian unitaries in the general N-mode. In the following sections the theory will be applied to the single mode and to the two-mode.

11.14 Gaussian Transformations in the Single Mode

In this section we develop for the single mode the Gaussian transformations introduced in Sect. 11.11, arriving at more specific results. We note that, having followed the unified form of Ma and Rhodes, the expressions for the single mode are practically identical to the ones of the N-mode, with slight simplifications.

As an exercise, a few results established for the general N-mode are re-established for the single mode.

11.14.1 Bogoliubov and Symplectic Transformations

In the single mode the Bogoliubov transformation (11.147) takes the form

$$\tilde{a} = U^* a U = E a + F a^* + y \qquad (11.198)$$

where E, F, and y are complex scalars. The commutation condition (11.148) reads as

$$|E|^2 - |F|^2 = 1. \qquad (11.199)$$

(see Problem 11.17 for a proof).

The symplectic transformation (11.149) becomes explicitly

$$\begin{bmatrix} \tilde{q} \\ \tilde{p} \end{bmatrix} = \begin{bmatrix} S_{11} & S_{12} \\ S_{21} & S_{22} \end{bmatrix} \begin{bmatrix} q \\ p \end{bmatrix} + \begin{bmatrix} d_1 \\ d_2 \end{bmatrix} \tag{11.200}$$

and the commutation condition (11.150) becomes

$$S\,\Omega\,S^{\mathrm{T}} = \begin{bmatrix} 0 & \det S \\ -\det S & 0 \end{bmatrix} = \begin{bmatrix} 0 & 1 \\ -1 & 0 \end{bmatrix}$$

with $\det S = 1$.

To find the relations between the parameters of the two transformations, in (11.198) we express the bosonic operators a and a^* in terms of the quadrature operators q and p, using (11.55), and we write the complex variables in the form $E = E_1 + iE_2$, $F = F_1 + iF_2$, and $y = y_1 + iy_2$. Then we find

$$\tilde{q} + \mathrm{i}\,\tilde{p} = (E_1 + \mathrm{i}\,E_2)(q + \mathrm{i}\,p) + (F_1 + \mathrm{i}\,F_2)(q - \mathrm{i}\,p) + y_1 + \mathrm{i}\,y_2$$

and hence

$$\begin{aligned} \tilde{q} &= (E_1 + F_1)\,q + (-E_2 + F_2)\,p + y_1 \\ \tilde{p} &= (E_2 + F_2)\,q + (E_1 - F_1)\,p) + y_2, \end{aligned} \tag{11.201}$$

which can be written in the symplectic form (11.200) with

$$S = \begin{bmatrix} \Re(E + F) & \Im(-E + F) \\ \Im(E + F) & \Re(E - F) \end{bmatrix}, \qquad d = \begin{bmatrix} y_1 \\ y_2 \end{bmatrix}. \tag{11.202}$$

11.14.2 *Williamson's Theorem and Thermal Decomposition*

For the single-mode, Williamson's theorem reads as

Theorem 11.5 (Williamson) *A single-mode covariance matrix can be decomposed in the form*

$$V = S_w\,\sigma^2\,I_2\,S_w^{\mathrm{T}} = \sigma^2\,S_w\,S_w^{\mathrm{T}} \tag{11.203}$$

where σ^2 is the symplectic eigenvalue of V.

Explicitly, with

$$V = \begin{bmatrix} v_{11} & v_{12} \\ v_{12} & v_{22} \end{bmatrix}, \qquad S_w = \begin{bmatrix} s_{11} & s_{12} \\ s_{21} & s_{22} \end{bmatrix}$$

the theorem states

$$\sigma^2 S_w S_w^{\mathrm{T}} = \sigma^2 \begin{bmatrix} s_{11}^2 + s_{12}^2 & s_{11}s_{21} + s_{12}s_{22} \\ s_{11}s_{21} + s_{12}s_{22} & s_{21}^2 + s_{22}^2 \end{bmatrix} = \begin{bmatrix} v_{11} & v_{12} \\ v_{12} & v_{22} \end{bmatrix}. \tag{11.204}$$

The symplectic condition for S_w gives

$$S_w \Omega S_w^{\mathrm{T}} = \begin{bmatrix} 0 & s_{11}s_{22} - s_{12}s_{21} \\ s_{12}s_{21} - s_{11}s_{22} & 0 \end{bmatrix} = \begin{bmatrix} 0 & 1 \\ -1 & 0 \end{bmatrix} = \Omega$$

which leads to the condition $\det(S_w) = s_{11}s_{22} - s_{12}s_{21} = 1$. The symplectic eigenvalue σ^2 is given by the positive (ordinary) eigenvalue of the matrix $i\Omega V$ and results in $\sigma^2 = \sqrt{v_{11}v_{22} - v_{12}^2} = \sqrt{\det(V)}$.

Now we can check that the matrix S_w is given by

$$\boxed{S_w = \sigma^{-1} V^{1/2}.} \tag{11.205}$$

Then, in general, to find S_w we have to evaluate the eigendecomposition $V = U \Lambda U^*$, which gives $V^{1/2} = U \Lambda^{1/2} U^*$.

As an example, consider the covariance matrix of a squeezed state, given by (11.238). In this case it is immediate to find that $\sigma = 1$ and

$$S_w = \begin{bmatrix} e^r & 0 \\ 0 & e^{-r} \end{bmatrix}.$$

11.14.3 Fundamental Gaussian Unitaries and Transformations

From the general definitions of Sect. 11.11 one gets:

(1) single-mode displacement operator

$$D(\alpha) = e^{\alpha a^* - \alpha^* a}, \qquad \alpha \in \mathbb{C}. \tag{11.206}$$

(2) single-mode rotation operator

$$R(\phi) = e^{i\phi a^* a}, \qquad \phi \in \mathbb{R}. \tag{11.207}$$

(3) single-mode squeeze operator

$$Z(z) = e^{\frac{1}{2}[z(a^*)^2 - z^* a^2]}, \qquad z = re^{i\theta} \in \mathbb{C}. \tag{11.208}$$

The combination of these operators allows us to get the most general single-mode Gaussian transformations.

As seen in Proposition 11.12, the **displacement operator** $U = D(\alpha)$ provides the Bogoliubov transformation

$$D^*(\alpha)\, a\, D(\alpha) = a + \alpha, .$$

The parameters of the symplectic map are

$$S_{\text{disp}} = \begin{bmatrix} 1 & 0 \\ 0 & 1 \end{bmatrix}, \qquad d_{\text{disp}} = \begin{bmatrix} \Re\alpha \\ \Im\alpha \end{bmatrix}. \tag{11.209}$$

Hence the displacement operator does not change the covariance matrix.

The **rotation operator** provides the Bogoliubov transformation (see Problem 11.18)

$$R^*(\phi)\, a\, R(\phi) = e^{i\phi}\, a \tag{11.210}$$

and the symplectic map $q\,p \to S_{\text{rot}}\, q\,p$ with

$$S_{\text{rot}}(\phi) = \begin{bmatrix} \cos\phi & -\sin\phi \\ \sin\phi & \cos\phi \end{bmatrix}, \qquad d_{\text{rot}} = \begin{bmatrix} 0 \\ 0 \end{bmatrix}. \tag{11.211}$$

From the **squeeze operator** with $z = re^{i\theta}$ one gets the Bogoliubov transformation (see Appendix section "Squeezed States. Proof of Bogoliubov Transformation (11.171)")

$$Z^*(z)\, a\, Z(z) = \cosh r\, a + \sinh r e^{i\theta}\, a^* \tag{11.212}$$

and the symplectic matrix is

$$S_{\text{sq}}(re^{i\theta}) = \begin{bmatrix} \cosh(r) + \cos(\theta)\sinh(r) & \sin(\theta)\sinh(r) \\ \sin(\theta)\sinh(r) & \cosh(r) - \cos(\theta)\sinh(r) \end{bmatrix} \tag{11.213}$$

while $d_{\text{sq}} = 0$. In particular for $\theta = 0$ the symplectic matrix is

$$S_{\text{sq}}(r) = \begin{bmatrix} e^r & 0 \\ 0 & e^{-r} \end{bmatrix}. \tag{11.214}$$

11.14.4 The Most General Single-Mode Gaussian Unitary

Theorem 11.3 gives for the single mode:

Theorem 11.6 *The most general Gaussian unitary in the single mode is given by the combination of the three fundamental Gaussian unitaries $D(\alpha)$, $Z(z)$, and $R(\phi)$,* *cascaded in an arbitrary order. Thus there are six possibilities:* $D(\alpha)\,Z(z)\,R(\phi)$, $Z(z)\,D(\alpha)\,R(\phi)$, *etc.*

In Appendix section "The Most General Gaussian Unitary in the Single Mode" we give a specific proof for the single mode, based on the SVD, which allows us to prove that the most general 2×2 symplectic matrix can be decomposed in the form

$$S = S_{\mathrm{rot}}(\theta)\,S_{\mathrm{sq}}(r)\,S_{\mathrm{rot}}(\phi). \tag{11.215}$$

Problem 11.17 ★   Prove in the single bosonic mode condition (11.148), which states that the commutation relation is preserved after a Bogoliubov transformation.

Hint Use the bilinearity of the commutator

$$[u_1 H_1 + u_2 H_2, v_1 K_1 + v_2 K_2] = u_1 v_1 [H_1, K_1] + u_1 v_2 [H_1, K_2] + u_2 v_1 [H_2, K_1] + u_2 v_2 [H_2, K_2] \tag{11.216}$$

where $u_i,\, v_j$ are complex numbers and $H_i,\, K_j$ are operators.

Problem 11.18 ★★   Prove that the rotation operator (11.207) produces the Bogoliubov transformation (11.210).

Problem 11.19 ★★   Prove that a squeezing followed by a displacement is equivalent to a displacement followed by a squeezing with the change of the displacement amount indicated in (11.176).

11.15 Single-Mode Gaussian States and Their Statistics

The general theory of the previous sections is particularized to the single mode with the target to find explicit results.

11.15.1 The Most General Gaussian State in the Single Mode

Theorem 11.4 formulates the most general Gaussian state in the N-mode. For the single mode the theorem is simplified as follows:

Theorem 11.7 *Any* **one-mode Gaussian state** *can be generated from the thermal state* $\rho_{\mathrm{th}}(\mathcal{N}) := \rho(0, \sigma^2)$ *by a squeezing* $Z(z)$ *followed by a displacement* $D(\alpha)$

$$\rho = D(\alpha)\,Z(z)\,\rho_{\mathrm{th}}(\mathcal{N})\,Z^*(z)\,D^*(\alpha) \tag{11.217}$$

where $\mathcal{N}$ *is the number thermal photons and* $\sigma^2 = 2\mathcal{N} + 1$ *is the variance of the thermal state.*

For the proof we start from the general form of Theorem 11.4, where, considering that the order of the fundamental operators is arbitrary, we choose rotation followed by squeezing, followed by displacement, to get

$$\rho = D(\alpha)\, Z(z)\, R(\phi)\, \rho_{\mathrm{th}}\,(\mathcal{N})\, R^*(\phi)\, Z^*(z)\, D^*(\alpha) := \rho_{\mathcal{N}}(z, \alpha).$$

Next we prove that the presence of the rotation is irrelevant (in the single mode). To see this, we consider the last relation in the phase space, where it modifies the covariance matrix of the thermal state $\sigma^2\, I_2$ in the form (recall that $S_{\mathrm{disp}}(\alpha)$ is the identity)

$$V = S_{\mathrm{sq}}(z)\, S_{\mathrm{rot}}(\phi)\, \sigma^2\, I_2\, S_{\mathrm{rot}}^{\mathrm{T}}(\phi)\, S_{\mathrm{sq}}^{\mathrm{T}}(z)$$

where $S_{\mathrm{rot}}(\phi)\, S_{\mathrm{rot}}^{\mathrm{T}}(\phi) = I_2$ and therefore $V = S_{\mathrm{sq}}(z)\, \sigma^2\, I_2\, S_{\mathrm{sq}}^{\mathrm{T}}(z)$. In words, the rotation is absorbed by the thermal state. The above result with the thermal state replaced by the vacuum state, is in agreement with Corollary 11.1:

Corollary 11.2 *The most general single-mode pure Gaussian state is a squeezed–displaced state.*

$$|z, \alpha\rangle = D(\alpha)\, Z(z)\, |0\rangle. \tag{11.218}$$

With $z, \alpha \in \mathbb{C}$ one can generate the whole class of pure Gaussian states.

11.15.2 Generality on Pure Gaussian States in the Single Mode

For convenience we recall the symbols introduced in Sect. 11.12, rewritten for the single mode:

- $|\alpha\rangle = D(\alpha)\, |0\rangle$: displaced (or coherent) state
- $|z\rangle = Z(z)\, |0\rangle$: squeezed state
- $|\alpha, \phi\rangle = R(\phi)\, D(\alpha)\, |0\rangle$: displaced–rotated state
- $|z, \alpha\rangle = D(\alpha)\, Z(z)\, |0\rangle$: squeezed–displaced state
- $|z, \alpha, \phi\rangle = R(\phi)\, D(\alpha)\, Z(z)|0\rangle$: squeezed–displaced–rotated state,

We want to find explicit results for each class of states, as the Fock representation, the mean value and the covariance matrix. We can do this for the class of squeezed–displaces states, which includes the other classes, but sometimes we prefer to begin with the simplest cases.

For the Fock expansion we use the general theory developed in the previous section, starting from the normal-ordered form, which reads for the single mode as

$$D(\alpha)\, Z(z) = K_0\, B(\beta, a)\, C(a)\, F(\beta, a) \tag{11.219}$$

where α and $z = re^{i\theta}$ are complex scalars,

$$\beta = \cosh r \, \alpha - e^{i\theta} \sinh r \, \alpha^* , \qquad \lambda = e^{i\theta} \tanh r$$
$$K_0 = (\operatorname{sech} r)^{1/2} \exp\left[-\tfrac{1}{2}(|\beta|^2 + \beta^2 \lambda)\right] \tag{11.219a}$$

and

$$B(\beta, a) := \exp\left[\beta \operatorname{sech} r \, a^* + \tfrac{1}{2}\lambda(a^*)^2\right]$$
$$C(a) := \exp[\log(e^{i\theta} \sinh r)\, a^* a] \tag{11.219b}$$
$$F(\beta, a) := \exp\left[-(\beta\lambda^* + \beta^*)\, a - \tfrac{1}{2}\lambda^* a^2\right].$$

To find the Fourier coefficients $|z, \alpha\rangle_n$ it is sufficient to get the expansion of the operator $B(\beta, a)$, which in the single mode has the form

$$B(\beta, a) = \exp\left[\beta \operatorname{sech} r \, a^* + \tfrac{1}{2}\lambda(a^*)^2\right] = \sum_{n=0}^{\infty} b(n)\, (a^*)^n. \tag{11.220}$$

The Fourier coefficients are then given by

$$|z, \alpha\rangle_n = K_0\, b(n)\, \sqrt{n!}.$$

The explicit form of (11.220) will be seen starting from for the particular cases and finally for the general case.

From the Fourier coefficients one gets the statistical parameters of interest, such as the probability distribution of the photon number, given by $p_n(i)|_{z, \alpha} := |\, |z, \alpha\rangle_i \,|^2$ and the average photon number, given by the two expressions

$$\bar{n}_{|z, \alpha\rangle} = \langle z, \alpha | a^* a | z, \alpha \rangle = \sum_{n=0}^{\infty} i\, p_n(i)|_{z, \alpha}. \tag{11.221}$$

From the symplectic transformation generated by the Gaussian unitary one obtains the symplectic matrix S and then the mean and the covariance matrix using (11.146).

11.15.3 Coherent States

The coherent states $|\alpha\rangle$, $\alpha \in \mathbb{C}$ are generated from the vacuum state $|0\rangle$ by application of the *displacement operator* (see Sect. 11.9.4), that is,

$$|\alpha\rangle = D(\alpha)\, |0\rangle = e^{\alpha a^* - \alpha^* a}\, |0\rangle . \tag{11.222}$$

Using the normal-ordered form of $D(\alpha)$, we can seen that the Fock representation has Fourier coefficients

$$|\alpha\rangle_n = e^{-\frac{1}{2}|\alpha|^2} \frac{\alpha^n}{\sqrt{n!}} \ |n\rangle \tag{11.223}$$

and that the photon number has a Poisson distribution with mean $\bar{n}_{|\alpha\rangle} = |\alpha|^2$. The distribution has been shown in Fig. 4.17.

We suggest the reader to evaluate (11.223) using the above general theory.

11.15.4 Rotated States and Displaced–Rotated States

The rotation operator applied to the vacuum state gives back the vacuum state

$$R(\phi)\,|0\rangle = e^{i\,a^* a}|0\rangle. \tag{11.224}$$

The rotation operator applied to a coherent state gives (see (11.210))

$$|\alpha, \phi\rangle = R(\phi)\,|\alpha\rangle = R(\phi)D(\alpha)|0\rangle = |e^{i\phi}\alpha\rangle.$$

The new state is still a coherent state identified by the complex number $e^{i\phi}\alpha$. Thus the Fock representation is given by (11.223) with α replaced by $e^{i\phi}\alpha$.

A phase rotation changes the covariance matrix in the form

$$V \quad \rightarrow \quad S_{\mathrm{rot}}(\phi)\, V\, S_{\mathrm{rot}}^{\mathrm{T}}(\phi)$$

where the symplectic matrix $S_{\mathrm{rot}}(\phi)$ is given by (11.170). But with a coherent state at the input (having $V = I_2$), the covariance does not change.

11.15.5 Squeezed States

Squeezed states have a fundamental role in the framework of continuous variables and therefore they will be seen in detail. The application of the squeeze operator to the vacuum state generates the **squeezed vacuum state**, briefly *squeezed state*, according to

$$|z\rangle = Z(z)\,|0\rangle = e^{\frac{1}{2}(z(a^*)^2 - z^* a^2)}\,|0\rangle \tag{11.225}$$

where $z = re^{i\theta} \in \mathbb{C}$ is the *squeeze factor*.

The evaluation of the Fock representation of the state $|z\rangle$ was developed by Yuen [31]. Here we follow the above general theory, which requires to find the expansion of $B(\beta, a) = \exp[\beta\,\mathrm{sech}\,r a^* + \frac{1}{2}\lambda(a^*)^2$ with $\alpha = \beta = 0$. We easily get

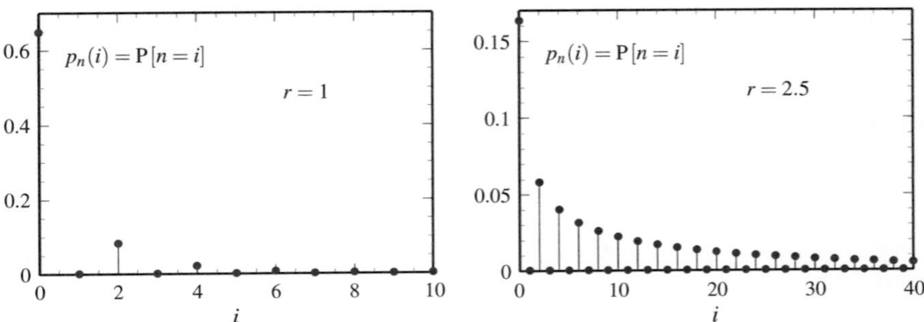

Fig. 11.12 Probability distribution of photon number in vacuum squeezed states, $p_n(i) = \mathrm{P}[n = i \mid z]$, for two real values of the squeeze factor z

$$B(0, a) = e^{\frac{1}{2}\lambda(a^*)^2} = \sum_{n=0}^{\infty} \frac{\lambda^n}{2^n n!}(a^*)^{2n}$$

so that $b(2n) = \lambda^n/(2^n n!)$ and $b(2n + 1) = 0$. Then

$$|z\rangle_{2n} = \sqrt{\operatorname{sech} r}\,\frac{\sqrt{(2n)!}}{2^n n!}\lambda^n \qquad \lambda = e^{i\theta}\tanh r \tag{11.226}$$

so that the squeezed state $|z\rangle$ is given by a linear combination of *even photon number states*. The photon statistics is described by the probability distribution

$$p_n(2i) = \mathrm{P}[n = 2i \mid z] = \operatorname{sech} r\,\frac{(2i)!}{(2^i i!)^2}\tanh^{2i} r$$

$$p_n(2i + 1) = \mathrm{P}[n = 2i + 1 \mid z] = 0, \tag{11.227}$$

as illustrated in Fig. 11.12.

11.15.6 Squeezed–Displaced States

We finally develop the general single-mode Gaussian states, given by

$$|z, \alpha\rangle = D(\alpha)\,Z(z)\,|0\rangle = e^{\alpha a^* - \alpha^* a}\,e^{\frac{1}{2}(z(a^*)^2 - z^* a^2)}\,|0\rangle \tag{11.228}$$

where now $z := re^{i\theta}$ is a scalar. The Fock representation of a squeezed–displaced state was evaluated by Yuen in a seminal paper [32], where the Fock coefficients are expressed by the Hermite polynomials $H_n(x)$, specifically[22]

---

[22] Yuen considered the state $D(\alpha)\,Z(-z)\,|0\rangle$ instead of $D(\alpha)\,Z(z)\,|0\rangle$.

$$|-z, \alpha\rangle_n = \frac{1}{\sqrt{\mu\, n!}} \left(\frac{\nu}{2\mu}\right)^{n/2} H_n\left(\frac{\beta}{\sqrt{2\mu\nu}}\right) \exp\left(-\frac{1}{2}|\beta|^2 + \frac{\nu^*}{2\mu}\beta^2\right) \quad (11.229)$$

where

$$\mu = \cosh r\,, \qquad \nu = \sinh r\,\exp(i\theta)\,, \qquad \beta = \mu\alpha - \nu\alpha^*. \qquad (11.229a)$$

Here, following our general formulation, we obtain the alternative expression for the Fock coefficients

$$|z, \alpha\rangle_n = \frac{\sqrt{n!}}{\mu} \left(\frac{\beta}{\mu}\right)^n \mathcal{H}_n\left(\frac{\mu\nu}{\beta^2}\right) \exp\left(-\frac{1}{2}|\beta|^2 - \beta^2\frac{\nu^*}{2\mu}\right) \qquad (11.230)$$

where $\mathcal{H}_n(x)$ are the polynomials (of degree $\lfloor n/2 \rfloor$)

$$\mathcal{H}_n(x) := \sum_{j=0}^{\lfloor n/2 \rfloor} \frac{1}{(n-2j)!\, j!}\, x^j. \qquad (11.231)$$

The proof of (11.230) is given in Appendix section "Alternative to Yuen's Formula for Squeezed–Displaced States". The advantage of the new formulation is that the polynomials $\mathcal{H}_n(x)$ are simpler than the Hermite polynomials $H_n(x)$.

The photon number distribution of the state (11.228) is given by $p_n(i) = \big|\, |z, \alpha\rangle_i\,\big|^2$ and it was illustrated in Fig. 7.42. Note that for $r = 0$, absence of squeezing, $p_n(i)$ becomes a Poisson distribution. For $\alpha = 0$, absence of displacement, the distribution becomes (11.227), in agreement with the property $H_i(0) = 0$ for i odd.

The mean photon number in a displaced squeezed state is given by [31, 39]

$$\bar{n}_{|z,\alpha\rangle} = |\alpha|^2 + \sinh^2 r \qquad (11.232)$$

with a clear separation of the contribution of the displacement and of the squeezing.

About the polynomials. The Hermite polynomials $H_n(x)$ are universally known, but also the polynomials $\mathcal{H}_n(x)$, defined by (11.231), are known in field of discrete probability distributions [33] since they are related to the so called *Hermite distribution*. The Hermite distribution is defined starting from two independent Poisson variables X and Y with means a and b. The probability distribution of the random variable $Z = X + 2Y$, which defines the Hermite distribution, is given by

$$P[Z = n] = e^{-a+b} \sum_{j=0}^{\lfloor n/2 \rfloor} \frac{a^{n-2j}b^j}{(n-2j)!\, j!} = e^{-a+b} a^n \mathcal{H}_n\left(\frac{b}{a^2}\right).$$

Note that here the polynomials $\mathcal{H}_n(x)$ appear in a *probability distribution*, while in (11.230) they express an *amplitude distribution*.

Problem 11.20 $\star\star$ Prove that in a cascade of three symplectic transformations $\tilde{X}_i = S_i X_i + d_i$, $i = 1, 2, 3$, the covariance matrix at the output is given by

$$V_{123} = S_3 S_2 S_1 V_0 S_1^{\mathrm{T}} S_2^{\mathrm{T}} S_3^{\mathrm{T}}$$

where V_0 is the covariance matrix at the input.

Problem 11.21 $\star\star\star$ Prove that the covariance matrix of a squeezed–displaced–rotated state $|z, \alpha, \phi\rangle$ is given by

$$V_{\mathrm{sq,rot}}\,(z, \phi) = \begin{bmatrix} \cosh(2r) + \cos(2\phi + \theta)\sinh(2r) & \sin(2\phi + \theta)\sinh(2r) \\ \sin(2\phi + \theta)\sinh(2r) & \cosh(2r) - \cos(2\phi + \theta)\sinh(2r) \end{bmatrix}.$$
$$(11.233)$$

11.16 More on Single-Mode Gaussian States

In the previous section we have seen that the class of Gaussian states coincides with the class of squeezed–displaced states. Then the general pure Gaussian state $|z, \alpha\rangle$ depends on two complex parameters

$$z = r e^{i\theta}, \qquad \alpha = \Delta\, e^{i\varepsilon}. \qquad (11.234)$$

By particularization of these parameters we get the subclasses

- $z = 0$, $\alpha = 0$: the vacuum state
- $z = 0$, $\alpha \neq 0$: the coherent states
- $z \neq 0$, $\alpha = 0$: the squeezed vacuum states.

The class of single-mode mixed states depends on a further parameter, $\sigma^2 = 2\mathcal{N} + 1$, where $\mathcal{N}$ is the number of thermal photons present in the state. For convenience, we will call these mixed states *noisy squeezed–displaced states*.

In this section we investigate these states more in depth because they represent the carriers of information in Quantum Communications systems.

11.16.1 Covariance Matrix and Wigner Function

We examine in detail the covariance matrix V and the Wigner function $W(x, y)$ of a general Gaussian state $|z, \alpha\rangle$. Now, in $|z, \alpha\rangle$ the squeezing part does not give any contribution to the mean value, so we have

$$\begin{bmatrix} \overline{q} \\ \overline{p} \end{bmatrix} = \begin{bmatrix} \Re\alpha \\ \Im\alpha \end{bmatrix} = \begin{bmatrix} \Delta\cos\varepsilon \\ \Delta\sin\varepsilon \end{bmatrix}. \qquad (11.235)$$

On the other hand the covariance matrix of the displacement component is the identity, so that the covariance matrix is simply given by $V = S_{sq}(z)S_{sq}^T(z)$, specifically (see Problem 11.22)

$$V = \begin{bmatrix} V_{11} & V_{12} \\ V_{12} & V_{22} \end{bmatrix} = \begin{bmatrix} \cosh 2r + \cos\theta \sinh 2r & \sin\theta \sinh 2r \\ \sin\theta \sinh 2r & \cosh 2r - \cos\theta \sinh 2r \end{bmatrix}.$$

$$(11.236)$$

Considering that $\det V = 1$, the Wigner function, given by (11.107) for the single mode, results in

$$W(x, y) = \frac{1}{2\pi} e^{-\frac{1}{2}\left[V_{22}(x - \overline{q})^2 + V_{11}(y - \overline{p})^2 - 2V_{12}(x - \overline{q})(y - \overline{p})\right]}$$

$$(11.237)$$

with $\overline{q}, \overline{p}$ given by (11.235) and the V_{ij} given by (11.236).

A convenient representation of the Wigner function $W(x, y)$ in the x, y plane is given by a *contour plot*, which represents the curves of equal levels, given by the relation $W(x, y) = L$, with L real. In general, these curves are *tilted* ellipses, as shown on the left of Fig. 11.13. The ellipses have the common center given by the displacement α, and the main axis is tilted by the angle $\frac{1}{2}\theta$. The lengths of the main axis and of the minor axis are proportional to e^{2r} and to e^{-2r}, respectively, and so they are independent of the squeeze phase θ (see Problem 11.21).

Absence of squeeze phase. With $\theta = 0$ we obtain a great simplification. The covariance matrix becomes

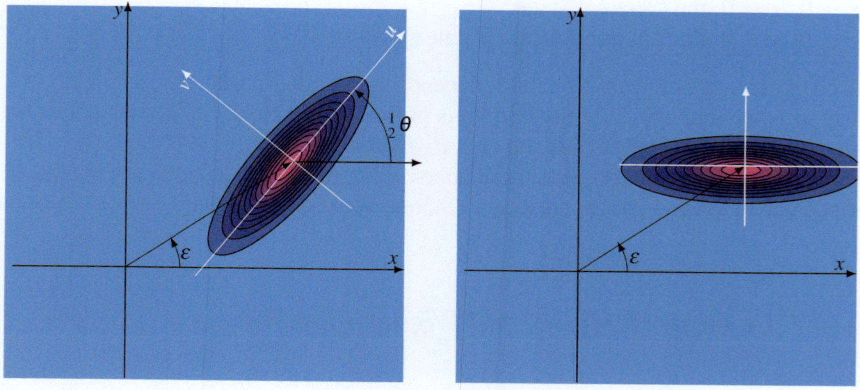

Fig. 11.13 On the *left*. Contour plot of the Wigner function $W(x, y)$ of a general Gaussian state $|z, \alpha\rangle$ with $z = r\,e^{i\theta}$ and $\alpha = \Delta\,e^{i\varepsilon}$. The mean vector $(\overline{q}, \overline{p}) = (\Delta\cos\varepsilon, \Delta\sin\varepsilon)$ gives the displacement amount and determines the center of the elliptic countours. The main axis of the ellipse is tilted with respect to the x-axis of the angle $\frac{1}{2}\theta$, independently of the displacement. On the *right*. Contour plot of the Wigner function $W(x, y)$ with a zero squeeze phase ($\theta = 0$)

$$V = \begin{bmatrix} \cosh^2 r + \sinh^2 r & 0 \\ 0 & \cosh^2 r - \sinh^2 r \end{bmatrix} = \begin{bmatrix} e^{2r} & 0 \\ 0 & e^{-2r} \end{bmatrix}. \tag{11.238}$$

Then in the contour plot of the Wigner function the ellipses turn out to be horizontally displayed, as shown on the right of Fig. 11.13.

11.16.2 Uncertainty Principle

The Uncertainty Principle states that the variances $V_q = V_{11}$ and $V_p = V_{22}$ of the canonic variables q and p are constrained as

$$V_q V_p \geq 1. \tag{11.239}$$

For a pure Gaussian state, considering that $\det V = V_{11} V_{22} - V_{12}^2 = 1$, we have

$$V_q V_p = 1 + V_{12}^2 = 1 + \sin^2 \theta \cosh^2 2r.$$

This is in agreement with the uncertainty principle stated by (11.186) in the N-mode. Then, in general, a Gaussian state (squeezed–displaced state or simply squeezed state) has different noise variances, $V_q = \cosh 2r + \cos \theta \sinh 2r$ and $V_p = \cosh 2r - \cos \theta \sinh 2r$, and also their product $V_q V_p$ is not the minimum established by the Uncertainty Principle. This is illustrated in Fig. 11.14, where V_q, V_p and $V_q V_p$ are plotted as a function of r for three different values of θ (recall that the variances do not depend on the displacement α).

For $\theta = 0$ (squeezed state with $z = r$ real as on the right of Fig. 11.13) the noise variances are still different, $V_q = e^{2r}$ and $V_p = e^{-2r}$, but their product keeps the minimum uncertainty $V_q V_p = 1$.

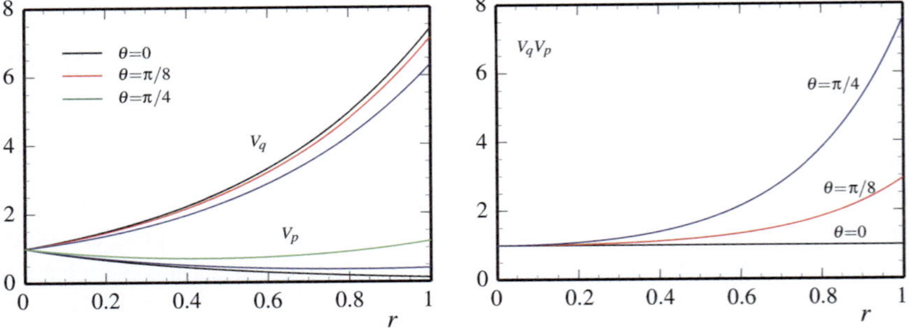

Fig. 11.14 Squeezed states have different noise variances $V_q = V_{11}$ and $V_p = V_{22}$. The product $V_q V_p$ keeps to the minimum uncertainty $V_q V_p = 1$ when the squeeze phase $\theta = 0$, while $V_q V_p > 1$ for $\theta \neq 0$

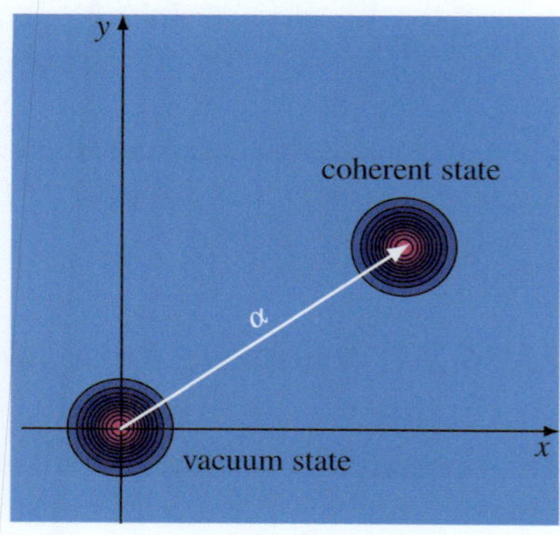

Fig. 11.15 Countour plot of the Wigner function of the vacuum state $|0\rangle$ and of a coherent state $|\alpha\rangle$

Finally for $z = 0$, corresponding to displaced states or to the vacuum state, the covariance matrix is the identity and therefore the position and the momentum have variance equal to one, $V_q = V_p = 1$, that is, the minimum variance established by the Uncertainty Principle. In the phase space these states have a contour plot given by circles, as shown in Fig. 11.15.

11.16.3 Noisy Gaussian States

Above we have considered pure Gaussian states, generated from the vacuum state. For mixed Gaussian states, generated from a thermal state with a given number of thermal photons $\mathcal{N}$, as stated by Theorem 11.7, we have the same classification as for pure Gaussian states, starting form the most general mixed state $\rho_{\mathcal{N}}(z, \alpha)$, which will be called *noisy squeezed–displaced state*.

The covariance matrix V of a noisy state is obtained by multiplying the covariance matrix of the corresponding pure state by the factor $\sigma^2 = 2\mathcal{N} + 1$, so that

$$V = (2\mathcal{N} + 1) \begin{bmatrix} \cosh 2r + \cos\theta \sinh 2r & \sin\theta \sinh 2r \\ \sin\theta \sinh 2r & \cosh 2r - \cos\theta \sinh 2r \end{bmatrix}. \qquad (11.240)$$

A useful interpretation of noisy states is to think that they are pure states corrupted by noise, as in Classical Communications. To this end the adequate model is a *quantum channel*, just called additive-noise Gaussian channel. Channels are *open systems*, which will be considered in Sect. 12.8. Now, if we send a pure Gaussian state $|z, \alpha\rangle$ with covariance matrix V_{in} through an additive-noise Gaussian channel specified by the number of thermal photons $\mathcal{N}$, at the output the state is still Gaussian and the covariance matrix becomes $(2\mathcal{N} + 1)V_{\text{in}}$, that is, we find the state denoted above by $\rho_{\mathcal{N}}(z, \alpha)$ (Fig. 11.16).

We realize that in the phase space the effect of noise is simply an increase of the variances with a consequent increase of the uncertainty. In the contour plot of the

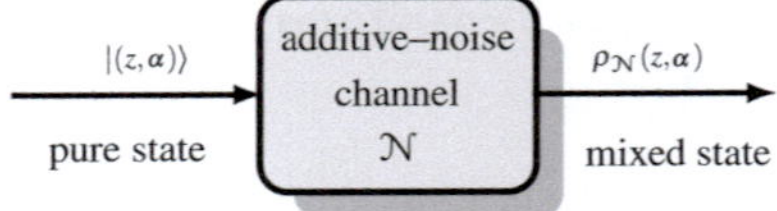

Fig. 11.16   A pure Gaussian state $|z, \alpha\rangle$ is sent through a Gaussian channel specified by the number of thermal photons $\mathcal{N}$. At the output the state is still Gaussian but becomes the mixed state $\rho_{\mathcal{N}}(z, \alpha)$

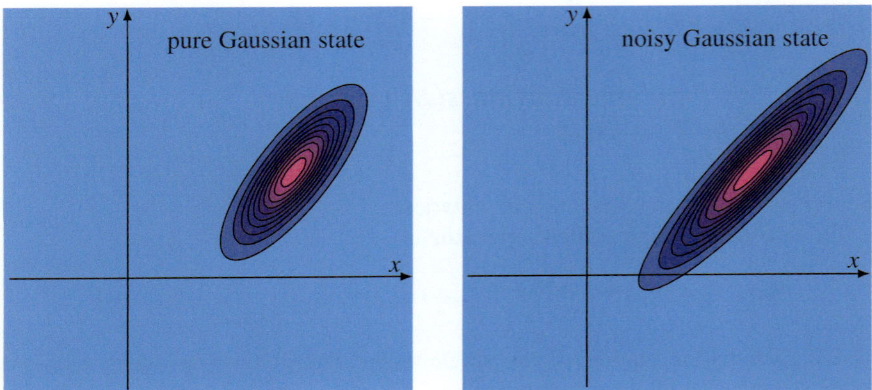

Fig. 11.17   Contour plot of a pure Gaussian state $|z, \alpha\rangle$ and of a noisy Gaussian state $\rho_{\mathcal{N}}(z, \alpha)$. The effect of noise is to increase the axes of the ellipses

Wigner function the noise increases the axes of the ellipses, as shown in Fig. 11.17 in the general case of a squeezed–displaced state.

Problem 11.22 ⋆   Prove that the covariance matrix of the single-mode Gaussian state $|z, \alpha\rangle$ is given by (11.236).

Problem 11.23 ⋆⋆   Consider the Wigner function $W(x, y)$ of a general Gaussian state given by (11.237) and introduce the change of coordinate (see the left of Fig. 11.13)

$$x = u \cos \tfrac{1}{2}\theta - v \sin \tfrac{1}{2}\theta, \qquad y = u \sin \tfrac{1}{2}\theta + v \cos \tfrac{1}{2}\theta$$

which provides a rotation of the angle $\tfrac{1}{2}\theta$. Prove that the new Wigner function $\tilde{W}(u, v)$ is obtained with the covariance matrix

$$V = \begin{bmatrix} e^{2r} & 0 \\ 0 & e^{-2r} \end{bmatrix}.$$

In words, the rotation of $\tfrac{1}{2}\theta$ removes the squeeze phase in $z = r e^{i\theta}$.

11.17 Gaussian States and Transformations
in the Two-Mode

The general theory for the N-mode developed in Sects. 11.10–11.12 is applied to
the two-mode. The application will be not systematic as done for the single mode,
but it is limited to a few important cases, in particular to evidence the presence of
entanglement, not possible in the single mode.

11.17.1 The Fundamental Gaussian Unitaries
in the Two-Mode

We write explicitly the fundamental unitaries.

The **two-mode displacement operator** is

$$D_2(\alpha) = e^{\alpha_1 a_1^* + \alpha_2 a_2^* - \alpha_1^* a_1 - \alpha_2^* a_2} = D(\alpha_1)\,D(\alpha_2)\,, \qquad \alpha = [\alpha_1, \alpha_2]^{\mathrm{T}}$$

so it is given by the product of two single-mode displacement operators, each one
acting separately on the corresponding mode.

The **two-mode rotation operator** is specified by a 2×2 Hermitian matrix ϕ and
reads explicitly as

$$R_2(\phi) = e^{i\,(a_1^* \phi_{11} a_1 + a_1^* \phi_{12} a_2 + a_2^* \phi_{21} a_1 + a_2^* \phi_{22} a_2)}$$

where $\phi_{21} = \phi_{12}^*$. This operator will be considered in the case of the beam splitter.

The **two-mode squeeze operator** is specified by a 2×2 symmetric matrix z and
reads explicitly as

$$Z_2(z) = e^{-\frac{1}{2}\left\{ [a_1^*, a_2^*] \begin{bmatrix} z_{11} & z_{12} \\ z_{12} & z_{22} \end{bmatrix} \begin{bmatrix} a_1^* \\ a_2^* \end{bmatrix} - [a_1, a_2] \begin{bmatrix} z_{11}^* & z_{12}^* \\ z_{12}^* & z_{22}^* \end{bmatrix} \begin{bmatrix} a_1 \\ a_2 \end{bmatrix} \right\}}.$$

The Bogoliubov and symplectic transformations corresponding to the fundamental operators are obtained by the general formulas of the N-mode, but in the general
case their explicit form becomes cumbersome for the presence of exponentials of
matrices (see Sect. 11.6.2). Then we will develop only the cases of interest.

11.17.2 Caves–Schumaker Two-Mode Squeezed–Displaced States

According to Corollary 11.1 the most general pure Gaussian states in the two-mode is a squeezed–displaced state $|z, \alpha\rangle$, obtained by the two-mode vacuum state $|0_2\rangle = |0\rangle_1 |0\rangle_2$ as

$$|z, \alpha\rangle = D_2(\alpha)\, Z_2(z)\, |0\rangle_1 |0\rangle_2 \tag{11.241}$$

where z is a 2×2 symmetric matrix. Here we develop the case usually considered in the literature as two-mode squeeze operator [8, 28, 34], where the matrix z has the simple form

$$z = \begin{bmatrix} 0 & z_0 \\ z_0 & 0 \end{bmatrix} \quad \rightarrow \quad Z_2(z_0) = e^{\frac{1}{2}\,(z_0\, a_1^* a_2^* - z_0^* a_1 a_2)} \tag{11.242}$$

where $z_0 = r_0\, e^{i\theta_0}$ with $r_0, \theta_0 \in \mathbb{R}$. Then the Gaussian unitary (11.241) becomes

$$D(\alpha_1) D(\alpha_2) Z_2(z_0) = e^{\alpha_1 a_1^* - \alpha_1^* a_1}\, e^{\alpha_2 a_2^* - \alpha_2^* a_2}\, e^{\frac{1}{2}\,(z_0\, a_1^* a_2^* - z_0^* a_1 a_2)} \tag{11.243}$$

and the squeezed–displaced state is generated in the form

$$|z_0, \alpha\rangle := D(\alpha_1) D(\alpha_2) Z_2(z_0)\, |0\rangle_1 |0\rangle_2 \,. \tag{11.244}$$

The covariance matrix of the state (11.244) is given by (see Problem 11.25)

$$V = \begin{bmatrix} \cosh 2r_0\; I_2 & \cos\theta_0 \sinh 2r_0\; Y_2 + \sin\theta_0 \sinh 2r_0\; W_2 \\ \cos\theta_0 \sinh 2r_0\; Y_2 + \sin\theta_0 \sinh 2r_0\; W_2 & \cosh 2r_0\; I_2 \end{bmatrix} \tag{11.245}$$

where

$$W_2 = \begin{bmatrix} 0 & 1 \\ 1 & 0 \end{bmatrix}, \qquad Y_2 = \begin{bmatrix} 1 & 0 \\ 0 & -1 \end{bmatrix}.$$

In particular, for $\theta_0 = 0$ we find

$$V = \begin{bmatrix} \cosh 2r_0\; I_2 & \sinh 2r_0\; Y_2 \\ \sinh 2r_0\; Y_2 & \cosh 2r_0\; I_2 \end{bmatrix}.$$

The Fock representation of the state (11.244) was evaluated by Caves et al. [28], who obtained the following expression for the Fock coefficients:

$$\boxed{\; |z_0, \alpha\rangle_{n_1, n_2} = K_0\, \frac{1}{q!}\, u_{12}^p\, u_1^{n_1 - p}\, u_2^{n_2 - p}\, L_p^{q - p}\!\left(-\frac{u_1 u_2}{u_{12}} \right) \;} \tag{11.246}$$

where $L_p^{q-p}(z)$ are the generalized Laguerre polynomials, $p = \min(n_1, n_2)$, $q = \max(n_1, n_2)$ and

$$K_0 = \operatorname{sech} r_0 \, \exp\left[-\tfrac{1}{2}(|\beta_1|^2 + |\beta_2|^2 + 2\beta_1\beta_2\lambda)\right]$$

$$\beta_1 = \cosh r_0 \, \alpha_1 - \mathrm{e}^{\mathrm{i}\theta_0} \, \sinh r_0 \, \alpha_1^*, \qquad \beta_2 = \cosh r_0 \, \alpha_2 - \mathrm{e}^{\mathrm{i}\theta_0} \, \sinh r_0 \, \alpha_2^*$$

$$u_1 = \operatorname{sech} r_0 \, \beta_2, \qquad u_2 = \operatorname{sech} r_0 \, \beta_1, \qquad u_{12} = \lambda = \mathrm{e}^{\mathrm{i}\theta_0} \, \tanh r_0.$$

$$(11.246a)$$

A simpler formula can be obtained following the general theory of Sect. 11.13, based on the the normal-ordered form of the operator (11.243). In this form the parameters of interest are K_0 and

$$B(\beta, a) = \mathrm{e}^L \quad \text{with} \quad L = \operatorname{sech} r_0(\beta_1 a_2^* + \beta_2 a_1^*) + \lambda \, a_1^* a_2^*. \tag{11.247}$$

Then we get the following expression of the Fock coefficients (see Appendix Section "Proof of Fock Expansion (11.248) of Caves–Schumaker States")

$$|z_0, \alpha\rangle_{n_1,n_2} = K_0 \sqrt{n_1! n_2!} \, u_1^{n_1} u_2^{n_2} \mathcal{L}_{n_1 n_2}\left(\frac{u_{12}}{u_1 u_2}\right) \tag{11.248}$$

where $\mathcal{L}_{n_1 n_2}(x)$ are the polynomials

$$\mathcal{L}_{n_1 n_2}(x) := \sum_{k=0}^{\min(n_1, n_2)} \frac{1}{(n_1 - k)!(n_2 - k)!k!} \, x^k. \tag{11.248a}$$

From the Fock coefficients one gets the distribution

$$p_{n_1,n_2}(i_1, i_2) := \mathrm{P}[n_1 = i_1, n_2 = i_2] = \left| \, |z_0, \alpha\rangle_{n_1,n_2} \, \right|^2 \tag{11.249}$$

giving the probability of the presence of i_1 photons in the first mode and of i_2 photons in the second mode. This distribution is illustrated in Fig. 11.18 with $\theta_0 = 0$ for $\alpha_1 = 3$, $\alpha_2 = 3$ and $r_0 = 1.5$.

About the polynomials. Here too, we have a competition between the celebrated Laguerre polynomials $L_p^{q-p}(x)$ and the polynomials $\mathcal{L}_{n_1 n_2}(x)$, practically unknown. In the theory of discrete probability distributions [33] the polynomials $\mathcal{L}_{n_1 n_2}(x)$ are related to bivariate Poisson distributions, which are defined starting form three independent Poisson variables Y_1, Y_2, Y_3 with means $\lambda_1, \lambda_2, \lambda_3$. Then we let $X_1 = Y_1 + Y_3$, $X_2 = Y_2 + Y_3$ and the bivariate Poisson distribution is defined as the probability $\mathrm{P}[X_1 = n_1, X_2 = n_2]$, which is given by

$$\mathrm{P}[X_1 = n_1, X_2 = n_2] = \exp\left(-\lambda_1 - \lambda_2 - \lambda_3\right) \lambda_1^{n_1} \lambda_2^{n_2} \mathcal{L}_{n_1 n_2}\left(\frac{\lambda_3}{\lambda_1 \lambda_2}\right).$$

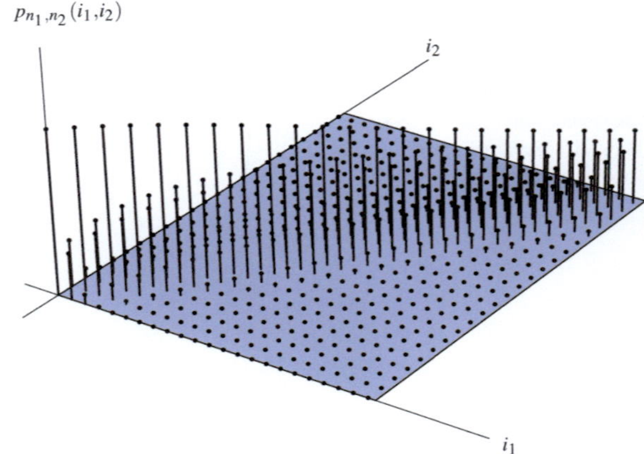

Fig. 11.18 Probability distribution $p_{n_1,n_2}(i_1,i_2)$ of pairs of photon numbers (n_1, n_2) in a **two-mode squeezed–displayed** state, for $\alpha_1 = 3$, $\alpha_2 = 3$ and $r = 1.5$

But here the polynomials $\mathcal{L}_{n_1 n_2}(x)$ express a *probability distribution*, while in (11.248) they express a *probability amplitude*.

11.17.3 Einstein–Podolsky–Rosen (EPR) States

A very important class (also for historic reasons) of two-mode states is obtained by omitting the displacement in the previous class of states, that is,

$$|z_0\rangle_{\text{epr}} := Z_2(r)\,|0\rangle_1\,|0\rangle_2 = e^{\frac{1}{2}\left(z_0\,a_1^* a_2^* - z_0^* a_1 a_2\right)}\,|0\rangle_1\,|0\rangle_2 \qquad z_0 = r_0 e^{i\theta_0} \quad (11.250)$$

which are known as Einstein-Podolsky–Rosen (EPR) states.

The covariance matrix of the EPR states is the same as that of squeezed–displaced states (see (11.245)). The parameters to evaluate the Fock representation are obtained from (11.247) and read

$$K_0 = \operatorname{sech} r_0 = \sqrt{1 - \tanh^2 r_0} \qquad\qquad (11.251)$$

$$B(0, a) = e^L \quad \text{with} \quad L = \lambda\, a_1^* a_2^* \qquad \lambda = e^{i\theta_0} \tanh r_0.$$

The expansion of the exponential is immediate and gives

$$B(0, a) = \sum_{n=0}^{\infty} \frac{(\lambda\, a_1^* a_2^*)^n}{n!}.$$

Hence the Fock expansion of EPR states is

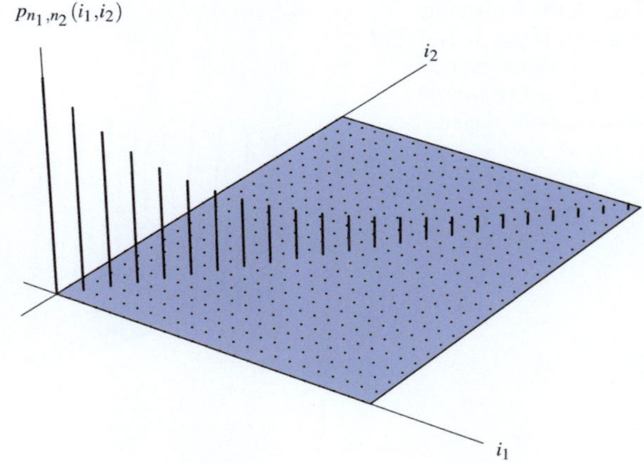

Fig. 11.19 Probability distribution $p_{n_1,n_2}(i_1, i_2)$ of pairs of photon numbers (n_1, n_2) in a **two-mode squeezed** state, for $r = 1.6$. The pairs have the same photon number in each mode

$$|z_0\rangle_{\text{epr}} = K_0\, B(0, a)\, |0\rangle_2 = \sqrt{1 - |\lambda|^2}\, \sum_{n=0}^{\infty} \lambda^n |n\rangle_1 |n\rangle_2 \qquad |\lambda| = \tanh r_0 \in (0, 1).$$

(11.252)

The statistical description of photon numbers is

$$p_{n_1 n_2}(i_1, i_2) = (1 - |\lambda|^2)\, |\lambda|^{2 i_1}\, \delta_{i_1 i_2}.$$

(11.253)

Hence, we have a diagonal distribution, which means that in the EPR states the photon numbers in the two modes are always equal. This is illustrated in Fig. 11.19.

11.17.4 *Fock Expansion of a General Two-Mode Gaussian State* ⇓

We mention the possibility of evaluating the Fock representation of a two-mode Gaussian state, given by

$$|z, \alpha\rangle = D_2(z) Z_2(\alpha) |0\rangle_2$$

(11.254)

where in general $z = r \mathrm{e}^{\mathrm{i}\theta}$ is an arbitrary 2×2 symmetric matrix. Then the squeeze operator takes the general form

$$Z_2(z) = \exp\left[\tfrac{1}{2}[a_1^*, a_2^*]\, z\, [a_1^*, a_2^*]^{\mathrm{T}} - \tfrac{1}{2}[a_1, a_2]\, z^*\, [a_1, a_2]^{\mathrm{T}} \right].$$

(11.255)

Using the normal-ordered form of the unitary $D_2(z) Z_2(\alpha)$ (see (11.195)) the state becomes

$$|z, \alpha\rangle = K_0\, B(\alpha, a) |0\rangle_2$$

(11.256)

where

$$K_0 = |\det S|^{1/2} \exp\left[-\tfrac{1}{2}(\alpha^*\alpha + \alpha^{\mathrm{T}} T^*\alpha)\right] \in \mathbb{C}$$

$$B(\alpha, a) := \exp\left[\alpha^{\mathrm{T}} S^{\mathrm{T}} a_\star + \tfrac{1}{2}a^* T a_\star\right].$$

with

$$S := \operatorname{sech} r = \begin{bmatrix} s_{11} & s_{12} \\ s_{21} & s_{22} \end{bmatrix}, \qquad T := \tanh z = \tanh r \, \mathrm{e}^{\mathrm{i}\theta} = \begin{bmatrix} t_{11} & t_{12} \\ t_{21} & t_{22} \end{bmatrix}.$$

For the explicit evaluation we have to find the *polar decomposition* of the squeeze matrix z, given by $r\mathrm{e}^{\mathrm{i}\theta}$, where $r = \sqrt{z\,z^*}$ and $\mathrm{e}^{\mathrm{i}\theta} = z\,r^{-1}$ (see Theorem 2.7). Hence, the preliminary step is the evaluation of the square root r of $z\,z^*$, using the EID, say $z\,z^* = U\operatorname{diag}[\sigma_1^2, \sigma_2^2]U^*$, where the eigenvalues are real and nonegative because $z\,z^*$ is PSD. Then $r = U\operatorname{diag}[\sigma_1, \sigma_2]U^*$. Next, we can evaluate the matrices S and T as

$$S = U\operatorname{diag}[\operatorname{sech}\sigma_1, \operatorname{sech}\sigma_2]U^*, \qquad T = U\operatorname{diag}[\tanh\sigma_1, \tanh\sigma_2]U^*\,\mathrm{e}^{\mathrm{i}\theta}.$$

Then the scalar K_0 is given by

$$K_0 = |\det S|^{1/2}\mathrm{e}^{-\tfrac{1}{2}\left(|\alpha_1|^2+|\alpha_2|^2+t_{11}\alpha_1^2+t_{12}\alpha_2^2+(t_{12}+t_{21})\alpha_1\alpha_2\right)} \tag{11.257}$$

and the operator $B(\alpha, a)$ reads as

$$B(\alpha, a) = \mathrm{e}^{u_1 a_1^* + u_2 a_2^* + u_{12} a_1^* a_2^* + v_1 (a_1^*)^2 + v_2 (a_2^*)^2} \tag{11.258}$$

where

$$u_1 = s_{11}\alpha_1 + s_{21}\alpha_2, \qquad u_2 = s_{22}\alpha_2 + s_{12}\alpha_1, \qquad u_{12} = \tfrac{1}{2}(t_{12} + t_{21})$$
$$v_1 = \tfrac{1}{2}t_{11}, \qquad v_2 = \tfrac{1}{2}t_{22}. \tag{11.259}$$

To proceed we have to distinguish the following cases: (a) $u_{12} \neq 0$, $v_1 = v_2 = 0$, (b) $u_{12} = 0$, and (c) $u_{12} \neq 0$, $v_1, v_2 \neq 0$. In the solution to problems we develop the three cases.

Problem 11.24 ★★   Prove that the symplectic transformation of the Gaussian unitary (11.243) for $\theta = 0$ is given by

$$S_{\mathrm{sq}}(z_0) = \begin{bmatrix} \cosh r_0 \, I_2 & \sinh r_0 \, Y_2 \\ \sinh r_0 \, Y_2 & \cosh r_0 \, I_2 \end{bmatrix}, \qquad Y_2 := \begin{bmatrix} 0 & 1 \\ -1 & 0 \end{bmatrix}. \tag{11.260}$$

Problem 11.25 ★★   Prove that the covariance matrix of the state (11.244) is given by (11.245).

Problem 11.26 ⋆⋆⋆　Develop the Fock expansion of the general two-mode Gaussian state (11.254), considering the three case listed above.

11.18 Beam Splitter

The beam splitter is one of the most important components in Classical and Quantum Optics. Its implementation was described in Sect. 9.2. Here we develop the quantum theory.

11.18.1 The Quantum Model

A beam splitter can be modeled by a two-mode rotation operator obtained with the phase matrix

$$\phi_{\text{bs}} = \begin{bmatrix} 0 & -i\beta \\ i\beta & 0 \end{bmatrix} \quad \rightarrow \quad R_2(\phi_{\text{bs}}) := e^{\beta(a_1^* a_2 - a_2 a_1^*)} \tag{11.261}$$

where β determines the transmissivity and the reflectivity given by $\tau = \cos^2 \beta$ and $\sqrt{1-\tau} = \sin\beta$ respectively (for $\tau = \frac{1}{2}$ the beam splitter is said to be *balanced*). We can use the general relation (11.169) to get the corresponding Bogoliubov transformation, namely

$$\tilde{a} = e^{i\phi_{\text{bs}}} a = \exp \begin{bmatrix} 0 & \beta \\ -\beta & 0 \end{bmatrix} a = u_{\text{bs}}(\beta) a \tag{11.262}$$

where (see (11.74))

$$u_{\text{bs}}(\beta) := \exp \begin{bmatrix} 0 & \beta \\ -\beta & 0 \end{bmatrix} = \begin{bmatrix} \cos\beta & \sin\beta \\ -\sin\beta & \cos\beta \end{bmatrix}. \tag{11.262a}$$

The corresponding symplectic matrix is obtained using Proposition 11.11 and reads (see Problem 11.27) as

$$S_{\text{bs}}(\beta) = \begin{bmatrix} \cos\beta\, I_2 & \sin\beta\, I_2 \\ -\sin\beta\, I_2 & \cos\beta\, I_2 \end{bmatrix} = u_{\text{bs}}(\beta) \otimes I_2. \tag{11.263}$$

Relation (11.263) has been obtained from the general relation (11.169).

11.18.2 The Beam Splitter as Two-Input–Two-Output Device

We apply at the input of the beam splitter a general two-mode Gaussian state $|z, \alpha\rangle$ (Fig. 11.20).

We know that at the output the state $|z', \alpha'\rangle$ is still Gaussian, but we want to find the new parameters. To this end we consider the generation of the input state $|z, \alpha\rangle$ from the vacuum $|0_2\rangle$, that is, applying a squeezing $Z_2(z)$ followed by a displacement $D_2(\alpha_0)$, as shown in Fig. 11.21. The beam splitter, modeled by $R_2(\phi_{\text{bs}})$, can be moved at the beginning of the cascade and finally dropped because $R_2(\phi_{\text{bs}})|0_2\rangle = |0_2\rangle$. The change of parameters $(z, \alpha) \;\rightarrow\; (z', \alpha')$ is provided by relations (11.176), which give the output state $|z', \alpha'\rangle$ with

$$z' = \mathrm{e}^{\mathrm{i}\phi_{\text{bs}}}\, z\, \mathrm{e}^{\mathrm{i}\phi_{\text{bs}}^{\mathrm{T}}} = u_{\text{bs}}(\beta)\, z\, u_{\text{bs}}^{\mathrm{T}}(\beta)\,, \qquad \alpha' = \mathrm{e}^{\mathrm{i}\phi_{\text{bs}}}\, \alpha = u_{\text{bs}}(\beta)\, \alpha. \qquad (11.264)$$

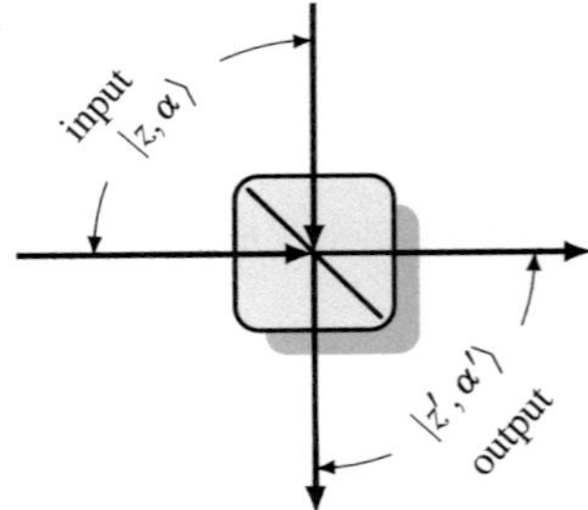

Fig. 11.20 Beam splitter with input a two-mode Gaussian state $|z, \alpha\rangle$ and output a new two-mode Gaussian state $|z', \alpha'\rangle$. Note that the states may be entangled (not separable)

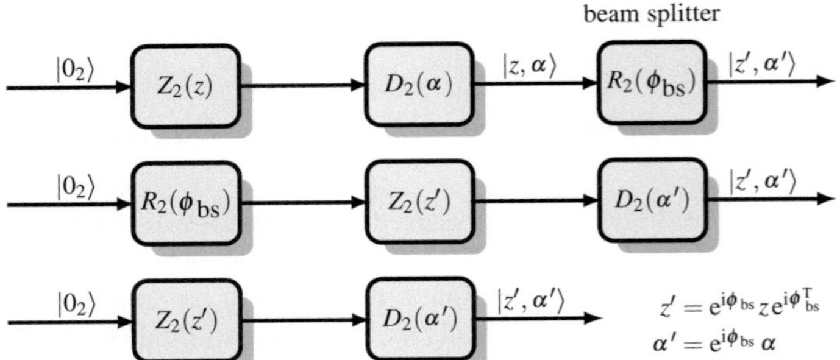

Fig. 11.21 Evaluation of the response of a beam splitter to a general two-mode Gaussian state. The input state $|z, \alpha\rangle$ is generated from the vacuum $|0_2\rangle$ with the application of the unitary $Z_2(z)$ followed by $D_2(\alpha)$; the beam splitter is modeled by a two-mode rotation operator $R_2(\phi_{\text{bs}})$ with phase matrix given by (11.261). To evaluate the output state $|z', \alpha'\rangle$, the rotation operator is moved at the beginning of the cascade and finally dropped, because $R_2(\phi_{\text{bs}})|0_2\rangle = |0_2\rangle$

Thus the beam splitter modifies both the squeezing and the displacement parameters.

To get a more explicit result we suppose that the input state is the Caves–Schumacher squeezed–displaced state considered above, where the squeeze matrix z has the form of (11.242). Then we find

	Input	Output
State	$\lvert z, \alpha \rangle$	$\lvert z', \alpha' \rangle$
Squeeze matrix	$z = \begin{bmatrix} 0 & z_0 \\ z_0 & 0 \end{bmatrix}, z_0 = r_0 e^{i\theta_0}$	$z' = \begin{bmatrix} z_0 \sin 2\beta & z_0 \cos 2\beta \\ z_0 \cos 2\beta & -z_0 \sin 2\beta \end{bmatrix}$
Displacement	$\alpha = \begin{bmatrix} \alpha_1 \\ \alpha_2 \end{bmatrix}$	$\begin{bmatrix} \alpha'_1 \\ \alpha'_2 \end{bmatrix} = \begin{bmatrix} \alpha_1 \cos \beta - \alpha_2 \sin \beta \\ \alpha_1 \sin \beta + \alpha_2 \cos \beta \end{bmatrix}$

Note that the average numbers of photons in the two modes at the input are given by

$$\bar{n}_1 = \operatorname{sech}^2(r_0) + |\alpha_1|^2, \qquad \bar{n}_2 = \operatorname{sech}^2(r_0) + |\alpha_2|^2 \tag{11.265}$$

and are modified by the beam splitter as (see Problem 11.28)

$$\begin{aligned}
\bar{n}'_1 &= \cos^2(\beta)\, \bar{n}_1 + \sin^2(\beta)\, \bar{n}_2 + \Delta n\, \sin 2\beta \\
\bar{n}'_2 &= \sin^2(\beta)\, \bar{n}_1 + \cos^2(\beta)\, \bar{n}_2 - \Delta n\, \sin 2\beta
\end{aligned} \tag{11.266}$$

where

$$\Delta n = \Re\left[(\alpha_2 \sin(\beta) + \alpha_1 \cos(\beta))(\alpha_2 \cos(\beta) - \alpha_1 \sin(\beta))^* \right] \tag{11.266a}$$

while the global number does not change

$$\bar{n}'_1 + \bar{n}'_2 = \bar{n}_1 + \bar{n}_2. \tag{11.266b}$$

11.18.3 The Beam Splitter as One-Input–Two-Output Device

In several applications the beam splitter is used as one-input–two-output device in the sense that the second input is not activated. The trick to preserve the above quantum model is to feed the first input with a single-mode state $\lvert z, \alpha \rangle$ and the second input with the ground state $\lvert 0 \rangle_2$, so that globally we have the two-mode state $\lvert z, \alpha \rangle \otimes \lvert 0 \rangle_2$ (Fig. 11.22).

To find the input–output relation we can still use the approach of Fig. 11.21 with the replacements

$$z = \begin{bmatrix} z_{11} & z_{12} \\ z_{12} & z_{22} \end{bmatrix} \rightarrow \begin{bmatrix} z_0 & 0 \\ 0 & 0 \end{bmatrix}, \qquad \alpha = \begin{bmatrix} \alpha_1 \\ \alpha_2 \end{bmatrix} \rightarrow \begin{bmatrix} \alpha_0 \\ 0 \end{bmatrix}$$

Fig. 11.22 Beam splitter as one-input–two-output device. At the first input there is a single-mode Gaussian state $|z_0, \alpha_0\rangle$; the second input is nullified with the ground state $|0\rangle_2$. The output is a two-mode Gaussian state $|z', \alpha'\rangle$

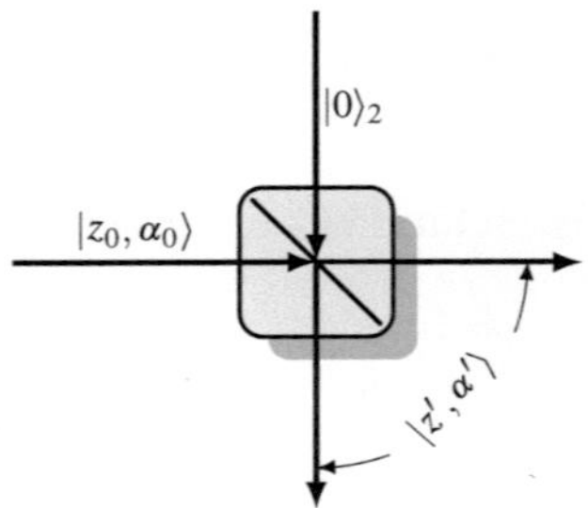

so that the two-mode operators degenerate as $Z_2(z) \rightarrow Z(z_0) \otimes I_{\mathcal{H}}$, $D_2(\alpha_0) \rightarrow D(\alpha_0) \otimes I_{\mathcal{H}}$, where $Z(z_0)$ and $D(\alpha_0)$ are single-mode. Then the input–output relation (11.264) becomes

$$
\begin{aligned}
z' &= u_{\mathrm{bs}}(\beta) \begin{bmatrix} z_0 & 0 \\ 0 & 0 \end{bmatrix} u_{\mathrm{bs}}^{\mathrm{T}}(\beta) = z_0 \begin{bmatrix} \cos^2 \beta & -\cos \beta \sin \beta \\ -\cos \beta \sin \beta & \sin^2 \beta \end{bmatrix} \\
\alpha' &= u_{\mathrm{bs}}(\beta) \begin{bmatrix} \alpha_0 \\ 0 \end{bmatrix} = \begin{bmatrix} \alpha_0 \cos \beta \\ \alpha_0 \sin \beta \end{bmatrix}
\end{aligned}
\tag{11.267}
$$

where the output $|z', \alpha'\rangle$ represents a quite general two-mode squeezed–displaced state.

Problem 11.27 ⋆ Prove that the symplectic matrix of the beam splitter is given by (11.263).

Problem 11.28 ⋆⋆⋆ Consider the beam splitter with a Caves-Schumacher state at the input. Prove that the average numbers of photons in the two modes are given by (11.265) at the input and by (11.266) at the output.

11.19 Entanglement in Two-Mode Gaussian States

Two-mode Gaussian states are usually entangled, that is, in general $|z, \alpha\rangle$ cannot be expressed as the tensor product of two single-mode states. In this section we consider the EPR states, as the most venerable example of continuous-variable entangled states, and then we develop a test of separability based on the Fock expansion. Finally, we will see that the beam splitter may act as an entangler.

11.19.1 Entanglement in EPR States

We consider, as an example of the entropic evaluation of the entanglement measure, the EPR states (see (11.252))

$$|z\rangle_{\mathrm{epr}} = \sum_{n=0}^{\infty} f(n)\,|n\rangle_1 \otimes |n\rangle_2\,, \qquad f(n) := \sqrt{1 - |\lambda|^2}\,\lambda^n\,, \quad \lambda = \tanh |z_0|\,.$$

From this Fock expansion we form the corresponding density operator

$$\rho_{\mathrm{epr}} = |z\rangle_{\mathrm{epr}} \langle z|_{\mathrm{epr}} = \sum_{m,n=0}^{\infty} f(m)\,f^*(n)|m\rangle_1 \otimes |m\rangle_2\,\,_2\langle n| \otimes\,_1\langle n|\,.$$

Then, tracing out with respect to mode 2, we find the mode 1 density operator

$$\rho_{\mathrm{epr},1} = \sum_{n=0}^{\infty} |f(n)|^2 |n\rangle_1\,_1\langle n|$$

where $|f(n)|^2 = (1 - |\lambda|^2)\,|\lambda|^{2n}$, which represents a thermal state (see Sect. 11.9.5) with number of thermal photons given by $\mathcal{N} = |\lambda|^2/(1 - |\lambda|^2)$.

Considering that the thermal state has the maximum von Neumann entropy, we conclude that the EPR states are maximally entangled (according to the entropic measure of entanglement) [8].

11.19.2 Separability Tests for Two-Mode Pure Gaussian States

A general two-mode Gaussian state is given by

$$|z, \alpha\rangle = D_2(\alpha) Z_2(z)|0\rangle_2 \tag{11.268}$$

where the displacement operator is always separable as $D_2(\alpha) = D(\alpha_1)D(\alpha_2)$, while $Z_2(z)$ is separable when the squeeze matrix z is diagonal, say

$$z = \begin{bmatrix} z_1 & 0 \\ 0 & z_2 \end{bmatrix} \tag{11.269}$$

giving

$$|z, \alpha\rangle = |z_1, \alpha_1\rangle \otimes |z_2, \alpha_2\rangle \tag{11.270}$$

where the factors $|z_i, \alpha_i\rangle$ are single-mode Gaussian states. This allows us to formulate a trivial test of separability for two-mode Gaussian states, which can be easily extended to the N-mode. However, the condition of a diagonal squeeze matrix is not a necessary condition, as we will see now.

We consider the separability criterion of Proposition 10.4, which is based on the Fourier expansion of the given bipartite state. In the bosonic space the natural expansion is given by the Fock expansion. Now, in Sect. 11.17.4, we have seen that the Fock expansion can be obtained starting from the expression

$$|z, \alpha\rangle = K_0 \, e^L \, |0\rangle_2, \qquad |0\rangle_2 = |0\rangle_1 |0\rangle_2 \tag{11.271}$$

where we have used the normal ordered form of the unitary $D_2(\alpha) Z_2(z)$. In (11.271) K_0 is a complex scalar and the exponent has the form

$$L = u_1 a_1^* + u_2 a_2^* + u_{12} a_1^* a_2^* + v_1 (a_1^*)^2 + v_2 (a_2^*)^2. \tag{11.272}$$

where the coefficients u and v depend on the matrix z and on the displacement $\alpha = [\alpha_1, \alpha_2]^{\mathrm{T}}$. In particular the coefficient u_{12} depends only on the squeeze matrix z through the hyperbolic tangent, specifically

$$u_{12} = \tfrac{1}{2}(t_{12} + t_{21}) \quad \text{with} \quad T = (\tanh r)\, e^{\mathrm{i}\theta} = \begin{bmatrix} t_{11} & t_{12} \\ t_{21} & t_{22} \end{bmatrix}.$$

Now the separability criterion is just based on this coefficient. In fact if and only if

$$\boxed{u_{12} = \tfrac{1}{2}(t_{12} + t_{21}) = 0} \tag{11.273}$$

the exponential becomes separable and the two-mode Gaussian state turns out to be

$$|z, \alpha\rangle = K_0 \, e^{u_1 a_1^* + v_1 (a_1^*)^2} \, |0\rangle_1 \; \otimes \; e^{u_2 a_2^* + v_2 (a_2^*)^2} \, |0\rangle_2. \tag{11.274}$$

It can be shown that the factors in (11.274) are single-mode Gaussian states, say $|z_1', \alpha_1'\rangle$ and $|z_2', \alpha_2'\rangle$, whose parameters can be calculated from the two-mode parameters z and α.

We now see an application of this criterion in a beam splitter.

11.19.3 The Beam Splitter as an Entangler

The possibility of using the beam splitter as an entangler has been long recognized [35, 36]. Here we give a simple example to illustrate this possibility.

We consider the beam splitter driven by two single-mode squeezed–displaced states, $|z_1, \alpha_1\rangle$ and $|z_2, \alpha_2\rangle$, which may be regarded as a separable two-mode squeezed–displaced state

$$|z, \alpha\rangle = |z_1, \alpha_1\rangle \otimes |z_2, \alpha_2\rangle.$$

The output of the beam splitter gives the state $|z', \alpha'\rangle$ with the squeeze matrix $z' = u_{\mathrm{bs}} \, z \, u_{\mathrm{bs}}^{\mathrm{T}}$, where u_{bs} is the beam splitter matrix defined by (11.262a). Considering that $z = \begin{bmatrix} z_1 & 0 \\ 0 & z_2 \end{bmatrix} = \begin{bmatrix} r_1 e^{\mathrm{i}\theta_1} & 0 \\ 0 & r_2 e^{\mathrm{i}\theta_2} \end{bmatrix}$, we find

$$z' = \begin{bmatrix} z_1 \cos^2 \beta + z_2 \sin^2 \beta & \cos \beta \sin \beta (z_2 - z_1) \\ \cos \beta \sin \beta (z_2 - z_1) & z_2 \cos^2 \beta + z_1 \sin^2 \beta \end{bmatrix}.$$

For the evaluation of the matrix $T := \tanh r' \, \mathrm{e}^{\mathrm{i}\theta'}$ we need the polar decomposition $r' \, \mathrm{e}^{\mathrm{i}\theta'}$ of the output squeeze matrix z'. Following the procedure outlined in Sect. 11.17.4, we first evaluate the square root r' of the PSD matrix

$$z' \, z'^* = \begin{bmatrix} r_1^2 \cos^2 \beta + r_2^2 \sin^2 \beta & \cos \beta (r_2^2 - r_1^2) \sin \beta \\ \cos \beta (r_2^2 - r_1^2) \sin \beta & r_2^2 \cos^2 \beta + r_1^2 \sin^2 \beta \end{bmatrix}.$$

This matrix has eigenvalues $\sigma_1^2 = r_1^2$ and $\sigma_2^2 = r_2^2$ and hence $r' = \sqrt{z' \, z'^*}$ is simply obtained by replacing r_1^2 and r_2^2 with their square roots r_1 and r_2 in $z' \, z'^*$, that is,

$$r' = \begin{bmatrix} r_1 \cos^2 \beta + r_2 \sin^2 \beta & \cos \beta (r_2 - r_1) \sin \beta \\ \cos \beta (r_2 - r_1) \sin \beta & r_2 \cos^2 \beta + r_1 \sin^2 \beta \end{bmatrix}.$$

Hence we find the matrix $\mathrm{e}^{\mathrm{i}\theta'}$ as $(r')^{-1} z'$. The matrix $\tanh r$ is obtained by replacing in r' its eigenvalues r_1 and r_2 with $\tanh r_1$ and $\tanh r_2$, respectively. Finally we have $T = \tanh r \, \mathrm{e}^{\mathrm{i}\theta'}$, namely

$$T = \begin{bmatrix} \mathrm{e}^{\mathrm{i}2\theta_1} \tanh r_1 \cos^2 \beta + \mathrm{e}^{\mathrm{i}2\theta_2} \sin^2 \beta \tanh r_2 & -\cos \beta \sin \beta \left[\mathrm{e}^{\mathrm{i}2\theta_1} \tanh r_1 - \mathrm{e}^{\mathrm{i}2\theta_2} \tanh r_2 \right] \\ -\cos \beta \sin \beta \left[\mathrm{e}^{\mathrm{i}2\theta_1} \tanh r_1 - \mathrm{e}^{\mathrm{i}2\theta_2} \tanh r_2 \right] & \mathrm{e}^{\mathrm{i}2\theta_2} \tanh r_2 \cos^2 \beta + \mathrm{e}^{\mathrm{i}2\theta_1} \sin^2 \beta \tanh r_1 \end{bmatrix}.$$

From the nondiagonal entries of T we find

$$u_{12} = \tfrac{1}{2}(t_{12} + t_{21}) = -\sin \beta \cos \beta \left[\mathrm{e}^{\mathrm{i}2\theta_1} \tanh r_1 - \mathrm{e}^{\mathrm{i}2\theta_2} \tanh r_2 \right]$$

which states that the output is entangled for $z_1 \neq z_2$ and $\sin \beta \cos \beta \neq 0$. To summarize:

Proposition 11.15 *A beam splitter driven by two squeezed–displaced states produces at the output a two-mode entangled state. Hence the beam splitter acts as an entangler.*

11.20 Gaussian States and Geometrically Uniform Symmetry

In this section we investigate the possibility that a constellation of Gaussian states have the GUS. We consider the problem in a general form.

Let $\mathcal{S} = \{|\psi(p)\rangle, \, p \in P\}$ be a class of quantum states dependent on a parameter p. The class is **closed with respect to rotations** if $R_N(\phi)|\psi(p)\rangle \in \mathcal{S}$, where $R_N(\phi)$

is the rotation operator. In words, the class is closed if, given a reference value of the parameter p_0, one can find a value $p_\phi \in P$, such that

$$R_N(\phi)|\psi(p_0)\rangle = |\psi(p_\phi)\rangle. \qquad (11.275)$$

With such a class we can construct constellations of any order K with the GUS property. In practice, in the single mode we get K-ary PSK constellations, by choosing an arbitrary reference state $|\psi(p_0)\rangle$ in S and using as symmetry operator $S = R(2\pi/K)$. In the multimode a relevant application is given by the PPM, where the "phase" ϕ becomes an $N \times N$ Hermitian matrix with the property $\exp(iK\phi) = I_N$, where K is the order of PPM and N is the dimension of the PPM Hilbert space (see the end of this section).

11.20.1 Rotated Gaussian States

The class of pure Gaussian states is closed with respect to rotations. In fact, we have seen that the most general pure Gaussian state can be generated in the form (with the notations introduced in Sect. 11.12.2)

$$|z, \alpha\rangle = D_N(\alpha)\, Z_N(z)\, |0_N\rangle, \qquad \alpha, z \in \mathbb{C}^N. \qquad (11.276)$$

In words, $|z, \alpha\rangle$ is a squeezed–displaced state. The application of a rotation gives a squeezed–displaced–rotated state

$$|z, \alpha, \phi\rangle = R_N(\phi)\, D_N(\alpha)\, Z_N(z)\, |0_N\rangle. \qquad (11.277)$$

Now we apply relation (11.176c) to get $R_N(\phi)\, D_N(\alpha) = D_N(e^{i\phi}\alpha)\, R_N(\phi)$. Next we apply (11.176b) to get $R_N(\phi)\, Z_N(z) = Z_N(e^{i\phi}\, z\, e^{i\phi^{\mathrm{T}}})\, R_N(\phi)$. Hence

$$|z, \alpha, \phi\rangle = D_N(e^{i\phi}\alpha)\, Z_N(e^{i\phi}z\, e^{i\phi^{\mathrm{T}}})\, R_N(\phi)\, |0_N\rangle$$

where $R_N(\phi)|0_N\rangle = |0_N\rangle$, so that the rotation can be dropped. In conclusion the rotation modifies the parameters in the form

$$z = \rightarrow\ e^{i\phi}z\, e^{i\phi^{\mathrm{T}}}, \qquad \alpha \rightarrow\ e^{i\phi}\alpha; . \qquad (11.278)$$

Theorem 11.8 *The class of squeezed–displaced states is closed under rotations. A squeezed–displaced–rotated state can be obtained from a squeezed–displaced state by modification of the squeeze matrix and of the displacement amount as*

$$\boxed{|z, \alpha, \phi\rangle = \left|e^{i\phi}z\, e^{i\phi^{\mathrm{T}}}, e^{i\phi}\alpha\right\rangle.} \qquad (11.279)$$

Fig. 11.23 A Gaussian state $|\psi(p)\rangle$ is sent through an additive-noise Gaussian channel. At the output the state is still Gaussian but it becomes a mixed state $\rho_{\mathcal{N}}(\psi(p))$

11.20.2 Geometrically Uniform Symmetry for Mixed States

The previous result obtained for pure Gaussian states cannot be extended to the whole class of mixed Gaussian states. In fact the critical point in the proof was the relation $R_N(\phi)|0_N\rangle = |0_N\rangle$, in which the ground state $|0_N\rangle$ "absorbs the rotation". This property does not hold in the N-mode when the ground state is replaced by a general thermal state.

To get useful results we have to limit the class of mixed Gaussian states to an appropriate subclass, obtained in the following way (but this reduced class seems to be the one of interest for the applications). We suppose that a pure Gaussian state $|\psi(p)\rangle$ of the class $\mathcal{S}$ is sent through an additive-noise channel specified by the number of thermal photons $\mathcal{N}$ (see Sect. 12.8). As seen in Sect. 11.16 for the single mode, at the output the noisy state is still Gaussian but specified by a density operator $\rho_{\mathcal{N}}(\psi(p))$ (Fig. 11.23). We denote by $\mathcal{S}_{\mathcal{N}} = \{\rho_{\mathcal{N}}(\psi(p)), p \in P\}$ this restricted subclass of Gaussian mixed states.

Theorem 11.9 *If the class of pure Gaussian states $\mathcal{S} = \{|\psi(p)\rangle, p \in P\}$, is closed under rotations, also the class of noisy states $\mathcal{S}_{\mathcal{N}} = \{\rho_{\mathcal{N}}(\psi(p)), p \in P\}$, obtained at the output of an additive-noise channel, is closed under rotations.*[23]

Proof The mean vector is not modified by an additive-noise channel, while the covariance matrix is modified as (see Sect. 12.8)[24]

$$V \quad \rightarrow \quad V + 2\mathcal{N}\, I_{2N}.$$

Let $V(p_\phi)$ be the covariance matrix corresponding to the rotation of ϕ in the class $\mathcal{S}$, so that we have to prove that, if $V(p_\phi) = S_{\mathrm{rot}}(\phi)V(p_0)S_{\mathrm{rot}}^{\mathrm{T}}(\phi)$, then

$$S_{\mathrm{rot}}(\phi)(V(p_0) + 2\mathcal{N}\, I_{2N})S_{\mathrm{rot}}^{\mathrm{T}}(\phi) = V(p_\phi) + 2\mathcal{N}\, I_{2N}.$$

---

[23] Recently this result was extended to other Gaussian channels, as attenuation channels [37].

[24] The restriction in this assumption is that in all the N-modes the average number of photons is the same. In general, denoting by $\mathcal{N}_i$ the average number of photons in the i-mode, the covariance relation should be modified as

$$V \quad \rightarrow \quad V + \bigoplus_{i=1}^{N} 2\mathcal{N}_i\, I_2.$$

In fact $S_{\text{rot}}(\phi)(V(p_0) + 2N\, I_{2N})S_{\text{rot}}^{\text{T}}(\phi) = S_{\text{rot}}(\phi)V(p_0)S_{\text{rot}}^{\text{T}}(\phi) + 2N\, S_{\text{rot}}(\phi)$
$S_{\text{rot}}^{\text{T}}(\phi)$, where $S_{\text{rot}}(\phi)$ verifies the condition $S_{\text{rot}}(\phi)S_{\text{rot}}^{\text{T}}(\phi) = I_{2N}$.

Note that the alternative proof in the bosonic space, based or the relation

$$R(\phi)\rho_{\mathcal{N}}(|\psi(p_0)\rangle R^*(\phi) = \rho_{\mathcal{N}}\left(\left|\psi(p_\phi)\right\rangle\right)$$

is very difficult for the complicated expressions of $\psi(p)$ and of $\rho_{\mathcal{N}}(\psi(p))$ (see [32]). Moreover, expressions of $\rho_{\mathcal{N}}(\psi(p))$ are not known explicitly for some classes, e.g. for squeezed states, and notwithstanding that it is possible to prove that their class is closed under rotations.

11.20.3 *Application to Pulse Position Modulation (PPM)*

The application of GUS to the PSK constellation is trivial, being based on the single mode. The application to PPM is less trivial because in this format the states become multimode. The symmetry operator S of K-ary PPM was defined in Proposition 7.3 of Sect. 7.13 and works in the Hilbert space $\mathcal{H} = \mathcal{H}_0^{\otimes K}$, where $\mathcal{H}_0$ has dimension n and $\mathcal{H}$ has dimension $N = n^K$. The expression of S is given by

$$S = \sum_{k=0}^{n-1} w_n(k) \otimes I_{n^{K-1}} \otimes w_n^{\text{T}}(k), \tag{11.280}$$

where $\otimes$ is Kronecker's product, $w_n(k)$ is a column vector of length n, with null elements except for one unitary element at position k. Then S is a matrix of dimension $N = n^K$, having the property $S^K = I_N$.

Now, it is not immediate to see that S is a rotation operator, that is, of the form $R_N(\phi) = e^{i\phi}$, with ϕ an $N \times N$ Hermitian matrix. To find the "phase" ϕ we use the EID of S, given by (see Theorem 2.2)

$$S = \sum_{m=0}^{K-1} \lambda_m P_m, \qquad \lambda_m = e^{i\,2\pi m/K} := W_K^m \tag{11.281}$$

where λ_m are the K distinct eigenvalues and P_m are K projectors. In this EID the eigenvalues are known, while the projectors should be evaluated from the expression (11.280), which defines a complicated permutation matrix. The alternative is the evaluation through the powers of S, given by

$$S^k = \sum_{m=0}^{K-1} W_K^{mk} P_m.$$

According to this relation, $[S^0, S^1, \ldots, S^{K-1}]$ turns out to be the DFT of $[P_0, P_1, \ldots, P_{K-1}]$. Thus, taking the inverse DFT one gets

$$P_m = \frac{1}{K} \sum_{k=0}^{K-1} W_K^{-mk} S^k$$

which is easy to evaluate. Next we recall that S is unitary and therefore it can be written in the form (see Theorem 2.5) $S = e^{i\phi}$. Then, by comparison we find that the EID of ϕ is given by

$$\phi = \sum_{m=0}^{K-1} \frac{2\pi m}{K} P_m. \tag{11.282}$$

where the eigenvalues become $2\pi m/K$ and the projectors are the same as in the EID (11.281).

Example 11.2 We give two examples of evaluation:

- $n = 3 \quad K = 2$

The symmetry operator is

$$S = \begin{bmatrix} 1 & 0 & 0 & 0 & 0 & 0 & 0 & 0 & 0 \\ 0 & 0 & 0 & 1 & 0 & 0 & 0 & 0 & 0 \\ 0 & 0 & 0 & 0 & 0 & 0 & 1 & 0 & 0 \\ 0 & 1 & 0 & 0 & 0 & 0 & 0 & 0 & 0 \\ 0 & 0 & 0 & 0 & 1 & 0 & 0 & 0 & 0 \\ 0 & 0 & 0 & 0 & 0 & 0 & 0 & 1 & 0 \\ 0 & 0 & 1 & 0 & 0 & 0 & 0 & 0 & 0 \\ 0 & 0 & 0 & 0 & 0 & 1 & 0 & 0 & 0 \\ 0 & 0 & 0 & 0 & 0 & 0 & 0 & 0 & 1 \end{bmatrix}$$

and the projectors are

$$P_0 = \frac{1}{2} \begin{bmatrix} 2 & 0 & 0 & 0 & 0 & 0 & 0 & 0 & 0 \\ 0 & 1 & 0 & 1 & 0 & 0 & 0 & 0 & 0 \\ 0 & 0 & 1 & 0 & 0 & 0 & 1 & 0 & 0 \\ 0 & 1 & 0 & 1 & 0 & 0 & 0 & 0 & 0 \\ 0 & 0 & 0 & 0 & 2 & 0 & 0 & 0 & 0 \\ 0 & 0 & 0 & 0 & 0 & 1 & 0 & 1 & 0 \\ 0 & 0 & 1 & 0 & 0 & 0 & 1 & 0 & 0 \\ 0 & 0 & 0 & 0 & 0 & 1 & 0 & 1 & 0 \\ 0 & 0 & 0 & 0 & 0 & 0 & 0 & 0 & 2 \end{bmatrix}, \quad P_1 = \frac{1}{2} \begin{bmatrix} 0 & 0 & 0 & 0 & 0 & 0 & 0 & 0 & 0 \\ 0 & 1 & 0 & -1 & 0 & 0 & 0 & 0 & 0 \\ 0 & 0 & 1 & 0 & 0 & 0 & -1 & 0 & 0 \\ 0 & -1 & 0 & 1 & 0 & 0 & 0 & 0 & 0 \\ 0 & 0 & 0 & 0 & 0 & 0 & 0 & 0 & 0 \\ 0 & 0 & 0 & 0 & 0 & 1 & 0 & -1 & 0 \\ 0 & 0 & -1 & 0 & 0 & 0 & 1 & 0 & 0 \\ 0 & 0 & 0 & 0 & 0 & -1 & 0 & 1 & 0 \\ 0 & 0 & 0 & 0 & 0 & 0 & 0 & 0 & 0 \end{bmatrix}$$

The "phase" matrix is

$$\phi = \frac{\pi}{2}
\begin{bmatrix}
0 & 0 & 0 & 0 & 0 & 0 & 0 & 0 & 0 \\
0 & 1 & 0 & -1 & 0 & 0 & 0 & 0 & 0 \\
0 & 0 & 1 & 0 & 0 & 0 & -1 & 0 & 0 \\
0 & -1 & 0 & 1 & 0 & 0 & 0 & 0 & 0 \\
0 & 0 & 0 & 0 & 0 & 0 & 0 & 0 & 0 \\
0 & 0 & 0 & 0 & 0 & 1 & 0 & -1 & 0 \\
0 & 0 & -1 & 0 & 0 & 0 & 1 & 0 & 0 \\
0 & 0 & 0 & 0 & 0 & -1 & 0 & 1 & 0 \\
0 & 0 & 0 & 0 & 0 & 0 & 0 & 0 & 0
\end{bmatrix}$$

- $n = 2 \quad K = 4$

The symmetry operator is

$$S =
\begin{bmatrix}
1&0&0&0&0&0&0&0&0&0&0&0&0&0&0&0 \\
0&0&0&0&0&0&0&0&1&0&0&0&0&0&0&0 \\
0&1&0&0&0&0&0&0&0&0&0&0&0&0&0&0 \\
0&0&0&0&0&0&0&0&0&1&0&0&0&0&0&0 \\
0&0&1&0&0&0&0&0&0&0&0&0&0&0&0&0 \\
0&0&0&0&0&0&0&0&0&0&1&0&0&0&0&0 \\
0&0&0&1&0&0&0&0&0&0&0&0&0&0&0&0 \\
0&0&0&0&0&0&0&0&0&0&0&1&0&0&0&0 \\
0&0&0&0&1&0&0&0&0&0&0&0&0&0&0&0 \\
0&0&0&0&0&0&0&0&0&0&0&0&1&0&0&0 \\
0&0&0&0&0&1&0&0&0&0&0&0&0&0&0&0 \\
0&0&0&0&0&0&0&0&0&0&0&0&0&1&0&0 \\
0&0&0&0&0&0&1&0&0&0&0&0&0&0&0&0 \\
0&0&0&0&0&0&0&0&0&0&0&0&0&0&1&0 \\
0&0&0&0&0&0&0&1&0&0&0&0&0&0&0&0 \\
0&0&0&0&0&0&0&0&0&0&0&0&0&0&0&1
\end{bmatrix}$$

We have evaluated the four 16×16 projectors P_0, P_1, P_2, P_2 and then the "phase" matrix, which results in

$$\phi = \frac{\pi}{4}
\begin{bmatrix}
0&0&0&0&0&0&0&0&0&0&0&0&0&0&0&0 \\
0&3&a^*&0&-1&0&0&0&a&0&0&0&0&0&0&0 \\
0&a&3&0&a^*&0&0&0&-1&0&0&0&0&0&0&0 \\
0&0&0&3&0&0&a^*&0&0&a&0&0&-1&0&0&0 \\
0&-1&a&0&3&0&0&0&a^*&0&0&0&0&0&0&0 \\
0&0&0&0&0&2&0&0&0&0&-2&0&0&0&0&0 \\
0&0&0&a&0&0&3&0&0&-1&0&0&a^*&0&0&0 \\
0&0&0&0&0&0&0&3&0&0&0&a&0&-1&a^*&0 \\
0&a^*&-1&0&a&0&0&0&3&0&0&0&0&0&0&0 \\
0&0&0&a^*&0&0&-1&0&0&3&0&0&a&0&0&0 \\
0&0&0&0&0&-2&0&0&0&0&2&0&0&0&0&0 \\
0&0&0&0&0&0&0&a^*&0&0&0&3&0&a&-1&0 \\
0&0&0&-1&0&0&a&0&0&a^*&0&0&3&0&0&0 \\
0&0&0&0&0&0&0&-1&0&0&0&a^*&0&3&a&0 \\
0&0&0&0&0&0&0&a&0&0&0&-1&0&a^*&3&0 \\
0&0&0&0&0&0&0&0&0&0&0&0&0&0&0&0
\end{bmatrix}, \qquad a = -1 + i.$$

In both cases we have verified (with Mathematica) that $e^{i\phi} = S$ and that $e^{iK\phi} = I_N$.

Problem 11.29 ★★★ Consider the Fock representation of a pure state in the single mode

$$|\psi(p)\rangle = \sum_{n=0}^{\infty} f_n(p) |n\rangle.$$

Prove that the application of the rotation operator $R(\phi)$ to $|\psi(p)\rangle$ modifies the Fourier coefficients as

$$f_n(p) \quad \rightarrow \quad e^{in\phi} f_n(p). \tag{11.283}$$

Problem 11.30 ★★★ Apply the statement of the previous problem to prove that the class of squeezed–displaced states is closed under rotations.

Problem 11.31 ★ Prove that the class of coherent states is closed with respect to rotations, using the Fock representation (11.191).

Appendix

Proof of Baker–Campbell–Hausdorff Identity

For the proof of (11.71) we let

$$E(x) = e^{xH} K e^{-xH}$$

so that $E(0) = K$ and

$$E'(x) = e^{xH} H K e^{-xH} - e^{xH} K H e^{-xH} = e^{xH} [H - K] e^{-xH}$$

from which $D_1 = E'(0) = [H, K]$ and

$$E''(x) = e^{xH} H [H, K] e^{-xH} - e^{xH} [H, K] H e^{-xH} = e^{xH} [H, [H, K]] e^{-xH}.$$

Hence $D_2 = E''(0) = [H, [H, K]]$, and so on. In conclusion, (11.71) is obtained as the Taylor expansion of $E(x)$.

To prove (11.70), note that, if H and $[H, K]$ commute, this relation reduces to $e^{xH} K e^{-xH} = K + [H, K]x$. Now, if we let

$$F(x) := e^{xH} e^{xK},$$

by derivation we get the differential equation

$$F'(x) = He^{xH}e^{xK} + e^{xH}e^{xK}K$$
$$= (H + e^{xH}Ke^{-xH})F(x) = (H + K + [H, K])F(x)$$

whose solution is

$$F(x) = e^{(H+K)x+[H,K]x^2/2} = e^{(H+K)x}e^{[H,K]x^2/2}$$

where we take into account the fact that $F(0) = I$ and that functions of the commuting operator $H + K$ and $[H, K]$ commute as well. Finally, setting $x = 1$ gives the conclusion.

Proof of Fock Representation of Weyl Operator (Proposition 11.7)

Use of (11.126) in (11.129) gives

$$D_{mn}(\xi) = e^{-\frac{1}{2}|\xi|^2} \langle m, \xi | n, \xi^* \rangle$$
$$= e^{-\frac{1}{2}|\xi|^2} \sum_{s=0}^{m} \sum_{r=0}^{n} \mu_{ms}\mu_{nr}\xi^s(-\xi^*)^r \langle m - s | n - r \rangle \tag{11.284}$$
$$= e^{-\frac{1}{2}|\xi|^2} \sum_{s=0}^{m} \sum_{r=0}^{n} \mu_{ms}\mu_{nr}\xi^s(-\xi^*)^r \delta_{m-s,n-r}$$

where the orthonormality of the Fock states has been used. Then

$$D_{mn}(\xi) = e^{-\frac{1}{2}|\xi|^2} \sum_{r=0}^{n} \mu_{m,m-n+r}\mu_{nr}\xi^{m-n+r}(-\xi^*)^r$$
$$= e^{-\frac{1}{2}|\xi|^2} \sum_{r=0}^{n} \frac{1}{(m-n+r)!r!}\sqrt{\frac{m!n!}{(n-r)!(n-r)!}}\xi^{m-n+r}(-\xi^*)^r$$
$$= e^{-\frac{1}{2}|\xi|^2} \sqrt{\frac{n!}{m!}} \sum_{r=0}^{n} \frac{m!}{(m-n+r)!r!(n-r)!}\xi^{m-n+r}(-\xi^*)^r$$
$$= e^{-\frac{1}{2}|\xi|^2} \sqrt{\frac{n!}{m!}} \sum_{r=0}^{n} \binom{m}{n-r}\frac{1}{r!}\xi^{m-n+r}(-\xi^*)^r.$$

and (11.130) follows at once.

Proof of Proposition 11.9 on the Orthogonality of the D_{mn}

We have to prove that the coefficients D_{mn} giving the Fock representation of the Weyl operator are orthogonal functions in $\mathbb{C}$, that is,

$$I_{mn}^{rs} := \frac{1}{\pi} \int_{\mathbb{C}} \mathrm{d}\xi \; D_{mn}^{*}(\xi)\, D_{rs}(\xi) = \delta_{mr}\delta_{ns}. \qquad (11.285)$$

To this end in the integrand we apply relation (11.130) twice to get

$$D_{mn}^{*}(\xi)\, D_{rs}(\xi) = \mathrm{e}^{-|\xi|^2}\, Z\, (\xi^{*})^{m-n}\xi^{r-s} L_n^{(m-n)}(|\xi|^2) L_s^{(r-s)}(|\xi|^2)$$

where $Z = \sqrt{n!s!/m!r!}$. Then we use polar coordinates letting $\xi = \sigma\, \mathrm{e}^{\mathrm{i}\phi}$

$$D_{mn}^{*}(\xi)\, D_{rs}(\xi) = Z\, \mathrm{e}^{-\sigma^2}\sigma^{m-n+r-s}\mathrm{e}^{\mathrm{i}\phi(n-m+r-s)} L_n^{(m-n)}(\sigma^2) L_s^{(r-s)}(\sigma^2)$$

and the integral becomes

$$I_{mn}^{rs} = Z\frac{1}{\pi}\int_0^{2\pi} \mathrm{e}^{\mathrm{i}\phi(n-m+r-s)}\mathrm{d}\phi \int_0^{\infty} \mathrm{e}^{-\sigma^2}\sigma^{m-n+r-s} L_n^{(m-n)}(\sigma^2) L_s^{(r-s)}(\sigma^2)\, \sigma\, \mathrm{d}\sigma$$

$$= Z\, 2\delta_{m-n,r-s}\int_0^{\infty} \mathrm{e}^{-\sigma^2}\sigma^{m-n+r-s} L_n^{(m-n)}(\sigma^2) L_s^{(r-s)}(\sigma^2)\, \sigma\, \mathrm{d}\sigma$$

$$= Z\, \delta_{m-n,r-s}\int_0^{\infty} \mathrm{e}^{-\sigma^2}\sigma^{2(m-n)} L_n^{(m-n)}(\sigma^2) L_s^{(m-n)}(\sigma^2)\, 2\sigma\, \mathrm{d}\sigma$$

$$= Z\, \delta_{m-n,r-s}\int_0^{\infty} \mathrm{e}^{-x} x^{m-n} L_n^{(m-n)}(x) L_s^{(m-n)}(x)\, \mathrm{d}x$$

where we can use the orthogonality of the generalized Laguerre polynomials, given by (11.128) (see also [25, p.8097.414–3]), to get

$$I_{mn}^{rs} = Z\, \delta_{m-n,r-s}\frac{\Gamma(m+1)}{n!}\delta_{n,s} = \sqrt{\frac{n!s!}{m!r!}}\delta_{m-n,r-s}\frac{m!}{n!}\delta_{n,s} = \delta_{mr}\delta_{ns}.$$

About Symplectic and Bogoliubov Transformations

Relations Between the Parameters of the Two Transformations

We prove Proposition 11.11. For brevity we develop the case $N = 2$, where

$$a = \begin{bmatrix} a_1 \\ a_2 \end{bmatrix}, \qquad a_{\star} = \begin{bmatrix} a_1^{*} \\ a_2^{*} \end{bmatrix}, \qquad q = \begin{bmatrix} q_1 \\ q_2 \end{bmatrix}, \qquad p = \begin{bmatrix} p_1 \\ p_2 \end{bmatrix},$$

$$q_p = X = \begin{bmatrix} X_1 \\ X_2 \\ X_3 \\ X_4 \end{bmatrix} = \begin{bmatrix} q_1 \\ p_1 \\ q_2 \\ p_2 \end{bmatrix}, \qquad Y = \begin{bmatrix} Y_1 \\ Y_2 \\ Y_3 \\ Y_4 \end{bmatrix} = \begin{bmatrix} q_1 \\ q_2 \\ p_1 \\ p_2 \end{bmatrix} = \begin{bmatrix} q \\ p \end{bmatrix}.$$

Note that

$$X = \Pi \, Y, \qquad Y = \Pi^{-1} X = \Pi \, X \tag{11.286}$$

where Π is the $2N \times 2N$ permutation matrix defined in Proposition 11.11 and having the property $\Pi^{-1} = \Pi^{\mathrm{T}}$.

It is easier to relate the vector of operators Y to the vectors a and $a_\star$ because

$$\Re a = q, \qquad \Im a = p, \qquad \Re a_\star = q, \qquad \Im a_\star = -p.$$

Now, we introduce the real and the imaginary parts in the relation $\tilde{a} = E\,a + F\,a_\star + y$ to get

$$\tilde{q} + i\tilde{p} = (\Re E + i\Im E)(q + ip) + (\Re F + i\Im F)(q - ip) + (\Re y + i\Im y)$$

Hence

$$\tilde{q} = (\Re E + \Re F)\,q + (-\Im E + \Im F)\,p + \Re y$$
$$\tilde{p} = (\Im E + \Im F)\,q + (\Re E - \Re F)\,p + \Im y$$

which can be written in the form

$$Y = T\,X + d_y \tag{11.287}$$

where

$$T = \begin{bmatrix} \Re(E + F) & \Im(-E + F) \\ \Im(E + F) & \Re(E - F) \end{bmatrix}, \qquad d_y = \begin{bmatrix} \Re y \\ \Im y \end{bmatrix} \tag{11.288}$$

Considering (11.286) and (11.150), relation (11.287) allows us to relate $\tilde{X}$ to X as $\tilde{X} = S\,X + d$ with $S = \Pi\,T\,\Pi^{\mathrm{T}}$, $T = \Pi^{\mathrm{T}} S\,\Pi$ and $d = \Pi\,d_y$. This completes the proof.

Commutation Conditions

Commutation conditions are given by (11.148) for Bogoliubov transformations and by (11.150) for symplectic transformations. We first prove (11.148), that is,

$$E E^* - F F^* = I, \qquad E F^{\mathrm{T}} = F E^{\mathrm{T}}. \tag{11.289}$$

We write the Bogoliubov relation (11.147) in scalar form

$$\tilde{a}_i = \sum_k E_{ik}\, a_k + \sum_l F_{il}\, a_l^* + z_i.$$

Then the Hermitian conjugate becomes

$$\tilde{a}_j^* = \sum_r E_{jr}^*\, a_r^* + \sum_s F_{js}^*\, a_s + z_j.$$

Next, using the bilinearity of the commutator, one gets

$$\tilde{a}_i, \tilde{a}_j^* t = \left[\sum_k E_{ik}\, a_k + \sum_l F_{il}\, a_l^* + z_i,\ \sum_r E_{jr}^*\, a_r^* + \sum_s F_{js}^*\, a_s + z_j \right]$$

$$= \sum_{k,r} E_{ik}\, E_{jr}^*\, [a_k, a_r^*] + \sum_{k,s} E_{ik}\, F_{js}^*\, [a_k, a_s]$$

$$+ \sum_{l,r} F_{il}\, E_{jr}^*\, [a_l^*, a_r^*] + \sum_{l,s} F_{il}\, F_{js}^*\, [a_l^*, a_s]$$

where (see (11.64)) $[a_k, a_r^*] = I\,\delta_{k,r}$, $[a_l^*, a_s] = -I\,\delta_{s,s}$, $[a_k, a_s] = [a_l^*, a_r^*] = 0$.
Hence

$$[\tilde{a}_i, \tilde{a}_j^*] = \sum_k E_{ik}\, E_{jk}^*\, I - \sum_l F_{il}\, F_{jl}^*\, I.$$

If $EE^* - FF^* = I$ the latter expression gives $[\tilde{a}_i, \tilde{a}_j^*] = I$. Analogously we find

$$[\tilde{a}_i, \tilde{a}_j] = \sum_k E_{ik}\, F_{jk}\, I - \sum_l F_{il}\, E_{jl}\, I$$

which gives $[\tilde{a}_i, \tilde{a}_j] = 0$ as soon as $EF^{\mathrm{T}} = FE^{\mathrm{T}}$.

Next we consider the symplectic condition (11.150), that is,

$$S\,\Omega\, S^{\mathrm{T}} = \Omega, \tag{11.290}$$

and, to establish the equivalence between the two commutation conditions, we prove that (11.290) implies the first of (11.289). To this end we write (11.290) in terms of the matrix T, namely

$$\Pi T \Pi^{\mathrm{T}}\, \Omega \Pi T^{\mathrm{T}} \Pi^{\mathrm{T}} = \Omega$$

that is,

$$T\Omega_0\, T^{\mathrm{T}} = \Omega_0 \quad \text{with} \quad \Omega_0 = \Pi^{\mathrm{T}}\Omega\,\Pi \quad \rightarrow \quad \Omega = \Pi\,\Omega_0\,\Pi^{\mathrm{T}}. \tag{11.291}$$

Now we note that for $N = 2$

$$\Omega = \begin{bmatrix} 0 & 1 & 0 & 0 \\ -1 & 0 & 0 & 0 \\ 0 & 0 & 0 & 1 \\ 0 & 0 & -1 & 0 \end{bmatrix}, \qquad \Omega_0 = \begin{bmatrix} 0 & 0 & 1 & 0 \\ 0 & 0 & 0 & 1 \\ -1 & 0 & 0 & 0 \\ 0 & -1 & 0 & 0 \end{bmatrix}.$$

For $N = 3$

$$\Omega = \begin{bmatrix} 0 & 1 & 0 & 0 & 0 & 0 \\ -1 & 0 & 0 & 0 & 0 & 0 \\ 0 & 0 & 0 & 1 & 0 & 0 \\ 0 & 0 & -1 & 0 & 0 & 0 \\ 0 & 0 & 0 & 0 & 0 & 1 \\ 0 & 0 & 0 & 0 & -1 & 0 \end{bmatrix}, \qquad \Omega_0 = \begin{bmatrix} 0 & 0 & 0 & 1 & 0 & 0 \\ 0 & 0 & 0 & 0 & 1 & 0 \\ 0 & 0 & 0 & 0 & 0 & 1 \\ -1 & 0 & 0 & 0 & 0 & 0 \\ 0 & -1 & 0 & 0 & 0 & 0 \\ 0 & 0 & -1 & 0 & 0 & 0 \end{bmatrix}.$$

Note that in general the matrix Ω_0 has the block form

$$\Omega_0 = \begin{bmatrix} 0 & I_N \\ -I_N & 0 \end{bmatrix} \tag{11.292}$$

where I_N is the identity matrix of order N. Then condition (11.291) becomes explicitly

$$\begin{bmatrix} \Re(E + F) & \Im(-E + F) \\ \Im(E + F) & \Re(E - F) \end{bmatrix} \begin{bmatrix} 0 & I_N \\ -I_N & 0 \end{bmatrix} \begin{bmatrix} \Re(E^\mathrm{T} + F^\mathrm{T}) & \Im(E^\mathrm{T} + F^\mathrm{T}) \\ \Im(-E^\mathrm{T} + F^\mathrm{T}) & \Re(E^\mathrm{T} - F^\mathrm{T}) \end{bmatrix} = \begin{bmatrix} 0 & I_N \\ -I_N & 0 \end{bmatrix}.$$

With $c = u + v$ and $d = u - v$ we get the four matrix equations

$$\Im(d)\Re(c^\mathrm{T}) - \Re(c)\Im(d^\mathrm{T}) = 0, \qquad \Im(d)\Im(c^\mathrm{T}) + \Re(c)\Re(d^\mathrm{T}) = I$$
$$\Im(c)\Im(d^\mathrm{T}) + \Re(d)\Re(c^\mathrm{T}) = I, \qquad \Im(c)\Re(d^\mathrm{T}) - \Re(d)\Im(c^\mathrm{T}) = 0.$$

The sum of the first multiplied by i with the second and the sum of the third with the fourth multiplied by i give, respectively,

$$i\,\Im(d)[\Re(c^\mathrm{T}) - i\,\Re(c^\mathrm{T})] + \Re(c)[\Re(d^\mathrm{T}) - i\,\Im(d^\mathrm{T})] = i\,\Im(d)\,c^* + \Re(c)\,d^* = I$$
$$i\,\Im(c)[\Re(d^\mathrm{T}) - i\,\Re(d^\mathrm{T})] + \Re(d)[\Re(c^\mathrm{T}) - i\,\Im(c^\mathrm{T})] = i\,\Im(c)\,d^* + \Re(d)\,c^* = I.$$

Hence

$$c\,d^* + d\,c^* = 2I \qquad \rightarrow \qquad E\,E^* - F\,F^* = I.$$

Proof of the Gaussianity Preservation Theorem (Theorem 11.1)

For brevity we let $t = u_v \in \mathbb{R}^{2N}$ and $x = q_p$, so that the Weyl operator reads as $D(t) = \mathrm{e}^{\mathrm{i}x^{\mathrm{T}}\Omega t}$. We develop the proof with the following steps:

(1) The symplectic transformation modifies the density operator and the Weyl operator as

$$\tilde{\rho} = U^*_{S,d}\,\rho\,U_{S,d}\,, \qquad \tilde{D}(t) = U^*_{S,d}\,D(t)\,U_{S,d},\tag{11.293}$$

where t denotes the pairs of vector (x, p) or their interlaced form x_p.
(2) Considering that the map for the quadrature operators is $\tilde{x} = Sx + d$, the new Weyl operator is more explicitly given by

$$\tilde{D}(t) = \mathrm{e}^{\mathrm{i}(Sx+d)^{\mathrm{T}}\Omega t}.\tag{11.294}$$

(3) We evaluate the new characteristic function, which is given by

$$\begin{aligned}
\tilde{\chi}(t) = \mathrm{Tr}[\tilde{\rho}\,D(t)] &= \mathrm{Tr}\left[U_{S,d}\,\rho\,U^*_{S,d}\,D(t)\right]\\
&= \mathrm{Tr}\left[\rho\,U^*_{S,d}\,D(t)\,U_{S,d}\right] = \mathrm{Tr}\left[\rho\,\tilde{D}(t)\right].
\end{aligned}\tag{11.295}$$

(4) We use the Gaussianity of the input state ρ, according to which the characteristic function is given by

$$\chi(t) = \exp\left[-\tfrac{1}{2}\left(\Omega\,V\Omega^{\mathrm{T}}\right)t - \mathrm{i}(\Omega\bar{x})^{\mathrm{T}}t\right].\tag{11.296}$$

Now we develop the above steps, taking into account the following properties of the matrices S and Ω (see (11.67) and (11.150)):

$$\Omega^{\mathrm{T}} = -\Omega\,, \qquad S^{\mathrm{T}}\Omega S = \Omega \quad \rightarrow \quad S^{\mathrm{T}}\Omega = \Omega S^{-1}.$$

Then (11.294) becomes explicitly

$$\begin{aligned}
\tilde{D}(t) &= \mathrm{e}^{\mathrm{i}(Sx+d)^{\mathrm{T}}\Omega t} = \mathrm{e}^{\mathrm{i}(Sx)^{\mathrm{T}}\Omega t}\mathrm{e}^{\mathrm{i}d^{\mathrm{T}}\Omega t}\\
&= \mathrm{e}^{\mathrm{i}x^{\mathrm{T}}S^{\mathrm{T}}\Omega t}\,\mathrm{e}^{\mathrm{i}d^{\mathrm{T}}\Omega t} = \mathrm{e}^{\mathrm{i}x^{\mathrm{T}}\Omega S^{-1}t}\mathrm{e}^{\mathrm{i}d^{\mathrm{T}}\Omega t}\\
&= D(S^{-1}t)\,\mathrm{e}^{\mathrm{i}d^{\mathrm{T}}\Omega t}
\end{aligned}\tag{11.297}$$

and the new characteristic function (11.295) reads as

$$\begin{aligned}
\tilde{\chi}(t) = \mathrm{Tr}\left[\rho\,\tilde{D}(t)\right] &= \mathrm{Tr}\left[\rho\,D_x(S^{-1}t)\,\mathrm{e}^{\mathrm{i}d^{\mathrm{T}}\Omega t}\right]\\
&= \chi(S^{-1}t)\,\mathrm{e}^{\mathrm{i}d^{\mathrm{T}}\Omega t}.
\end{aligned}\tag{11.298}$$

Finally we use the Gaussianity (11.296) for $\chi(t)$ to get

$$\tilde{\chi}(t) = \chi(S^{-1}t)\, e^{i\,d^{\mathrm{T}}\Omega\, t}$$

$$= \exp\left[-\frac{1}{2}(S^{-1}t)^{\mathrm{T}}\left(\Omega\, V\Omega^{\mathrm{T}}\right) S^{-1}t - i(\Omega\bar{x})^{\mathrm{T}}S^{-1}t + i\,d^{\mathrm{T}}\Omega\, t\right]$$

where

$$(S^{-1}t)^{\mathrm{T}}\Omega = t^{\mathrm{T}}S^{-1^{\mathrm{T}}}\Omega = t^{\mathrm{T}}\Omega\, S\,, \qquad \Omega^{\mathrm{T}}S^{-1} = S^{\mathrm{T}}\,\Omega^{\mathrm{T}}$$

$$-i(\Omega\bar{x})^{\mathrm{T}}S^{-1}t + i\,d^{\mathrm{T}}\Omega t = -i\left[\bar{x}^{\mathrm{T}}\Omega^{\mathrm{T}}S^{-1} - d^{\mathrm{T}}\Omega\right]t$$

$$= -i\left\{\bar{x}^{\mathrm{T}}S^{\mathrm{T}}\Omega^{\mathrm{T}} + d^{\mathrm{T}}\Omega^{\mathrm{T}}\right\}t = -i\{\Omega(S\bar{x}+d)\}^{\mathrm{T}}\,.$$

Hence

$$\tilde{\chi}(t) = \exp\left[-\frac{1}{2}\left(t^{\mathrm{T}}\Omega\, S\, V S^{\mathrm{T}}\,\Omega^{\mathrm{T}}\, t\right) - i\{\Omega(S\bar{x}+d)\}^{\mathrm{T}}\right]$$

which is the characteristic function of a Gaussian state with the parameters indicated in (11.146).

Squeezed States. Proof of Bogoliubov Transformation (11.171)

The relation to develop is

$$Z^*(z)\, a\, Z(z) = e^{\frac{1}{2}(z^*a^2 - z\,b^2)}\, a\, e^{\frac{1}{2}(zb^2 - z^*a^2)}\,, \qquad (b = a^*).$$

We apply BCH formula (11.71) with

$$x = 1\,, \qquad H = \tfrac{1}{2}(z^*a^2 - z\,b^2)\,, \qquad K = a.$$

Then

$$Z^*(z)\, a\, Z(z) = \sum_{n=0}^{\infty}\frac{1}{n!}\, D_n$$

where $D_0 = K = a$ and we have the recursion $D_n = \tfrac{1}{2}[(z^*a^2 - z\,b^2), D_{n-1}]$. We prove that

$$D_{2n} = (zz^*)^n a\,, \qquad D_{2n+1} = (zz^*)^n z\, b. \tag{11.299}$$

In this case the technique is the introduction of the commutation relations

$$[b, a] = ba - ab = -I\,, \qquad [a, b] = ab - ba = I$$

which give

$$[b^2, a] = -2b\,, \qquad [a^2, b] = -2a\,, \qquad [b^2, b] = 0\,, \qquad [a^2, a] = 0. \qquad (11.300)$$

In fact

$$[b^2, a] = b^2 a - ab^2 = b^2 a - bab + bab - ab^2 = b(ba - ab) + (ba - ab)b = -2b$$
$$[a^2, b] = a^2 b - ba^2 = a^2 b - aba + aba - ba^2 = a(ab - ba) + (ab - ba)a = 2a.$$

Then

$$D_1 = \frac{1}{2}[(z^* a^2 - z\, b^2), a] = \frac{1}{2}z^*[a^2, a] - \frac{1}{2}z[b^2, a] = b$$

and so on. In conclusion

$$Z^*(z)\, a\, Z(z) = a + z\, a^* + \frac{1}{2!}\, zz^*\, a + \frac{1}{3!}\, (zz^*)z\, a^* + \frac{1}{4!}\, (zz^*)^2\, a + \cdots \qquad (11.301)$$

Letting $z = r\, e^{i\theta}$, the even and odd terms in (11.301) give respectively

$$E = (1 + \frac{1}{2!}\, r^2 + \frac{1}{4!}\, r^4 + \cdots)\, a = \cosh r\, a$$
$$O = (r + \frac{1}{3!}\, r^3 + \frac{1}{5!}\, r^5 \cdots)e^{i\theta}\, b = \sinh r e^{i\theta}\, b$$

and (11.171) follows.

The Most General Gaussian Unitary in the Single Mode

In the phase space a Gaussian unitary is represented by the pair (S, d) with

$$S = \begin{bmatrix} s_{11} & s_{12} \\ s_{21} & s_{22} \end{bmatrix}, \qquad d = \begin{bmatrix} d_1 \\ d_2 \end{bmatrix}. \qquad (11.302)$$

We have to find the most general pair (S, d) with the appropriate constraints. In (11.302) all the six parameters are real and the displacement has the only constraint $d \in \mathbb{R}^2$. The matrix S should be symplectic, $S\,\Omega\,S^{\mathrm{T}} = \Omega$, which gives the condition $s_{11}s_{22} - s_{12}s_{21} = \det S = 1$. Now it is possible to find which Gaussian unitaries ensure these conditions. This is provided by the SVD, which gives

$$S = S_{\mathrm{rot}}(\theta)\, S_{\mathrm{sq}}(r)\, S_{\mathrm{rot}}(\phi) \qquad (11.303)$$

where

$$S_{\text{rot}}(\theta) = \begin{bmatrix} \cos\theta - \sin\theta \\ \sin\theta \ \ \cos\theta \end{bmatrix}, \quad S_{\text{sq}}(r) = \begin{bmatrix} e^{-r} \ 0 \\ 0 \ \ e^{r} \end{bmatrix}, \quad S_{\text{rot}}(\phi) = \begin{bmatrix} \cos\phi - \sin\phi \\ \sin\phi \ \ \cos\phi \end{bmatrix}.$$

It is not easy to get the SVD (11.303).[25] We have used `Mathematica`, calling $\{U, \Lambda, V\} = $ `SingularValueDecomposition[S]`, where

$$S = \begin{bmatrix} a \ b \\ c \ d \end{bmatrix}.$$

`Mathematica` gives complicated formulas for the factors, but imposing that S be real and symplectic ($ad - bc = \det S = 1$), it was possible (with a great effort) to reduce the formulas in the following form. The singular values are

$$\lambda_{\pm} = \frac{\sqrt{A^2 \pm \sqrt{A^4 - 4}}}{\sqrt{2}}, \qquad A^2 = a^2 + b^2 + c^2 + d^2$$

and are each other reciprocal, $\lambda_{-}\lambda_{+} = 1$. This justifies the structure of the matrix $S_{\text{sq}}(r)$ with $e^{-r} = \lambda_{-}$ and $e^{r} = \lambda_{+}$. The left rotation matrix is

$$U = S_{\text{rot}}(\theta) = \begin{bmatrix} \cos\theta \ -\sin\theta \\ \sin\theta \ \ \cos\theta \end{bmatrix} = \begin{bmatrix} \dfrac{f_1}{\sqrt{|f_1|^2 + |f_2|^2}} & -\dfrac{f_2}{\sqrt{|f_1|^2 + |f_2|^2}} \\ \dfrac{f_2}{\sqrt{|f_1|^2 + |f_2|^2}} & \dfrac{f_1}{\sqrt{|f_1|^2 + |f_2|^2}} \end{bmatrix}$$

where

$$f_1 = a^3 + ab^2 - aB + ac^2 + 2bcd - ad^2, \qquad B := \sqrt{A^4 - 4}$$
$$f_2 = a^2 c - b^2 c - Bc + c^3 + 2abd + cd^2.$$

The right rotation matrix is

$$V = S_{\text{rot}}(\phi) = \begin{bmatrix} \cos\phi \ -\sin\phi \\ \sin\phi \ \ \cos\phi \end{bmatrix} = \begin{bmatrix} \dfrac{T}{\sqrt{T^2+1}} & -\dfrac{1}{\sqrt{T^2+1}} \\ \dfrac{1}{\sqrt{T^2+1}} & \dfrac{T}{\sqrt{T^2+1}} \end{bmatrix}$$

where

$$T = \frac{a^2 - b^2 + c^2 - d^2 + \sqrt{\left[(b+c)^2 + (a-d)^2\right]\left[(b-c)^2 + (a+d)^2\right]}}{2(ab + cd)}.$$

Now we search for the most general Gaussian unitary, starting from (11.303). We have the cascade of three symplectic transformations, which corresponds to

---

[25] A similar decomposition is given in [38], but not with the explicit evaluation of the factors.

the cascade of unitaries $U_S = R(\theta)Z(r)R(\phi)$. But these unitaries lead to a zero mean vector and to get a mean vector d we add the displacement operator $D(d)$, which does not modify the global symplectic matrix. Then we get the cascade $U = D(d)R(\theta)Z(r)R(\phi)$. Next we apply the switching condition (11.176b) to get $U = D(d)R(\theta)R(\phi)Z(\mathrm{e}^{\mathrm{i}\phi} r\, \mathrm{e}^{\mathrm{i}\phi})$. Also, in the single mode, $R(\theta)R(\phi) = R(\theta + \phi)$. In conclusion we have

$$U = D(d)R(\phi_0)Z(z) \qquad (11.304)$$

where d, $\phi_0 = \theta + \phi$, and $z = \mathrm{e}^{\mathrm{i}\phi} r\, \mathrm{e}^{\mathrm{i}\phi}$ are arbitrary.

On the other hand, use of relations (11.176) allows to get an arbitrary order of the three fundamental unitaries. This complete the proof of Theorem 11.6.

Two-Mode Squeezed States and EPR States

For the proof of Bogoliubov transformation we write explicitly (11.147) in the two-mode

$$U^* a_1 U = E_{11} a_1 + E_{12} a_2 + F_{11} a_1^* + F_{12} a_2^*$$
$$U^* a_2 U = E_{21} a_1 + E_{22} a_2 + F_{21} a_1^* + F_{22} a_2^*.$$

We recognize that in the first we can apply the BCH identity with

$$x = \tfrac{1}{2} r_0, \qquad H = a_1^* a_2^* - a_1 a_2, \qquad K = a_1.$$

Hence

$$\tilde{a}_1 = \sum_{n=0}^{\infty} \frac{r_0^n}{2^n n!} D_n(1) \qquad (11.305)$$

where $D_0(1) = a_1$ and $D_n(1) = [a, D_{n-1}(1)]$, $n \geq 1$. Then, considering that operators of different modes commute, we find

$$D_1(1) = [A, a_1] = a_1^* a_2^* a_1 - a_1 a_1^* a_2^* - a_1 a_2 a_1 + a_1 a_2 a_1$$
$$= a_2^* a_1^* a_1 - a_2^* a_1 a_1^*$$
$$= a_2^*(a_1^* a_1 - a_1 a_1^*) = -a_2^*.$$

Analogously

$$D_2(1) = [a_1^* a_2^* - a_1 a_2 - a_2^*] = a_1(a_2 a_2^* - a_2^* a_2) = a_1$$
$$D_3(1) = [a_1^* a_2^* - a_1 a_2 a_1] = -a_2^*$$

and in general

$$D_{2n}(1) = a_1, \qquad D_{2n+1}(1) = -a_2^*.$$

Hence (11.305) becomes

$$\tilde{a}_1 = \sum_{n=0}^{\infty} \frac{r_0^{2n}}{2^{2n}(2n)!} D_{2n}(1) + \sum_{n=0}^{\infty} \frac{r_0^{2n+1}}{2^{2n+1}(2n+1)!} D_{2n+1}(1)$$

$$= \sum_{n=0}^{\infty} \frac{r_0^{2n}}{2^{2n}(2n)!} a_1 - \sum_{n=0}^{\infty} \frac{r_0^{2n+1}}{2^{2n+1}(2n+1)!} a_2.$$

Alternative to Yuen's Formula for Squeezed–Displaced States

We begin with the evaluation of the expansion of the state (11.228). The parameters needed for the expansion are (see Sect. 11.15.2)

$$\beta = \cosh r\, \alpha - e^{i\theta}\, \sinh r\, \alpha^*$$

$$K_0 = (\operatorname{sech} r)^{1/2} \exp\left[-\tfrac{1}{2}(|\beta|^2 + \beta^2 e^{i\theta}\, \tanh r)\right]$$

$$B(\beta, a) := \exp\left[\beta \operatorname{sech} r\, a^* + \tfrac{1}{2}e^{i\theta}\, \tanh r\, (a^*)^2\right] = \exp\left[u\, a^* + v\, (a^*)^2\right]$$

where $u = \beta \operatorname{sech} r$ and $v = \tfrac{1}{2}e^{i\theta}\, \tanh r$. The expansion of the exponential reads as

$$B(\beta, a) = \exp\left[u\, a^* + v\, (a^*)^2\right] = \sum_{n=0}^{\infty} b(n)\, (a^*)^n \tag{11.306}$$

and the Fock coefficients are then given by

$$|z, \alpha\rangle_n = K_0\, b(n) \sqrt{n!}.$$

Now we find that the coefficients $b(n)$ of the expansion are explicitly given by

$$b(n) = \sum_{j=0}^{\lfloor n/2 \rfloor} \frac{1}{(n-2j)!\, j!} u^{n-2j} v^j = u^n \mathcal{H}_n\left(\frac{v}{u^2}\right)$$

where $\mathcal{H}_n(x)$ is the polynomial defined by (11.231). Considering (11.229a), this completes the proof of (11.230).

To see the equivalence of formula (11.229) with Yuen's formula (7.148), we compare the polynomials $\mathcal{H}_n(x)$ with the (physicists') Hermite polynomials, which are given by

$$H_n(x) = n! \sum_{m=0}^{\lfloor n/2 \rfloor} \frac{(-1)^m}{m!(n-2m)!} (2x)^{n-2m}.$$

The comparison gives

$$\mathcal{H}_n(x) = H_n\left(\frac{1}{2\sqrt{-x}}\right)\frac{(-x)^{n/2}}{n!}.$$

Using this relation we can see the equivalence.

The polynomials $\mathcal{H}_n(x)$ are related to the confluent hypergeometric function ${}_1F_1$ as

$$\mathcal{H}_{2n}(x) = \frac{1}{n!}\,x^{-n}\,{}_1F_1\left[-n;\frac{1}{2};-\frac{1}{4x}\right]$$

$$\mathcal{H}_{2n+1}(x) = \frac{1}{n!}\,x^{-n-1}\,{}_1F_1\left[-n;\frac{3}{2};-\frac{1}{4x}\right].$$

Proof of Fock Expansion (11.248) of Caves–Schumaker States

The coefficients $b(n_1, n_2)$ are obtained from the expansion

$$e^{u_1t_1+u_2t_2+u_{12}t_1t_2} = \sum_{n_1=0}^{\infty}\sum_{n_2=0}^{\infty} b(n_1, n_2)\, t_1^{n_1} t_2^{n_2}. \tag{11.307}$$

The exponential has the following expansion:

$$e^{u_1t_1+u_2t_2+u_{12}t_1t_2} = \sum_{\ell=0}^{\infty}\frac{1}{\ell!}(u_1t_1 + u_2t_2 + u_{12}t_1t_2)^{\ell}$$

$$= \sum_{ijk}\frac{1}{i!\,j!\,k!}(u_1t_1)^i (u_2t_2)^j (u_{12}t_1t_2)^k$$

$$= \sum_{ijk}\frac{1}{i!\,j!\,k!}u_1^i u_2^j u_{12}^k\, t_1^{i+k} t_2^{j+k}$$

where the summation is extended to the naturals $i,\, j,\, k$ such that $i + j + k = \ell$. With $n_1 = i + k$ and $n_2 = j + k$ the coefficients read as

$$b(n_1, n_2) = \sum_{k=0}^{\min(n_1,n_2)}\frac{1}{(n_1 - k)!(n_2 - k)!k!}u_1^{n_1-k}u_2^{n_2-k}u_{12}^k$$

$$u_1^{n_1}u_2^{n_2}\sum_{k=0}^{\min(n_1,n_2)}\frac{1}{(n_1 - k)!(n_2 - k)!k!}\left(\frac{u_{12}}{u_1u_2}\right)^k$$

and can be expressed in the form

$$b(n_1, n_2) = u_1^{n_1} u_2^{n_2} \mathcal{L}_{n_1 n_2} \left(\frac{u_{12}}{u_1 u_2} \right)$$

where $\mathcal{L}_{n_1 n_2}(x)$ are the polynomials defined by (11.248a).

In [28] the Fock expansion of the state CS is formulated through the generalized Laguerre polynomials, which verify the identity

$$\sum_{j=0}^{p} \frac{1}{j!(n_1 - j)(n_2 - j)!} (-x)^{-j} = L_p^{q-p}(x) \frac{1}{q!(-x)^p}. \qquad (11.308)$$

Then we have the relation

$$\mathcal{L}_{n_1 n_2}(x) = L_p^{q-p} \left(-\frac{1}{x} \right) \frac{x^p}{q!}.$$

The polynomial $\mathcal{L}_{n_1 n_2}(x)$ can be also expressed through confluent hypergeometric function $U(a, b, c)$ as

$$\mathcal{L}_{n_1 n_2}(x) = \frac{1}{n_1! n_2!} \left(-\frac{1}{x} \right)^{n_1} U \left(-n_1, -n_1 + n_2 + 1, -\frac{1}{x} \right).$$

References

1. C. Weedbrook, S. Pirandola, R. García-Patrón, N.J. Cerf, T.C. Ralph, J.H. Shapiro, S. Lloyd, Gaussian quantum information. Rev. Mod. Phys. **84**, 621–669 (2012)
2. P.A.M. Dirac, *The Principles of Quantum Mechanics* (Oxford University Press, Oxford, 1958)
3. S.L. Braunstein, P. van Loock, Quantum information with continuous variables. Rev. Mod. Phys. **77**, 513–577 (2005)
4. U. Andersen, G. Leuchs, C. Silberhorn, Continuous-variable quantum information processing. Laser Photon. Rev. **4**(3), 337–354 (2010)
5. S.L. Braunstein, A.K. Pati, *Quantum Information with Continuous Variables* (Kluwer Academic Publishers, Dordrecht, 2003)
6. N. Cerf, G. Leuchs, E. Polzik, *Quantum Information with Continuous Variables of Atoms and Light* (Imperial College Press, London, 2007)
7. X.B. Wang, T. Hiroshima, A. Tomita, M. Hayashi, Quantum information with Gaussian states. Phys. Rep. **448**(1–4), 1–111 (2007)
8. A. Ferraro, S. Olivares, M. Paris, *Gaussian States in Continuous Variable Quantum Information*, Napoli Series on Physics and Astrophysics (2005). (ed. Bibliopolis, Napoli, 2005)
9. S. Olivares, Quantum optics in the phase space. Eur. Phys. J. Spec. Top. **203**(1), 3–24 (2012)
10. J. Eisert, M.B. Plenio, Introduction to the basics of entanglement theory in continuous-variable systems. Int. J. Quantum Inf. **01**(04), 479–506 (2003)
11. W.H. Louisell, *Radiation and Noise in Quantum Electronics* (McGraw-Hill, New York, 1964)
12. A. Papoulis, *Probability, Random Variables, and Stochastic Processes* (McGraw-Hill, New York, 1965)

13. R.J. Glauber, The quantum theory of optical coherence. Phys. Rev. **130**, 2529–2539 (1963)
14. K.E. Cahill, R.J. Glauber, Ordered expansions in Boson amplitude operators. Phys. Rev. **177**, 1857–1881 (1969)
15. R. Simon, N. Mukunda, B. Dutta, Quantum-noise matrix for multimode systems: u(n) invariance, squeezing, and normal forms. Phys. Rev. A **49**, 1567–1583 (1994)
16. W. Rudin, *Fourier Analysis on Groups* (Interscience Publishers, New York, 1962)
17. G. Cariolaro, *Unified Signal Theory* (Springer, London, 2011)
18. S. Wolfram, *Mathematica: A System for Doing Mathematics by Computer*, 2nd edn. (Addison-Wesley, Redwood, 1991)
19. E. Wigner, On the quantum correction for thermodynamic equilibrium. Phys. Rev. **40**, 749–759 (1932)
20. K.E. Cahill, R.J. Glauber, Density operators and quasiprobability distributions. Phys. Rev. **177**, 1882–1902 (1969)
21. M.G.A. Paris, The modern tools of quantum mechanics. Eur. Phys. J. Spec. Top. **203**(1), 61–86 (2012)
22. R. Simon, E.C.G. Sudarshan, N. Mukunda, Gaussian-Wigner distributions in quantum mechanics and optics. Phys. Rev. A **36**, 3868–3880 (1987)
23. M. Abramowitz, I. Stegun, *Handbook of Mathematical Functions* (Dover Publications, New York, 1970)
24. G. Cariolaro, G. Pierobon, The Weyl operator as an eigenoperator of the symplectic Fourier transform, to be published
25. I.S. Gradshteyn, I.M. Ryzhik, *Tables of Integrals, Series, and Products*, 7th edn. (Elsevier, Amsterdam, 2007)
26. X. Ma, W. Rhodes, Multimode squeeze operators and squeezed states. Phys. Rev. A **41**, 4625–4631 (1990)
27. D.F. Walls, G.J. Milburn, *Quantum Optics* (Springer, Berlin, 2008)
28. C.M. Caves, C. Zhu, G.J. Milburn, W. Schleich, Photon statistics of two-mode squeezed states and interference in four-dimensional phase space. Phys. Rev. A **43**, 3854–3861 (1991)
29. B.L. Schumaker, Quantum mechanical pure states with Gaussian wave functions. Phys. Rep. **135**(6), 317–408 (1986)
30. X. Ma, Time evolution of stable squeezed states. J. Mod. Opt. **36**(8), 1059–1064 (1989)
31. H.P. Yuen, R. Kennedy, M. Lax, Optimum testing of multiple hypotheses in quantum detection theory. IEEE Trans. Inf. Theory **21**(2), 125–134 (1975)
32. H.P. Yuen, Two-photon coherent states of the radiation field. Phys. Rev. A **13**, 2226–2243 (1976)
33. N.L. Johnson, S. Kotz, N. Balakrishnan, *Discrete Multivariate Distributions* (Wiley, New York, 1997)
34. C.M. Caves, B.L. Schumaker, New formalism for two-photon quantum optics. I. Quadrature phases and squeezed states. Phys. Rev. A **31**, 3068–3092 (1985)
35. M.S. Kim, W. Son, V. Bužek, P.L. Knight, Entanglement by a beam splitter: nonclassicality as a prerequisite for entanglement. Phys. Rev. A **65**, paper no. 032323 (2002). http://link.aps.org/doi/10.1103/PhysRevA.65.032323
36. M.M. Wolf, J. Eisert, M.B. Plenio, Entangling power of passive optical elements. Phys. Rev. Lett. **90**, paper no. 047904 (2003)
37. G. Cariolaro, R. Corvaja, G. Pierobon, Gaussian states and geometrically uniform symmetry. Phys. Rev. A **90**(4) (2014)
38. R.A. Horn, C.R. Johnson, *Matrix Analysis* (Cambridge University Press, Cambridge, 1998)
39. M.S. Kim, F.A.M. de Oliveira, P.L. Knight, Properties of squeezed number states and squeezed thermal states. Phys. Rev. A **40**, 2494–2503 (1989)

Chapter 12
Classical and Quantum Information Theory

List of Main Symbols

$\mathcal{D}(\mathcal{H})$	class of density operators acting in the Hilbert space $\mathcal{H}$
$\mathcal{A}$	alphabet
$p_A(a), a \in \mathcal{A}$	probability distribution over $\mathcal{A}$
A	symbol (synonymous: random variable, information source)
$A \sim (\mathcal{A}, p_A)$	random variable A generated by the ensemble $\{\mathcal{A}, p_A\}$
$A^L := (A_1, \ldots, A_L)$	word of length L
$\rho_A \sim (\mathcal{A}, p_A)$	density operator ρ_A generated by the ensemble $\{\mathcal{A}, p_A\}$, with $\mathcal{A}$ alphabet of states
$\rho_A \sim (\mathcal{A}, p_A)_{\text{ort}}$	density operator ρ_A generated by the ensemble $\{\mathcal{A}, p_A\}$, with $\mathcal{A}$ orthonormal basis of states
$H(A) \quad [H(\{p_A\})]$	classical entropy of A [generated by the distribution $\{p_A\}$]
$S(\rho) \quad [S(\{p_A\})]$	quantum entropy of ρ [generated by the eigenvalues $\{p_A\}$]
$\chi(\{p_a, \rho_a\}) = \chi(\mathcal{L})$	Holevo χ of the ensemble $\mathcal{L} = (\{p_a, \rho_a\})$

12.1 Introduction

Information Theory is a mathematical discipline, born within the field of Telecommunications in 1948, with the revolutionary ideas developed by Shannon [1]. Its purposes are mainly: (1) to define *information* mathematically and quantitatively, (2) to represent information in an efficient way (through data compression) for storage and transmission, and (3) to ensure the protection of information (through encoding) in the presence of noise and other impairments. As Mechanics is conventionally subdivided into Classical and Quantum Mechanics, also Information Theory has an analogous subdivision into **Classical Information Theory**, where the above goals are accomplished in accordance to the laws of Classical Physics, and **Quantum Information Theory**, which is based on quantum mechanical principles. As we

© Springer International Publishing Switzerland 2015

G. Cariolaro, *Quantum Communications*, Signals and Communication Technology,
DOI 10.1007/978-3-319-15600-2_12

shall see, Quantum Information Theory is intrinsically richer than its Classical counterpart, because of intriguing resources, such as entanglement, and therefore it is more interesting and challenging.

The purpose of this chapter is to provide an overview of Quantum Information Theory starting from Classical Information Theory. As a matter of fact, the quantum version cannot be developed without a robust preliminary introduction on the classical version. For this reason, each of the three items listed above will be developed starting from the classical case. Moreover, there is a deep interplay between the two disciplines deriving from the fact that the purpose of Quantum Communications is usually the transmission of **classical information** through **quantum states**.

12.1.1 Measure of Information

The information contained in a message, such as a written text, a speech, a music, a piece of an image, a string of data, is related to its randomness or uncertainty: if a message is known in advance, it does not bring any information. As a matter of fact, in its simplest version, the classical theory formulates a source of information as a *random variable A*, specified by a probability distribution p_A over a given alphabet $\mathcal{A}$. The classical (or Shannon) entropy of such a source is defined as

$$H(A) := - \sum_a p_A(a) \, \log_2 p_A(a) \tag{12.1}$$

and gives the *average information content* of the source A. With the choice of the binary logarithm, in (12.1) the measure of information is expressed in binary digits or *bits*.

In the quantum version, the information environment becomes a Hilbert space $\mathcal{H}$ and the source of information lies on quantum states, and, more precisely, the randomness required for the presence of information is provided by a *mixed state*. In this case, the average information content is provided by the quantum (or von Neumann) entropy, which is defined by

$$S(\rho_A) := -\mathrm{Tr}[\rho_A \, \log_2 \rho_A] \tag{12.2}$$

where ρ_A is the density operator describing the mixed state. This definition seems to depart considerably from the classical counterpart $H(A)$, but we shall see that $S(\rho)$ can be written in the same form as the classical entropy $H(A)$ (the differences being others).

Now we spend a few words on the measure unit for information. The term **bit** denotes a classical notion having two distinct meanings: as a binary system with two possible states, say 0 and 1, and as a measure unit for information, corresponding to the information of a binary random variable, where 0 and 1 are equally likely. The quantum counterpart of the bit (intended as binary system) is the *quantum bit* or **qubit**. We have seen in Sect. 3.2 that a qubit is a quantum system represented by a Hilbert space of dimension 2, with an orthonormal basis $\{|0\rangle, |1\rangle\}$. Then $|0\rangle$ and $|1\rangle$ are possible states corresponding to 0 and 1 of the classical bit. The fundamental difference between the bit and the qubit is that a qubit can be in any state obtained as a linear combination $\alpha|0\rangle + \beta|1\rangle$ with $|\alpha|^2 + |\beta|^2 = 1$, and when a measurement is performed it gives 0 with probability $|\alpha|^2$ and 1 with probability $|\beta|^2$.

The qubit, too can be considered as unit of measure, but in a subtler sense related to the dimensionality in the context of compression. Usually, the compressed space $\mathcal{H}_c$ is given by the tensor product of L qubits. Now, if a quantum state can be correctly represented in $\mathcal{H}_c$, one says that the state has L qubits. Also the quantum entropy is expressed in qubit per symbol, but we will reconsider this point in Sect. 12.4 after we have acquired the concept of quantum entropy.

Finally we discuss the terms **classical information** and **quantum information**. The distinction is not easy although both terms are currently used with the meaning obtained from the context. In principle, classical information refers to messages, as text, speech, etc., and the mathematical models are random variables, random vectors and also random processes. Quantum information refers to quantum states, as seen above. But in very important cases, as in Quantum Communications (see Part II of this book), classical information is sent through a quantum system. Then we encounter conversions of the form "classical" $\rightarrow$ "quantum" and "quantum" $\rightarrow$ "classical"; this will be seen in a precise form in Sect. 12.8.

12.1.2 Compression of Information. Data Compression

In practice, sources of information are often *redundant*, in the sense that they produce some messages more frequently than others. On the other hand, the ideal source (having no redundancy) has a uniform probability distribution over the symbols. Then the goal is to transform a redundant source into a uniform source to achieve the minimum amount of bits per message. This operation is called *data compression* and also *source coding*.

The limits of data compression are established by *source coding theorems*. In these theorems, the reference model for source coding is represented by an *independent identically distributed* (IID) sequence of random variables, say $(A_1, \ldots, A_L)$. The first Shannon coding theorem states that for $L \gg 1$, the minimum number of bits to represent such a sequence is given by the entropy of the sequence.

The quantum counterpart of an IID sequence is obtained starting from a density operator $\rho \in \mathcal{D}(\mathcal{H})$, and is given by the tensor product $\rho^L = \rho^{\otimes L} \in \mathcal{D}(\mathcal{H}^{\otimes L})$. In this context, the target of data compression is the *reduction of the dimensionality*. The dimension is meaningfully expressed by the number of qubits. The minimum compression rate is established by Schumacher's noiseless quantum channel coding theorem, according to which there exists an asymptotically reliable compression protocol, which represents a state ρ in a space of $S(\rho)$ qubits.

The proofs of Shannon's and Schumacher's theorems are based on an entropic application of the law of large numbers and related *typical sequences*. They establish asymptotic results (valid for $L \to \infty$) and for this reason the related topic is called *Asymptotic* Information Theory.

12.1.3 Reliable Transmission of Information over Noisy Channels

In the presence of noise, the communication channel distorts the messages and, in order to counteract this effect, a *channel coding* is introduced before the transmission.

In the classical case, a source message of a given length L_0 is encoded into an L-long codeword W from a prefixed codebook and sent to the noisy channel, which provides a stochastic mapping $W \to W'$ (stochastic for the randomness in noise). The parameter to optimize is the *rate R* given by the ratio of the length of the messages to the length of the codewords. In his second (channel coding) theorem (believed to be the most important result of Information Theory) Shannon established that, providing that an appropriate message codebook is selected, there is a threshold, called **channel capacity** C, below which any rate R can be achieved with an arbitrarily small error probability.

In the quantum case, a first goal is to establish an adequate model for a noisy channel. The problem is that a real system, with the interaction with external impairments that damage the information, is an *open system*, while the Postulates of Quantum Mechanics consider only closed systems. The solution is to operate with a larger, composite closed system. As we shall see in Sect. 12.8, the literature has recently consolidated a satisfactory (and beautiful) model for noisy quantum channels [2]. A second goal is that the decoding at the receiver side (Bob) be performed by a quantum measurement. This leads to the concept of *accessible information I_{acc}*, given by the maximum amount of information that Bob can gain through any possible measurement. An upper bound for the accessible information was established by Holevo in 1973 and is called Holevo-χ or χ-information. The Holevo-χ allows us to define the so-called *product-state capacity* and to establish the corresponding coding theorem, known as Holevo–Schumacher–Westmoreland (HSW) Theorem. But, differently from the classical case, where the channel capacity is unique, in the quantum case a noisy channel has various capacities in dependence of the presence or absence of entanglement. This is presently a hot topic in Quantum Information Theory.

Organization of the Chapter

In Sects. 12.2 and 12.3, we introduce the fundamental definitions and properties of Classical Information Theory in adequate probabilistic environments, and in Sect. 12.4 this is done for Quantum Information Theory.

In Sects. 12.5 and in 12.6, classical and quantum data compression are developed, respectively.

Section 12.7 deals with classical channel coding and the fundamental theorem of Information Theory.

Section 12.8 deals with open systems and quantum channels and Sect. 12.9 with accessible information and its bound. Finally, Sect. 12.10 deals with quantum channel coding and capacities.

Throughout the chapter each topic is developed in the simplest case, as is common practice in Information Theory, e.g., information sources with IID variables and memoryless channels. The proofs of statements are often omitted, but adequate references are given.

12.2 Messages of Classical Information

The objects of Information Theory, as messages and symbols, are random quantities and therefore they are formulated in terms of Probability Theory. We consider only *digital messages*, which take values from a countable set, called *alphabet*. As happens with written messages (texts), it is convenient to view a message as a sequence of *symbols*. Sometimes symbols are collected in *words*.

We now introduce the statistical description of symbols, words, and infinite sequences, which in Probability Theory are called random variables, random vectors, and random processes, respectively.

12.2.1 Symbols as Random Variables

A symbol A may be defined as a *finite random variable*, which takes values from a finite-sized alphabet, say $\mathcal{A} = \{1, \ldots, K\}$ or $\mathcal{A} = \{0, 1, \ldots, K-1\}$, with probability

$$p_A(a) := \mathrm{P}[A = a], \qquad a \in \mathcal{A}. \tag{12.3}$$

The function $p_A(a)$ forms a *probability distribution* and verifies the conditions

$$\forall a \in \mathcal{A} \qquad p_A(a) \geq 0, \qquad \sum_{a \in \mathcal{A}} p_a(a) = 1. \tag{12.3a}$$

Then a symbol A is described by the ensemble $(\mathcal{A}, p_A)$, briefly symbolized as $A \sim (\mathcal{A}, p_A)$. We often identify A as an *information source*.

Binary source. It is the simplest information source: $A \sim (\mathcal{A}, p_A)$, where $\mathcal{A} = \{0, 1\}$ and $p_A(0) = q$, $p_A(1) = p$, with $p + q = 1$. The graphical representation is given in Fig. 12.1.

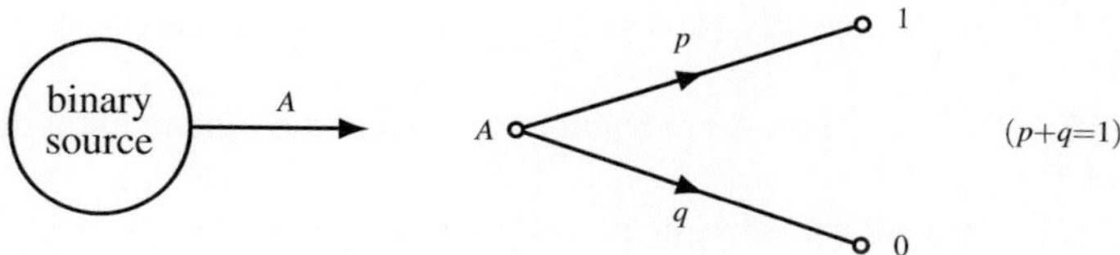

Fig. 12.1   Graphical representation of a binary source

12.2.2 Words as Random Vectors

An L-length word consists of a sequence of L symbols

$$A^L := (A_1, A_2, \ldots, A_L) \tag{12.4}$$

described by the probabilities

$$p_{A^L}(a_1, \ldots, a_L) = \mathrm{P}[A_1 = a_1, \ldots, A_L = a_L], \qquad a_i \in \mathcal{A}. \tag{12.5}$$

In general, the probability distribution of each symbol is not sufficient for the statistical description of a word, but all the joint probabilities (12.5) are needed.

Words with independent symbols. When the L symbols in a word $A^L = (A_1, \ldots, A_L)$ are *statistically independent*, the joint probabilities (12.5) are factored in the form

$$p_{A_1 \ldots A_L}(a_1, \ldots, a_L) = p_{A_1}(a_1) \ldots p_{A_L}(a_L).$$

In this case, the statistical description of the component symbols becomes sufficient.

Realizations. It is important to clearly bear in mind the difference between a random object (here denoted in uppercase) and its realizations (denoted in lower case). For instance, a binary word of length 3, $A^3 = (A_1, A_2, A_3)$ is a random vector, while $a^3 = (a_1, a_2, a_3)$ is a generic realization of A^3, which can explicitly take the values $(000), (001), \ldots, (111)$.

Example 12.1   Figure 12.2 shows the tree representation of a binary word of length 3 with $\mathrm{P}[A_i = 1] = p$ and $\mathrm{P}[A_i = 0] = q = 1 - p$. From the probabilities of the realizations indicated in the figure, one can deduce that the symbols are statistically independent.

12.2.3 Messages as Random Processes

The sequence $\{A_\infty\} = A_1, A_2, \ldots$, where A_n is the symbol at the normalized time n, is modeled as a *discrete-time random process*. The statistical description of $\{A_\infty\}$ is given by all the distributions of the form

Fig. 12.2 Graphical representation of a binary word of length $L = 3$ consisting of independent symbols. On the *right* the probabilities of each realization are indicated

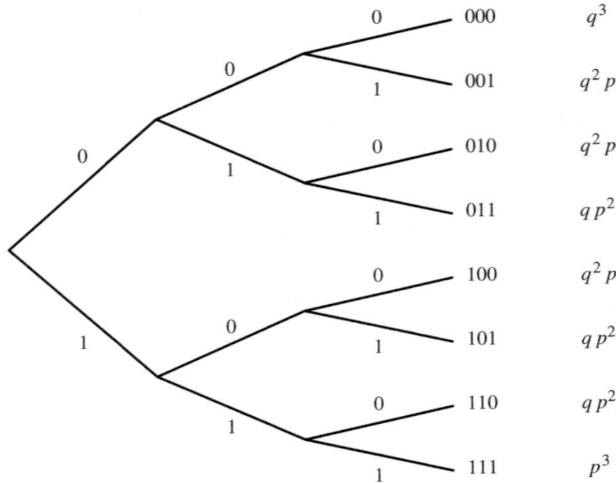

$$p_A(a_1, \ldots, a_L; n_1, \ldots, n_L) = P[A_{n_1} = a_1, \ldots, A_{n_L} = a_L] \qquad (12.6)$$

where L is an arbitrary positive integer, $n_1, n_2, \ldots, n_L$ are arbitrary integers and $a_1, \ldots, a_L$ are values of a same alphabet $\mathcal{A}$. Distributions (12.6) are often abbreviated in the form $p(A_{n_1}, \ldots, A_{n_K})$.

From a message one can extract words of any length. It is easy to prove that the complete statistical description, given by (12.6), can be limited to words consisting of L consecutive symbols, say $(A_n, A_{n+1}, \ldots, A_{n+L-1})$, that is, by the probabilities $p(A_n, A_{n+1}, \ldots, A_{n+L-1})$ for every initial time n and every length L. When such probabilities are independent of the time n, the message is said to be *stationary*. If this is the case, all the words of length L, as $(A_n, \ldots, A_{n+L-1})$ and $(A_1, \ldots, A_L)$, have the same statistical description. Hereafter, we will consider only stationary messages.

Note that, by using the definition of *conditional* probability, the probability of the word $(A_1, \ldots, A_L)$ can be factored in the form (chain rule)

$$p(A_1, \ldots, A_L) = p(A_1)p(A_2|A_1) \ldots p(A_L|A_1, \ldots, A_{L-1}) \qquad (12.7)$$

where $p(|)$ are conditional probabilities.

Messages with independent symbols. If the symbols of a message are *statistically independent*, the probabilities (12.6) are factorized as

$$p(A_n, A_{n+1}, \ldots, A_{n+L-1}) = p(A_n)\, p(A_{n+1}) \cdots p(A_{n+L-1}) \qquad (12.8)$$

and then the description is limited to the probabilities of each symbol $p(A_n) = P[A_n = a]$.

12.3 Measure of Information and Classical Entropy

We begin with the evaluation of the amount of information (briefly *information*) of a symbol. Then we consider a pair of symbols and finally a sequence of symbols of arbitrary length. The statistical average (expectation) of the information gives the entropy.

12.3.1 The Logarithmic Law

We make a few intuitive considerations on a symbol $A \sim (\mathcal{A}, p_A)$. A first remark is that the information does not depend on the nature of the alphabet, but only on its size. Thus the binary alphabets $\{0, 1\}$ and $\{$head, tail$\}$ are equivalent. A second remark is that the information is related do the *degree of uncertainty* of a symbol: a symbol that occurs with certainty does not carry information and, on the other hand, a symbol that occurs very seldom carries a lot of information. Technically speaking, this means that the information $i(a)$ of a specific symbol a (realization of A) is a function of the probability $p_A(a)$, say $i(a) = f[p(a)]$, where $f[x]$ must be a decreasing function of x with $f[x] = 0$ when $x = 1$, that is, when the symbol a occurs with certainty.

It remains to choose the function f. A convenient choice, made by Nyquist in 1928 [3] and reconsidered by Shannon in his masterpiece, is the logarithmic function[1]

$$i(a) := -\log_2 p(a) \qquad\qquad (12.9)$$

shown in Fig. 12.3. The choice of the base 2 logarithm leads to the binary digit (*bit*) as the unit of information. Letting $p(a) = 1/2$ in (12.9) one gets $i(a) = 1$ bit. Hereafter, $\log_2$ will be simply denoted as "log."

It can be shown that the logarithm is the only "smooth" function having the property that the information carried by two *independent* symbols is the sum of the information carried separately by the two symbols (see Problem 12.1).

Problem 12.1 ★★   Consider a pair (A, B) of statistically independent symbols. Prove that, by imposing the condition $i(a, b) = i(a) + i(b)$, the unique function $f[\cdot]$ defining the information $i(a) = f[p(a]$ is the logarithm.

---

[1] Considering that $\lim_{p \to 0^+} p \log p = 0$, the convention is assuming $0 \log 0 = 0$.

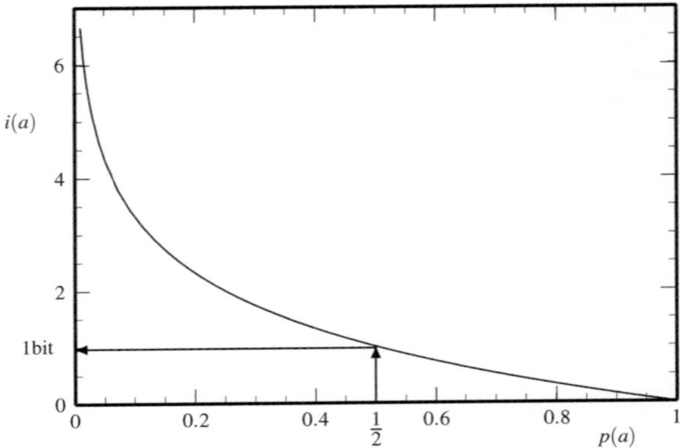

Fig. 12.3 Information of a symbol as a function of its probability. For $p = \frac{1}{2}$ the symbol carries the information of 1 bit

12.3.2 *Entropy of a Symbol*

If in the logarithmic law $i(a) = -\log p(a)$ the argument a is replaced by the (discrete) random variable A, also $i(A)$ becomes a random variable. The statistical average (expectation) of $i(A)$, that is,

$$H(A) := \mathrm{E}[i(A)] = \mathrm{E}[-\log p(A)] \tag{12.10}$$

and explicitly

$$H(A) = \sum_{a \in \mathcal{A}} p_A(a)\, i(a) = -\sum_{a \in \mathcal{A}} p_A(a) \log p_A(a) \tag{12.11}$$

gives the *average information of the symbol A*, or entropy of the source A. It is also called Shannon entropy or classical entropy. The unit of entropy is the *bit*.

Entropy of a binary source. With $A \in \mathcal{A} = \{0,\, 1\}$, $p_A(1) = p$, $p_A(0) = 1 - p$, $0 \le p \le 1$, the entropy is given by

$$H(A) = -p \log p - (1 - p) \log(1 - p) \tag{12.12}$$

and is shown in Fig. 12.4 as a function of $p = \mathrm{P}[A = 1]$.

One can check that $H(A)$ has its maximum for $p = 1/2$, where it takes the value of 1 bit/symbol. Moreover, $H(A)$ is zero for $p = 0$ and $p = 1$, in agreement with the intuitive considerations made above.

The entropy of a K-ary symbol verifies the conditions

$$0 \le H(A) \le \log K \tag{12.13}$$

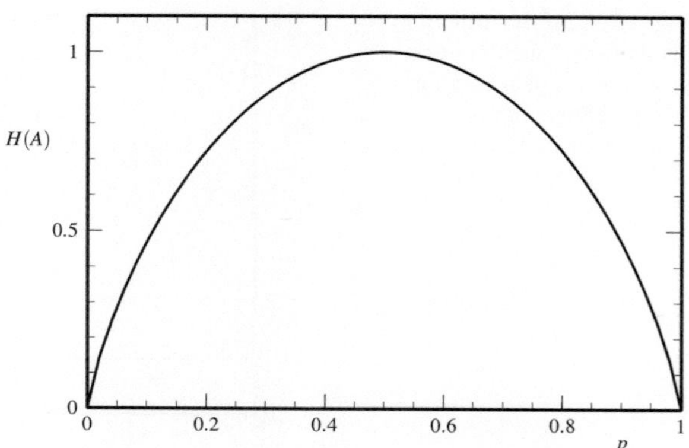

Fig. 12.4 Entropy $H(A)$ of a binary symbol A as a function of the probability $p = \mathrm{P}[A = 1]$

which can be articulated in the following properties:

(1) $H(A) \geq 0$,
(2) $H(A) = 0$, if one value of A occurs with probability 1 (almost deterministic source),
(3) $H(A) \leq \log K$,
(4) $H(A) = \log K$ if A has equiprobable values (uniform source).

Property (1) is a consequence of the fact that $i(a) \geq 0$ (see Fig. 12.3). Property (2) follows from the fact that the equation $p \log p = 0$, with $0 \leq p \leq 1$, has the unique solutions $p = 0$ and $p = 1$. For (3) and (4), we use the inequality

$$\log_e x \leq x - 1, \qquad x > 0 \tag{12.14}$$

where the equality holds only for $x = 1$.

Notations. A symbol A is specified by an alphabet $\mathcal{A}$ and by a probability distribution p_A, which globally forms an *ensemble* $\mathcal{E} = (\mathcal{A}, p_A)$. Then for the entropy, we find alternative notations as $H(A) = H(\mathcal{E}) = H((\mathcal{A}, p_A))$. But, considering that the entropy only depends on the distribution, frequently we find also the notation $H(\{p_A\})$.

12.3.3 Classical Entropies in a Bipartite System

Let A and B be two symbols with alphabets of size K and M, respectively, which are statistically related, e.g., A is the input and B is the output in a communication channel. For the pair (A, B), which forms a word of length 2, one can consider several informations and entropies, which are very useful to consolidate the meaning of the information measure. This multiplicity is due to the fact that for a symbol pair one has several probability systems: marginal probabilities $p(A)$ and $p(B)$,

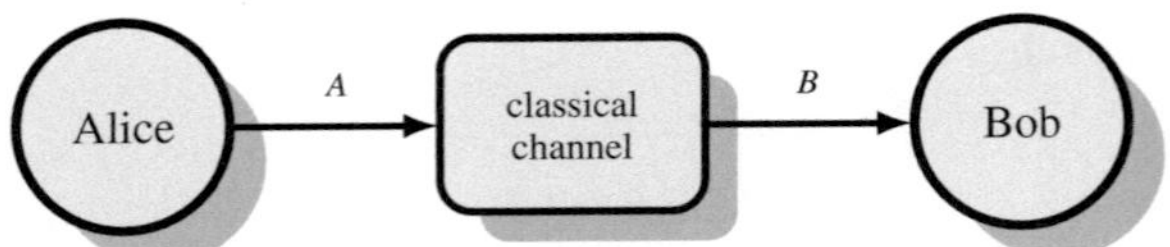

Fig. 12.5 Source of information A (Alice) connected to Bob through a classical channel. Alice sends the symbol A to the channel and at the output Bob receives the symbol B

joint probabilities $p(A, B)$, and conditional probabilities $p(A|B)$, $p(B|A)$. Then we have the informations $i(A)$, $i(B)$, $i(A, B)$, $i(A|B)$, $i(B|A)$, and the corresponding entropies $H(A)$, $H(B)$, $H(A, B)$, $H(A|B)$, $H(B|A)$. We focus our attention only on $H(A, B)$ and $H(A|B)$, having $H(B|A)$ the specular properties of $H(A|B)$.

We first consider the joint information $i(A, B) = -\log p(A, B)$, which measures the total uncertainty one has about the pair (A, B). The corresponding entropy is given by the expectation $H(A, B) = \mathrm{E}[i(A, B)]$, and explicitly

$$
\begin{aligned}
H(A, B) &= \sum_a \sum_b p(a, b) i(a, b) \\
&= -\sum_a \sum_b p(a, b) \log p(a, b).
\end{aligned}
\tag{12.15}
$$

Next, suppose that we know the symbol B, then the remaining uncertainty about the symbol A is given by the conditional information $i(A|B) = -\log p(A|B)$. For the interpretation of this quantity it is convenient to think that Alice transmits the symbol A through a classical channel and at the output Bob reads the symbol B, as shown in Fig. 12.5. Then Bob has some uncertainty about the symbol A, given by $i(A|B)$. The corresponding entropy is $H(A|B) = \mathrm{E}[i(A|B)]$ and explicitly

$$
H(A|B) = -\sum_a \sum_b p(a, b) \log p(a|b).
$$

An important property is obtained from the relation between joint and conditional probabilities: $p(A, B) = p(A|B)\, p(B)$, which gives $i(A, B) = i(A|B) + i(B)$ and hence, considering the linearity of expectation,

$$
H(A, B) = H(A|B) + H(B)
\tag{12.16}
$$

which reads: the average information on the pair (A, B) is given by the information on B plus the information on A, when B is known.

For a clear interpretation, two limit cases are particularly useful:

(1) when the symbols A and B are statistically independent, $H(A, B) = H(A) + H(B)$,

(2) when the knowledge of B implies the full knowledge of A: $H(A, B) = H(B)$, $H(A|B) = 0$.

An important inequality is

$$\boxed{H(A|B) \leq H(A)} \tag{12.17}$$

which states that, on average, the residual uncertainty on A, when B is known, cannot be greater than the uncertainty on A when B is not known. The proof of (12.17) is based on the logarithm inequality (12.14).[2] As a consequence of (12.17), we have the other inequality

$$H(A) \leq H(A, B). \tag{12.18}$$

Using the previous results, inequality (12.17) can be articulated as:

(1) $H(A|B) \geq 0$,
(2) $H(A|B) = 0$ if and only if knowledge of B implies full knowledge of A,
(3) $H(A|B) \leq H(A)$,
(4) $H(A|B) = H(A)$ if and only if A and B are independent.

Moreover, the combination of (12.17) and (12.16) gives $H(A, B) \leq H(A) + H(B)$.

Another important entropic quantity in bipartite systems is *mutual information*, which will be introduced in Sect. 12.7 in the context of channels.

12.3.4 Informations and Entropies of a Sequence

The information of a sequence (word) of L symbols $A^L = (A_1, \ldots, A_L)$ is given by

$$i(A_1, \ldots, A_L) = -\log p_{A^L}(A_1, \ldots, A_L)$$

where p_{A^L} are the joint probabilities defined by (12.5). The expectation gives the entropy of the sequence

$$H(A_1, \ldots, A_L) = \mathrm{E}[i(A_1, \ldots, A_L)].$$

The chain rule on the joint probabilities, given by (12.7), allows us to write this entropy in the form

$$H(A_1, \ldots, A_L) = H(A_1) + H(A_2|A_1) + H(A_3|A_1A_2) + \cdots + H(A_L|A_1 \cdots A_{L-1}) \tag{12.19}$$

which generalizes formula (12.16) seen for bipartite systems and has a similar interpretation.

---

[2] An alternative prove is based on the concept of *relative entropy* (see [4]).

The entropy of a word satisfies several inequalities. In particular,

$$H(A_1, \ldots, A_L) \leq H(A_1) + H(A_2) + \cdots + H(A_L) \qquad (12.20)$$

where the equality holds when the symbols in the word are independent.

For an unlimited sequence (random process) $\{A_\infty\} = A_1, A_2, A_3, \ldots$, the entropy is in general infinite and it becomes convenient to introduce the *entropy rate*

$$R = \lim_{L \to \infty} \frac{1}{L} H(A_1, \ldots, A_L) \qquad (12.21)$$

which is expressed in *bits/symbol*. For a stationary process, in which $H(A_n)$ is independent of n, using inequality (12.19) we find that the entropy rate verifies the inequality

$$R \leq H(A_1) \qquad (12.22)$$

where the equality holds when the symbols of the sequence are independent.

Final Comment

In this section, we have introduced the elementary concepts on information and entropy in a minimal form. Some entropic quantities, as the relative entropy, have been omitted. Also important properties, as the concavity of entropy, were not considered. A good "source of information" to complete the topic is the book by Cover and Thomas [4].

12.4 Quantum Entropy

In a quantum system described by a Hilbert space $\mathcal{H}$, the information is contained in quantum states. To understand why, it is convenient to recall the considerations leading to the concept of mixed states in Sect. 3.3.

12.4.1 Quantum States as Sources of Information

According to Postulate 1 of Quantum Mechanics, a closed quantum system, described by a Hilbert space, at each time of its evolution, is completely specified by a quantum state $s = |\psi\rangle$ (pure state), but, if the observer only has a probabilistic knowledge of the system, the state s must be regarded as a *random state* (mixed state), which can take its values in a set $\mathcal{S} = \{|\psi_1\rangle, |\psi_2\rangle, \ldots\}$, with probabilities $p_k := \mathrm{P}[s = |\psi_k\rangle]$. Then we get an alphabet-probability distribution ensemble $\mathcal{E}(\rho) = (\mathcal{S}, p)$, as seen for a classical symbol A. We recall that in the evaluation of information the nature of the alphabet is irrelevant, as it can contain literal symbols as well as quantum states.

The ensemble is encoded into a density operator, in the form

$$\mathcal{E}(\rho) = \quad \rightarrow \quad \rho = \sum_k p_k \, |\psi_k\rangle\langle\psi_k| \tag{12.23}$$

which contains both the information of the states, through the elementary operators $|\psi_k\rangle\langle\psi_k|$, and of the probability distribution p. In conclusion, a mixed quantum state described by a density operator may be viewed as a source of quantum information. To stress the analogy with the classical case, where a symbol A is a **random variable**, described be the ensemble $\mathcal{E} = (\mathcal{A}, p_A)$, we can think of a density operator ρ as generating a **quantum random state** $|\Lambda\rangle$, described by an ensemble $\mathcal{E}(\rho)$.

However, while an ensemble identifies a unique density operator, from a density operator one can obtain infinitely many ensembles, even with different cardinalities, as we saw in Sect. 3.11. An ensemble is also provided by the EID of ρ

$$\rho = U \Lambda U^* = \sum_k \lambda_k \, |\lambda_k\rangle\langle\lambda_k| \tag{12.24}$$

where λ_k are the eigenvalues and $|\lambda_k\rangle$ are the corresponding eigenvectors. In fact, the eigenvalues λ_k form a probability distribution, with $\lambda_k \geq 0$ and $\sum_k \lambda_k = 1$, as a direct consequence of the properties of a density operator: $\rho \geq 0$ and $\mathrm{Tr}[\rho] = 1$. Moreover, the eigenvectors are orthonormal, $\langle\lambda_i|\lambda_j\rangle = \delta_{ij}$. Then we get the ensemble, written in compact form

$$\rho \quad \rightarrow \quad \mathcal{E}_\perp(\rho) = (\{\lambda_k, |\lambda_k\rangle\}) \tag{12.25}$$

where the subscript $\perp$ emphasizes the orthonormality of the eigenvectors.

As we shall see now in the evaluation of quantum entropy, *only the orthogonal ensemble* $\mathcal{E}_\perp(\rho)$ obtained by the EID is correct, while the other ensembles $\mathcal{E}(\rho)$ will lead to wrong entropic evaluations.

12.4.2 Definition of Quantum Entropy

The quantum (or von Neumann) entropy refers to a quantum state described by a density operator ρ, and is defined as

$$\boxed{S(\rho) := -\mathrm{Tr}[\rho \log \rho].} \tag{12.26}$$

The explicit evaluation of $S(\rho)$ can be done from the EID of ρ, given by (12.24). Using the definition of a function of an operator (see Sect. 2.12) in (12.26) with $f(z) = -z \log z$, we get

$$S(\rho) = \mathrm{Tr}[U \, U^* f(D)] = \mathrm{Tr}[f(D)] = \sum_k f(\lambda_k)$$

which gives

$$S(\rho) = -\sum_k \lambda_k \log \lambda_k. \tag{12.27}$$

Thus, formally we get the same expression, given by (12.11), as the classical entropy $H(A)$ of a symbol A. Note that $S(\rho)$ depends only on the probability distribution provided by the eigenvalues. Relation (12.26) is often written in the form

$$S(\rho) = H(\{\lambda_k\}) \tag{12.27a}$$

which states that the quantum entropy of the state ρ is given by the entropy of a classical source with probability distribution $\{\lambda_k\}$, with λ_k the eigenvalues of ρ.

Example 12.2 (Qubit states) Consider a qubit system with standard basis $\{|0\rangle, |1\rangle\}$ and the state given by

$$\rho_0 = \tfrac{1}{2}|0\rangle\langle 0| + \tfrac{1}{2}|1\rangle\langle 1|.$$

This expression is just the EID of ρ_0, which gives the ensemble $\mathcal{E}_\perp(\rho_0) = \{(\tfrac{1}{2}, |0\rangle),$ $(\tfrac{1}{2}, |1\rangle)\}$ and therefore the quantum entropy is $S(\rho_0) = 1$. This is just the classical entropy of a symbol, with alphabet $\{|0\rangle, |1\rangle\}$, where $|0\rangle$ and $|1\rangle$ are equiprobable.

Next, consider the qubit state (used in the B92 protocol for quantum key distribution)

$$\rho_1 = \tfrac{1}{2}|0\rangle\langle 0| + \tfrac{1}{2}|+\rangle\langle +|$$

where $|+\rangle\langle +| = \frac{1}{\sqrt{2}}(|0\rangle\langle 0| + |1\rangle\langle 1|)$. Now, if we calculate the classical entropy of the ensemble $\mathcal{E}(\rho_1) = \{(\tfrac{1}{2}, |0\rangle), (\tfrac{1}{2}, |+\rangle)\}$ we find again $H(\mathcal{E}) = 1$ bit. But this is not the von Neumann entropy because $\{\tfrac{1}{2}, \tfrac{1}{2}\}$ are not the eigenvalues of ρ_1. In fact, the EID of ρ_1 reads

$$\left\{\lambda_0 = \cos^2 \tfrac{\pi}{8} \ , \ \lambda_1 = \sin^2 \tfrac{\pi}{8}\right\}$$

$$\left\{|\lambda_0\rangle = \cos \tfrac{\pi}{8}|0\rangle + \sin \tfrac{\pi}{8}|1\rangle \ , \ |\lambda_1\rangle = \sin \tfrac{\pi}{8}|0\rangle - \cos \tfrac{\pi}{8}|1\rangle\right\}$$

and therefore the quantum entropy is given by

$$S(\rho_1) = -\cos^2 \tfrac{\pi}{8} \log \cos^2 \tfrac{\pi}{8} - \sin^2 \tfrac{\pi}{8} \log \sin^2 \tfrac{\pi}{8} \simeq 0.6009 \text{ qubits.}$$

This second case allows us to understand a first difference between the classical and the quantum entropy and more generally between classical and quantum information. The difference is due to the fact that classical entropy does not take into account the geometry of ensembles.

12.4.3 Properties of Quantum Entropy

Considering that $S(\rho)$ is equal to the classical entropy $H(\{\lambda_k\})$, although expressed in a different language, we have the following properties (see Sect. 12.3.3):

(1) $S(\rho) \geq 0$,
(2) $S(\rho) = 0$, if and only if the state is pure,
(3) $S(\rho) \leq \log K$, where K is the rank of ρ,
(4) $S(\rho) = \log K$ if and only if $\rho = I_{\mathcal{H}}/K$. This state is called *completely mixed state* or *chaotic state*.
(5) $S(\rho)$ is invariant with respect to unitary transformations.

Property (2): when $\rho = |\lambda\rangle\langle\lambda|$ we have $\mathrm{P}[\Lambda = |\lambda\rangle] = 1$, that is the state is known with certainty. Property (4): if all the eigenvalues ρ are equal, $\lambda_k = 1/K$, from (12.24) we have $\rho = (1/K)\sum_k |\lambda_k\rangle\langle\lambda_k|$, where the sum gives $I_{\mathcal{H}}$ by the completeness of kets $|\lambda_k\rangle$. Property (5) follows from the fact that ρ and $U\rho U^*$, with U unitary, have the same eigenvalues.

Problem 12.2 ⋆⋆   (*Thermal states*) A thermal state may be defined as the bosonic state that maximizes the von Neumann entropy for a given mean number of photons $\mathcal{N}$ [5]. It has the following Fock representation (see Sect. 11.9)

$$\rho_{\text{th}} = \sum_{n=0}^{\infty} \frac{\mathcal{N}^n}{(\mathcal{N}+1)^{n+1}} \, |n\rangle\langle n|. \tag{12.28}$$

Find its quantum entropy.

12.4.4 Entropy of Gaussian States

For an N-mode Gaussian state, the quantum entropy can be easily evaluated starting from the N symplectic eigenvalues σ_k^2 of the covariance matrix V (see Sect. 11.10). The expression is [6]

$$S(\rho) = \sum_{k=1}^{N} g(\sigma_k^2) \tag{12.29}$$

where $g(x)$ is the function (Fig. 12.6)

$$g(x) := (x + \tfrac{1}{2}) \log(x + \tfrac{1}{2}) - (x - \tfrac{1}{2}) \log(x - \tfrac{1}{2}).$$

For the single mode we have one symplectic eigenvalue given by $\sigma_1^2 = \sqrt{\det V} = 2\mathcal{N} + 1$, where $\mathcal{N}$ is the number of thermal photon of the state.

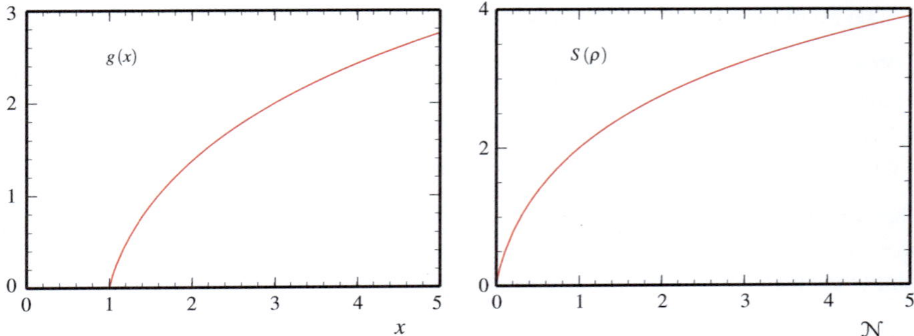

Fig. 12.6 The function $g(x)$ for the evaluation of the quantum entropy of Gaussian states from the symplectic eigenvalues σ_k^2 and the quantum entropy $S(\rho) = g(2\mathcal{N}+1)$ of a single mode Gaussian state as a function of the number of thermal photons $\mathcal{N}$

Then, $S(\rho) = g(2\mathcal{N}+1)$. Note that for $\mathcal{N} = 0$, corresponding to a pure Gaussian state, the quantum entropy is zero.

For multimode Gaussian states, the evaluation of the symplectic eigenvalues is not so simple [7].

12.4.5 Quantum Entropies in a Bipartite Quantum System

Let $A\,B$ be a bipartite quantum system described by the Hilbert space $\mathcal{H} = \mathcal{H}_A \otimes \mathcal{H}_B$, and let ρ_{AB} be a density operator of $\mathcal{H}$, having the "marginal" density operators $\rho_A = \mathrm{Tr}_B[\rho_{AB}]$ and $\rho_B = \mathrm{Tr}_A[\rho_{AB}]$. From these operators, we introduce the following quantum entropies[3]

$$
\begin{aligned}
S(A, B) &= -\mathrm{Tr}[\rho_{AB} \log \rho_{AB}] && \text{joint quantum entropy of } AB, \\
S(A) &= -\mathrm{Tr}[\rho_A \log \rho_A] && \text{quantum entropy of } A, \\
S(B) &= -\mathrm{Tr}[\rho_B \log \rho_B] && \text{quantum entropy of } B.
\end{aligned}
\tag{12.30}
$$

The quantum conditional entropy is defined as the difference

$$
S(A|B) = S(A, B) - S(B) \quad \text{quantum conditional entropy}
\tag{12.31}
$$

in agreement with the relation $H(A|B) = H(A, B) - H(B)$ seen for classical entropies. But, while $H(A|B)$ is defined from the corresponding conditional distribution, $S(A|B)$ is not related to a density operator (a conditional density operator does not exist).

---

[3] The change of notation $S(\rho_{AB}) \rightarrow S(A, B)$, $S(\rho_A) \rightarrow S(A)$, etc., are frequently used in the literature, where the density operator are replaced by the label of the system.

The main properties of entropies (12.30) and (12.31) are [8]

- $S(A, B) = S(A) + S(B)$ if $\rho_{AB} = \rho_A \otimes \rho_B$ (additivity of the quantum entropy under the tensor product),
- $S(A, B) \le S(A) + S(B)$ (subadditivity inequality),
- $S(A, B) \ge |S(A) - S(B)|$ (triangle inequality).

The novelty is that the quantum conditional entropy $S(A|B)$ may be negative for the presence of entanglement in contrast with the classical case, where $H(A|B) \ge 0$. A simple example to see this possibility is provided by:

Example 12.3 (*Bell state*) The Bell state

$$|\psi\rangle_{\text{Bell}} = \frac{1}{\sqrt{2}}(|00\rangle + |11\rangle) \tag{12.32}$$

is maximally entangled. The corresponding density operator is

$$\rho_{AB} = \frac{1}{2}\left(|00\rangle\langle 00| + |00\rangle\langle 11| + |11\rangle\langle 00| + |11\rangle\langle 11|\right) \tag{12.33}$$

and therefore

$$\rho_A = \frac{1}{2}I_{\mathcal{H}_A}, \qquad \rho_B = \frac{1}{2}I_{\mathcal{H}_B}.$$

Considering that ρ_{AB} corresponds to a pure state, we have $S(A, B) = 0$, whereas $S(B) = 1$ and $S(A|B) = -1$.

12.4.6 About the Difference Between Classical and Quantum Entropies

In this paragraph, we wish to explain where the difference between classical and quantum entropies arises from a bipartite system. In both cases, the entropies depend only on probability distributions, but there is a deep difference concerning the way the marginal distributions are derived from the joint distribution.

Let us consider the evaluation of classical entropies, for simplicity in the case of two symbols with binary alphabets, where we have the ensembles $(\mathcal{A} \times \mathcal{B}, p_{AB})$, $(\mathcal{A}, p_A)$, and $(\mathcal{B}, p_B)$, with $\mathcal{A} = \mathcal{B} = \{0, 1\}$. The distributions can be displayed in the form

$$\begin{array}{cc} p_A(0) & p_A(1) \\ \begin{array}{c} p_B(0) \\ p_B(1) \end{array} \left[\begin{array}{cc} p_{AB}(0, 0) & p_{AB}(0, 1) \\ p_{AB}(1, 0) & p_{AB}(1, 1) \end{array} \right]. \end{array}$$

and verify the usual conditions and in particular the marginal laws

$$p_A(i) = \sum_j p_{AB}(i, j), \qquad p_B(j) = \sum_i p_{AB}(i, j) \tag{12.34}$$

where p_A is simply obtained by summing the elements of the columns of the matrix of the joint probabilities p_{AB}. Pictorially, we can describe this operation as a projection along the columns, which collects and sums the probabilities encountered. The same consideration holds for the marginal distribution p_B, where the projection is along the rows of the matrix.

In the quantum case, the probability ensembles are obtained from the density operators ρ_{AB}, ρ_A, and ρ_B. We must consider this problem in great detail because it can easily generate confusion. To begin with, suppose that the bipartite operator ρ_{AB} has the decomposition

$$\rho_{AB} = \sum_i \sum_j q_{AB}(i, j)\, |\psi_i\rangle \otimes |\phi_j\rangle\langle\psi_i| \otimes \langle\phi_j| \qquad (12.35)$$

where $|\psi_i\rangle \in \mathcal{H}_A$, $|\phi_j\rangle \in \mathcal{H}_B$, and q_{AB} is a joint probability distribution having the meaning $q_{AB}(i, j) = \mathrm{P}[s_{AB} = |\psi_i\rangle \otimes |\phi_j\rangle]$. For the marginal operators we find

$$\begin{aligned}
\rho_A &= \mathrm{Tr}_B[\rho_{AB}] = \sum_i q_A(i)\, |\psi_i\rangle\langle\psi_i| \\
\rho_B &= \mathrm{Tr}_A[\rho_{AB}] = \sum_j q_B(j)\, |\phi_i\rangle\langle\phi_i|
\end{aligned} \qquad (12.36)$$

where $q_A(i) = \sum_j q_{AB}(i, j)$ and $q_B(j) = \sum_i q_{AB}(i, j)$. Thus we have obtained probability distributions that verify conditions (12.34), exactly, as in the classical case. If we evaluate the quantum entropies with these distributions, we obtain the same properties as for the classical entropies, and in particular $S(A, B) \geq S(B)$.

Why? The reason is that in the evaluation of von Neumann entropies the **probability distributions must be obtained as eigenvalues of the density operators**, as seen in the above remark. Now the EID of the bipartite operator has the form

$$\rho_{AB} = \sum_k \lambda_{AB}(k)\, |\lambda_{AB}(k)\rangle\langle\lambda_{AB}(k)|$$

where the $|\lambda_{AB}(k)\rangle$ form an orthonormal basis of $\mathcal{H}_{AB}$ and are not separable and may be entangled in general, whereas in (12.35) they are separable. For the marginal operator $\rho_A = \mathrm{Tr}_B[\rho_{AB}]$, we have to perform a separate EID to get

$$\rho_A = \sum_i \lambda_A(i)\, |\lambda_A(i)\rangle\langle\lambda_A(i)|$$

where $p_A(i) = \lambda_A(i)$ and the kets $|\lambda_A(i)\rangle$ form an orthonormal basis of $\mathcal{H}_A$, whereas in (12.36) the kets are not orthonormal in general. Analogous considerations hold for the EID of ρ_B. The critical point that makes all the difference is that **the EID**

of ρ_A **cannot be obtained by tracing out the EID of** ρ_{AB}, but must be carried out independently. Analogous considerations hold for the EID of ρ_B. In general, we will find

$$\lambda_A(i) \neq \sum_j \lambda_{AB}(i, j), \qquad \lambda_B(j) \neq \sum_i \lambda_{AB}(i, j). \tag{12.37}$$

Evaluation of EID in a composite quantum system. We recall from Chap. 2 that the EID of an operator is evaluated on the basis of its matrix representation, which is assumed to be an ordinary 2D square matrix. But in a composite quantum system AB described by the Hilbert space $\mathcal{H}_A \otimes \mathcal{H}_B$ the matrix becomes 4D and the EID, or the equivalent singular value decomposition (SVD), are developed in multilinear algebra (see, e.g., Kolda, Tamara G.; Bader, Brett W. "Tensor Decompositions and Application". SIAM Rev. 51), but the practical evaluation is cumbersome and ultimately not useful. If the operator is separable, $\rho_{AB} = \rho_A \otimes \rho_B$, the EID is obtained through the ordinary EID of the component operators. What to do in the nonseparable case? The solution is to represent the 4D matrix ρ_{AB} by a 2D matrix through the lexicographical order (see Sect. 2.13 and the forthcoming example).

12.4.7 Example of Bipartite Quantum System

The example of negative conditional quantum entropy given by the Bell state is too extreme to understand the paradox due to the presence of entanglement. Now we consider a more articulated case, where the bipartite density operator consists of a convex combination of a Bell state and a separable state

$$\rho_{AB} = \alpha \rho_{\text{Bell}} + \beta \rho_{\text{sep}}, \qquad \alpha + \beta = 1$$

where ρ_{Bell} is given by (12.33) and

$$\rho_{\text{sep}} = (\varepsilon|0\rangle\langle 0| + \mu|1\rangle\langle 1|) \otimes (\varepsilon|0\rangle\langle 0| + \mu|1\rangle\langle 1|), \qquad \varepsilon + \mu = 1.$$

The parameter α may be regarded as the degree of entanglement, with $\alpha = 1$ corresponding to maximally entangled and $\alpha = 0$ to unentangled. The marginal density operators are given by

$$\rho_A = \alpha \tfrac{1}{2} I_{\mathcal{H}_A} + \beta(\varepsilon|0\rangle\langle 0| + \mu|1\rangle\langle 1|), \qquad \rho_B = \alpha \tfrac{1}{2} I_{\mathcal{H}_B} + \beta(\varepsilon|0\rangle\langle 0| + \mu|1\rangle\langle 1|).$$

The matrix representations are, respectively,[4]

---

[4] Note the 2D matrix representing ρ_{AB} through the lexicographical order.

$$
\rho_{AB} = \begin{array}{c} \\ \langle 00| \\ \langle 01| \\ \langle 10| \\ \langle 11| \end{array}
\begin{array}{cccc}
|00\rangle & |01\rangle & |10\rangle & |11\rangle
\end{array}
\left[\begin{array}{cccc}
\frac{1}{2}\alpha + \beta\varepsilon^2 & 0 & 0 & \frac{1}{2}\alpha \\
0 & \beta\varepsilon\mu & 0 & 0 \\
0 & 0 & \beta\varepsilon\mu & 0 \\
\frac{1}{2}\alpha & 0 & 0 & \frac{1}{2}\alpha + \beta\mu^2
\end{array}\right]
$$

$$
\rho_A = \rho_B = \begin{array}{c} \\ \langle 0| \\ \langle 1| \end{array}
\begin{array}{cc}
|0\rangle & |1\rangle
\end{array}
\left[\begin{array}{cc}
\frac{1}{2}\alpha + \beta\varepsilon & 0 \\
0 & \frac{1}{2}\alpha + \beta\mu
\end{array}\right].
$$

Inside ρ_{AB} and ρ_A we find the probabilities, e.g., $q_{AB}(00) = \mathrm{P}[s_{AB} = |00\rangle] = \frac{1}{2}\alpha + \beta\varepsilon^2$, $q_A(0) = \mathrm{P}[s_A = |0\rangle] = \frac{1}{2}\alpha + \beta\varepsilon$, which form classical probability distributions as in (12.35) and (12.36). But we have to evaluate the distributions through EIDs.

The eigenvalues of the matrix ρ_{AB} are $\lambda_\pm = \frac{1}{2}\left[\beta\varepsilon^2 + \beta\mu^2 + \alpha \pm \sqrt{\Delta}\right]$ and $\lambda_3 = \lambda_4 = \beta\varepsilon\mu$, where $\Delta = \alpha^2 + \beta^2(\varepsilon^2 - \mu^2)$, and give the following joint probability matrix

$$
\lambda_{AB} = \left[\begin{array}{cc}
\frac{1}{2}\left(\beta\varepsilon^2 + \beta\mu^2 + \alpha + \sqrt{\Delta}\right) & \beta\varepsilon\mu \\
\beta\varepsilon\mu & \frac{1}{2}\left(\beta\varepsilon^2 + \beta\mu^2 + \alpha - \sqrt{\Delta}\right)
\end{array}\right]. \tag{12.38}
$$

Now, by summing the entries along the columns and the rows, we get the vectors of the **classical** marginal probabilities, say $\tilde{\lambda}_A(i) = \sum_j \lambda_{AB}(i, j)$ and $\tilde{\lambda}_B = \sum_i \lambda_{AB}(i, j)$, respectively. But the **quantum** marginal probabilities, obtained as the eigenvalues of ρ_A and ρ_B, are

$$
\lambda_A = \lambda_B = \left[\tfrac{1}{2}\alpha + \beta\varepsilon , \ \tfrac{1}{2}\alpha + \beta\mu\right]. \tag{12.39}
$$

The difference between classical and quantum marginal probabilities is illustrated in Fig. 12.7 as a function of the degree of entanglement α. Note that for $\alpha = 0$ (absence of entanglement), the classical and the quantum probabilities coincide, while for $\alpha = 1$ (maximum entanglement) the difference reaches its maximum value.

Now we can evaluate the quantum entropies $S(A, B)$ and $S(B)$ from the corresponding (quantum) probabilities, λ_{AB} given by (12.38) and λ_B given by (12.39), that is,

$$
S(A, B) = -\sum_{i=0}^{1}\sum_{j=0}^{1} \lambda_{AB}(i, j) \log \lambda_{AB}(i, j), \qquad S(B) = -\sum_{i=0}^{1} \lambda_B(i) \log \lambda_B(i).
$$

Both $S(A, B)$ and $S(B)$ depend on the parameters α and ε. The quantum entropies are illustrated in Fig. 12.8 as a function of the degree of entanglement α for two

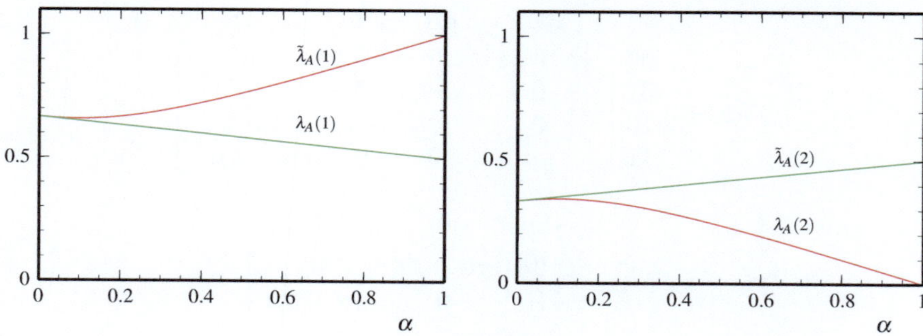

Fig. 12.7 Comparison of marginal probabilities in the example of a bipartite state as a function of the entanglement factor α for $\varepsilon = 1/3$. In *red* the **classical** probabilities $\tilde{\lambda}_A(i)$ obtained as the sum $\sum_j \lambda_{AB}(i, j)$. In *green* the **quantum** probabilities $\lambda_A(i)$ obtained as eigenvalues of ρ_A. Note that for $\alpha = 0$ (absence of entanglement) classical and quantum probabilities coincide

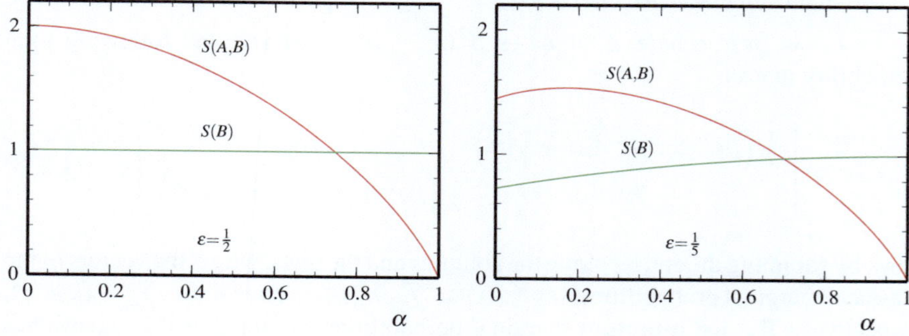

Fig. 12.8 Comparison of quantum entropies $S(A, B)$ and $S(B)$ in the example of a bipartite state, as a function of the entanglement factor α for $\varepsilon = 1/2$ (*left*) and $\varepsilon = 1/5$ (*right*). For $\alpha = 0$ (absence of entanglement) the entropies are given as in the classical case with $S(B) = H(B) = H(A)$ and $S(A, B) = H(A, B) = 2S(B)$. For $\alpha = 1$ (maximum entanglement) $S(B) = 1$ and $S(A, B) = 0$

values of ε. Note that for $\alpha = 0$ (absence of entanglement) the entropies are given as in the classical case, while for $\alpha = 1$ (maximum entanglement) $S(B) = 1$ and $S(A, B) = 0$.

12.4.8 About the Bit and the Qubit

Let us consider the discussion made in the introduction about the bit and the qubit as units of measurement.

In Quantum Information theory, the **logarithm of the dimensionality** of the space is the measure of the information content of a system and plays the role of the logarithm of the size of a codebook in Classical Information Theory [2].

More specifically, a codebook $\mathcal{C} = \mathcal{A}^L$ with $|\mathcal{A}| = d$ contains d^L codewords and $\log |\mathcal{C}| = L \log d$, so that each codeword is represented by $L \log d$ bits. In the quantum case we consider L-long sequences of states of the form $|\Psi\rangle = |\psi_{i_1}\rangle \otimes \ldots \otimes |\psi_{i_L}\rangle$, where $|\psi_i\rangle \in \mathcal{H}$, so that $|\Psi\rangle \in \mathcal{H}^{\otimes L}$. If $\dim \mathcal{H} = d$, then $\log \dim \mathcal{H}^{\otimes L} = L \log d$. If we call *qudit* the basic Hilbert space $\mathcal{H}$, then the sequence $|\Psi\rangle$ is represented by L qudits, or equivalently by 1 qudit/symbol. If the basic Hilbert space is a qubit, $\mathcal{H} = \mathcal{Q}$, with $\dim \mathcal{Q} = 2$, the sequence is represented by L qubits, that is by 1 qubit per symbol.

In these considerations, "bit" and "qubit" are regarded as binary systems, without entering into the information contents of the codewords and of the quantum sequences. But we may ask what is the minimum number of bit per symbol or qubit per symbol needed to represent a codeword or a state sequence. The answer is given by the source coding theorems. Suppose for simplicity that codewords and quantum sequences are produced in a memoryless way, that is, independently of each other and with a given probability distribution $\{p_i\}$. Then, the Shannon entropy is given by $LH(A)$, with $H(A)$ the entropy of a symbol (determined by the distribution). Shannon's source coding theorem states that in the limit ($L \to \infty$), the minimum number is given by the entropy $H(A)$, expressed in bit per symbol. In the quantum case, the probability distribution is encoded in a density operator as $\rho = \sum_i p_i |\psi_i\rangle\langle\psi_i|$ and determines the quantum entropy of a state $S(\rho)$, which becomes $L\,S(\rho)$ for the L-long sequence. Schumacher's source coding theorem states that in the limit the minimum number is given by a quantum entropy $S(\rho)$, expressed in qubit per symbol.

12.5   Classical Data Compression (Source Coding)

One of the fundamental problems in Information Theory is the efficient representation of messages produced by a source of information, which is of interest both for the storage and the transmission of information. In general, a source produces a redundant message, as quantified by the entropy, and the problem is the reduction or, even, the suppression of all the redundancy, thus allowing for a reduction of the message length.

The first and main result on data compression was stated by Shannon under the name *Noiseless channel coding theorem*, where the sender (Alice) and the receiver (Bob) are connected by an ideal (noiseless) channel, as shown in Fig. 12.9. The topic is also known as *source coding*, whereas the coding in the presence of a noisy connection is known as *channel coding*, which will be developed in Sect. 12.7.

Shannon's theorem on source coding is based on a brilliant and imaginative application of the law of large numbers and essentially establishes that the length of a message can be reduced to its entropy, that is, a message m with entropy $H(m)$ can be represented by a string of $H(m)$ bits.

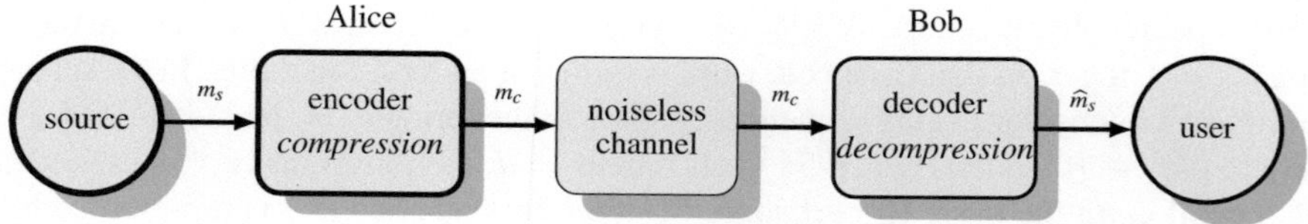

Fig. 12.9 Operations in compression and decompression of an information message: m_s = source message, m_c = coded message, $\widehat{m}_s$ = decoded message

12.5.1 IID Classical Source and the Law of Large Numbers

In this topic the reference source is given by a sequence of *independent identically distributed* (IID) random variables. To define this source, we start from a random variable $A \sim (\mathcal{A}, p_A)$, where $\mathcal{A} = \{1, \dots, K\}$ is a K-ary alphabet.[5] From A, we form a word $A^L := (A_1, \dots, A_L)$ of random variables by imposing that the A_k are statistically independent and have the common distribution p_A. Then the possible realizations $a^L := (a_1, \dots, a_L) \in \mathcal{A}^L$ of A^L have probability

$$P[A^L = a^L] = p_A(a_1) \cdots p_A(a_L). \tag{12.40}$$

The law of large numbers is concerned with the mean $m(A) = E[A]$ of the random variable A and its *estimator* is

$$\mathcal{E}(A^L) := \frac{1}{L} \sum_{n=1}^{L} A_n \tag{12.41}$$

which is a random quantity with expectation $E[\mathcal{E}(A^L)] = E[A]$. The law of large numbers claims that the mean estimator $\mathcal{E}(A^L)$ converges to the mean $m(A) = E[A]$ as L diverges.

Theorem 12.1 (Law of large numbers) *In an IID source, the mean estimator $\mathcal{E}(A^L)$, defined by (12.41), converges in probability to the mean $m(A) = E[A]$:*

$$\lim_{L \to \infty} \mathcal{E}(A^L) = m(A) \quad \text{(in probability)} \tag{12.42}$$

that is, for any $\varepsilon > 0$, $P[|\mathcal{E}(A^L) - m(A)| > \varepsilon] \to 0$ as $L \to \infty$.

---

[5] Equivalently, an IID source may be viewed as a **stationary** random process $\{A_\infty\} = (A_1, A_2, \dots)$ with **independent symbols** and therefore completely specified by an ensemble $(\mathcal{A}, p_A)$, where $p_A(a), a \in \mathcal{A}$ is the common probability distribution, giving $p_A(a) = P[A_n = a]$ for any n. From the random process one can extract words of any length, $(A_1, \dots, A_L)$, which, by the stationarity of the random process and the independence of its symbols, turn out to be L-tuples of IID random variables.

Remark We have introduced the abbreviations, which we will be often used hereafter,

$$A^L := (A_1, \ldots, A_L), \qquad a^L := (a_1, \ldots, a_L) \in \mathcal{A}^L$$

to denote, respectively, the sequence generated by the random variable A and a specific realization of A^L. The number of possible realizations is $|\mathcal{A}|^L = K^L$. In this context, the IID sequence A^L plays the role of *sourceword* (which will be encoded into a *codeword*).

12.5.2 Entropic Application of the Law of Large Numbers

In an IID source, the information of a source word A^L is given by

$$i(A^L) = -\log[p_A(A_1) \cdots p_A(A_L)] \tag{12.43}$$

and the corresponding entropy is given by $H(A^L) = L\, H(A)$.

We apply Theorem 12.1 considering, instead of the mean $m(A) = \mathrm{E}[A]$, the *entropy* $H(A) = \mathrm{E}[i(A)]$. In this case, the estimator is given by

$$\mathcal{I}(A^L) := \frac{1}{L} i(A_1, \ldots, A_L) = \frac{1}{L} \sum_{n=1}^{L} i(A_n) = -\frac{1}{L} \sum_{n=1}^{L} \log p_A(A_n) \tag{12.44}$$

whose expectation is $\mathrm{E}[\mathcal{I}_L(A)] = H(A)$. Then we can apply Theorem 12.1 with the replacements $\mathcal{E}(A^L) \to \mathcal{I}(A^L)$ and $m(A) \to H(A)$ to get:

Theorem 12.2 *In an IID source, the entropy estimator $\mathcal{I}(A^L)$ converges to the entropy*

$$\lim_{L \to \infty} \mathcal{I}(A^L) = H(A) \qquad \text{(in probability)}. \tag{12.45}$$

12.5.3 Preview of Shannon's Protocol

In the asymptotic theory of information, Theorem 12.2 is a central result and establishes the so-called *asymptotic equipartition property*, where the set of the possible sequences emitted by an information source is subdivided into two distinct classes, called *typical sequences* and *atypical sequences*. To introduce these concepts, we consider an IID binary source, which emits 1 with probability $p = \mathrm{P}[A = 1]$ and 0 with probability $q = 1 - p$. As the length L of the symbol sequence emitted by the source increases, we expect to see in the sequence approximately $L\,p$ 1s, and $L\,q$ 0s. A realization a^L for which this assumption is true is a *typical* sequence. Its probability is given by

$$p(a^L) = p_A(a_1) \cdots p_A(a_L) \simeq p^{Lp} q^{Lq}$$

and the corresponding information is

$$i(a^L) \simeq -Lp \log p - Lq \log q = L H(A)$$

where $H(A)$ is the entropy of the binary source (see (12.12)). By combination of the above relations, we find

$$p(a^L) \simeq 2^{-L H(A)}$$

which states two things: (1) the typical sequences are equally likely (uniformly distributed), and (2) their number is at most $2^{L H(A)}$ (because the probability of all the sequences is unitary).

Using the above considerations, it is possible to outline the original protocol of Shannon for data compression. Since there are at most $N \simeq 2^{L H(A)}$ typical sequences, it is sufficient to create a codebook, or an index set, of cardinality N, where each codeword can be represented by a string of $\log N \simeq L H(A)$ bits, that is, **using $H(A)$ bits/symbol**. When Alice realizes that a sequence is typical, she encodes the sequence and sends it to Bob, who knows the codebook and therefore is able to decompress correctly the codeword. As the length L of the sequence becomes larger, this procedure works correctly with probability close to one.

We now formalize the concept of typical sequence. We have a classical IID source, which emits the random sequence A^L with realizations a^L. The typical sequences refer to the entropy estimator $\mathfrak{I}(A^L)$, evaluated for each realization a^L, that is,

$$\mathfrak{I}(a^L) = -\frac{1}{L} \log p_A(a_1) \cdots p_A(a_L) = -\frac{1}{L} \sum_{n=1}^{L} \log p_A(a_n). \tag{12.46}$$

Then, $\mathfrak{I}(a^L)$, which is called *sample entropy* or empirical entropy, is compared with the entropy $H(A)$.

Definition 12.1 A realization a^L is ε-typical if its sample entropy $\mathfrak{I}(a^L)$ is ε-close to the entropy $H(A)$, that is,

$$\boxed{H(A) - \varepsilon \le \mathfrak{I}(a^L) \le H(A) + \varepsilon.} \tag{12.47}$$

The ε-typical set $A_\varepsilon^{(L)}$ is the set of all ε-typical sequences. Considering (12.46), relation (12.47) is equivalent to

$$2^{-L(H(A)+\varepsilon)} \le p(a_1) \cdots p(a_L) \le 2^{-L(H(A)-\varepsilon)}. \tag{12.48}$$

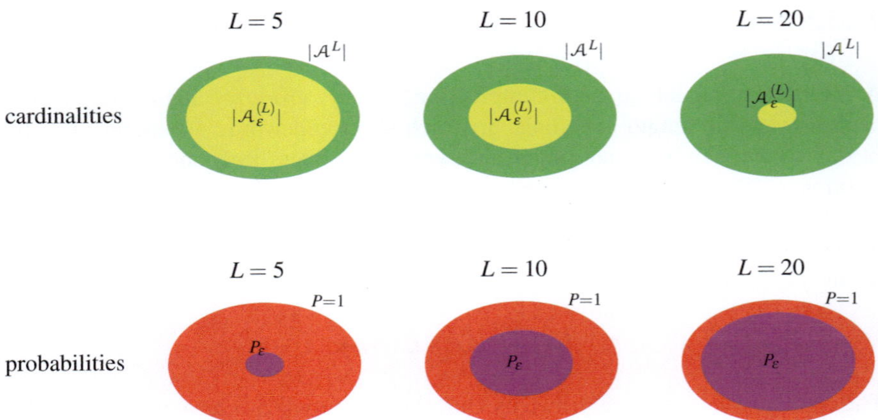

Fig. 12.10 Cardinality and probabilities of the ε-typical set for three values of the sequence length L. The fraction of ε-typical sequences decreases with L. The probability P_ε of ε-typical sequences increases with L

For the typical sequences, it is not difficult to prove [4] the facts illustrated in Fig. 12.10.

12.5.4 Shannon's Compression Protocol

A compression or coding protocol $\mathcal{C}_L$ maps the sourceword $A^L = (A_1, \ldots, A_L) \in \mathcal{A}^L$ into binary codeword $C^{LR} = (C_1, \ldots, C_{LR})$ of $L\,R$ bits, and more precisely of $\lceil L\,R \rceil$ if $L\,R$ is not an integer. A decompression or decoding protocol $\mathcal{D}_L$ maps the $L\,R$ bits back into a sequence $\tilde{A}^L = \mathcal{D}(C^{LR})$ of the original alphabet. Then the maps have the structure

$$\mathcal{C}_L : \ \mathcal{A}^L \ \to \ \{0, 1\}^{LR}, \quad \mathcal{D}_L : \ \{0, 1\}^{LR} \ \to \ \mathcal{A}^L. \tag{12.49}$$

The protocol is *reliable* if

$$\mathcal{D}_L(\mathcal{C}_L(A^L) = A^L \tag{12.50}$$

and it is *asymptotically reliable* if the probability of (12.50) approaches one as $L \to \infty$. The parameter R is called the *compression rate* and represents the number of bits/symbol used in the encoding.

Theorem 12.3 (Shannon's noiseless channel coding) *Consider an IID source with entropy $H(A)$ and a compression rate of R bit/symbol. Then, if $R > H(A)$, there exists an asymptotically reliable compression protocol. Conversely, if $R < H(A)$ any compression protocol is not reliable.*

The proof is a simple application of the properties of typical sequences [4].

12.5.5 *Source Coding Evolution After Shannon's Theorem*

A different approach to data compression, called *zero-error data compression*, exploits variable length encoding, in which short codewords are assigned to the most frequent symbols and longer codewords to the less frequent ones. A typical example of variable length coding is given by the Morse telegraph alphabet. A little later after the proof of the Shannon coding theorem, it was proved that, given a source with entropy $H(A)$, there exists a variable length uniquely decodable code such that the average length L of the codewords satisfies the inequality

$$H(A) \leq L < H(A) + 1.$$

The coding procedure, established by Huffman [9], provides a zero-error probability compression, with compression rate approaching the entropy bound for long symbol sequences.

12.6 Quantum Data Compression

Quantum data compression is similar to Classical data compression but with a few relevant differences. The target is to find an encoding protocol that maps the states of a Hilbert space into a new Hilbert space of reduced dimensionality, with the possibility of recovering the original states. As in the classical compression, where a sequence of symbols is considered, in quantum compression a sequence of states, rather than a single state, is considered. Hence, we start from a Hilbert space $\mathcal{H}$ with a given dimension $d = \dim \mathcal{H}$ and consider its L-extension $\mathcal{H}^{\otimes L}$ of dimension d^L, where the source state sequences (quantum sourcewords) are defined, and we map these sequences into quantum codewords belonging to a new Hilbert space $\mathcal{H}_c$ with the target to get

$$d_c := \dim \mathcal{H}_c < \dim \mathcal{H}^{\otimes L} = d^L. \tag{12.51}$$

12.6.1 *IID Quantum Source*

Usually, an IID quantum source is defined starting from a density operator ρ acting on a Hilbert space $\mathcal{H}$ and considering the L-replica $\rho^{\otimes L}$ obtained by the tensor product. Here we prefer an alternative, but perfectly equivalent, approach which is closer to the classical approach. The given density operator identifies a **random state** $|A\rangle$, which is described by an ensemble $(\mathcal{A}, p_A)$, where $\mathcal{A} = \{|1\rangle, \ldots, |K\rangle\}$ is an alphabet of states in $\mathcal{H}$ and p_A is a probability distribution over $\mathcal{A}$, with the meaning $p_A(a) = \mathrm{P}[|A\rangle = |a\rangle]$.

From the random state $|A\rangle \sim (\mathcal{A}, p_A)$, we form a sequence of L random states by the tensor product, namely $|A^L\rangle := |A_1\rangle \otimes \cdots \otimes |A_L\rangle \in \mathcal{H}^{\otimes L}$, with $|A_i\rangle \sim (\mathcal{A}, p_A)$ and we impose that the kets $|A_k\rangle$ be statistically independent. The possible realizations of random state sequences $|A^L\rangle$ are $|a^L\rangle := |a_1\rangle \otimes \cdots \otimes |a_L\rangle$ and have probability

$$P[|A_L\rangle = |a^L\rangle] = p_A(a_1) \cdots p_A(a_L). \tag{12.52}$$

We add the assumption that **the states of $\mathcal{A}$ form an orthonormal basis of $\mathcal{H}$**, so that dim $\mathcal{H} = |\mathcal{A}| = K$. This assumption remarks a first difference with respect to the classical case, where no geometrical property has been assumed for the alphabet of the symbols.

In the standard approach to define an IID quantum sequence, one starts from a density operator $\rho \in \mathcal{D}(\mathcal{H})$ generated by an ensemble $(\mathcal{A}, p_A)$, where $\mathcal{A}$ is an orthonormal basis of $\mathcal{H}$. Then the density operator reads

$$\rho = \sum_{a \in \mathcal{A}} p_A(a) |a\rangle\langle a| \tag{12.53}$$

which represents an EID, with $p_A(a)$ the eigenvalues and $|a\rangle$ the corresponding eigenvectors (by the orthonormality of $\mathcal{A}$). In the mixed-state ρ given by (12.53) $p_A(a)$ has just the meaning that $|a\rangle$ is present with probability $p_A(a)$. Next, consider the EID of $\rho^{\otimes L}$, which is given by

$$\rho^{\otimes L} = \sum_{a_1} \cdots \sum_{a_L} p_A(a_1) \cdots p_A(a_L) |a_1\rangle\langle a_1| \otimes \cdots \otimes |a_L\rangle\langle a_L|. \tag{12.54}$$

Again, we can see that in this decomposition the sequences of states have just the probabilities given by (12.52).

12.6.2 Typical Quantum Sequences and Typical Subspace

In a quantum IID source we have, for the additivity of the quantum entropy under the tensor product (see Sect. 12.4.4),

$$S(\rho^{\otimes L}) = L\, S(\rho) = L\, H(A)$$

where $S(\rho) = H(A)$ is the classical entropy of the ensemble $(\mathcal{A}, p_A)$ (for the orthonormality of the basis $\mathcal{A}$). Now we can proceed as in the classical case with the entropic application of the law of large numbers (see (12.45) and Theorem 12.2). We introduce the estimator of the quantum entropy $S(\rho)$ as

$$\mathfrak{I}(|A^L\rangle) := \frac{1}{L}i(|A_1\rangle, \ldots, |A_L\rangle) = \frac{1}{L}\sum_{n=1}^{L} i(A_n) = -\frac{1}{L}\sum_{n=1}^{L} \log p_A(A_n). \quad (12.55)$$

Then

Theorem 12.4 *In an IID quantum source, the entropy estimator $\mathfrak{I}(|A^L\rangle)$ converges to the entropy*

$$\lim_{L\to\infty} \mathfrak{I}(|A^L\rangle) = S(\rho) \qquad \textit{(in probability)}. \quad (12.56)$$

Next, in analogy with the classical case, we introduce the *sample entropy* as

$$\mathfrak{I}(|a^L\rangle) = -\frac{1}{L}\log p_A(a_1)\cdots p_A(a_L) = -\frac{1}{L}\sum_{n=1}^{L} \log p_A(a_n). \quad (12.57)$$

Definition 12.2 A realization $|a^L\rangle$ of the state sequence $|A^L\rangle$ is ε typical if its sample entropy $\mathfrak{I}(|a^L\rangle)$ is ε close to the quantum entropy $S(\rho)$, that is,

$$\boxed{S(\rho) - \varepsilon \leq \mathfrak{I}(|a^L\rangle) \leq S(\rho) + \varepsilon.} \quad (12.58)$$

The ε-typical **subspace** $\mathcal{H}_\varepsilon^{(L)}$ is defined as the subspace of $\mathcal{H}^{\otimes L}$ spanned by all ε-typical state sequences (Fig. 12.11). $\square$

The dimension of the space $\mathcal{H}^{\otimes L}$ is given by (see (12.51)) dim $\mathcal{H}^{\otimes L} = (\dim \mathcal{H})^L$. To find the dimension of the reduced space $\mathcal{H}_\varepsilon^{(L)}$, the projector onto the ε-typical subspace (*typical projector*) is introduced

$$\Pi_\varepsilon^{(L)} = \sum_{\varepsilon\text{-typical}} |a_1\rangle\langle a_1| \otimes |a_2\rangle\langle a_2| \otimes \cdots \otimes |a_L\rangle\langle a_L| \quad (12.59)$$

Fig. 12.11 Hilbert space $\mathcal{H}^{\otimes L}$ of quantum sourcewords and ε-typical subspace $\mathcal{H}_\varepsilon^{(L)}$ of quantum codewords

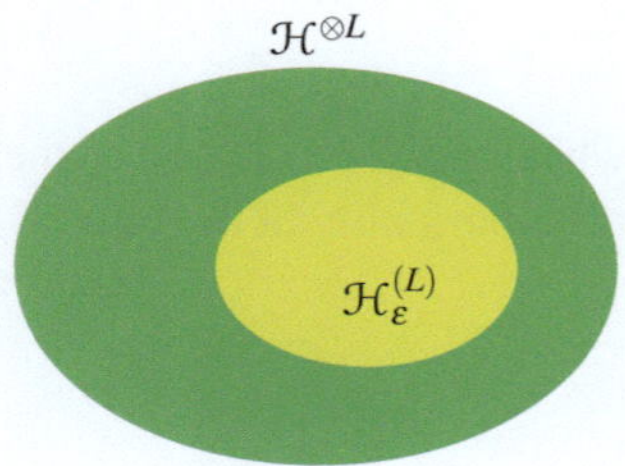

where the summation is limited to the ε-typical sequences defined by (12.58). Then

$$\dim \mathcal{H}_\varepsilon^{(L)} = \mathrm{Tr}\,\Pi_\varepsilon^{(L)}.$$

12.6.3 Schumacher's Compression Protocol

Quantum compression is similar to classical compression, with the adequate modifications. Now, the classical encoding and decoding maps seen in (12.49) take the form (Fig. 12.12)

$$\mathcal{C}_L : \ \mathcal{H}^{\otimes L} \ \to \ \mathcal{H}_c^{(L)}, \qquad \mathcal{D}_L : \ \mathcal{H}_c^{(L)} \ \to \ \mathcal{H}^{\otimes L}. \qquad (12.60)$$

where $\mathcal{H}_c^{(L)}$ is the compressed Hilbert space (in practice given by the ε-typical subspace $\mathcal{H}_\varepsilon^{(L)}$). The *compression rate* is defined as

$$R := \frac{\log \dim \mathcal{H}_c^{(L)}}{\log \dim \mathcal{H}^{\otimes L}} = \frac{\log \dim \mathcal{H}_c^{(L)}}{L \log d}.$$

In particular, if $\mathcal{H}^{\otimes L}$ consists of L qubits ($d = 2$), the compression rate becomes $R = \log \dim \mathcal{H}^{(L)}/L$ and allows one to express the dimension of the reduced space in the form

$$\dim \mathcal{H}_c^{(L)} = 2^{L R}$$

and R reads in **qubits/symbol**. For the reliability of the protocol, the criterion is based on the concept of *fidelity* F_L, which compares the decompressed state to the original state. This parameter verifies the condition $0 \leq F_L \leq 1$ and becomes $F_L = 1$ only when the two states coincide [8].

Theorem 12.5 (Schumacher's noiseless quantum channel coding) *Consider an IID quantum source with entropy $S(\rho)$ and rate R. Then, if $R > S(\rho)$ there exists an*

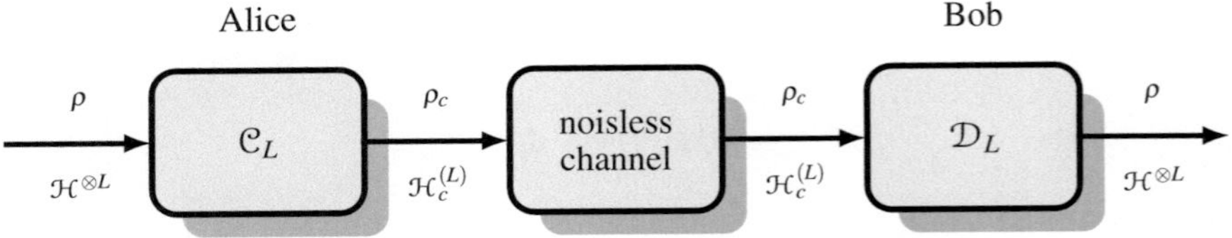

Fig. 12.12 Quantum compression and decompression. The map $\mathcal{C}_L$ compresses a state ρ of $L \log d$ qubits (space $\mathcal{H}^{\otimes L}$) into a state ρ_c of LR qubits (space $\mathcal{H}_c^{(L)}$), and the map $\mathcal{D}_L$ decompresses ρ_c into the original state ρ

asymptotically reliable compression protocol (where $F_L \rightarrow 1$ as $L \rightarrow \infty$). Conversely, if $R < H(A)$ any compression protocol is not reliable.

The proof is based on an application of the properties of typical sequences.

12.6.4 Difference Between Quantum and State Compression

In the final part of Chap. 5, we have developed the **state compression** as a representation of quantum states in a Hilbert space of reduced dimensionality. Thus the target was similar to the one of **quantum compression** developed in this section. However, the two techniques are completely different because they are based on different properties of the states. The state compression is based only on the geometric and algebraic properties of vector spaces (where quantum states live). The quantum compression is concerned with the statistical properties encoded in quantum states, whose randomness represents a source of information.

Anyway, it is interesting to remark that state compression preserves information, as expected by the fact that it is a reversible transformation.

Proposition 12.1 *The entropy of a quantum state does not change after a state compression*

$$S(\overline{\rho}) = S(\rho) \quad \text{with} \quad \overline{\rho} = U_r^* \rho \, U_r. \tag{12.61}$$

Proof Let $n = \dim \mathcal{H}$, $r = \dim \overline{\mathcal{H}}$, and let $\sigma(\rho) = \{\lambda_1, \ldots, \lambda_r, 0, \ldots, 0\}$ be the spectrum of ρ, where the last $n - r$ eigenvalues are zero. We have to prove that

$$\sigma(\overline{\rho}) = \{\lambda_1, \ldots, \lambda_r\}. \tag{12.62}$$

Note the zero eigenvalues do not give a contribution to quantum entropy because we assume $0 \log 0 = 0$. We start from the reduced EID of ρ, say $\rho = V_r \, \Lambda_r \, V_r^*$ and use the compressor U_r^* to get

$$\overline{\rho} = U_r^* \rho \, U_r = U_r^* V_r \, \Lambda_r \, V_r^* \, U_r = Z \Lambda Z^*. \tag{12.63}$$

If we prove that the matrix $Z := U_r^* V_r$ is unitary, then we get that (12.63) is an EID of $\overline{\rho}$ with Λ_r containing the spectrum (12.62) on the diagonal. We have

$$Z^* Z = V_r^* U_r \, U_r^* V_r = V_r^* P_{\mathcal{U}} V_r$$

where $P_{\mathcal{U}} = U_r \, U_r^*$ is the projector from $\mathcal{H}$ onto the compressed space $\overline{\mathcal{H}}$ with the property $P_{\mathcal{U}}|u\rangle = |u\rangle$, $\forall |u\rangle \in \mathcal{U}$ (see (5.140) and Fig. 5.14). Considering that the kets of V_r belong to $\mathcal{U}$, we have that $P_{\mathcal{U}} V_r = P_{\mathcal{U}}[|v_1\rangle, \cdots, |v_r\rangle] = V_r$. Then, $Z^* Z = V_r^* V_r = I_r$ from the orthonormality of the kets $V_r = [|v_1\rangle, \cdots, |v_r\rangle]$, and Z^* turns out to be a unitary matrix, and the same applies to Z.

12.7 Classical Channels and Channel Encoding

In this section, we consider the transmission of information from Alice to Bob through a classical noisy channel.

In data compression, the transmission channel is noiseless and therefore preserves the message and its information. In the presence of noise, the communication channel distorts the messages and to counteract this effect a *channel coding* is introduced before the transmission. While in source encoding the redundancy is removed, in channel encoding a suitable redundancy is introduced with the target that, although the message is corrupted, at the reception, the redundancy makes it possible to retrieve the original message with a small error probability (reliable communication).

The central result will be the second Shannon's theorem, according to which the transmission can be made highly reliable, with a probability of error approaching zero, provided that the information message is appropriately encoded and the transmission rate R is less than the channel capacity C, a parameter that characterizes a noisy channel. The scenario is illustrated in Fig. 12.13, where Alice wants to transmit to Bob a message A. The message is mapped into a *codeword* C by a *channel encoder* and sent through a noisy channel. At the channel output, Bob receives the word $\hat{C}$, and tries to get an estimate $\hat{A}$ of the original message A.

12.7.1 Probabilities and Information in a Discrete Channel

We consider only memoryless channels, which can be defined by a triplet $(\mathcal{A}, \mathcal{B}, p_c)$ where $\mathcal{A}$ and $\mathcal{B}$ are respectively the input and output alphabets, while p_c are the transition probabilities $p_c(b|a) = P[B = b|A = a], a \in \mathcal{A}, b \in \mathcal{B}$. Figure 12.14 illustrates the graphical representation of a ternary–quaternary channel, together with the representation of the ternary source.

Given the source probabilities $p_A(a)$ and the transition probabilities $p_c = p_{B|A}(b|a)$ one can evaluate the other probabilities in the bipartite system $A \to B$, that is, $p_B, p_{AB}, p_{A|B}$ and then evaluate the corresponding informations and entropies, seen in Sect. 12.3. Here we introduce a new entropic quantity: the mutual information, which finds the following motivation. The target of a channel is the transmission of the information carried by the symbol A, when Bob reads the received symbol B. In the ideal case, from B Bob should know A, but in practice (for the presence of

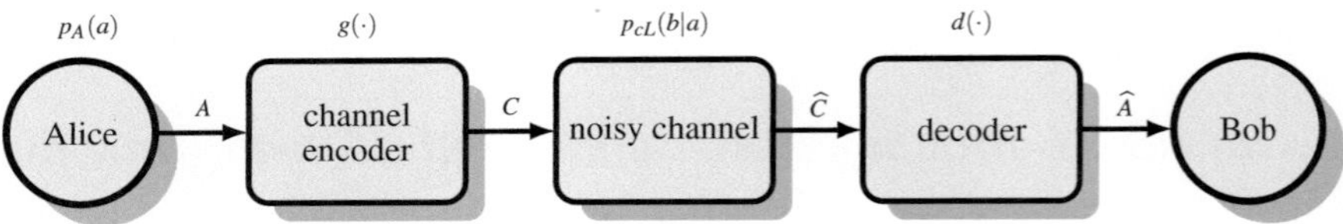

Fig. 12.13 Communication through a noisy channel

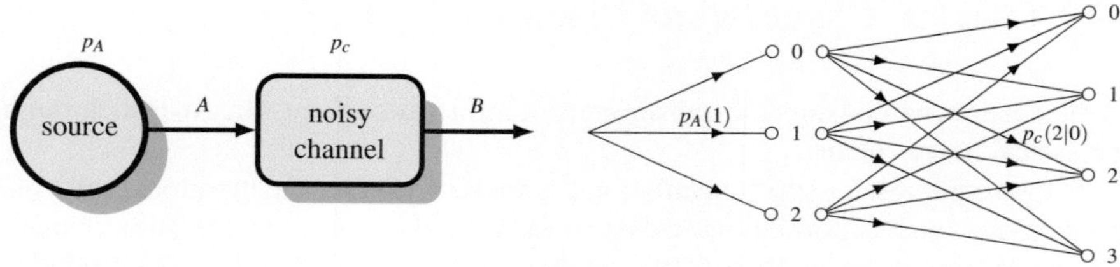

Fig. 12.14 Noisy ternary–quaternary channel driven by a ternary source: A source (or transmitted) symbol and B received symbol. The source is specified by the *prior probabilities* $p_A(a)$ and the channel by the *transition probabilities* $p_c(b|a)$

noise), he has an uncertainty on A. Specifically, $i(A)$ is the *a priori* uncertainty on A, while $i(A|B)$ is the *a posteriori* uncertainty on A, or residual uncertainty, when b is known. The difference

$$\Delta i(A; B) = i(A) - i(A|B) = \log \frac{p(A|B)}{p(A)} \tag{12.64}$$

may be regarded as the information transmitted by the channel according to the budget

$$\underbrace{\text{a priori uncertainty}}_{i(A)} = \underbrace{\text{a posteriori uncertainty}}_{i(A|B)} + \underbrace{\text{transmitted information}}_{\Delta i(A; B)} \tag{12.65}$$

The average of $\Delta i(A; B)$

$$\boxed{I(A; B) = \mathrm{E}[\Delta i(A; B)] = H(A) - H(A|B)} \tag{12.66}$$

is called **mutual information** of the pair (A, B). Note the symmetry $\Delta i(A; B) = \Delta i(B; A)$, which gives $I(A; B) = I(B; A)$, that is, the mutual informations of (A, B) and (B, A) are the same. Finally, from inequality (12.17), we have $I(A; B) \geq 0$.

In the ideal case (noiseless channel), all the source information is transmitted by the channel because $H(A|B) = 0$, and we have $I(A; B) = H(A)$. In a noisy channel $H(A|B) > 0$, and then $I(A; B) < H(A)$ and in (12.66), the conditional entropy $H(A|B)$ may be regarded as the loss of information due to the presence of noise. For this reason, $H(A|B)$ is called **equivocation** and also **information loss**.

Note the alternative expression of the mutual entropy

$$I(A; B) = H(A) - H(A|B) = H(B) - H(B|A) = H(A) - H(B) - H(A, B). \tag{12.67}$$

Example 12.4 (*Binary symmetric channel*) This channel is specified by the *cross* transition probabilities, which are assumed to be equal $p_c(0|1) = p_c(1|0) = \varepsilon$, as

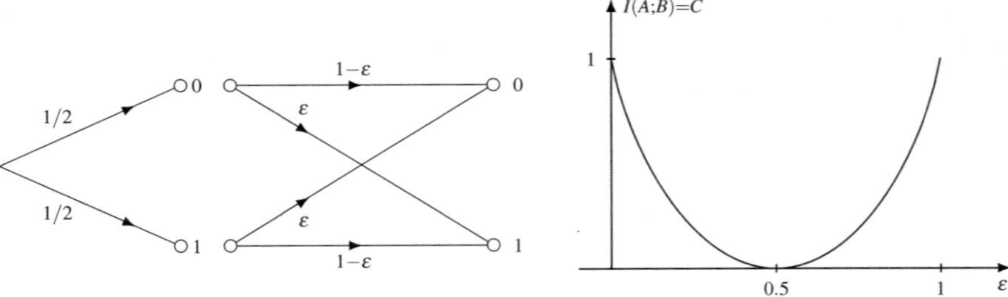

Fig. 12.15 Binary symmetric channel driven by a symmetric source ($p_A(0) = p_A(1) = 1//2$). The mutual information $I(A; B)$ gives also the channel capacity C (owing to the source symmetry)

shown in Fig. 12.15. Assuming that also the a priori probabilities are equal, $p_A(0) = p_A(1) = \frac{1}{2}$, the source entropy is $H(A) = 1$ bit/symbol. The equivocation results (see Problem 12.3)

$$H(A|B) = -(1 - \varepsilon)\log(1 - \varepsilon) - \varepsilon \log \varepsilon \, .$$

Hence the mutual information in a binary symmetric channel is given by

$$I(A; B) = H(A) - H(A|B) = 1 + (1 - \varepsilon)\log(1 - \varepsilon) + \varepsilon \log \varepsilon \qquad (12.68)$$

and is illustrated in Fig. 12.15 as a function of ε. Note the extreme cases, $\varepsilon = 0$ and $\varepsilon = 1$, in which the channel is noiseless and we have $I(A; B) = 1$. The case $\varepsilon = 1/2$ gives $I(A; B) = 0$ and corresponds to a *useless* channel.

12.7.2 Channel Capacity

The mutual information $I(A; B)$ depends both on the source, through the a priori probabilities, and on the channel, through the transition probabilities. Now, for a given channel, we can vary $I(A; B)$ by changing the a priori probabilities p_A. The maximum that one obtains gives the capacity of the channel.

Definition 12.3 The **capacity** of a discrete memoryless channel is given by the maximum of the mutual information taken over all possible a priori distributions:

$$\boxed{C = \max_{p_A} I(A; B).}\qquad (12.69)$$

Then, by definition

$$0 \le I(A; B) \le C. \qquad (12.70)$$

For the evaluation of the capacity, it will be convenient to express $I(A\,;\,B) = H(A) - H(A|B)$ in terms of the a priori distribution and the transition probabilities. The expression is (see Problem 12.4)

$$I(A\,;\,B) = \sum_{a,b} p_A(a)\, p_c(b|a) \log \frac{p_c(b|a)}{\sum_{a'} p_A(a')\, p_c(b|a')}. \tag{12.71}$$

In the evaluation, the channel transition probabilities are assumed as given and we have to perform the maximization with respect to the a priori probabilities p_A. The problem is far from being easy because the p_A have the constraint $p_A(a) \geq 0$ and $\sum_a p_A(a) = 1$, and in general we do not achieve a closed-form result. Only in the presence of symmetries, explicit results are obtained. In particular, for the binary symmetric channel, a symmetry consideration leads to the conclusion that the capacity is obtained when the source is symmetric, that is, $p_A(0) = p_B(1) = 1/2$. Hence, the formula of the mutual information (12.68), obtained in such condition, gives the channel capacity of the binary symmetric channel.

12.7.3 *Coding in the Presence of Noise*

The scenario has been illustrated in Fig. 12.13. The source message is encoded into *codewords* of a given length L, which are transmitted by the L-extension of a discrete channel. The received words are decoded into the original source format. The presence of noise corrupts the codewords, but if the redundancy introduced by encoding is sufficiently high, the decoder can recover the original message with a small error probability. Note that there is a special kind of duality between the source coding and the channel coding: in the first the redundancy is removed, while in the second the redundancy is intentionally introduced to make the recovery reliable.

To describe the system, we introduce the following notations:

$$\begin{aligned}
&A \in \mathcal{A} = \{1, \ldots, K\} &&\text{symbol (or message) produced by the source,}\\
&X^L = (X_1 X_2 \ldots X_L) \in \mathcal{C} &&\text{codeword sent to the channel,}\\
&Y^L = (Y_1 Y_2 \ldots Y_L) \in \mathcal{Y}^L &&\text{received codeword,}\\
&\widehat{A} \in \mathcal{A} &&\text{decoded symbol (or message).}
\end{aligned}$$

The set $\mathcal{C}$ of the codewords is called *codebook* and has the same cardinality as the source alphabet, $|\mathcal{C}| = |\mathcal{A}| = K$; in general it is given by a small subset of the L-extension of an alphabet $\mathcal{X} = \{x_1, \ldots, x_N\}$, often given by the binary alphabet $\mathcal{X} = \{0, 1\}$. At the channel output, for the presence of noise, the received word Y^L may be different from the codeword X^L and potentially it may be any L-tuple, that is, $Y^L \in \mathcal{Y}^L$. Usually, the channel is the L-extension of a binary symmetric channel, so that the codewords are given *only by K binary L-tuples*, while at the channel output the words may be *any binary L-tuple*.[6]

Now we write the relations of each operation in the scheme of Fig. 12.13. The channel encoder maps, with a one-to-one correspondence, the source symbols onto the codewords:

$$x^L = g(a), \qquad a \in \mathcal{A}, x^L \in \mathcal{C} \qquad (12.72)$$

where $g(a)$ is the *encoding function*. The channel is the L-extension $(\mathcal{X}^L, \mathcal{Y}^L, p_{cL})$ of a discrete channel $(\mathcal{X}, \mathcal{Y}, p_c)$, where the transition probabilities work as follows (assuming a memoryless channel)

$$\boxed{p_{cL}(y_l y_2 \ldots y_L | x_1 x_2 \ldots x_L) = \prod_{n=1}^{L} p_c(y_n | x_n).} \qquad (12.73)$$

The decoder cannot realize the inverse map of (12.72) (because $\mathcal{Y}^L \supset \mathcal{C}$), but, observing the received word y^L, it makes a *decision* about the source symbol, according to a *decision function* $d : \mathcal{Y}^L \to \mathcal{A}$, where

$$\hat{a} = d(y^L), \qquad y^L \in \mathcal{Y}^L, \hat{a} \in \mathcal{A}. \qquad (12.74)$$

The global system is equivalent to a discrete channel $(\mathcal{A}, \mathcal{A}, p_g)$, where the transition probabilities $p_g(a'|a) = \mathrm{P}[\hat{A} = a' | A = a]$ can be calculated from the previous rules. In particular, we can obtain the *error probability* as

$$P_e = \mathrm{P}[\hat{A} \neq A] = 1 - \sum_{a \in \mathcal{A}} p_g[a|a] \, p_A(a). \qquad (12.75)$$

12.7.4 An Elementary Example

A simple example of channel encoding is obtained by transmitting each source binary data three times. This is known as a $(3,1)$ *repetition code* and is illustrated in Fig. 12.16. Then the codeword is simply $C = (AAA)$, with codebook $\mathcal{C} = \{(000), (111)\}$. Using the 3-extension of a binary symmetric channel, the dictionary of the received words $\mathcal{Y}^3$ consists of all the possible 8 binary triplets. The transition probabilities of the channel can be calculated using (12.73). For instance, with the transmission of the codeword $x^3 = (000)$, the probability that the received word is $y^3 = (001)$ is given by

---

[6] We continue with the convention of denoting random quantities by upper case, as A and X^L, and their realizations by the corresponding lower case letters, as a and x^L.

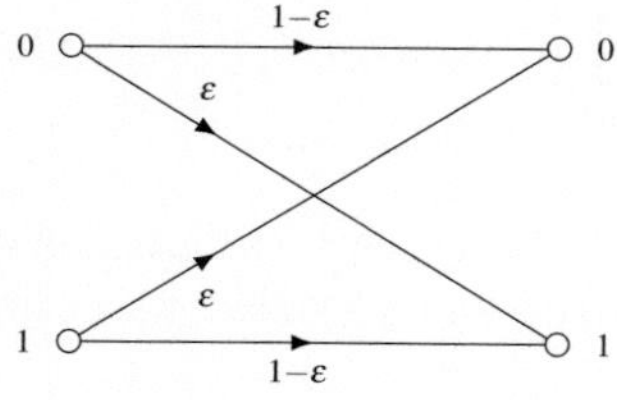

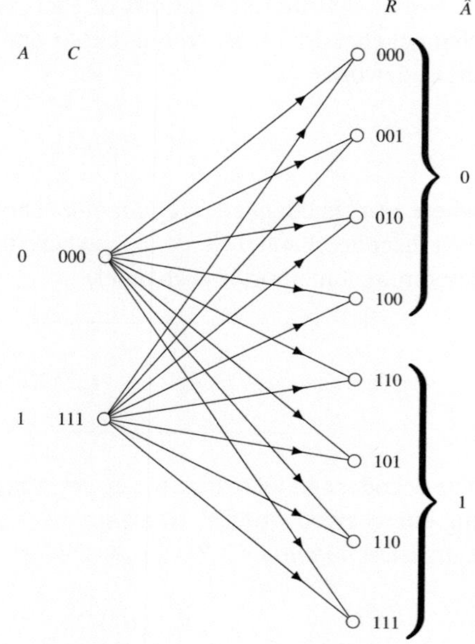

Fig. 12.16 Example of channel encoding for the transmission of binary symbols, using the 3-extension of a binary symmetric channel. The table gives the source alphabet $\mathcal{A}$, the codebook $\mathcal{C}$, and the set $\mathcal{R}$ of the possible words at the channel output. In the graph, at the *right hand side*, the decision rule is specified through the *curly brackets*

$$p_{c3}(001|000) = p_c(0|0)\,p_c(0|0)\,p_c(1|0) = (1-\varepsilon)(1-\varepsilon)\varepsilon\,.$$

In general, we have

$$\boxed{p_{cL}(r|c) = \varepsilon^{D(r,c)}(1-\varepsilon)^{L-D(r,c)}} \tag{12.76}$$

where $D(r, c)$ is the *Hamming distance* between the received word r and the codeword c (given by the number of different bits between the two words).

The decoding function is based on the majority rule with a partition of the dictionary of the received words $\mathcal{Y}^3$ into two parts

$$\mathcal{Y}_0^3 = \{(000),\,(001),\,(010),\,(100)\}\,, \qquad \mathcal{Y}_1^3 = \{(101),\,(110),\,(011),\,(111)\}$$

The rule gives $\widehat{A} = i$ if $y^3 \in \mathcal{Y}_i^3$, $i = 0, 1$. Then, the error probability is given by

$$P_e = \sum_{r_i \notin \mathcal{R}_i} p_{cL}(r_i|c_i) p_A(i).$$

Since we are considering a memoryless channel, the errors occurring in the symbols of words are statistically independent. Then, by inspection we find

$$P_{e3} = P[\widehat{A} \neq A] = \binom{3}{2} \varepsilon^2 (1 - \varepsilon) + \varepsilon^3.$$

To fix the ideas and to be anchored to the real world, we consider a specific case: a binary source with equally likely symbols, so that $H(A) = 1$ bit/symbol and we suppose that our binary channel allows the transmission of $f_c = 1000$ bit/s with the error probability $\varepsilon = 0.1$. With these data, we compare the transmission without the encoding and in the presence of encoding.

In the absence of coding, the information sent to the channel is $H' = H(A) f_c = 1000$ bit/s. and at the channel output the error probability is $P_e = \varepsilon = 0.1$, that is, on average, about one error every ten bits.

Now the introduction of the $(3, 1)$ repetition code allows for the reduction of the error probability by encoding.

$$P_e := P[\hat{a} \neq a] = \binom{3}{2} \varepsilon^2 (1 - \varepsilon) + \varepsilon^3 = 0.028.$$

with an appreciable improvement. But the penalty to pay is the reduction of the entropy rate because we repeat the same symbol three times and the channel guarantees the error probability $\varepsilon = 0.1$ with a rate of $f_c = 1000$ symbol/s, where now $f_c = 3 f_s$. Hence the entropy rate should be reduced to $H' \cong 333$ bit/s.

Following this line, we can increase the number of repetitions to improve the reliability of the transmission with a corresponding reduction of the entropy rate, according to the paradigm (for $L \to \infty$).

$$P_e \to 0 \quad \text{provided that} \quad H' = f_s H \to 0.$$

This seems the intuitive conclusion. But Shannon established a counter intuitive conclusion: there exists a channel encoding such that

$$P_e \to 0 \quad \text{provided that} \quad H' < C'.$$

In the above example, $C' = f_c[-\varepsilon \log \varepsilon - (1 - \varepsilon) \log(1 - \varepsilon)] \cong 529$ bit/s, so that we can realize a completely (asymptotically) reliable transmission ($P_e \sim 0$), with an entropy rate of 529 bit/s.

12.7.5 The Fundamental Theorem of Information Theory

Theorem 12.6 (The second Shannon theorem) *Given a noisy channel with capacity C and a source with K messages, provided that*

$$\frac{1}{L} \log K < C,$$

there exists a coding and encoding system that guarantees an arbitrarily small error probability. Conversely, if

$$\frac{1}{L} \log K > C$$

no encoding scheme guarantees arbitrarily small error probability.

The proof of the theorem is based on a very sophisticated coding and decoding procedure excogitated by Shannon [1], which can be summarized in the following steps.

1. A third party, say Charlie, on the basis of the channel transition probabilities $p_c(y|x)$ and of a source probability distribution $p_X(x)$, computes the joint input–output distribution $p_{XY}(x, y) = p_X(c)p_c(y|x)$ with the corresponding joint entropy $H(A, B)$.

2. For a fixed integer length L and a real number $\varepsilon > 0$, Charlie computes the joint ε-typical set, i.e., the set of the input and output realizations (x^L, y^L), such that the entropy estimator $\mathcal{J}(x^L, y^L)$ satisfies the inequalities

$$H(X, Y) - \varepsilon \leq \mathcal{J}(X^L, Y^L) \leq H(X, Y) + \varepsilon$$

(see the corresponding Definition 12.1).

3. For each source message $a \in \mathcal{A}$, Charlie generates a *random codeword* $x^L(a) = (x_1(a), x_2(a), \ldots, x_L(a))$ with IID symbols having common probability distribution $p_X(x)$. The K codewords $x_L(1), \ldots, x^L(K)$ are chosen independently.

4. Charlie sends to Alice and Bob the random code generated at step 3 and the joint typical set computed at step 2.

5. Alice, on the basis of the message a emitted by the source, sends through the channel the random codeword $x^L(a)$. Note that this word is doubly random, because of the randomness of both the source symbol a and of the code generated by Charlie.

6. After receiving the word y^L emitted by the noisy channel, Bob searches an input codeword $x^L(\hat{a})$ such that the pair $(x^L(\hat{a}), y^L)$ belongs to the joint typical set, according to the so-called *typical set decoding*. If such a codeword exists and is unique, the corresponding message $\hat{a}$ is chosen as the decided message. If such a codeword either does not exist or it is not unique, a random message is chosen.

The brilliant Shannon's idea resides in the random coding of step 3, that became a standard paradigm in the evolution of both Classical and Quantum Information Theory. The philosophy supporting the proof of the theorem is as follows. For any

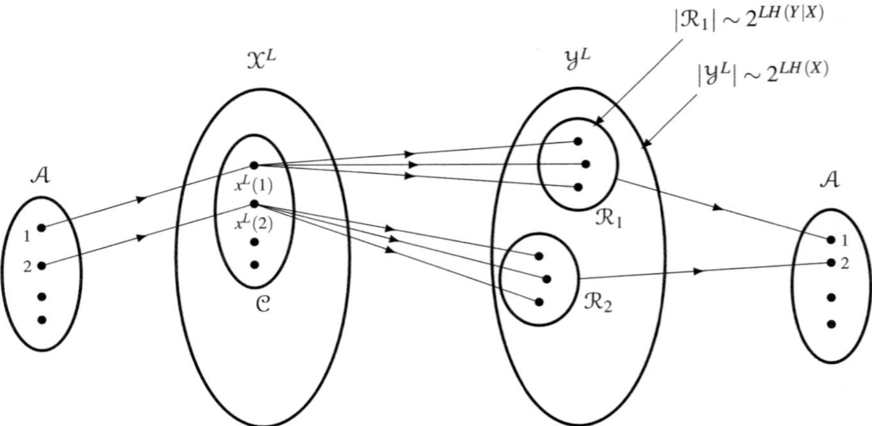

Fig. 12.17 Sets in random coding: $\mathcal{A}$ source alphabet, $\mathcal{X}^L$ set of possible input sequences, $\mathcal{C}$ codebook, $\mathcal{Y}^L$ set of possible output sequences, $\mathcal{R}_1$ set of received sequences for codeword $x^L(1)$, $\mathcal{R}_2$ set of received sequences for codeword $x^L(2)$, ...

realization of the code, the error probability could be evaluated, at least in principle. Owing to the randomness of the code, the error probability P_e turns out to be a random variable. The evaluation of the mean error probability $\overline{P_e}$ is long and cumbersome, but by no means difficult (see [4]). In particular, it may be shown that if L is large enough, namely if it satisfies the inequality

$$\frac{1}{L} \log K < I(X; Y),$$

one gets $\overline{P_e} < \varepsilon$.

Now, if the mean error probability $\overline{P_e}$ is computed, there exists at least a code realization with error probability not greater than $\overline{P_e}$ and, in conclusion, less than ε. Finally, if Charlie chooses as $p_X(x)$ the input probability distribution corresponding to the maximum of $I(X, Y)$, i.e., to the capacity C, the proof is complete.

The result can be explained in terms of typical sequences as in Fig. 12.17. For L large enough there are approximately $2^{LH(X)}$ distinct typical output sequences that could be assigned to the source messages as codewords. Then for a uniquely decodable coding it must be $K \leq 2^{LH(X)}$. On the other hand, this is not sufficient to guarantee that distinct typical input codewords produce distinct output words and we must choose a subset of K typical input words as codewords. To evaluate the cardinality of this subset consider that, given an input typical sequence x^L, for L large enough the are $2^{LH(Y|X)}$ output typical sequences. In order that the output typical words produced by the K codewords cover the set of the $2^{LH(Y)}$, we must have $K2^{LH(Y|X)} \leq 2^{LH(Y)}$, i.e.,

$$K \leq 2^{L[H(Y)-H(Y|X)]} = 2^{LI(X;Y)}.$$

Note that $I(X; Y) \leq H(X)$, so that the condition is more stringent than $K \leq 2^{LH(X)}$.

Problem 12.3 ⋆   Prove that in a binary symmetric channel with cross transition probability ε and equal a priori probabilities (see Fig. 12.15), the equivocation is given by

$$H(A|B) = -(1 - \varepsilon) \log(1 - \varepsilon) - \varepsilon \log \varepsilon.$$

Problem 12.4 ⋆   Prove formula (12.71) giving the mutual information in terms of the a priori probabilities and the transition probabilities.

12.8 Quantum Channels and Open Systems

In this section, we develop the theory of quantum channels as a preparation for the reliable transmission of information, which will be formulated in the next section. In this review, we follow closely a recent paper by Holevo and Giovannetti [2].

Quantum channels are a key part of quantum communication systems. In particular, consider the following scenario: classical data belonging to a finite-size alphabet $\mathcal{A}$ are transmitted through a physical line which employs a quantum carrier to convey information (e.g., an optical fiber). In this case it is required that the data are encoded into quantum states of the carrier and finally the quantum states are mapped back to the original classical format. This scenario is illustrated in Fig. 12.18.

The initial encoding is formulated as a *classical-quantum* (c $\rightarrow$ q) mapping

$$A \;\rightarrow\; \rho_A, \qquad A \in \mathcal{A} \tag{12.77}$$

for each symbol A of the data message. In the decoding stage we find a *quantum-classical* (q $\rightarrow$ c) mapping, which is performed by a quantum measurement using a projector system or more generally a POVM system $\{Q_b\,,\ b \in \mathcal{B}\}$ with an alphabet $\mathcal{B}$. In general, $\mathcal{B}$ may be different from $\mathcal{A}$, but hereafter we suppose that $\mathcal{B} = \mathcal{A}$. Then the outcome m of the measurement is $B = b \in \mathcal{A}$ with probability

$$p_{B|A}(b|a) = \mathrm{Tr}\left[\rho_a' \, Q_b\right], \qquad b, a \in \mathcal{A}. \tag{12.78}$$

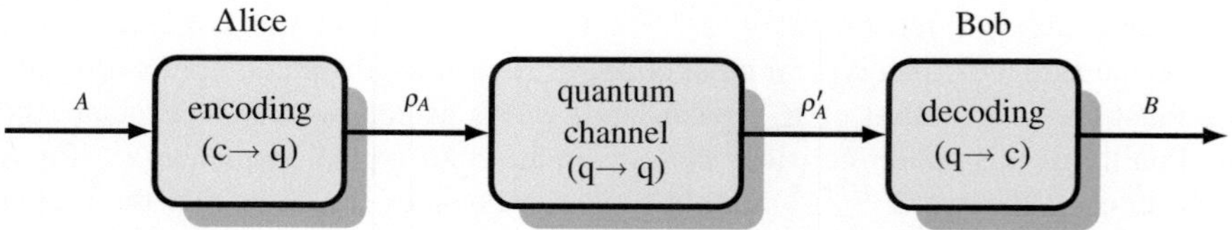

Fig. 12.18   Environment of a quantum channel. Alice encodes the incoming classical symbol $A \in \mathcal{A}$ into a quantum state ρ_A. The quantum channel transfers ρ_A reproducing a distorted replica ρ_A'. Bob decodes ρ_A' into a classical symbol $B \in \mathcal{B}$ by a quantum measurement

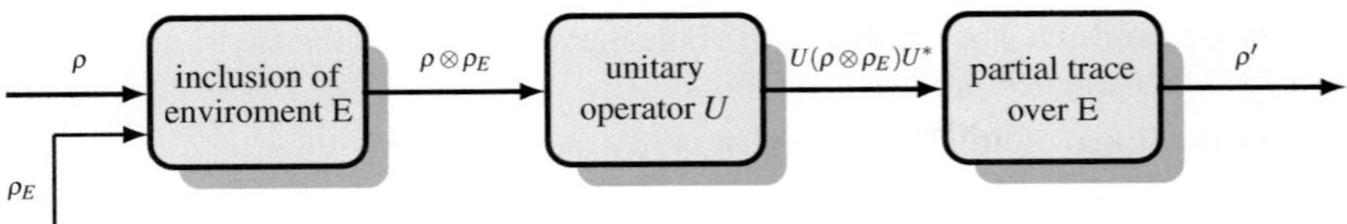

Fig. 12.19 Unitary representation of a noisy quantum channel through the introduction of ancillary density operator ρ_E describing the environment interaction

The central operation in Fig. 12.18 is a q $\to$ q mapping

$$\Phi: \quad \rho \quad \to \quad \rho' \tag{12.79}$$

where $\rho = \rho_A$ is the input quantum state and $\rho' = \rho'_A$ is the output quantum state and represents the **quantum channel**. In practice the output $\rho' \neq \rho$ because of the errors introduced by the physical channel, mainly due to noise. The noise can be classical but also intrinsically quantum, i.e., introduced by the interaction of the system with the environment. Note that the map Φ in general maps operators to operators, possibly in a non-unitary way. It describes in fact the dynamics of an **open system**, and therefore is not directly covered by the elementary postulates of Quantum Mechanics (which are concerned with closed or isolated quantum systems, evolving unitarily). The technique for dealing with open quantum system to regain Quantum Mechanics postulates consists in adding an ancillary Hilbert space $\mathcal{H}_E$ to the input Hilbert space $\mathcal{H}$ to describe the environment interaction and assigning to it an appropriate initial state ρ_E. In such a way, the composite Hilbert space $\mathcal{H} \otimes \mathcal{H}_E$ forms an isolated quantum system, where the evolution is described by a unitary operator U, as shown in Fig. 12.19. The final output is obtained by tracing out over E the state of $\mathcal{H}$ produced by the unitary operator, that is,

$$\rho' = \Phi[\rho] = \mathrm{Tr}_E \left[U \left(\rho \otimes \rho_E \right) U^* \right]. \tag{12.80}$$

This model is very general to include *noiseless* channels as well as *noisy* channels, and also, with some artifacts, the hybrid c $\to$ q and q $\to$ c transformations seen in Fig. 12.19 [2]. The model may include also quantum measurements. A noiseless channel is obtained by omitting the E part and therefore it is simply governed by a unitary transformation, that is,

$$\rho' = \Phi[\rho] = U \rho U^*. \tag{12.81}$$

A noiseless channel is reversible, whereas a noisy channel is not.

A map Φ derived as in (12.80) satisfies the following properties [2, 8]:

A1 **Trace preservation**: Both the input ρ and the output $\rho' = \Phi[\rho]$ are density operators and therefore $\mathrm{Tr}\rho' = \mathrm{Tr}\rho = 1$.

A2 Tensor product: The combination of the input state ρ with the environment state ρ_E is made by a tensor product.

A3 Convex linear map: The map must verify the condition

$$\Phi\left[\sum_i \pi_i\,\rho_i\right] = \sum_i \pi_i\,\Phi[\rho_i] \tag{12.82}$$

for every probability distribution $\{\pi_i\}$ and every set $\{\rho_i\}$ of density operators.

A4 Complete positivity: An obvious condition comes from the fact that the channel mapping Φ transforms a quantum state $\rho \geq 0$ into a quantum state $\rho' \geq 0$, which implies the preservation of *positivity*. But there is a subtle and stronger condition, called *complete positivity*, which means that the extension of a channel Φ with the parallel of an ideal channel Id should be again positive.

The above assumptions are usually abbreviated as *completely positivity and trace preserving* (CPTP).

12.8.1 Kraus Representation

Assumptions CPTP allow us to reformulate the mapping (12.80) in the form

$$\rho' = \Phi[\rho] = \sum_{k=1}^{d_E} V_k\,\rho\,V_k^* \tag{12.83}$$

which is known as a *Kraus representation* of a noisy quantum channel and also *operation-sum representation*. In this expression, the V_k are operators acting on the input Hilbert space $\mathcal{H}$ that verify the completeness condition

$$\sum_{k=1}^{d_E} V_k\,V_k^* = I_{\mathcal{H}} \tag{12.83a}$$

where d_E is the dimension of the environment space $\mathcal{H}_E$. The Kraus representation is not unique, and it becomes important to find minimal representations (the ones with the minimal dimension d_E).

Note, e.g., that the trace is preserved both in (12.80) and in (12.83), where $\mathrm{Tr}\rho' = \mathrm{Tr}\rho$. Also the positivity is ensured in Kraus representation because $\rho \geq 0$ implies $V_k\,\rho\,V_k^* \geq 0$. Less evident is the requirement of complete positivity.

In the case of noiseless channels, Kraus' representation with $d_E = 1$ gives the standard relation of closed systems, that is, $\rho' = U\,\rho\,U^*$.

Example 12.5 (Bit-flip channel) In Sect. 3.12 we have seen that the bit-flip operation in a qubit system is provided by Pauli's matrix σ_x. That operation must be regarded

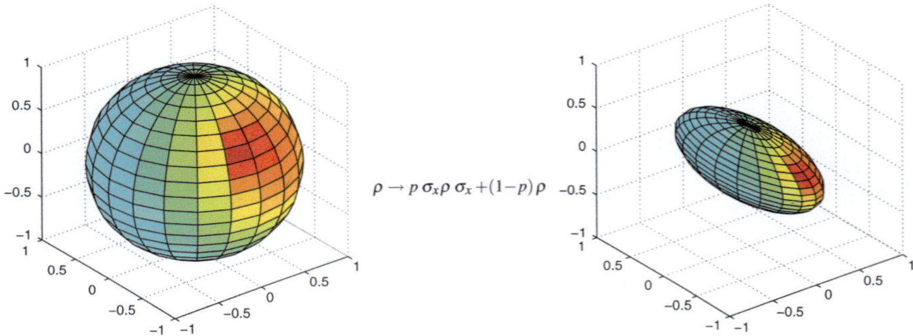

Fig. 12.20 Interpretation of the bit-flip channel with the *Bloch sphere*. On the *left*, the sphere represents the set of all pure states in a qubit system. On the *right*, the sphere is deformed into an ellipsoid by the noisy bit-flip channel with $p = 0.32$. An input pure state $\rho = |\psi\rangle\langle\psi|$ is transformed into a mixed-state (noisy state)

as a noiseless channel because σ_x is a unitary operator. In the noisy version. the channel provides a bit-flip with probability p and leaves the qubit unchanged with the probability $1 - p$. The relation reads

$$\Phi(\rho) = p\,\sigma_x\,\rho\,\sigma_x + (1 - p)\,\rho. \tag{12.84}$$

To verify that this relation describes a quantum noisy channel, it is sufficient to prove that it is a Kraus operation (12.83) with specific V_k. In fact, with

$$d_E = 2, \qquad V_1 = \sqrt{1 - p}\,I_2, \qquad V_2 = \sqrt{p}\,\sigma_x$$

the general formula (12.83) gives (12.84). It remains to check the completeness condition (12.83a). We have, recalling that $\sigma_x^2 = I_2$,

$$V_1 V_1^* + V_2 V_2^* = (1 - p)I_2 + p\sigma_x^2 = I_2.$$

The effect of the transit over a noisy bit-flip channel is illustrated in Fig. 12.20 through the Bloch sphere, which is distorted into an ellipsoid, so that an input pure-state qubit becomes a mixed-state qubit.

Example 12.6 (Depolarizing channel) This channel is a convex combination of the identity channel $Id : \rho \to \rho$ and a completely depolarizing channel, which transforms any ρ into a chaotic state, $\rho \to (1/d)I_{\mathcal{H}}$, with probability p; d is the dimension of the Hilbert space. Then

$$\Phi(\rho) = (1 - p)\rho + p\,(1/d)I_{\mathcal{H}}. \tag{12.85}$$

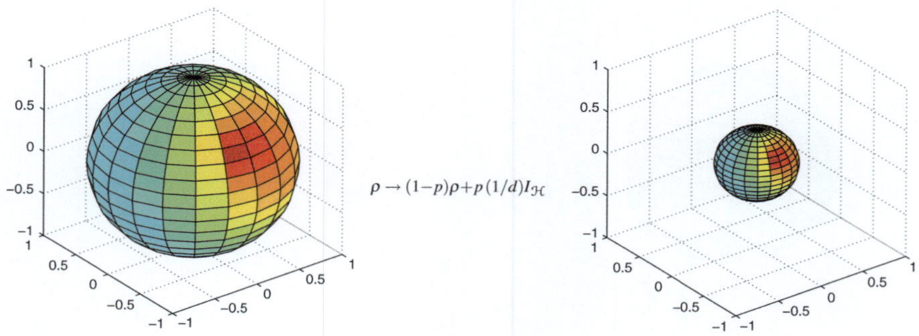

Fig. 12.21 Interpretation of the depolarizing channel with the *Bloch sphere*. On the *left*, the sphere represents the set of all pure states in a qubit system. On the *right*, the sphere is uniformly contracted by the depolarizing channel in dependence of the probability p (here $p = 0.5$). An input pure-state $\rho = |\psi\rangle\langle\psi|$ is transformed into a mixed-state (noisy state)

The Kraus representation can be obtained for a general finite-dimensional Hilbert space [2]. In a qubit space, $\mathcal{H} = \mathcal{Q} = \mathbb{C}^2$, it is obtained using the following relation, which expresses the identity $I_{\mathcal{H}}$ through Pauli's matrices, given by (3.91), in the form (see Problem 12.5)

$$I_{\mathcal{H}} = I_2 = \tfrac{1}{2} \sum_{i=0,x,y,z} \sigma_i\,\rho\,\sigma_i^*. \tag{12.86}$$

The effect of this channel is illustrated in Fig. 12.21 with the Bloch sphere, which is uniformly contracted in dependence of the probability p. Also in this case an input pure-state qubit becomes a mixed-state qubit.

12.8.2 Gaussian Channels

Gaussian channels play a fundamental role in quantum-optical communications, where the information carrier is carried by Gaussian states (coherent and also squeezed states). By definition a Gaussian channel is a quantum channel that preserves the Gaussianity.

We recall that a Gaussian state in the N bosonic mode is completely specified by its mean vector $\overline{X} = \overline{q_{\mathrm{p}}}$ and its covariance matrix V. Then it is sufficient to establish how these parameters are changed in a general Gaussian channel. We have [2]

Proposition 12.2 *In the N bosonic mode, a Gaussian channel is specified by a triplet* (S, B, d), *where S and B are real $2N \times 2N$ matrices with the constraint (uncertainty relation)*

$$B \geq \tfrac{1}{2}\,\mathrm{i}(\Omega - S\,\Omega\,S^{\mathrm{T}}). \tag{12.87}$$

and d is a vector in $\mathbb{R}^{2N}$; Ω is the $2N \times 2N$ matrix defined by (11.67). The mean value $\overline{X} = \overline{q_p}$ and the covariance matrix V of the input state $\rho(\overline{X}, V)$ are transformed by the Gaussian channel, as

$$\overline{X} \rightarrow S\overline{X} + d, \qquad V \rightarrow S V S^{\mathrm{T}} + B. \tag{12.88}$$

The matrix B is the noise parameter. For $B = 0$ the Gaussian channel is noiseless, and therefore governed by a unitary transformation as in (12.81). In this case, the matrix S becomes symplectic, that is, $S\Omega S^{\mathrm{T}} = \Omega$, and the relation between the input and output quadrature operators is governed by a linear unitary symplectic transformation

$$q_p \rightarrow q_p S + d.$$

Then we find the relation of a symplectic transformation seen in Sect. 11.10, and consequently we can classify such transformations as a noiseless channel.

For $B \neq 0$, the Gaussian channel becomes noisy. In this case, the unitary representation of the quantum channel, given by (12.80), is obtained by modeling the environment by a multimode bosonic mode, say q_{p_E}, with input–output relations of the form

$$q_p \rightarrow q_p S + q_{p_E} S_E + d, \qquad q_{p_E} \rightarrow q_p M + q_{p_E} M_E + d_E$$

where S, S_E, M, and M_E are real matrices, such that

$$\begin{bmatrix} S & M \\ S_E & M_E \end{bmatrix}$$

is a symplectic matrix.

12.8.3 Examples of Gaussian Channels

We illustrate the main examples of Gaussian channels in the single bosonic mode $(N = 1)$:

Attenuation and amplification channels. The channel matrices S and B have the forms

$$S = k \begin{bmatrix} 1 & 0 \\ 0 & 1 \end{bmatrix}, \qquad B = \left[\mathcal{N}_0 + \tfrac{1}{2}|1 - k^2| \right] \begin{bmatrix} 1 & 0 \\ 0 & 1 \end{bmatrix}$$

where $\mathcal{N}_0 \geq 0$ and the parameter k expresses an attenuation for $0 < k < 1$ and an amplification for $k > 1$.

It can be shown [10] that in the case of attenuation, the environment of the general scheme of Fig. 12.19 is provided by a beam splitter of transmissivity k (see Sect. 11.17), driven by a thermal state $\rho_E = \rho_{\mathrm{th}}$ with mean photon number

$\mathcal{N} = \mathcal{N}_0/(1-k^2)$. In the case of amplification, the environment is provided by a two-mode squeezing driven by a thermal state with mean photon number $\mathcal{N} = \mathcal{N}_0/(k^2-1)$ (see Sect. 11.17).

Additive noise channels. The channel matrices S and B have the simple forms

$$S = \begin{bmatrix} 1 & 0 \\ 0 & 1 \end{bmatrix}, \qquad B = 2\mathcal{N}\begin{bmatrix} 1 & 0 \\ 0 & 1 \end{bmatrix}$$

where $\mathcal{N} \geq 0$ is the number of thermal photons. The new covariance matrix becomes

$$V \quad \to \quad V + 2\mathcal{N}\, I_2.$$

Problem 12.5 ⋆⋆   Find the Kraus representation of a depolarizing channel in a qubit system, using identity (12.86).

12.9 Accessible Information and Holevo Bound

In the transmission of **classical information** through a channel, we have a bipartite system Alice $\rightarrow$ Bob, where Alice at the transmission side handles a classical source $A \simeq (\mathcal{A}, p_A)$ with p_A the a priori probabilities and at the reception side Bob sees a classical source $B \simeq (\mathcal{A}, p_{A|B})$, with a posteriori probabilities $p_{A|B}$. The entropic connection is provided by the *mutual information*

$$I(A; B) = H(A) - H(A|B) \tag{12.89}$$

introduced in Sect. 12.7 (see (12.66)), which quantifies how much information A and B have in common and can be calculated from the probabilities $\{p_A(a)\}$ and $\{p_{B|A}(b|a)\}$ (see (12.66)). The interpretation of the a posteriori probabilities is that Bob makes

- a correct inference on $A = a$ with probability $p_{B|A}(a|a)$,
- a wrong inference with probability $1 - p_{B|A}(a|a)$.

This holds both with a classical and with a quantum channel because we are considering classical information.

   In the next section, we will consider the transmission through a **noisy channel**; but, as a preliminary, it is important to deal with a **noiseless channel** to remark the difference between the classical and the quantum case. In the classical case, Bob has no problem to get a right inference on the transmitted symbol; since, with a noiseless channel,

$$p_{B|A}(b|a) = \begin{cases} 1 & \text{if } b = a \\ 0 & \text{if } b \neq a \end{cases} \tag{12.90}$$

Fig. 12.22 Transmission through a noiseless quantum channel. Alice encodes the incoming classical symbol $A \in \mathcal{A}$ into a quantum state ρ_A. The quantum channel transfers ρ_A uncorrupted. Bob decodes ρ_A into a classical symbol $B \in \mathcal{B}$ by a quantum measurement

and then

$$I(A; B) = H(A) \qquad \text{(classical case: noiseless channel).} \qquad (12.91)$$

We now consider the quantum case with the required detail, continuing with the assumption of a noiseless channel. The scenario is depicted in Fig. 12.22, where the source emits a symbol $A \simeq (\mathcal{A}, p_A)$, and Alice wants to transmit A to Bob. If the source emits the message $A = a$, she encodes the message into a quantum state of some physical system, according to the *classical-quantum* (c $\to$ q) mapping

$$a \; \to \; \rho_a \,, \qquad a \in \mathcal{A}. \qquad (12.92)$$

Note that for the encoding Alice should have the availability of a constellation of K distinct density operators $\{\rho_1, \rho_2, \ldots, \rho_K\}$, where $K = |\mathcal{A}|$ is the size of the source alphabet. The scenario is the one we considered in quantum communications in Chap. 7 (where we assumed pure states instead of density operators).

Since the channel is noiseless, Bob receives exactly ρ_A and performs a quantum measurement using a POVM system $\{Q_b \,, \; b \in \mathcal{A}\}$. The outcome B of the measurement is $B = b \in \mathcal{A}$ with probability

$$p_{B|A}(b|a) = \text{Tr}[\rho_a \, Q_b] \,, \qquad b, a \in \mathcal{A}. \qquad (12.93)$$

Now, can Bob reach a condition (12.90) to get a right inference with certainty on the symbol transmitted by Alice? The answer is that (12.90) is verified only in very special cases, which essentially requires that the **states be orthogonal**. Then to make a comparison with (12.91), we have

$$I(A; B) = H(A) \quad \text{(quantum case: noiseless channel, orthogonal states).} \quad (12.94)$$

But in practice, as we have seen in Chap. 7 with coherent states (and also with squeezed states), the states are never orthogonal, and, although the channel is noiseless, Bob cannot infer with certainty on the transmitted symbol. This is an important difference of quantum information transmission with respect to the classical case.

Then in the quantum case, we have to consider the full expression of the mutual information $I(A; B) = H(A) - H(A|B)$, which depends on the measurement performed by Bob. But, with the limit imposed by quantum mechanics, Bob should do his best, as stated by the following definition

Definition 12.4 The maximum of mutual information obtained through any possible measurement

$$I_{\text{acc}} := \max_{\{\text{POVM}\}} I(A; B) \tag{12.95}$$

is called **accessible information**.

But before proceeding we have to study an important entropic quantity.

12.9.1 Holevo Entropy with a Constellation of Quantum State

In a constellation of K quantum states we have the ensemble

$$\mathcal{L} = \{(p_1, \rho_1), \ldots, (p_K, \rho_K)\} = \{p_a, \rho_a\}$$

to which the following entropic quantity can be associated

$$\boxed{\chi(\{p_a, \rho_a\}) := S(\rho) - \sum_a p_a S(\rho_a)} \tag{12.96}$$

which is called χ-information or Holevo-χ. In (12.96) ρ is the average density operator: $\rho = \mathrm{E}[\rho_A] = \sum_a p_a \rho_a$.

Holevo-χ has several properties

(1) It depends only on the ensemble $\mathcal{L}$ and, for this reason, it is often indicated as $\chi(\mathcal{L})$. In some respects, it can be interpreted as the "entropy of the constellation".
(2) It is nonnegative, $\chi(\mathcal{L}) \geq 0$, for the concavity of quantum entropy [8].
(3) For pure states, for which $S(\rho_a) = 0$, it results

$$\chi(\{p_a, \rho_a\}) = S(\rho). \tag{12.97}$$

(4) A quantum operation Φ can never increase the Holevo-χ, that is, if $\mathcal{L}' = \{(p_a, \Phi(\rho_a))\}$, then $\chi(\mathcal{L}') \leq \chi(\mathcal{L})$ (from Lindblad–Uhlmann monotonicity of quantum relative entropy [11]).

But the most important property is

Theorem 12.7 (Holevo's bound) *For any measurement Bob can do, the mutual information is limited by the Holevo-χ*

$$I(A; B) \leq \chi(\{p_a, \rho_a\}). \tag{12.98}$$

This theorem, proved by Holevo in 1973, is one of the most important results of quantum information theory. The proof [11] is based on an elegant formulation of a tripartite system formed by: (1) The source $(\mathcal{A}, p_A)$, seen as a fictitious quantum system, (2) the encoder, and (3) the measurement device used by Bob. It is an extraordinary application of several statements: (i) Properties of the quantum entropies, as the strong subadditivity of the von Neumann entropy, (ii) the fact that quantum operations can never increase the χ-information, and (iii) the theory of quantum open system seen in the previous section.

The conditions for the equality in (12.98) are: (1) The density operators ρ_a commute (are simultaneously diagonalizable) and (2) Bob perform the POVM measurement with the common eigenbasis of the $\{\rho_a\}$ (see Problem 12.8).

Example 12.7 We consider a qubit system where Alice prepares the states

$$|\psi_0\rangle = |0\rangle, \qquad |\psi_1\rangle = \cos\theta \, |0\rangle + \sin\theta \, |1\rangle$$

with probability $p_0 = p$ and $p_1 = 1 - p$, respectively, θ being a parameter. Note that the corresponding density operators are respectively

$$\rho_0 = |\psi_0\rangle\langle\psi_0| = \begin{bmatrix} 1 & 0 \\ 0 & 0 \end{bmatrix}, \qquad \rho_1 = |\psi_1\rangle\langle\psi_1| = \begin{bmatrix} \cos^2\theta & \cos\theta \sin\theta \\ \cos\theta \sin\theta & \sin^2\theta \end{bmatrix}.$$

The average density operator is

$$\rho = p\,\rho_0 + (1-p)\,\rho_1 = \begin{bmatrix} (1-p)\cos^2\theta + p & (1-p)\cos\theta \sin\theta \\ (1-p)\cos\theta \sin\theta & (1-p)\sin^2\theta \end{bmatrix}.$$

Note that the inner product is $\langle\psi_1|\psi_0\rangle = \cos\theta$, so that the states are orthogonal only for $\theta = \pi/2$.

The quantum entropies of the two (pure) states are zero, $S(\rho_0) = S(\rho_1) = 0$. For the evaluation of the quantum entropy of the average density operator ρ, we calculate the eigenvalues which result in

$$\lambda_\pm = \frac{1}{2}\left[1 \pm \sqrt{2p^2 - 2(p-1)\cos 2\theta p - 2p + 1}\right].$$

Then $S(\rho)$ is given by the classical entropy of a binary source with probabilities $\{\lambda_-, \lambda_+\}$

$$\chi(\{p_a, \rho_a\}) = S(\rho) = H(\{\lambda_-, \lambda_+\}).$$

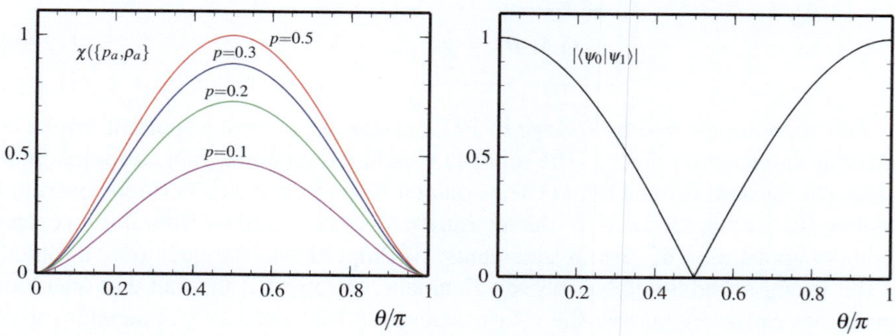

Fig. 12.23 Holevo bound as a function of the parameter θ in a qubit system when the states $|\psi_0\rangle$ and $|\psi_1\rangle$ are generated , respectively, with probability p and $1 - p$. For each value of p, the maximum is given by the classical entropy $H(\{p, 1 - p\})$ and is obtained when the states are orthogonal, that is, for $\theta = \pi/2$ (see the inner product $|\langle\psi_0|\psi_1\rangle|$ as function of θ on the *right*)

The Holevo bound χ is illustrated in Fig. 12.23 as function of θ for four values of the probability p. Note that the maximum is obtained for $\theta = \pi/2$, that is, when the two states are orthogonal, and in particular that for $p = 1/2$ the maximum is given by 1 bit. In correspondence to the maximum, it is possible for Bob to establish with certainty which state Alice has prepared. For all the other values of θ, the state orthogonality does not hold and Bob cannot establish with certainty which state Alice has prepared.

Problem 12.6 ⋆ Consider the following ensemble in a qubit system

$$\mathcal{L}: \quad p_0 = \tfrac{1}{2}, \quad \rho_0 = \begin{bmatrix} 0.8 & 0.25 \\ 0.25 & 0.2 \end{bmatrix}, \qquad p_1 = \tfrac{1}{2}, \quad \rho_1 = \begin{bmatrix} 0.1 & 0.3 \\ 0.3 & 0.9 \end{bmatrix}$$

Evaluate the Holevo χ.

Problem 12.7 ⋆⋆ With the ensemble $\mathcal{L}$ specified in the previous problem, evaluate the mutual information, assuming that Bob uses the measurement operators provided by the Helstrom theory. Then verify the Holevo bound $I(A, B) \leq \chi(\mathcal{L})$.

Problem 12.8 ⋆⋆⋆ Prove that the Holevo bound holds with the equality sign if (1) the density operators $\{\rho_a\}$ commute, that is, they are *simultaneously diagonalizable* and (2) the POVM measurement is performed with the common eigenbasis of the $\{\rho_a\}$.

Problem 12.9 ⋆ Prove that in a constellation of distinct pure states $\{\rho_a = |\psi_a\rangle\langle\psi_a| , \ a \in \mathcal{A}\}$, the density operators commute if and only if the states are orthogonal.

12.10 Transmission Through a Noisy Quantum Channel

As the subject is very complex and still debated, we will limit ourselves to giving an overview, without delving into the details of the various formulations.

We now suppose that the quantum channel is noisy (Fig. 12.24) and then we have to change the target and the methodology. The change with respect to the noiseless case is that the state received by Bob is distorted according to a quantum operation Φ, specific of the channel, say

$$\rho'_A = \Phi(\rho_A).\tag{12.99}$$

An intuitively obvious remark is that *the amount of classical information that Bob can receive will be reduced* with respect to the noiseless case. This is mathematically proved by Lindblad–Uhlmann monotonicity of quantum relative entropy [11], which states that **a quantum operation Φ can never increase the χ-information**, namely $\chi(\{p_a, \Phi(\rho_a)\}) \le \chi(\{p_a, \rho_a\})$.

Now the problem is to establish the maximum amount of classical information that can be reliably transmitted through a noisy channel, specified by the quantum operation Φ. In other words, we have to define and evaluate the *classical capacity* of the quantum channel Φ. The answer to this problem is based upon a multiple use of the noisy channel, similarly to the classical channel encoding. But, while a classical channel has a unique capacity, for a quantum channel we may define **several capacities**, in dependence of the presence or absence of entanglement and of the type of quantum measurement. Specifically, the capacity of quantum channel may depend on whether:

- the information to be transmitted is classical or quantum,
- the channel is memoryless or not,
- the encoding uses separable states or entangled states,
- the quantum measurement is individual or collective (global),
- Alice and Bob share or do not share entanglement resources.

Hereafter, we suppose that the information to be transmitted is classical and that the channel is memoryless.

Fig. 12.24 Transmission through a noisy quantum channel. Alice encodes the incoming classical symbol $A \in \mathcal{A}$ into a quantum state ρ_A. The quantum channel modifies the state as $\Phi(\rho_A)$. Bob decodes $\Phi(\rho_A)$ into a classical symbol $B \in \mathcal{B}$ by a quantum measurement

12.10.1 General Formulation with Classical Information

We now consider a quite general formulation and introduce the main concepts, as reliability and achievability, but without arriving at explicit results, which will be obtained in specific cases.

The scenario is the following: Alice has a classical information source, which emits a symbol $A \simeq (\mathcal{A}, p_A)$, and wants to transmit A to Bob. Assume that Φ represents a qubit channel. Then

- Alice encodes the messages a of $\mathcal{A}$ into a quantum state of L qubits: $a \rightarrow \rho_a^{(L)}$, where $\rho_a^{(L)}$ acts in the L-qubit space $\mathcal{H}^{\otimes L}$,
- Alice sends $\rho_a^{(L)}$ to Bob through L uses of the qubit channel Φ,
- Bob receives the state $\sigma_a^{(L)} := \Phi^{\otimes L}(\rho_a^{(L)})$,
- Bob makes a collective measurement with a POVM system $\{Q_a\}$ in the L-qubit space $\mathcal{H}^{\otimes L}$.

The error probability is

$$P_e^{(L)} = \frac{1}{|\mathcal{A}|} \sum_a \left(1 - \mathrm{Tr}\left[\sigma_a^{(L)} Q_a\right]\right).$$

The transmission is reliable if $P_e^{(L)} \rightarrow 0$ as $L \rightarrow \infty$.

We now proceed with the general formulation of a quantum channel capacity, following Holevo and Giovannetti [2] The authors begin with the *Shannon capacity*, defined as

$$C_{\mathrm{Shan}} = \max_{\{p_a\}} I(A; B) \tag{12.100}$$

where $I(A; B)$ is the mutual information and the maximization is performed with respect to all possible input distribution (This is the same definition of capacity we gave in Sect. 12.7 for a memoryless discrete classical channel). Then they extend the definition to the L-block channel as

$$C_{\mathrm{Shan}}^{(L)} = L \, C_{\mathrm{Shan}} \tag{12.101}$$

where the memoryless nature is reflected by the additive property. The maximization is now taken over all input joint distributions of the sequence A^L, including the correlated ones.

Then the authors consider the transfer of classical information into a quantum sequence of separable states, $A^L \rightarrow \rho_{a_1} \otimes \cdots \otimes \rho_{a_L}$, and define $C^{(L)}$ as the Shannon capacity, obtained by maximizing over all possible measurements on $\mathcal{H}^{\otimes L}$. Because of **possibly entangled measurements**, $C^{(L)}$ may be strictly superadditive, $C^{(L)} > L \, C^{(1)}$. Then, the definition of capacity needs a regularization

$$C_\chi := \lim_{L \to \infty} \frac{1}{L} C^{(L)}. \tag{12.102}$$

Note that, in general, calculations as (12.102) represent a formidable task; but in this case the result is very simple

$$C_\chi = \max_{\{p_a\}} \chi(\{p_a, \rho_a\}) \tag{12.103}$$

where $\chi(\{p_a, \rho_a\})$ is Holevo-χ and the maximum is **over all possible input probabilities, with the** ρ_a **fixed**. Relation (12.103) is the HSW theorem (according to Holevo himself!). We will come back to this statement below.

12.10.2 Capacity of the Input $c \to q$ Channel

We follow closely Holevo and Giovannetti's paper [2], where they evaluate the capacity of the $c \to q$ quantum channel. The scenario is the following (Fig. 12.25): The input $A \simeq (\mathcal{A}, p)$ is mapped into a fixed family of quantum states $\{\rho_a\}$. If the letters of the message $a^L = (a_1, \ldots a_L)$ are transmitted independently of each other, at the output of the composite channel one has the separable state $\rho^L = \rho_{a_1} \otimes \cdots \otimes \rho_{a_L}$. The direct decoding of this sequence (without the presence of a noisy channel) requires a quantum measurement in $\mathcal{H}^{\otimes L}$. Then we have two classical random variables: The random variable A^L at the input (of which a^L is realization), and the random variable $\hat{A}^L$ given by the result of the measurement. From these random variables, we can evaluate the classical mutual information $I(A^L; \hat{A}^L)$. Then the

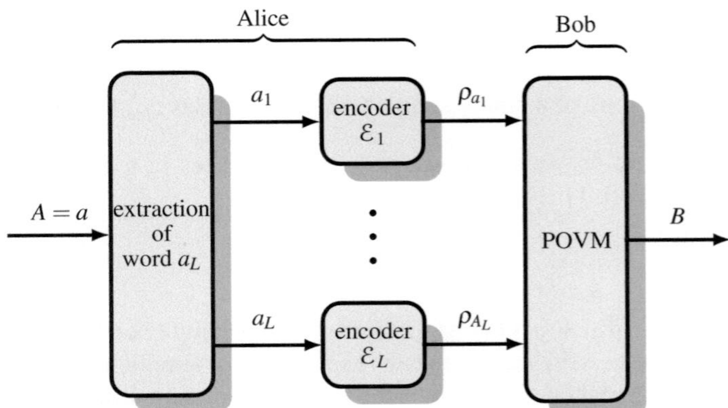

Fig. 12.25 Evaluation of the capacity of the input $c \to q$ channel. If the source of information emits the message $A = a$, Alice chooses the word $a^L = (a_1, \ldots, a_L)$, encodes it in the quantum word $\rho_{a_1} \otimes \cdots \otimes \rho_{a_L}$, and sends separately the L factors ρ_{a_i} by L uses of the channel Φ. Bob decodes the factors ρ_{a_i} by a POVM system

Shannon capacity $C^{(L)}$ is obtained by maximizing **over all the measurements on** $\mathcal{H}^{\otimes L}$. Because of **possibly entangled measurements**, $C^{(L)}$ may be strictly super-additive, $C^{(L)} > L\,C^{(1)}$. Then, the definition of capacity needs a regularization, as in (12.102)

$$C_{\mathrm{q}\to\mathrm{c}} := \lim_{L\to\infty} \frac{1}{L} C^{(L)}. \tag{12.104}$$

Again, calculations according to (12.104) are quite challenging in general, because we have to evaluate $C^{(L)}$ for an arbitrary length and then take the limit. But in this case, the result is very simple

Theorem 12.8 (HSW theorem I) *The capacity of the input $c \to q$ channel (12.104) can be obtained by the maximization of the χ-information*

$$C_{\mathrm{q}\to\mathrm{c}}(\{\rho_a\}) = \max_{p_a} \chi(\{p_a, \rho_a\}) \tag{12.105}$$

where the maximization is performed over the probabilities $\{p_a\}$, with $\{\rho_a\}$ fixed.

Note that in this formulation the capacity depends on the input states, as indicated in the symbol in (12.105).

Example 12.8 (*BPSK system*) Consider a BPSK system where the input consists of the two coherent states $|\pm A\rangle$ of amplitude $\pm A$. In [12, 13] the classical capacity C_χ defined by (12.105) is calculated. The result is

$$C_\chi = h_2(\tfrac{1}{2}(1 + |X|^2))$$

where $|X|^2 = \exp(-4A^2)$ is the quadratic overlap of the two coherent states and

$$h_2(p) = -p \log p - (1-p)\log(1-p) \tag{12.106}$$

is the classical entropy of a binary source with probabilities $(p, 1-p)$.

Example 12.9 Consider an input ensemble of pure states in a qubit system $\mathcal{L} = \{p_a, |\psi_a\rangle\langle\psi_a|, a = 0, 1\}$. Then

$$\chi(\mathcal{L}) = S(\rho)$$

where $\rho = p_0|\psi_0\rangle\langle\psi_0| + p_1|\psi_1\rangle\langle\psi_1|$ is the average density operator. For the evaluation of ρ we have to write the pure states explicitly. For simplicity, we assume that they are given by a real combination of the basis vector; so that they can be written in the trigonometric form

$$|\psi_0\rangle = \cos\phi|0\rangle + \sin\phi|1\rangle\,, \qquad |\psi_1\rangle = \cos\beta|0\rangle + \sin\beta|1\rangle$$

The corresponding density operators are

$$\rho_0 = \begin{bmatrix} \cos^2\phi & \cos\phi\sin\phi \\ \cos\phi\sin\phi & \sin^2\phi \end{bmatrix}, \qquad \rho_1 = \begin{bmatrix} \cos^2\beta & \cos\beta\sin\beta \\ \cos\beta\sin\beta & \sin^2\beta \end{bmatrix}$$

and the average density operator is given by

$$\rho = \begin{bmatrix} \cos^2\phi\, p_0 - \cos^2\beta(p_0-1) & \cos\phi\sin\phi\, p_0 - \cos\beta\sin\beta(p_0-1) \\ \cos\phi\sin\phi\, p_0 - \cos\beta\sin\beta(p_0-1) & \sin^2\phi\, p_0 - \sin^2\beta(p_0-1) \end{bmatrix}$$

and the corresponding eigenvalues are

$$\lambda_\pm = \frac{1}{2}\left[1 \pm \sqrt{2p_0^2 - 2\cos(2(\beta-\phi))(p_0-1)p_0 - 2p_0 + 1}\right].$$

From these eigenvalues we can evaluate the Holevo-χ as $\chi(\mathcal{L}) = S(\rho) = H(\{\lambda_-, \lambda_+\}) = h_2(\lambda_-)$, where H is the Shannon entropy of a binary source. To find the capacity we have to maximize with respect to the input probabilities. For reasons of symmetry, the optimum is obtained with $p_0 = p_1 = \frac{1}{2}$. Then

$$\rho = \begin{bmatrix} \frac{1}{2}\left[\cos^2\beta + \cos^2\phi\right] & \frac{1}{4}(\sin 2\beta + \sin 2\phi) \\ \frac{1}{4}(\sin 2\beta + \sin 2\phi) & \frac{1}{2}\left[\sin^2\beta + \sin^2\phi\right] \end{bmatrix}$$

and

$$\lambda_\pm = \frac{1}{2}\left[1 \pm \sqrt{\cos^2(\beta-\phi)}\right] = \frac{1}{2}[1 \pm |X|]$$

where $X = \cos(\beta - \phi)$ is the inner product of the two pure states.

In conclusion, the capacity of the input $c \to q$ channel is given by

$$C_{c\to q} = h_2\left(\tfrac{1}{2}\left[1 - |X|\right]\right)$$

where X is the inner product between the input states. Figure 12.26 shows the plot as a function of $|X|$.

12.10.3 Product-State Capacity

The simplest approach to transmit classical information over a noisy quantum channel is similar to the one used in Shannon's noisy channel coding theorem, that is, a random classical code is selected according to a given distribution $p(x)$. Alice encodes the message to be sent into a *quantum codeword* obtained as the tensor product from an

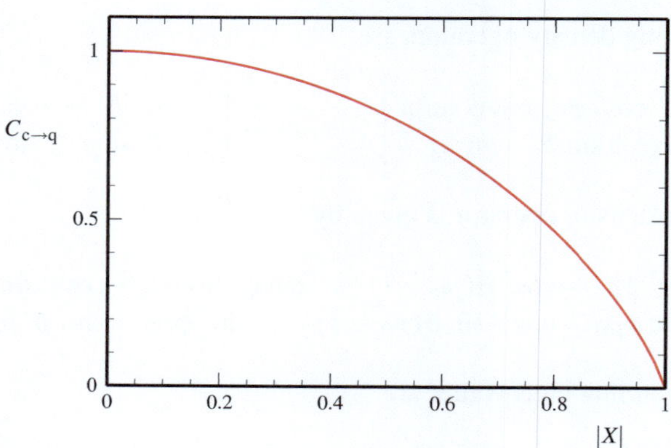

Fig. 12.26 Capacity $C_{c \to q}$ of an input qubit source as a function of the superposition parameter $|X|$ of the input states

alphabet of quantum states and transmits it by multiuse of a noise quantum channel. At the reception, Bob performs *individual* POVM measurements and determines in such a way a conditional probability distribution $p_{Y|X}(y|x)$, as in a classical noisy channel. The corresponding mutual information $I(X; Y)$ represents an achievable rate. The classical capacity is obtained by maximization of $I(X; Y)$ over Alice's encoding possible choices and Bob's possible measurements. The scenario is illustrated in Fig. 12.27.

We now give the details. Let $A \simeq (\mathcal{A}, p_A)$ be the information source and let $p(x)$ be a probability distribution, according to which a codebook of IID words is generated

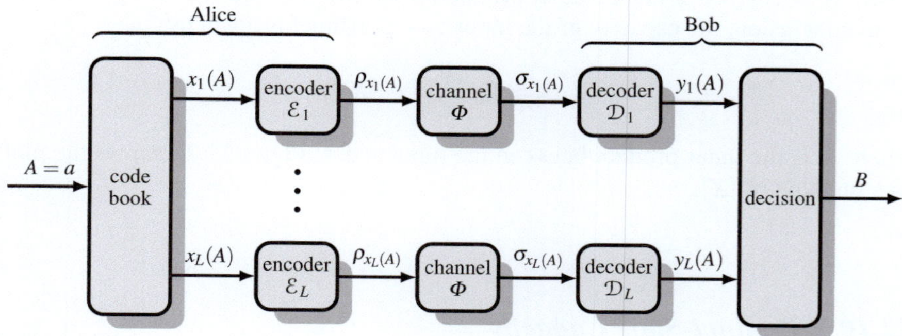

Fig. 12.27 Evaluation of product-state capacity of a noisy channel. Alice and Bob share a codebook of L words generated by a probability distribution $p_X(x)$. If the source of information emits the message $A = a$, Alice chooses the codeword $x^L(a) = (x_1(a), \ldots, x_L(a))$, encodes it in the quantum word $\rho_{x_1(a)} \otimes \cdots \otimes \rho_{x_L(a)}$, and sends separately the L factors $\rho_{x_i(a)}$ by L uses of the channel Φ. Bob receives the corrupted versions $\sigma_{x_i(a)}$ and decodes them by a POVM system

$$
\mathcal{C} = \begin{bmatrix} x^L(1) \\ \vdots \\ x^L(L) \end{bmatrix} = \begin{bmatrix} x_1(1) & x_2(1) & \cdots & x_L(1) \\ \vdots & \vdots & \ddots & \vdots \\ x_1(L) & x_2(L) & \cdots & x_L(L) \end{bmatrix}.
$$

where L is the number of possible messages, that is, $L = |\mathcal{A}|$. The distribution $p(x)$ and the codebook must be known by both Alice and Bob. If the source emits the message $A = a$, Alice chooses the corresponding codeword $x^L(a)$ in the codebook and prepares the quantum codeword with the format

$$
\rho_{x^L(a)} = \rho_{x_1(a)} \otimes \rho_{x_2(a)} \otimes \cdots \otimes \rho_{x_L(a)} \tag{12.107}
$$

where the ρ_x are density operators to act as input of the quantum channel Φ. The factors in (12.107) are individually sent to the channel and Bob receives a corrupted replica of each factor

$$
\sigma_{x_i(a)} = \Phi(\rho_{x_i(a)}), \qquad i = 1, 2, \ldots, L.
$$

Bob makes L individual measurements using a POVM system $\{Q_i\}$. These measurements induce the probability distributions

$$
p_{Y_i|X_i}(y_i|x_i(a)) = \mathrm{Tr}\left[Q_{y_i}\Phi(\rho_{x_i(a)})\right], \qquad i = 1, 2, \ldots, L
$$

and, considering the independence and the IID, the L-fold conditional distribution is given by $p_{Y_1\cdots Y_L|X_1\cdots X_L}(y_1 \cdots y_L|x_i(a)\cdots x_L(a)) = \prod_i p_{Y_i|X_i}(y_i|x_i(a))$. From this distribution Bob can evaluate the mutual information $I(X;Y)$, which depends on the distribution $p(x)$, on the quantum alphabet $\{\rho_x\}$, on the channel Φ, and on the POVM system Q_{y_i}. For a fixed channel, the maximization over the other parameters gives the **classical channel capacity**, also called **product-state capacity**.

The HSW theorem states that the product-state capacity can be obtained by the Holevo-χ (see [8, 11, 14]). Note that the procedure is essentially the same as in the second Shannon theorem.

Theorem 12.9 (HSW theorem II) *The product-state capacity can be obtained by the maximization of the χ-information*

$$
C(\Phi)_{\mathrm{prod}} = \max_{\mathcal{L}} \chi(\mathcal{L}) \tag{12.108}
$$

over all the input ensembles $\mathcal{L} = \{p_a, \rho_a\}$.

Example 12.10 Consider the depolarizing qubit channel with relation (see Example 12.6)

$$
\Phi(\rho) = (1 - p)\rho + \tfrac{1}{2}p\, I_2
$$

with the input ensemble of pure states $\mathcal{L} = \{p_a, |\psi_a\rangle\langle\psi_a|, a = 0, 1\}$. Then the output χ results in

$$\chi(\{p_a, \Phi(\rho_a)\}) = S(\Phi(\rho)) - p_0\, S(\Phi(\rho_0)) - p_1\, S(\Phi(\rho_1))$$

where $\rho = p_0|\psi_0\rangle\langle\psi_0| + p_1|\psi_1\rangle\langle\psi_1|$ and

$$\Phi(\rho) = (1 - p)\,[p_0|\psi_0\rangle\langle\psi_0| + p_1|\psi_1\rangle\langle\psi_1|] + \tfrac{1}{2}p\, I_2.$$

The eigenvalues of $\Phi(\rho_i) = \Phi(|\psi_i\rangle\langle\psi_i|)$ are independent of the states and given by $\frac{1}{2}(1 \pm p)$. Therefore

$$S(\Phi(\rho_i)) = h_2(\tfrac{1}{2}(1 - p)) \tag{12.109}$$

with $h_2(x)$ the entropy of a classical binary source given by (12.106).

For the evaluation of $\Phi(\rho)$, we have to write the pure-states explicitly. As done in Example 12.9 we assume that they are given by a real combination of the basis vector so that they can be written in the trigonometric form

$$|\psi_0\rangle = \cos\phi|0\rangle + \sin(\phi)|1\rangle, \qquad |\psi_1\rangle = \cos\beta|0\rangle + \sin\beta|1\rangle.$$

The corresponding density operators are

$$\rho_0 = \begin{bmatrix} \cos^2\phi & \cos\phi\sin\phi \\ \cos\phi\sin\phi & \sin^2\phi \end{bmatrix}, \qquad \rho_1 = \begin{bmatrix} \cos^2\beta & \cos\beta\sin\beta \\ \cos\beta\sin\beta & \sin^2\beta \end{bmatrix}$$

The average density operator is $\rho = p_0\rho_0 + p_1\rho_1$, which becomes at the channel output $\Phi(\rho) = (1 - p)\rho + pI_2/2$. Explicitly, we find

$$\Phi(\rho) = \begin{bmatrix} qp_1\cos^2\beta + \tfrac{1}{2}p + qp_0\cos^2\phi & -q(p_0\cos\phi\sin\phi + p_1\cos\beta\sin\beta) \\ -q(p_0\cos\phi\sin\phi + p_1\cos\beta\sin\beta) & qp_1\sin^2\beta + \tfrac{1}{2}p + qp_0\sin^2\phi \end{bmatrix}$$

with $p_1 = 1 - p_0$ and $q = 1 - p$. The corresponding eigenvalues are

$$\lambda_\pm = \left\{ \frac{1}{2}\left[1 \pm \sqrt{-(p - 1)^2\left[-2p_0^2 + 2\cos(2(\beta - \phi))(p_0 - 1)p_0 + 2p_0 - 1\right]}\right]\right\}$$

and the corresponding entropy is $S(\rho) = h_2(\lambda_-)$. But we have to find the optimization with respect to the input probabilities $p_0, 1 - p_0$. Considering the symmetry, the optimization is obtained with $p_0 = \frac{1}{2}$. Then we have the simplifications

$$\Phi(\rho) = \begin{bmatrix} \tfrac{1}{4}(p\cos 2\beta + p\cos 2\phi + 2) & \tfrac{1}{4}p(\sin 2\beta + \sin 2\phi) \\ \tfrac{1}{4}p(\sin 2\beta + \sin 2\phi) & \tfrac{1}{2}\left[p\sin^2\beta + p\sin^2\phi + p\right] \end{bmatrix}$$

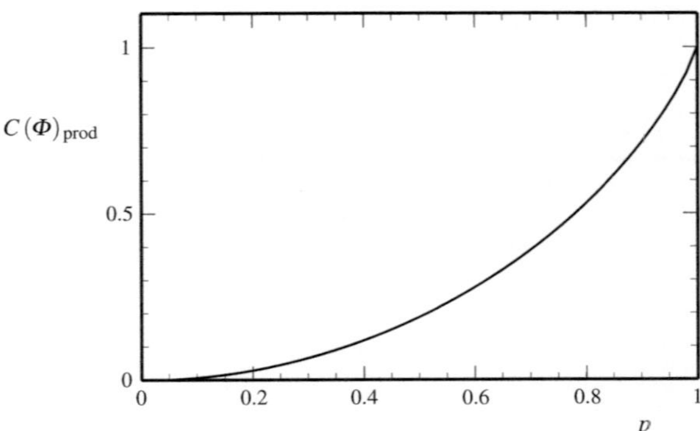

Fig. 12.28 Product-state capacity of a depolarizing qubit channel as a function of the probability p. The relation of the channel is $\Phi(\rho) = (1 - p)\rho + \frac{1}{2}p\, I_2$ and the expression of the capacity is $C(\Phi)_{\mathrm{prod}} = 1 - h_2(\frac{1}{2}(1 - p))$

and

$$\lambda_\pm = \frac{1}{2}\left[1 \pm \sqrt{p^2 X^2}\right] \text{ with } X^2 = \cos^2(\beta - \phi) \tag{12.110}$$

where $X = \langle\psi_0|\psi_1\rangle$ is the inner product. Thus the Holevo-χ obtained with the optimal input probabilities $p_0 = p_1 = \frac{1}{2}$ is given by

$$\chi = h_2(\tfrac{1}{2}(1 - pX)) - h_2(\tfrac{1}{2}(1 - p))$$

where X is the inner product between the input states and p is the probability of the depolarizing channel. Now, to find the product-state capacity we have to make the optimization also with respect to the input states. Clearly this leads to the orthogonality of the states, $X = 0$. Then $h_2(\frac{1}{2}(1 - pX)) = h_2(\frac{1}{2}) = 1$ bit. In conclusion the product-state capacity is

$$C(\Phi)_{\mathrm{prod}} = 1 - h_2(\tfrac{1}{2}(1 - p)).$$

Figure 12.28 shows the plot of the capacity as a function of the probability p.

12.10.4 A More General Approach with Input Entanglement

The previous scenario depicted in Fig. 12.27 is essentially a classical scheme because it makes no use of quantum mechanical features such as entanglement. More generally, we can encode the classical information into a general (possibly entangled) density operator acting on a composite Hilbert space $\mathcal{H}^{\otimes L}$ (Fig. 12.29). If $\rho^L(a)$ is the quantum codeword corresponding to the message a (in general not given by the tensor product of L density operators as in the previous scheme), it is assumed [2]

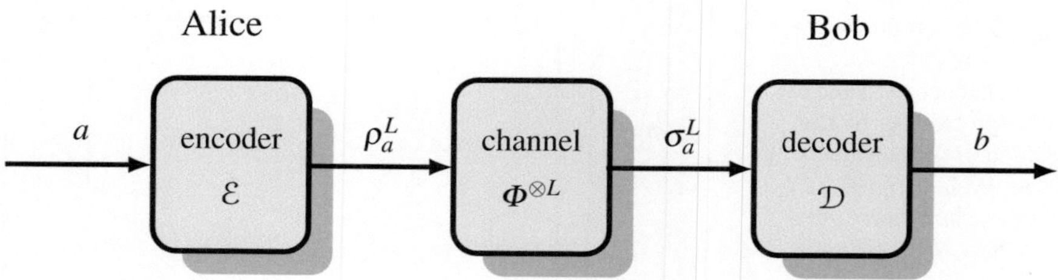

Fig. 12.29 General transmission of classical information through L independent uses of a quantum channel Φ. The information source emits the message a. Alice prepares a composite quantum state $\rho^L(a)$ acting on $\mathcal{H}^{\otimes L}$ (possibly entangled) for input to L independent uses of a noisy quantum channel Φ. At the output Bob receives a corrupted version $\sigma_a^L = \Phi^{\otimes L}(\rho^L(a))$ and decodes it with a collective measurement

that each component is individually affected by the same noise channel Φ, producing the transformation

$$\rho_a^L \;\rightarrow\; \Phi^{\otimes L}[\rho_a^L]$$

where $\Phi^{\otimes L}$ is the L-fold tensor product of the channel map.[7]

The corresponding capacity can be defined as

$$C_\chi(\Phi^{\otimes L}) = \max_{\{p_a, \rho_a^L\}} \chi(\{p_a\}, \{\Phi^{\otimes L}(\rho_a^L)\})$$

where the maximum is taken over the ensembles $\{p_a\}$ and $\{\rho_a^L\}$. This capacity gives the rate of classical bits using blocks of size L and therefore $C_\chi(\Phi^{\otimes L})/L$ represents the bit rate per individual use of the channel Φ. Asymptotically one has the capacity

$$C(\Phi) := \lim_{L\to\infty} \frac{1}{L} C_\chi(\Phi^{\otimes L}).$$

Now, if the *additivity* holds

$$C_\chi(\Phi^{\otimes L}) = L\, C_\chi(\Phi) \tag{12.111}$$

---

[7] Intuitively, the tensor product of two channel maps Φ_1 and Φ_2 acts as the parallel of the two channels in a composite Hilbert space $\mathcal{H}_1 \otimes \mathcal{H}_2$. Specifically, one has [2]

$$\Phi_1 \otimes \Phi_2 = (\Phi_1 \otimes Id_2) \circ (Id_1 \otimes \Phi_2)$$

where $\circ$ is the concatenation ($\Phi_1 \circ \Phi_2$ is obtained by application of Φ_2 at the output of Φ_1), Id_1 and Id_2 are the identity channels in $\mathcal{H}_1$ and $\mathcal{H}_2$, respectively. The interpretation becomes clear when $\rho_{12} = \rho_1 \otimes \rho_2$, where

$$(\Phi_1 \otimes \Phi_2)[\rho_1 \otimes \rho_2] = \Phi_1[\rho_1] \otimes \Phi_2[\rho_2].$$

Roughly speaking, we can say that in the channel $\Phi^{\otimes L}$ each "component" of the input state sees the channel Φ.

one gets

$$C(\Phi) = C_\chi(\Phi).$$

Relation (12.111) is known as the **additive conjecture**, which has been proved for several quantum channels, but for some other it does not hold. This is still an open question, as discussed in [2].

12.10.5 Capacity with Entropy Exchange

The previous scenarios are essentially based on the Holevo-χ, which does not account for a possible cooperation between the input and output of the quantum noisy channel. To investigate this opportunity, we need further generalizations.

The new scenario considered in [2] is based on the procedure of purification, as depicted in Fig. 12.30. Of course, the central role is played by the noisy channel Φ, which maps the input state $\rho = \rho_A$ into the output state $\rho_B = \Phi(\rho_A)$, both acting in the Hilbert space $\mathcal{H}_A$. But the newcomer is a reference system $\mathcal{H}_R$, which allows us to purify ρ_A as $|\psi_{AR}\rangle\langle\psi_{AR}|$. From the theory of purification of Sect. 10.4 we have that the density operator $\rho_R = \mathrm{Tr}_A|\psi_{AR}\rangle\langle\psi_{AR}|$ has the same spectrum as ρ_A and hence $S(\rho_R) = S(\rho_A)$. Also we have that the transmission of $|\psi_{AR}\rangle\langle\psi_{AR}|$ through the channel $\Phi \otimes Id_R$ gives at the output the state

$$\rho_{ER} = (\Phi \otimes Id_R)(|\psi_{AR}\rangle\langle\psi_{AR}|).$$

The above density operator allows us to define the **quantum mutual information** as

$$I(\rho, \Phi) := S(\rho_R) + S(\rho_B) - S(\rho_{BR})$$

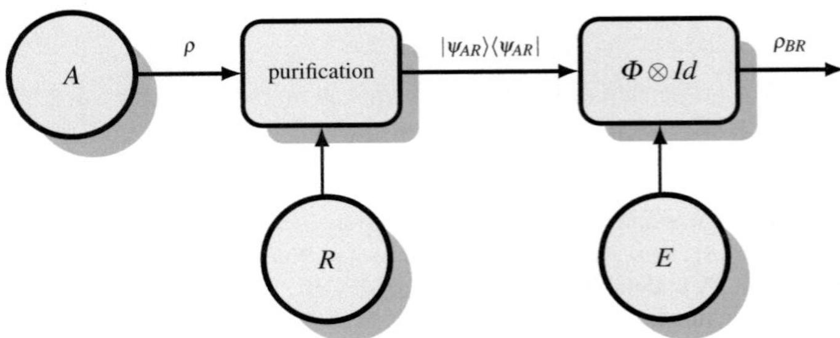

Fig. 12.30 Scenario for entropy exchange. The source A, the reference system R for the purification of the input state, and the channel environment E form a tripartite system, which is the basis for the entropy exchange

which is formed by: the input entropy $S(\rho_R) = S(\rho_A)$, the output entropy $S(\rho_B) = S(\Phi(\rho_A))$, and the joint entropy $S(\rho_{BR})$. We have to recall that the channel includes also an environment system E, so that we have a tripartite system $\mathcal{H}_A$, $\mathcal{H}_R$, and $\mathcal{H}_E$. It can be shown that $S(\rho_{BR}) = S(\rho_E)$. This allows us to define the **entropy exchange**

$$S(\rho, \Phi) := S(\rho_{BR}) = S(\rho_E)$$

and the quantum mutual information can now be written in the meaningful form

$$\boxed{I(\rho, \Phi) = S(\rho_A) + S(\Phi(\rho_A)) - S(\rho, \Phi).}$$
(12.112)

12.10.6 Concluding Remarks

In this last section of the chapter, we have considered the transmission of (classical) information through a noisy quantum channel. The topic is not concluded, in agreement with the fact that several questions, related to the definition of quantum capacity, are still open. A deeper investigation goes beyond the scope of this book. We have only tried to formulate a few of the many scenarios offered by quantum information theory, in sharp contrast with classic information theory, where channel capacity is defined in an unique way.

A further indication that the topic is not consolidated is given by the number of different formulations we find in the recent literature. Compare, e.g., the seminal paper by Holevo and Giovannetti [2], the formulation of Datta in [11], the books by Nielsend and Chuang [8], and the book by Wilde [14].

References

1. C.E. Shannon, A mathematical theory of communication. Bell Syst. Tech. J. **27**(3), 379–423 (1948)
2. A.S. Holevo, V. Giovannetti, Quantum channels and their entropic characteristics. Rep. Prog. Phys. **75**(4), 046001 (2012)
3. R.W. Hartley, Transmission of information. Bell Syst. Tech. J. **7**, 535–564 (1928)
4. T. Cover, J. Thomas, *Elements of Information Theory* (Wiley, New York, 1991)
5. C. Weedbrook, S. Pirandola, R. García-Patrón, N.J. Cerf, T.C. Ralph, J.H. Shapiro, S. Lloyd, Gaussian quantum information. Rev. Mod. Phys. **84**, 621–669 (2012)
6. A.S. Holevo, M. Sohma, O. Hirota, Capacity of quantum Gaussian channels. Phys. Rev. A **59**, 1820–1828 (1999)
7. S. Olivares, Quantum optics in the phase space. Eur. Phys. J. Spec. Top. **203**(1), 3–24 (2012)
8. M.A. Nielsen, I.L. Chuang, *Quantum Computation and Quantum Information* (Cambridge University Press, Cambridge, 2000)

9. D.A. Huffman, A method for the construction of minimum redundancy codes. Proc. IRE **40**, 1098–1101 (1952)
10. B.L. Schumaker, C.M. Caves, New formalism for two-photon quantum optics. II. Mathematical foundation and compact notation. Phys. Rev. A **31**, 3093–3111 (1985)
11. N. Datta, Quantum entropy and information, in *Quantum Information, Computation and Cryptography*, Lecture Notes in Physics, vol. 808, ed. by F. Benatti, M. Fannes, R. Floreanini, D. Petritis (Springer, Berlin, 2010), pp. 175–214
12. A.S. Holevo, Complementary channels and the additivity problem. Theory Probab. Appl. **51**(1), 92–100 (2007)
13. L.B. Levitin, Optimal quantum measurements for two pure and mixed states, in *Quantum Communications and Measurement*, ed. by V. Belavkin, O. Hirota, R. Hudson (Springer, US, 1995), pp. 439–448
14. M.M. Wilde, *Quantum Information Theory* (Cambridge University Press, Cambridge, 2013). Cambridge Books Online

Chapter 13
Applications of Quantum Information

Main Acronyms

PTNG   Pseudo-random number generation
QRNG   Quantum random number generation
LCG   Linear congruential generators
QKD   Quantum key distribution
DV–QKD   QKD with discrete variables
CV–QKD   QKD with continuous variables

13.1 Introduction

Besides the problem of reliably transmitting classical information through quantum means, which is the focus of this book, Quantum Information has seen an impressive diversity of applications, ranging from quantum computing to quantum cryptography, and from quantum teleportation to quantum metrology (for an extensive review see [1]). In this chapter we briefly present some examples of application, with the sole purpose of illustrating the many potential uses of Quantum Information.

In fact, the inherent randomness in quantum measurements lends itself to devising methods for the fast automatic generation of *true random numbers* with quantum devices. Similarly, the possibility (granted by Postulate 3) of detecting that some measurement operation has been performed on a single quantum system by employing a different measurement on the same system, has opened the way to *quantum cryptography*. This constitutes an unconditionally secure replacement for the schemes that currently lie at the core of many protocols for securing the transmission and storing of information from a rational attacker. Eventually, we devote a paragraph to the topic of *quantum teleportation*, that is, the transfer of an unknown quantum state between two different locations that is achieved by making use of entanglement and only transmitting classical information.

© Springer International Publishing Switzerland 2015
G. Cariolaro, *Quantum Communications*, Signals and Communication Technology,
DOI 10.1007/978-3-319-15600-2_13

13.2 Quantum Random Number Generation

One of the most striking applications of Quantum Mechanics in the field of Quantum Information is the generation of true random numbers. Random numbers represent a resource in many areas of science and technology. They provide the main ingredient of Monte Carlo methods and cryptographic protocols. In particular, for what concerns the former, whenever it is too difficult to solve a problem analytically, numerical simulations provide the most viable solution.

Regarding cryptographic applications, random numbers are fundamental to the ciphering of information. At the time of writing, most of the random numbers used in the cited fields, are obtained by means of *pseudo-random number generators* (PRNG). The adjective *pseudo* stands for *false* because PRNGs can only mimic the task of a generator, that is, to yield an identical and independent distribution of random variables. PRNGs are indeed nothing more that algorithms recursively executed by computers, which output a number at every operation. Unfortunately, these numbers seem random if one does not know the initial state, the so-called *seed*, of the generator, or if one has not exceeded its *period*, that is, the number of times the algorithm can be run before it goes back to outputting the same numbers. Clearly, when PRNGs are used in cryptography one has to take all the precautions to prevent a possible eavesdropper from predicting the generated number and then getting a copy of the key. In addition, a third problem is related to the way RNG algorithms are often engineered. More in detail, it happens that only after many years of use some widely employed PRNGs reveal dramatic nonrandom features, as was the case for the RAND-U generator, which belongs to the class of Linear Congruential Generators (LCG). In this generator a random number s_n is obtained according to the algorithm $s_n = (65,539\, s_{n-1})_{\mathrm{mod}\,231}$ with the initial state s_0 being an arbitrary seed. The dangerous feature of this generator is that it lacks randomness in a subtle way: indeed, if one maps consecutive triplets $\{s_n, s_{n-1}, s_{n-2}\}$ in 3D space, one can see that the numbers *mainly fall on parallel planes*, as shown in Fig. 13.1.[1] It is then clear that PRNGs not only represent a very weak point in cryptographic protocols but may also be the cause for erroneous results in simulations. Indeed John Von Neumann, one of the fathers of modern Computer Science and one of the first to employ random numbers in simulations, pointed out that *anyone who attempts to generate random numbers by deterministic means is, of course, living in a state of sin.*

Now, we will present two recipes showing how Quantum Mechanics can solve the problem of the generation of random numbers impossible to be forecast in any way.

---

[1] Citing the paper of Marsaglia [2] the mathematician that was the first to discover this weird behavior.

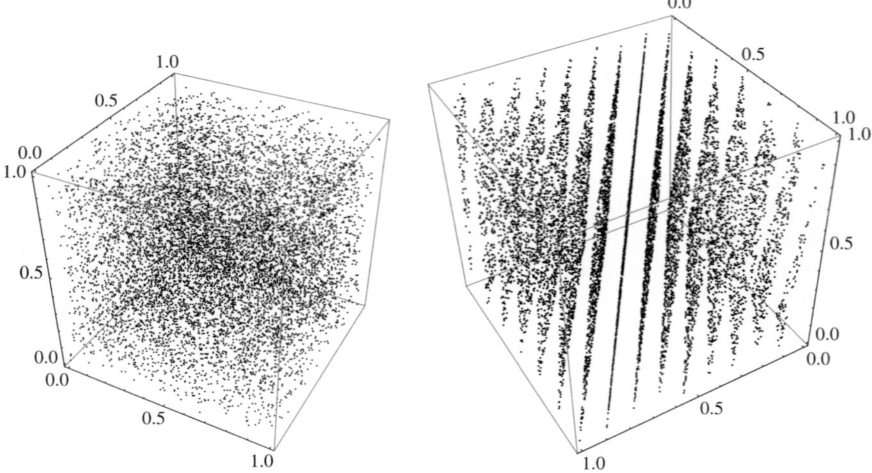

Fig. 13.1 *Left* Triplets of random numbers produced by employing the Linear Congruential Generator RAND-U are mapped in the space. *Right* If the point of sight is conveniently tilted, one can see that the points have the tendency to distribute along planes, a clear mark of lack of spatial uniformity

13.2.1 A Discrete Variable Quantum RNG

A solution to the issue of predictability is given by considering a physical quantum system. The latest step in the technology of random number generation devices is indeed the quantum random number generation (QRNG). The underlying principle of a QRNG is the impossibility of predicting the outcome of a measurement on a quantum system S prepared in a proper state ρ_S. As a simple example to understand how a QRNG works, one can consider a single photon state $|1\rangle$ impinging on a 50:50 beam-splitter. Let us suppose that the photon enters through input arm 1, whereas the unused port 2 carries the vacuum state $|0\rangle$. The overall input state is then given by

$$\psi = |1, 0\rangle_{1,2} = |1\rangle_1 \otimes |0\rangle_2 \tag{13.1}$$

which is equivalent to

$$\psi = a_1^* |0, 0\rangle_{1,2} \tag{13.2}$$

having introduced the field creation operator a_1^* for the mode of input 1. Considering that the beam splitter is modeled as a unitary transformation $U_{\text{b.s}}(\theta)$ on the field operators (see Sects. 9.2.2 and 11.18), one has that in the balanced case the field creation operator transforms according to

$$a_1^* = \tfrac{1}{\sqrt{2}} \left(i\, a_1^* + a_4^* \right) \tag{13.3}$$

so that the output state is then given by

$$\psi' = \tfrac{1}{\sqrt{2}} \left(i\, |1, 0\rangle_{3,4} + |0, 1\rangle_{3,4} \right). \tag{13.4}$$

The state ψ' is an entangled state of the modes 3 and 4: the photon is at the same time in both and none of the output arms of the beam splitter. By placing in front of the two outputs a pair of single-photon detectors, one realizes the following measurement operators:

$$P_{\text{out}}^{\text{no}} = |0\rangle \langle 0|_{\text{out}}, \qquad P_{\text{out}}^{\text{yes}} = |1\rangle \langle 1|_{\text{out}} \tag{13.5}$$

which measure, respectively, the absence or the presence of the photon in the respective output arm with out $\in \{3, 4\}$. Since the two measurements are independent, after the interaction of the single photon with the beam-splitter, the detectors perform the two possible bit-generating measurements

$$\Pi_0 = P_3^{\text{no}} \otimes P_4^{\text{yes}}, \qquad \Pi_1 = P_4^{\text{no}} \otimes P_3^{\text{yes}}. \tag{13.6}$$

By computing the outcome probability from (13.6) on the state $\rho_S = |\psi'\rangle \langle \psi'|$,

$$\text{Tr}\,[\Pi_0 \rho_S] = \text{Tr}[\Pi_1 \rho_S] = \frac{1}{2} \tag{13.7}$$

one sees that in a completely unpredictable way, as stated by the Born probability rule, it is possible to get 0 or 1 with exactly the same probability.

This approach was suggested and realized for the first time in [3] and it superseded the first attempts to generate random numbers by employing radioactive sources. Indeed, by using controllable optical photon sources as LEDs or lasers, one can easily prepare the state to be measured and fit both the sources and the detectors into compact and small devices (see for example [4] or [5]).

A drawback of these QRNGs is that they are limited by the count rate of the single photon counters, which at the present time do not allow one to extract random numbers at a rate higher than tens of gigabits per second. A way to overcome this limit is by changing the paradigm from discrete to continuous variables.

13.2.2 A Continuous Variable QRNG

The vacuum state of the electromagnetic field represents a source of entropy which has recently been employed to extract random numbers. When the quadratures of a pure vacuum state of the electromagnetic state are measured, one can collect a set of unpredictable random variables distributed according to the normal distribution. This becomes evident when one considers the Wigner function of the vacuum state

$$W(q, p) = \frac{1}{2\pi} \exp\left(-\frac{1}{2}(q^2 + p^2)\right) \tag{13.8}$$

where q and p are the eigenvalues relative to the momentum and position operators, respectively. When the state is measured along a given quadrature q, the possible measurement outcomes are distributed as follows:

$$w(q) = \int_{-\infty}^{+\infty} \mathrm{d}p\, W(q, p) = \frac{1}{\sqrt{2\pi}} \exp\left(-\frac{1}{2}q^2\right). \tag{13.9}$$

In the experiment, quadrature measurements are performed by means of homodyne detection, according to the scheme of Fig. 13.2. A coherent electromagnetic field, the so called *local oscillator* is mixed to the vacuum field entering through the unused port of a 50:50 beam splitter. More specifically, with respect to the single-photon discrete-variable approach, here the local oscillator is so intense that it can be treated as a classical field with amplitude $\alpha = |\alpha|e^{i\theta}$, playing the role of a vacuum fluctuations *amplifier*. The mixed fields exiting from the beam splitter outputs are intercepted by a couple of large bandwidth photodiodes which generate a current signal ΔI proportional to the light intensity hitting them. The two currents are respectively subtracted, so that one is left with a signal whose fluctuations are proportional to the quantum fluctuations of the field Fig. 13.3. In addition, local oscillator noise of classical origin, which would affect both incoming beams, is thus eliminated. In particular, if we denote by A and B the detectors intercepting the fields at the output of arms 3 and 4, respectively, we have that the output current of the

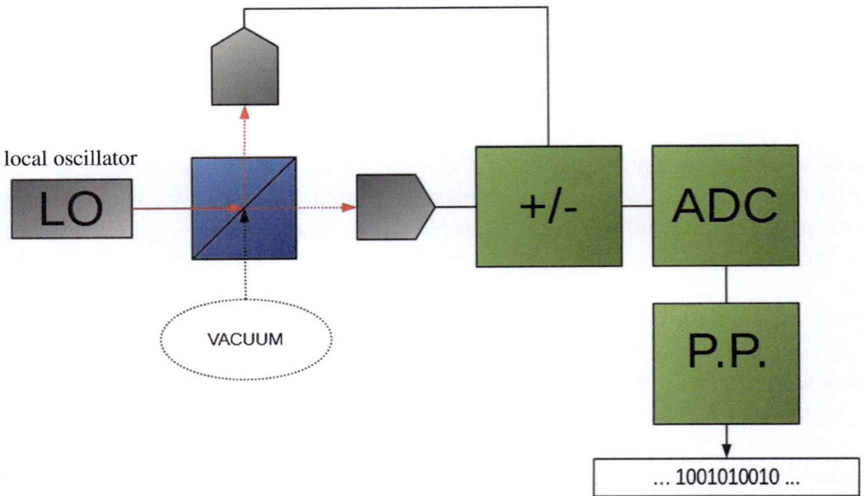

Fig. 13.2 Generic scheme to generate random numbers by homodyning the vacuum

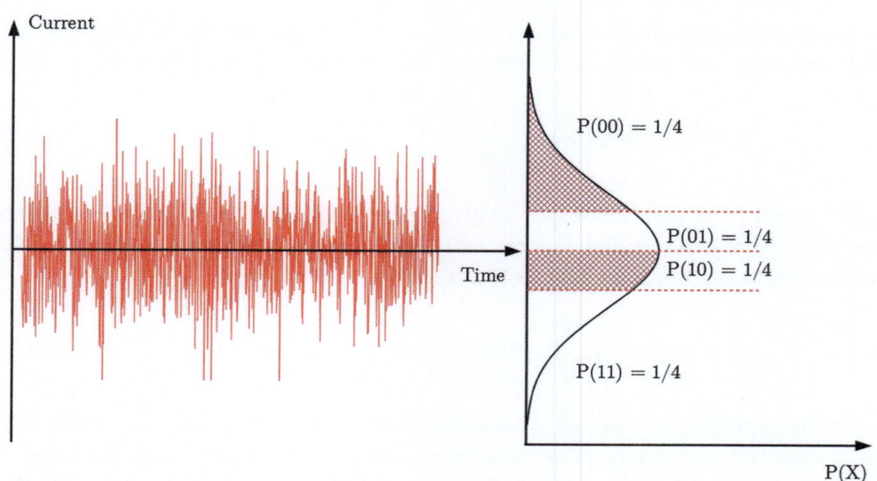

Fig. 13.3 On the *left*, the fluctuating current signal obtained by subtracting the outputs of the two photodiodes. On the *right*, the amplitude distribution of the signal is shown: in order to obtain numbers with a uniform distribution rather than a Gaussian one, the range of possible outcomes is split into a series of equal probability intervals. In the example of the picture, one has four possible intervals, each one with probability $\frac{1}{4}$ for a number in the range $[0, 4]$

setup is proportional to the difference of photon numbers given by the homodyne measurement operator $\hat{\Delta} = \hat{n}_A - \hat{n}_B$, where $\hat{n}_A = a\,a_3^*\,a\,a_3$ and $\hat{n}_B = a\,a_4^*a\,a_4$. By expressing the output operators as functions of the input ones, and considering the local oscillator classically, one has explicitly

$$
\begin{aligned}
\Delta &= n_A - n_B \\
&= \tfrac{1}{2}\left((\alpha^* + a\,a_1^*)(\alpha + a\,a_1) - (a\,a_1^* - \alpha^*)(a\,a_1 - \alpha)\right) \\
&= \tfrac{1}{2}\left((a\,a_1^*\alpha) + (a\,a_1\alpha^*)\right) \\
&= \tfrac{1}{2}|\alpha|\left(a\,a_1\mathrm{e}^{\mathrm{i}\theta} + a\,a_1\mathrm{e}^{-\mathrm{i}\theta}\right).
\end{aligned}
\tag{13.10}
$$

At this point it is easy to see that if the local oscillator is in- (out) phase, $\theta = 0$ (respectively $\theta = \frac{\pi}{2}$), with the field entering at input 1 it is possible to measure its q (respectively p) quadrature. For example if $\theta = 0$, and the input state at arm 1 is the vacuum, one will get a ΔI proportional to $\Delta = \sqrt{2}|\alpha|q$. Random numbers are then obtained by sampling the ΔI signal with an analog-to-digital converter (ADC). However, since the quadrature values are normally distributed according to (13.9), it is necessary to make equal the appearance probability of every number. For this purpose, a post-processing algorithm splits the range of possible current values into *equal probability* intervals, as shown in Fig. 13.3, and then outputs a given number according to the interval within which the measured value falls. This approach for random number generation was presented in [6] and for further details see [7].

13.3 Introduction to Quantum Cryptography

Nowadays, cryptography represents the general instrument for protecting information against a rational adversary. Cryptographic algorithms lie at the core of most security protocols and mechanisms, such as: encryption of data to ensure confidentiality, data authentication to detect forged messages, or integrity protection against illegitimate modification of messages in transit.

The majority of classic cryptographic algorithms can only offer computational security, that is, they guarantee that an adversary with limited computational capabilities has a low probability of success in attacking the security protocol within a reasonable amount of time. Such is the case, for instance, of all public-key cryptography, e.g., RSA encryption [8] and DSA signatures [9], as well as most symmetric schemes, e.g., AES encryption [10], and deterministic hashing, e.g., SHA [11]. If the amount of computational time that is needed by any adversary to break the security scheme considerably exceeds the useful life span of the relevant information, the scheme can be deemed properly secure. However, such schemes do not offer long-term protection of secured information from possible future technological or algorithmic breakthroughs. In particular, some public-key schemes, such as the above-mentioned RSA and DSA, have already been proven vulnerable to quantum computing attacks, since Shor's quantum algorithm [12] allows one to solve the task of finding the periodicity of a function with limited error probability and in polynomial time. In fact, that task is crucial in solving the integer factorization and the finite logarithm problems, the hardness of which (for classical computers) ensures the computational security of RSA and DSA, respectively.

Other classical schemes offer unconditional security (also known as information theoretic security), where the limit to the success probability of the attacker is no longer set by his/her computational capabilities, but rather by the information that is available to him/her. However, this is typically done at the expense of requiring the legitimate users to preshare a large quantity of secret material, as in the one-time-pad scheme, where the encrypted message is obtained by summing the secret message with a random secret key with the same entropy as the message. Alternatively, some information is required at the legitimate terminals about the attacker channel, as in designing wiretap coding schemes, and in this case the diversity between the legitimate and the attacker channel is leveraged to provide the required security. However, it should be noted that the assumption of knowing the attacker channel is unrealistic in general, since it cannot rely on any collaboration from the adversary.

By contrast, quantum cryptography can offer unconditional, information theoretic, security, as it is based on:

- the inherent randomness in the outcomes of quantum measurements,
- the possibility of statistically bounding the amount of measurements taken by the adversary, from the statistics of nonorthogonal measurements by the legitimate parties.

From the above two properties, one can state that there is no such thing as a purely passive, undetectable attacker in the realm of quantum information.

Starting from the pioneering work of Wiesner [13] who, as early as 1970 (even if his paper was only published many years later), set forth the possibility to create unforgettable quantum money, quantum counterparts have been subsequently developed for many cryptographic primitives, such as bit commitment, oblivious transfer, coin flipping, and random number generation (as was seen in Sect. 13.2.1). In the following sections, however, we will limit ourselves to describing the quantum cryptographic primitive that has been the earliest and most successfully implemented, that is, quantum key agreement (aka quantum key distribution).

13.4 Quantum Key Distribution (QKD)

A *key agreement protocol* is a security mechanism upon which two parties, Alice and Bob, jointly generate a common random variable or string (the *key*) $K \in \mathcal{K}$ that is uniformly distributed and unknown to any other party. Thus, K can securely be used as a cryptographic key for symmetric algorithms between them, e.g., for encryption or message authentication. To this purpose Alice and Bob can locally process separate secret random variables A and B, respectively, at each terminal, and exchange messages m_A, m_B over a public and authenticated channel (*public discussion*), where all transmissions can be observed, but not forged or altered, by any third party.

The most widely known and adopted key agreement scheme is the Diffie-Hellman protocol [14], which allows for separate and independent generation of the initial random variables A and B at Alice and Bob, and offers computational security based on the hardness of the discrete logarithm problem.

On the other hand, *information theoretic* key agreement schemes offer unconditional security, but require that some randomness is shared beforehand between Alice and Bob, that is to say, their initial random variables A and B must be correlated. This can be obtained either by separate noisy observations of the same random quantity (in the so-called *source model*), or by generating a random signal at one end (say, A at Alice) and transmitting it to the other end (say, Bob) through a noisy channel (in the *channel model*). However, when such interaction is allowed for the legitimate terminals, the same must be granted to a generic eavesdropper Eve, who will therefore have access to a third variable C, itself correlated with A and B.

The performance measure of an information theoretic key agreement scheme is given by the *secret key rate* R_k, that is, the information rate (in bit/s) of the final output key under the asymptotic constraints

$$
\begin{aligned}
&\mathrm{P}[K_A \neq K_B] < \varepsilon && \text{(correctness)} \\
&\log_2 |\mathcal{K}| - H(K) < \varepsilon && \text{(uniformity)} \\
&I(K; C, m_A, m_B) < \varepsilon && \text{(secrecy)}.
\end{aligned}
\tag{13.11}
$$

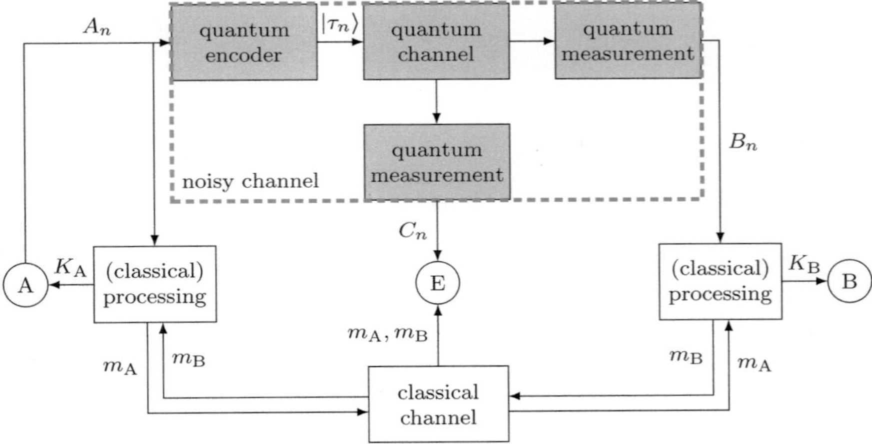

Fig. 13.4 Quantum cryptographic implementation of information theoretic key agreement in the channel model via a prepare-and-measure QKD system

If the random variables initially available to Alice, Bob, and Eve are symbol sequences, denoted by A_n, B_n, and C_n, respectively, generated by a memoryless source or noisy channel with symbol rate R_s, and joint symbol probability distribution $p(A, B, C)$, it can be shown that the maximum achievable secret key rate satisfies the bounds

$$R_s\left[I(A; B) - \min\{I(A; C), I(B; C)\}\right] \le R_k \le R_s \min\{I(A; B), I(A; B|C)\}$$

$$(13.12)$$

Quantum cryptography allows for an effective implementation of information theoretic key agreement schemes,[2] leading to the development of quantum key distribution (QKD) protocols. In particular, channel model schemes can be implemented through *prepare-and-measure* protocols as illustrated in Fig. 13.4, while source model schemes find a proper embodiment in *entanglement-based* protocols, as shown in Fig. 13.5.

When considering a quantum environment, the secrecy notion in (13.11) should be stated in quantum information terms, e.g., by bounding the accessible information at Eve, as $I_{acc} < \varepsilon$, since, in general Eve may optimize her measurement after Alice and Bob have performed their agreement protocol.

Traditionally, QKD protocols are divided into *discrete variable* (DV-) and *continuous variable* (CV-) QKD, according to the nature of the initial random variables A_n, B_n and of the quantum states that represent them. In the following, we shall examine an example of both prepare-and-measure and entanglement-based, DV-QKD. Eventually, we shall also briefly outline a QKD protocol with continuous variables.

---

[2] Historically, the first formulation of a QKD protocol [15] preceded that of general information theoretic key agreement schemes [16].

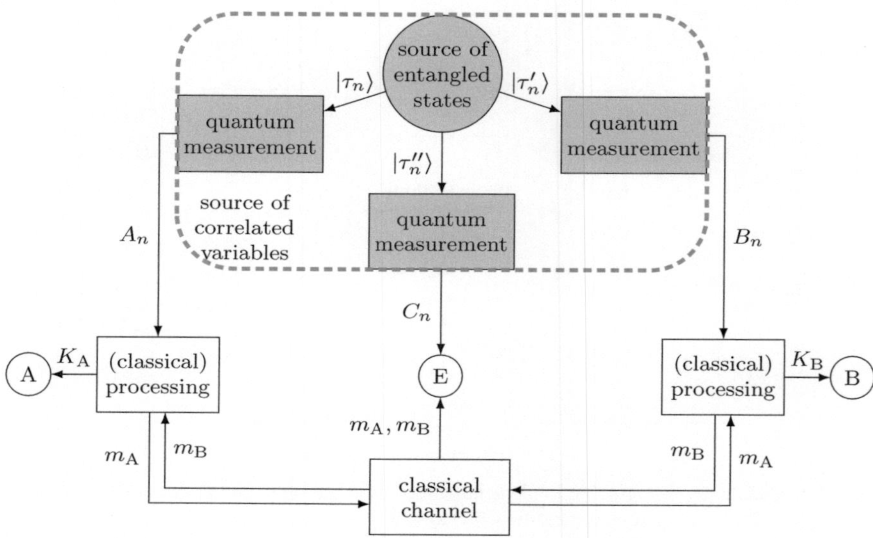

Fig. 13.5 Quantum cryptographic implementation of information theoretic key agreement in the source model via an entanglement-based QKD system

13.4.1 A Discrete-Variable-QKD Prepare-and-Measure Protocol

In this section we describe a *prepare-and-measure* protocol for DV-QKD that was proposed in [17] and is known as *efficient BB84*. It represents a variation of the original *BB84* protocol, the first to be proposed for DV-QKD in [15], and lends itself to a compact description and a precise security analysis [18, 19].

Transmission and Detection

According to this protocol, four states $|\gamma_0^+\rangle, |\gamma_1^+\rangle, |\gamma_0^\times\rangle, |\gamma_1^\times\rangle \in \mathcal{H}$ are used for transmission along the qubit channel. They are chosen to be pairwise orthogonal with $\langle \gamma_0^+|\gamma_1^+\rangle = 0$ and $\langle \gamma_0^\times|\gamma_1^\times\rangle = 0$ and hence make up two distinct bases for $\mathcal{H}$. The basis $\mathcal{B}^+ = \{|\gamma_0^+\rangle, |\gamma_1^+\rangle\}$ is called the *majority basis* (or *bit basis*), and is used to share a common binary string between the two legitimate terminals, whereas the *minority basis* $\mathcal{B}^\times = \{|\gamma_0^\times\rangle, |\gamma_1^\times\rangle\}$ (sometimes called *phase basis*) is used to detect any eavesdropping on the qubit channel. In fact, if eavesdropping is detected, the protocol aborts, and the eavesdropped key is discarded.

The transmitter (Alice) generates a sequence of independent–identically distributed binary symbols $\{A_n\}$ with equally likely 0 and 1 and encodes each bit randomly and independently into either basis, so that the transmitted state at the nth symbol period is

$$|\tau_n\rangle = \begin{cases} |\gamma_{A_n}^+\rangle & \text{with probability } p \\ |\gamma_{A_n}^\times\rangle & \text{with probability } 1 - p \end{cases}$$

for some fixed probability p.

On the other side of the channel, the receiver (Bob) measures each incoming state $|\tau_n\rangle$ with a POVM M_n, that is composed of a pair of orthogonal rank-1 projectors along the states that make up either basis. In fact, the measurement operators are chosen randomly and independently at each symbol period, and independently of the encoding choices made by Alice, with

$$M_n = \begin{cases} \{\Pi_0^+, \Pi_1^+\} & \text{with probability } p' \\ \{\Pi_0^\times, \Pi_1^\times\} & \text{with probability } 1 - p' \end{cases}$$

where $\Pi_0^+ = |\gamma_0^+\rangle\langle\gamma_0^+|$ and analogously for $\Pi_1^+, \Pi_0^\times, \Pi_1^\times$, and for some fixed p'. We denote by $B_n \in \{0, 1\}$ the corresponding outcome.

Hence, from (3.29) the channel transition probabilities are

$$p_c(i|j) = \begin{cases} \left|\left\langle\gamma_i^+|\gamma_j^+\right\rangle\right|^2 & \text{when both Alice and Bob use } \mathcal{B}^+ \\ \left|\left\langle\gamma_i^\times|\gamma_j^\times\right\rangle\right|^2 & \text{when both Alice and Bob use } \mathcal{B}^\times \\ \left|\left\langle\gamma_i^\times|\gamma_j^+\right\rangle\right|^2 & \text{when Alice uses } \mathcal{B}^+ \text{ and Bob uses } \mathcal{B}^\times \\ \left|\left\langle\gamma_i^+|\gamma_j^\times\right\rangle\right|^2 & \text{when Alice uses } \mathcal{B}^\times \text{ and Bob uses } \mathcal{B}^+. \end{cases}$$

In particular, observe that, due to the orthogonality between states in the same basis, whenever Alice and Bob choose the same basis they have a correct transition with probability 1, whereas when the chosen bases differ, there will be a bit error with probability $\delta = |\langle\gamma_0^+|\gamma_1^\times\rangle|^2 = |\langle\gamma_1^+|\gamma_0^\times\rangle|^2$.

Eavesdropping

Now, consider that an eavesdropper (Eve) sitting along the Alice-Bob channel has observed (measured) each single qubit coming from Alice. Her best chance is to always use the $\{\Pi_0^+, \Pi_1^+\}$ measurement operators, as this will give her full information on the secret bits that will be shared between Alice and Bob. Let C_n denote the outcome of her measurement; because of the no-cloning theorem, in order to share the same information with Bob, she must re-encode it as

$$|\tilde{\tau}_n\rangle = |\gamma_{C_n}^+\rangle$$

and transmit it along the channel to Bob. Whenever both Alice and Bob choose the majority basis, both measurements by Eve and Bob will yield a correct transition, and it will be $A_n = B_n = C_n$. However, if both Alice and Bob choose the minority basis, it will be

$$p_{C_n|A_n}(i|j) = \left|\left\langle \gamma_i^+ | \gamma_j^\times \right\rangle\right|^2$$

and

$$p_{B_n|C_n}(i|j) = \left|\left\langle \gamma_i^\times | \gamma_j^+ \right\rangle\right|^2 .$$

By conditioning on C_n, and applying the total probability theorem, we then obtain

$$
\begin{aligned}
p_{\mathrm{c}}(i|j) &= \sum_{\ell=0}^{1} p_{B_n|A_nC_n}(i|j,\ell)\, p_{C_n|A_n}(\ell|j) \\
&= \sum_{\ell=0}^{1} p_{B_n|C_n}(i|\ell)\, p_{C_n|A_n}(\ell|j) \\
&= |\langle \gamma_i^\times | \gamma_0^+ \rangle|^2 \left|\left\langle \gamma_0^+ | \gamma_j^\times \right\rangle\right|^2 + |\langle \gamma_i^\times | \gamma_1^+ \rangle|^2 \left|\left\langle \gamma_1^+ | \gamma_j^\times \right\rangle\right|^2 \\
&= \begin{cases} \delta^2 + (1-\delta)^2 & \text{for } i = j \\ 2\delta(1-\delta) & \text{for } i \neq j. \end{cases}
\end{aligned}
$$

$$(13.13)$$

Therefore, when both Alice and Bob choose the minority basis and Eve performs the measurement and re-encoding on the transmitted qubit using the majority basis, Alice and Bob will experience a bit error with probability $\delta' = 2\delta(1-\delta)$.

Sifting and Eavesdropping Detection

After the transmission is completed, Alice and Bob can share the following information along the public channel, that is, by m_A and m_B

1. In m_A, Alice tells Bob the subset of indices $N_\mathrm{A} = \{n \mid \tau_n \in \mathcal{B}^+\}$ in which she used the majority basis;
2. In m_B, Bob tells Alice the subset of indices $N_\mathrm{B} = \{n \mid M_n = \{\Pi_0^+, \Pi_1^+\}\}$ in which he used the majority basis;

so that each of them can infer the subset of indices $N = N_\mathrm{A} \cap N_\mathrm{B}$ in which they have both used the majority basis, and $N' = N_\mathrm{A}^\mathrm{c} \cap N_\mathrm{B}^\mathrm{c}$ in which they have both used the minority basis. Then:

- the bits A_n, B_n for $n \notin N \cup N'$ are discarded (*sifting*);
- the bits A_n, B_n for $n \in N$ are kept undisclosed and will be used to build the secret key;
- the bits A_n, B_n for $n \in N'$ are exchanged by Alice and Bob over the public channel, so that by comparing their values they can detect any errors.

Assume that n_{tot} total qubits have been transmitted, and that Alice and Bob declare that eavesdropping has been detected if for some $n \in N'$, $A_n \neq B_n$. The probability that Eve observing all qubits goes undetected is the probability that there are no errors in all the bits where both Alice and Bob use the minority basis, that is,

$$P_{\text{md}} = \prod_{n=1}^{n_{\text{tot}}} \left(\mathrm{P}[n \notin N'] + (1 - \delta')\mathrm{P}[n \in N'] \right) = \left[1 - \delta'(1 - (p + p') + pp') \right]^{n_{\text{tot}}}.$$

Choice of the Parameters p', δ, δ'

So far, we have left the values of parameters p, p', δ, δ' unspecified, subject to system design choices. We will now show that some optimal choice can be made straight away, with the aim of maximizing the number of bits that can be used to build the secret key, and at the same time of minimizing the probability that an attack by Eve goes undetected.

Consider the probability that a particular bit A_n (and correspondingly, B_n) is used to build the secret key, called the *sifted key rate*, which is given by $\mathrm{P}[n \in N] = pp'$. This is clearly maximized by the choice $p = p' = 1$ (always using the majority basis), which, unfortunately, would eliminate the possibility of detecting Eve's attack, and yield $P_{\text{md}} = 1$. Therefore, a tradeoff must be sought between increasing the sifted key rate and the attack detection probability. However, one can notice that for any fixed value of sifted rate pp', the value of P_{md} is minimized by making the sum $p + p'$ as small as possible, that is, by choosing $p' = p$, and by maximizing δ'.

As δ' is a quadratic function of δ, it is easily seen that its maximum is achieved at $\delta = 1/2$ yielding $\delta' = 1/2$. Observe that this choice corresponds to having the inner products

$$\left| \langle \gamma_i^+ | \gamma_j^\times \rangle \right| = \frac{1}{\sqrt{2}}, \quad i, j = 0, 1 \tag{13.14}$$

that are obtained by choosing the two bases in a symmetric fashion in the qubit space, which intuitively justifies our $\mathcal{B}^+$, $\mathcal{B}^\times$ notation. For instance, if the information is encoded into the polarization state τ_n of a single photon, one may choose horizontal, vertical, and diagonal polarization states as follows:

$$\gamma_0^+ = |\uparrow\rangle, \gamma_1^+ = |\rightarrow\rangle, \gamma_0^\times = |\nearrow\rangle, \gamma_1^\times = |\nwarrow\rangle$$

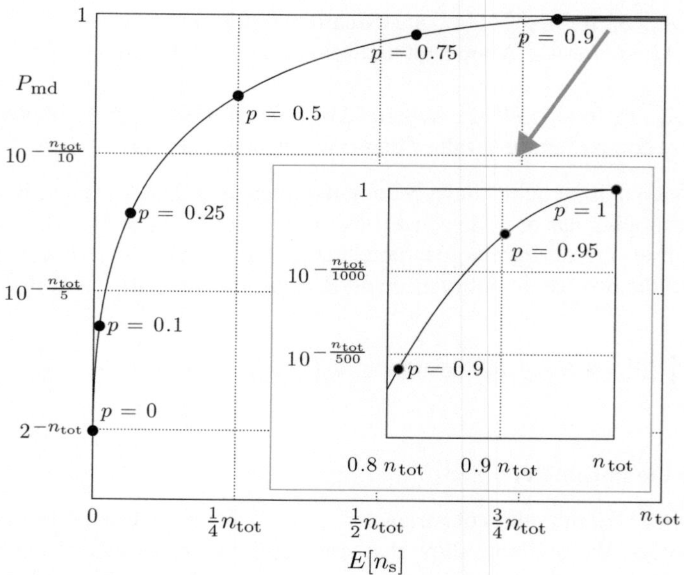

Fig. 13.6 Illustration of the tradeoff between the expected length of the sifted key E $[n_{\mathrm{s}}]$ and the missed detection probability P_{md}, depending on the value of the probability p of the majority basis. In the *lower right* corner an expanded view of the *upper right* corner, which is typically the region of practical interest. For instance observe that, with $n_{\mathrm{tot}} = 10^4$ transmitted qubits, if it is required to keep $P_{\mathrm{md}} < 10^{-20}$, one has to choose $p \leq 0.9$, and hence obtain no more than 8 200 sifted bits, on average

From now on, we will therefore assume that (13.14) holds and $p' = p$, thus yielding the missed detection probability and the expected sifted key length

$$P_{\mathrm{md}} = \left(\frac{1}{2} + p - \frac{1}{2}p^2\right)^{n_{\mathrm{tot}}}, \quad \mathrm{E}\,[n_{\mathrm{s}}] = n_{\mathrm{tot}}\,p^2$$

where the value of p allows us to trade the sifted key length for the attack detection capabilities of the scheme. The tradeoff is illustrated in Fig. 13.6.

Note that in this ideal setting the sifted keys $A' = [A'_1, \ldots, A'_{n_{\mathrm{s}}}] = [A_n]_{n \in N}$ and $B' = [B'_1, \ldots, B'_{n_{\mathrm{s}}}] = [B_n]_{n \in N}$ can be directly used as a secret cryptographic key pair, since they are identical with unit probability, and provided P_{md} is sufficiently low, any eavesdropping would have been detected with high probability.[3]

---

[3] A somewhat subtle point should be made here. The security of the protocol does not guarantee that eavesdropping is unlikely, given that no errors have been detected in the minority basis. Rather, it states that if eavesdropping takes place, it will be detected with high probability. In symbols, let E denote the event that eavesdropping has taken place and D the event that no errors have been detected, we can only upper bound $P_{\mathrm{md}} = \mathrm{P}[D|E]$, but nothing can be said about $\mathrm{P}[E|D]$, since no assumption can be made on the probability of event E which is totally under the control of the attacker.

13.4.2 A DV-QKD Entanglement-Based Protocol

In this section we present a QKD protocol which makes use of entangled particles to share a secret key between two parties. The BBM92 protocol (Fig. 13.7), described here, was first proposed by Bennett, Brassard and Mermin in [20], as a simpler version of the Ekert protocol [21].

In the BBM92 scheme the channel consists of a source, called Charlie, that emits entangled particles and sends them to opposite directions.

The particles are received by two users, Alice and Bob, who perform measurements M_n^A and M_n^B. Both Alice and Bob choose their measurement operators randomly, independently and with equal probability between $\{\Pi_0^+, \Pi_1^+\}$ and $\{\Pi_0^\times, \Pi_1^\times\}$. Similar to BB84, we have that $\{\Pi_0^+, \Pi_1^+\}$ and $\{\Pi_0^\times, \Pi_1^\times\}$ should be selected symmetrically in the qubit space. Usually the bases are chosen to be the Pauli operator, i.e., $\Pi_0^\times = \sigma_z$ and $\Pi_0^+ = \sigma_x$, which satisfy the nonorthogonality condition. After a sequence of n_{tot} entangled particles are received and measured, Alice and Bob publicly announce which basis they used for each particle, but not the outcomes of the measurements. During sifting, Alice and Bob discard the events in which they measured in different bases, or in which the measurement failed because of imperfect detection. The remaining instances, in which both measured in the same basis, should be perfectly correlated if they actually measured entangled pairs. In order to verify this, Alice and Bob publicly compare their outcomes A_n, B_n in a subset $n \in N'$ of the undiscarded events. If Alice and Bob find perfect correlation on the tested set N', they can state that the transmission was secure and no eavesdropping was performed, and keep the remaining A_n, B_n, $n \in N = N_{AB} \setminus N'$ to produce the secret key K.

Security Proof

In examining the security of the BBM92 protocol, it is interesting to notice that this protocol bears many analogies to the BB84 presented in the previous section. We consider the most common attacks, i.e., *intercept and resend*, and *source substitution*. The discussion of the protocol robustness against the former kind of attack is analogous to the one given for the BB84 protocol and we refer to the previous section. The source substitution attack happens when an eavesdropper Eve sends Alice and Bob

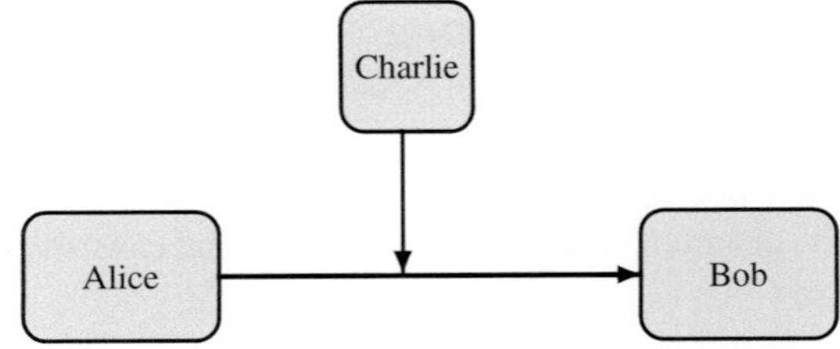

Fig. 13.7 BBM92 scheme. Alice and Bob: receiving users. Charlie: source of entangled particles

pairs that are somehow entangled with systems available to her. The most general entangled state Eve can prepare is equal to

$$|\Phi\rangle = |11\rangle|e_0\rangle + |00\rangle|e_1\rangle + |10\rangle|e_2\rangle + |01\rangle|e_3\rangle,$$

where $|1\rangle$ and $|0\rangle$ form an orthonormal qubit basis and $|e_0\rangle, |e_1\rangle, |e_2\rangle$ and $|e_3\rangle$ are the states of Eve's system. We can notice that in general Eve does not even have to decide her measurements until Alice and Bob have published theirs. Eve's aim is to be completely invisible, therefore, if Alice and Bob measure in the $\mathcal{B}^+$ basis, in order that they have fully correlated outcomes, the state $|\Phi\rangle$ must be an eigenstate of $\sigma_z^a \sigma_z^b$ with eigenvalue -1. This implies that $|\Phi\rangle$ must assume the form:

$$|\Phi\rangle = |10\rangle|e_2\rangle + |01\rangle|e_3\rangle.$$

At the same time, if Alice and Bob measure in the $\mathcal{B}^\times$ bases, the state $|\Phi\rangle$ must be an eigenstate of $\sigma_x^a \sigma_x^b$ with eigenvalue -1. This further restricts $|\Phi\rangle$ as follows:

$$|\Phi\rangle = (|10\rangle - |01\rangle)\,|e_2\rangle.$$

From this, the only Eve's source that will surely be undetected by Alice's and Bob's test is the one in which Eve's system is completely uncorrelated with the entangled particles. Thus, every measurement gives her no information about Alice's and Bob's outcomes.

Out of Curiosity

In 1992 there was a heated discussion between Ekert and Bennett, Brassard and Mermin, about the described protocol. Ekert stated that the security proof must be based on "non-locality" and "non-reality" tests given by Bell's theorem. Bennett et al. demonstrated their protocol without these assumptions using a simpler scheme. Recently, Vallone et al. [22] proposed a new protocol which uses a simple scheme as Bennett et al. and bases its security on Bell's theorem thanks to the use of non-maximally entangled particles.

13.4.3 Key Processing

In introducing the above protocols, we have ideally supposed that, provided Alice and Bob choose the same basis and Eve does not interfere, the sharing of a bit through the quantum channel is error-free. In that case, the sifted keys are also the final secret keys.

In a more realistic environment, however, distortion introduced by the quantum channel, temporal or spatial misalignment between the two terminals, and quantum noise in the receiver may introduce some errors, even for bits A_n, B_n with $n \in N \cup N'$, and without any attack. This has two potentially fatal consequences:

1. errors $A_n \neq B_n$ for $n \in N$ will propagate to the distilled secret key;
2. errors $A_n \neq B_n$ for $n \in N'$ will make Alice and Bob abort the protocol, even in the absence of an attacker.

The two problems above can be solved with techniques for the processing of random signals in the classical domain, to yield the final secret keys where both the mismatch between Alice and Bob's keys and the information leaked to the attacker have been removed with high probability.

In the following we assume that errors in each basis are symmetric and independent across symbols, so that the transformation linking A_n to B_n for $n \in N$ (respectively, $n \in N'$) is a binary symmetric channel with error rate ε^+ (respectively, $\varepsilon^\times$).

Information Reconciliation

In order to solve problem 1, techniques similar to traditional forward error correction coding for the binary symmetric channel can be used, providing they are suitably adapted to the secrecy requirement in the QKD framework. In fact, since the binary sifted sequence $A' = [A'_1, \ldots, A'_{n_s}] = [A_n]_{n \in N}$, (and the analogous B') is only known after sifting, the redundancy bits that allow error correction must be transmitted later, along the public classical channel, and can be observed by any attacker. The amount of redundancy must therefore be kept to a minimum, not for efficiency reasons, but to limit as far as possible the amount of information that leaks to an eavesdropper. As is well known, a lower bound on the amount of redundancy that must be transmitted in order to have reliable error correction is given by $r = n_s h_2(\varepsilon^+)$, with $h_2(\cdot)$ denoting the binary entropy function (see Chap. 12) $h_2(\varepsilon) = -\varepsilon \log_2 \varepsilon - (1 - \varepsilon) \log_2(1 - \varepsilon)$.

One possibility is to generate the redundancy bits m'_A by using systematic encoding for a block channel code, where properly sized blocks taken from the sifted sequence make up the information words, that is, $m'_A = GA'$, where A' and m'_A are seen as columns vectors, and G denotes the nonidentity portion of the systematic generating matrix. Thus, upon receiving $\tilde{m}'_A$ over the public channel, Bob can perform minimum distance decoding, that is replace B' by

$$B'' = \arg \min_{\beta \in \{0,1\}^{n_s}} d_H([\beta, G\beta], [B', \tilde{m}_A])$$

with $d_H(\cdot, \cdot)$ representing the Hamming distance between two binary strings.[4]

However, typically the public channel is assumed error-free and authenticated (that is, each message can be verified to actually come from Alice and not having been altered in transit), so that $\tilde{m}'_A = m'_A$ and there is no need to protect the redundancy bits from channel errors. In this case, it is more efficient in terms of error correction capability to obtain the redundancy bits as a hash of the sifted sequence, given, for

---

[4] That is, the number of positions at which they differ.

instance, through the parity check matrix G' of a linear code, yielding $m'_\mathrm{A} = G'A'$. Thus, upon receiving $\tilde{m}'_\mathrm{A}$ over the public channel, Bob can perform minimum distance decoding, that is, replace B' by

$$B'' = \arg \min_{\beta \in \{0,1\}^{n_\mathrm{s}}} d_\mathrm{H}([\beta, G'\beta], [B', \tilde{m}_\mathrm{A}]).$$

The latter approach is currently the most widely used in the QKD literature, typically by employing LDPC codes (see [23]), especially with stable channels and long processing blocks, where the code parameters can be precisely tuned to require an amount of redundancy that is close to the lower bound. On the other hand, when the channel conditions are varying, and/or shorter blocks need to be used, other ad hoc solutions are considered that require more interaction between the terminals along the public channel and intrinsically adapt to the channel conditions (see [19]).

In a symmetric fashion, one can have Alice correct A' to match Bob's sifted sequence B' based on a public message m_B sent by Bob. Alternatively, one can use a two-way reconciliation scheme where both Alice and Bob send public messages and each one partially correct their sifted keys.

Privacy Amplification

The obvious solution to problem 2 above is to allow for some errors in the bits with $n \in N'$ without aborting the protocol, as long as the number of errors n_err is below some specified threshold θ. The threshold is typically chosen depending on the cardinality of N' and the channel error rate.

However, this would introduce a vulnerability in the protocol. It makes it possible for the eavesdropper to perform a *selective intercept and resend* attack on a limited, yet significant, fraction of the qubits shared between Alice and Bob, by retransmitting them through an error-free channel. In this way, Eve's observations may not be detected, as Alice and Bob will attribute the errors to the channel and tolerate them, whereas they were actually induced by the eavesdropper measurements.

Therefore, a conservative countermeasure requires Alice and Bob to remove the partial information that Eve may have acquired through undetected qubit observations or by accessing the redundancy transmitted over the public channel for the purpose of reconciliation. *Privacy amplification* is the process of removing any information available to the attacker from the reconciled keys to yield the final *secret key K*.

This is done through the application of a common hashing function $f : \{0,1\}^{n_\mathrm{s}} \to \{0,1\}^{\ell}$ at each reconciled sequence A''_n and B''_n, where $\ell < n_\mathrm{s}$ represents the length of the final key. Clearly, applying the same function allows to maintain correctness. In fact, since the sequences A''_n and B''_n are supposedly identical with very high probability, so will be the corresponding outputs K_A, K_B. On the other hand, compressing the sequence with a function that is surjective, but not injective, makes it possible to remove bits that have been learnt by Eve, and the redundancy that has been inserted for reconciliation purposes, to obtain a key that is as uniform and independent of the eavesdropper observations as possible.

Typically, the hash function is simply a multiplication by a matrix $F \in \{0, 1\}^{\ell \times n_s}$ on the binary field. Also, a potential eavesdropper knowledge (C, m_A, m_B) about the reconciled sequence $A'' = B''$ can itself be described as a matrix function $M \in \{0, 1\}^{t \times n_s}$. For instance, suppose that Eve has performed selective intercept and resend so that she knows a subset $C'' = \{A_n'', n \in N_E\}$ of the reconciled sequence, for some N_E, and that she has observed the bits $m_A' = G A''$ transmitted along the public channel for reconciliation. Then, we can write

$$M = \begin{bmatrix} I_{N_E} \\ G \end{bmatrix}$$

where I_{N_E} is made of the rows from the $n_s \times n_s$ identity matrix with indices in N_E, and $t = r + |N_E|$.

If the eavesdrop matrix M were known to Alice and Bob, it would in principle be possible to choose the privacy amplification matrix F to yield a perfectly secret key. In fact, in this case, since A'' is uniform over $\{0, 1\}^{n_s}$ (as a consequence of the fact that A_n and B_n are assumed to be iid uniform sequences), it can be easily seen that the final key K is uniform in as $\{0, 1\}^{\ell}$ and independent of the eavesdropper observations if and only if the null spaces of F and M satisfy

$$\dim \mathcal{N}(M) - \dim \left(\mathcal{N}(M) \cap \mathcal{N}(F) \right) = \ell \, . \tag{13.15}$$

On the other hand, if M is not known, but the value of t is (or can at least be upper bounded), Alice and Bob can choose the hashing function f randomly after sifting (so that Eve can not tailor her observations to it) and communicate the choice over the public channel. It was shown in [24] that the average of the mutual information in (13.11) over the choice of f can be upper bounded as

$$I(K; C, m_A, m_B, f) < \frac{1}{\log 2} \frac{1}{2^{n_s - t - \ell}} \tag{13.16}$$

by choosing f uniformly within a universal hashing class,[5] such as that of all $\ell \times n_s$ binary Toeplitz matrices.

In general, however, it is more realistic to assume that neither the exact position, nor even the exact amount of the qubits observed by the eavesdropper are known to the legitimate parties.

Therefore, privacy amplification is usually performed in two steps. First, since the matrix G of information reconciliation is perfectly known, a matrix $F_1 \in \{0, 1\}^{\ell_1 \times n_s}$

---

[5] A class $\mathcal{F}$ of functions mapping the same domain X to the same range Y is called *universal hashing* if it maps inputs to outputs "uniformly", that is,

$$\begin{cases} |\{f \mid f(x) = y\}| = |\mathcal{F}|/|Y| & \text{for all } x \in X, y \in Y \\ |\{f \mid f(x_1) = f(x_2)\}| = |\mathcal{F}|/|Y| & \text{for all } x_1, x_2 \in X. \end{cases}$$

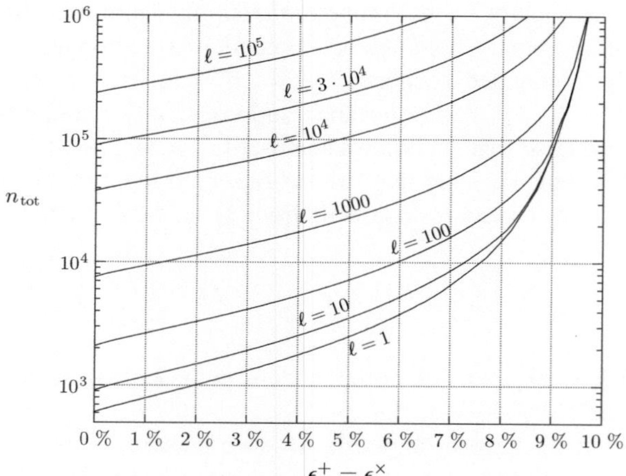

Fig. 13.8 Minimum amount of qubits n_{tot} that need to be transmitted in the efficient BB84 protocol, as a function of the quantum BER (assumed equal on both bases, $\varepsilon^+ = \varepsilon^\times$), for different target values of the final secret key length ℓ. The plot is based on an optimization of the finite key bound in [18]

that satisfies (13.15) with F_1 replacing F and G replacing M is applied. Then, the amount of information that is available to Eve from undetected qubit observations is upper bounded probabilistically in terms of the abort threshold on the number of detected errors, that is, t_{ub} is chosen so that $\text{P}[n_{\text{err}} < \theta | t > t_{\text{ub}}]$ is acceptably low. Eventually, a matrix $F_2 \in \{0, 1\}^{\ell \times \ell_1}$ chosen randomly from a universal hashing class is applied so that (13.16) is satisfied with very high probability.

Clearly, in the limit of $n_{\text{tot}}, |N'| \to \infty$ the rate of information that is available to Eve can be precisely estimated. On the other hand, when n_{tot} is limited (in the so-called *finite key* regime) such estimates have a large amount of uncertainty, and significant margins must be allowed when choosing t_{ub} and ℓ. Several bounds for ℓ have been formulated in the finite key regime [18, 19]. Figure 13.8 shows a contour plot of the final key length as a function of the total transmitted qubits and the error rates in the channel for the efficient BB84 protocol, according to the bound provided in [18] and for optimal choices of the threshold θ and the majority basis rate p. Observe that ℓ decreases rapidly following n_{tot}.

13.4.4 A Continuous Variable QKD Protocol

As an example of CV-QKD, we consider the GG02 protocol [25], as introduced in [26], a *prepare-and-measure* scheme, which makes use of coherent states with Gaussian displacements.

In fact, the transmitter Alice generates a sequence A_n of iid complex Gaussian random variables with circular symmetry.[6] Then she encodes each variable A_n into the coherent state with displacement given by the corresponding realization of A_n, that is,

$$A_n = \alpha \Rightarrow |\tau_n\rangle = |\alpha\rangle.$$

Alternatively, this can be viewed as encoding $\Re A_n$ into the position and $\Im A_n$ into the momentum displacement of $|\tau_n\rangle$.

The protocol is based on the fact that the uncertainty principle prevents measuring both quadratures with full accuracy. On the other side of the quantum channel, Bob measures either the position or momentum of each incoming state $|\tau_n\rangle$, by randomly and independently choosing each measurement observable as

$$M_n = \begin{cases} q & \text{with probability } 1/2 \\ p & \text{with probability } 1/2 \end{cases}$$

and we denote by $B_n \in \mathbb{R}$ the corresponding continuous-valued outcome.

Thus, B_n will be a Gaussian random variable correlated with either $\Re A_n$ of $\Im A_n$ according to whether $M_n = q$ or $M_n = p$. After the transmission is completed, Bob tells Alice via the message m_B the sequence of measurements $\{M_n\}$ so that Alice can sift her sequence and obtain

$$A_n' = \begin{cases} \Re A_n & \text{if } M_n = q \\ \Im A_n & \text{if } M_n = p. \end{cases}$$

The mutual information between A_n (or equivalently A_n') and B_n is therefore given by

$$I(A; B) = \frac{1}{2} \log_2 \left(1 + \frac{\sigma_A^2}{\sigma_0^2} \right)$$

where σ_0^2 represents the fluctuations of the coherent state around its displacement.

The possibility of detecting an intercept and resend attack by Eve lies in the impossibility for Eve to perform both a position and momentum observation on τ_n, analogously to what was shown for discrete variable protocols.

13.5 Teleportation

Quantum teleportation is one of the many important applications of entanglement. It allows an *unknown* quantum state to be transported from Alice to Bob by transmitting only classical information. In particular, a qubit can be teleported by using two

---

[6] A complex-valued random variable X is called circular symmetric Gaussian if $\Re X$ and $\Im X$ are independent Gaussian variables with zero mean and the same variance σ_X^2.

classical bits. Let us consider the simplest example, given by a single qubit in a generic state unknown to Alice

$$|\varphi\rangle_c = \alpha|0\rangle_c + \beta|1\rangle_c. \tag{13.17}$$

Due to the no-cloning theorem, Alice cannot clone such state and cannot know all the quantum information by measuring the qubit. Indeed, the parameters α and β can only be obtained if Alice has many copies of the state $|\varphi\rangle_c$ and performs several measurements on them. It is important to note that the state $|\varphi\rangle_c$ contains an infinite amount of classical information, parameterized by the complex (continuous) parameters α and β.

Quantum teleportation [27] allows Alice to send such qubit by sending Bob only two bits of classical information. The key resource to achieve such goal is a maximally entangled state between Alice and Bob. We recall the four Bell states, which are maximally entangled states forming a basis in the Hilbert space of two qubits

$$|\phi^{\pm}\rangle = \tfrac{1}{\sqrt{2}} \left(|00\rangle_{AB} \pm |11\rangle_{AB}\right), \qquad |\psi^{\pm}\rangle = \tfrac{1}{\sqrt{2}} \left(|01\rangle_{AB} \pm |10\rangle_{AB}\right). \tag{13.18}$$

Any Bell state can be used for quantum teleportation. Here we show how to achieve it with the state $|\psi^-\rangle_{AB}$. Alice holds the unknown qubit $|\varphi\rangle_c$ and her part of the entangled state, A. The total state shared by Alice and Bob can be written as

$$|\Psi\rangle_{CAB} = |\varphi\rangle_c \otimes |\psi^-\rangle_{AB} \tag{13.19}$$

By expanding the state we obtain

$$|\Psi\rangle_{CAB} = \tfrac{1}{\sqrt{2}} \left(\alpha|001\rangle_{CAB} - \alpha|010\rangle_{CAB} + \beta|101\rangle_{CAB} - \beta|110\rangle_{CAB}\right) \tag{13.20}$$

with the easy notation $|001\rangle_{CAB} \equiv |0\rangle_C \otimes |0\rangle_A \otimes |1\rangle_B$. From the definition of the Bell states, it is possible to show that the following equalities hold

$$|00\rangle_{CA} = \tfrac{1}{\sqrt{2}} \left(|\phi^+\rangle + |\phi^-\rangle\right), \qquad |11\rangle_{CA} = \tfrac{1}{\sqrt{2}} \left(|\phi^+\rangle - |\phi^-\rangle\right)$$
$$|01\rangle_{CA} = \tfrac{1}{\sqrt{2}} \left(|\psi^+\rangle + |\psi^-\rangle\right), \qquad |10\rangle_{CA} = \tfrac{1}{\sqrt{2}} \left(|\psi^+\rangle - |\psi^-\rangle\right).$$

Thus the total state can be written in the following form:

$$|\Psi\rangle_{CAB} = \tfrac{1}{2} \left(|\phi^+\rangle_{CA} \otimes \sigma_x\sigma_z|\varphi\rangle_B + |\phi^-\rangle_{CA} \otimes \sigma_x|\varphi\rangle_B - |\psi^+\rangle_{CA} \otimes \sigma_z|\varphi\rangle_B - |\psi^-\rangle_{CA} \otimes |\varphi\rangle_B\right). \tag{13.21}$$

The above equation contains all the information needed to understand quantum teleportation. To complete the protocol, Alice needs to perform a measurement on the two qubits C and A. Indeed, she performs a *Bell measurement*, consisting in a projective measurement that distinguishes between the four orthogonal Bell states

$\{|\phi^+\rangle, |\phi^-\rangle, |\psi^+\rangle, |\psi^-\rangle\}$. If she obtains $|\phi^+\rangle$, relation (13.21) indicates that Bob obtains the state $\sigma_x\sigma_z|\varphi\rangle_B$. If she obtains $|\phi^-\rangle$, $|\psi^+\rangle$ or $|\psi^-\rangle$, Bob is left with the state $\sigma_x|\varphi\rangle_B$, $\sigma_z|\varphi\rangle_B$, or $|\varphi\rangle_B$, respectively. Then Alice communicates to Bob which state she has measured (since she has four possibilities, two classical bits are sufficient). Bob performs a different unitary transformation $\mathcal{U}$ depending on the outcomes obtained by Alice to recover the input state (unknown to both Alice and Bob). The operation performed by Bob is summarized in the following table:

Alice outcome	Bob Operation ($\mathcal{U}$)
$|\phi^+\rangle$	$\sigma_z\sigma_x$
$|\phi^-\rangle$	σ_x
$|\psi^+\rangle$	σ_z
$|\psi^-\rangle$	$\mathbf{1}$

It is important to underline that the Bell measurement gives no information on the input state $|\varphi\rangle_C$ and that for any input state Alice has equal probability, 1/4, of obtaining each of the four Bell states.

In the quantum teleportation protocol (Fig. 13.9), the input quantum state is not traveling between Alice and Bob: what is "traveling" is the *quantum information* contained in the parameters α and β. Indeed, it is worth noticing that the input and the teleported qubits can be implemented in different physical systems. For instance, the input qubit can be encoded in the polarization of a photon, while the teleported qubit can be represented by a two-energy-level atom system. Moreover, there is no

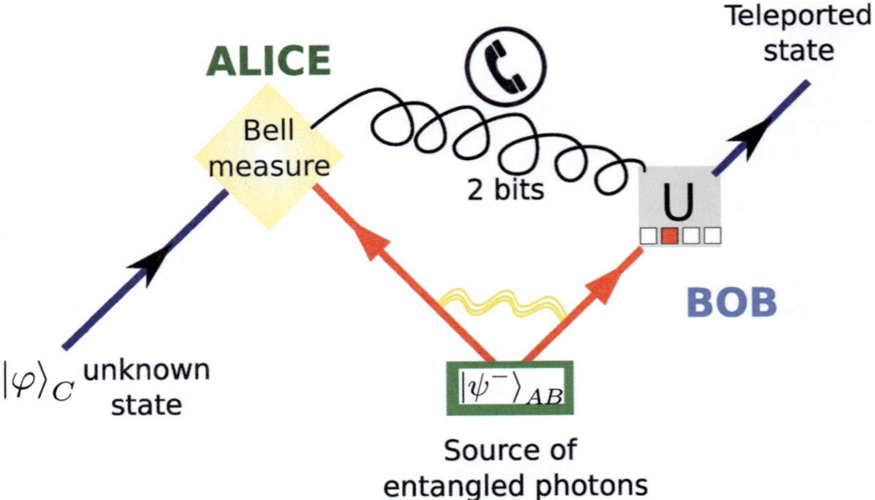

Fig. 13.9 Teleportation environment

contradiction with the no-cloning theorem: indeed, the unknown state vanishes at Alice's side and appears at Bob's location. Then there is no "cloning" of the input state.

Finally, even if the collapse of the wave function is instantaneous (at the moment in which Alice obtains her outcome), Bob's state immediately collapses to $\sigma_x\sigma_z|\varphi\rangle_B$, $\sigma_x|\varphi\rangle_B$, $\sigma_z|\varphi\rangle_B$, or $|\varphi\rangle_B$, a classical communication is necessary between Alice and Bob to correctly recover the input qubit. Then teleportation does not violate the "no faster than light" communication principle.

The first experimental demonstrations were performed with photons in Rome and Vienna in 1997 [28, 29]. Further experiments were realized with coherent states [30] and nuclear magnetic resonance [31]. Recent experiments reported quantum teleportation of photons along distances of more than 100 km [32, 33].

References

1. C. Weedbrook, S. Pirandola, R. García-Patrón, N.J. Cerf, T.C. Ralph, J.H. Shapiro, S. Lloyd, Gaussian quantum information. Rev. Mod. Phys. **84**, 621–669 (2012)
2. G. Marsaglia, Random numbers fall mainly in the planes. Proc. Natl. Acad. Sci. **61**(1), 25–28 (1968)
3. T. Jennewein, U. Achleitner, G. Weihs, H. Weinfurter, A. Zeilinger, A fast and compact quantum random number generator. Rev. Sci. Instrum. **71**(4), paper no. 1675 (2000)
4. M. Fürst, H. Weier, S. Nauerth, D.G. Marangon, C. Kurtsiefer, H. Weinfurter, High speed optical quantum random number generation. Opt. Express **18**(12), 13029–13037 (2010)
5. M. Stipčević, B.M. Rogina, Quantum random number generator based on photonic emission in semiconductors. Rev. Sci. Instrum. **78**(4), paper no. 045104 (2007)
6. C. Gabriel, C. Wittmann, D. Sych, R. Dong, W. Mauerer, U.L. Andersen, C. Marquardt, G. Leuchs, A generator for unique quantum random numbers based on vacuum states. Nat. Photonics **4**(10), 711–715 (2010)
7. T. Symul, S.M. Assad, P.K. Lam, Real time demonstration of high bitrate quantum random number generation with coherent laser light. Appl. Phys. Lett. **1**, 2–5 (2011)
8. R.L. Rivest, A. Shamir, L. Adleman, A method for obtaining digital signatures and public-key cryptosystems. Commun. ACM **21**(2), 120–126 (1978)
9. T. Elgamal, A public key cryptosystem and a signature scheme based on discrete logarithms. IEEE Trans. Inf. Theory **31**(4), 469–472 (1985)
10. National Institute of Standards and Technology (NIST), Advanced Encryption Standard (AES). *Federal Information Processing Standards*, Publication 197 (FIPS PUB 197), November 2001
11. National Institute of Standards and Technology (NIST), Secure Hash Standard (SHS), *Federal Information Processing Standards*, Publication 180-4 (FIPS PUB 180-4), March 2012
12. P.W. Shor, Polynomial-time algorithms for prime factorization and discrete logarithms on a quantum computer. SIAM J. Comput. **26**(5), 1484–1509 (1997)
13. S. Wiesner, Conjugate coding. ACM SIGACT News **15**(1), 78–88 (1983)
14. W. Diffie, M.E. Hellman, New directions in cryptography. IEEE Trans. Inf. Theory **22**(6), 644–654 (1976)
15. C.H. Bennett, G. Brassard, in *Quantum cryptography: public-key distribution and coin tossing*. IEEE International Conference on Computers, Systems and Signal Processing (IEEE Computer Society, Bangalore, 1984), pp. 175–179
16. U.M. Maurer, Secret key agreement by public discussion from common information. J. IEEE Trans. Inf. Theory **39**(3), 733–742 (1993)

17. H.K. Lo, H. Chau, M. Ardehali, Efficient quantum key distribution scheme and a proof of its unconditional security. J. Cryptol. **18**(2), 133–165 (2004)
18. M. Tomamichel, C.C.W. Lim, N. Gisin, R. Renner, Tight finite-key analysis for quantum cryptography. Nat. Commun. **3**, 634 (2012)
19. D. Bacco, M. Canale, N. Laurenti, G. Vallone, P. Villoresi, Experimental quantum key distribution with finite-key security analysis for noisy channels. Nat. Commun. **4**, paper no. 2363, (2013)
20. C.H. Bennett, G. Brassard, N.D. Mermin, Quantum cryptography without Bell's theorem. Phys. Rev. Lett. **68**(5), 557–559 (1992)
21. A.K. Ekert, Quantum cryptography based on Bell's theorem. Phys. Rev. Lett. **67**, 661–663 (1991)
22. G. Vallone, A. Dall'Arche, M. Tomasin, P. Villoresi, Loss tolerant device-independent quantum key distribution: a proof of principle. New J. Phys. **16**(6), paper no. 063064 (2014)
23. D. Elkouss, J. Martinez-Mateo, V. Martin, Information reconciliation for quantum key distribution. Quantum Inf. Comput. **11**(3&4), 226–238 (2011)
24. C.H. Bennett, G. Brassard, C. Crepeau, U.M. Maurer, Generalized privacy amplification. IEEE Trans. Inf. Theory **41**(6), 1915–1923 (1995)
25. F. Grosshans, P. Grangier, Continuous variable quantum cryptography using coherent states. Phys. Rev. Lett. **88**(5), paper n. 057902 (2002)
26. F. Grosshans, G. Van Assche, J. Wenger, R. Brouri, N.J. Cerf, P. Grangier, Quantum key distribution using Gaussian-modulated coherent states. Nature **421**(6920), 41–238 (2003)
27. C.H. Bennett, G. Brassard, C. Crépeau, R. Jozsa, A. Peres, W.K. Wootters, Teleporting an unknown quantum state via dual classical and Einstein-Podolsky-Rosen channels. Phys. Rev. Lett. **70**, 1895–1899 (1993)
28. D. Bouwmeester, J.W. Pan, K. Mattle, M. Eibl, H. Weinfurter, A. Zeilinger, Experimental quantum teleportation. Nature **390**, 575–579 (1997)
29. D. Boschi, S. Branca, F. De Martini, L. Hardy, S. Popescu, Experimental realization of teleporting an unknown pure quantum state via dual classical and Einstein-Podolsky-Rosen channels. Phys. Rev. Lett. **80**, 1121–1125 (1998)
30. A. Furusawa, J.L. Srensen, S.L. Braunstein, C.A. Fuchs, H.J. Kimble, E.S. Polzik, Unconditional quantum teleportation. Science **282**(5389), 706–709 (1998)
31. M.A. Nielsen, E. Knill, R. Laflamme, Complete quantum teleportation using nuclear magnetic resonance. Nature **396**(6706), 52–55 (1998)
32. J. Yin, J.G. Ren, H. Lu, Y. Cao, H.L. Yong, Y.P. Wu, C. Liu, S.K. Liao, F. Zhou, Y. Jiang, X.D. Cai, P. Xu, G.S. Pan, J.J. Jia, Y.M. Huang, H. Yin, J.Y. Wang, Y.A. Chen, C.Z. Peng, J.W. Pan, Quantum teleportation and entanglement distribution over 100-kilometre free-space channels. Nature **488**(7410), 185–188 (2012)
33. X.S. Ma, T. Herbst, T. Scheidl, D. Wang, S. Kropatschek, W. Naylor, B. Wittmann, A. Mech, J. Kofler, E. Anisimova, V. Makarov, T. Jennewein, R. Ursin, A. Zeilinger, Quantum teleportation over 143 kilometres using active feed-forward. Nature **489**(7415), 269–273 (2012)

Index

© Springer International Publishing Switzerland 2015
G. Cariolaro, *Quantum Communications*, Signals and Communication Technology,
DOI 10.1007/978-3-319-15600-2